Multiple Scattering of Light by Particles

This volume provides a thorough and up-to-date treatment of multiple scattering of light and other electromagnetic radiation in media composed of randomly and sparsely positioned particles. For the first time in monographic literature, the radiative transfer theory (RTT) is systematically and consistently presented as a branch of classical macrosopic electromagnetics. The book traces the fundamental link between the RTT and the effect of coherent backscattering (CB) and explains their place in the context of a comprehensive hierarchy of electromagnetic scattering problems. Dedicated sections present a thorough discussion of the physical meaning and range of applicability of the radiative transfer equation (RTE) and compare the self-consistent microphysical and the traditional phenomenological approaches to radiative transfer. The work describes advanced techniques for solving the RTE and gives examples of physically based applications of the RTT and CB in noninvasive particle characterization and remote sensing. This thorough and self-contained book will be valuable for science professionals, engineers, and graduate students working in a wide range of disciplines including optics, electromagnetics, remote sensing, atmospheric radiation, astrophysics, and biomedicine.

Michael I. Mishchenko is a senior scientist at the NASA Goddard Institute for Space Studies in New York City. After gaining a Ph.D. in physics in 1987, he has been principal investigator on several NASA and DoD projects and has served as topical editor and editorial board member of several leading scientific journals. Dr. Mishchenko is a recipient of the Henry G. Houghton Award of the American Meteorological Society and is an elected Fellow of the American Geophysical Union, the Optical Society of America, and the Institute of Physics. His research interests include electromagnetic scattering, radiative transfer, and remote sensing.

Larry D. Travis is presently Associate Chief of the NASA Goddard Institute for Space Studies. He gained a Ph.D. in astronomy at Pennsylvania State University in 1971. Dr. Travis has acted as principal investigator on several NASA projects and was awarded a NASA Exceptional Scientific Achievement Medal. His research interests include the theoretical interpretation of remote sensing measurements of polarization, planetary atmospheres, atmospheric dynamics, and radiative transfer.

Andrew A. Lacis is a senior scientist at the NASA Goddard Institute for Space Studies, and teaches radiative transfer at Columbia University. He gained a Ph.D. in physics at the University of Iowa in 1970 and has acted as principal investigator on numerous NASA and DoE projects. His research interests include radiative transfer in planetary atmospheres, the absorption of solar radiation by the Earth's atmosphere, and climate modeling.

Among the numerous scientific publications by these authors is the monograph on *Scattering, Absorption, and Emission of Light by Small Particles* published by Cambridge University Press in 2002.

Multiple Scattering of Light by Particles

Radiative Transfer and Coherent Backscattering

Michael I. Mishchenko
Larry D. Travis
Andrew A. Lacis
NASA Goddard Institute for Space Studies, New York

CAMBRIDGE
UNIVERSITY PRESS

University Printing House, Cambridge CB2 8BS, United Kingdom

One Liberty Plaza, 20th Floor, New York, NY 10006, USA

477 Williamstown Road, Port Melbourne, VIC 3207, Australia

4843/24, 2nd Floor, Ansari Road, Daryaganj, Delhi - 110002, India

79 Anson Road, #06-04/06, Singapore 079906

Cambridge University Press is part of the University of Cambridge.

It furthers the University's mission by disseminating knowledge in the pursuit of education, learning and research at the highest international levels of excellence.

www.cambridge.org
Information on this title: www.cambridge.org/9780521158015

First published 2006
First paperback edition 2017

A catalogue record for this publication is available from the British Library

ISBN 978-0-521-83490-2 Hardback
ISBN 978-0-521-15801-5 Paperback

Contents

Preface xi
Dedication and acknowledgments xv

Chapter 1 **Introduction** 1

1.1 Electromagnetic scattering by a fixed finite object 1
1.2 Actual observables 5
1.3 Foldy–Lax equations 6
1.4 Dynamic and static scattering by random groups of particles 7
1.5 Ergodicity 9
1.6 Single scattering by random particles 10
1.7 Multiple scattering by a large random group of particles 12
1.8 Coherent backscattering 14
1.9 Classification of electromagnetic scattering problems 16
1.10 Notes and further reading 18

Chapter 2 **Maxwell equations, electromagnetic waves, and Stokes parameters** 20

2.1 Maxwell equations and constitutive relations 20
2.2 Boundary conditions 23
2.3 Time-harmonic fields 26
2.4 The Poynting vector 28
2.5 Plane-wave solution 31
2.6 Coherency matrix and Stokes parameters 37
2.7 Ellipsometric interpretation of the Stokes parameters 41

2.8 Rotation transformation rule for the Stokes parameters 47
2.9 Quasi-monochromatic light 48
2.10 Measurement of the Stokes parameters 54
2.11 Spherical-wave solution 58
2.12 Coherency dyad of the electric field 62
2.13 Historical notes and further reading 64

Chapter 3 **Basic theory of electromagnetic scattering** 66

3.1 Volume integral equation and Lippmann–Schwinger equation 67
3.2 Scattering in the far-field zone 71
3.3 Scattering dyadic and amplitude scattering matrix 78
3.4 Reciprocity 80
3.5 Scale invariance rule 84
3.6 Electromagnetic power and electromagnetic energy density 87
3.7 Phase matrix 93
3.8 Extinction matrix 99
3.9 Extinction, scattering, and absorption cross sections 102
3.10 Coherency dyad of the total electric field 105
3.11 Other types of illumination 109
3.12 Variable scatterers 110
3.13 Thermal emission 112
3.14 Historical notes and further reading 114

Chapter 4 **Scattering by a fixed multi-particle group** 115

4.1 Vector form of the Foldy–Lax equations 115
4.2 Far-field version of the vector Foldy–Lax equations 118

Chapter 5 **Statistical averaging** 123

5.1 Statistical averages 124
5.2 Configurational averaging 126
5.3 Averaging over particle states 126

Chapter 6 **Scattering by a single random particle** 131

6.1 Scattering in the far-field zone of the trap volume 131
6.2 "Near-field" scattering 136

Chapter 7 **Single scattering by a small random particle group** 140

7.1 Single-scattering approximation for a fixed group of particles 141
7.2 Far-field single-scattering approximation for a fixed particle group 142
7.3 Far-field uncorrelated single-scattering approximation and modified uncorrelated single-scattering approximation 145

7.4 Forward-scattering interference 147
7.5 Energy conservation 151
7.6 Conditions of validity of the far-field modified uncorrelated single-scattering approximation 151
7.7 First-order-scattering approximation 158
7.8 Discussion 163

Chapter 8 **Radiative transfer equation** 165

8.1 The Twersky approximation 166
8.2 The Twersky expansion of the coherent field 171
8.3 Coherent field 173
8.4 Transfer equation for the coherent field 180
8.5 Dyadic correlation function in the ladder approximation 181
8.6 Integral equation for the ladder specific coherency dyadic 191
8.7 Integro-differential equation for the diffuse specific coherency dyadic 195
8.8 Integral and integro-differential equations for the diffuse specific coherency matrix 197
8.9 Integral and integro-differential equations for the diffuse specific coherency column vector 198
8.10 Integral and integro-differential equations for the specific intensity column vector 199
8.11 Summary of assumptions and approximations 200
8.12 Physical meaning of the diffuse specific intensity column vector and the coherent Stokes column vector 203
8.13 Energy conservation 208
8.14 External observation points 209
8.14.1 Coherent field 210
8.14.2 Ladder coherency dyadic 211
8.14.3 Specific intensity column vector 213
8.14.4 Discussion 214
8.14.5 Illustrative example: first-order scattering 216
8.15 Other types of illumination 217
8.16 Phenomenological approach to radiative transfer 218
8.17 Scattering media with thermal emission 224
8.18 Historical notes and further reading 225

Chapter 9 **Calculations and measurements of single-particle characteristics** 227

9.1 Exact theoretical techniques 227
9.2 Approximations 234
9.3 Measurement techniques 237

9.4 Further reading 239

Chapter 10 **Radiative transfer in plane-parallel scattering media** 240

10.1 The standard problem 240
10.2 The propagator 243
10.3 The general problem 245
10.4 Adding equations 247
10.5 Invariant imbedding equations 255
10.6 Ambarzumian equation 258
10.7 Reciprocity relations for the reflection and transmission matrices 259
10.8 Notes and further reading 260

Chapter 11 **Macroscopically isotropic and mirror-symmetric scattering media** 261

11.1 Symmetries of the Stokes scattering matrix 262
11.2 Macroscopically isotropic and mirror-symmetric scattering medium 265
11.3 Phase matrix 266
11.4 Forward-scattering direction and extinction matrix 270
11.5 Backward scattering 273
11.6 Scattering cross section and asymmetry parameter 275
11.7 Thermal emission 276
11.8 Spherically symmetric particles 277
11.9 Effects of nonsphericity and orientation 278
11.10 Normalized scattering and phase matrices 279
11.11 Expansion in generalized spherical functions 282
11.12 Circular-polarization representation 286
11.13 Illustrative examples 291

Chapter 12 **Radiative transfer in plane-parallel, macroscopically isotropic and mirror-symmetric scattering media** 302

12.1 The standard problem 302
12.2 The general problem 304
12.3 Adding equations 306
12.4 Invariant imbedding and Ambarzumian equations 311
12.5 Successive orders of scattering 313
12.6 Symmetry relations 315
12.6.1 Phase matrix 315
12.6.2 Reflection and transmission matrices 316
12.6.3 Matrices describing the internal field 317
12.6.4 Perpendicular directions 317
12.7 Fourier decomposition 318

12.7.1 Fourier decomposition of the VRTE 318
12.7.2 Fourier components of the phase matrix 319
12.8 Scalar approximation 321
12.9 Notes and further reading 322

Chapter 13 **Illustrative applications of radiative transfer theory** 324

13.1 Accuracy of the scalar approximation 324
13.1.1 Rayleigh-scattering slabs 325
13.1.2 Polydisperse spherical particles and spheroids 337
13.2 Directional reflectance and spherical and plane albedos 347
13.3 Polarization as an effect and as a particle characterization tool 357
13.4 Depolarization 362
13.5 Further reading 362

Chapter 14 **Coherent backscattering** 365

14.1 Specific coherency dyadic 366
14.2 Reflected light 371
14.3 Exact backscattering direction 373
14.4 Other types of illumination 379
14.5 Photometric and polarimetric characteristics of coherent backscattering 380
14.5.1 Unpolarized incident light 380
14.5.2 Linearly polarized incident light 381
14.5.3 Circularly polarized incident light 382
14.5.4 General properties of the enhancement factors and polarization ratios 383
14.5.5 Spherically symmetric particles 385
14.5.6 Benchmark results for Rayleigh scattering 386
14.6 Numerical results for polydisperse spheres and polydisperse, randomly oriented spheroids 386
14.7 Angular profile of coherent backscattering 395
14.8 Further discussion of theoretical and practical aspects of coherent backscattering 402
14.9 Applications and further reading 404

Appendix A **Dyads and dyadics** 407

Appendix B **Spherical wave expansion of a plane wave in the far-field zone** 409

Appendix C **Euler rotation angles** 411

Appendix D **Integration quadrature formulas** 413

Appendix E **Stationary phase evaluation of a double integral** 416

Appendix F **Wigner functions, Jacobi polynomials, and generalized spherical functions** 418

F.1 Wigner *d*-functions 418
F.2 Jacobi polynomials 422
F.3 Orthogonality and completeness 422
F.4 Recurrence relations 423
F.5 Legendre polynomials and associated Legendre functions 424
F.6 Generalized spherical functions 425
F.7 Wigner *D*-functions, addition theorem, and unitarity 426
F.8 Further reading 428

Appendix G **Système International units** 429

Appendix H **Abbreviations** 431

Appendix I **Glossary of symbols** 433

References 442
Index 469

Preface

Since the seminal papers by Lommel (1887), Chwolson (1889), and Schuster (1905), the radiative transfer equation (RTE) has been widely used in diverse areas of science and engineering to describe multiple scattering of light and other electromagnetic radiation in media composed of randomly and sparsely distributed particles. Analytical studies of the RTE have formed a separate branch of mathematical physics. However, despite the importance and the widespread use of the radiative transfer theory (RTT), its physical basis had not been established firmly until quite recently.

Indeed, the traditional "phenomenological" way to introduce the RTE has been to invoke an eclectic combination of principles borrowed from classical radiometry (i.e., intuitively appealing arguments of energy balance and the simple heuristic concepts of light rays and ray pencils), classical electromagnetics (electromagnetic scattering, Stokes parameters, and phase and extinction matrices), and even quantum electrodynamics ("photons"). Furthermore, the phenomenological approach has always relied on an illusive concept of an "elementary (or differential) volume element" of the discrete scattering medium. To sew together these motley concepts, one needs a set of postulates that appear to be plausible at first sight but turn out to be artificial upon close examination.

This inconsistent approach to radiative transfer is quite deceptive since it implies that in order to derive the RTE for media composed of elastically scattering particles one needs postulates other than those already contained in classical electromagnetics. The phenomenological "derivation" becomes especially questionable when one attempts to include the effects of polarization described by the so-called vector RTE and/or to take into account the effects of particle nonsphericity and orientation. Furthermore, it does not allow one to determine the range of applicability of the RTE and trace the fundamental link between the RTT and the effect of coherent backscattering.

During the past few decades, there has been significant progress in studies of the statistical wave content of the RTT. This research has resulted in a much improved understanding of the basic assumptions leading to the RTE and has indeed demonstrated it to be a corollary of the Maxwell equations. Hence, the main goal of this monograph is to consistently present the RTT as a branch of classical electromagnetics as applied to discrete random media and to clarify the relationship between radiative transfer and coherent backscattering.

Another motivation for writing this book was the recognition of the scarcity of comprehensive monographs describing the fundamentals of polarized radiative transfer and its applications in a way intelligible to graduate students and non-expert scientists.[1] This factor has significantly impeded the development and wide dissemination of physically-based remote sensing and particle characterization techniques. Hence, the additional purpose of this volume is to present a broad and coherent outline of the subject and to make the technical material accessible to a larger audience than those specializing in this research area. Consistent with this purpose, our presentation assumes minimal prior knowledge of the subject matter and the relevant theoretical approaches. We expect, therefore, that the book will be useful to science professionals, engineers, and graduate students working in a broad range of disciplines: optics, electromagnetics, atmospheric radiation and remote sensing, radar meteorology, oceanography, climate research, astrophysics, optical engineering and technology, particle characterization, and biomedical optics.

This volume is a natural continuation of our recent monograph on *Scattering, Absorption, and Emission of Light by Small Particles* (Mishchenko *et al.*, 2002; hereinafter referred to as MTL[2]) in that it consistently uses the same general methodology and notation system while applying them to multiple scattering by random particle ensembles. However, the present book contains all the necessary background material and is self-contained.

As in MTL, we usually denote vectors using the Times bold font and matrices using the Arial bold font. Unit vectors are denoted by a caret, whereas dyads and dyadics are denoted by the symbol ↔. The Times italic font is reserved for scalar quantities, important exceptions being the square root of minus one, the differential sign, and the base of natural logarithms, which are denoted by Times roman characters i, d, and e, respectively. Another exception is the relative refractive index, which is denoted by a sloping sans serif *m*. For the reader's convenience, a glossary listing the symbols used, their meaning and dimension, and the section where they first appear is provided at the end of the book (Appendix I). Appendix H contains a list of abbreviations.

---

[1] The recent book by Hovenier *et al.* (2004) is a notable exception.

[2] By agreement with Cambridge University Press, MTL is now publicly available in the .pdf format at http://www.giss.nasa.gov/~crmim/books.html.

We did not try to compile a comprehensive and detailed reference list. Instead, preference was given to seminal publications as well as to relevant books and reviews where further references can be found.

We mention several relevant computer programs made publicly available on-line. These programs have been thoroughly tested and are expected to generate reliable results provided that they are implemented as instructed. It is not inconceivable, however, that some of these programs contain errors and/or are not platform-independent. Also, it is possible that users could specify input parameter values that are outside the intended range for which accurate results can be expected. For these reasons the authors of this book and the publisher disclaim all liability for any damage that may result from the use of the programs. Although the authors and the publisher have used their best endeavors to ensure that the URLs for external Internet sites referred to in this book are correct and active at the time of this book going to press, they cannot guarantee that a site will remain live or that its content is or will remain appropriate.

Michael I. Mishchenko
Larry D. Travis
Andrew A. Lacis

New York
September 2005

Dedication and acknowledgments

The phenomenological theory of radiative transfer in discrete random media had been widely used for many decades in numerous research and engineering disciplines despite the fact that its physical origin had not been established. This uncomfortable situation has finally changed, and the RTT has become a legitimate branch of classical electromagnetics. It was very exciting for us to be able to write this entire monograph on radiative transfer without ever having to leave the firm grounds of electromagnetic theory. We, therefore, appreciatively dedicate this book to James Clerk Maxwell, whose monumental contribution to physics can be compared only to that of Sir Isaac Newton and whose equations of electromagnetism have been voted by scientists to be the greatest equations ever (Crease, 2004).

Several prominent scientists have made important contributions to the evolving subject of multiple wave scattering by small particles and microphysical justification of the RTT. Our own research has been most influenced by the publications of Yuri Barabanenkov, Anatoli Borovoi, Akira Ishimaru, Leung Tsang, Victor Twersky, and Hendrik van de Hulst to whom we express sincere appreciation.

We are deeply indebted to Joop Hovenier, Michael Kahnert, and Cornelis van der Mee for numerous discussions, continued encouragement, and valuable comments on a preliminary version of this book that resulted in a much improved manuscript. We also gratefully acknowledge illuminating discussions with Yuri Barabanenkov, Anatoli Borovoi, Oleg Bugaenko, Brian Cairns, Barbara Carlson, Zhanna Dlugach, Helmut Domke, James Hansen, Vsevolod Ivanov, Nikolai Khlebtsov, Kuo-Nan Liou, Daniel Mackowski, Bart van Tiggelen, Victor Tishkovets, Gorden Videen, Edgard Yanovitskij, and many other colleagues.

We thank Joop Hovenier for several invitations to visit him at de Vrije Universiteit te Amsterdam and de Universiteit van Amsterdam and for generous travel support. MIM thanks Helmut Domke for kind hospitality during a short stay at Zentralinstitut für Astrophysik in Potsdam.

Our research endeavors have been generously funded over the years by grants from the United States Government. We thankfully acknowledge the continuing support from the NASA Glory Mission Project, the NASA Earth Observing System Program, and the Department of Energy Atmospheric Radiation Measurement Program. The preparation of this book was sponsored by a grant from the NASA Radiation Sciences Program managed by Donald Anderson and Hal Maring.

We thank Zoe Wai and Josefina Mora for their help in tracing some of the less accessible publications and acknowledge the fine cooperation that we received from the staff of Cambridge University Press at all stages of the production of this book. We are particularly grateful to Jacqueline Garget for her patience and encouragement and to Jo Tyszka for thorough copy-editing work.

Our special thanks go to Nadia Zakharova who provided invaluable assistance during the preparation of the manuscript and the proofreading stage. She also contributed many numerical results and almost all computer graphics.

Chapter 1

Introduction

Natural and man-made environments provide countless examples of diverse scattering media composed of particles. The varying complexity of these media suggests multiple ways of using electromagnetic scattering for particle characterization and gives rise to a distinctive hierarchy of theoretical models that can be used to simulate specific remote-sensing or laboratory measurements. Hence the objective of this introductory chapter is to present a simple classification of scattering problems involving small particles and to briefly outline solution approaches described in detail in later chapters.

1.1 Electromagnetic scattering by a fixed finite object

A parallel monochromatic beam of light propagates in a vacuum without a change in its intensity or polarization state. However, inserting an object into the beam (see Fig. 1.1.1) causes several distinct effects. First, the object extracts some of the incident energy and spreads it in all directions at the frequency of the incident beam. This phenomenon is called elastic *scattering* and, in general, gives rise to light with a polarization state different from that of the incident beam. Second, the object may convert some of the energy contained in the beam into other forms of energy such as heat. This phenomenon is called *absorption*. The energy contained in the incident beam is accordingly reduced by the amount equal to the sum of the scattered and absorbed energy. This reduction is called *extinction*. The extinction rates for different polarization components of the incident beam can be different, which is called *dichroism*.

In electromagnetic terms, the parallel monochromatic beam of light is represented by a harmonically oscillating plane electromagnetic wave. The latter propagates in a

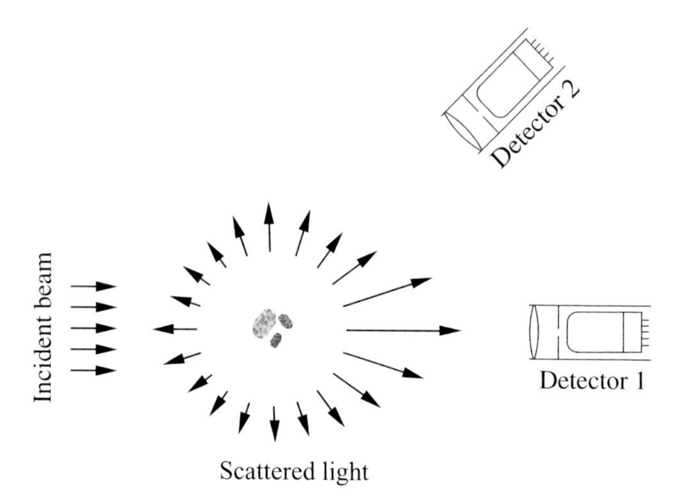

Figure 1.1.1. Scattering by a fixed finite object. In this case the object consists of three disjoint, heterogeneous, stationary bodies.

vacuum without a change in its intensity or polarization state (see Fig. 1.1.2(a)). However, the presence of a finite object, as illustrated in Fig. 1.1.2(b), changes both the electric, $\mathcal{E}$, and the magnetic, $\mathcal{H}$, field that would otherwise exist in an unbounded homogeneous space. The difference between the total fields in the presence of the object, $\mathcal{E}(\mathbf{r},t)$ and $\mathcal{H}(\mathbf{r},t)$, and the original fields that would exist in the absence of the object, $\mathcal{E}^{\text{inc}}(\mathbf{r},t)$ and $\mathcal{H}^{\text{inc}}(\mathbf{r},t)$, can be thought of as the fields scattered by the object, $\mathcal{E}^{\text{sca}}(\mathbf{r},t)$ and $\mathcal{H}^{\text{sca}}(\mathbf{r},t)$, where $\mathbf{r}$ is the position (radius) vector and t is time (Fig. 1.1.2(b)). In other words, the total electric and magnetic fields in the presence of the object are equal to vector sums of the respective incident (original) and scattered fields:

$$\mathcal{E}(\mathbf{r},t) = \mathcal{E}^{\text{inc}}(\mathbf{r},t) + \mathcal{E}^{\text{sca}}(\mathbf{r},t), \tag{1.1.1}$$

$$\mathcal{H}(\mathbf{r},t) = \mathcal{H}^{\text{inc}}(\mathbf{r},t) + \mathcal{H}^{\text{sca}}(\mathbf{r},t). \tag{1.1.2}$$

The origin of the scattered electromagnetic field can be understood by recalling that in terms of microscopic electrodynamics, the object is an aggregation of a large number of discrete elementary electric charges. The oscillating electromagnetic field of the incident wave excites these charges to vibrate with the same frequency and thereby radiate secondary electromagnetic waves. The superposition of all the secondary waves gives the total elastically scattered field. If the charges do not oscillate exactly in phase or exactly in anti-phase with the incident field then there is dissipation of electromagnetic energy into the object. This means that the object is absorbing and scatters less total energy than it extracts from the incident wave.

Electromagnetic scattering is an exceedingly complex phenomenon because a secondary wave generated by a vibrating charge also stimulates vibrations of all other charges forming the object and thus modifies their respective secondary waves. As a

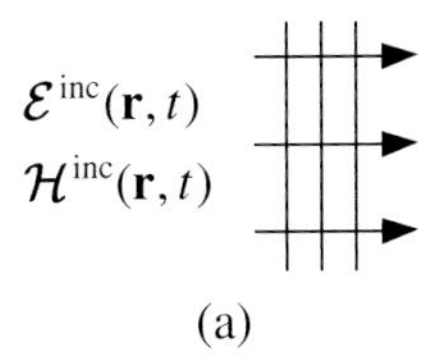

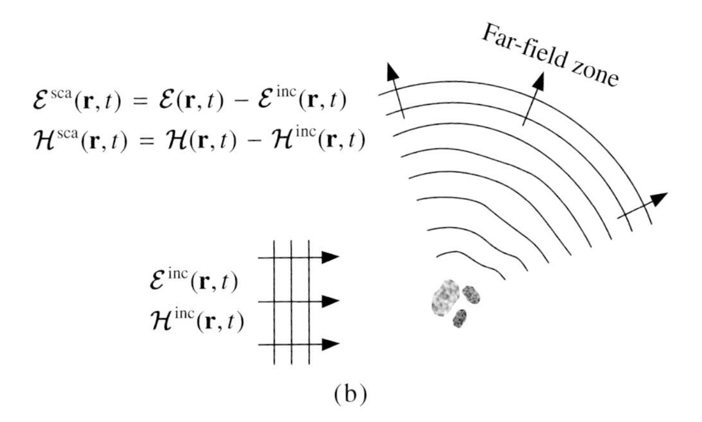

Figure 1.1.2. Schematic representation of the electromagnetic scattering problem.

result, all the secondary waves become interdependent. Furthermore, the computation of the total scattered field by superposing the secondary waves must take account of their phase differences, which change every time the incidence and/or the scattering direction is changed. Therefore, the total scattered field depends on the way the charges are arranged to form the object with respect to the incidence and scattering directions.

Since the number of elementary charges forming an object can be extremely large, solving the scattering problem directly by computing and superposing all secondary waves is impracticable even with the aid of modern computers. Fortunately, the scattering problem can also be solved using the concepts of macroscopic electromagnetics, which treat the large collection of charges as one or several macroscopic bodies with a specific distribution of the refractive index. Consequently, the scattered field can be computed by solving the Maxwell equations for the macroscopic electromagnetic field subject to appropriate boundary conditions. It is this approach that forms the basis of the modern theory of electromagnetic scattering by macroscopic objects.

To simplify the solution of the scattering problem, we will adhere throughout the book to the following five well-defined restrictions:

1. We will always assume that the unbounded host medium surrounding the scattering object is homogeneous, linear, isotropic, and nonabsorbing.

2. We will always assume that the scattering object is illuminated by either:

   (i) a time-harmonic plane electromagnetic wave given, in the complex-field representation, by

$$\begin{cases} \mathbf{E}(\mathbf{r},t) = \mathbf{E}_0 \exp(\mathrm{i}\mathbf{k}\cdot\mathbf{r} - \mathrm{i}\omega t), \\ \mathbf{H}(\mathbf{r},t) = \mathbf{H}_0 \exp(\mathrm{i}\mathbf{k}\cdot\mathbf{r} - \mathrm{i}\omega t), \end{cases} \quad \mathbf{r} \in \mathfrak{R}^3, \tag{1.1.3}$$

with constant amplitudes $\mathbf{E}_0$ and $\mathbf{H}_0$, where ω is the angular frequency, $\mathbf{k}$ is the real-valued wave vector, $\mathrm{i} = (-1)^{1/2}$, and $\mathfrak{R}^3$ denotes the entire three-dimensional space, or

(ii) a quasi-monochromatic parallel beam of light given by

$$\begin{cases} \mathbf{E}(\mathbf{r},t) = \mathbf{E}_0(t) \exp(\mathrm{i}\mathbf{k}\cdot\mathbf{r} - \mathrm{i}\omega t), \\ \mathbf{H}(\mathbf{r},t) = \mathbf{H}_0(t) \exp(\mathrm{i}\mathbf{k}\cdot\mathbf{r} - \mathrm{i}\omega t), \end{cases} \quad \mathbf{r} \in \mathfrak{R}^3, \tag{1.1.4}$$

where fluctuations in time of the complex amplitudes of the electric and magnetic fields, $\mathbf{E}_0(t)$ and $\mathbf{H}_0(t)$, around their respective mean values occur much more slowly than the harmonic oscillations of the time factor $\exp(-\mathrm{i}\omega t)$.

This restriction excludes other types of illumination such as a focused laser beam of finite lateral extent or a pulsed beam.

3. We will exclude nonlinear optics effects by assuming that the conductivity, permeability, and electric susceptibility of both the scattering object and the surrounding medium are independent of the electric and magnetic fields.

4. We will assume that electromagnetic scattering occurs without frequency redistribution, i.e., the scattered light has the same frequency as the incident light. This restriction excludes inelastic scattering phenomena such as Raman and Brillouin scattering and fluorescence. It also excludes the specific consideration of the small Doppler shift of frequency of the scattered light relative to that of the incident light due the movement of the scatterer with respect to the source of illumination.

5. We will largely exclude from consideration the phenomenon of thermal emission. The latter is caused by electron transitions from one energy level to a lower level in macroscopic bodies with absolute temperature different from zero. A macroscopic object is a complex system of molecules with a large number of degrees of freedom. Therefore, many different electron transitions produce spectral emission lines so closely spaced that the resulting radiation spectrum becomes effectively continuous and includes emitted energy at all frequencies. By neglecting thermal emission, we will implicitly assume that the temperature of the object is low enough that the intensity of the emitted radiation at the frequency of the incident light is much smaller that the elastically scattered intensity. This assumption is usually valid for objects at room or lower temperature and for near-infrared and shorter wavelengths.

The theoretical and numerical techniques for computing the electromagnetic field elastically scattered by a finite fixed object composed of one or several physical bodies are many and are reviewed thoroughly in Mishchenko *et al.* (2000a), MTL, and Kahnert (2003). Since all of these techniques have certain limitations in terms of the object morphology and object size relative to the incident wavelength, a practitioner should analyze carefully the relative strengths and weaknesses of the available solution techniques before attempting to address the specific problem in hand.

1.2 Actual observables

Because of high frequency of time-harmonic oscillations, traditional optical instruments cannot measure the electric and magnetic fields associated with the incident and scattered waves. Indeed, accumulating and averaging a signal proportional to the electric or the magnetic field over a time interval long compared with the period of oscillations would result in a zero net result:

$$\frac{1}{T}\int_{t}^{t+T} \mathrm{d}t' \exp(-\mathrm{i}\omega t') \underset{T \gg 2\pi/\omega}{=} 0. \tag{1.2.1}$$

Therefore, the majority of optical instruments measure quantities which have the dimension of energy flux and are defined in such a way that the time-harmonic factor $\exp(-\mathrm{i}\omega t)$ vanishes upon multiplication by its complex-conjugate counterpart: $\exp(-\mathrm{i}\omega t)[\exp(-\mathrm{i}\omega t)]^* \equiv 1$. This means that in order to make the theory applicable to analyses of actual optical observations, the scattering process must be characterized in terms of carefully chosen derivative quantities that can be measured directly. This explains why the concept of an actual *observable* is central to the discipline of light scattering by particles.

Although one can always define the magnitude and the direction of the electromagnetic energy flux at any point in space in terms of the Poynting vector, the latter carries no information about the polarization state of the incident and scattered fields. The conventional approach to ameliorate this problem dates back to Sir George Gabriel Stokes. He proposed using four real-valued quantities which have the dimension of monochromatic energy flux and fully characterize a *transverse* electromagnetic wave[1] inasmuch as it is subject to practical optical analysis (Stokes, 1852). These quantities, called the Stokes parameters, form the so-called four-component Stokes column vector and carry information about both the intensity and the polarization state of the wave.

In the so-called far-field zone of a fixed object, the propagation of the scattered electromagnetic wave is away from the object (Fig. 1.1.2(b)). Furthermore, the elec-

[1] By definition, the electric and magnetic field vectors of a transverse electromagnetic wave vibrate in the plane perpendicular to the propagation direction.

tric and magnetic field vectors vibrate in the plane perpendicular to the propagation direction and their amplitudes decay inversely with distance from the object. The tranversality of both the incident plane wave and the scattered spherical wave allows one to define the corresponding sets of Stokes parameters and to describe the response of a well-collimated polarization-sensitive detector of light in terms of the 4×4 so-called phase and extinction matrices. Specifically, detector 2 in Fig. 1.1.1 collects only the scattered light, and its response is fully characterized by the product of the phase matrix and the Stokes column vector of the incident wave. Thus the phase matrix realizes the transformation of the Stokes parameters of the incident wave into the Stokes parameters of the scattered wave. The response of detector 1 consists of three parts:

1. The one due to the incident light.
2. The one due to the forward-scattered light.
3. The one due to the interference of the incident wave and the wave scattered by the object in the exact forward direction.

The third part is described by minus the product of the extinction matrix and the Stokes vector of the incident wave and accounts for both the total attenuation of the detector signal due to extinction of light by the object and the effect of dichroism.

The phase and extinction matrices depend on object characteristics such as size, shape, refractive index, and orientation and can be readily computed provided that the scattered field is known from the solution of the Maxwell equations.

The main convenience of the far-field approximation is that it allows one to treat the object essentially as a point source of scattered radiation. However, the criteria defining the far-field zone are rather stringent and are often violated in practice. A good example is remote sensing of water clouds in the terrestrial atmosphere using detectors of electromagnetic radiation mounted on aircraft or satellite platforms. Such detectors typically measure radiation coming from a small part of a cloud and do not "perceive" the entire cloud as a single point-like scatterer. Furthermore, the notion of the far-field zone of the cloud becomes completely meaningless if a detector is placed inside the cloud. It is thus clear that to characterize the response of such "near-field" detectors one must define quantities other than the Stokes parameters and the extinction and phase matrices. Still the actual observables must be defined in such a way that they can be measured by an optical device ultimately recording the flux of electromagnetic energy.

1.3 Foldy–Lax equations

Many theoretical techniques based on directly solving the differential Maxwell equations or their integral counterparts are applicable to an arbitrary fixed finite object, be it a single physical body or a cluster consisting of several distinct components, either touching or spatially separated. These techniques are based on treating the object as a single scatterer and yield the total scattered electric and magnetic fields. However, if the object is a multi-particle cluster then it is often convenient to represent the total scattered field as a vector superposition of partial fields scattered by the individual cluster components. This means, for example, that the total electric field at a point $\mathbf{r}$ is written as follows:

$$\mathbf{E}(\mathbf{r},t) = \mathbf{E}^{\mathrm{inc}}(\mathbf{r},t) + \sum_{i=1}^{N} \mathbf{E}_i^{\mathrm{sca}}(\mathbf{r},t), \qquad \mathbf{r} \in \Re^3, \tag{1.3.1}$$

where N is the number of particles in the cluster and $\mathbf{E}_i^{\mathrm{sca}}(\mathbf{r},t)$ is the ith partial scattered electric field. The total magnetic field is given by a similar expression. The partial scattered fields can be found by solving vector so-called Foldy–Lax equations which follow directly from the volume integral equation counterpart of the Maxwell equations and are exact. By iterating the Foldy–Lax equations, one can derive an order-of-scattering expansion of the scattered field which, in combination with statistical averaging, forms the basis of the modern theory of multiple scattering by random particle ensembles.

1.4 Dynamic and static scattering by random groups of particles

Solving the Maxwell equations yields the field scattered by a fixed object. This approach can be used directly in analyses of microwave analog measurements (e.g., Gustafson, 2000; Section 8.2 of MTL), in which the scattering object is held fixed relative to the source of electromagnetic radiation during the measurement cycle. However, it is inapplicable in the majority of laboratory and remote-sensing observations. Even if the scattering object is a single microparticle trapped inside an electrostatic or optical levitator (e.g., Chapter 2 of Davis and Schweiger, 2002), it rapidly changes its position and orientation during the time necessary to take a measurement. Furthermore, one often encounters situations in which light is scattered by a very large group of particles forming a constantly varying spatial configuration. A typical example is a cloud of water droplets or ice crystals in which the particles are constantly moving, spinning, and even changing their shapes and sizes due to oscillations of the droplet surface, evaporation, condensation, sublimation, and melting. Although such a particle collection can be treated at each given moment as a fixed cluster, a typical measurement of light scattering takes a finite amount of time over which the spatial configuration of the component particles and their sizes, orientations, and/or shapes continuously and randomly change. Therefore, the registered signal is in effect an average over a large number of distinct clusters.

When a fixed group of particles is illuminated by a monochromatic, spatially co-

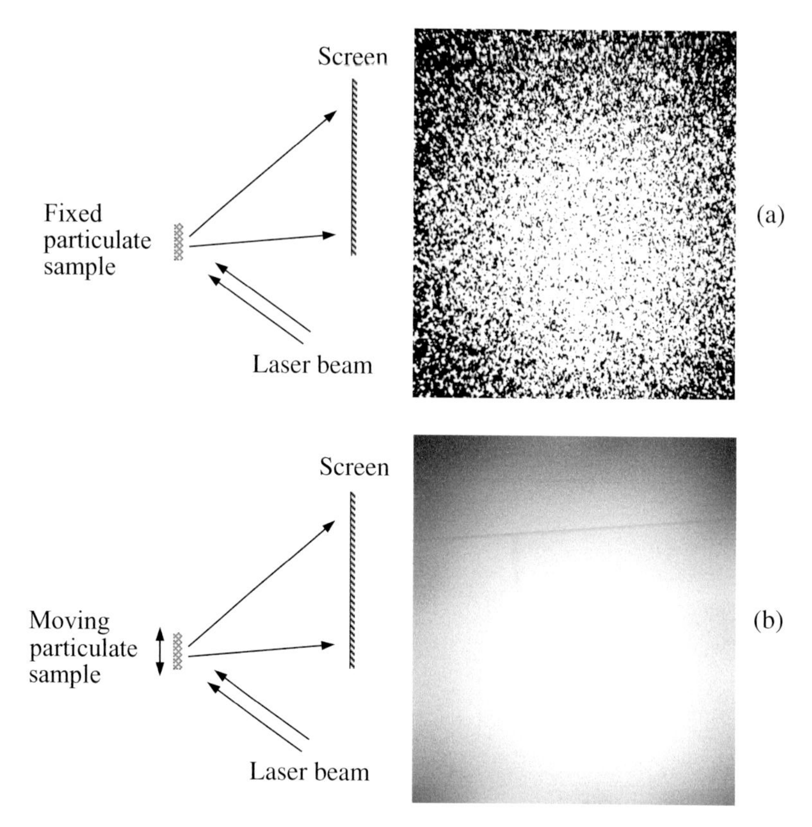

Figure 1.4.1. (a) Speckle pattern produced by laser light reflected by a fixed particulate sample. (b) Moving the scattering sample during the measurement averages the speckle pattern out. (After Lenke and Maret, 2000a.)

herent plane wave (e.g., laser light), the light scattered by the group onto a distant screen generates a characteristic speckle pattern consisting of randomly located bright spots of various sizes and shapes (see Fig. 1.4.1(a)). This pattern is the result of constructive or destructive interference of the partial waves scattered by different particles towards a point on the screen. When the particles move, the phase relations between the partial waves constantly change, thereby causing rapid fluctuations of the speckle pattern. Accumulating the signal over a sufficiently long period of time averages the speckle pattern out and results in a rather smooth "incoherent" distribution of the scattered intensity (Fig. 1.4.1(b)).

It has been shown that measurements of the temporal and/or spatial fluctuations of the speckle pattern contain useful information about the particles, in particular about their motion. Statistical analyses of light scattered by dilute and dense particle suspensions, respectively, are the subject of the disciplines called photon correlation spectroscopy (PCS) and diffusing wave spectroscopy (DWS) and form the basis of many well-established experimental techniques for the measurement of various particle characteristics such as velocity, size, and dispersity (e.g., Berne and Pecora, 1976;

Pine *et al.*, 1990). The recent extension of PCS to account for particles changing the polarization state of the incident coherent beam, so-called polarization fluctuation spectroscopy, enables the shapes in addition to the sizes of particles to be sensed (Hopcraft *et al.*, 2004).

Photon correlation spectroscopy and diffusing wave spectroscopy study *dynamic* aspects of light scattering by groups of randomly moving particles and as such will not be discussed in this volume. Instead, we will assume that the effect of temporal fluctuations is eliminated by averaging the speckle pattern over a period of time much longer than the typical period of the fluctuations. In other words, we will deal with the average, static component of the scattering pattern. Therefore, the subject of this book can be called *static* light scattering.

1.5 Ergodicity

Quantitative analyses of static scattering measurements require the use of a theoretical averaging procedure. Let us consider, for example, the measurement of a scattering characteristic A of a cloud of spherical water droplets. This characteristic depends on time implicitly by being a function of time-dependent physical parameters of the cloud such as the coordinates and sizes of all the constituent particles. The full set of particle positions and sizes will be denoted collectively by ψ and determines the state of the entire cloud at a moment in time. In order to interpret the measurement of $A[\psi(t)]$ accumulated over a period of time extending from $t = t_0$ to $t = t_0 + T$, one needs a way of predicting theoretically the average value

$$\overline{A} = \frac{1}{T} \int_{t_0}^{t_0+T} \mathrm{d}t\, A[\psi(t)]. \tag{1.5.1}$$

Quite often the temporal evolution of a complex scattering object such as the cloud of water droplets is controlled by several physical processes and is described by an intricate system of equations. To incorporate the solution of this system of equations for each moment of time into the theoretical averaging procedure (1.5.1) can be a formidable task and is rarely, if ever, done. Instead, averaging over time is replaced by ensemble averaging based on the following rationale.

Although the coordinates and sizes of water droplets in the cloud change with time in a specific way, the range of instantaneous states of the cloud captured by the detector during the measurement becomes representative of that captured over an infinite period of time provided that T is sufficiently large. We thus have

$$\overline{A} \approx \lim_{\mathcal{T}\to\infty} \frac{1}{\mathcal{T}} \int_{t_0}^{t_0+\mathcal{T}} \mathrm{d}t\, A[\psi(t)] = \langle A \rangle_t. \tag{1.5.2}$$

Notice now that the infinite integral in Eq. (1.5.2) can be expected to "sample" every physically realizable state ψ of the cloud. Furthermore, this sampling is statistically representative in that the number of times each state is sampled is large and tends to infinity in the limit $\tau \to \infty$. Most importantly, the cumulative contribution of a cloud state ψ to $\langle A \rangle_t$ is independent of the specific moments of time when this state actually occurred in the process of the temporal evolution of the cloud. Rather, it depends on how many times this state was sampled. Therefore, this cumulative contribution can be thought of as being proportional to the probability of occurrence of the state ψ at *any* moment of time. This means that instead of specifying the state of the cloud at each moment t and integrating over all t, one can introduce an appropriate time-independent probability density function $p(\psi)$ and integrate over the entire physically realizable range of cloud states:

$$\langle A \rangle_t \approx \int \mathrm{d}\psi \, p(\psi) A(\psi) = \langle A \rangle_\psi, \tag{1.5.3}$$

where

$$\int \mathrm{d}\psi \, p(\psi) = 1. \tag{1.5.4}$$

The assumption that averaging over time for a "sufficiently random" object can be replaced by ensemble averaging is called the ergodic hypothesis. Although it has not been possible to establish mathematically the full ergodicity of real dynamical systems, more restricted versions of the ergodic theorem have been proven. Physical processes such as Brownian motion and turbulence often help to establish a significant degree of randomness of particle positions and orientations, which seems to explain why many theoretical predictions based on the ergodic hypothesis have agreed very well with experimental data (e.g., Berne and Pecora, 1976). Therefore, we will assume throughout this book that the scattering system in question is ergodic and, thus, Eq. (1.5.3) is applicable.

1.6 Single scattering by random particles

The simplest stochastic scattering object is a single particle undergoing random changes of position, orientation, size, and/or shape during the measurement. A good example is a solid or liquid particle trapped inside an electrostatic or optical levitator. In this case particle positions are confined to a small volume with diameter often much smaller than the distance from the volume center to the detector (Fig. 1.6.1). It is then rather straightforward to show that the detector signal accumulated over a period of time is independent of particle positions and can be described in terms of phase and extinction matrices averaged over appropriate ranges of particle orientations, sizes, and shapes. The formalism remains largely the same as in the case of far-field scattering by a fixed object.

A more difficult case is the scattering by a small random group of particles (Fig.

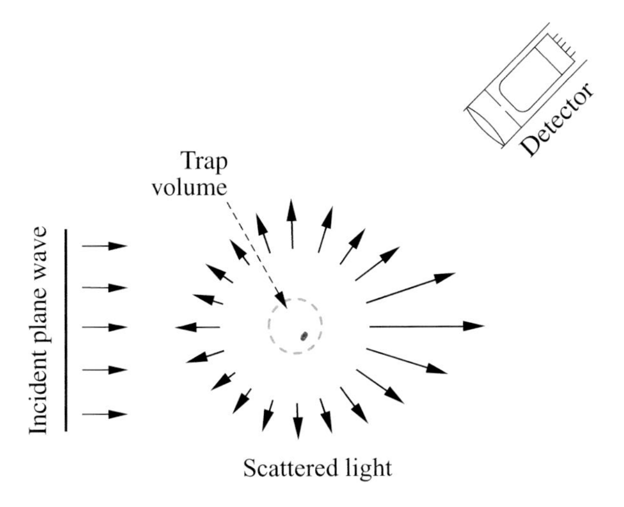

Figure 1.6.1. Scattering by a single random particle.

1.6.2). Still most of the far-field-scattering formalism can be preserved if the group is observed from a large distance and is sufficiently tenuous. Specifically, if the number of particles is sufficiently small and the separation between them is sufficiently large then one can neglect the response of each particle to the fields scattered by all other particles and assume that each particle is excited only by the external field. This is the essence of the so-called single-scattering approximation, which leads to a significant simplification of the Foldy–Lax equations. Another assumption is that particle positions are uncorrelated and sufficiently random and are independent of particle states (i.e., combinations of particle sizes, refractive indices, shapes, and orientations). One can then show that the signal accumulated by a distant detector over a period of time

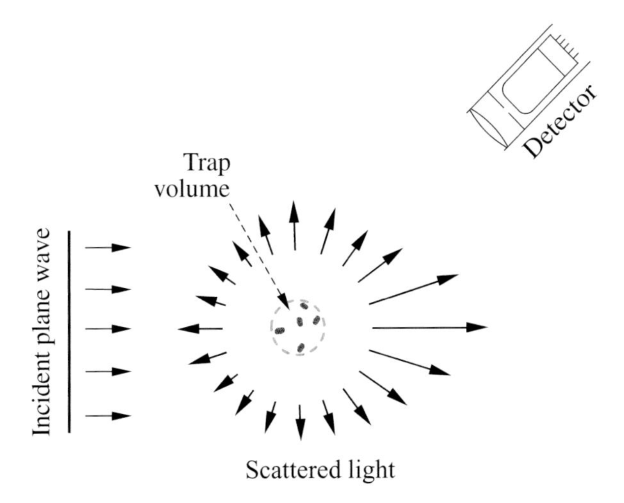

Figure 1.6.2. Scattering by a small random particle group.

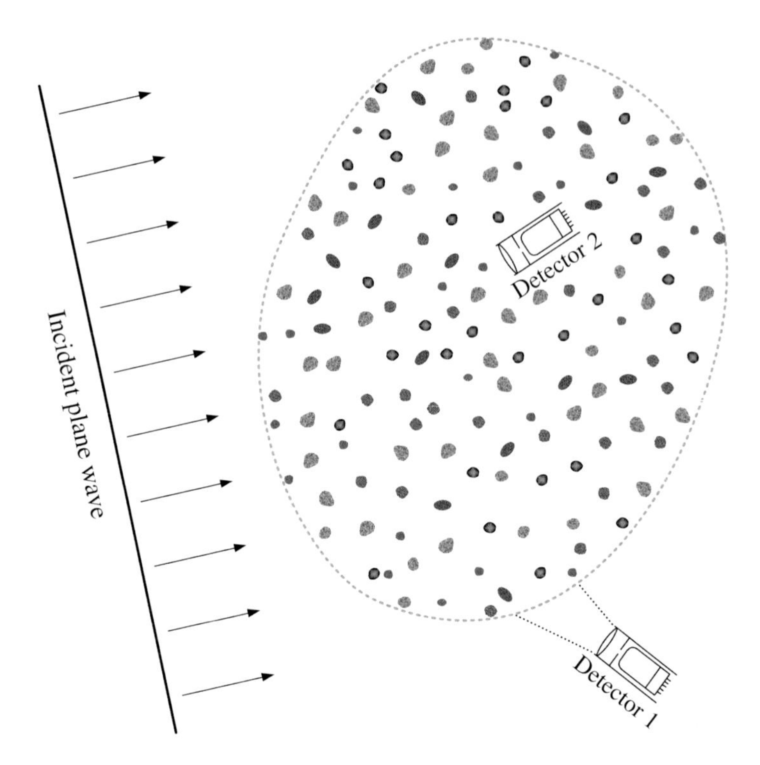

Figure 1.7.1. Scattering by a large random particle group.

can be directly described in terms of single-particle phase and extinction matrices averaged over the states (but not the positions!) of all the particles and multiplied by the number of particles.

1.7 Multiple scattering by a large random group of particles

The problem of utmost complexity is electromagnetic scattering by a very large random group of particles occupying a large volume of space (Fig. 1.7.1). The far-field-scattering formalism becomes totally inapplicable since the angular aperture of an external detector may subtend only a small fraction of the scattering volume (detector 1) or, worse, the detector may be placed inside the scattering medium (detector 2). Furthermore, the field created by a particle in response to the fields scattered by all the other particles forming the medium can be comparable to or even greater than that created in response to the incident field, which means that the single-scattering approximation is no longer valid.

To deal with this problem, one has to make several crucial assumptions. The first is to assume that each particle is located in the far-field zones of all the other particles and that the observation point is also located in the far-field zones of all the particles forming the scattering medium. This assumption leads to a dramatic simplification of the Foldy–Lax equations wherein the latter are converted from a system of volume integral equations into a system of linear algebraic equations. However, it limits the applicability of the final result by requiring that the particles in the scattering medium are not closely spaced, a condition that is nonetheless met in many natural circumstances.

The algebraic system of the far-field Foldy–Lax equations can be cast into an order-of-scattering form, in which the total electric field at a point in space is represented as a sum of contributions arising from light-scattering paths going through all possible particle sequences. The second major assumption, called the Twersky approximation, is that all paths going through the same particle more than once can be neglected. It can be demonstrated that doing this is justified provided that the total number of particles in the scattering volume is very large.

The third major assumption is that of full ergodicity, which allows one to replace averaging over time by averaging over particle positions and states.

The fourth major assumption is that (i) the position and state of each particle are statistically independent of each other and of those of all the other particles and (ii) the spatial distribution of the particles throughout the medium is random and statistically uniform. As one might expect, this assumption leads to a major simplification of all analytical derivations.

The next major step is the characterization of the multiply scattered radiation by the coherency dyadic

$$\ddot{C}(\mathbf{r}) = \langle \mathbf{E}(\mathbf{r}, t) \otimes \mathbf{E}^*(\mathbf{r}, t) \rangle_t \tag{1.7.1}$$

followed by the angular decomposition

$$\ddot{C}(\mathbf{r}) = \int_{4\pi} \mathrm{d}\hat{\mathbf{q}}\, \ddot{\Sigma}(\mathbf{r}, \hat{\mathbf{q}}) \tag{1.7.2}$$

in terms of the so-called specific coherency dyadic $\ddot{\Sigma}(\mathbf{r}, \hat{\mathbf{q}})$, where $\otimes$ denotes the dyadic product of two vectors and the integration is performed over all propagation directions as specified by the unit vector $\hat{\mathbf{q}}$. The introduction of these quantities offers three decisive benefits. First, one can sum the so-called ladder diagrams appearing in the diagrammatic representation of the coherency dyadic and show that the ladder component of the specific coherency dyadic satisfies a radiative transfer equation. Second, the ladder component of the specific coherency dyadic can be used to define the so-called specific intensity column vector which also satisfies an RTE. Third, one can use the integral form of the RTE to show that the specific intensity column vector directly describes the radiometric and polarimetric response of detectors 1 and 2 in Fig. 1.7.1 averaged over a period of time.

The fact that the specific intensity column vector can be both computed theoretically by solving the RTE and measured with a suitable optical device explains the

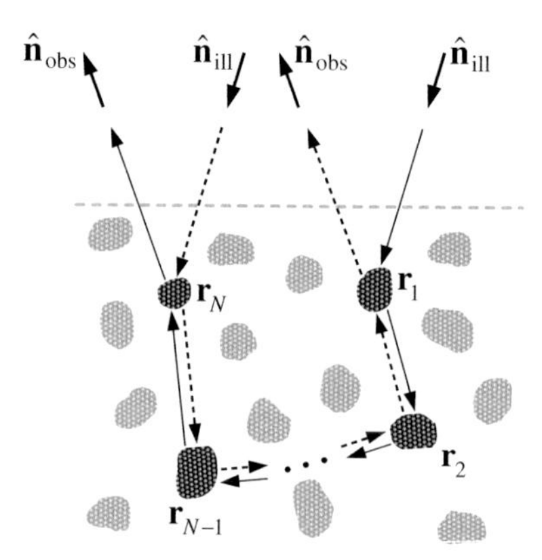

Figure 1.8.1. Schematic explanation of coherent backscattering.

practical usefulness of the radiative transfer theory in countless applications in various branches of science and engineering. Furthermore, the microphysical derivation of the RTE outlined above and described in detail in Chapter 8 gives the RTT the firm footing that it had needed for many decades in order to refute the criticism on the part of physicists (Apresyan and Kravtsov, 1996).

1.8 Coherent backscattering

Despite the restrictions of the RTT, it provides a powerful and reasonably general prescription for the treatment of the interaction of light with particulate media and is accordingly applicable to a broad range of practical situations. However, owing to some of the basic assumptions in the derivation of the RTE, there are circumstances for which it is not sufficient. An important example is the so-called coherent backscattering (CB) effect (otherwise known as weak localization of electromagnetic waves).

To trace the physical origin of this effect, let us consider a layer composed of randomly positioned particles and illuminated by a plane electromagnetic wave incident in the direction $\hat{\mathbf{n}}_{\text{ill}}$ (Fig. 1.8.1). The (infinitely) distant observer measures the intensity of light reflected by the layer in the direction $\hat{\mathbf{n}}_{\text{obs}}$. The reflected signal is composed of the contributions made by waves scattered along various paths inside the layer involving different combinations of particles. Let us consider the two conjugate scattering paths shown in Fig. 1.8.1 by solid and broken lines. These paths go through the same group of N particles, denoted by their positions $\mathbf{r}_1$, $\mathbf{r}_2$, ..., $\mathbf{r}_N$, but in opposite directions. The waves scattered along the two conjugate paths interfere, the interference being constructive or destructive depending on the phase difference

$$\varDelta = k_1(\mathbf{r}_N - \mathbf{r}_1)\cdot(\hat{\mathbf{n}}_{\text{ill}} + \hat{\mathbf{n}}_{\text{obs}}), \tag{1.8.1}$$

where k_1 is the wave number in the surrounding medium. If the observation direction

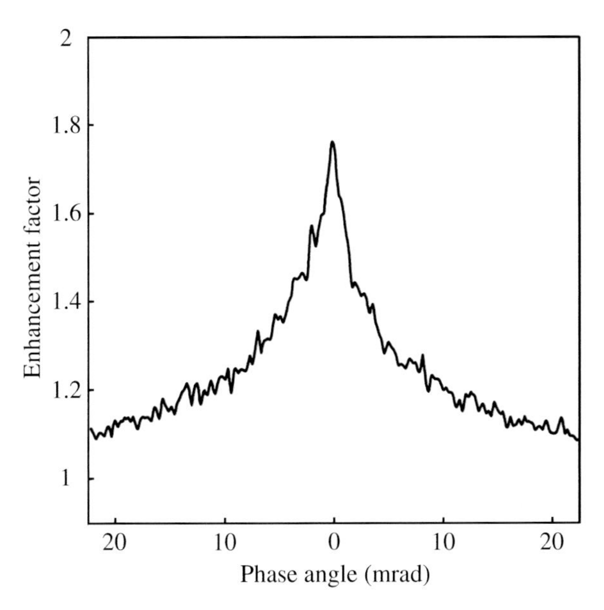

Figure 1.8.2. Angular profile of the coherent backscattering peak produced by a 1500 μm-thick slab of 9.6 vol% of 0.215 μm-diameter polystyrene spheres suspended in water. The slab was illuminated by a linearly polarized laser beam propagating normally to the slab surface. The incident wavelength was 633 nm. The scattering plane (i.e., the plane through the vectors $\hat{\mathbf{n}}_{\text{obs}}$ and $\hat{\mathbf{n}}_{\text{ill}}$) was fixed in such a way that the electric field vector of the incident beam vibrated in this plane. The detector measured the component of the backscattered intensity polarized parallel to the scattering plane. The curve shows the profile of the backscattered intensity normalized by the intensity of the incoherent background as a function of the phase angle. The latter is defined as the angle between the vectors $\hat{\mathbf{n}}_{\text{obs}}$ and $-\hat{\mathbf{n}}_{\text{ill}}$. (After van Albada *et al.*, 1987.)

is far from the exact backscattering direction given by $-\hat{\mathbf{n}}_{\text{ill}}$, then the waves scattered along conjugate paths involving different groups of particles interfere in different ways, and the average effect of the interference is zero owing to the randomness of particle positions. Consequently, the observer measures some average, incoherent intensity that is well described by the RTE. However, at exactly the backscattering direction ($\hat{\mathbf{n}}_{\text{obs}} = -\hat{\mathbf{n}}_{\text{ill}}$), the phase difference between conjugate paths involving any group of particles is identically equal to zero, Eq. (1.8.1), and the interference is always constructive, thereby resulting in a coherent intensity peak superposed on the incoherent background (Fig. 1.8.2).

The failure of the RTE to reproduce the CB peak is explained by the fact that of all kinds of diagrams in the diagrammatic representation of the coherency dyadic it keeps only the ladder diagrams, whereas CB is caused by so-called cyclical (or maximally crossed) diagrams. The inclusion of the cyclical diagrams makes the computation of the coherency dyadic much more involved and limits the range of problems that can be solved accurately. However, the reciprocal nature of each single-scattering event leads to an interesting exact result: the characteristics of the CB effect at the exact backscattering direction can be rigorously expressed in terms of the solution of the RTE. This result will be discussed in detail in Chapter 14.

The ladder and cyclical diagrams are the dominant but not the only types of diagrams in the diagrammatic representation of the coherency dyadic. However, to include all the other diagrams in the calculation of multiply scattered radiation is a formidable task that goes beyond the scope of this book.

1.9 Classification of electromagnetic scattering problems

To develop a comprehensive and universal classification of electromagnetic scattering problems borders on being impossible. This chapter provides only an outline tailored to the specifics of radiative transfer and coherent backscattering, whereas those working on another aspect of electromagnetic scattering might prefer a modified classification with somewhat different emphases. We hope, however, that our outline, summarized graphically in Fig. 1.9.1, fulfills its limited objective and explains adequately the place of the RTT and CB within the broader context of classical macroscopic electromagnetics.

As is obvious from the diagram in Fig. 1.9.1, there are two broad classes of problems that we have not touched upon so far and which, in fact, will not be discussed specifically in this book. We have emphasized several times that the main theme of this book is multiple scattering by randomly positioned discrete particles with refractive index distinctly different from that of the surrounding medium. However, one can also consider multiple scattering in continuous media with random fluctuations of the refractive index. This class of problems requires special solution approaches that are beyond the scope of this book. The reader can find relevant information in the monographs by Fabelinskii (1968), Crosignani *et al.* (1975), Kuz'min *et al.* (1994), Apresyan and Kravtsov (1996), and Tsang and Kong (2001) as well as in the recent reviews by van Tiggelen and Stark (2000) and Klyatskin (2004).

Another important problem is electromagnetic scattering by an infinite random rough surface separating two half-spaces with different refractive indices. Although some rough surfaces, such as the ocean surface, indeed change randomly in time, many rough interfaces do not change and are deterministic rather than random. However, quite often their position relative to the source of light and/or the detector is not fixed during the measurement and their vertical profile is described by a highly irregular function of lateral coordinates. Even minute displacements of the source of light and/or the detector change phase differences entirely, thereby destroying the speckle pattern. Furthermore, the detector may view different parts of the surface at different moments in time, thereby in effect recording an average over a temporally varying surface profile. These two factors make the concept of a random rough surface a good model for describing the results of many actual static measurements. Detailed information on this subject can be found in the books by Fung (1994), Tsang

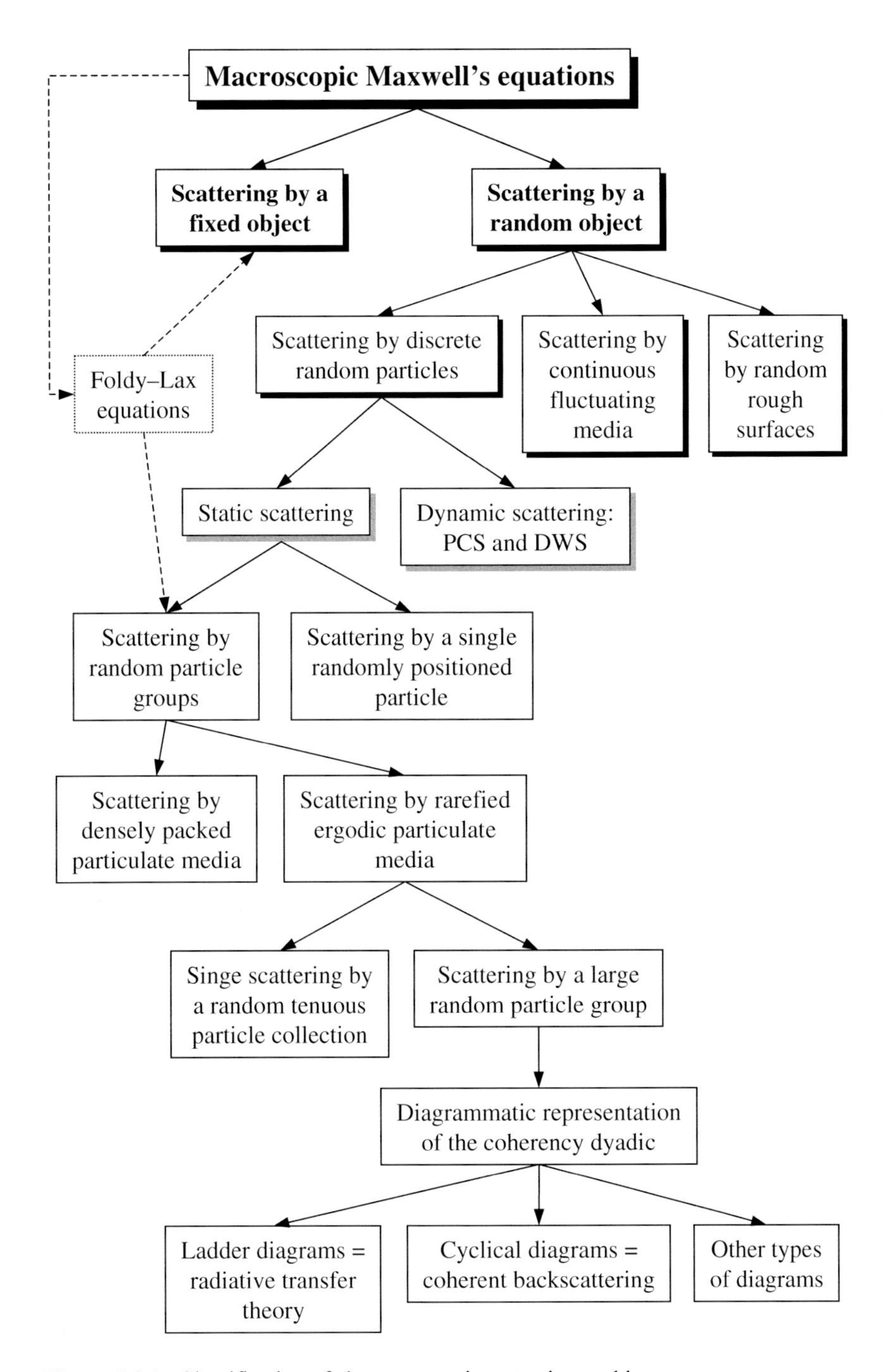

Figure 1.9.1. Classification of electromagnetic scattering problems.

and Kong (2001), and Tsang *et al.* (2001) as well as in the recent reviews by Saillard and Sentenac (2001), Elfouhaily and Guérin (2004), and Shchegrov *et al.* (2004) and in the numerous publications cited therein.

One can also think of more complex problems involving different types of volume and/or surface scattering. A good example is electromagnetic scattering by a layer of continuous fluctuating medium comprising randomly positioned discrete particles and bounded by random rough surfaces. Although problems like this one are important in practice and have been treated using various phenomenological approaches, microphysical treatments based on consistent application of the Maxwell equations have been extremely scarce.

1.10 Notes and further reading

A useful analytical modification of the Foldy–Lax approach to split the total field scattered by a cluster into partial fields scattered by the individual components is the so-called superposition *T*-matrix method. This technique, pioneered by Bruning and Lo (1971a,b) and Peterson and Ström (1973), is based on expanding the fields entering the Foldy–Lax equations in appropriate sets of vector spherical wave functions (see, e.g., Section 5.9 of MTL and Borghese *et al.*, 2003). The fundamentals of the superposition *T*-matrix method and its applications to various multiple-scattering problems are described in Part 2 of Varadan and Varadan (1980), Chapter 6 of Tsang and Kong (2001), and Section 10.4 of Tsang *et al.* (2001). This method becomes especially useful when one considers multiple scattering by densely packed media in which particles are not located in each other's far-field zones. A detailed list of relevant publications can be found in the database compiled by Mishchenko *et al.* (2004a). This database also cites publications in which the *T*-matrix method has been applied to electromagnetic scattering by configurations involving a particle and an infinite interface, either plane or rough.

The optics of laser speckles is discussed in the book edited by Dainty (1984). Diffusing wave spectroscopy was pioneered by Maret and Wolf (1987) and Pine *et al.* (1988). Detailed information about PCS and DWS and their diverse applications can be found in the books by Cummins and Pike (1974, 1977), Crosignani *et al.* (1975), Pecora (1985), Schmitz (1990), Chu (1991), Brown (1993, 1996), Pike and Abbiss (1997), Sebbah (2001), Tuchin (2002, 2004), Albrecht *et al.* (2003), and van Tiggelen and Skipetrov (2003) as well as in the recent feature issues of *Applied Optics* edited by Meyer *et al.* (1997, 2001). An overview of the polarization-sensitive speckle spectroscopy can be found in Zimnyakov *et al.* (2004). The particle-shape sensitivity of polarization PCS measurements has been studied by Pitter *et al.* (1999), Jakeman (2000), Smith *et al.* (2001), Kusmartseva and Smith (2002), and Chang *et al.* (2002).

A review of inelastic scattering processes and their applications to optical particle characterization can be found in Chapter 8 of Davis and Schweiger (2002). A good number of particle characterization techniques are based on measurements of frequency shifts due to the Doppler effect caused by nonzero particle velocities. Since the scattering of the laser beam in the particle reference frame is still elastic and since the frequency changes measured in the laboratory reference frame are usually minute, the underlying phenomenon is often referred to as quasi-elastic light scattering. A comprehensive overview of various laser Doppler methods (LDMs) for measuring particle velocities, sizes, and concentrations was published recently by Albrecht *et al.* (2003).

Unlike the PCS, DWS, RTT, CB, and LDMs, which deal with groups of randomly moving particles, the technique of optical coherence tomography (OCT) was specifically designed as a means of noninvasive optical characterization of stationary objects with complex morphology such as biological tissues (see, e.g., the reviews by Schmitt (1999) and Fercher *et al.* (2003) as well as the recent books edited by Bouma and Tearney (2002) and Tuchin (2004)). This technique uses a spectrally broadband source of light and a two-beam Michelson interferometer with the mirror in one arm replaced by a turbid sample. By measuring the interference between the beams backscattered from the sample and from the moving reference mirror, one can measure the depth and magnitude of optical scattering within the sample with micrometer-scale precision (limited by the coherence length of the source). Scanning the light beam across the sample produces a two-dimensional representation of the optical backscattering of the sample's cross section, which is often displayed as a gray-scale or false-color image. Polarization-sensitive OCT uses the information contained in the polarization state of the recorded interference fringe intensity to provide additional contrast in the sample cross-sectional images (Schmitt and Xiang, 1998; de Boer and Milner, 2002).

The ergodic hypothesis was introduced by James Clerk Maxwell and Ludwig Boltzmann as a basic underlying principle of statistical mechanics. The details of the ergodic theory, its relation to the famous Poincaré recurrence theorem (Poincaré, 1890), and its applications to statistical mechanics and kinetic theory are described by Khinchin (1949), Uhlenbeck and Ford (1963), and Farquhar (1964). Interesting discussions of the ergodic hypothesis and specific examples of nonergodic scattering media can be found in Pusey and van Megen (1989), Joosten *et al.* (1990), Xue *et al.* (1992), Nisato *et al.* (2000), and Scheffold *et al.* (2001).

Chapter 2

Maxwell equations, electromagnetic waves, and Stokes parameters

The theoretical basis for describing single and multiple scattering of light by particles is formed by classical electromagnetics. In order to make the book sufficiently self-contained, this chapter provides a summary of those concepts and equations of electromagnetic theory that will be used extensively in later chapters and introduces the necessary notation. We start by formulating the macroscopic Maxwell equations and constitutive relations and discuss the fundamental time-harmonic plane-wave solution that underlies the basic optical idea of a monochromatic parallel beam of light. This is followed by the introduction of the Stokes parameters and a discussion of their ellipsometric content. Then we consider the concept of a quasi-monochromatic beam of light and its implications and briefly discuss how the Stokes parameters of monochromatic and quasi-monochromatic light can be measured in practice. In the final two sections, we discuss another fundamental solution of the Maxwell equations in the form of time-harmonic outgoing and incoming spherical waves and introduce the concept of the coherency dyad of the electric field. The latter plays a vital role in the theory of single and multiple light scattering by random particle ensembles.

2.1 Maxwell equations and constitutive relations

The theory of classical optics phenomena is based on the set of four Maxwell equations for the macroscopic electromagnetic field at interior points in matter, which in SI units read:

$$\nabla \cdot \boldsymbol{\mathcal{D}}(\mathbf{r}, t) = \rho(\mathbf{r}, t), \tag{2.1.1}$$

$$\nabla \times \mathcal{E}(\mathbf{r}, t) = -\frac{\partial \mathcal{B}(\mathbf{r}, t)}{\partial t}, \tag{2.1.2}$$

$$\nabla \cdot \mathcal{B}(\mathbf{r}, t) = 0, \tag{2.1.3}$$

$$\nabla \times \mathcal{H}(\mathbf{r}, t) = \mathcal{J}(\mathbf{r}, t) + \frac{\partial \mathcal{D}(\mathbf{r}, t)}{\partial t}, \tag{2.1.4}$$

where $\mathcal{E}$ is the electric and $\mathcal{H}$ the magnetic field, $\mathcal{B}$ the magnetic induction, $\mathcal{D}$ the electric displacement, and ρ and $\mathcal{J}$ the macroscopic (free) charge density and current density, respectively. All quantities entering Eqs. (2.1.1)–(2.1.4) are functions of time, t, and spatial coordinates, $\mathbf{r}$. Implicit in the Maxwell equations is the continuity equation

$$\frac{\partial \rho(\mathbf{r}, t)}{\partial t} + \nabla \cdot \mathcal{J}(\mathbf{r}, t) = 0, \tag{2.1.5}$$

which is obtained by combining the time derivative of Eq. (2.1.1) with the divergence of Eq. (2.1.4) and taking into account the vector identity $\nabla \cdot (\nabla \times \mathbf{a}) = 0$. The vector fields entering Eqs. (2.1.1)–(2.1.4) are related by

$$\mathcal{D}(\mathbf{r}, t) = \epsilon_0 \mathcal{E}(\mathbf{r}, t) + \mathcal{P}(\mathbf{r}, t), \tag{2.1.6}$$

$$\mathcal{H}(\mathbf{r}, t) = \frac{1}{\mu_0} \mathcal{B}(\mathbf{r}, t) - \mathcal{M}(\mathbf{r}, t), \tag{2.1.7}$$

where $\mathcal{P}$ is the electric polarization (average electric dipole moment per unit volume), $\mathcal{M}$ is the magnetization (average magnetic dipole moment per unit volume), and ϵ_0 and μ_0 are the electric permittivity and the magnetic permeability of free space, respectively.

Equations (2.1.1)–(2.1.7) are insufficient for a unique determination of the electric and magnetic fields from a given distribution of charges and currents and must be supplemented with so-called constitutive relations:

$$\mathcal{P}(\mathbf{r}, t) = \epsilon_0 \chi(\mathbf{r}) \mathcal{E}(\mathbf{r}, t), \tag{2.1.8}$$

$$\mathcal{B}(\mathbf{r}, t) = \mu(\mathbf{r}) \mathcal{H}(\mathbf{r}, t), \tag{2.1.9}$$

$$\mathcal{J}(\mathbf{r}, t) = \sigma(\mathbf{r}) \mathcal{E}(\mathbf{r}, t), \tag{2.1.10}$$

where χ is the electric susceptibility, μ the magnetic permeability, and σ the conductivity. Equations (2.1.6) and (2.1.8) yield

$$\mathcal{D}(\mathbf{r}, t) = \epsilon(\mathbf{r}) \mathcal{E}(\mathbf{r}, t), \tag{2.1.11}$$

where

$$\epsilon(\mathbf{r}) = \epsilon_0 [1 + \chi(\mathbf{r})] \tag{2.1.12}$$

is the electric permittivity. For linear and isotropic media, χ, μ, σ, and ϵ are scalars independent of the fields. The microphysical derivation and the range of validity of the macroscopic Maxwell equations are discussed in detail by Jackson (1998).

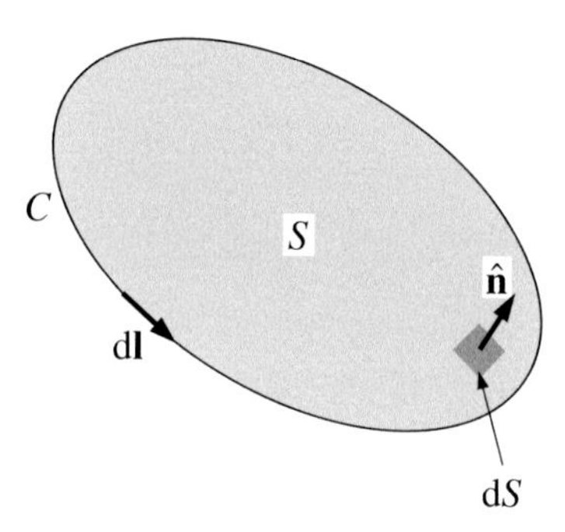

Figure 2.1.1. A finite surface S bounded by a closed contour C.

The constitutive relations (2.1.9)–(2.1.11) connect the field vectors at the same moment of time t and are valid for electromagnetic fields in a vacuum and also for electromagnetic fields in macroscopic material media provided that the fields are constant or vary in time rather slowly. For a rapidly varying field in a material medium, the state of the medium depends not only on the current value of the field but also on the values of the field at all previous times. Therefore, for a linear, time-invariant medium, the constitutive relations (2.1.9)–(2.1.11) must be replaced by the following general causal relations that take into account the effect of the prior history on the electromagnetic properties of the medium:

$$\mathcal{D}(\mathbf{r},t) = \int_{-\infty}^{t} \mathrm{d}t' \tilde{\epsilon}(\mathbf{r},t-t')\mathcal{E}(\mathbf{r},t'), \tag{2.1.13}$$

$$\mathcal{B}(\mathbf{r},t) = \int_{-\infty}^{t} \mathrm{d}t' \tilde{\mu}(\mathbf{r},t-t')\mathcal{H}(\mathbf{r},t'), \tag{2.1.14}$$

$$\mathcal{J}(\mathbf{r},t) = \int_{-\infty}^{t} \mathrm{d}t' \tilde{\sigma}(\mathbf{r},t-t')\mathcal{E}(\mathbf{r},t'). \tag{2.1.15}$$

The medium characterized by the constitutive relations (2.1.13)–(2.1.15) is called time-dispersive.

It is straightforward to rewrite the Maxwell equations and the continuity equation in an integral form. Specifically, integrating Eqs. (2.1.2) and (2.1.4) over a surface S bounded by a closed contour C (see Fig. 2.1.1) and applying the Stokes theorem,

$$\int_S \mathrm{d}S\,(\nabla\times\mathbf{A})\cdot\hat{\mathbf{n}} = \oint_C \mathrm{d}\mathbf{l}\cdot\mathbf{A}, \tag{2.1.16}$$

yields

$$\oint_C \mathrm{d}\mathbf{l}\cdot\mathcal{E} = -\frac{\partial}{\partial t}\int_S \mathrm{d}S\,\mathcal{B}\cdot\hat{\mathbf{n}}, \tag{2.1.17}$$

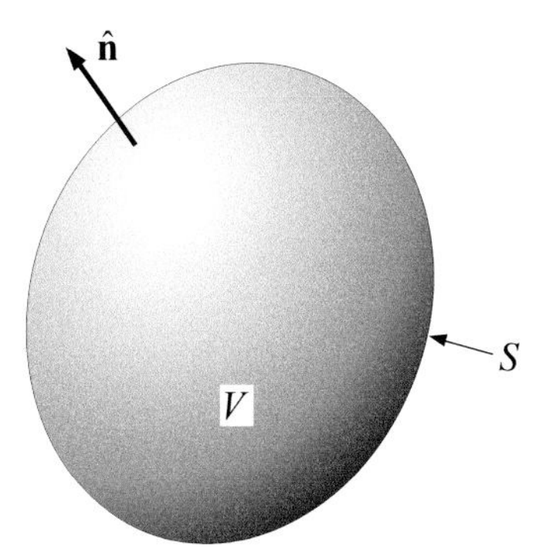

Figure 2.1.2. A finite volume V bounded by a closed surface S.

$$\oint_C \mathrm{d}\mathbf{l} \cdot \boldsymbol{\mathcal{H}} = \int_S \mathrm{d}S\, \boldsymbol{\mathcal{J}} \cdot \hat{\mathbf{n}} + \frac{\partial}{\partial t} \int_S \mathrm{d}S\, \boldsymbol{\mathcal{D}} \cdot \hat{\mathbf{n}}, \tag{2.1.18}$$

where we employ the usual convention that the direction of the differential length vector $\mathbf{dl}$ is related to the direction of the unit vector along the local normal to the surface $\hat{\mathbf{n}}$ according to the right-hand rule.

Similarly, integrating Eqs. (2.1.1), (2.1.3), and (2.1.5) over a finite volume V bounded by a closed surface S (see Fig. 2.1.2) and using the Gauss theorem,

$$\int_V \mathrm{d}\mathbf{r}\, \nabla \cdot \mathbf{A} = \oint_S \mathrm{d}S\, \mathbf{A} \cdot \hat{\mathbf{n}}, \tag{2.1.19}$$

we derive

$$\oint_S \mathrm{d}S\, \boldsymbol{\mathcal{D}} \cdot \hat{\mathbf{n}} = \int_V \mathrm{d}\mathbf{r}\, \rho, \tag{2.1.20}$$

$$\oint_S \mathrm{d}S\, \boldsymbol{\mathcal{B}} \cdot \hat{\mathbf{n}} = 0, \tag{2.1.21}$$

$$\oint_S \mathrm{d}S\, \boldsymbol{\mathcal{J}} \cdot \hat{\mathbf{n}} = -\frac{\partial}{\partial t} \int_V \mathrm{d}\mathbf{r}\, \rho, \tag{2.1.22}$$

where the unit vector $\hat{\mathbf{n}}$ is directed along the outward local normal to the surface.

2.2 Boundary conditions

The Maxwell equations are strictly valid only for points in whose neighborhood the physical properties of the medium, as characterized by the constitutive parameters χ, μ, and σ, vary continuously. However, across an interface separating one medium from another the constitutive parameters may change abruptly, and one may expect

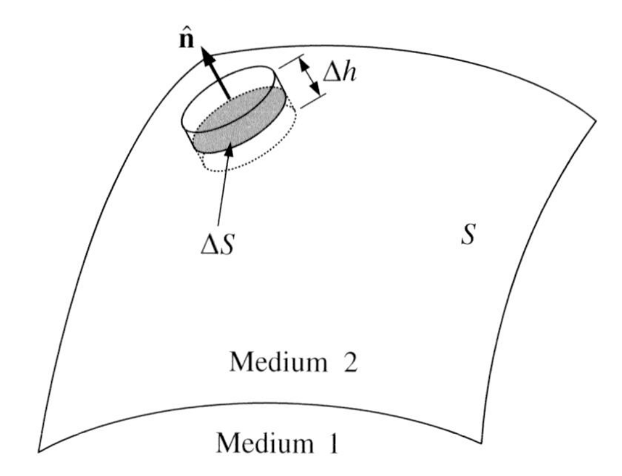

Figure 2.2.1. Pillbox used in the derivation of the boundary conditions for $\mathcal{B}$ and $\mathcal{D}$.

similar discontinuous behavior of the field vectors $\mathcal{E}$, $\mathcal{D}$, $\mathcal{H}$, and $\mathcal{B}$. The boundary conditions at such an interface can be derived from the integral form of the Maxwell equations as described below.

Consider two different continuous media separated by an interface S as shown in Fig. 2.2.1. Let $\hat{\mathbf{n}}$ be a unit vector along the local normal to the interface, pointing from medium 1 toward medium 2. Let us take the integral in Eq. (2.1.21) over the closed surface of a small cylinder with bases parallel to a small surface element ΔS such that half of the cylinder is in medium 1 and half in medium 2. The contribution from the curved surface of the cylinder vanishes in the limit $\Delta h \to 0$, and we thus obtain

$$(\mathcal{B}_2 - \mathcal{B}_1) \cdot \hat{\mathbf{n}} = 0, \tag{2.2.1}$$

which means that the normal component of the magnetic induction is continuous across the interface.

Similarly, evaluating the integrals on the left- and right-hand sides of Eq. (2.1.20) over the surface and volume of the cylinder, respectively, we derive

$$(\mathcal{D}_2 - \mathcal{D}_1) \cdot \hat{\mathbf{n}} = \lim_{\Delta h \to 0} \Delta h \rho = \rho_S, \tag{2.2.2}$$

where ρ_S is the surface charge density (the charge per unit area) measured in coulombs per square meter. Thus, there is a discontinuity in the normal component of $\mathcal{D}$ if the interface carries a layer of surface charge density.

Let us now consider a small rectangular loop of area ΔA formed by sides of length Δl perpendicular to the local normal and ends of length Δh parallel to the local normal, as shown in Fig. 2.2.2. The surface integral on the right-hand side of Eq. (2.1.17) vanishes in the limit $\Delta h \to 0$,

$$\lim_{\Delta h \to 0} \int_{\Delta A} \mathrm{d}S\, \mathcal{B} \cdot (\hat{\mathbf{n}} \times \hat{\mathbf{l}}) = \lim_{\Delta h \to 0} \Delta l \Delta h \mathcal{B} \cdot (\hat{\mathbf{n}} \times \hat{\mathbf{l}}) = 0,$$

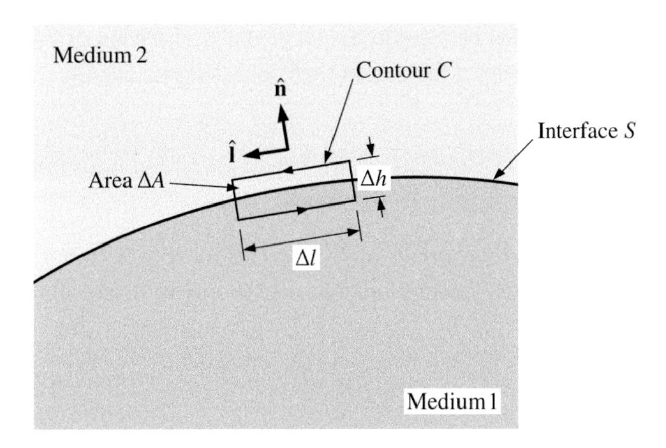

Figure 2.2.2. Rectangular loop used in the derivation of the boundary conditions for $\mathcal{E}$ and $\mathcal{H}$.

so that

$$\hat{\mathbf{l}} \cdot (\mathcal{E}_2 - \mathcal{E}_1) = 0. \tag{2.2.3}$$

Since the orientation of the rectangle – and hence also of $\hat{\mathbf{l}}$ – is arbitrary, Eq. (2.2.3) means that the vector $\mathcal{E}_2 - \mathcal{E}_1$ must be perpendicular to the interface. Thus,

$$\hat{\mathbf{n}} \times (\mathcal{E}_2 - \mathcal{E}_1) = \mathbf{0}, \tag{2.2.4}$$

where $\mathbf{0}$ is a zero vector. This implies that the tangential component of $\mathcal{E}$ is continuous across the interface.

Similarly, Eq. (2.1.18) yields

$$\hat{\mathbf{l}} \cdot (\mathcal{H}_2 - \mathcal{H}_1) = \lim_{\Delta h \to 0} \Delta h (\hat{\mathbf{n}} \times \hat{\mathbf{l}}) \cdot \mathcal{J} = (\hat{\mathbf{n}} \times \hat{\mathbf{l}}) \cdot \mathcal{J}_S, \tag{2.2.5}$$

where $\mathcal{J}_S$ is the surface current density measured in amperes per meter. Since

$$\hat{\mathbf{l}} = (\hat{\mathbf{n}} \times \hat{\mathbf{l}}) \times \hat{\mathbf{n}}, \tag{2.2.6}$$

we can use the vector identity

$$(\mathbf{a} \times \mathbf{b}) \cdot \mathbf{c} = \mathbf{a} \cdot (\mathbf{b} \times \mathbf{c}) \tag{2.2.7}$$

to derive

$$[(\hat{\mathbf{n}} \times \hat{\mathbf{l}}) \times \hat{\mathbf{n}}] \cdot (\mathcal{H}_2 - \mathcal{H}_1) = (\hat{\mathbf{n}} \times \hat{\mathbf{l}}) \cdot [\hat{\mathbf{n}} \times (\mathcal{H}_2 - \mathcal{H}_1)] = (\hat{\mathbf{n}} \times \hat{\mathbf{l}}) \cdot \mathcal{J}_S. \tag{2.2.8}$$

Taking into account that this equality must be valid for any orientation of the rectangle and, thus, of the tangent unit vector $\hat{\mathbf{l}}$ finally yields

$$\hat{\mathbf{n}} \times (\mathcal{H}_2 - \mathcal{H}_1) = \mathcal{J}_S, \tag{2.2.9}$$

which means that there is a discontinuity in the tangential component of $\mathcal{H}$ if the interface can carry a surface current. Media with finite conductivity cannot support surface currents, so that

$$\hat{\mathbf{n}} \times (\mathcal{H}_2 - \mathcal{H}_1) = \mathbf{0} \quad \text{(finite conductivity).} \tag{2.2.10}$$

The boundary conditions (2.2.1), (2.2.2), (2.2.4), (2.2.9), and (2.2.10) are useful in solving the differential Maxwell equations in different adjacent regions with continuous physical properties and then linking the partial solutions to determine the fields throughout all space.

2.3 Time-harmonic fields

Let us now assume that all fields and sources are time harmonic (or monochromatic), which means that their time dependence can be fully described by expressing them as sums of terms proportional to either $\cos\omega t$ or $\sin\omega t$, where ω is the angular frequency. It is standard practice to represent real monochromatic fields as real parts of the respective complex fields, e.g.,

$$\begin{aligned} \mathcal{E}(\mathbf{r},t) &= \operatorname{Re}\mathbf{E}(\mathbf{r},t) = \operatorname{Re}[\mathbf{E}(\mathbf{r})\exp(-\mathrm{i}\omega t)] \\ &= \tfrac{1}{2}[\mathbf{E}(\mathbf{r})\exp(-\mathrm{i}\omega t) + \mathbf{E}^*(\mathbf{r})\exp(\mathrm{i}\omega t)] \end{aligned} \tag{2.3.1}$$

and analogously for $\mathcal{D}$, $\mathcal{H}$, $\mathcal{B}$, $\mathcal{J}$, ρ, $\mathcal{P}$, and $\mathcal{M}$, where $\mathbf{E}(\mathbf{r})$ is complex, and the asterisk denotes a complex-conjugate value.[1] Equations (2.1.1)–(2.1.5) then yield the following frequency-domain Maxwell equations and continuity equation for the time-independent components of the complex fields:

$$\nabla \cdot \mathbf{D}(\mathbf{r}) = \mathrm{\rho}(\mathbf{r}), \tag{2.3.2}$$

$$\nabla \times \mathbf{E}(\mathbf{r}) = \mathrm{i}\omega\mathbf{B}(\mathbf{r}), \tag{2.3.3}$$

$$\nabla \cdot \mathbf{B}(\mathbf{r}) = 0, \tag{2.3.4}$$

$$\nabla \times \mathbf{H}(\mathbf{r}) = \mathbf{J}(\mathbf{r}) - \mathrm{i}\omega\mathbf{D}(\mathbf{r}), \tag{2.3.5}$$

$$-\mathrm{i}\omega\mathrm{\rho}(\mathbf{r}) + \nabla \cdot \mathbf{J}(\mathbf{r}) = 0, \tag{2.3.6}$$

where we emphasize the typographical distinction between the symbols for the real quantities $\mathcal{E}$, $\mathcal{D}$, $\mathcal{H}$, $\mathcal{B}$, $\mathcal{J}$, and ρ and for their complex counterparts **E**, **D**, **H**, **B**, **J**, and $\mathrm{\rho}$.

[1] A complex vector is formally defined as $\mathbf{a} = \mathbf{b} + \mathrm{i}\mathbf{c}$, where **b** and **c** are usual real vectors. All operations with complex vectors are defined in a way analogous to the definition of operations with complex numbers and real vectors. For example, the complex conjugate $\mathbf{a}^*$ is defined as the complex vector $\mathbf{b} - \mathrm{i}\mathbf{c}$, the scalar product of two complex vectors $\mathbf{a} = \mathbf{b} + \mathrm{i}\mathbf{c}$ and $\mathbf{d} = \mathbf{e} + \mathrm{i}\mathbf{f}$ with real **b**, **c**, **e**, and **f** is defined as $\mathbf{a} \cdot \mathbf{d} = \mathbf{b} \cdot \mathbf{e} - \mathbf{c} \cdot \mathbf{f} + \mathrm{i}(\mathbf{b} \cdot \mathbf{f} + \mathbf{c} \cdot \mathbf{e})$, etc.

The constitutive relations remain unchanged in the frequency domain for a non-dispersive medium:

$$\mathbf{D}(\mathbf{r}) = \epsilon(\mathbf{r})\mathbf{E}(\mathbf{r}), \tag{2.3.7}$$

$$\mathbf{B}(\mathbf{r}) = \mu(\mathbf{r})\mathbf{H}(\mathbf{r}), \tag{2.3.8}$$

$$\mathbf{J}(\mathbf{r}) = \sigma(\mathbf{r})\mathbf{E}(\mathbf{r}). \tag{2.3.9}$$

For a time-dispersive medium, we can substitute the monochromatic fields of the form (2.3.1) into Eqs. (2.1.13)–(2.1.15), which yields

$$\mathbf{D}(\mathbf{r}) = \mathrm{\epsilon}(\mathbf{r},\omega)\mathbf{E}(\mathbf{r}), \tag{2.3.10}$$

$$\mathbf{B}(\mathbf{r}) = \mathrm{\mu}(\mathbf{r},\omega)\mathbf{H}(\mathbf{r}), \tag{2.3.11}$$

$$\mathbf{J}(\mathbf{r}) = \mathrm{\sigma}(\mathbf{r},\omega)\mathbf{E}(\mathbf{r}), \tag{2.3.12}$$

where

$$\mathrm{\epsilon}(\mathbf{r},\omega) = \int_0^\infty \mathrm{d}t\, \tilde{\epsilon}(\mathbf{r},t)\exp(\mathrm{i}\omega t), \tag{2.3.13}$$

$$\mathrm{\mu}(\mathbf{r},\omega) = \int_0^\infty \mathrm{d}t\, \tilde{\mu}(\mathbf{r},t)\exp(\mathrm{i}\omega t), \tag{2.3.14}$$

$$\mathrm{\sigma}(\mathbf{r},\omega) = \int_0^\infty \mathrm{d}t\, \tilde{\sigma}(\mathbf{r},t)\exp(\mathrm{i}\omega t) \tag{2.3.15}$$

are complex functions of the angular frequency. Note that we use sloping Greek letters in Eqs. (2.3.7)–(2.3.9) and upright Greek letters in Eqs. (2.3.10)–(2.3.12) to differentiate between the frequency-independent and the frequency-dependent constitutive parameters, respectively. Equations (2.3.2) and (2.3.5) can be rewritten in the form

$$\nabla \cdot [\varepsilon(\mathbf{r},\omega)\mathbf{E}(\mathbf{r})] = 0, \tag{2.3.16}$$

$$\nabla \times \mathbf{H}(\mathbf{r}) = -\mathrm{i}\omega\varepsilon(\mathbf{r},\omega)\mathbf{E}(\mathbf{r}), \tag{2.3.17}$$

where

$$\varepsilon(\mathbf{r},\omega) = \mathrm{\epsilon}(\mathbf{r},\omega) + \mathrm{i}\frac{\mathrm{\sigma}(\mathbf{r},\omega)}{\omega} \tag{2.3.18}$$

is the so-called complex permittivity. Again, the reader should note the typographical distinction between the frequency-dependent electric permittivity $\mathrm{\epsilon}$ (which can, in principle, be complex-valued for a dispersive medium) and the complex permittivity ε. We will show later that a direct consequence of a complex-valued ε and/or $\mathrm{\mu}$ is a non-zero imaginary part of the refractive index (Eq. (2.5.19)), which causes absorption of electromagnetic energy (Eq. (2.5.20)) by converting it into other forms of energy, e.g., heat.

It is straightforward to verify that the frequency-domain form of the integral counterparts of the Maxwell equations (2.1.17), (2.1.18), (2.1.20), and (2.1.21) is as follows:

$$\oint_C \mathrm{d}\mathbf{l} \cdot \mathbf{E}(\mathbf{r}) = \mathrm{i}\omega \int_S \mathrm{d}S\, \mu(\mathbf{r},\omega)\mathbf{H}(\mathbf{r}) \cdot \hat{\mathbf{n}}, \tag{2.3.19}$$

$$\oint_C \mathrm{d}\mathbf{l} \cdot \mathbf{H}(\mathbf{r}) = -\mathrm{i}\omega \int_S \mathrm{d}S\, \varepsilon(\mathbf{r},\omega)\mathbf{E}(\mathbf{r}) \cdot \hat{\mathbf{n}}, \tag{2.3.20}$$

$$\oint_S \mathrm{d}S\, \varepsilon(\mathbf{r},\omega)\mathbf{E}(\mathbf{r}) \cdot \hat{\mathbf{n}} = 0, \tag{2.3.21}$$

$$\oint_S \mathrm{d}S\, \mu(\mathbf{r},\omega)\mathbf{H}(\mathbf{r}) \cdot \hat{\mathbf{n}} = 0. \tag{2.3.22}$$

The linearity of these equations with respect to $\mathbf{E}(\mathbf{r})$ and $\mathbf{H}(\mathbf{r})$ leads to the fundamental *principle of superposition*: if the electromagnetic fields $[\mathbf{E}_1(\mathbf{r}), \mathbf{H}_1(\mathbf{r})]$ and $[\mathbf{E}_2(\mathbf{r}), \mathbf{H}_2(\mathbf{r})]$ are solutions of the Maxwell equations, then the electromagnetic field $[\mathbf{E}_1(\mathbf{r}) + \mathbf{E}_2(\mathbf{r}), \mathbf{H}_1(\mathbf{r}) + \mathbf{H}_2(\mathbf{r})]$ is also a solution.

Neither the scalar nor the vector product of two real vector fields is equal to the real part of the respective product of the corresponding complex vector fields. Instead,

$$\begin{aligned}\mathcal{C}(\mathbf{r},t) &= \mathcal{A}(\mathbf{r},t)\cdot\mathcal{G}(\mathbf{r},t)\\ &= \tfrac{1}{4}[\mathbf{A}(\mathbf{r})\exp(-\mathrm{i}\omega t) + \mathbf{A}^*(\mathbf{r})\exp(\mathrm{i}\omega t)]\\ &\quad\cdot[\mathbf{G}(\mathbf{r})\exp(-\mathrm{i}\omega t) + \mathbf{G}^*(\mathbf{r})\exp(\mathrm{i}\omega t)]\\ &= \tfrac{1}{2}\mathrm{Re}[\mathbf{A}(\mathbf{r})\cdot\mathbf{G}^*(\mathbf{r}) + \mathbf{A}(\mathbf{r})\cdot\mathbf{G}(\mathbf{r})\exp(-2\mathrm{i}\omega t)],\end{aligned} \tag{2.3.23}$$

and similarly for a vector product. Usually the angular frequency ω is so high that traditional optical measuring devices are not capable of following the rapid oscillations of the instantaneous product values but rather respond to a time average

$$\langle\mathcal{C}(\mathbf{r},t)\rangle_t = \frac{1}{T}\int_t^{t+T} \mathrm{d}t'\,\mathcal{C}(\mathbf{r},t'), \tag{2.3.24}$$

where T is a time interval long compared with the period of the time-harmonic oscillations, $2\pi/\omega$. Therefore, Eqs. (2.3.23) and (2.3.24) imply that the time average of a product of two real fields is equal to one half of the real part of the respective product of one complex field with the complex conjugate of the other, e.g.,

$$\langle\mathcal{C}(\mathbf{r},t)\rangle_t = \tfrac{1}{2}\mathrm{Re}[\mathbf{A}(\mathbf{r})\cdot\mathbf{B}^*(\mathbf{r})]. \tag{2.3.25}$$

2.4 The Poynting vector

Both the value and the direction of the electromagnetic energy flow are described by the so-called Poynting vector $\mathcal{S}$ (Jackson, 1998). The expression for $\mathcal{S}$ can be derived by considering conservation of energy and taking into account that the magnetic field does no work and that for a local charge q the rate of doing work by the electric field is $q(\mathbf{r},t)\,\mathbf{v}(\mathbf{r},t)\cdot\mathcal{E}(\mathbf{r},t)$, where $\mathbf{v}$ is the velocity of the charge.

Indeed, the total rate of work done by the electromagnetic field in a finite volume V is given by

$$\int_V \mathrm{d}\mathbf{r}\,\mathcal{J}(\mathbf{r},t)\cdot\mathcal{E}(\mathbf{r},t) \tag{2.4.1}$$

and represents the rate of conversion of electromagnetic energy into mechanical or thermal energy. This power must be balanced by the corresponding rate of decrease of the electromagnetic field energy within the volume V. Using Eqs. (2.1.2) and (2.1.4) and the vector identity

$$\nabla\cdot(\mathbf{a}\times\mathbf{b}) = \mathbf{b}\cdot(\nabla\times\mathbf{a}) - \mathbf{a}\cdot(\nabla\times\mathbf{b}), \tag{2.4.2}$$

we derive

$$\begin{aligned}\int_V \mathrm{d}\mathbf{r}\,\mathcal{J}\cdot\mathcal{E} &= \int_V \mathrm{d}\mathbf{r}\,\mathcal{E}\cdot\left(\nabla\times\mathcal{H} - \frac{\partial\mathcal{D}}{\partial t}\right)\\ &= -\int_V \mathrm{d}\mathbf{r}\left[\nabla\cdot(\mathcal{E}\times\mathcal{H}) + \mathcal{E}\cdot\frac{\partial\mathcal{D}}{\partial t} + \mathcal{H}\cdot\frac{\partial\mathcal{B}}{\partial t}\right].\end{aligned} \tag{2.4.3}$$

Let us first consider a linear medium without dispersion and introduce the total electromagnetic energy density,

$$\mathcal{U}(\mathbf{r},t) = \tfrac{1}{2}[\mathcal{E}(\mathbf{r},t)\cdot\mathcal{D}(\mathbf{r},t) + \mathcal{H}(\mathbf{r},t)\cdot\mathcal{B}(\mathbf{r},t)], \tag{2.4.4}$$

and the Poynting vector,

$$\mathcal{S}(\mathbf{r},t) = \mathcal{E}(\mathbf{r},t)\times\mathcal{H}(\mathbf{r},t). \tag{2.4.5}$$

The latter represents electromagnetic energy flow and has the dimension [energy/(area $\times$ time)]. Using also the Gauss theorem (2.1.19), we finally obtain

$$\int_V \mathrm{d}\mathbf{r}\,\mathcal{J}\cdot\mathcal{E} + \oint_S \mathrm{d}S\,\mathcal{S}\cdot\hat{\mathbf{n}} + \int_V \mathrm{d}\mathbf{r}\frac{\partial\mathcal{U}}{\partial t} = 0, \tag{2.4.6}$$

where the closed surface S bounds the volume V and $\hat{\mathbf{n}}$ is a unit vector in the direction of the local outward normal to the surface. Equation (2.4.6) manifests the conservation of energy by requiring that the rate of the total work done by the fields on the sources within the volume, the rate of change of electromagnetic energy within the volume, and the electromagnetic energy flowing out through the volume boundary per unit time add up to zero. Since the volume V is arbitrary, Eq. (2.4.3) can also be written in the form of a differential continuity equation:

$$\frac{\partial \mathcal{U}}{\partial t} + \nabla \cdot \boldsymbol{\mathcal{S}} = -\boldsymbol{\mathcal{J}} \cdot \boldsymbol{\mathcal{E}}. \tag{2.4.7}$$

Owing to the vector identity $\nabla \cdot (\nabla \times \mathbf{a}) = 0$, it is clear from Eq. (2.4.7) that adding the curl of a vector field to the Poynting vector will not change the energy balance. This seems to suggest that there is a degree of arbitrariness in the definition of the Poynting vector. However, relativistic considerations discussed in Section 12.10 of Jackson (1998) show that the definition (2.4.5) is, in fact, unique.

Let us now allow the medium to be dispersive. Instead of Eq. (2.4.1), we now consider the integral

$$\frac{1}{2}\int_V \mathrm{d}\mathbf{r}\, \mathbf{J}^*(\mathbf{r}) \cdot \mathbf{E}(\mathbf{r}) \tag{2.4.8}$$

whose real part gives the time-averaged rate of work done by the electromagnetic field (cf. Eq. (2.3.25)). Using Eqs. (2.3.3), (2.3.5), and (2.4.2), we derive

$$\begin{aligned}\frac{1}{2}\int_V \mathrm{d}\mathbf{r}\, \mathbf{J}^*(\mathbf{r}) \cdot \mathbf{E}(\mathbf{r}) &= \frac{1}{2}\int_V \mathrm{d}\mathbf{r}\, \mathbf{E}(\mathbf{r}) \cdot [\nabla \times \mathbf{H}^*(\mathbf{r}) - \mathrm{i}\omega \mathbf{D}^*(\mathbf{r})] \\ &= -\frac{1}{2}\int_V \mathrm{d}\mathbf{r}\, \{\nabla \cdot [\mathbf{E}(\mathbf{r}) \times \mathbf{H}^*(\mathbf{r})] \\ &\qquad + \mathrm{i}\omega [\mathbf{E}(\mathbf{r}) \cdot \mathbf{D}^*(\mathbf{r}) - \mathbf{B}(\mathbf{r}) \cdot \mathbf{H}^*(\mathbf{r})]\}.\end{aligned} \tag{2.4.9}$$

If we now define the complex Poynting vector by

$$\mathbf{S}(\mathbf{r}) = \tfrac{1}{2}[\mathbf{E}(\mathbf{r}) \times \mathbf{H}^*(\mathbf{r})] \tag{2.4.10}$$

and the complex electric and magnetic energy densities by

$$w_{\mathrm{e}}(\mathbf{r}) = \tfrac{1}{4}[\mathbf{E}(\mathbf{r}) \cdot \mathbf{D}^*(\mathbf{r})], \tag{2.4.11}$$

$$w_{\mathrm{m}}(\mathbf{r}) = \tfrac{1}{4}[\mathbf{B}(\mathbf{r}) \cdot \mathbf{H}^*(\mathbf{r})], \tag{2.4.12}$$

respectively, and apply the Gauss theorem, we then have

$$\frac{1}{2}\int_V \mathrm{d}\mathbf{r}\, \mathbf{J}^*(\mathbf{r}) \cdot \mathbf{E}(\mathbf{r}) + \oint_S \mathrm{d}S\, \mathbf{S}(\mathbf{r}) \cdot \hat{\mathbf{n}} + 2\mathrm{i}\omega \int_V \mathrm{d}\mathbf{r}[w_{\mathrm{e}}(\mathbf{r}) - w_{\mathrm{m}}(\mathbf{r})] = 0. \tag{2.4.13}$$

Obviously, the real part of Eq. (2.4.13) manifests the conservation of energy for the corresponding time-averaged quantities. In particular, the time-averaged Poynting vector $\langle \boldsymbol{\mathcal{S}}(\mathbf{r}, t) \rangle_t$ is equal to the real part of the complex Poynting vector,

$$\langle \boldsymbol{\mathcal{S}}(\mathbf{r}, t) \rangle_t = \mathrm{Re}[\mathbf{S}(\mathbf{r})]. \tag{2.4.14}$$

The net rate W at which the electromagnetic energy crosses the closed surface S is given by

$$W = -\oint_S \mathrm{d}S \, \langle \boldsymbol{\mathcal{S}}(\mathbf{r}, t) \rangle_t \cdot \hat{\mathbf{n}}. \tag{2.4.15}$$

The rate is defined such that it is positive if there is a net transfer of electromagnetic energy into the volume V and is negative otherwise.

2.5 Plane-wave solution

Consider an infinite homogeneous medium without sources. The use of the formulas

$$\nabla \cdot (f\mathbf{a}) = f\nabla \cdot \mathbf{a} + \mathbf{a} \cdot \nabla f, \tag{2.5.1}$$

$$\nabla \times (f\mathbf{a}) = f\nabla \times \mathbf{a} + (\nabla f) \times \mathbf{a}, \tag{2.5.2}$$

$$\nabla \exp(\mathrm{i}\mathbf{k} \cdot \mathbf{r}) = \mathrm{i}\mathbf{k} \exp(\mathrm{i}\mathbf{k} \cdot \mathbf{r}) \tag{2.5.3}$$

in Eqs. (2.3.3), (2.3.4), (2.3.16), and (2.3.17) shows that the complex field vectors

$$\mathbf{E}(\mathbf{r}, t) = \mathbf{E}_0 \exp(\mathrm{i}\mathbf{k} \cdot \mathbf{r} - \mathrm{i}\omega t), \tag{2.5.4}$$

$$\mathbf{H}(\mathbf{r}, t) = \mathbf{H}_0 \exp(\mathrm{i}\mathbf{k} \cdot \mathbf{r} - \mathrm{i}\omega t), \tag{2.5.5}$$

where $\mathbf{E}_0$, $\mathbf{H}_0$, and $\mathbf{k}$ are constant complex vectors, are a solution of the Maxwell equations provided that

$$\mathbf{k} \cdot \mathbf{E}_0 = 0, \tag{2.5.6}$$

$$\mathbf{k} \cdot \mathbf{H}_0 = 0, \tag{2.5.7}$$

$$\mathbf{k} \times \mathbf{E}_0 = \omega\mu\mathbf{H}_0, \tag{2.5.8}$$

$$\mathbf{k} \times \mathbf{H}_0 = -\omega\varepsilon\mathbf{E}_0. \tag{2.5.9}$$

The so-called wave vector $\mathbf{k}$ is usually expressed as

$$\mathbf{k} = \mathbf{k}_\mathrm{R} + \mathrm{i}\mathbf{k}_\mathrm{I}, \tag{2.5.10}$$

where $\mathbf{k}_\mathrm{R}$ and $\mathbf{k}_\mathrm{I}$ are real vectors. Thus

$$\mathbf{E}(\mathbf{r}, t) = \mathbf{E}_0 \exp(-\mathbf{k}_\mathrm{I} \cdot \mathbf{r}) \exp(\mathrm{i}\mathbf{k}_\mathrm{R} \cdot \mathbf{r} - \mathrm{i}\omega t), \tag{2.5.11}$$

$$\mathbf{H}(\mathbf{r}, t) = \mathbf{H}_0 \exp(-\mathbf{k}_\mathrm{I} \cdot \mathbf{r}) \exp(\mathrm{i}\mathbf{k}_\mathrm{R} \cdot \mathbf{r} - \mathrm{i}\omega t). \tag{2.5.12}$$

$\mathbf{E}_0 \exp(-\mathbf{k}_\mathrm{I} \cdot \mathbf{r})$ and $\mathbf{H}_0 \exp(-\mathbf{k}_\mathrm{I} \cdot \mathbf{r})$ are the complex amplitudes of the electric and magnetic fields, respectively, and $\phi = \mathbf{k}_\mathrm{R} \cdot \mathbf{r} - \omega t$ is their phase.

The vector $\mathbf{k}_\mathrm{R}$ is normal to the surfaces of constant phase, whereas $\mathbf{k}_\mathrm{I}$ is normal to the surfaces of constant amplitude. Indeed, a plane surface normal to a real vector $\mathbf{K}$ is described by $\mathbf{r} \cdot \mathbf{K} = \text{constant}$, where $\mathbf{r}$ is the radius vector drawn from the origin of the reference frame to any point in the plane (see Fig. 2.5.1). Also, it is easy to see that surfaces of constant phase propagate in the direction of $\mathbf{k}_\mathrm{R}$ with the phase velocity

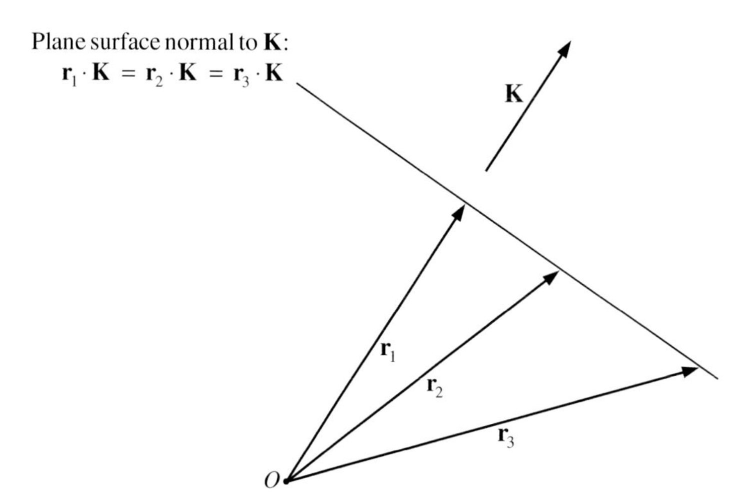

Figure 2.5.1. Plane surface normal to a real vector $\mathbf{K}$.

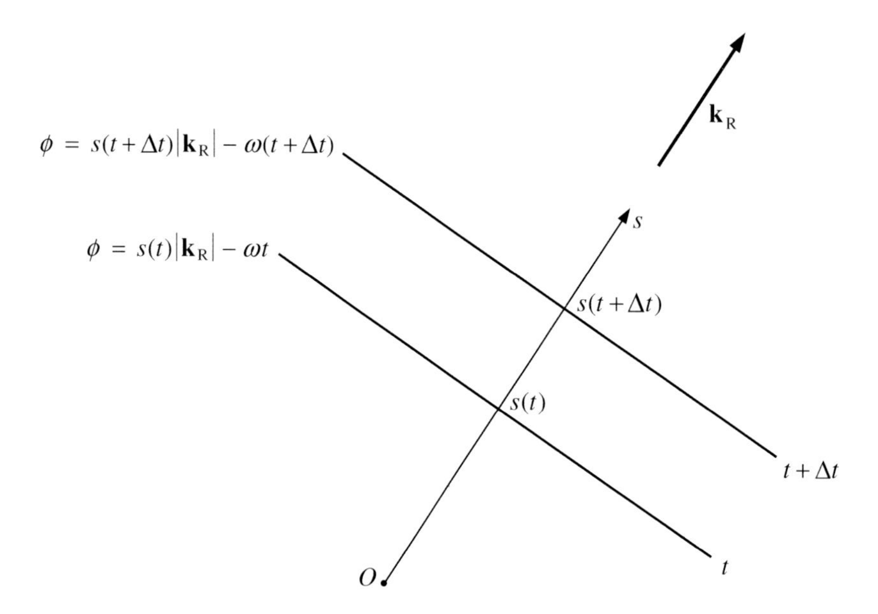

Figure 2.5.2. The plane of constant phase ϕ = constant travels a distance Δs over the time period Δt. The s-axis is drawn from the origin of the coordinate system along the vector $\mathbf{k}_{\mathrm{R}}$.

$$v = \frac{\omega}{|\mathbf{k}_{\mathrm{R}}|}. \tag{2.5.13}$$

Indeed, the planes corresponding to the instant times t and $t + \Delta t$ are separated by the distance $\Delta s = \omega \Delta t / |\mathbf{k}_{\mathrm{R}}|$ (see Fig. 2.5.2), which gives Eq. (2.5.13). Thus Eqs. (2.5.4) and (2.5.5) describe a plane electromagnetic wave propagating in a homogeneous

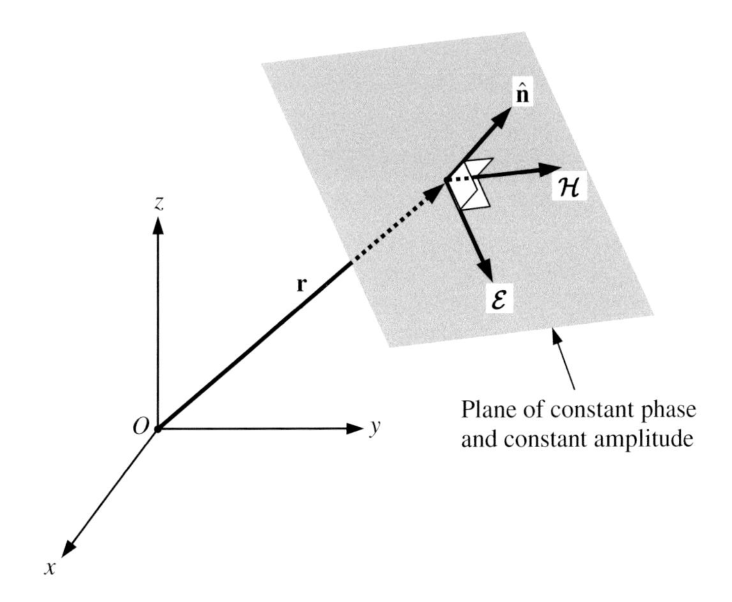

Figure 2.5.3. Plane wave propagating in a homogeneous medium with no dispersion and losses.

medium without sources. This is a very important solution of the Maxwell equations because it embodies the concept of a perfectly monochromatic parallel beam of light of infinite lateral extent and represents the transport of electromagnetic energy from one point to another.

Equations (2.5.4), (2.5.5), and (2.5.8) yield

$$\mathbf{H}(\mathbf{r},t) = \frac{1}{\omega\mu}\mathbf{k}\times\mathbf{E}(\mathbf{r},t). \tag{2.5.14}$$

Therefore, a plane electromagnetic wave always can be considered in terms of only the electric (or only the magnetic) field.

The electromagnetic wave is called homogeneous if $\mathbf{k}_{\mathrm{R}}$ and $\mathbf{k}_{\mathrm{I}}$ are parallel (including the case $\mathbf{k}_{\mathrm{I}} = \mathbf{0}$); otherwise it is called inhomogeneous. When $\mathbf{k}_{\mathrm{R}} \| \mathbf{k}_{\mathrm{I}}$, the complex wave vector can be expressed as $\mathbf{k} = (k_{\mathrm{R}} + \mathrm{i}k_{\mathrm{I}})\hat{\mathbf{n}}$, where $\hat{\mathbf{n}}$ is a real unit vector in the direction of propagation and both k_{R} and k_{I} are real and nonnegative.

According to Eqs. (2.5.6) and (2.5.7), the plane electromagnetic wave is transverse: both $\mathbf{E}_0$ and $\mathbf{H}_0$ are perpendicular to $\mathbf{k}$. Furthermore, it is evident from either Eq. (2.5.8) or Eq. (2.5.9) that $\mathbf{E}_0$ and $\mathbf{H}_0$ are mutually perpendicular: $\mathbf{E}_0 \cdot \mathbf{H}_0 = 0$. Since $\mathbf{E}_0$, $\mathbf{H}_0$, and $\mathbf{k}$ are, in general, complex vectors, the physical interpretation of these facts can be far from obvious. It becomes most transparent when both ε, μ, and $\mathbf{k}$ are real. The reader can verify that in this case the real field vectors $\mathcal{E}$ and $\mathcal{H}$ are mutually perpendicular and lie in a plane normal to the direction of wave propagation $\hat{\mathbf{n}}$ (see Fig. 2.5.3).

Taking the vector product of **k** with the left-hand side and the right-hand side of Eq. (2.5.8) and using Eq. (2.5.9) and the vector identity

$$\mathbf{a}\times(\mathbf{b}\times\mathbf{c}) = \mathbf{b}(\mathbf{a}\cdot\mathbf{c}) - \mathbf{c}(\mathbf{a}\cdot\mathbf{b}) \tag{2.5.15}$$

together with Eq. (2.5.6) yields

$$\mathbf{k}\cdot\mathbf{k} = \omega^2\varepsilon\mu. \tag{2.5.16}$$

In the practically important case of a homogeneous plane wave, we obtain from Eq. (2.5.16)

$$k = k_R + \mathrm{i}k_I = \omega\sqrt{\varepsilon\mu} = \frac{\omega m}{c}, \tag{2.5.17}$$

where k is the wave number,

$$c = \frac{1}{\sqrt{\epsilon_0\mu_0}} \tag{2.5.18}$$

is the speed of light in a vacuum, and

$$m = \frac{ck}{\omega} = m_R + \mathrm{i}m_I = \sqrt{\frac{\varepsilon\mu}{\epsilon_0\mu_0}} = c\sqrt{\varepsilon\mu} \tag{2.5.19}$$

is the complex refractive index with a nonnegative real part m_R and a nonnegative imaginary part m_I. Thus, the complex electric field vector of the homogeneous plane wave has the form

$$\mathbf{E}(\mathbf{r},t) = \mathbf{E}_0\exp\left(-\frac{\omega}{c}m_I\hat{\mathbf{n}}\cdot\mathbf{r}\right)\exp\left(\mathrm{i}\frac{\omega}{c}m_R\hat{\mathbf{n}}\cdot\mathbf{r} - \mathrm{i}\omega t\right). \tag{2.5.20}$$

If the imaginary part of the refractive index is nonzero then it determines the decay of the amplitude of the wave as it propagates through the medium, which is thus absorbing. On the other hand, a medium is nonabsorbing if it is nondispersive ($\epsilon = \epsilon$ and $\mu = \mu$) and lossless ($\sigma = 0$), which causes the refractive index $m = m_R = c(\epsilon\mu)^{1/2}$ to be real-valued. The real part of the refractive index determines the phase velocity of the wave:

$$v = \frac{c}{m_R}. \tag{2.5.21}$$

In a vacuum, $m = m_R = 1$ and $v = c$.

As follows from Eqs. (2.4.10), (2.4.14), (2.5.4), (2.5.5), (2.5.8), and (2.5.15), the time-averaged Poynting vector of a plane wave is

$$\langle\boldsymbol{\mathcal{S}}(\mathbf{r},t)\rangle_t = \mathrm{Re}\left(\frac{\mathbf{k}^*[\mathbf{E}(\mathbf{r})\cdot\mathbf{E}^*(\mathbf{r})] - \mathbf{E}^*(\mathbf{r})[\mathbf{k}^*\cdot\mathbf{E}(\mathbf{r})]}{2\omega\mu^*}\right). \tag{2.5.22}$$

If the wave is homogeneous then $\mathbf{k}\cdot\mathbf{E}(\mathbf{r}) = 0$ and so $\mathbf{k}^*\cdot\mathbf{E}(\mathbf{r}) = 0$. Therefore,

$$\langle \mathcal{S}(\mathbf{r},t)\rangle_t = \tfrac{1}{2}\mathrm{Re}\left(\sqrt{\frac{\varepsilon}{\mu}}\right)|\mathbf{E}_0|^2 \exp\left(-2\frac{\omega}{c}m_\mathrm{I}\hat{\mathbf{n}}\cdot\mathbf{r}\right)\hat{\mathbf{n}}. \tag{2.5.23}$$

Thus, $\langle \mathcal{S}(\mathbf{r},t)\rangle_t$ is in the direction of propagation and its absolute value, called the intensity, is attenuated exponentially provided that the medium is absorbing:

$$I(\mathbf{r}) = |\langle \mathcal{S}(\mathbf{r},t)\rangle_t| = I_0 \exp(-\alpha\hat{\mathbf{n}}\cdot\mathbf{r}), \tag{2.5.24}$$

where I_0 is the intensity at $\mathbf{r} = \mathbf{0}$. The absorption coefficient α is

$$\alpha = 2\frac{\omega}{c}m_\mathrm{I} = \frac{4\pi m_\mathrm{I}}{\lambda_0}, \tag{2.5.25}$$

where

$$\lambda_0 = \frac{2\pi c}{\omega} \tag{2.5.26}$$

is the free-space wavelength. The intensity has the dimension of monochromatic energy flux, [energy/(area$\times$time)], and is equal to the amount of electromagnetic energy crossing a unit surface element normal to $\hat{\mathbf{n}}$ per unit time.

The expression for the time-averaged energy density of a time-harmonic electromagnetic field existing in a medium without dispersion follows from Eqs. (2.3.7), (2.3.8), (2.3.25), and (2.4.4):

$$\langle \mathcal{U}(\mathbf{r},t)\rangle_t = \tfrac{1}{4}[\epsilon\mathbf{E}(\mathbf{r})\cdot\mathbf{E}^*(\mathbf{r}) + \mu\mathbf{H}(\mathbf{r})\cdot\mathbf{H}^*(\mathbf{r})]. \tag{2.5.27}$$

Assuming further that the medium is lossless and recalling Eqs. (2.5.6), (2.5.8), and (2.5.16) as well as the vector identity

$$(\mathbf{a}\times\mathbf{b})\cdot(\mathbf{c}\times\mathbf{d}) = (\mathbf{a}\cdot\mathbf{c})(\mathbf{b}\cdot\mathbf{d}) - (\mathbf{a}\cdot\mathbf{d})(\mathbf{b}\cdot\mathbf{c}), \tag{2.5.28}$$

we derive for a plane electromagnetic wave

$$\langle \mathcal{U}(\mathbf{r},t)\rangle_t = \tfrac{1}{2}\epsilon|\mathbf{E}_0|^2. \tag{2.5.29}$$

Comparison of Eqs. (2.5.23), (2.5.24), and (2.5.29) shows that

$$I(\mathbf{r}) = \frac{1}{\sqrt{\epsilon\mu}}\langle \mathcal{U}(\mathbf{r},t)\rangle_t = v\langle \mathcal{U}(\mathbf{r},t)\rangle_t, \tag{2.5.30}$$

where v is the speed of light in the nonabsorbing material medium. The physical interpretation of this result is quite clear: the amount of electromagnetic energy crossing a surface element of unit area normal to the direction of propagation per unit time is equal to the product of the speed of light and the amount of electromagnetic energy per unit volume.

Figure 2.5.4 gives a simple example of a plane electromagnetic wave propagating along the y-axis in a nonabsorbing homogeneous medium and described by the following real electric and magnetic field vectors:

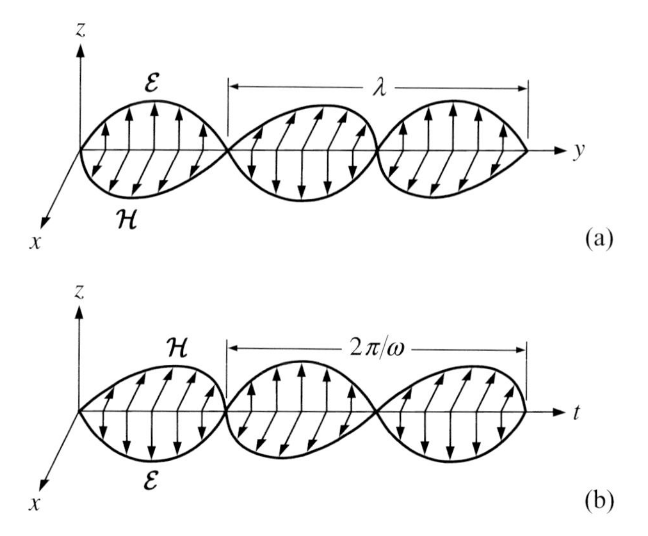

Figure 2.5.4. Plane electromagnetic wave described by Eqs. (2.5.31) and (2.5.32).

$$\boldsymbol{\mathcal{E}}(\mathbf{r},t) = \mathcal{E}\cos(ky-\omega t-\pi/2)\hat{\mathbf{z}}, \tag{2.5.31}$$

$$\boldsymbol{\mathcal{H}}(\mathbf{r},t) = \mathcal{H}\cos(ky-\omega t-\pi/2)\hat{\mathbf{x}}, \tag{2.5.32}$$

where $\mathcal{E}$, $\mathcal{H}$, and k are real and $\hat{\mathbf{x}}$ and $\hat{\mathbf{z}}$ are the unit vectors along the x-axis and the z-axis, respectively. Panel (a) shows the electric and magnetic fields as a function of y at the moment $t = 0$, while panel (b) depicts the fields as a function of time at any point in the plane $y = 0$. The period of the sinusoids in panel (a) is given by

$$\lambda = \frac{2\pi}{k} \tag{2.5.33}$$

and defines the wavelength of light in the nonabsorbing material medium, whereas the period of the sinusoids in panel (b) is equal to $2\pi/\omega$.

It is straightforward to verify that a choice of the time dependence $\exp(\mathrm{i}\omega t)$ rather than $\exp(-\mathrm{i}\omega t)$ in the complex representation of time-harmonic fields in Eq. (2.3.1) would have led to $m = m_{\mathrm{R}} - \mathrm{i}m_{\mathrm{I}}$ with a nonnegative m_{I}. The $\exp(-\mathrm{i}\omega t)$ time-factor convention adopted here has been used in many books on optics and light scattering (e.g., Born and Wolf, 1999; Bohren and Huffman, 1983; Barber and Hill, 1990; MTL), electromagnetics (e.g., Stratton, 1941; Jackson, 1998; Tsang *et al.*, 2000; Kong, 2000), and solid-state physics (e.g., Kittel, 1963). However, van de Hulst (1957), Kerker (1969), and Hovenier *et al.* (2004) used the time factor $\exp(\mathrm{i}\omega t)$, which implies a nonpositive imaginary part of the complex refractive index. It does not matter in the final analysis which convention is chosen because all measurable quantities of practical interest are always real. However, it is important to remember that once a choice of the time factor has been made, its consistent use throughout all derivations is imperative.

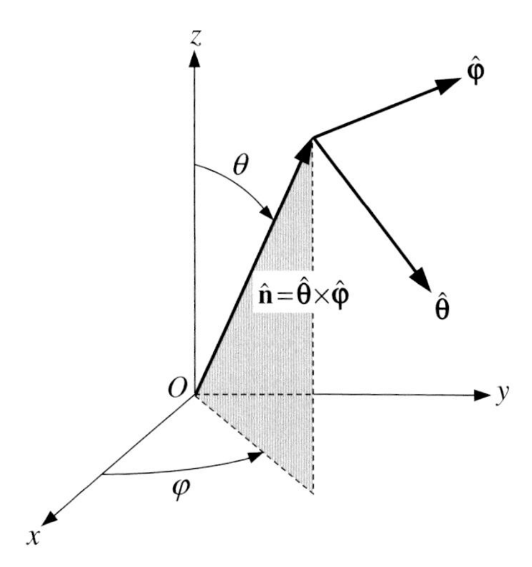

Figure 2.6.1. Local coordinate system used to describe the direction of propagation and the polarization state of a plane electromagnetic wave at the observation point O.

2.6 Coherency matrix and Stokes parameters

Traditional optical devices cannot measure the electric and magnetic fields associated with a beam of light; rather they measure quantities that are time averages of real-valued linear combinations of products of field vector components and have the dimension of the intensity. In order to define these quantities, we use polar spherical coordinates associated with the local right-handed Cartesian coordinate system with origin at the observation point, as shown in Fig. 2.6.1. Assuming that the medium is homogeneous and has no dispersion and losses, we specify the direction of propagation of a plane electromagnetic wave by a unit vector $\hat{\mathbf{n}}$ or, equivalently, by a couplet $\{\theta, \varphi\}$, where $\theta \in [0, \pi]$ is the polar (zenith) angle measured from the positive z-axis and $\varphi \in [0, 2\pi)$ is the azimuth angle measured from the positive x-axis in the clockwise direction when looking in the direction of the positive z-axis. Since the component of the electric field vector along the direction of propagation $\hat{\mathbf{n}}$ is equal to zero, the electric field at the observation point can be expressed as

$$\mathbf{E} = \mathbf{E}_\theta + \mathbf{E}_\varphi,$$

where $\mathbf{E}_\theta$ and $\mathbf{E}_\varphi$ are the θ- and φ-components of the electric field vector, respectively. The component

$$\mathbf{E}_\theta = E_\theta \hat{\boldsymbol{\theta}}$$

lies in the meridional plane (i.e., the plane through $\hat{\mathbf{n}}$ and the z-axis), whereas the component

$$\mathbf{E}_\varphi = E_\varphi \hat{\boldsymbol{\varphi}}$$

is perpendicular to this plane.[2] $\hat{\boldsymbol{\theta}}$ and $\hat{\boldsymbol{\varphi}}$ are the corresponding unit vectors such that

$$\hat{\mathbf{n}} = \hat{\boldsymbol{\theta}} \times \hat{\boldsymbol{\varphi}}.$$

The specification of a unit vector $\hat{\mathbf{n}}$ uniquely determines the meridional plane of the propagation direction except when $\hat{\mathbf{n}}$ is oriented along the positive or negative direction of the z-axis. Although it may seem redundant to specify φ in addition to θ when $\theta = 0$ or π, the unit θ- and φ-vectors and, thus, the electric-field vector components $\mathbf{E}_\theta$ and $\mathbf{E}_\varphi$ still depend on the orientation of the meridional plane. Therefore, we will always assume that the specification of $\hat{\mathbf{n}}$ implicitly includes the specification of the appropriate meridional plane in cases when $\hat{\mathbf{n}}$ is parallel to the z-axis. To minimize confusion, we often will specify explicitly the direction of propagation using the angles θ and φ; the latter uniquely defines the meridional plane when $\theta = 0$ or π.

Consider a plane electromagnetic wave propagating in a homogeneous medium without dispersion and losses and given by

$$\mathbf{E}(\mathbf{r}, t) = \mathbf{E}_0 \exp(\mathrm{i}k\hat{\mathbf{n}} \cdot \mathbf{r} - \mathrm{i}\omega t) \tag{2.6.1}$$

with a real k. The simplest complete set of linearly independent quadratic combinations of the electric field vector components with nonzero time averages consists of the following four quantities:

$$E_\theta(\mathbf{r}, t)[E_\theta(\mathbf{r}, t)]^* = E_{0\theta} E_{0\theta}^*, \qquad E_\theta(\mathbf{r}, t)[E_\varphi(\mathbf{r}, t)]^* = E_{0\theta} E_{0\varphi}^*,$$

$$E_\varphi(\mathbf{r}, t)[E_\theta(\mathbf{r}, t)]^* = E_{0\varphi} E_{0\theta}^*, \qquad E_\varphi(\mathbf{r}, t)[E_\varphi(\mathbf{r}, t)]^* = E_{0\varphi} E_{0\varphi}^*.$$

The products of these quantities and $\frac{1}{2}(\epsilon/\mu)^{1/2}$ have the dimension of monochromatic energy flux and form the 2×2 coherency (or density) matrix $\boldsymbol{\rho}$ (Born and Wolf, 1999):

$$\boldsymbol{\rho} = \begin{bmatrix} \rho_{11} & \rho_{12} \\ \rho_{21} & \rho_{22} \end{bmatrix} = \frac{1}{2}\sqrt{\frac{\epsilon}{\mu}} \begin{bmatrix} E_{0\theta} E_{0\theta}^* & E_{0\theta} E_{0\varphi}^* \\ E_{0\varphi} E_{0\theta}^* & E_{0\varphi} E_{0\varphi}^* \end{bmatrix}. \tag{2.6.2}$$

The completeness of the set of the four coherency matrix elements means that any plane-wave characteristic directly observable with a traditional optical instrument is a real-valued linear combination of these quantities.

Since ρ_{12} and ρ_{21} are, in general, complex, it is convenient to introduce an alternative complete set of four real, linearly independent quantities called Stokes parameters. We first group the elements of the 2×2 coherency matrix into a 4×1 coherency column vector:

---

[2] In the microwave remote-sensing literature, $\mathbf{E}_\theta$ and $\mathbf{E}_\varphi$ are often denoted as $\mathbf{E}_\mathrm{v}$ and $\mathbf{E}_\mathrm{h}$ and called the vertical and horizontal electric-field vector components, respectively (e.g., Ulaby and Elachi, 1990; Tsang *et al.*, 2000).

$$\mathbf{J} = \begin{bmatrix} \rho_{11} \\ \rho_{12} \\ \rho_{21} \\ \rho_{22} \end{bmatrix} = \frac{1}{2}\sqrt{\frac{\epsilon}{\mu}} \begin{bmatrix} E_{0\theta} E_{0\theta}^* \\ E_{0\theta} E_{0\varphi}^* \\ E_{0\varphi} E_{0\theta}^* \\ E_{0\varphi} E_{0\varphi}^* \end{bmatrix}. \tag{2.6.3}$$

The Stokes parameters I, Q, U, and V are then defined as the elements of a 4×1 Stokes column vector $\mathbf{I}$ as follows:

$$\mathbf{I} = \begin{bmatrix} I \\ Q \\ U \\ V \end{bmatrix} = \mathbf{DJ} = \frac{1}{2}\sqrt{\frac{\epsilon}{\mu}} \begin{bmatrix} E_{0\theta} E_{0\theta}^* + E_{0\varphi} E_{0\varphi}^* \\ E_{0\theta} E_{0\theta}^* - E_{0\varphi} E_{0\varphi}^* \\ -E_{0\theta} E_{0\varphi}^* - E_{0\varphi} E_{0\theta}^* \\ \mathrm{i}(E_{0\varphi} E_{0\theta}^* - E_{0\theta} E_{0\varphi}^*) \end{bmatrix}$$

$$= \frac{1}{2}\sqrt{\frac{\epsilon}{\mu}} \begin{bmatrix} E_{0\theta} E_{0\theta}^* + E_{0\varphi} E_{0\varphi}^* \\ E_{0\theta} E_{0\theta}^* - E_{0\varphi} E_{0\varphi}^* \\ -2\,\mathrm{Re}(E_{0\theta} E_{0\varphi}^*) \\ 2\,\mathrm{Im}(E_{0\theta} E_{0\varphi}^*) \end{bmatrix}, \tag{2.6.4}$$

where

$$\mathbf{D} = \begin{bmatrix} 1 & 0 & 0 & 1 \\ 1 & 0 & 0 & -1 \\ 0 & -1 & -1 & 0 \\ 0 & -\mathrm{i} & \mathrm{i} & 0 \end{bmatrix}. \tag{2.6.5}$$

Conversely,

$$\mathbf{J} = \mathbf{D}^{-1}\mathbf{I}, \tag{2.6.6}$$

where the inverse of $\mathbf{D}$ is given by

$$\mathbf{D}^{-1} = \frac{1}{2} \begin{bmatrix} 1 & 1 & 0 & 0 \\ 0 & 0 & -1 & \mathrm{i} \\ 0 & 0 & -1 & -\mathrm{i} \\ 1 & -1 & 0 & 0 \end{bmatrix}. \tag{2.6.7}$$

By virtue of being real-valued and having the dimension of energy flux, the Stokes parameters form a complete set of quantities needed to characterize a plane electromagnetic wave, inasmuch as it is subject to practical analysis. This means that:

- Any other observable quantity is a linear combination of the four Stokes parameters.
- It is impossible to distinguish between two plane waves with the same values of the Stokes parameters using a traditional optical device (the so-called prin-

ciple of optical equivalence).

Indeed, the two complex amplitudes $E_{0\theta} = a_\theta \exp(\mathrm{i}\Delta_\theta)$ and $E_{0\varphi} = a_\varphi \exp(\mathrm{i}\Delta_\varphi)$ are characterized by four real numbers: the nonnegative amplitudes a_θ and a_φ and the phases Δ_θ and $\Delta_\varphi = \Delta_\theta - \Delta$. The Stokes parameters carry information about the amplitudes and the phase difference Δ, but not about Δ_θ. The latter is the only quantity that could be used to distinguish different waves with the same a_θ, a_φ, and Δ (and thus the same Stokes parameters), but it vanishes when a field vector component is multiplied by the complex conjugate value of the same or another field vector component.

The first Stokes parameter, I, is the intensity introduced in the previous section, with the explicit definition here applicable to a homogeneous, nonabsorbing medium (cf. Eqs. (2.5.23), (2.5.24), and (2.6.4)). The Stokes parameters Q, U, and V describe the polarization state of the wave. The ellipsometric interpretation of the Stokes parameters will be the subject of the next section. It is easy to verify that the Stokes parameters of a plane monochromatic wave are not completely independent but rather are related by the quadratic so-called Stokes identity

$$I^2 = Q^2 + U^2 + V^2. \tag{2.6.8}$$

We will see later, however, that this identity may not hold for a quasi-monochromatic beam of light.

The coherency matrix and the Stokes column vector are not the only representations of polarization and not always the most convenient ones. Two other frequently used representations are the real so-called modified Stokes column vector given by

$$\mathbf{I}^{\mathrm{MS}} = \begin{bmatrix} I_\mathrm{v} \\ I_\mathrm{h} \\ U \\ V \end{bmatrix} = \mathbf{B}\mathbf{I} = \begin{bmatrix} \frac{1}{2}(I+Q) \\ \frac{1}{2}(I-Q) \\ U \\ V \end{bmatrix} \tag{2.6.9}$$

and the complex circular-polarization column vector defined as

$$\mathbf{I}^{\mathrm{CP}} = \begin{bmatrix} I_2 \\ I_0 \\ I_{-0} \\ I_{-2} \end{bmatrix} = \mathbf{A}\mathbf{I} = \frac{1}{2}\begin{bmatrix} Q+\mathrm{i}U \\ I+V \\ I-V \\ Q-\mathrm{i}U \end{bmatrix}, \tag{2.6.10}$$

where

$$\mathbf{B} = \begin{bmatrix} 1/2 & 1/2 & 0 & 0 \\ 1/2 & -1/2 & 0 & 0 \\ 0 & 0 & 1 & 0 \\ 0 & 0 & 0 & 1 \end{bmatrix}, \tag{2.6.11}$$

$$\mathbf{A} = \frac{1}{2}\begin{bmatrix} 0 & 1 & \mathrm{i} & 0 \\ 1 & 0 & 0 & 1 \\ 1 & 0 & 0 & -1 \\ 0 & 1 & -\mathrm{i} & 0 \end{bmatrix}. \tag{2.6.12}$$

It is easy to verify that

$$\mathbf{I} = \mathbf{B}^{-1}\mathbf{I}^{\mathrm{MS}} \tag{2.6.13}$$

and

$$\mathbf{I} = \mathbf{A}^{-1}\mathbf{I}^{\mathrm{CP}}, \tag{2.6.14}$$

where

$$\mathbf{B}^{-1} = \begin{bmatrix} 1 & 1 & 0 & 0 \\ 1 & -1 & 0 & 0 \\ 0 & 0 & 1 & 0 \\ 0 & 0 & 0 & 1 \end{bmatrix} \tag{2.6.15}$$

and

$$\mathbf{A}^{-1} = \begin{bmatrix} 0 & 1 & 1 & 0 \\ 1 & 0 & 0 & 1 \\ -\mathrm{i} & 0 & 0 & \mathrm{i} \\ 0 & 1 & -1 & 0 \end{bmatrix}. \tag{2.6.16}$$

The usefulness of the modified Stokes and circular-polarization column vectors will be illustrated in the following section.

We conclude this section with a caution. It is important to remember that whereas the Poynting vector can be defined for an arbitrary electromagnetic field, the Stokes parameters can only be defined for transverse fields such as plane waves discussed in the previous section or spherical waves discussed in Section 2.11. Quite often the electromagnetic field at an observation point is not a well-defined transverse electromagnetic wave, in which case the Stokes vector formalism cannot be applied directly.

2.7 Ellipsometric interpretation of the Stokes parameters

In this section we show how the Stokes parameters can be used to derive the ellipsometric characteristics of the plane electromagnetic wave given by Eq. (2.6.1). Writing

$$E_{0\theta} = a_\theta \exp(\mathrm{i}\Delta_\theta), \tag{2.7.1}$$

$$E_{0\varphi} = a_\varphi \exp(\mathrm{i}\Delta_\varphi) \tag{2.7.2}$$

with real nonnegative amplitudes a_θ and a_φ and real phases Δ_θ and Δ_φ and recalling the definition (2.6.4), we obtain for the Stokes parameters

$$I = \frac{1}{2}\sqrt{\frac{\epsilon}{\mu}}\,(a_\theta^2 + a_\varphi^2), \tag{2.7.3}$$

$$Q = \frac{1}{2}\sqrt{\frac{\epsilon}{\mu}}\,(a_\theta^2 - a_\varphi^2), \tag{2.7.4}$$

$$U = -\sqrt{\frac{\epsilon}{\mu}}\,a_\theta a_\varphi \cos\Delta, \tag{2.7.5}$$

$$V = \sqrt{\frac{\epsilon}{\mu}}\,a_\theta a_\varphi \sin\Delta, \tag{2.7.6}$$

where

$$\Delta = \Delta_\theta - \Delta_\varphi. \tag{2.7.7}$$

Substituting Eqs. (2.7.1) and (2.7.2) in Eqs. (2.3.1) and (2.6.1), we have for the real electric field vector components

$$\mathcal{E}_\theta(\mathbf{r},t) = a_\theta \cos(\delta_\theta - \omega t), \tag{2.7.8}$$

$$\mathcal{E}_\varphi(\mathbf{r},t) = a_\varphi \cos(\delta_\varphi - \omega t), \tag{2.7.9}$$

where

$$\delta_\theta = \Delta_\theta + k\hat{\mathbf{n}}\cdot\mathbf{r}, \tag{2.7.10}$$

$$\delta_\varphi = \Delta_\varphi + k\hat{\mathbf{n}}\cdot\mathbf{r}. \tag{2.7.11}$$

At any fixed point O in space, the endpoint of the real electric field vector given by Eqs. (2.7.8)–(2.7.11) describes an ellipse with specific major and minor axes and orientation (see the top panel of Fig. 2.7.1). The major axis of the ellipse makes an angle ζ with the positive direction of the φ-axis such that $\zeta \in [0, \pi)$. By definition, this orientation angle is obtained by rotating the φ-axis in the *clockwise* direction when looking in the direction of propagation, until it is directed along the major axis of the ellipse. The ellipticity is defined as the ratio of the minor to the major axes of the ellipse and is usually expressed as $|\tan\beta|$, where $\beta \in [-\pi/4, \pi/4]$. By definition, β is positive when the real electric field vector at O rotates clockwise, as viewed by an observer looking in the direction of propagation (Fig. 2.7.1(a)). The polarization for positive β is called right-handed, as opposed to the left-handed polarization corresponding to the anti-clockwise rotation of the electric field vector.

To express the orientation ζ of the ellipse and the ellipticity $|\tan\beta|$ in terms of the Stokes parameters, we first write the equations representing the rotation of the real

(a) Polarization ellipse

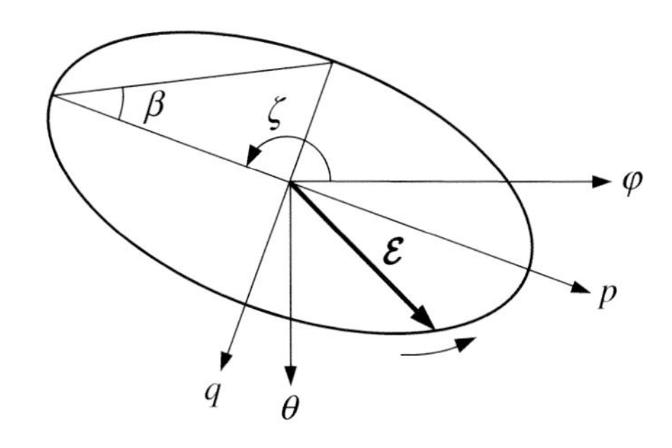

(b) Elliptical polarization $(V \neq 0)$

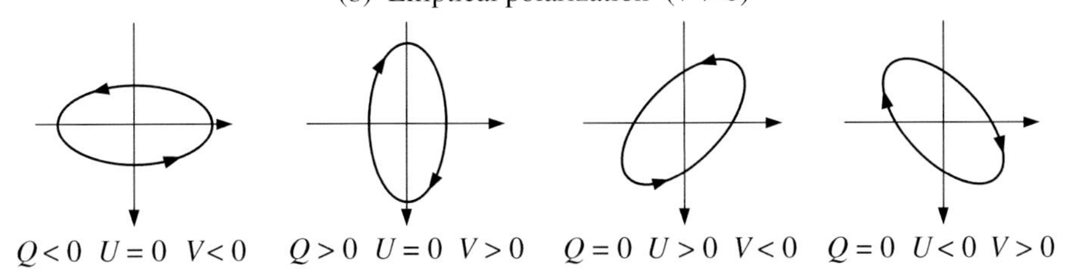

(c) Linear polarization $(V = 0)$

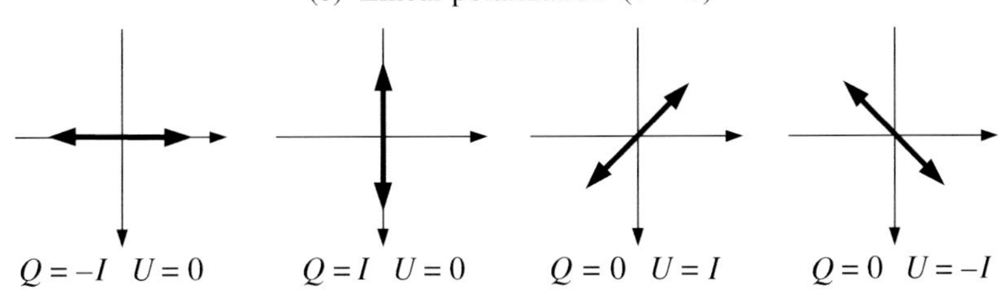

(d) Circular polarization $(Q = U = 0)$

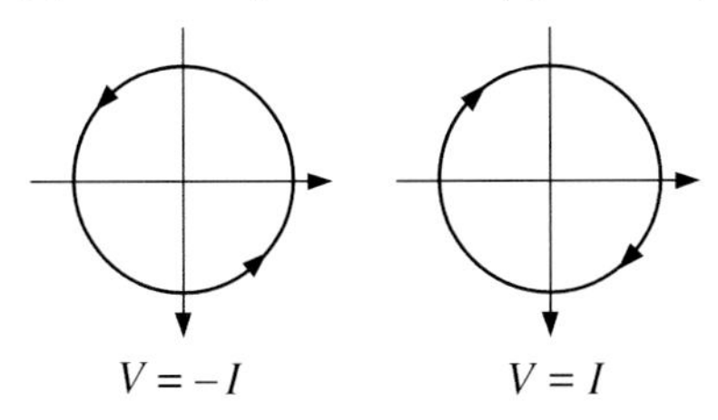

Figure 2.7.1. Ellipse described by the tip of the real electric vector at a fixed point O in space (top panel) and particular cases of elliptical, linear, and circular polarization. The plane electromagnetic wave propagates towards the reader.

electric field vector at O in the form

$$\mathcal{E}_q(\mathbf{r},t) = a\sin\beta\sin(\delta-\omega t), \tag{2.7.12}$$

$$\mathcal{E}_p(\mathbf{r},t) = a\cos\beta\cos(\delta-\omega t), \tag{2.7.13}$$

where $\mathcal{E}_p$ and $\mathcal{E}_q$ are the electric field components along the major and minor axes of the ellipse, respectively, Fig. 2.7.1(a). One easily verifies that a positive (negative) β indeed corresponds to the right-handed (left-handed) polarization. The connection between Eqs. (2.7.8)–(2.7.11) and Eqs. (2.7.12)–(2.7.13) can be established by using the simple transformation rule for rotation of a two-dimensional coordinate system:

$$\mathcal{E}_\theta(\mathbf{r},t) = -\mathcal{E}_q(\mathbf{r},t)\cos\zeta + \mathcal{E}_p(\mathbf{r},t)\sin\zeta, \tag{2.7.14}$$

$$\mathcal{E}_\varphi(\mathbf{r},t) = -\mathcal{E}_q(\mathbf{r},t)\sin\zeta - \mathcal{E}_p(\mathbf{r},t)\cos\zeta. \tag{2.7.15}$$

By equating the coefficients of $\cos\omega t$ and $\sin\omega t$ in the expanded Eqs. (2.7.8) and (2.7.9) with those in (2.7.14) and (2.7.15), we obtain

$$a_\theta\cos\delta_\theta = -a\sin\beta\sin\delta\cos\zeta + a\cos\beta\cos\delta\sin\zeta, \tag{2.7.16}$$

$$a_\theta\sin\delta_\theta = a\sin\beta\cos\delta\cos\zeta + a\cos\beta\sin\delta\sin\zeta, \tag{2.7.17}$$

$$a_\varphi\cos\delta_\varphi = -a\sin\beta\sin\delta\sin\zeta - a\cos\beta\cos\delta\cos\zeta, \tag{2.7.18}$$

$$a_\varphi\sin\delta_\varphi = a\sin\beta\cos\delta\sin\zeta - a\cos\beta\sin\delta\cos\zeta. \tag{2.7.19}$$

Squaring and adding Eqs. (2.7.16) and (2.7.17) and Eqs. (2.7.18) and (2.7.19) gives

$$a_\theta^2 = a^2(\sin^2\beta\cos^2\zeta + \cos^2\beta\sin^2\zeta), \tag{2.7.20}$$

$$a_\varphi^2 = a^2(\sin^2\beta\sin^2\zeta + \cos^2\beta\cos^2\zeta). \tag{2.7.21}$$

Multiplying Eqs. (2.7.16) and (2.7.18) and Eqs. (2.7.17) and (2.7.19) and adding yields

$$a_\theta a_\varphi\cos\Delta = -\tfrac{1}{2}a^2\cos2\beta\sin2\zeta. \tag{2.7.22}$$

Similarly, multiplying Eqs. (2.7.17) and (2.7.18) and Eqs. (2.7.16) and (2.7.19) and subtracting gives

$$a_\theta a_\varphi\sin\Delta = -\tfrac{1}{2}a^2\sin2\beta. \tag{2.7.23}$$

Comparing Eqs. (2.7.3)–(2.7.6) with Eqs. (2.7.20)–(2.7.23), we finally derive

$$I = \frac{1}{2}\sqrt{\frac{\epsilon}{\mu}}\,a^2, \tag{2.7.24}$$

$$Q = -I\cos2\beta\cos2\zeta, \tag{2.7.25}$$

$$U = I\cos2\beta\sin2\zeta, \tag{2.7.26}$$

$$V = -I\sin2\beta. \tag{2.7.27}$$

The parameters of the polarization ellipse are thus expressed in terms of the Stokes parameters as follows. The major and minor axes are given by

$$\sqrt{2I\sqrt{\mu/\epsilon}}\,\cos\beta$$

and

$$\sqrt{2I\sqrt{\mu/\epsilon}}\,|\sin\beta|,$$

respectively (cf. Eqs. (2.7.12) and (2.7.13)). Equations (2.7.25) and (2.7.26) yield

$$\tan 2\zeta = -\frac{U}{Q}. \tag{2.7.28}$$

Because $|\beta| \le \pi/4$, we have $\cos 2\beta \ge 0$ so that $\cos 2\zeta$ has the same sign as $-Q$. Therefore, from the different values of ζ that satisfy Eq. (2.7.28) but differ by $\pi/2$, we must choose the one that makes the sign of $\cos 2\zeta$ to be the same as that of $-Q$. The ellipticity and handedness follow from

$$\tan 2\beta = -\frac{V}{\sqrt{Q^2 + U^2}}. \tag{2.7.29}$$

Thus, the polarization is left-handed if V is positive and is right-handed if V is negative (Fig. 2.7.1(b)).

The electromagnetic wave becomes linearly polarized when β vanishes; then the electric field vector vibrates along the line making the angle ζ with the φ-axis (Fig. 2.7.1(a)) and $V = 0$. Furthermore, if $\zeta = 0$ or $\zeta = \pi/2$ then U vanishes as well. This explains the usefulness of the modified Stokes representation of polarization given by Eq. (2.6.9) in situations involving linearly polarized light as follows. The modified Stokes column vector has only one nonzero element and is equal to $[I\ 0\ 0\ 0]^{\mathrm{T}}$ if $\zeta = \pi/2$ (the electric field vector vibrates along the θ-axis, i.e., in the meridional plane) or $[0\ I\ 0\ 0]^{\mathrm{T}}$ if $\zeta = 0$ (the electric field vector vibrates along the φ-axis, i.e., in the plane perpendicular to the meridional plane), where T indicates the transpose of a matrix (see Fig. 2.7.1(c)).

If, however, $\beta = \pm\pi/4$, then both Q and U vanish, and the electric field vector describes a circle in the clockwise ($\beta = \pi/4$, $V = -I$) or anti-clockwise ($\beta = -\pi/4$, $V = I$) direction, as viewed by an observer looking in the direction of propagation (Fig. 2.7.1(d)). In this case the electromagnetic wave is circularly polarized; the circular-polarization column vector $\mathbf{I}^{\mathrm{CP}}$ has only one nonzero element and takes the values $[0\ 0\ I\ 0]^{\mathrm{T}}$ and $[0\ I\ 0\ 0]^{\mathrm{T}}$, respectively (see Eq. (2.6.10)).

The polarization ellipse, along with a designation of the rotation direction (right- or left-handed), fully describes the temporal evolution of the real electric field vector at a fixed point in space. This evolution can also be visualized by plotting the curve in (θ, φ, t) coordinates described by the tip of the electric field vector as a function of time. For example, in the case of an elliptically polarized plane wave with right-

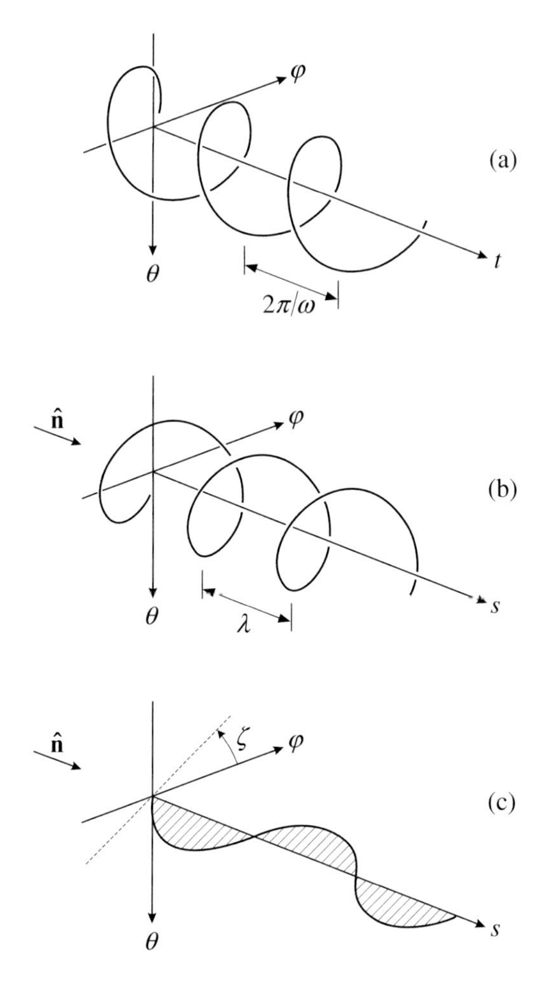

Figure 2.7.2. (a) The helix described by the tip of the real electric field vector of a plane electromagnetic wave with right-handed polarization in the (θ, φ, t) coordinates at a fixed point in space. (b) As in (a), but in the (θ, φ, s) coordinates at a fixed moment in time. (c) As in (b), but for a linearly polarized wave.

handed polarization the curve is a right-handed helix with an elliptical projection onto the $\theta\varphi$-plane centered around the t-axis (see Fig. 2.7.2(a)). The pitch of the helix is simply $2\pi/\omega$, where ω is the angular frequency of the wave.

Another way to visualize a plane wave is to fix a moment in time and draw a three-dimensional curve in (θ, φ, s) coordinates described by the tip of the electric field vector as a function of a spatial coordinate $s = \hat{\mathbf{n}} \cdot \mathbf{r}$ oriented along the direction of propagation $\hat{\mathbf{n}}$. According to Eqs. (2.7.8)–(2.7.11), the electric field is the same for all position–time combinations with constant $ks - \omega t$. Therefore, at any instant of time (say, $t = 0$) the locus of the points described by the tip of the electric field vector originating at different points on the s axis is also a helix, with the same projection onto the $\theta\varphi$-plane as the respective helix in the (θ, φ, t) coordinates but with

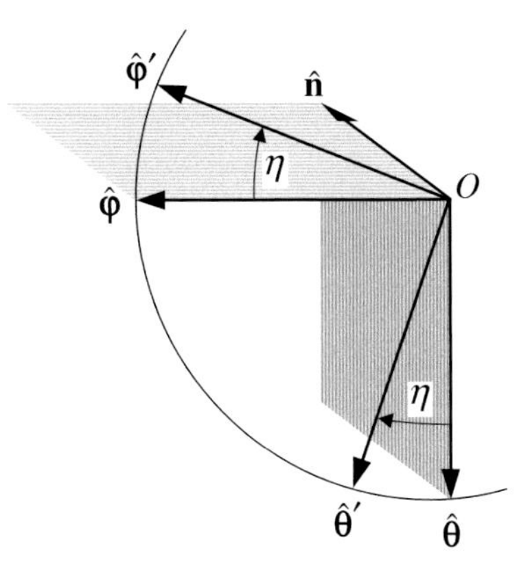

Figure 2.8.1. Rotation of the θ- and φ-axes through an angle $\eta \geq 0$ around $\hat{\mathbf{n}}$ in the clockwise direction when looking in the direction of propagation.

opposite handedness. For example, for the wave with right-handed elliptical polarization shown in Fig. 2.7.2(a), the respective curve in the (θ, φ, s) coordinates is a left-handed elliptical helix, shown in Fig. 2.7.2(b). The pitch of this helix is the wavelength λ. It is now clear that the propagation of the wave in time and space can be represented by progressive movement in time of the helix shown in Fig. 2.7.2(b) in the direction of $\hat{\mathbf{n}}$ with the speed of light. With increasing time, the intersection of the helix with any plane s = constant describes a right-handed vibration ellipse.

In the case of a circularly polarized wave, the elliptical helix becomes a helix with a circular projection onto the $\theta\varphi$-plane. If the wave is linearly polarized, then the helix degenerates into a simple sinusoidal curve in the plane making an angle ζ with the φ-axis (Fig. 2.7.2(c)).

2.8 Rotation transformation rule for the Stokes parameters

The Stokes parameters of a plane electromagnetic wave are always defined with respect to a reference plane containing the direction of wave propagation. If the reference plane is rotated about the direction of propagation then the Stokes parameters are modified according to a rotation transformation rule, which can be derived as follows. Consider a rotation of the coordinate axes θ and φ through an angle $0 \leq \eta < 2\pi$ in the *clockwise* direction when looking in the direction of propagation (Fig. 2.8.1). The transformation rule for rotation of a two-dimensional coordinate system yields

$$E'_{0\theta} = E_{0\theta}\cos\eta + E_{0\varphi}\sin\eta, \tag{2.8.1}$$

$$E'_{0\varphi} = -E_{0\theta}\sin\eta + E_{0\varphi}\cos\eta, \tag{2.8.2}$$

where the primes denote the electric-field vector components with respect to the new reference frame. It then follows from Eq. (2.6.4) that the rotation transformation rule for the Stokes column vector is

$$\mathbf{I}' = \mathbf{L}(\eta)\mathbf{I}, \tag{2.8.3}$$

where

$$\mathbf{L}(\eta) = \begin{bmatrix} 1 & 0 & 0 & 0 \\ 0 & \cos 2\eta & -\sin 2\eta & 0 \\ 0 & \sin 2\eta & \cos 2\eta & 0 \\ 0 & 0 & 0 & 1 \end{bmatrix} \tag{2.8.4}$$

is the so-called Stokes rotation matrix for angle η. It is obvious that a $\eta = \pi$ rotation does not change the Stokes parameters.

Because

$$(\mathbf{I}^{\mathrm{MS}})' = \mathbf{B}\mathbf{I}' = \mathbf{B}\mathbf{L}(\eta)\mathbf{I} = \mathbf{B}\mathbf{L}(\eta)\mathbf{B}^{-1}\mathbf{I}^{\mathrm{MS}}, \tag{2.8.5}$$

the rotation matrix for the modified Stokes column vector is given by

$$\mathbf{L}^{\mathrm{MS}}(\eta) = \mathbf{B}\mathbf{L}(\eta)\mathbf{B}^{-1} = \begin{bmatrix} \cos^2\eta & \sin^2\eta & -\frac{1}{2}\sin 2\eta & 0 \\ \sin^2\eta & \cos^2\eta & \frac{1}{2}\sin 2\eta & 0 \\ \sin 2\eta & -\sin 2\eta & \cos 2\eta & 0 \\ 0 & 0 & 0 & 1 \end{bmatrix}. \tag{2.8.6}$$

Similarly, for the circular-polarization representation,

$$(\mathbf{I}^{\mathrm{CP}})' = \mathbf{A}\mathbf{I}' = \mathbf{A}\mathbf{L}(\eta)\mathbf{I} = \mathbf{A}\mathbf{L}(\eta)\mathbf{A}^{-1}\mathbf{I}^{\mathrm{CP}}, \tag{2.8.7}$$

and the corresponding rotation matrix is diagonal:

$$\mathbf{L}^{\mathrm{CP}}(\eta) = \mathbf{A}\mathbf{L}(\eta)\mathbf{A}^{-1} = \begin{bmatrix} \exp(\mathrm{i}2\eta) & 0 & 0 & 0 \\ 0 & 1 & 0 & 0 \\ 0 & 0 & 1 & 0 \\ 0 & 0 & 0 & \exp(-\mathrm{i}2\eta) \end{bmatrix} \tag{2.8.8}$$

(Hovenier and van der Mee, 1983).

2.9 Quasi-monochromatic light

The definition of a monochromatic plane electromagnetic wave given by Eqs. (2.5.4) and (2.5.5) implies that the complex amplitudes $\mathbf{E}_0$ and $\mathbf{H}_0$ are constant. In reality, both amplitudes often fluctuate in time, albeit much more slowly than the time-harmonic factor $\exp(-\mathrm{i}\omega t)$. The fluctuations of the complex amplitudes include, in general, fluctuations of both the amplitude and the phase of the real electric and magnetic field vectors.

It is straightforward to verify that the electromagnetic field given by

$$\mathbf{E}(\mathbf{r},t) = \mathbf{E}_0(t)\exp(\mathrm{i}\mathbf{k}\cdot\mathbf{r} - \mathrm{i}\omega t), \tag{2.9.1}$$

$$\mathbf{H}(\mathbf{r},t) = \mathbf{H}_0(t)\exp(\mathrm{i}\mathbf{k}\cdot\mathbf{r} - \mathrm{i}\omega t) \tag{2.9.2}$$

still satisfies the Maxwell equations (2.1.1)–(2.1.4) at any moment in time provided that the medium is homogeneous and source-free and that

$$\mathbf{k}\cdot\mathbf{E}_0(t) = 0, \tag{2.9.3}$$

$$\mathbf{k}\cdot\mathbf{H}_0(t) = 0, \tag{2.9.4}$$

$$\mathbf{k}\times\mathbf{E}_0(t) = \omega\mu\mathbf{H}_0(t), \tag{2.9.5}$$

$$\mathbf{k}\times\mathbf{H}_0(t) = -\omega\varepsilon\mathbf{E}_0(t), \tag{2.9.6}$$

$$\left|\frac{\partial\mathbf{E}_0(t)}{\partial t}\right| \ll \omega|\mathbf{E}_0(t)|, \tag{2.9.7}$$

$$\left|\frac{\partial\mathbf{H}_0(t)}{\partial t}\right| \ll \omega|\mathbf{H}_0(t)|. \tag{2.9.8}$$

Equations (2.9.1)–(2.9.8) collectively define a parallel *quasi-monochromatic* beam of light. The latter can be thought of as a superposition of a large number of monochromatic plane electromagnetic waves which propagate in the same direction and are randomly distributed over a range of angular frequencies $[\omega - \Delta\omega, \omega + \Delta\omega]$ such that

$$\frac{\Delta\omega}{\omega} \ll 1 \tag{2.9.9}$$

(see, e.g., Subsection 7.3.3 of Born and Wolf, 1999).

Although the typical frequency of the fluctuations of the complex electric and magnetic field amplitudes is much smaller than the angular frequency ω, it is still so high that most optical instruments are incapable of tracing the instantaneous values of the Stokes parameters but rather respond to an average of the Stokes parameters over a relatively long period of time. Therefore, the definition of the Stokes parameters for a quasi-monochromatic beam of light propagating in a homogeneous nonabsorbing medium must be modified as follows:

$$I = \frac{1}{2}\sqrt{\frac{\epsilon}{\mu}}\,\{\langle E_{0\theta}(t)[E_{0\theta}(t)]^*\rangle_t + \langle E_{0\varphi}(t)[E_{0\varphi}(t)]^*\rangle_t\}, \tag{2.9.10}$$

$$Q = \frac{1}{2}\sqrt{\frac{\epsilon}{\mu}}\,\{\langle E_{0\theta}(t)[E_{0\theta}(t)]^*\rangle_t - \langle E_{0\varphi}(t)[E_{0\varphi}(t)]^*\rangle_t\}, \tag{2.9.11}$$

$$U = -\frac{1}{2}\sqrt{\frac{\epsilon}{\mu}}\,\{\langle E_{0\theta}(t)[E_{0\varphi}(t)]^*\rangle_t + \langle E_{0\varphi}(t)[E_{0\theta}(t)]^*\rangle_t\}, \tag{2.9.12}$$

$$V = \mathrm{i}\frac{1}{2}\sqrt{\frac{\epsilon}{\mu}}\,\{\langle E_{0\varphi}(t)[E_{0\theta}(t)]^*\rangle_t - \langle E_{0\theta}(t)[E_{0\varphi}(t)]^*\rangle_t\}, \tag{2.9.13}$$

where

$$\langle f(t)\rangle_t = \frac{1}{T}\int_t^{t+T} \mathrm{d}t' f(t') \tag{2.9.14}$$

denotes the average over a time interval T long compared with the typical period of fluctuation.

Equations (2.9.10)–(2.9.14) illustrate the usefulness of the concept of quasi-monochromatic light. Indeed, quasi-monochromatic light can be considered as monochromatic over time intervals long compared with the period of time-harmonic oscillations, $2\pi/\omega$, but short compared with the typical period of fluctuation. Therefore, the corresponding electric and magnetic field vectors at any moment in time can still be found by solving the time-harmonic Maxwell equations. Any observable characteristic of the quasi-monochromatic light can then be found by assuming that the corresponding monochromatic characteristic is "slowly varying" and averaging it over a sufficiently long time interval.

The Stokes identity (2.6.8) is not valid, in general, for a quasi-monochromatic beam. Indeed, now we have

$$\begin{aligned}
&I^2 - Q^2 - U^2 - V^2 \\
&= \frac{\epsilon}{\mu}[\langle a_\theta^2\rangle_t \langle a_\varphi^2\rangle_t - \langle a_\theta a_\varphi \cos\Delta\rangle_t^2 - \langle a_\theta a_\varphi \sin\Delta\rangle_t^2] \\
&= \frac{\epsilon}{\mu}\frac{1}{T^2}\int_t^{t+T} \mathrm{d}t' \int_t^{t+T} \mathrm{d}t'' \{[a_\theta(t')]^2[a_\varphi(t'')]^2 \\
&\qquad - a_\theta(t')a_\varphi(t')\cos[\Delta(t')]a_\theta(t'')a_\varphi(t'')\cos[\Delta(t'')] \\
&\qquad - a_\theta(t')a_\varphi(t')\sin[\Delta(t')]a_\theta(t'')a_\varphi(t'')\sin[\Delta(t'')]\} \\
&= \frac{\epsilon}{\mu}\frac{1}{T^2}\int_t^{t+T} \mathrm{d}t' \int_t^{t+T} \mathrm{d}t'' \{[a_\theta(t')]^2[a_\varphi(t'')]^2 \\
&\qquad - a_\theta(t')a_\varphi(t')a_\theta(t'')a_\varphi(t'')\cos[\Delta(t') - \Delta(t'')]\} \\
&= \frac{\epsilon}{\mu}\frac{1}{2T^2}\int_t^{t+T} \mathrm{d}t' \int_t^{t+T} \mathrm{d}t'' \{[a_\theta(t')]^2[a_\varphi(t'')]^2 + [a_\theta(t'')]^2[a_\varphi(t')]^2 \\
&\qquad - 2a_\theta(t')a_\varphi(t')a_\theta(t'')a_\varphi(t'')\cos[\Delta(t') - \Delta(t'')]\} \\
&\geq \frac{\epsilon}{\mu}\frac{1}{2T^2}\int_t^{t+T} \mathrm{d}t' \int_t^{t+T} \mathrm{d}t'' \{[a_\theta(t')]^2[a_\varphi(t'')]^2 + [a_\theta(t'')]^2[a_\varphi(t')]^2 \\
&\qquad - 2a_\theta(t')a_\varphi(t')a_\theta(t'')a_\varphi(t'')\} \\
&= \frac{\epsilon}{\mu}\frac{1}{2T^2}\int_t^{t+T} \mathrm{d}t' \int_t^{t+T} \mathrm{d}t'' [a_\theta(t')a_\varphi(t'') - a_\theta(t'')a_\varphi(t')]^2 \\
&\geq 0,
\end{aligned} \tag{2.9.15}$$

thereby yielding

$$I^2 \geq Q^2 + U^2 + V^2. \tag{2.9.16}$$

The equality holds only if the ratio $a_\theta(t)/a_\varphi(t)$ of the real amplitudes and the phase difference $\Delta(t)$ are independent of time, which means that $E_{0\theta}(t)$ and $E_{0\varphi}(t)$ are completely correlated. In this case the beam is said to be fully (or completely) polarized. This definition includes a monochromatic plane wave, but is, of course, more general. However, if $a_\theta(t)$, $a_\varphi(t)$, $\Delta_\theta(t)$, and $\Delta_\varphi(t)$ are totally uncorrelated and $\langle a_\theta^2 \rangle_t = \langle a_\varphi^2 \rangle_t$ then $Q = U = V = 0$, and the quasi-monochromatic beam of light is said to be unpolarized (or natural).

One way to visualize quasi-monochromatic light is to assume that Eqs. (2.9.1) and (2.9.2) describe an instantaneous polarization ellipse with ellipticity, handedness, orientation, and size fluctuating in time. This means that for unpolarized light, the parameters of the vibration ellipse traced by the endpoint of the electric field vector fluctuate in such a way that there is no preferred vibration ellipse. For a completely polarized beam, the ellipticity, handedness, and orientation of the ellipse remain constant, and only the size of the ellipse may change in time. In all other cases, the quasi-monochromatic beam is partially polarized with certain "amounts of preference" for ellipticity, handedness, and orientation; these amounts of preference are not equal to 100% for all of the three ellipse parameters.

Thus, quasi-monochromatic light can be partially polarized and even completely unpolarized, whereas a plane electromagnetic wave is always fully polarized. The realization of this fact was the main motivation for the introduction of the Stokes parameters as descriptors of the polarization state of a light beam (Stokes, 1852).

When two or more quasi-monochromatic beams propagating in the same direction are mixed incoherently, which means that there is no permanent phase relation between the separate beams, then the Stokes column vector of the mixture is equal to the sum of the Stokes column vectors of the individual beams:

$$\mathbf{I} = \sum_n \mathbf{I}_n, \tag{2.9.17}$$

where n numbers the beams. Indeed, inserting Eqs. (2.7.1) and (2.7.2) in Eq. (2.9.10), we obtain for the total intensity

$$\begin{aligned} I &= \frac{1}{2}\sqrt{\frac{\epsilon}{\mu}} \sum_n \sum_m \langle a_{\theta n} a_{\theta m} \exp[\mathrm{i}(\Delta_{\theta n} - \Delta_{\theta m})] \\ &\qquad + a_{\varphi n} a_{\varphi m} \exp[\mathrm{i}(\Delta_{\varphi n} - \Delta_{\varphi m})] \rangle_t \\ &= \frac{1}{2}\sqrt{\frac{\epsilon}{\mu}} \left\{ \sum_n I_n + \sum_n \sum_{m \neq n} \langle a_{\theta n} a_{\theta m} \exp[\mathrm{i}(\Delta_{\theta n} - \Delta_{\theta m})] \right. \\ &\qquad \left. + a_{\varphi n} a_{\varphi m} \exp[\mathrm{i}(\Delta_{\varphi n} - \Delta_{\varphi m})] \rangle_t \right\}. \end{aligned} \tag{2.9.18}$$

Since the phases of different beams are uncorrelated, the second term on the right-hand side of the relation above vanishes. Hence

$$I = \sum_n I_n, \tag{2.9.19}$$

and similarly for Q, U, and V. Of course, this additivity rule also applies to the coherency matrix $\boldsymbol{\rho}$, the modified Stokes column vector $\mathbf{I}^{\mathrm{MS}}$, and the circular-polarization column vector $\mathbf{I}^{\mathrm{CP}}$.

The additivity of the Stokes parameters allows us to generalize the principle of optical equivalence (Section 2.6) to quasi-monochromatic light as follows: it is impossible by means of a traditional optical instrument to distinguish between various incoherent mixtures of quasi-monochromatic beams that form a beam with the same Stokes parameters (I, Q, U, V). For example, there is only one kind of unpolarized light, although it can be composed of quasi-monochromatic beams in an infinite variety of optically indistinguishable ways.

According to Eqs. (2.9.16) and (2.9.17), it is always possible *mathematically* to decompose any quasi-monochromatic beam into two incoherent parts, one unpolarized, with a Stokes column vector

$$[I - \sqrt{Q^2 + U^2 + V^2} \quad 0 \quad 0 \quad 0]^{\mathrm{T}},$$

and one fully polarized, with a Stokes column vector

$$[\sqrt{Q^2 + U^2 + V^2} \quad Q \quad U \quad V]^{\mathrm{T}}.$$

Thus, the intensity of the fully polarized component is $(Q^2 + U^2 + V^2)^{1/2}$, so that the degree of (elliptical) polarization of the quasi-monochromatic beam is

$$P = \frac{\sqrt{Q^2 + U^2 + V^2}}{I}. \tag{2.9.20}$$

The degree of linear polarization is defined as

$$P_{\mathrm{L}} = \frac{\sqrt{Q^2 + U^2}}{I} \tag{2.9.21}$$

and the degree of circular polarization as

$$P_{\mathrm{C}} = V/I. \tag{2.9.22}$$

P vanishes for unpolarized light and is equal to unity for fully polarized light. For a partially polarized beam $(0 < P < 1)$ with $V \neq 0$, the sign of V indicates the preferential handedness of the vibration ellipses described by the endpoint of the electric field vector. Specifically, a positive V indicates left-handed polarization and a negative V indicates right-handed polarization. By analogy with Eqs. (2.7.28) and (2.7.29), the quantities $-U/Q$ and $|V|/(Q^2 + U^2)^{1/2}$ can be interpreted as specifying the preferential orientation and ellipticity of the vibration ellipse. Unlike the Stokes parameters, these quantities are not additive. According to Eqs. (2.8.3) and (2.8.4), P, P_{L},

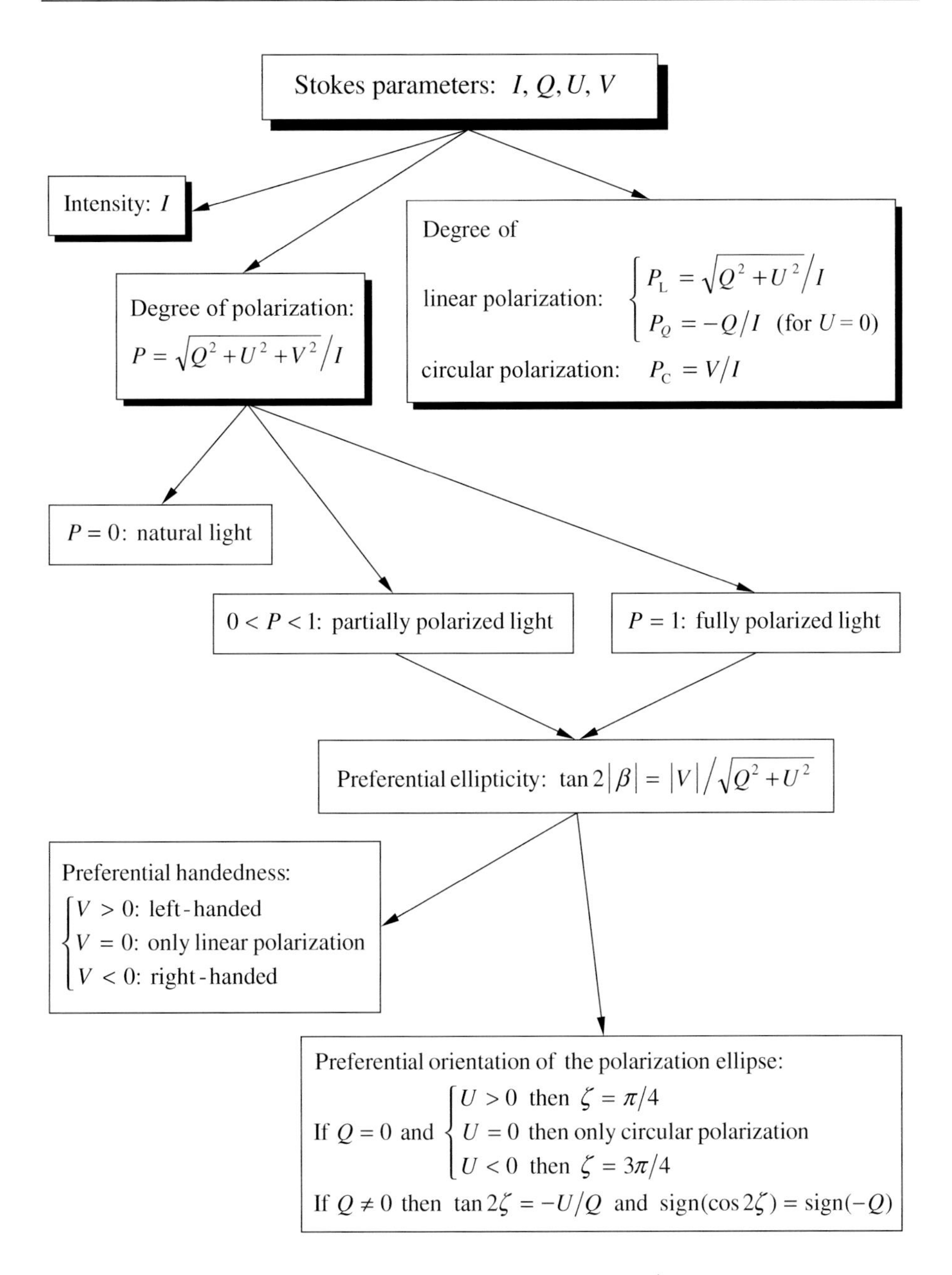

Figure 2.9.1. Analysis of a quasi-monochromatic beam with Stokes parameters I, Q, U, and V.

and P_C are invariant with respect to rotations of the reference frame around the direction of propagation.

When $U = 0$, the ratio

$$P_Q = -Q/I \tag{2.9.23}$$

is also called the degree of linear polarization (or the *signed* degree of linear polarization). P_Q is positive when the vibrations of the electric field vector in the φ-direction (i.e., the direction perpendicular to the meridional plane of the beam) dominate those in the θ-direction and is negative otherwise.

The standard polarimetric analysis of a general quasi-monochromatic beam with Stokes parameters I, Q, U, and V is summarized in Fig. 2.9.1 (after Hovenier *et al.*, 2004).

2.10 Measurement of the Stokes parameters

Most detectors of electromagnetic radiation, especially those in the visible and infrared spectral range, are insensitive to the polarization state of the beam impinging on the detector surface and can measure only the first Stokes parameter of the beam, viz., the intensity. Therefore, to measure the entire Stokes column vector of the beam, one has to insert between the source of light and the detector one or several optical elements that modify the beam so that the new first Stokes parameter of the radiation reaching the detector now contains information about the second, third, and fourth Stokes parameters of the original beam. This is usually done with so-called polarizers and retarders, and typically entails a succession of several such measurements to fully characterize the Stokes column vector.

A polarizer is an optical element that attenuates the orthogonal components of the electric field vector of an electromagnetic wave unevenly. Let us denote the corresponding attenuation coefficients as p_θ and p_φ and consider first the situation when the two orthogonal transmission axes of a polarizer coincide with the θ- and φ-axes of the laboratory coordinate system (see Fig. 2.10.1). This means that after the electromagnetic wave goes through the polarizer, the orthogonal components of the electric field change as follows:

$$E'_\theta = p_\theta E_\theta, \quad 0 \le p_\theta \le 1, \tag{2.10.1}$$

$$E'_\varphi = p_\varphi E_\varphi, \quad 0 \le p_\varphi \le 1. \tag{2.10.2}$$

It then follows from the definition of the Stokes parameters, Eq. (2.6.4), that the Stokes column vector of the wave modifies according to

$$\mathbf{I}' = \mathbf{P}\mathbf{I}, \tag{2.10.3}$$

where

$$\mathbf{P} = \frac{1}{2}\begin{bmatrix} p_\theta^2 + p_\varphi^2 & p_\theta^2 - p_\varphi^2 & 0 & 0 \\ p_\theta^2 - p_\varphi^2 & p_\theta^2 + p_\varphi^2 & 0 & 0 \\ 0 & 0 & 2p_\theta p_\varphi & 0 \\ 0 & 0 & 0 & 2p_\theta p_\varphi \end{bmatrix} \tag{2.10.4}$$

is the so-called Mueller matrix representing the effect of the polarizer.

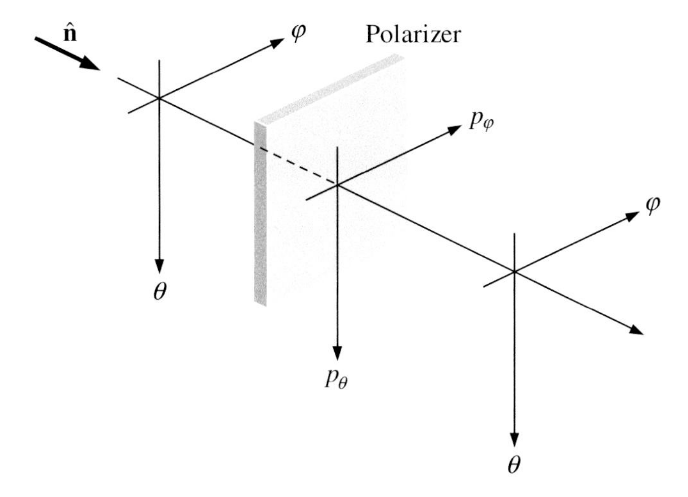

Figure 2.10.1. The transmission axes of the polarizer coincide with those of the laboratory reference frame.

An important example of a polarizer is a neutral filter with $p_\theta = p_\varphi = p$, which equally attenuates the orthogonal components of the electric field vector and does not change the polarization state of the wave:

$$\mathbf{P} = p^2 \begin{bmatrix} 1 & 0 & 0 & 0 \\ 0 & 1 & 0 & 0 \\ 0 & 0 & 1 & 0 \\ 0 & 0 & 0 & 1 \end{bmatrix}. \tag{2.10.5}$$

In contrast, an ideal linear polarizer transmits only one orthogonal component of the wave (say, the θ-component) and completely blocks the other one ($p_\varphi = 0$):

$$\mathbf{P} = \frac{p_\theta^2}{2} \begin{bmatrix} 1 & 1 & 0 & 0 \\ 1 & 1 & 0 & 0 \\ 0 & 0 & 0 & 0 \\ 0 & 0 & 0 & 0 \end{bmatrix}. \tag{2.10.6}$$

An ideal perfect linear polarizer does not change one orthogonal component ($p_\theta = 1$) and completely blocks the other one ($p_\varphi = 0$):

$$\mathbf{P} = \frac{1}{2} \begin{bmatrix} 1 & 1 & 0 & 0 \\ 1 & 1 & 0 & 0 \\ 0 & 0 & 0 & 0 \\ 0 & 0 & 0 & 0 \end{bmatrix}. \tag{2.10.7}$$

If the transmission axes of a polarizer are rotated relative to the laboratory co-

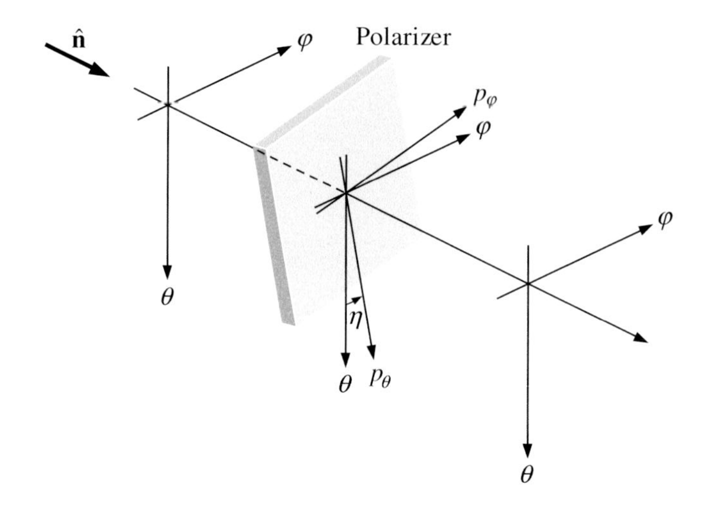

Figure 2.10.2. The polarizer transmission axes are rotated through an angle $\eta \geq 0$ around $\hat{\mathbf{n}}$ in the clockwise direction when looking in the direction of propagation.

ordinate system (Fig. 2.10.2) then its Mueller matrix with respect to the laboratory coordinate system also changes. To obtain the resulting Stokes column vector with respect to the laboratory coordinate system, we need to:

- "Rotate" the initial Stokes column vector through the angle η in the clockwise direction in order to obtain the Stokes parameters of the original beam with respect to the polarizer axes.
- Multiply the "rotated" Stokes column vector by the original (nonrotated) polarizer Mueller matrix.
- "Rotate" the Stokes column vector thus obtained through the angle $-\eta$ in order to calculate the Stokes parameters of the resulting beam with respect to the laboratory coordinate system.

The final result is as follows:

$$\mathbf{I}' = \mathbf{L}(-\eta)\mathbf{P}\mathbf{L}(\eta)\mathbf{I}. \tag{2.10.8}$$

Hence the Mueller matrix of the rotated polarizer computed with respect to the laboratory coordinate system is given by

$$\mathbf{P}(\eta) = \mathbf{L}(-\eta)\mathbf{P}\mathbf{L}(\eta) \tag{2.10.9}$$

with $\mathbf{P}(0) = \mathbf{P}$.

A retarder is an optical element that changes the phase of the beam by causing a phase shift of $+\zeta/2$ along the θ-axis and a phase shift of $-\zeta/2$ along the φ-axis (Fig. 2.10.3). We thus have

$$E'_\theta = \exp(+\mathrm{i}\zeta/2)E_\theta, \tag{2.10.10}$$

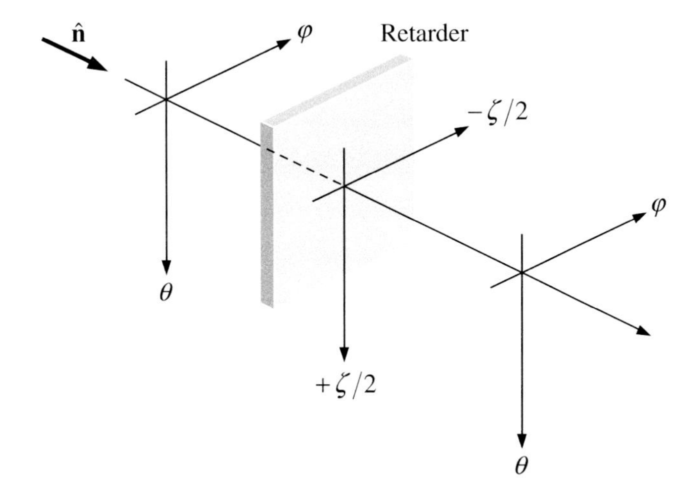

Figure 2.10.3. Propagation of a beam through a retarder.

$$E'_\varphi = \exp(-\mathrm{i}\zeta/2)E_\varphi, \tag{2.10.11}$$

which yields

$$\mathbf{I}' = \mathbf{R}(\zeta)\mathbf{I}, \tag{2.10.12}$$

where

$$\mathbf{R}(\zeta) = \begin{bmatrix} 1 & 0 & 0 & 0 \\ 0 & 1 & 0 & 0 \\ 0 & 0 & \cos\zeta & \sin\zeta \\ 0 & 0 & -\sin\zeta & \cos\zeta \end{bmatrix} \tag{2.10.13}$$

is the Mueller matrix of the retarder.

Consider now the optical path shown in Fig. 2.10.4. The beam of light goes through a retarder and a rotated ideal perfect linear polarizer and then impinges on the surface of a polarization-insensitive detector. The Stokes column vector of the resulting beam impinging on the detector surface is given by

$$\mathbf{I}' = \mathbf{P}(\eta)\mathbf{R}(\zeta)\mathbf{I}, \tag{2.10.14}$$

where the polarizer Mueller matrix is

$$\mathbf{P}(\eta) = \frac{1}{2}\begin{bmatrix} 1 & \cos 2\eta & -\sin 2\eta & 0 \\ \cos 2\eta & \cos^2 2\eta & -\cos 2\eta \sin 2\eta & 0 \\ -\sin 2\eta & -\cos 2\eta \sin 2\eta & \sin^2 2\eta & 0 \\ 0 & 0 & 0 & 0 \end{bmatrix} \tag{2.10.15}$$

(cf. Eqs. (2.10.7) and (2.10.9)). Hence the intensity of the resulting beam as a function

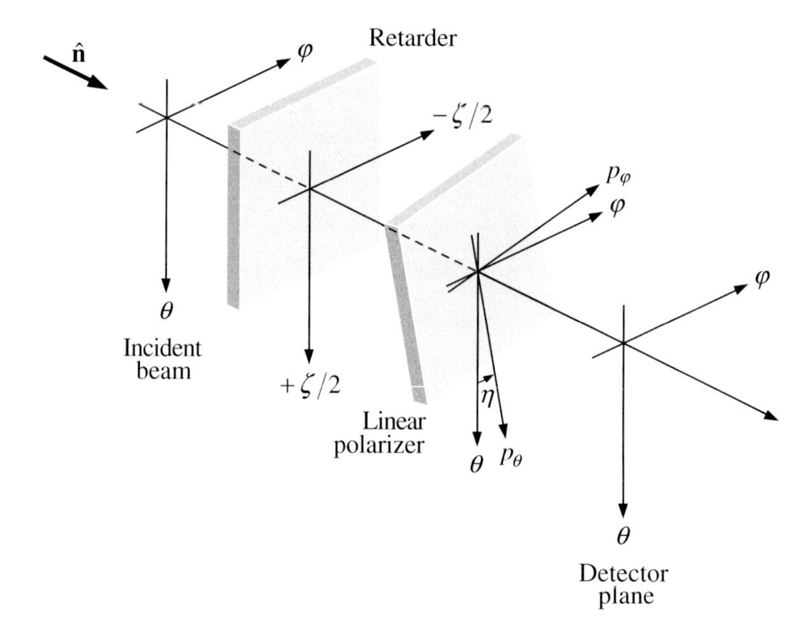

Figure 2.10.4. Measurement of the Stokes parameters with a retarder and an ideal perfect linear polarizer rotated with respect to the laboratory reference frame.

of η and ζ is given by

$$I'(\eta,\zeta) = \tfrac{1}{2}(I + Q\cos 2\eta - U\sin 2\eta\cos\zeta - V\sin 2\eta\sin\zeta). \tag{2.10.16}$$

This formula suggests a simple way to determine the Stokes parameters of the original beam by measuring the intensity of the resulting beam using four different combinations of η and ζ:

$$I = I'(0°, 0°) + I'(90°, 0°), \tag{2.10.17}$$

$$Q = I'(0°, 0°) - I'(90°, 0°), \tag{2.10.18}$$

$$U = -2I'(45°, 0°) + I, \tag{2.10.19}$$

$$V = I - 2I'(45°, 90°). \tag{2.10.20}$$

Other methods for measuring the Stokes parameters and practical aspects of polarimetry are discussed in detail in the books by Shurcliff (1962), Clarke and Grainger (1971), Azzam and Bashara (1977), Kliger *et al.* (1990), and Collett (1992).

2.11 Spherical-wave solution

As we have seen, plane electromagnetic waves represent a fundamental solution of the Maxwell equations underlying the concept of a monochromatic parallel beam of light. Another fundamental solution representing the outward propagation of electromagnetic energy from a point-like source is a transverse spherical wave. To derive this solution, we need Eqs. (2.3.3), (2.3.4), (2.3.11), (2.3.16), (2.3.17), (2.5.1), and (2.5.2) as well as the following formulas:

$$\nabla \frac{\exp(\pm \mathrm{i}kr)}{r} = \left(\pm \mathrm{i}k - \frac{1}{r} \right) \frac{\exp(\pm \mathrm{i}kr)}{r} \hat{\mathbf{r}}, \tag{2.11.1}$$

$$\nabla \cdot \mathbf{a} = \frac{\partial a_r}{\partial r} + \frac{2a_r}{r} + \frac{1}{r}\frac{\partial a_\theta}{\partial \theta} + \frac{a_\theta}{r \tan\theta} + \frac{1}{r \sin\theta}\frac{\partial a_\varphi}{\partial \varphi}, \tag{2.11.2}$$

$$\nabla \times \mathbf{a} = \left(\frac{1}{r}\frac{\partial a_\varphi}{\partial \theta} + \frac{a_\varphi}{r\tan\theta} - \frac{1}{r\sin\theta}\frac{\partial a_\theta}{\partial \varphi} \right)\hat{\mathbf{r}} + \left(\frac{1}{r\sin\theta}\frac{\partial a_r}{\partial \varphi} - \frac{\partial a_\varphi}{\partial r} - \frac{a_\varphi}{r} \right)\hat{\boldsymbol{\theta}} + \left(\frac{\partial a_\theta}{\partial r} + \frac{a_\theta}{r} - \frac{1}{r}\frac{\partial a_r}{\partial \theta} \right)\hat{\boldsymbol{\varphi}}, \tag{2.11.3}$$

where $\hat{\mathbf{r}} = \mathbf{r}/|\mathbf{r}|$ is the unit vector in the direction of the position vector $\mathbf{r}$. It is then straightforward to verify that the complex field vectors

$$\mathbf{E}(\mathbf{r},t) = \frac{\exp(\mathrm{i}kr)}{r}\mathbf{E}_1(\hat{\mathbf{r}})\exp(-\mathrm{i}\omega t), \tag{2.11.4}$$

$$\mathbf{H}(\mathbf{r},t) = \frac{\exp(\mathrm{i}kr)}{r}\mathbf{H}_1(\hat{\mathbf{r}})\exp(-\mathrm{i}\omega t) \tag{2.11.5}$$

are a solution of the Maxwell equations in the limit $kr \to \infty$ provided that the medium is homogeneous and that

$$\hat{\mathbf{r}} \cdot \mathbf{E}_1(\hat{\mathbf{r}}) = 0, \tag{2.11.6}$$

$$\hat{\mathbf{r}} \cdot \mathbf{H}_1(\hat{\mathbf{r}}) = 0, \tag{2.11.7}$$

$$k\hat{\mathbf{r}} \times \mathbf{E}_1(\hat{\mathbf{r}}) = \omega\mu\mathbf{H}_1(\hat{\mathbf{r}}), \tag{2.11.8}$$

$$k\hat{\mathbf{r}} \times \mathbf{H}_1(\hat{\mathbf{r}}) = -\omega\varepsilon\mathbf{E}_1(\hat{\mathbf{r}}), \tag{2.11.9}$$

where the wave number $k = k_{\mathrm{R}} + \mathrm{i}k_{\mathrm{I}} = \omega(\varepsilon\mu)^{1/2} = \omega m/c$ may be complex and the $\mathbf{E}_1(\hat{\mathbf{r}})$ and $\mathbf{H}_1(\hat{\mathbf{r}})$ are independent of the distance r from the origin.

Equations (2.11.4)–(2.11.9) describe an *outgoing* transverse spherical wave propagating radially with the phase velocity $v = \omega/k_{\mathrm{R}}$ and having mutually perpendicular complex electric and magnetic field vectors. The wave is homogeneous in that the real and imaginary parts of the local complex wave vector $k\hat{\mathbf{r}}$ are parallel. The surfaces of constant phase coincide with the surfaces of constant amplitude and are spherical. It is obvious that

$$\mathbf{H}(\mathbf{r},t) = \frac{k}{\omega\mu}\hat{\mathbf{r}} \times \mathbf{E}(\mathbf{r},t), \tag{2.11.10}$$

which allows one to consider the spherical wave in terms of the electric (or the mag-

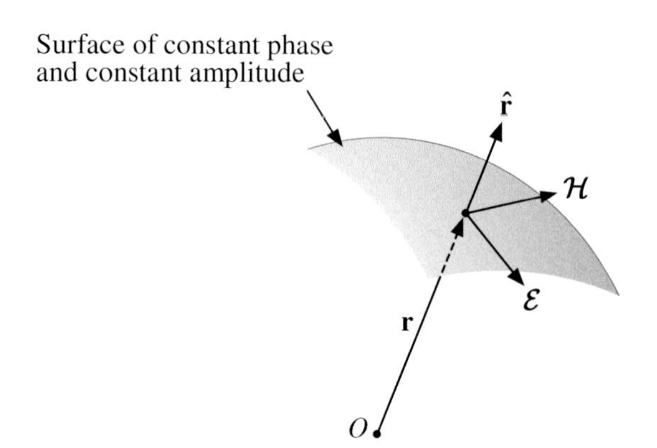

Figure 2.11.1. Spherical electromagnetic wave propagating in a homogeneous medium with no dispersion and losses.

netic) field only. The time-averaged Poynting vector of the wave is given by

$$\langle \boldsymbol{\mathcal{S}}(\mathbf{r},t)\rangle_t = \tfrac{1}{2}\mathrm{Re}\left(\sqrt{\frac{\varepsilon}{\mu}}\right)\frac{|\mathbf{E}_1(\hat{\mathbf{r}})|^2}{r^2}\exp\left(-2\frac{\omega}{c}m_{\mathrm{I}}r\right)\hat{\mathbf{r}}, \tag{2.11.11}$$

where, as before, $m_{\mathrm{I}} = ck_{\mathrm{I}}/\omega$. Thus, the local direction of the electromagnetic energy flow is away from the origin. The intensity of the spherical wave is defined as the absolute value of the time-averaged Poynting vector,

$$I(\mathbf{r}) = |\langle \boldsymbol{\mathcal{S}}(\mathbf{r},t)\rangle_t| = \tfrac{1}{2}\mathrm{Re}\left(\sqrt{\frac{\varepsilon}{\mu}}\right)\frac{|\mathbf{E}_1(\hat{\mathbf{r}})|^2}{r^2}\exp\left(-2\frac{\omega}{c}m_{\mathrm{I}}r\right). \tag{2.11.12}$$

The intensity has the dimension of monochromatic energy flux and specifies the amount of electromagnetic energy crossing a unit surface element normal to $\hat{\mathbf{r}}$ per unit time. The intensity is attenuated exponentially by absorption and in addition decreases as the inverse square of the distance from the origin.

In the case of a medium with no dispersion and losses, the real electric and magnetic field vectors are mutually orthogonal and are normal to the direction of propagation $\hat{\mathbf{r}}$ (Fig. 2.11.1). The energy conservation law takes the form

$$\begin{aligned}\oint_S \mathrm{d}S\,\langle \boldsymbol{\mathcal{S}}(\mathbf{r},t)\rangle_t\cdot\hat{\mathbf{r}} &= \oint_S \mathrm{d}S\,I(\mathbf{r})\\ &= \frac{1}{2}\sqrt{\frac{\epsilon}{\mu}}\frac{1}{r^2}\oint_S \mathrm{d}S\,|\mathbf{E}_1(\hat{\mathbf{r}})|^2\\ &= \frac{1}{2}\sqrt{\frac{\epsilon}{\mu}}\int_{4\pi}\mathrm{d}\hat{\mathbf{r}}\,|\mathbf{E}_1(\hat{\mathbf{r}})|^2\\ &= \text{constant},\end{aligned} \tag{2.11.13}$$

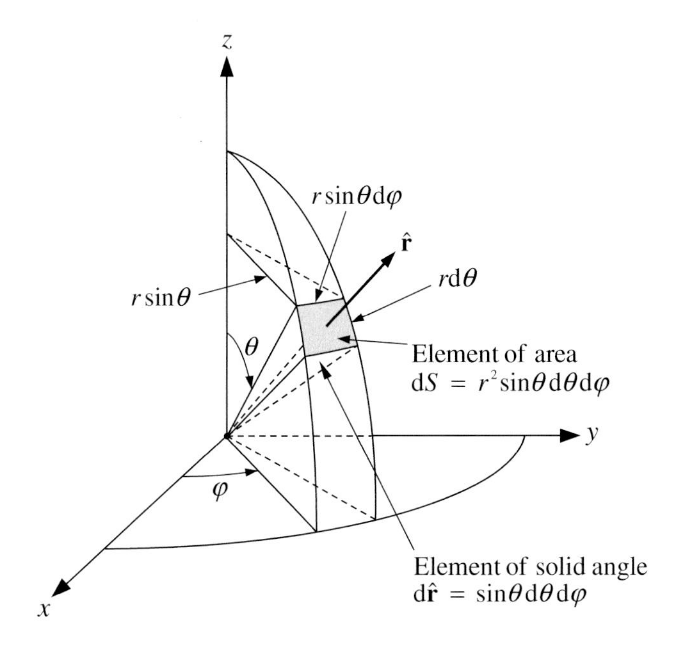

Figure 2.11.2. Differential solid angle in polar spherical coordinates.

where S is the sphere of radius r and

$$d\hat{\mathbf{r}} = \frac{dS}{r^2} = \sin\theta d\theta d\varphi \tag{2.11.14}$$

is a differential solid angle element around the direction $\hat{\mathbf{r}}$ (see Fig. 2.11.2). It is also easy to show that in the case of a nonabsorbing medium, the time-averaged energy density of a spherical wave is given by

$$\langle \mathcal{U}(\mathbf{r},t)\rangle_t = \tfrac{1}{2}\epsilon\frac{|\mathbf{E}_1(\hat{\mathbf{r}})|^2}{r^2}. \tag{2.11.15}$$

Equations (2.11.12) and (2.11.15) show that

$$I(\mathbf{r}) = v\langle \mathcal{U}(\mathbf{r},t)\rangle_t, \tag{2.11.16}$$

where $v = 1/(\epsilon\mu)^{1/2}$ is the speed of light in the material medium. The same result was obtained previously for a plane wave propagating in a nonabsorbing medium (cf. Eq. (2.5.30)).

In complete analogy with the case of a plane wave, the coherency matrix, the coherency column vector, and the Stokes column vector of a spherical wave propagating in a homogeneous medium with no dispersion and losses can be defined as

$$\boldsymbol{\rho}(\mathbf{r}) = \begin{bmatrix} \rho_{11}(\mathbf{r}) & \rho_{12}(\mathbf{r}) \\ \rho_{21}(\mathbf{r}) & \rho_{22}(\mathbf{r}) \end{bmatrix} = \frac{1}{2}\sqrt{\frac{\epsilon}{\mu}}\frac{1}{r^2}\begin{bmatrix} E_{1\theta}(\hat{\mathbf{r}})[E_{1\theta}(\hat{\mathbf{r}})]^* & E_{1\theta}(\hat{\mathbf{r}})[E_{1\varphi}(\hat{\mathbf{r}})]^* \\ E_{1\varphi}(\hat{\mathbf{r}})[E_{1\theta}(\hat{\mathbf{r}})]^* & E_{1\varphi}(\hat{\mathbf{r}})[E_{1\varphi}(\hat{\mathbf{r}})]^* \end{bmatrix}, \tag{2.11.17}$$

$$\mathsf{J}(\mathbf{r}) = \begin{bmatrix} \rho_{11}(\mathbf{r}) \\ \rho_{12}(\mathbf{r}) \\ \rho_{21}(\mathbf{r}) \\ \rho_{22}(\mathbf{r}) \end{bmatrix} = \frac{1}{2}\sqrt{\frac{\epsilon}{\mu}}\frac{1}{r^2} \begin{bmatrix} E_{1\theta}(\hat{\mathbf{r}})[E_{1\theta}(\hat{\mathbf{r}})]^* \\ E_{1\theta}(\hat{\mathbf{r}})[E_{1\varphi}(\hat{\mathbf{r}})]^* \\ E_{1\varphi}(\hat{\mathbf{r}})[E_{1\theta}(\hat{\mathbf{r}})]^* \\ E_{1\varphi}(\hat{\mathbf{r}})[E_{1\varphi}(\hat{\mathbf{r}})]^* \end{bmatrix}, \tag{2.11.18}$$

$$\mathsf{I}(\mathbf{r}) = \begin{bmatrix} I(\mathbf{r}) \\ Q(\mathbf{r}) \\ U(\mathbf{r}) \\ V(\mathbf{r}) \end{bmatrix} = \mathsf{D}\,\mathsf{J}(\mathbf{r}) = \frac{1}{2}\sqrt{\frac{\epsilon}{\mu}}\frac{1}{r^2} \begin{bmatrix} E_{1\theta}(\hat{\mathbf{r}})[E_{1\theta}(\hat{\mathbf{r}})]^* + E_{1\varphi}(\hat{\mathbf{r}})[E_{1\varphi}(\hat{\mathbf{r}})]^* \\ E_{1\theta}(\hat{\mathbf{r}})[E_{1\theta}(\hat{\mathbf{r}})]^* - E_{1\varphi}(\hat{\mathbf{r}})[E_{1\varphi}(\hat{\mathbf{r}})]^* \\ -E_{1\theta}(\hat{\mathbf{r}})[E_{1\varphi}(\hat{\mathbf{r}})]^* - E_{1\varphi}(\hat{\mathbf{r}})[E_{1\theta}(\hat{\mathbf{r}})]^* \\ \mathrm{i}\{E_{1\varphi}(\hat{\mathbf{r}})[E_{1\theta}(\hat{\mathbf{r}})]^* - E_{1\theta}(\hat{\mathbf{r}})[E_{1\varphi}(\hat{\mathbf{r}})]^*\} \end{bmatrix}, \tag{2.11.19}$$

respectively. All these quantities have the dimension of monochromatic energy flux. As before, the first Stokes parameter is the intensity (defined this time by Eq. (2.11.12)).

The reader is invited to verify that the complex field vectors

$$\mathbf{E}(\mathbf{r},t) = \frac{\exp(-\mathrm{i}kr)}{r}\mathbf{E}_1(\hat{\mathbf{r}})\exp(-\mathrm{i}\omega t), \tag{2.11.20}$$

$$\mathbf{H}(\mathbf{r},t) = \frac{\exp(-\mathrm{i}kr)}{r}\mathbf{H}_1(\hat{\mathbf{r}})\exp(-\mathrm{i}\omega t) \tag{2.11.21}$$

represent yet another solution of the Maxwell equations in the limit $kr \to \infty$ provided that the medium is homogeneous and that

$$\hat{\mathbf{r}}\cdot\mathbf{E}_1(\hat{\mathbf{r}}) = 0, \tag{2.11.22}$$

$$\hat{\mathbf{r}}\cdot\mathbf{H}_1(\hat{\mathbf{r}}) = 0, \tag{2.11.23}$$

$$k\hat{\mathbf{r}}\times\mathbf{E}_1(\hat{\mathbf{r}}) = -\omega\mu\mathbf{H}_1(\hat{\mathbf{r}}), \tag{2.11.24}$$

$$k\hat{\mathbf{r}}\times\mathbf{H}_1(\hat{\mathbf{r}}) = \omega\varepsilon\mathbf{E}_1(\hat{\mathbf{r}}). \tag{2.11.25}$$

These formulas describe an *incoming* transverse spherical wave with mutually perpendicular complex electric and magnetic field vectors. The spherical surfaces of constant phase and constant amplitude and the electromagnetic energy propagate radially in the direction of the local unit vector $-\hat{\mathbf{r}}$.

2.12 Coherency dyad of the electric field

The definition of the coherency and Stokes column vectors explicitly exploits the transverse character of an electromagnetic wave and requires the use of a local spherical coordinate system. However, in some cases it is convenient to introduce an alternative quantity that also provides a complete optical specification of a transverse electromagnetic wave but is defined without explicit use of a coordinate system. One such quantity is called the coherency dyad and, in the general case of an arbitrary electromagnetic field, is given by

$$\ddot{\rho}(\mathbf{r},t) = \mathbf{E}(\mathbf{r},t)\otimes[\mathbf{E}(\mathbf{r},t)]^*, \tag{2.12.1}$$

where $\otimes$ denotes the dyadic product of two vectors (see Appendix A for a discussion of dyads and dyadics). It is then clear that the coherency and Stokes column vectors of a transverse time-harmonic electromagnetic wave propagating in the direction $\hat{\mathbf{n}}$ through a homogeneous medium with no dispersion and losses can be expressed in terms of the coherency dyad as follows:

$$\mathbf{J} = \frac{1}{2}\sqrt{\frac{\epsilon}{\mu}}\begin{bmatrix} \hat{\boldsymbol{\theta}}\cdot\ddot{\rho}\cdot\hat{\boldsymbol{\theta}} \\ \hat{\boldsymbol{\theta}}\cdot\ddot{\rho}\cdot\hat{\boldsymbol{\varphi}} \\ \hat{\boldsymbol{\varphi}}\cdot\ddot{\rho}\cdot\hat{\boldsymbol{\theta}} \\ \hat{\boldsymbol{\varphi}}\cdot\ddot{\rho}\cdot\hat{\boldsymbol{\varphi}}) \end{bmatrix}, \tag{2.12.2}$$

$$\mathbf{I} = \frac{1}{2}\sqrt{\frac{\epsilon}{\mu}}\begin{bmatrix} \hat{\boldsymbol{\theta}}\cdot\ddot{\rho}\cdot\hat{\boldsymbol{\theta}} + \hat{\boldsymbol{\varphi}}\cdot\ddot{\rho}\cdot\hat{\boldsymbol{\varphi}} \\ \hat{\boldsymbol{\theta}}\cdot\ddot{\rho}\cdot\hat{\boldsymbol{\theta}} - \hat{\boldsymbol{\varphi}}\cdot\ddot{\rho}\cdot\hat{\boldsymbol{\varphi}} \\ -\hat{\boldsymbol{\theta}}\cdot\ddot{\rho}\cdot\hat{\boldsymbol{\varphi}} - \hat{\boldsymbol{\varphi}}\cdot\ddot{\rho}\cdot\hat{\boldsymbol{\theta}} \\ \mathrm{i}(\hat{\boldsymbol{\varphi}}\cdot\ddot{\rho}\cdot\hat{\boldsymbol{\theta}} - \hat{\boldsymbol{\theta}}\cdot\ddot{\rho}\cdot\hat{\boldsymbol{\varphi}}) \end{bmatrix}, \tag{2.12.3}$$

whereas the products $\ddot{\rho}\cdot\hat{\mathbf{n}}$ and $\hat{\mathbf{n}}\cdot\ddot{\rho}$ vanish. It follows from the definition of the coherency dyad that it is Hermitian:

$$\ddot{\rho}^{\mathrm{T}} = \rho^*, \tag{2.12.4}$$

where T denotes the transpose of a dyad(ic).

The coherency dyad is a more general quantity than the coherency and Stokes column vectors because it can be applied to any electromagnetic field and not just to a transverse electromagnetic wave. The simplest example of a situation in which the coherency dyad can be introduced, whereas the Stokes column vector cannot, involves the superposition of two plane electromagnetic waves propagating in different directions. The more general nature of the coherency dyad makes it very convenient in studies of random electromagnetic fields created by large stochastic groups of scatterers. For example, the additivity of the Stokes parameters (see Section 2.9) is a concept that can be applied only to transverse waves propagating in exactly the same direction, whereas a statistical average of the coherency dyad of a random electromagnetic field at an observation point can sometimes be reduced to an incoherent sum of coherency dyads of transverse waves propagating in various directions (see Section 8.6).

It is important to remember, however, that when the coherency dyad is applied to an arbitrary electromagnetic field, it may not always have as definite a physical meaning as, for example, the Poynting vector. The relationship between the coherency dyad and the actual physical observables may change depending on the problem in hand and must be established carefully whenever this quantity is used in a theoretical analysis of a specific measurement procedure. For example, the right-hand sides of Eqs. (2.12.2) and (2.12.3) may become rather meaningless if the products $\ddot{p} \cdot \hat{\mathbf{n}}$ and $\hat{\mathbf{n}} \cdot \ddot{p}$ do not vanish.

2.13 Historical notes and further reading

The equations of classical electromagnetics were written originally by James Clerk Maxwell (1831–79) in Cartesian component form (Maxwell, 1891) and were cast in the modern vector form by Oliver Heaviside (1850–1925). The subsequent experimental verification of Maxwell's theory by Heinrich Rudolf Hertz (1857–94) made it a well-established discipline. Since then classical electromagnetics has been a cornerstone of physics and has played a critical role in the development of a great variety of scientific, engineering, and biomedical disciplines. The fundamental nature of Maxwell's electromagnetics was ultimately asserted by the development of the relativity theory by Jules Henri Poincaré (1854–1912) and Hendrik Antoon Lorentz (1853–1928) (Whittaker, 1987).

The two-volume monograph by Sir Edmund Whittaker referenced above remains by far the most complete and balanced account of the history of electromagnetism from the time of William Gilbert (1544–1603) and René Descartes (1596–1650) to the relativity theory. This magnificent work should be read by everyone interested in a masterful and meticulously documented recreation of the actual sequence of events and publications that shaped the physical science.

Comprehensive modern accounts of classical electromagnetics and optics can be found in the monographs by Stratton (1941), Jackson (1998), Born and Wolf (1999), and Kong (2000).

Sir George Gabriel Stokes (1819–1903) was the first to discover that four quantities, now known as the Stokes parameters, could conveniently characterize the polarization state of any light beam, including partially polarized and unpolarized light (Stokes, 1852). Furthermore, he noted that unlike the quantities entering the amplitude formulation of the optical field, these parameters could be directly measured by a suitable optical instrument.

The fascinating subject of polarization attracted the attention of many other great scientists before and after Stokes, including Augustin Jean Fresnel (1788–1827), Dominique François Arago (1786–1853), Thomas Young (1773–1829), Subrahmanyan Chandrasekhar (1910–95), and Hendrik van de Hulst (1918–2000). Even Poincaré, who is rightfully considered to be one the greatest geniuses of all time, had found the time to contribute to this discipline by developing a useful polarization analysis tool known as the Poincaré sphere (Poincaré, 1892; see also Kliger *et al.*, 1990 and Collett, 1992).

Extensive treatments of theoretical and experimental aspects of polarimetry have been provided by Shurcliff (1962), Kliger *et al.* (1990), Collett (1992), Brosseau (1998), and Hovenier *et al.* (2004). In Pye (2001), numerous manifestations of polarization in science and nature are described.

Chapter 3

Basic theory of electromagnetic scattering

Although the main subject of this book is multiple scattering of light by groups of randomly positioned particles, many quantities used in the derivation of the radiative transfer equation and finally entering it originate in the electromagnetic theory of single scattering by a fixed object. Therefore, we will introduce in this chapter the relevant single-scattering concepts and definitions and summarize the theoretical results that will be necessary for understanding the material presented in the following chapters.

As we have indicated in Chapter 1, the presence of an object with a refractive index different from that of the surrounding medium changes the electromagnetic field that would otherwise exist in an unbounded homogeneous space. The difference of the total field in the presence of the object and the original field that would exist in the absence of the object can be thought of as the field *scattered* by the object. In other words, the total field is the vector sum of the *incident* (original) field and the *scattered* field.

The specific angular distribution and polarization state of the scattered field depend on the polarization and directional characteristics and wavelength of the incident field as well as on such properties of the scatterer as its size, shape, relative refractive index, and orientation. However, the principal objective of this chapter is to consider the general mathematical description of the scattering process without making detailed assumptions about the scattering object except that it is composed of a linear and isotropic material. As always, we begin with a field description of the scattering process and then proceed by introducing quantities that can be directly measured with a suitable optical instrument.

An important part of this chapter is the application of the concept of the coherency dyad of the total electric field. Although this quantity may not have as definite a physical meaning as, for example, the Poynting vector, it proves to be an essential

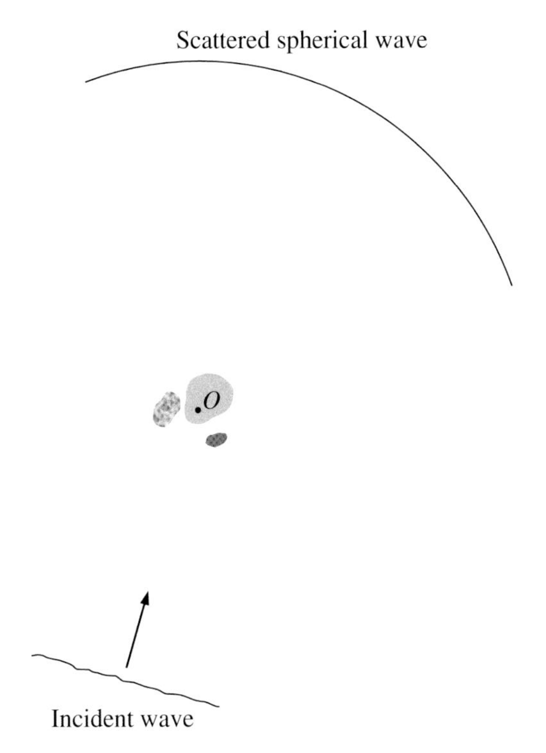

Figure 3.1.1. Schematic representation of the electromagnetic scattering problem. The unshaded exterior region V_{EXT} is unbounded in all directions, whereas the shaded areas collectively represent the interior region V_{INT}.

instrument in the solution of various electromagnetic scattering problems including the microphysical derivation of the RTE.

3.1 Volume integral equation and Lippmann–Schwinger equation

Consider a fixed scattering object embedded in an infinite, homogeneous, linear, isotropic, and nonabsorbing medium (see Fig. 3.1.1). The scatterer occupies a finite interior region V_{INT} and is surrounded by the infinite exterior region V_{EXT} such that $V_{\mathrm{INT}} \cup V_{\mathrm{EXT}} = \Re^3$, where, as before, $\Re^3$ denotes the entire three-dimensional space. The interior region is filled with an isotropic, linear, and possibly inhomogeneous material. The scatterer can be either a single body or a cluster with touching and/or separated components. Point O serves as the common origin of all position vectors and as the origin of the laboratory coordinate system.

It is well known that optical properties of bulk substances in solid or liquid phases are qualitatively different from those of their constituent atoms and molecules when the latter are isolated. This may cause a problem when one applies the concept of bulk optical constants to a very small particle because either the optical constants determined for bulk matter provide an inaccurate estimate or the particle is so small that the entire concept of optical constants loses its validity. We will therefore assume that the individual bodies forming the scattering object are sufficiently large so that they can still be characterized by optical constants appropriate to bulk matter. According to Huffman (1988), this implies that each body is larger than approximately 50 Å.

The frequency-domain monochromatic Maxwell curl equations (2.3.3) and (2.3.17) describing the scattering problem can be rewritten as follows:

$$\left.\begin{aligned} \nabla\times\mathbf{E}(\mathbf{r}) &= \mathrm{i}\omega\mu_1\mathbf{H}(\mathbf{r}) \\ \nabla\times\mathbf{H}(\mathbf{r}) &= -\mathrm{i}\omega\epsilon_1\mathbf{E}(\mathbf{r}) \end{aligned}\right\} \quad \mathbf{r}\in V_{\mathrm{EXT}}, \tag{3.1.1}$$

$$\left.\begin{aligned} \nabla\times\mathbf{E}(\mathbf{r}) &= \mathrm{i}\omega\mu_2(\mathbf{r},\omega)\mathbf{H}(\mathbf{r}) \\ \nabla\times\mathbf{H}(\mathbf{r}) &= -\mathrm{i}\omega\varepsilon_2(\mathbf{r},\omega)\mathbf{E}(\mathbf{r}) \end{aligned}\right\} \quad \mathbf{r}\in V_{\mathrm{INT}}, \tag{3.1.2}$$

where the subscripts 1 and 2 refer to the exterior and interior regions, respectively. The permeability and the complex permittivity for the interior region are functions of $\mathbf{r}$ to provide for the general case of the scattering object being inhomogeneous. Since the first relations in Eqs. (3.1.1) and (3.1.2) yield the magnetic field provided that the electric field is known everywhere, we will look for the solution of Eqs. (3.1.1) and (3.1.2) in terms of only the electric field.

Assuming that the host medium and the scattering object are nonmagnetic, i.e., $\mu_2(\mathbf{r},\omega) \equiv \mu_1 = \mu_0$, where μ_0 is the permeability of a vacuum, we derive the following vector wave equations:

$$\nabla\times\nabla\times\mathbf{E}(\mathbf{r}) - k_1^2\mathbf{E}(\mathbf{r}) = 0, \qquad \mathbf{r}\in V_{\mathrm{EXT}}, \tag{3.1.3}$$

$$\nabla\times\nabla\times\mathbf{E}(\mathbf{r}) - k_2^2(\mathbf{r},\omega)\mathbf{E}(\mathbf{r}) = 0, \qquad \mathbf{r}\in V_{\mathrm{INT}}, \tag{3.1.4}$$

where $k_1 = \omega(\epsilon_1\mu_0)^{1/2}$ and $k_2(\mathbf{r},\omega) = \omega[\varepsilon_2(\mathbf{r},\omega)\mu_0]^{1/2}$ are the wave numbers of the exterior and interior regions, respectively. Equations (3.1.3) and (3.1.4) can then be rewritten as a single inhomogeneous differential equation

$$\nabla\times\nabla\times\mathbf{E}(\mathbf{r}) - k_1^2\mathbf{E}(\mathbf{r}) = \mathbf{j}(\mathbf{r}), \qquad \mathbf{r}\in\Re^3, \tag{3.1.5}$$

where

$$\mathbf{j}(\mathbf{r}) = k_1^2[\tilde{m}^2(\mathbf{r},\omega) - 1]\,\mathbf{E}(\mathbf{r}), \tag{3.1.6}$$

$$\tilde{m}(\mathbf{r},\omega) = \begin{cases} 1, & \mathbf{r}\in V_{\mathrm{EXT}}, \\ m(\mathbf{r},\omega) = k_2(\mathbf{r},\omega)/k_1 = m_2(\mathbf{r},\omega)/m_1, & \mathbf{r}\in V_{\mathrm{INT}}, \end{cases} \tag{3.1.7}$$

and $m(\mathbf{r},\omega)$ is the refractive index of the interior relative to that of the exterior. From this point on, we will omit the argument ω for the sake of brevity, while still remembering that the relative refractive index may be frequency-dependent. It follows from Eq. (3.1.6) that the forcing function $\mathbf{j}(\mathbf{r})$ vanishes everywhere outside the interior region.

Any solution of an inhomogeneous linear differential equation can be divided into two parts: (i) a solution of the respective homogeneous equation with the right-hand side identically equal to zero and (ii) a particular solution of the inhomogeneous equation. The first part satisfies the equation

$$\nabla\times\nabla\times\mathbf{E}^{\mathrm{inc}}(\mathbf{r}) - k_1^2\mathbf{E}^{\mathrm{inc}}(\mathbf{r}) = 0, \qquad \mathbf{r}\in\Re^3, \tag{3.1.8}$$

and describes the field that would exist in the absence of the scattering object, i.e., the *incident field.* The physically appropriate particular solution of Eq. (3.1.5) must give the *scattered field* $\mathbf{E}^{\mathrm{sca}}(\mathbf{r})$ generated by the forcing function $\mathbf{j}(\mathbf{r})$. Obviously, of all possible particular solutions of Eq. (3.1.5), we must choose the one that vanishes at large distances from the scattering object and ensures energy conservation.

To find $\mathbf{E}^{\mathrm{sca}}(\mathbf{r})$, we first introduce the free space dyadic Green's function $\overleftrightarrow{G}(\mathbf{r},\mathbf{r}')$ as a dyadic satisfying the differential equation

$$\nabla\times\nabla\times\overleftrightarrow{G}(\mathbf{r},\mathbf{r}') - k_1^2\overleftrightarrow{G}(\mathbf{r},\mathbf{r}') = \overleftrightarrow{I}\,\delta(\mathbf{r}-\mathbf{r}'), \tag{3.1.9}$$

where $\overleftrightarrow{I}$ is the identity dyadic,

$$\delta(\mathbf{r}-\mathbf{r}') = \begin{cases} \delta(x-x')\,\delta(y-y')\,\delta(z-z') & \text{(Cartesian coordinates)} \\ \dfrac{1}{r^2}\delta(r-r')\,\delta(\cos\theta-\cos\theta')\,\delta(\varphi-\varphi') & \text{(spherical polar coordinates)} \end{cases} \tag{3.1.10}$$

is the three-dimensional delta function, and $\delta(x-x')$ is the usual Dirac delta function. Taking into account that

$$\nabla\times[\overleftrightarrow{G}(\mathbf{r},\mathbf{r}')\cdot\mathbf{j}(\mathbf{r}')] = [\nabla\times\overleftrightarrow{G}(\mathbf{r},\mathbf{r}')]\cdot\mathbf{j}(\mathbf{r}'), \tag{3.1.11}$$

we get

$$\nabla\times\nabla\times[\overleftrightarrow{G}(\mathbf{r},\mathbf{r}')\cdot\mathbf{j}(\mathbf{r}')] - k_1^2[\overleftrightarrow{G}(\mathbf{r},\mathbf{r}')\cdot\mathbf{j}(\mathbf{r}')] = \overleftrightarrow{I}\cdot\mathbf{j}(\mathbf{r}')\,\delta(\mathbf{r}-\mathbf{r}'). \tag{3.1.12}$$

We integrate both sides of this equation over the entire space to obtain

$$(\nabla\times\nabla\times\overleftrightarrow{I} - k_1^2\overleftrightarrow{I})\cdot\int_{\Re^3}\mathrm{d}\mathbf{r}'\,\overleftrightarrow{G}(\mathbf{r},\mathbf{r}')\cdot\mathbf{j}(\mathbf{r}') = \mathbf{j}(\mathbf{r}), \tag{3.1.13}$$

where the infinitesimal volume element is given by $\mathrm{d}\mathbf{r}' = \mathrm{d}x'\mathrm{d}y'\mathrm{d}z'$ in Cartesian coordinates and by $\mathrm{d}\mathbf{r}' = r'^2\mathrm{d}r'\sin\theta'\mathrm{d}\theta'\mathrm{d}\varphi'$ in spherical polar coordinates. Comparison with Eq. (3.1.5) now shows that

$$\mathbf{E}^{\mathrm{sca}}(\mathbf{r}) = \int_{V_{\mathrm{INT}}}\mathrm{d}\mathbf{r}'\,\overleftrightarrow{G}(\mathbf{r},\mathbf{r}')\cdot\mathbf{j}(\mathbf{r}'), \qquad \mathbf{r}\in\Re^3, \tag{3.1.14}$$

where we have taken into account that $\mathbf{j}(\mathbf{r})$ vanishes everywhere outside V_{INT}. We will see in the following section that this particular solution of Eq. (3.1.5) indeed vanishes at infinity and ensures energy conservation, thereby being the physically appropriate particular solution. Hence, the complete solution of Eq. (3.1.5) is

$$\mathbf{E}(\mathbf{r}) = \mathbf{E}^{\text{inc}}(\mathbf{r}) + \int_{V_{\text{INT}}} d\mathbf{r}' \ddot{G}(\mathbf{r}, \mathbf{r}') \cdot \mathbf{j}(\mathbf{r}'), \qquad \mathbf{r} \in \Re^3. \tag{3.1.15}$$

To find the free space dyadic Green's function, we first express it in terms of a scalar Green's function $g(\mathbf{r}, \mathbf{r}')$ as follows:

$$\ddot{G}(\mathbf{r}, \mathbf{r}') = \left(\ddot{I} + \frac{1}{k_1^2} \nabla \otimes \nabla \right) g(\mathbf{r}, \mathbf{r}'). \tag{3.1.16}$$

Inserting Eq. (3.1.16) into Eq. (3.1.9) and noticing that

$$\nabla \times [\nabla \times (\nabla \otimes \nabla)] = \nabla \times [(\nabla \times \nabla) \otimes \nabla] = \ddot{0}, \tag{3.1.17}$$

$$\nabla \times \nabla \times (\ddot{I} g) = \nabla \otimes \nabla g - \ddot{I} \nabla^2 g, \tag{3.1.18}$$

where $\ddot{0}$ is a zero dyad, we obtain the following differential equation for g:

$$(\nabla^2 + k_1^2) g(\mathbf{r}, \mathbf{r}') = -\delta(\mathbf{r} - \mathbf{r}'). \tag{3.1.19}$$

The well-known solution of this equation, which satisfies the condition

$$\lim_{k_1|\mathbf{r}-\mathbf{r}'| \to \infty} g(\mathbf{r}, \mathbf{r}') = 0$$

and represents *outgoing* waves, is

$$g(\mathbf{r}, \mathbf{r}') = \frac{\exp(\mathrm{i}k_1|\mathbf{r} - \mathbf{r}'|)}{4\pi|\mathbf{r} - \mathbf{r}'|} \tag{3.1.20}$$

(e.g., Jackson, 1998, p. 427). Hence, Eqs. (3.1.6), (3.1.7), (3.1.15), (3.1.16), and (3.1.20) finally yield (Saxon, 1955b; Shifrin, 1968)

$$\begin{aligned} \mathbf{E}(\mathbf{r}) &= \mathbf{E}^{\text{inc}}(\mathbf{r}) + k_1^2 \int_{V_{\text{INT}}} d\mathbf{r}' \ddot{G}(\mathbf{r}, \mathbf{r}') \cdot \mathbf{E}(\mathbf{r}') [m^2(\mathbf{r}') - 1] \\ &= \mathbf{E}^{\text{inc}}(\mathbf{r}) + k_1^2 \left(\ddot{I} + \frac{1}{k_1^2} \nabla \otimes \nabla \right) \\ &\quad \cdot \int_{V_{\text{INT}}} d\mathbf{r}' [m^2(\mathbf{r}') - 1] \mathbf{E}(\mathbf{r}') \frac{\exp(\mathrm{i}k_1|\mathbf{r} - \mathbf{r}'|)}{4\pi|\mathbf{r} - \mathbf{r}'|}, \qquad \mathbf{r} \in \Re^3. \end{aligned} \tag{3.1.21}$$

The scattered electric field is then given by

$$\mathbf{E}^{\text{sca}}(\mathbf{r}) = k_1^2 \int_{V_{\text{INT}}} d\mathbf{r}' \ddot{G}(\mathbf{r}, \mathbf{r}') \cdot \mathbf{E}(\mathbf{r}') [m^2(\mathbf{r}') - 1]$$

$$= k_1^2\left(\ddot{I} + \frac{1}{k_1^2}\nabla \otimes \nabla\right) \cdot \int_{V_{\rm INT}} d\mathbf{r}'[m^2(\mathbf{r}') - 1]\mathbf{E}(\mathbf{r}') \frac{\exp(ik_1|\mathbf{r} - \mathbf{r}'|)}{4\pi|\mathbf{r} - \mathbf{r}'|},$$

$$\mathbf{r} \in \Re^3. \tag{3.1.22}$$

Equation (3.1.21) is a volume integral equation expressing the total electric field everywhere in space in terms of the incident field and the total field inside the scatterer. Since the latter is not known in general, one must solve Eq. (3.1.21) either numerically or analytically. As a first step, the internal field can be approximated by the incident field. This is the gist of the so-called Rayleigh–Gans approximation (RGA) otherwise known as the Rayleigh–Debye or Born approximation (van de Hulst, 1957; Ishimaru, 1978). The total field computed in the RGA can be substituted in the integral on the right-hand side of Eq. (3.1.21) in order to compute an improved approximation, and this iterative process can be continued until the total field converges within a given numerical accuracy. Although this procedure can be rather involved, it shows that in the final analysis the scattered electric field can be expressed in terms of the incident field as follows:

$$\mathbf{E}^{\rm sca}(\mathbf{r}) = \int_{V_{\rm INT}} d\mathbf{r}'\ddot{G}(\mathbf{r}, \mathbf{r}') \cdot \int_{V_{\rm INT}} d\mathbf{r}''\ddot{T}(\mathbf{r}', \mathbf{r}'') \cdot \mathbf{E}^{\rm inc}(\mathbf{r}''), \quad \mathbf{r} \in \Re^3, \tag{3.1.23}$$

where $\ddot{T}$ is the so-called dyadic transition operator (Tsang *et al.*, 1985). Substituting Eq. (3.1.23) in Eq. (3.1.21) yields the following integral equation for $\ddot{T}$:

$$\ddot{T}(\mathbf{r}, \mathbf{r}') = k_1^2[m^2(\mathbf{r}) - 1]\,\delta(\mathbf{r} - \mathbf{r}')\ddot{I}$$

$$+ k_1^2[m^2(\mathbf{r}) - 1] \int_{V_{\rm INT}} d\mathbf{r}''\ddot{G}(\mathbf{r}, \mathbf{r}'') \cdot \ddot{T}(\mathbf{r}'', \mathbf{r}'), \qquad \mathbf{r}, \mathbf{r}' \in V_{\rm INT}. \tag{3.1.24}$$

Equations of this type appear in the quantum theory of scattering and are called Lippmann–Schwinger equations (Lippmann and Schwinger, 1950; Newton, 1982).

Equation (3.1.23) shows that if $\mathbf{E}_1^{\rm inc}(\mathbf{r})$ and $\mathbf{E}_2^{\rm inc}(\mathbf{r})$ are two different incident fields and $\mathbf{E}_1^{\rm sca}(\mathbf{r})$ and $\mathbf{E}_2^{\rm sca}(\mathbf{r})$ are the corresponding scattered fields, then $\mathbf{E}_1^{\rm sca}(\mathbf{r}) + \mathbf{E}_2^{\rm sca}(\mathbf{r})$ is the scattered field corresponding to the incident field $\mathbf{E}_1^{\rm inc}(\mathbf{r}) + \mathbf{E}_2^{\rm inc}(\mathbf{r})$. This result is, of course, a consequence of the principle of superposition discussed in Section 2.3.

3.2 Scattering in the far-field zone

Let us now subdivide the scattering object into a large number of elementary volume elements ΔV and rewrite Eq. (3.1.22) for an external observation point $\mathbf{r}$ in the discrete form:

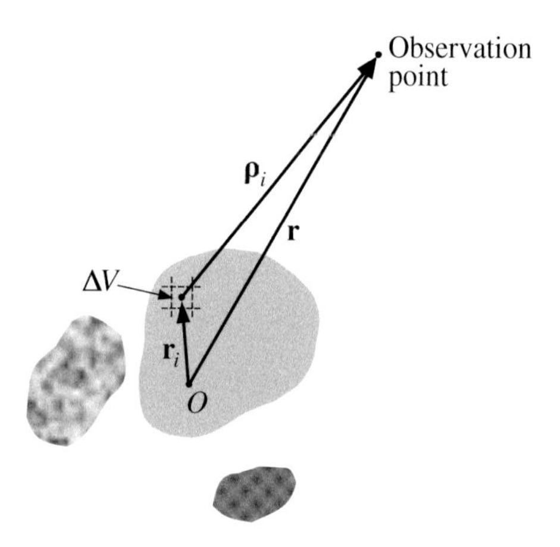

Figure 3.2.1. Derivation of Eq. (3.2.5).

$$\mathbf{E}^{\text{sca}}(\mathbf{r}) = \frac{k_1^2}{4\pi} \lim_{\Delta V \to 0} \sum_i \Delta V (m_i^2 - 1) \left(\ddot{I} + \frac{1}{k_1^2} \nabla \otimes \nabla \right) \cdot \mathbf{E}_i \frac{\exp(\mathrm{i}k_1 \rho_i)}{\rho_i},$$

$$\mathbf{r} \in V_{\text{EXT}}, \tag{3.2.1}$$

where the index i numbers the volume elements, $\mathbf{E}_i$ and m_i are the electric field and relative refractive index values, respectively, at the center of the ith volume element, $\rho_i = |\boldsymbol{\rho}_i|$ is the distance from the center of the ith volume element to the observation point, $\boldsymbol{\rho}_i = \mathbf{r} - \mathbf{r}_i$ is the vector connecting the center of the ith volume element and the observation point, and $\mathbf{r}_i$ is the radius vector of the center of the ith volume element (Fig. 3.2.1). Recall now that in spherical polar coordinates,

$$\nabla = \hat{\mathbf{r}} \frac{\partial}{\partial r} + \hat{\boldsymbol{\theta}} \frac{1}{r} \frac{\partial}{\partial \theta} + \hat{\boldsymbol{\varphi}} \frac{1}{r \sin\theta} \frac{\partial}{\partial \varphi}, \tag{3.2.2}$$

$$\frac{\partial}{\partial r} \hat{\mathbf{r}} = 0, \qquad \frac{\partial}{\partial \theta} \hat{\mathbf{r}} = \hat{\boldsymbol{\theta}}, \qquad \frac{\partial}{\partial \varphi} \hat{\mathbf{r}} = \hat{\boldsymbol{\varphi}} \sin\theta, \tag{3.2.3}$$

where the order of operator components relative to $\hat{\mathbf{r}}$, $\hat{\boldsymbol{\theta}}$, and $\hat{\boldsymbol{\varphi}}$ in Eq. (3.2.2) is essential because the unit basis vectors depend on θ and φ. The simplicity of these formulas makes it convenient to evaluate the contribution of each volume element to the sum on the right-hand side of Eq. (3.2.1) by using a local coordinate system originating at the center of this volume element and having the same orientation as the laboratory reference frame. This is done by making the substitution $\mathbf{r} \to \boldsymbol{\rho}_i$ for each new i. Recalling Eqs. (2.5.1) and (2.11.1) and assuming that

$$k_1 \rho_i \gg 1 \text{ for any } i \tag{3.2.4}$$

then yields

$$\mathbf{E}^{\text{sca}}(\mathbf{r}) = \frac{k_1^2}{4\pi} \lim_{\Delta V \to 0} \sum_i \Delta V\,(m_i^2 - 1)\frac{\exp(\mathrm{i}k_1\rho_i)}{\rho_i}(\ddot{I} - \hat{\boldsymbol{\rho}}_i \otimes \hat{\boldsymbol{\rho}}_i)\cdot \mathbf{E}_i,$$
$$\mathbf{r} \in V_{\text{EXT}}, \tag{3.2.5}$$

where $\hat{\boldsymbol{\rho}}_i = \boldsymbol{\rho}_i/\rho_i$ is the unit vector originating at the center of the ith volume element and directed towards the observation point. Finally,

$$\mathbf{E}^{\text{sca}}(\mathbf{r}) = \frac{k_1^2}{4\pi}\int_{V_{\text{INT}}} \mathrm{d}\mathbf{r}'[m^2(\mathbf{r}') - 1]\frac{\exp(\mathrm{i}k_1|\mathbf{r} - \mathbf{r}'|)}{|\mathbf{r} - \mathbf{r}'|}(\ddot{I} - \hat{\boldsymbol{\rho}}' \otimes \hat{\boldsymbol{\rho}}')\cdot \mathbf{E}(\mathbf{r}'),$$
$$\mathbf{r} \in V_{\text{EXT}}, \tag{3.2.6}$$

where

$$\hat{\boldsymbol{\rho}}' = \frac{\mathbf{r} - \mathbf{r}'}{|\mathbf{r} - \mathbf{r}'|}. \tag{3.2.7}$$

Equation (3.2.6) has two important implications. First, it shows that the scattered field at an external observation point is a vector superposition of partial scattered fields (wavelets) which are created by infinitesimal volume elements constituting the interior of the object. Second, it demonstrates that each wavelet is an outgoing transverse spherical wave (Fig. 3.2.2). Indeed, the identity dyadic in spherical polar coordinates is given by Eq. (A.12), so that the dyadic factor

$$\ddot{I} - \hat{\boldsymbol{\rho}}' \otimes \hat{\boldsymbol{\rho}}'$$

in Eq. (3.2.6) ensures that each wavelet is transverse, i.e., the electric field vector of the wavelet at the observation point is perpendicular to its propagation direction $\hat{\boldsymbol{\rho}}'$:

$$\hat{\boldsymbol{\rho}}' \cdot (\ddot{I} - \hat{\boldsymbol{\rho}}' \otimes \hat{\boldsymbol{\rho}}')\cdot \mathbf{E}(\mathbf{r}') = 0. \tag{3.2.8}$$

Furthermore, the electric field of the wavelet decays inversely with distance $|\mathbf{r} - \mathbf{r}'|$ from the center of the infinitesimal volume element.

Let us now assume that the origin of the laboratory coordinate system O is close to the geometrical center of the scattering object (Figs. 3.1.1 and 3.2.3). Usually one is interested in calculating the scattered field in the so-called far-field zone of the entire object. Specifically, assuming that the distance r from the origin to the observation point is much larger than any linear dimension of the scatterer,

$$r \gg r' \quad \text{for any } \mathbf{r}' \in V_{\text{INT}}, \tag{3.2.9}$$

we have

$$\ddot{I} - \hat{\boldsymbol{\rho}}' \otimes \hat{\boldsymbol{\rho}}' \approx \ddot{I} - \hat{\mathbf{r}} \otimes \hat{\mathbf{r}}, \tag{3.2.10}$$

$$|\mathbf{r} - \mathbf{r}'| = r\sqrt{1 - 2\frac{\hat{\mathbf{r}}\cdot\mathbf{r}'}{r} + \frac{r'^2}{r^2}}$$
$$\approx r - \hat{\mathbf{r}}\cdot\mathbf{r}' + \frac{r'^2}{2r}, \tag{3.2.11}$$

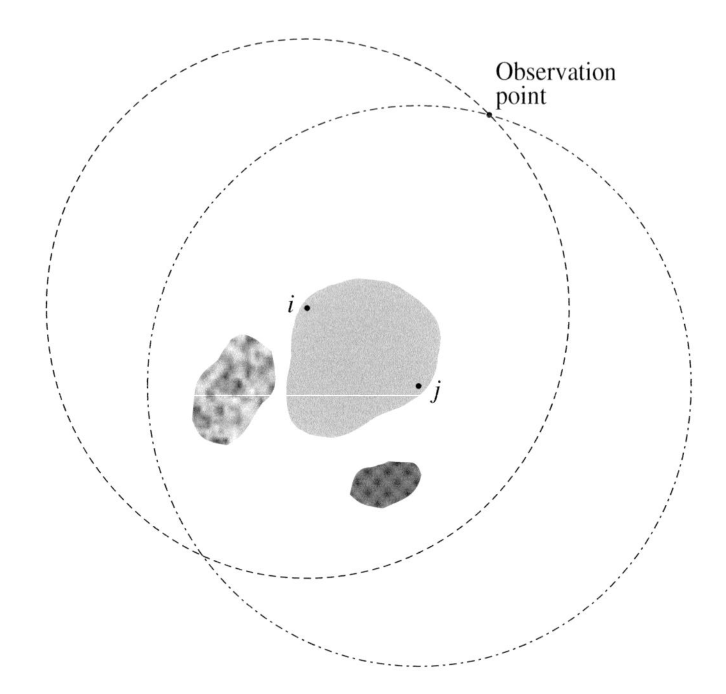

Figure 3.2.2. Spherical wavelets generated by infinitesimal volume elements centered at points i (broken line) and j (dot-dashed line).

where $\hat{\mathbf{r}} = \mathbf{r}/r$ is the unit vector in the direction of $\mathbf{r}$, Fig. 3.2.3. The last two terms on the right-hand side of Eq. (3.2.11) can be neglected in computing the slowly varying denominator in the expression on the right-hand side of Eq. (3.2.6), thereby yielding

$$\frac{1}{|\mathbf{r} - \mathbf{r}'|} \approx \frac{1}{r}, \tag{3.2.12}$$

but not in computing the rapidly oscillating factor $\exp(\mathrm{i}k_1|\mathbf{r} - \mathbf{r}'|)$. Assuming, however, that

$$\frac{k_1 r'^2}{2r} \ll 1 \quad \text{for any} \quad \mathbf{r}' \in V_{\mathrm{INT}} \tag{3.2.13}$$

we finally obtain

$$\mathbf{E}^{\mathrm{sca}}(\mathbf{r}) \approx \frac{\exp(\mathrm{i}k_1 r)}{r} \frac{k_1^2}{4\pi} (\ddot{I} - \hat{\mathbf{r}} \otimes \hat{\mathbf{r}}) \cdot \int_{V_{\mathrm{INT}}} \mathrm{d}\mathbf{r}' [m^2(\mathbf{r}') - 1] \mathbf{E}(\mathbf{r}') \times \exp(-\mathrm{i}k_1 \hat{\mathbf{r}} \cdot \mathbf{r}'). \tag{3.2.14}$$

This remarkable formula is the main result of the *far-field approximation* and demonstrates that the scattered electric field at a large distance from the object behaves as a single outgoing transverse spherical wave centered at O and propagating in the direction of the radial unit vector $\hat{\mathbf{r}}$. Indeed, the scattered field decays inversely

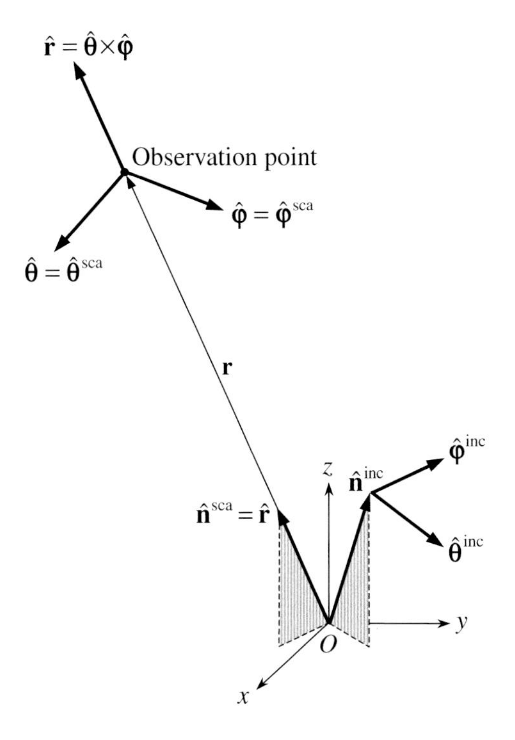

Figure 3.2.3. Scattering in the far-field zone of the entire object.

with distance r from the origin and

$$\hat{\mathbf{r}} \cdot \mathbf{E}^{\mathrm{sca}}(\mathbf{r}) = 0. \tag{3.2.15}$$

Thus, only the θ- and φ-components of the electric vector of the scattered field are nonzero. Equation (3.2.14) can be rewritten in the form

$$\mathbf{E}^{\mathrm{sca}}(\mathbf{r}) = \frac{\exp(\mathrm{i}k_1 r)}{r}\mathbf{E}_1^{\mathrm{sca}}(\hat{\mathbf{r}}), \qquad \hat{\mathbf{r}} \cdot \mathbf{E}_1^{\mathrm{sca}}(\hat{\mathbf{r}}) = 0, \tag{3.2.16}$$

where the vector $\mathbf{E}_1^{\mathrm{sca}}(\hat{\mathbf{r}})$ is independent of r and describes the angular distribution of the scattered radiation in the far-field zone.

Let a be the radius of the smallest circumscribing sphere of the scattering object centered at O. Then the criteria (3.2.4), (3.2.9), and (3.2.13) of the far-field approximation can be summarized as follows:

$$k_1(r - a) \gg 1, \tag{3.2.17}$$

$$r \gg a \quad \text{or} \quad k_1 r \gg k_1 a, \tag{3.2.18}$$

$$r \gg \frac{k_1 a^2}{2} \quad \text{or} \quad k_1 r \gg \frac{k_1^2 a^2}{2}. \tag{3.2.19}$$

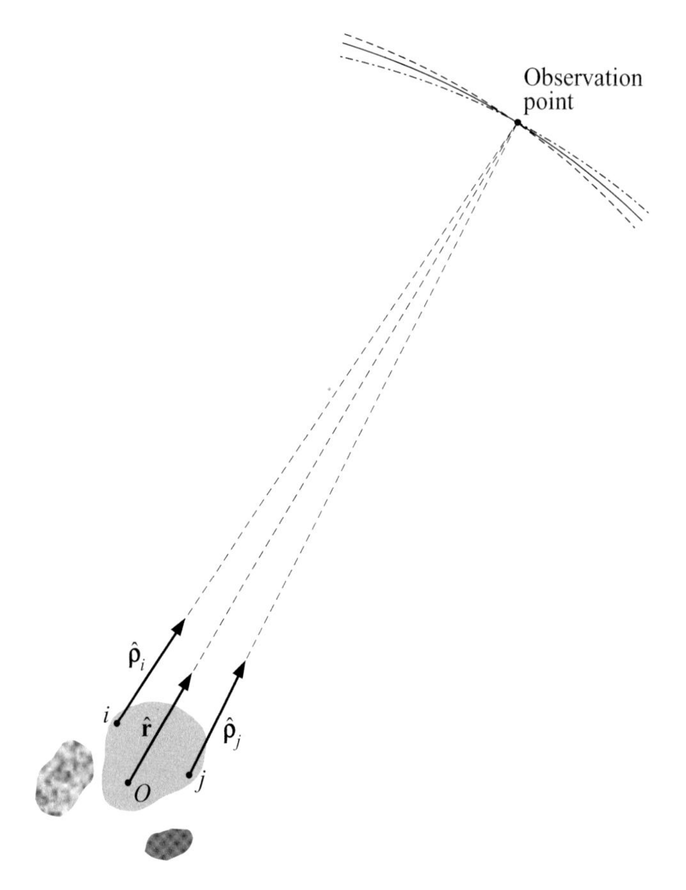

Figure 3.2.4. The individual spherical wavefronts generated by infinitesimal volume elements centered at points i (broken curve) and j (dot-dashed curve) nearly merge with increasing distance of the observation point from the scattering object and become locally indistinguishable from the unified spherical wavefront centered at the common origin (solid curve). The respective propagation directions at the observation point also become close and eventually coincide.

The inequality (3.2.17) means that the distance from any point inside the object to the observation point must be much greater than the wavelength. This ensures that at the observation point, the partial field scattered by any differential volume element develops into an outgoing spherical wavelet.

The inequality (3.2.18) requires the observation point to be located at a distance from the object much greater than the object size. This ensures that when the partial wavelets generated by the elementary volume elements constituting the object arrive at the observation point, they propagate in essentially the same scattering direction, Fig. 3.2.4, and are equally attenuated by the factor 1/distance:

$$\frac{\mathbf{r}-\mathbf{r}'}{|\mathbf{r}-\mathbf{r}'|} \approx \hat{\mathbf{r}} \quad \text{and} \quad \frac{1}{|\mathbf{r}-\mathbf{r}'|} \approx \frac{1}{r} \quad \text{for any } \mathbf{r}' \in V_{\mathrm{INT}}. \tag{3.2.20}$$

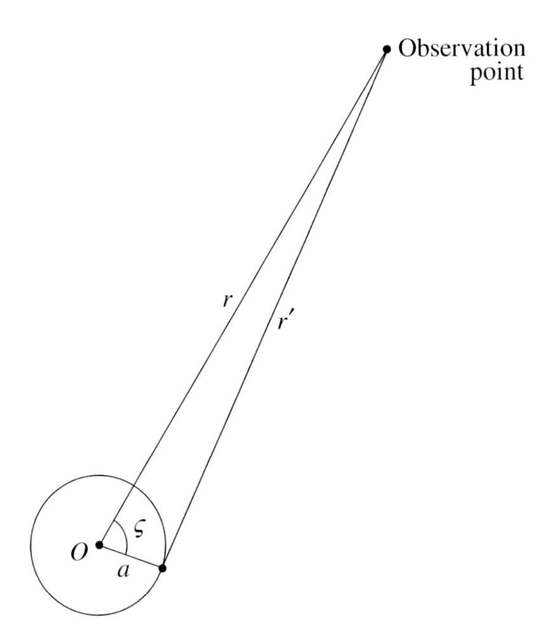

Figure 3.2.5. Interpretation of the inequality (3.2.19).

The meaning of the inequality (3.2.19) is a bit more subtle, but becomes clear from the inspection of Fig. 3.2.5, in which the observation point is shown relative to the smallest circumscribing sphere of the object. The phase difference between the straight path connecting the observation point and a point on the sphere surface and the path connecting the observation point and the origin is given by

$$k_1(r' - r) \approx \frac{k_1 a^2}{2r} - k_1 a \cos\varsigma. \tag{3.2.21}$$

The second term on the right-hand side of this expression is independent of r (for a fixed scattering direction), whereas the variation of the first term with changing r is significant unless $k_1 a^2/2r \ll 1$. Therefore, we can interpret the inequality (3.2.19) as the requirement that the observation point be so far from the scatterer that the phase difference between the paths connecting the observation point and any two points of the scatterer becomes independent of r for any fixed scattering direction. As a consequence, the surfaces of constant phase of the partial wavelets generated by the elementary volume elements constituting the object coincide locally when they reach an observation point situated in the far-field zone, and the wavelets form a single outgoing spherical wave (compare Figs. 3.2.2 and 3.2.4). This implies that the entire scatterer is effectively treated as a point-like body located at the origin of the laboratory coordinate system.

The relative importance of the far-field-zone criteria (3.2.17)–(3.2.19) changes with particle size relative to the wavelength. For particles much smaller than the wavelength ($k_1 a \ll 1$), the inequality (3.2.17) is the most restrictive of the three. When the size parameter $k_1 a$ is of order unity, all three criteria are roughly equivalent. For particles much greater than the wavelength ($k_1 a \gg 1$), the inequality (3.2.19) becomes the most demanding and can "move" the far-field zone much farther from the particle than the other two inequalities.

In view of the inequality (3.2.18), the inequality (3.2.17) can be simplified:

$$k_1 r \gg 1. \tag{3.2.22}$$

Furthermore, all three criteria of far-field scattering can be written as the following single inequality:

$$k_1 r \gg \max(1, \tfrac{1}{2}x^2), \tag{3.2.23}$$

where $x = k_1 a$ is the dimensionless so-called size parameter of the object.

3.3 Scattering dyadic and amplitude scattering matrix

Assuming that the incident field is a plane electromagnetic wave,

$$\mathbf{E}^{\text{inc}}(\mathbf{r}) = \mathbf{E}_0^{\text{inc}} \exp(\mathrm{i}k_1 \hat{\mathbf{n}}^{\text{inc}} \cdot \mathbf{r}), \tag{3.3.1}$$

and using Eq. (3.1.23), we have

$$\mathbf{E}^{\text{sca}}(r\hat{\mathbf{n}}^{\text{sca}}) = \frac{\exp(\mathrm{i}k_1 r)}{r} \overset{\leftrightarrow}{A}(\hat{\mathbf{n}}^{\text{sca}}, \hat{\mathbf{n}}^{\text{inc}}) \cdot \mathbf{E}_0^{\text{inc}}, \tag{3.3.2}$$

where $\hat{\mathbf{n}}^{\text{sca}} = \hat{\mathbf{r}}$ (see Fig. 3.2.3) and $\overset{\leftrightarrow}{A}$ is the so-called scattering dyadic. It follows from Eqs. (3.2.15) and (3.3.2) that

$$\hat{\mathbf{n}}^{\text{sca}} \cdot \overset{\leftrightarrow}{A}(\hat{\mathbf{n}}^{\text{sca}}, \hat{\mathbf{n}}^{\text{inc}}) = \mathbf{0}. \tag{3.3.3}$$

However, because the incident field given by Eq. (3.3.1) is a transverse wave with electric field vector perpendicular to the direction of propagation, the dot product $\overset{\leftrightarrow}{A}(\hat{\mathbf{n}}^{\text{sca}}, \hat{\mathbf{n}}^{\text{inc}}) \cdot \hat{\mathbf{n}}^{\text{inc}}$ is not defined by Eq. (3.3.2). To complete the definition, we take this product to be zero:

$$\overset{\leftrightarrow}{A}(\hat{\mathbf{n}}^{\text{sca}}, \hat{\mathbf{n}}^{\text{inc}}) \cdot \hat{\mathbf{n}}^{\text{inc}} = \mathbf{0}. \tag{3.3.4}$$

Therefore, the final expression for the scattering dyadic in terms of the dyadic transition operator is as follows:

$$\begin{aligned} \overset{\leftrightarrow}{A}(\hat{\mathbf{n}}^{\text{sca}}, \hat{\mathbf{n}}^{\text{inc}}) = {} & \frac{1}{4\pi}(\overset{\leftrightarrow}{I} - \hat{\mathbf{n}}^{\text{sca}} \otimes \hat{\mathbf{n}}^{\text{sca}}) \cdot \int_{V_{\text{INT}}} \mathrm{d}\mathbf{r}' \exp(-\mathrm{i}k_1 \hat{\mathbf{n}}^{\text{sca}} \cdot \mathbf{r}') \\ & \times \int_{V_{\text{INT}}} \mathrm{d}\mathbf{r}'' \overset{\leftrightarrow}{T}(\mathbf{r}', \mathbf{r}'') \exp(\mathrm{i}k_1 \hat{\mathbf{n}}^{\text{inc}} \cdot \mathbf{r}'') \cdot (\overset{\leftrightarrow}{I} - \hat{\mathbf{n}}^{\text{inc}} \otimes \hat{\mathbf{n}}^{\text{inc}}). \end{aligned} \tag{3.3.5}$$

The elements of the scattering dyadic have the dimension of length.

According to the above definition, the scattering dyadic describes far-field scattering of a plane electromagnetic wave. Although this may appear to suggest that the usefulness of this quantity is rather limited, its actual range of applicability is much wider. Indeed, it follows directly from the principle of superposition that the scattering dyadic can be used to compute far-field scattering of any incident field as long as the latter can be expanded in elementary plane waves.

Equations (3.3.3) and (3.3.4) show that only four out of the nine components of the scattering dyadic are independent in the spherical polar coordinate system centered at the origin, Fig. 3.2.3. It is therefore convenient to introduce the 2×2 so-called amplitude scattering matrix $\mathbf{S}$, which describes the transformation of the θ- and φ-components of the incident plane wave into the θ- and φ-components of the scattered spherical wave:

$$\mathbf{E}^{\text{sca}}(r\hat{\mathbf{n}}^{\text{sca}}) = \frac{\exp(\mathrm{i}k_1 r)}{r}\mathbf{S}(\hat{\mathbf{n}}^{\text{sca}}, \hat{\mathbf{n}}^{\text{inc}})\mathbf{E}_0^{\text{inc}}, \tag{3.3.6}$$

where $\mathbf{E}$ denotes a two-component column formed by the θ- and φ-components of the electric field vector:

$$\mathbf{E} = \begin{bmatrix} E_\theta \\ E_\varphi \end{bmatrix}. \tag{3.3.7}$$

The elements of the amplitude scattering matrix have the dimension of length and are expressed in terms of the scattering dyadic as follows:

$$S_{11} = \hat{\boldsymbol{\theta}}^{\text{sca}} \cdot \ddot{A} \cdot \hat{\boldsymbol{\theta}}^{\text{inc}}, \tag{3.3.8}$$

$$S_{12} = \hat{\boldsymbol{\theta}}^{\text{sca}} \cdot \ddot{A} \cdot \hat{\boldsymbol{\varphi}}^{\text{inc}}, \tag{3.3.9}$$

$$S_{21} = \hat{\boldsymbol{\varphi}}^{\text{sca}} \cdot \ddot{A} \cdot \hat{\boldsymbol{\theta}}^{\text{inc}}, \tag{3.3.10}$$

$$S_{22} = \hat{\boldsymbol{\varphi}}^{\text{sca}} \cdot \ddot{A} \cdot \hat{\boldsymbol{\varphi}}^{\text{inc}}. \tag{3.3.11}$$

The amplitude scattering matrix depends on the directions of incidence and scattering as well as on the size, morphology, composition, and orientation of the scattering object with respect to the coordinate system. It also depends on the choice of the origin of the coordinate system relative to the object. If known, the amplitude scattering matrix gives the scattered and thus the total field, thereby providing a complete description of the scattering pattern in the far-field zone.

We have pointed out in Section 2.6 that when a wave propagates along the z-axis, the θ- and φ-components of the electric field vector are determined by the specific choice of the meridional plane. Therefore, the amplitude scattering matrix explicitly depends on φ^{inc} and φ^{sca} even when $\theta^{\text{inc}} = 0$ or π and/or $\theta^{\text{sca}} = 0$ or π.

3.4 Reciprocity

A fundamental property of the scattering dyadic is the reciprocity relation, which is a manifestation of the symmetry of the scattering process with respect to an inversion of time (Saxon, 1955a). To derive the reciprocity relation, we first consider the scattering of a spherical incoming wave by an arbitrary finite object embedded in an infinite, homogeneous, nonabsorbing medium. In the far-field zone of the object, the total electric field is the sum of the incoming and scattered spherical waves:

$$\mathbf{E}(r\hat{\mathbf{r}}) = \frac{\exp(-\mathrm{i}k_1 r)}{r}\mathbf{E}^{\mathrm{inc}}(\hat{\mathbf{r}}) + \frac{\exp(\mathrm{i}k_1 r)}{r}\mathbf{E}^{\mathrm{sca}}(\hat{\mathbf{r}}), \tag{3.4.1}$$

where $\mathbf{E}^{\mathrm{inc}}(\hat{\mathbf{r}})$ and $\mathbf{E}^{\mathrm{sca}}(\hat{\mathbf{r}})$ are independent of r and

$$\hat{\mathbf{r}}\cdot\mathbf{E}^{\mathrm{inc}}(\hat{\mathbf{r}}) = 0, \tag{3.4.2}$$

$$\hat{\mathbf{r}}\cdot\mathbf{E}^{\mathrm{sca}}(\hat{\mathbf{r}}) = 0 \tag{3.4.3}$$

(cf. Eq. (2.11.6)).

Because of the linearity of the Maxwell equations and by analogy with Eq. (3.3.2), the outgoing spherical wave must be linearly related to the incoming spherical wave. Following Saxon (1955a), we express this relationship in terms of the so-called scattering tensor $\overleftrightarrow{S}$ as follows:

$$\mathbf{E}^{\mathrm{sca}}(\hat{\mathbf{r}}) = -\int_{4\pi} \mathrm{d}\hat{\mathbf{r}}'\,\overleftrightarrow{S}(\hat{\mathbf{r}},\hat{\mathbf{r}}')\cdot\mathbf{E}^{\mathrm{inc}}(-\hat{\mathbf{r}}'). \tag{3.4.4}$$

In view of Eq. (3.4.3), we have

$$\hat{\mathbf{r}}\cdot\overleftrightarrow{S}(\hat{\mathbf{r}},\hat{\mathbf{r}}') = \mathbf{0}. \tag{3.4.5}$$

Since $\mathbf{E}^{\mathrm{inc}}(\hat{\mathbf{r}})$ is transverse, the product $\overleftrightarrow{S}(\hat{\mathbf{r}},\hat{\mathbf{r}}')\cdot\hat{\mathbf{r}}'$ remains undefined by Eq. (3.4.4). As before, we will complete the definition of the scattering tensor by taking this product to be zero:

$$\overleftrightarrow{S}(\hat{\mathbf{r}},\hat{\mathbf{r}}')\cdot\hat{\mathbf{r}}' = \mathbf{0}. \tag{3.4.6}$$

As a consequence of Eqs. (3.4.5) and (3.4.6), $\overleftrightarrow{S}$ has only four independent components.

The derivation of the reciprocity relation for the scattering tensor starts from the statement that if $\mathbf{E}_1$ and $\mathbf{E}_2$ are any two solutions of the Maxwell equations (but with the same harmonic time-dependence), then

$$r^2\int_{4\pi} \mathrm{d}\hat{\mathbf{r}}\,\hat{\mathbf{r}}\cdot\{\mathbf{E}_2(r\hat{\mathbf{r}})\times[\nabla\times\mathbf{E}_1(r\hat{\mathbf{r}})] - \mathbf{E}_1(r\hat{\mathbf{r}})\times[\nabla\times\mathbf{E}_2(r\hat{\mathbf{r}})]\} \underset{r\to\infty}{=} 0. \tag{3.4.7}$$

Indeed, using Eqs. (2.4.2), (3.1.1), and (3.1.2), we easily establish that $\nabla\cdot(\mathbf{E}_2\times\mathbf{H}_1 - \mathbf{E}_1\times\mathbf{H}_2)$ vanishes identically everywhere in space. Integrating this quantity over all space and applying the Gauss theorem (2.1.19) then yields Eq. (3.4.7).

We now take $\mathbf{E}_1$ and $\mathbf{E}_2$ at infinity to be superpositions of incoming and outgoing spherical waves:

$$\mathbf{E}_j(r\hat{\mathbf{r}}) = \frac{\exp(-\mathrm{i}k_1 r)}{r}\mathbf{E}_j^{\mathrm{inc}}(\hat{\mathbf{r}}) + \frac{\exp(\mathrm{i}k_1 r)}{r}\mathbf{E}_j^{\mathrm{sca}}(\hat{\mathbf{r}}), \quad j = 1, 2. \tag{3.4.8}$$

Taking into account Eq. (2.5.2), (2.5.15), (3.4.2), (3.4.3), (2.11.1), and the formula (see Eq. (3.2.2))

$$\nabla \times \mathbf{E}_{1,2}^{\mathrm{inc,sca}}(\hat{\mathbf{r}}) = \boldsymbol{O}(r^{-1}), \tag{3.4.9}$$

where $\boldsymbol{O}(r^{-1})$ is a vector with components vanishing at infinity at least as r^{-1}, we derive the following after some algebra:

$$\int_{4\pi} \mathrm{d}\hat{\mathbf{r}} [\mathbf{E}_2^{\mathrm{inc}}(\hat{\mathbf{r}}) \cdot \mathbf{E}_1^{\mathrm{sca}}(\hat{\mathbf{r}}) - \mathbf{E}_1^{\mathrm{inc}}(\hat{\mathbf{r}}) \cdot \mathbf{E}_2^{\mathrm{sca}}(\hat{\mathbf{r}})] = 0. \tag{3.4.10}$$

Using Eq. (3.4.4) to express the outgoing waves in terms of the incoming waves, we then have

$$\int_{4\pi} \mathrm{d}\hat{\mathbf{r}} \int_{4\pi} \mathrm{d}\hat{\mathbf{r}}' [\mathbf{E}_2^{\mathrm{inc}}(\hat{\mathbf{r}}) \cdot \ddot{S}(\hat{\mathbf{r}}, \hat{\mathbf{r}}') \cdot \mathbf{E}_1^{\mathrm{inc}}(-\hat{\mathbf{r}}')$$
$$- \mathbf{E}_1^{\mathrm{inc}}(\hat{\mathbf{r}}) \cdot \ddot{S}(\hat{\mathbf{r}}, \hat{\mathbf{r}}') \cdot \mathbf{E}_2^{\mathrm{inc}}(-\hat{\mathbf{r}}')] = 0. \tag{3.4.11}$$

Replacing $\hat{\mathbf{r}}$ by $-\hat{\mathbf{r}}'$ and $\hat{\mathbf{r}}'$ by $-\hat{\mathbf{r}}$ in the last term and transposing the tensor product according to Eq. (A.6), we derive

$$\int_{4\pi} \mathrm{d}\hat{\mathbf{r}} \int_{4\pi} \mathrm{d}\hat{\mathbf{r}}' \, \mathbf{E}_2^{\mathrm{inc}}(\hat{\mathbf{r}}) \cdot \{\ddot{S}(\hat{\mathbf{r}}, \hat{\mathbf{r}}') - [\ddot{S}(-\hat{\mathbf{r}}', -\hat{\mathbf{r}})]^{\mathrm{T}}\} \cdot \mathbf{E}_1^{\mathrm{inc}}(-\hat{\mathbf{r}}') = 0. \tag{3.4.12}$$

Since $\mathbf{E}_1^{\mathrm{inc}}$ and $\mathbf{E}_2^{\mathrm{inc}}$ are arbitrary, we finally have $\ddot{S}(\hat{\mathbf{r}}, \hat{\mathbf{r}}') = [\ddot{S}(-\hat{\mathbf{r}}', -\hat{\mathbf{r}})]^{\mathrm{T}}$ or

$$\ddot{S}(-\hat{\mathbf{r}}', -\hat{\mathbf{r}}) = [\ddot{S}(\hat{\mathbf{r}}, \hat{\mathbf{r}}')]^{\mathrm{T}}. \tag{3.4.13}$$

This is the reciprocity condition for the scattering tensor.

It should be remarked that in deriving Eq. (3.4.7), we assumed, as almost everywhere else in this book, that the permeability, permittivity, and conductivity are scalars. However, it is easily checked that Eq. (3.4.7) and thus the reciprocity condition (3.4.13) remain valid even when the permeability, permittivity, and conductivity of the scattering object are tensors provided that all these tensors are symmetric. If any of these tensors is not symmetric, then Eq. (3.4.13) may become invalid (Dolginov *et al.*, 1995; Lacoste and van Tiggelen, 1999).

We now use Eq. (3.4.13) to derive the reciprocity relation for the scattering dyadic by considering the case in which the scattering object is illuminated by a plane wave incident along the direction $\hat{\mathbf{n}}^{\mathrm{inc}}$. As follows from Eqs. (3.2.16) and (3.3.1), the total electric field in the far-field zone is given by

$$\mathbf{E}(r\hat{\mathbf{n}}^{\text{sca}}) = \mathbf{E}_0^{\text{inc}} \exp(\mathrm{i}k_1 r\hat{\mathbf{n}}^{\text{inc}} \cdot \hat{\mathbf{n}}^{\text{sca}}) + \mathbf{E}_1^{\text{sca}}(\hat{\mathbf{n}}^{\text{sca}}) \frac{\exp(\mathrm{i}k_1 r)}{r}. \tag{3.4.14}$$

Representing the incident plane wave as a superposition of incoming and outgoing spherical waves,

$$\exp(\mathrm{i}k_1 r\hat{\mathbf{n}}^{\text{inc}} \cdot \hat{\mathbf{n}}^{\text{sca}}) \underset{k_1 r\to\infty}{=} \frac{\mathrm{i}2\pi}{k_1}\left[\delta(\hat{\mathbf{n}}^{\text{inc}} + \hat{\mathbf{n}}^{\text{sca}})\frac{\exp(-\mathrm{i}k_1 r)}{r} - \delta(\hat{\mathbf{n}}^{\text{inc}} - \hat{\mathbf{n}}^{\text{sca}})\frac{\exp(\mathrm{i}k_1 r)}{r}\right] \tag{3.4.15}$$

(see Appendix B), we derive

$$\mathbf{E}(r\hat{\mathbf{n}}^{\text{sca}}) \underset{k_1 r\to\infty}{=} \frac{\mathrm{i}2\pi}{k_1}\mathbf{E}_0^{\text{inc}}\,\delta(\hat{\mathbf{n}}^{\text{inc}} + \hat{\mathbf{n}}^{\text{sca}})\frac{\exp(-\mathrm{i}k_1 r)}{r} + \left[\mathbf{E}_1^{\text{sca}}(\hat{\mathbf{n}}^{\text{sca}}) - \frac{\mathrm{i}2\pi}{k_1}\delta(\hat{\mathbf{n}}^{\text{inc}} - \hat{\mathbf{n}}^{\text{sca}})\mathbf{E}_0^{\text{inc}}\right]\frac{\exp(\mathrm{i}k_1 r)}{r}. \tag{3.4.16}$$

Considering this a special form of Eq. (3.4.1) and recalling the definition of the scattering tensor, Eq. (3.4.4), we have

$$\mathbf{E}_1^{\text{sca}}(\hat{\mathbf{n}}^{\text{sca}}) = \frac{\mathrm{i}2\pi}{k_1}[\delta(\hat{\mathbf{n}}^{\text{inc}} - \hat{\mathbf{n}}^{\text{sca}})\mathbf{E}_0^{\text{inc}} - \vec{\vec{S}}(\hat{\mathbf{n}}^{\text{sca}}, \hat{\mathbf{n}}^{\text{inc}})\cdot\mathbf{E}_0^{\text{inc}}]. \tag{3.4.17}$$

It now follows from the definition of the scattering dyadic, Eqs. (3.3.2)–(3.3.4), that

$$\vec{\vec{A}}(\hat{\mathbf{n}}^{\text{sca}}, \hat{\mathbf{n}}^{\text{inc}}) = \frac{\mathrm{i}2\pi}{k_1}[(\vec{\vec{I}} - \hat{\mathbf{n}}^{\text{inc}} \otimes \hat{\mathbf{n}}^{\text{inc}})\delta(\hat{\mathbf{n}}^{\text{inc}} - \hat{\mathbf{n}}^{\text{sca}}) - \vec{\vec{S}}(\hat{\mathbf{n}}^{\text{sca}}, \hat{\mathbf{n}}^{\text{inc}})]. \tag{3.4.18}$$

Finally, from Eqs. (3.4.13) and (3.4.18) we derive the reciprocity relation for the scattering dyadic:

$$\vec{\vec{A}}(-\hat{\mathbf{n}}^{\text{inc}}, -\hat{\mathbf{n}}^{\text{sca}}) = [\vec{\vec{A}}(\hat{\mathbf{n}}^{\text{sca}}, \hat{\mathbf{n}}^{\text{inc}})]^{\mathrm{T}}. \tag{3.4.19}$$

It is easy to see that the reciprocity relation can be interpreted as follows: if the source of light and the detector are interchanged then the new scattering tensor is obtained by transposing the original scattering tensor (Fig. 3.4.1).

The reciprocity relation for the amplitude scattering matrix follows from Eqs. (3.3.8)–(3.3.11) and (3.4.19) and the unit vector identities

$$\hat{\boldsymbol{\theta}}(-\hat{\mathbf{n}}) = \hat{\boldsymbol{\theta}}(\hat{\mathbf{n}}), \qquad \hat{\boldsymbol{\varphi}}(-\hat{\mathbf{n}}) = -\hat{\boldsymbol{\varphi}}(\hat{\mathbf{n}}). \tag{3.4.20}$$

Simple algebra gives

$$\mathbf{S}(-\hat{\mathbf{n}}^{\text{inc}}, -\hat{\mathbf{n}}^{\text{sca}}) = \begin{bmatrix} S_{11}(\hat{\mathbf{n}}^{\text{sca}}, \hat{\mathbf{n}}^{\text{inc}}) & -S_{21}(\hat{\mathbf{n}}^{\text{sca}}, \hat{\mathbf{n}}^{\text{inc}}) \\ -S_{12}(\hat{\mathbf{n}}^{\text{sca}}, \hat{\mathbf{n}}^{\text{inc}}) & S_{22}(\hat{\mathbf{n}}^{\text{sca}}, \hat{\mathbf{n}}^{\text{inc}}) \end{bmatrix}. \tag{3.4.21}$$

An interesting consequence of reciprocity is the so-called backscattering theorem, which directly follows from Eq. (3.4.21) after substituting $\hat{\mathbf{n}}^{\text{inc}} = \hat{\mathbf{n}}$ and $\hat{\mathbf{n}}^{\text{sca}} = -\hat{\mathbf{n}}$:

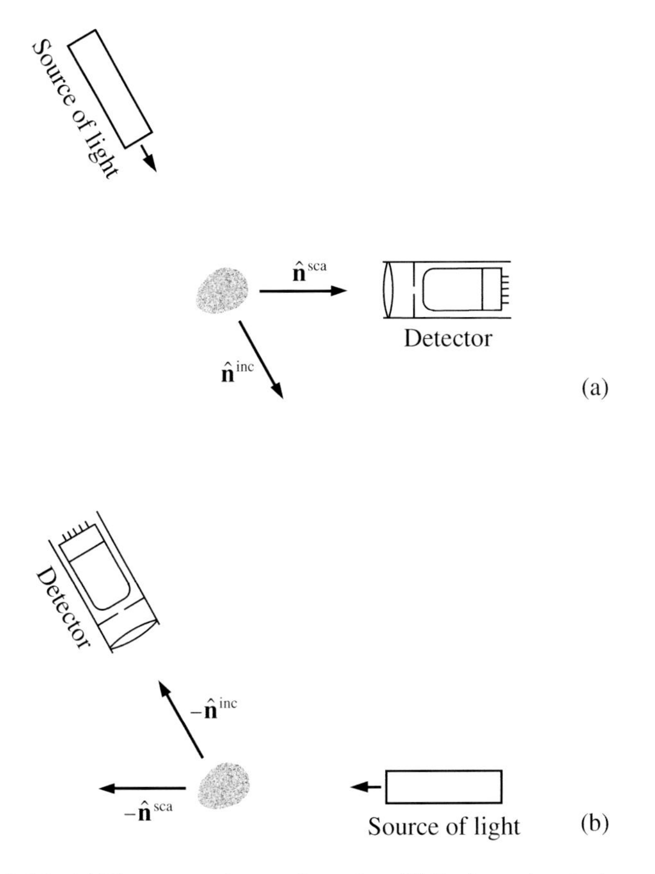

Figure 3.4.1. (a) Direct scattering configuration. (b) Reciprocal scattering configuration.

$$S_{21}(-\hat{\mathbf{n}}, \hat{\mathbf{n}}) = -S_{12}(-\hat{\mathbf{n}}, \hat{\mathbf{n}}) \tag{3.4.22}$$

(van de Hulst, 1957, Section 5.32).

Because of the universal nature of reciprocity, Eqs. (3.4.19), (3.4.21), and (3.4.22) are important tests in computations or measurements of light scattering by small particles: violation of reciprocity means that the computations or measurements are incorrect or inaccurate. Alternatively, the use of reciprocity can substantially shorten required computer time or reduce the measurement effort because one may calculate or measure light scattering for only half of the scattering geometries and then use Eqs. (3.4.19) and (3.4.21) for the reciprocal geometries. Reciprocity also plays a fundamental role in the effect of coherent backscattering of light from discrete random media discussed in Chapter 14.

As we have already indicated, Eqs. (3.4.19) and (3.4.21) are valid provided that the permeability, permittivity, and conductivity of the scattering object are symmetric tensors. If the scattering object and/or the surrounding medium consist of magneto-optic materials and are placed in a constant magnetic field $\mathcal{B}$, then Eqs. (3.4.19) and (3.4.21) must be replaced by

$$\overleftrightarrow{A}(-\hat{\mathbf{n}}^{\text{inc}}, -\hat{\mathbf{n}}^{\text{sca}}, -\mathcal{B}) = [\overleftrightarrow{A}(\hat{\mathbf{n}}^{\text{sca}}, \hat{\mathbf{n}}^{\text{inc}}, \mathcal{B})]^{\text{T}}, \tag{3.4.23}$$

$$\mathbf{S}(-\hat{\mathbf{n}}^{\text{inc}}, -\hat{\mathbf{n}}^{\text{sca}}, -\mathcal{B}) = \begin{bmatrix} S_{11}(\hat{\mathbf{n}}^{\text{sca}}, \hat{\mathbf{n}}^{\text{inc}}, \mathcal{B}) & -S_{21}(\hat{\mathbf{n}}^{\text{sca}}, \hat{\mathbf{n}}^{\text{inc}}, \mathcal{B}) \\ -S_{12}(\hat{\mathbf{n}}^{\text{sca}}, \hat{\mathbf{n}}^{\text{inc}}, \mathcal{B}) & S_{22}(\hat{\mathbf{n}}^{\text{sca}}, \hat{\mathbf{n}}^{\text{inc}}, \mathcal{B}) \end{bmatrix} \tag{3.4.24}$$

(Dolginov *et al.*, 1995).

3.5 Scale invariance rule

Another fundamental property of electromagnetic scattering is the so-called scale invariance rule (also referred to as the principle of electromagnetic similitude). The general derivation of this rule was given by Mishchenko (2006) and starts with the introduction of the following dimensionless quantities:

$$\breve{\mathbf{r}} = k_1 \mathbf{r}, \tag{3.5.1}$$

$$\breve{\overleftrightarrow{G}}(\breve{\mathbf{r}}, \breve{\mathbf{r}}') = (\overleftrightarrow{I} + \breve{\nabla} \otimes \breve{\nabla}) \frac{\exp(\mathrm{i}|\breve{\mathbf{r}} - \breve{\mathbf{r}}'|)}{4\pi|\breve{\mathbf{r}} - \breve{\mathbf{r}}'|}, \tag{3.5.2}$$

$$\breve{\nabla} = \frac{1}{k_1}\nabla, \tag{3.5.3}$$

$$\breve{\overleftrightarrow{T}}(\breve{\mathbf{r}}, \breve{\mathbf{r}}') = \frac{1}{k_1^5}\overleftrightarrow{T}(\mathbf{r}, \mathbf{r}'), \tag{3.5.4}$$

$$\breve{m}(\breve{\mathbf{r}}) = m(\mathbf{r}). \tag{3.5.5}$$

It is then rather obvious that the Lippmann–Schwinger equation (3.1.24) can be rewritten for a dimensionless dyadic transition operator as follows:

$$\breve{\overleftrightarrow{T}}(\breve{\mathbf{r}}, \breve{\mathbf{r}}') = [\breve{m}^2(\breve{\mathbf{r}}) - 1]\,\delta(\breve{\mathbf{r}} - \breve{\mathbf{r}}')\overleftrightarrow{I} + [\breve{m}^2(\breve{\mathbf{r}}) - 1] \int_{\breve{V}_{\text{INT}}} \mathrm{d}\breve{\mathbf{r}}''\, \breve{\overleftrightarrow{G}}(\breve{\mathbf{r}}, \breve{\mathbf{r}}'') \cdot \breve{\overleftrightarrow{T}}(\breve{\mathbf{r}}'', \breve{\mathbf{r}}'), \quad \breve{\mathbf{r}}, \breve{\mathbf{r}}' \in \breve{V}_{\text{INT}}, \tag{3.5.6}$$

where we have taken into account that

$$\delta(\mathbf{r}) = b^3 \delta(b\mathbf{r}), \tag{3.5.7}$$

and the dimensionless "volume" $\breve{V}_{\text{INT}} = k_1^3 V_{\text{INT}}$ is obtained from the actual volume V_{INT} by multiplying all dimensions of the latter by k_1. Solving Eq. (3.5.6) by iteration shows that the dimensionless dyadic transition operator depends on the dimensionless particle volume rather than on the actual volume and on the wave number separately.

The next step is to introduce the dimensionless scattering dyadic as follows:

$$\breve{\overleftrightarrow{A}}(\hat{\mathbf{n}}^{\text{sca}}, \hat{\mathbf{n}}^{\text{inc}}) = k_1 \overleftrightarrow{A}(\hat{\mathbf{n}}^{\text{sca}}, \hat{\mathbf{n}}^{\text{inc}}). \tag{3.5.8}$$

Then Eq. (3.3.5) takes the form

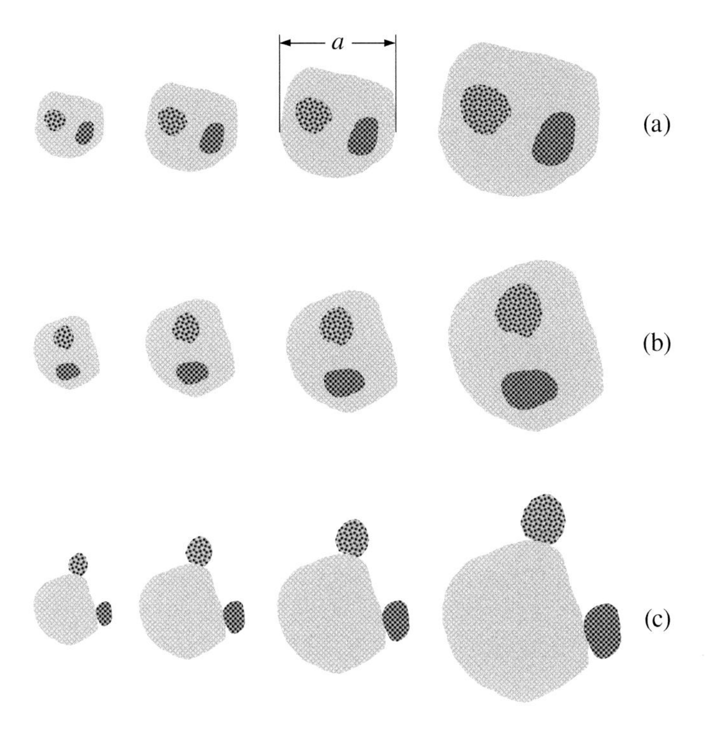

Figure 3.5.1. Three classes of electromagnetically similar objects. Note that the objects in a class have geometrically similar shapes and morphologies as well as identical orientations with respect to the laboratory reference frame.

$$\breve{\overleftrightarrow{A}}(\hat{\mathbf{n}}^{\text{sca}}, \hat{\mathbf{n}}^{\text{inc}}) = \frac{1}{4\pi}(\overleftrightarrow{I} - \hat{\mathbf{n}}^{\text{sca}} \otimes \hat{\mathbf{n}}^{\text{sca}}) \cdot \int_{\breve{V}_{\text{INT}}} \mathrm{d}\breve{\mathbf{r}}' \exp(-\mathrm{i}\hat{\mathbf{n}}^{\text{sca}} \cdot \breve{\mathbf{r}}')$$
$$\times \int_{\breve{V}_{\text{INT}}} \mathrm{d}\mathbf{r}'' \breve{\overleftrightarrow{T}}(\breve{\mathbf{r}}', \breve{\mathbf{r}}'') \exp(\mathrm{i}\hat{\mathbf{n}}^{\text{inc}} \cdot \breve{\mathbf{r}}'') \cdot (\overleftrightarrow{I} - \hat{\mathbf{n}}^{\text{inc}} \otimes \hat{\mathbf{n}}^{\text{inc}}), \qquad (3.5.9)$$

which shows again that the dimensionless scattering dyadic is a function of the dimensionless particle volume rather than a function of the actual particle volume as well as of the wave number. Of course, the same is true of the dimensionless amplitude scattering matrix

$$\breve{\mathbf{S}}(\hat{\mathbf{n}}^{\text{sca}}, \hat{\mathbf{n}}^{\text{inc}}) = k_1 \mathbf{S}(\hat{\mathbf{n}}^{\text{sca}}, \hat{\mathbf{n}}^{\text{inc}}).$$

The scale invariance rule is a direct consequence of these results and states the following. If one multiplies all linear dimensions of the scattering object by a constant factor f (thereby not changing the shape and morphology of the object and its orientation with respect to the coordinate system) and multiplies the wave number k_1 by a factor $1/f$, then the dimensionless scattering dyadic and the dimensionless amplitude scattering matrix of the object do not change.

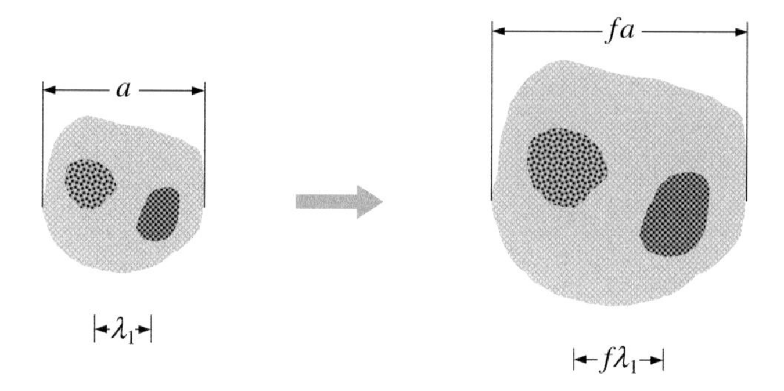

Figure 3.5.2. Scale invariance rule.

This rule can be reformulated as follows. Consider a class of geometrically similar objects with geometrically similar spatial distributions of the relative refractive index and the same orientation with respect to the laboratory reference frame (Fig. 3.5.1). It is clear that each object from the class can be uniquely identified by the value of a typical linear dimension a (for example, the largest or the smallest dimension of the object or the radius of the surface- or volume-equivalent sphere). Then the scale invariance rule implies that the dimensionless scattering characteristics of the objects do not depend on specific values of a and k_1, but rather depend on the product of a and k_1 traditionally called the size parameter x.

The size parameter can also be expressed in terms of the wavelength of the incident wave in the exterior region, $\lambda_1 = 2\pi/k_1$, as $x = 2\pi a/\lambda_1$. This means that multiplying the typical particle size and the wavelength by the same factor f (see Fig. 3.5.2) does not change the dimensionless scattering dyadic and the dimensionless amplitude scattering matrix.

The scale invariance rule can be very helpful in practice because it makes a single computation or measurement applicable to all couplets {size, wavelength} having the same ratio of size to wavelength, provided that the relative refractive index remains the same. In particular, the scale invariance rule is the basic physical principle of the so-called microwave analog technique. The latter involves measurements of microwave scattering by easily manufactured centimeter-sized objects followed by extrapolation to other wavelengths (e.g., visible or infrared) by keeping the ratio of size to wavelength fixed (e.g., Gustafson, 2000 and Section 8.2 of MTL).

The ratios $\overset{\leftrightarrow}{A}(\hat{\mathbf{n}}^{\rm sca}, \hat{\mathbf{n}}^{\rm inc})/a$ and $\mathbf{S}(\hat{\mathbf{n}}^{\rm sca}, \hat{\mathbf{n}}^{\rm inc})/a$ are also scale-invariant quantities. Indeed, since assuming

$$k_1 a = \text{constant}$$

yields

$$k_1 \overset{\leftrightarrow}{A}(\hat{\mathbf{n}}^{\rm sca}, \hat{\mathbf{n}}^{\rm inc}) = \text{constant}$$

and

$$k_1\mathbf{S}(\hat{\mathbf{n}}^{\text{sca}}, \hat{\mathbf{n}}^{\text{inc}}) = \text{constant},$$

dividing the latter two equalities by the first equality must also yield constants.

3.6 Electromagnetic power and electromagnetic energy density

Although the knowledge of the amplitude scattering matrix provides the complete description of the monochromatic scattering process in the far-field zone, the measurement of the amplitude scattering matrix is a very complex experimental problem involving the determination of both the amplitude and the phase of the incident and scattered waves. Measuring the phase is especially difficult, and only a handful of such experiments have been performed, all using the microwave analog technique (Gustafson, 2000). The majority of other experiments have dealt with quasi-monochromatic rather than monochromatic light and involved measurements of derivative quantities having the dimension of energy flux rather than the electric field itself. It is therefore more convenient to characterize the scattering process using quantities that are easier to measure and are encountered more often, even though they may provide a less complete description of the scattering pattern in some cases. Such quantities will be introduced in this and the following sections.

Consider the standard measurement configuration involving a well-collimated detector of electromagnetic radiation located at a distance r from the scattering object in the far-field zone, with its sensitive surface aligned normal to and centered on the position vector $\mathbf{r} = r\hat{\mathbf{r}}$ (see Fig. 3.6.1). The functional definition of a well-collimated detector suitable for our purposes is that of a sensitive plane surface of an area ΔS that registers the energy of monochromatic or quasi-monochromatic light impinging on any point of ΔS in directions confined to a small solid angle $\Delta\Omega$ (called the detector angular aperture) centered at the local normal to the detector surface. We will assume that the angular size of the sensitive surface of the detector as seen from the scattering object is smaller than the detector angular aperture:

$$\frac{\Delta S}{r^2} < \Delta\Omega. \tag{3.6.1}$$

This important inequality ensures that if the detector is centered on the scattering object then all radiation scattered by the object in radial directions and impinging on ΔS is detected. For well-collimated detectors with a small angular aperture, the condition (3.6.1) usually implies that the distance r from the scattering object to the detector is much greater than the diameter D of the sensitive surface of the detector:

$$r \gg D. \tag{3.6.2}$$

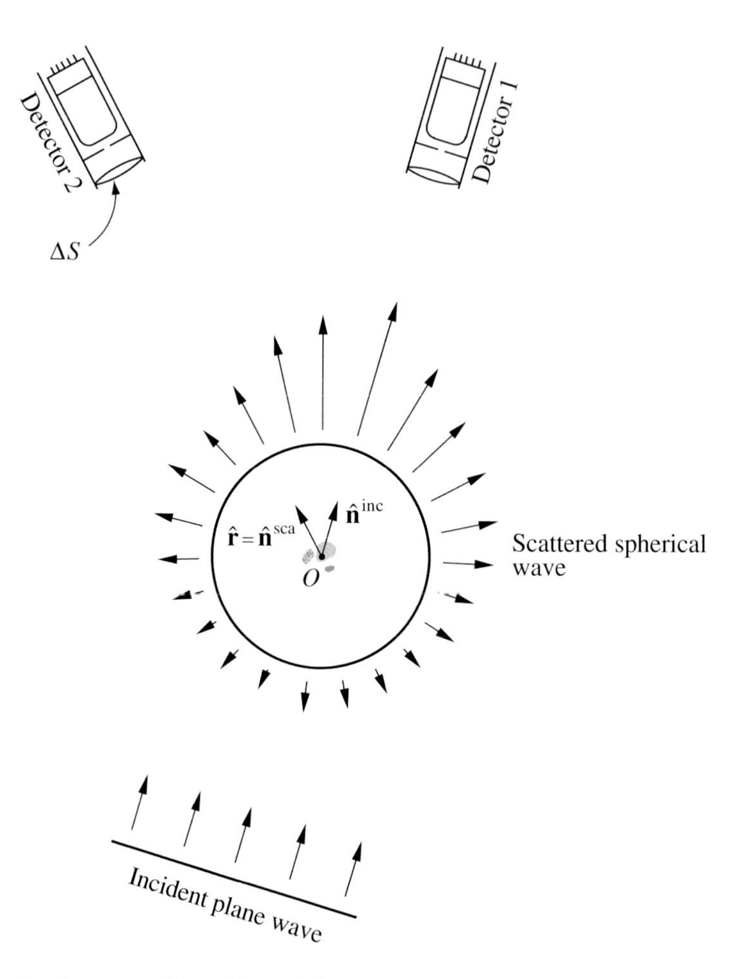

Figure 3.6.1. Response of the collimated detector depends on the line of sight.

We begin by writing the time-averaged Poynting vector $\langle \boldsymbol{\mathcal{S}}(\mathbf{r}', t) \rangle_t$ at any point of the sensitive surface of the detector located in the far-field zone as the sum of three terms:

$$\begin{aligned} \langle \boldsymbol{\mathcal{S}}(\mathbf{r}', t) \rangle_t &= \tfrac{1}{2} \mathrm{Re}\{\mathbf{E}(\mathbf{r}') \times [\mathbf{H}(\mathbf{r}')]^*\} \\ &= \langle \boldsymbol{\mathcal{S}}^{\text{inc}}(\mathbf{r}', t) \rangle_t + \langle \boldsymbol{\mathcal{S}}^{\text{sca}}(\mathbf{r}', t) \rangle_t + \langle \boldsymbol{\mathcal{S}}^{\text{ext}}(\mathbf{r}', t) \rangle_t, \end{aligned} \tag{3.6.3}$$

where $\mathbf{r}' = r'\hat{\mathbf{r}}'$ is the corresponding radius vector,

$$\langle \boldsymbol{\mathcal{S}}^{\text{inc}}(\mathbf{r}', t) \rangle_t = \tfrac{1}{2} \mathrm{Re}\{\mathbf{E}^{\text{inc}}(\mathbf{r}') \times [\mathbf{H}^{\text{inc}}(\mathbf{r}')]^*\} \tag{3.6.4}$$

and

$$\langle \boldsymbol{\mathcal{S}}^{\text{sca}}(\mathbf{r}', t) \rangle_t = \tfrac{1}{2} \mathrm{Re}\{\mathbf{E}^{\text{sca}}(\mathbf{r}') \times [\mathbf{H}^{\text{sca}}(\mathbf{r}')]^*\} \tag{3.6.5}$$

are the Poynting vectors associated with the incident and the scattered fields, respectively, and

$$\langle \mathcal{S}^{\text{ext}}(\mathbf{r}',t)\rangle_t = \tfrac{1}{2}\text{Re}\{\mathbf{E}^{\text{inc}}(\mathbf{r}')\times[\mathbf{H}^{\text{sca}}(\mathbf{r}')]^* + \mathbf{E}^{\text{sca}}(\mathbf{r}')\times[\mathbf{H}^{\text{inc}}(\mathbf{r}')]^*\} \tag{3.6.6}$$

can be interpreted as the term caused by interaction between the incident and the scattered fields.

Let us consider a scattering object illuminated by a plane electromagnetic wave. Recalling Eqs. (2.5.6), (2.5.8), (2.5.17), (3.3.1), and (B.10), we have for the incident wave in the far-field zone of the scattering particle:

$$\begin{aligned}\mathbf{E}^{\text{inc}}(\mathbf{r}') &= \mathbf{E}_0^{\text{inc}}\exp(\mathrm{i}k_1\hat{\mathbf{n}}^{\text{inc}}\cdot\mathbf{r}')\\ &\underset{k_1r'\to\infty}{=} \frac{\mathrm{i}2\pi}{k_1}\left[\delta(\hat{\mathbf{n}}^{\text{inc}}+\hat{\mathbf{r}}')\frac{\exp(-\mathrm{i}k_1r')}{r'} - \delta(\hat{\mathbf{n}}^{\text{inc}}-\hat{\mathbf{r}}')\frac{\exp(\mathrm{i}k_1r')}{r'}\right]\mathbf{E}_0^{\text{inc}},\end{aligned} \tag{3.6.7}$$

$$\mathbf{E}_0^{\text{inc}}\cdot\hat{\mathbf{n}}^{\text{inc}} = 0, \tag{3.6.8}$$

$$\begin{aligned}\mathbf{H}^{\text{inc}}(\mathbf{r}') &= \sqrt{\frac{\epsilon_1}{\mu_0}}\exp(\mathrm{i}k_1\hat{\mathbf{n}}^{\text{inc}}\cdot\mathbf{r}')\,\hat{\mathbf{n}}^{\text{inc}}\times\mathbf{E}_0^{\text{inc}}\\ &\underset{k_1r'\to\infty}{=} \frac{\mathrm{i}2\pi}{k_1}\left[\delta(\hat{\mathbf{n}}^{\text{inc}}+\hat{\mathbf{r}}')\frac{\exp(-\mathrm{i}k_1r')}{r'}\right.\\ &\qquad\left. - \delta(\hat{\mathbf{n}}^{\text{inc}}-\hat{\mathbf{r}}')\frac{\exp(\mathrm{i}k_1r')}{r'}\right]\sqrt{\frac{\epsilon_1}{\mu_0}}\,\hat{\mathbf{n}}^{\text{inc}}\times\mathbf{E}_0^{\text{inc}}.\end{aligned} \tag{3.6.9}$$

Equations (2.11.8) and (3.2.16) give for the scattered spherical wave:

$$\mathbf{E}^{\text{sca}}(\mathbf{r}') = \frac{\exp(\mathrm{i}k_1r')}{r'}\mathbf{E}_1^{\text{sca}}(\hat{\mathbf{r}}'), \qquad \mathbf{E}_1^{\text{sca}}(\hat{\mathbf{r}}')\cdot\hat{\mathbf{r}}' = 0, \tag{3.6.10}$$

$$\mathbf{H}^{\text{sca}}(\mathbf{r}') = \sqrt{\frac{\epsilon_1}{\mu_0}}\frac{\exp(\mathrm{i}k_1r')}{r'}\hat{\mathbf{r}}'\times\mathbf{E}_1^{\text{sca}}(\hat{\mathbf{r}}'). \tag{3.6.11}$$

One can now derive that the total electromagnetic power received by a well-collimated detector is

$$\begin{aligned}W_{\Delta S}(\hat{\mathbf{r}}) &= \int_{\Delta S}\mathrm{d}S\,\hat{\mathbf{r}}'\cdot\langle\mathcal{S}(\mathbf{r}',t)\rangle_t\\ &\approx \Delta S\frac{1}{2}\sqrt{\frac{\epsilon_1}{\mu_0}}\frac{1}{r^2}|\mathbf{E}_1^{\text{sca}}(\hat{\mathbf{r}})|^2\end{aligned} \tag{3.6.12}$$

when $\hat{\mathbf{r}} \neq \hat{\mathbf{n}}^{\text{inc}}$ (detector 2 in Fig. 3.6.1), whereas for the exact forward-scattering direction (detector 1),

$$W_{\Delta S}(\hat{\mathbf{n}}^{\text{inc}}) = \int_{\Delta S}\mathrm{d}S\,\mathbf{r}'\cdot\langle\mathcal{S}(\mathbf{r}',t)\rangle_t$$

$$\approx \Delta S \frac{1}{2}\sqrt{\frac{\epsilon_1}{\mu_0}} |\mathbf{E}_0^{\text{inc}}|^2 + \int_{\Delta S} \mathrm{d}S\,\hat{\mathbf{n}}^{\text{inc}} \cdot [\langle \boldsymbol{\mathcal{S}}^{\text{sca}}(\mathbf{r}',t)\rangle_t + \langle \boldsymbol{\mathcal{S}}^{\text{ext}}(\mathbf{r}',t)\rangle_t]$$

$$\approx \Delta S \frac{1}{2}\sqrt{\frac{\epsilon_1}{\mu_0}} \left[|\mathbf{E}_0^{\text{inc}}|^2 + \frac{1}{r^2}|\mathbf{E}_1^{\text{sca}}(\hat{\mathbf{n}}^{\text{inc}})|^2 \right]$$

$$+ r^2 \int_{\Omega'} \mathrm{d}\hat{\mathbf{r}}'\,\hat{\mathbf{n}}^{\text{inc}} \cdot \langle \boldsymbol{\mathcal{S}}^{\text{ext}}(r\hat{\mathbf{r}}',t)\rangle_t$$

$$\approx \Delta S \frac{1}{2}\sqrt{\frac{\epsilon_1}{\mu_0}} \left[|\mathbf{E}_0^{\text{inc}}|^2 + \frac{1}{r^2}|\mathbf{E}_1^{\text{sca}}(\hat{\mathbf{n}}^{\text{inc}})|^2 \right]$$

$$- \frac{2\pi}{k_1}\sqrt{\frac{\epsilon_1}{\mu_0}}\, \mathrm{Im}[\mathbf{E}_1^{\text{sca}}(\hat{\mathbf{n}}^{\text{inc}}) \cdot (\mathbf{E}_0^{\text{inc}})^*] \tag{3.6.13a}$$

$$= \Delta S \frac{1}{2}\sqrt{\frac{\epsilon_1}{\mu_0}} |\mathbf{E}_0^{\text{inc}}|^2 - \frac{2\pi}{k_1}\sqrt{\frac{\epsilon_1}{\mu_0}}\, \mathrm{Im}[\mathbf{E}_1^{\text{sca}}(\hat{\mathbf{n}}^{\text{inc}}) \cdot (\mathbf{E}_0^{\text{inc}})^*]$$

$$+ O(r^{-2}), \tag{3.6.13b}$$

where $\Omega' = \Delta S / r^2$ is the solid angle centered at the direction $\hat{\mathbf{n}}^{\text{inc}}$ and subtended by the detector surface at the distance r from the particle. Equations (3.6.13a) and (3.6.13b) are a particular case of the so-called optical theorem.

Note that the presence of the terms proportional to the delta function $\delta(\hat{\mathbf{n}}^{\text{inc}} + \hat{\mathbf{r}})$ on the right-hand sides of Eqs. (3.6.7) and (3.6.9) seems to indicate that there is interference of the incident field and the field scattered in the exact backscattering direction. It is easy to verify, however, that the contribution of the interference term $\langle \boldsymbol{\mathcal{S}}^{\text{ext}}(\mathbf{r}',t)\rangle_t$ to the signal measured by a detector facing the exact backscattering direction vanishes upon taking the real part of the signal according to Eq. (3.6.6).

The first term on the right-hand side of Eq. (3.6.13b) is proportional to the detector area ΔS and is equal to the electromagnetic power that would be received by detector 1 in the absence of the scattering object (cf. Eq. (2.5.23) with $m_1 = 0$), whereas the second term is independent of ΔS and describes attenuation caused by interposing the object between the light source and the detector. Thus, the detector centered at the exact forward-scattering direction measures the power of the incident light attenuated by the interference of the incident and the scattered fields plus a relatively small contribution from the scattered light, whereas the detector centered at any other direction registers only the scattered light. These are two fundamental features of electromagnetic scattering by a fixed object.

It is extremely important that in either case the detector reacts to a transverse electromagnetic wave, be it the scattered spherical wave propagating in a direction away from $\hat{\mathbf{n}}^{\text{inc}}$ (detector 2) or the superposition of the incident plane wave and the scattered spherical wave propagating in the incidence direction (detector 1). We will see in the following two sections that this allows one to describe the polarization response of detectors 1 and 2 using the Stokes vector formalism.

The formula for the time-averaged electromagnetic energy density of the total field at a point **r** follows from Eq. (2.5.27):

$$\langle \mathcal{U}(\mathbf{r},t)\rangle_t = \langle \mathcal{U}^{\rm inc}(\mathbf{r},t)\rangle_t + \langle \mathcal{U}^{\rm sca}(\mathbf{r},t)\rangle_t + \langle \mathcal{U}^{\rm ext}(\mathbf{r},t)\rangle_t, \tag{3.6.14}$$

where

$$\begin{aligned}\langle \mathcal{U}^{\rm inc}(\mathbf{r},t)\rangle_t &= \tfrac{1}{4}\{\epsilon_1\mathbf{E}^{\rm inc}(\mathbf{r})\cdot[\mathbf{E}^{\rm inc}(\mathbf{r})]^* + \mu_0\mathbf{H}^{\rm inc}(\mathbf{r})\cdot[\mathbf{H}^{\rm inc}(\mathbf{r})]^*\}\\ &= \tfrac{1}{2}\epsilon_1|\mathbf{E}_0^{\rm inc}|^2\end{aligned} \tag{3.6.15}$$

is the component due to the incident field,

$$\begin{aligned}\langle \mathcal{U}^{\rm sca}(\mathbf{r},t)\rangle_t &= \tfrac{1}{4}\{\epsilon_1\mathbf{E}^{\rm sca}(\mathbf{r})\cdot[\mathbf{E}^{\rm sca}(\mathbf{r})]^* + \mu_0\mathbf{H}^{\rm sca}(\mathbf{r})\cdot[\mathbf{H}^{\rm sca}(\mathbf{r})]^*\}\\ &= \frac{\epsilon_1|\mathbf{E}_0^{\rm inc}|^2}{2r^2}\end{aligned} \tag{3.6.16}$$

is that due to the scattered field, and

$$\begin{aligned}\langle \mathcal{U}^{\rm ext}(\mathbf{r},t)\rangle_t &= \tfrac{1}{2}{\rm Re}\{\epsilon_1\mathbf{E}^{\rm inc}(\mathbf{r})\cdot[\mathbf{E}^{\rm sca}(\mathbf{r})]^* + \mu_0\mathbf{H}^{\rm inc}(\mathbf{r})\cdot[\mathbf{H}^{\rm sca}(\mathbf{r})]^*\}\\ &= -\frac{2\pi\epsilon_1}{k_1 r^2}\delta(\hat{\mathbf{n}}^{\rm inc}-\hat{\mathbf{r}})\,{\rm Im}[\mathbf{E}_1^{\rm sca}(\hat{\mathbf{n}}^{\rm inc})\cdot(\mathbf{E}_0^{\rm inc})^*]\end{aligned} \tag{3.6.17}$$

is that due to the interference of the incident and the forward-scattered field. The latter term vanishes everywhere except along the straight line originating at the scattering object and extending in the incidence direction.

The ratio of minus the second term on the right-hand side of Eq. (3.6.13b) to the incident energy flux has the dimension of area and is called the extinction cross section, $C_{\rm ext}$ (see also Section 3.9). Therefore, Eq. (3.6.13b) can also be written as follows:

$$W_{\Delta S}(\hat{\mathbf{r}}=\hat{\mathbf{n}}^{\rm inc}) \approx (\Delta S - C_{\rm ext})\frac{1}{2}\sqrt{\frac{\epsilon_1}{\mu_0}}|\mathbf{E}_0^{\rm inc}|^2 + O(r^{-2}). \tag{3.6.18}$$

It is well known that the extinction cross section can exceed the area of the object's geometrical projection by a factor of several (e.g., Section 9.1 of MTL). Therefore, it is necessary to assume that the diameter D of the surface of detector 1 is significantly greater than any linear dimension of the scattering object:

$$D \gg 2a. \tag{3.6.19}$$

Indeed, this requirement ensures that the signal measured by detector 1 is positive and, thus, physically meaningful. In general, this requirement does not apply to detector 2 in Fig. 3.6.1.

There is another fundamental reason for imposing the requirement (3.6.19). As we have already emphasized, the far-field-zone approximation implies the treatment of the scatterer as a point-like object. This treatment is justified for the derivation of the above formulas, but becomes too crude when one attempts to describe the interaction

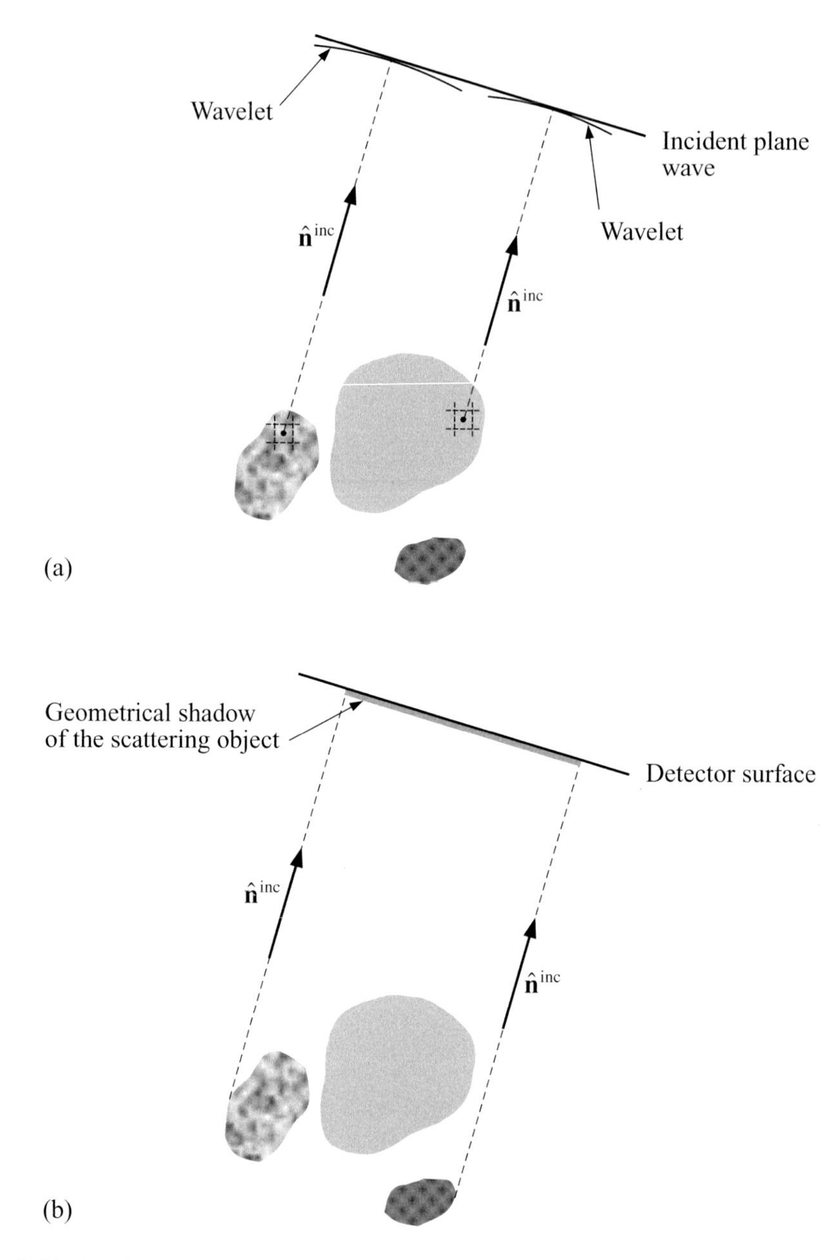

Figure 3.6.2. (a) The wavelets generated by different elementary volume elements interact with the incident plane wave along the respective straight lines parallel to $\hat{\mathbf{n}}^{\text{inc}}$. (b) To capture the interaction of all wavelets generated by different elementary volume elements with the incident plane wave, the detector surface must be greater than the object's geometrical shadow.

of the incident plane wave and the scattered field *across* the sensitive surface of the detector facing the incident light. To do that properly, we should recall that the actual scattered field is a superposition of spherical wavelets generated by elementary volume elements of the scattering object. It is clear from the previous discussion that each wavelet interacts with the incident plane wave only along the straight line drawn through the center of the corresponding volume element and parallel to the incidence direction $\hat{\mathbf{n}}^{\mathrm{inc}}$ (Fig. 3.6.2(a)). Therefore, to capture each individual interaction, the detector surface must exceed the area of the shadow cast by the scatterer onto the plane normal to $\hat{\mathbf{n}}^{\mathrm{inc}}$, Fig. 3.6.2(b).

Another important practical aspect of scattering measurements is that the angular scattering pattern for a particle comparable to and larger than the wavelength is known to vary dramatically with scattering direction. This angular variability can be traced back to the complex exponential factor $\exp(-\mathrm{i}k_1\hat{\mathbf{r}}\cdot\mathbf{r}')$ on the right-hand side of Eq. (3.2.14). Indeed, the electric field contributions from two arbitrary elementary volumes of the scattering object centered at $\mathbf{r}'$ and $\mathbf{r}''$ interfere in the far-field zone, the result of the interference being controlled by the product

$$\exp(-\mathrm{i}k_1\hat{\mathbf{r}}\cdot\mathbf{r}')[\exp(-\mathrm{i}k_1\hat{\mathbf{r}}\cdot\mathbf{r}'')]^* = \exp[-\mathrm{i}k_1\hat{\mathbf{r}}\cdot(\mathbf{r}'-\mathbf{r}'')]. \tag{3.6.20}$$

Obviously, depending on the angle between $\hat{\mathbf{r}}$ and $\mathbf{r}'-\mathbf{r}''$ and on $|\mathbf{r}'-\mathbf{r}''|$, this complex exponential can be a rapidly varying function of $\hat{\mathbf{r}}$. As a result, the angular scattering pattern in the far-field zone can be expected to be a superposition of multiple maxima and minima generated by different pairs of elementary volume elements of the scatterer. The most rapidly changing component of the scattering pattern should be caused by the pairs of elementary volume elements with $(\mathbf{r}'-\mathbf{r}'')\perp\hat{\mathbf{r}}$ and $|\mathbf{r}'-\mathbf{r}''|\approx 2a$. Therefore, the far-field angular pattern can be expected to vary quite significantly even when the scattering direction changes by as little as $\pi/(2k_1a)$ (rad) since this change corresponds to a change of the phase $k_1\hat{\mathbf{r}}\cdot(\mathbf{r}'-\mathbf{r}'')$ equal to π. An example of this strong angular variability can be seen in the upper left panel of Plate 11.13.1, which will be further discussed in Section 11.13.

Consequently, if a detector were to fully resolve this angular variability, the distance r from the scattering object to the detector must satisfy the following inequality:

$$r \gg \frac{Dk_1a}{\pi}. \tag{3.6.21}$$

If this requirement is not met, then the detector will record a convolution of the angular scattering pattern with the detector angular aperture. For particles greater than the wavelength, the condition (3.6.21) becomes much stronger than the condition (3.6.2).

3.7 Phase matrix

In the thought experiment discussed in the previous section, it was assumed that the detectors can measure only the total power of electromagnetic radiation impinging on their sensitive surfaces in directions within their angular apertures and that they make no distinction between electromagnetic waves with different states of polarization. Many detectors of electromagnetic energy are indeed polarization-insensitive. However, by interposing one or more optical elements such as polarizers and retarders (see Section 2.10) between the source of light and the scattering object, one can generate incident light with a specific state of polarization, whereas interposing one or more optical elements between the object and the detector enables the latter to measure the power corresponding to a particular polarization component of the scattered light. By repeating the measurement for a number of different combinations and/or orientations of the optical elements, one can, in principle, determine the specific prescription for the transformation of a complete set of polarization characteristics of the incident light into that of the scattered light *provided that both sets of characteristics have the same dimension of energy flux.* This prescription is usually formulated in terms of the so-called phase and extinction matrices.

As discussed in Section 2.6, convenient complete sets of polarization characteristics having the dimension of monochromatic energy flux are the coherency and Stokes column vectors. So we will now assume that a measurement device realizing the situation shown schematically in Fig. 3.6.1 can:

1. Generate incident light with different (but physically realizable) combinations of coherency or Stokes column vector components.
2. Measure the electromagnetic power associated with any component of the coherency or the Stokes column vector and equal to the integral of the component over the surface ΔS of the collimated detector aligned normal to the direction of propagation $\hat{\mathbf{r}}$. The component itself is then found by dividing the measured power by ΔS.

Let us first consider the situation when the scattering direction is *away from the incidence direction* ($\hat{\mathbf{r}} \neq \hat{\mathbf{n}}^{\text{inc}}$). According to the discussion of the previous section, detector 2 in Fig. 3.6.1 registers only the scattered radiation in the form of a transverse outgoing spherical wave. Therefore, one can express the polarization response of the detector in terms of the coherency column vector of the scattered wave as follows:

$$(\textbf{Signal 2})^{J} = \Delta S\, \mathbf{J}^{\text{sca}}(r\hat{\mathbf{n}}^{\text{sca}}), \qquad \hat{\mathbf{n}}^{\text{sca}} \neq \hat{\mathbf{n}}^{\text{inc}}, \tag{3.7.1}$$

where

$$\mathbf{J}^{\text{sca}}(r\hat{\mathbf{n}}^{\text{sca}}) = \frac{1}{2}\sqrt{\frac{\epsilon_1}{\mu_0}} \begin{bmatrix} E_\theta^{\text{sca}}(r\hat{\mathbf{n}}^{\text{sca}})[E_\theta^{\text{sca}}(r\hat{\mathbf{n}}^{\text{sca}})]^* \\ E_\theta^{\text{sca}}(r\hat{\mathbf{n}}^{\text{sca}})[E_\varphi^{\text{sca}}(r\hat{\mathbf{n}}^{\text{sca}})]^* \\ E_\varphi^{\text{sca}}(r\hat{\mathbf{n}}^{\text{sca}})[E_\theta^{\text{sca}}(r\hat{\mathbf{n}}^{\text{sca}})]^* \\ E_\varphi^{\text{sca}}(r\hat{\mathbf{n}}^{\text{sca}})[E_\varphi^{\text{sca}}(r\hat{\mathbf{n}}^{\text{sca}})]^* \end{bmatrix}$$

$$
= \frac{1}{r^2}\frac{1}{2}\sqrt{\frac{\epsilon_1}{\mu_0}}\begin{bmatrix} E_{1\theta}^{\text{sca}}(\hat{\mathbf{n}}^{\text{sca}})[E_{1\theta}^{\text{sca}}(\hat{\mathbf{n}}^{\text{sca}})]^* \\ E_{1\theta}^{\text{sca}}(\hat{\mathbf{n}}^{\text{sca}})[E_{1\varphi}^{\text{sca}}(\hat{\mathbf{n}}^{\text{sca}})]^* \\ E_{1\varphi}^{\text{sca}}(\hat{\mathbf{n}}^{\text{sca}})[E_{1\theta}^{\text{sca}}(\hat{\mathbf{n}}^{\text{sca}})]^* \\ E_{1\varphi}^{\text{sca}}(\hat{\mathbf{n}}^{\text{sca}})[E_{1\varphi}^{\text{sca}}(\hat{\mathbf{n}}^{\text{sca}})]^* \end{bmatrix}. \tag{3.7.2}
$$

Recalling that the coherency column vector of the incident plane wave is given by

$$
\mathbf{J}^{\text{inc}} = \frac{1}{2}\sqrt{\frac{\epsilon_1}{\mu_0}}\begin{bmatrix} E_{0\theta}^{\text{inc}}(E_{0\theta}^{\text{inc}})^* \\ E_{0\theta}^{\text{inc}}(E_{0\varphi}^{\text{inc}})^* \\ E_{0\varphi}^{\text{inc}}(E_{0\theta}^{\text{inc}})^* \\ E_{0\varphi}^{\text{inc}}(E_{0\varphi}^{\text{inc}})^* \end{bmatrix}, \tag{3.7.3}
$$

it is straightforward to derive the following relationship between the coherency column vectors of the incident and scattered light:

$$
\mathbf{J}^{\text{sca}}(r\hat{\mathbf{n}}^{\text{sca}}) = \frac{1}{r^2}\mathbf{Z}^J(\hat{\mathbf{n}}^{\text{sca}}, \hat{\mathbf{n}}^{\text{inc}})\mathbf{J}^{\text{inc}}, \tag{3.7.4}
$$

where the elements of the 4×4 coherency phase matrix $\mathbf{Z}^J(\hat{\mathbf{n}}^{\text{sca}}, \hat{\mathbf{n}}^{\text{inc}})$ have the dimension of area and are quadratic combinations of the elements of the amplitude scattering matrix $\mathbf{S}(\hat{\mathbf{n}}^{\text{sca}}, \hat{\mathbf{n}}^{\text{inc}})$:

$$
\mathbf{Z}^J = \begin{bmatrix} |S_{11}|^2 & S_{11}S_{12}^* & S_{12}S_{11}^* & |S_{12}|^2 \\ S_{11}S_{21}^* & S_{11}S_{22}^* & S_{12}S_{21}^* & S_{12}S_{22}^* \\ S_{21}S_{11}^* & S_{21}S_{12}^* & S_{22}S_{11}^* & S_{22}S_{12}^* \\ |S_{21}|^2 & S_{21}S_{22}^* & S_{22}S_{21}^* & |S_{22}|^2 \end{bmatrix}. \tag{3.7.5}
$$

In the Stokes-vector representation,

$$
\textbf{Signal 2} = \Delta S\,\mathbf{I}^{\text{sca}}(r\hat{\mathbf{n}}^{\text{sca}}), \qquad \hat{\mathbf{n}}^{\text{sca}} \neq \hat{\mathbf{n}}^{\text{inc}}, \tag{3.7.6}
$$

where the Stokes column vector of the scattered spherical wave is given by

$$
\mathbf{I}^{\text{sca}}(r\hat{\mathbf{n}}^{\text{sca}}) = \mathbf{D}\mathbf{J}^{\text{sca}}(r\hat{\mathbf{n}}^{\text{sca}}) = \frac{1}{r^2}\frac{1}{2}\sqrt{\frac{\epsilon_1}{\mu_0}}\begin{bmatrix} E_{1\theta}^{\text{sca}}(E_{1\theta}^{\text{sca}})^* + E_{1\varphi}^{\text{sca}}(E_{1\varphi}^{\text{sca}})^* \\ E_{1\theta}^{\text{sca}}(E_{1\theta}^{\text{sca}})^* - E_{1\varphi}^{\text{sca}}(E_{1\varphi}^{\text{sca}})^* \\ -E_{1\theta}^{\text{sca}}(E_{1\varphi}^{\text{sca}})^* - E_{1\varphi}^{\text{sca}}(E_{1\theta}^{\text{sca}})^* \\ \mathrm{i}[E_{1\varphi}^{\text{sca}}(E_{1\theta}^{\text{sca}})^* - E_{1\theta}^{\text{sca}}(E_{1\varphi}^{\text{sca}})^*] \end{bmatrix}. \tag{3.7.7}
$$

The corresponding scattering transformation now reads:

$$
\mathbf{I}^{\text{sca}}(r\hat{\mathbf{n}}^{\text{sca}}) = \frac{1}{r^2}\mathbf{Z}(\hat{\mathbf{n}}^{\text{sca}}, \hat{\mathbf{n}}^{\text{inc}})\mathbf{I}^{\text{inc}}, \tag{3.7.8}
$$

where $\mathbf{Z}(\hat{\mathbf{n}}^{\text{sca}}, \hat{\mathbf{n}}^{\text{inc}})$ is the 4×4 Stokes phase matrix, and the Stokes column vector of the incident plane wave is given by

$$\mathbf{I}^{\text{inc}} = \mathbf{D}\mathbf{J}^{\text{inc}} = \frac{1}{2}\sqrt{\frac{\epsilon_1}{\mu_0}} \begin{bmatrix} E_{0\theta}^{\text{inc}}(E_{0\theta}^{\text{inc}})^* + E_{0\varphi}^{\text{inc}}(E_{0\varphi}^{\text{inc}})^* \\ E_{0\theta}^{\text{inc}}(E_{0\theta}^{\text{inc}})^* - E_{0\varphi}^{\text{inc}}(E_{0\varphi}^{\text{inc}})^* \\ -E_{0\theta}^{\text{inc}}(E_{0\varphi}^{\text{inc}})^* - E_{0\varphi}^{\text{inc}}(E_{0\theta}^{\text{inc}})^* \\ \mathrm{i}[E_{0\varphi}^{\text{inc}}(E_{0\theta}^{\text{inc}})^* - E_{0\theta}^{\text{inc}}(E_{0\varphi}^{\text{inc}})^*] \end{bmatrix}. \tag{3.7.9}$$

Explicit formulas for the elements of the Stokes phase matrix in terms of the amplitude scattering matrix elements result from

$$\mathbf{Z}(\hat{\mathbf{n}}^{\text{sca}}, \hat{\mathbf{n}}^{\text{inc}}) = \mathbf{D}\mathbf{Z}^J(\hat{\mathbf{n}}^{\text{sca}}, \hat{\mathbf{n}}^{\text{inc}})\mathbf{D}^{-1} \tag{3.7.10}$$

(cf. Eqs. (2.6.5) and (2.6.7)) and are as follows:

$$Z_{11} = \tfrac{1}{2}(|S_{11}|^2 + |S_{12}|^2 + |S_{21}|^2 + |S_{22}|^2), \tag{3.7.11}$$

$$Z_{12} = \tfrac{1}{2}(|S_{11}|^2 - |S_{12}|^2 + |S_{21}|^2 - |S_{22}|^2), \tag{3.7.12}$$

$$Z_{13} = -\mathrm{Re}(S_{11}S_{12}^* + S_{22}S_{21}^*), \tag{3.7.13}$$

$$Z_{14} = -\mathrm{Im}(S_{11}S_{12}^* - S_{22}S_{21}^*), \tag{3.7.14}$$

$$Z_{21} = \tfrac{1}{2}(|S_{11}|^2 + |S_{12}|^2 - |S_{21}|^2 - |S_{22}|^2), \tag{3.7.15}$$

$$Z_{22} = \tfrac{1}{2}(|S_{11}|^2 - |S_{12}|^2 - |S_{21}|^2 + |S_{22}|^2), \tag{3.7.16}$$

$$Z_{23} = -\mathrm{Re}(S_{11}S_{12}^* - S_{22}S_{21}^*), \tag{3.7.17}$$

$$Z_{24} = -\mathrm{Im}(S_{11}S_{12}^* + S_{22}S_{21}^*), \tag{3.7.18}$$

$$Z_{31} = -\mathrm{Re}(S_{11}S_{21}^* + S_{22}S_{12}^*), \tag{3.7.19}$$

$$Z_{32} = -\mathrm{Re}(S_{11}S_{21}^* - S_{22}S_{12}^*), \tag{3.7.20}$$

$$Z_{33} = \mathrm{Re}(S_{11}S_{22}^* + S_{12}S_{21}^*), \tag{3.7.21}$$

$$Z_{34} = \mathrm{Im}(S_{11}S_{22}^* + S_{21}S_{12}^*), \tag{3.7.22}$$

$$Z_{41} = -\mathrm{Im}(S_{21}S_{11}^* + S_{22}S_{12}^*), \tag{3.7.23}$$

$$Z_{42} = -\mathrm{Im}(S_{21}S_{11}^* - S_{22}S_{12}^*), \tag{3.7.24}$$

$$Z_{43} = \mathrm{Im}(S_{22}S_{11}^* - S_{12}S_{21}^*), \tag{3.7.25}$$

$$Z_{44} = \mathrm{Re}(S_{22}S_{11}^* - S_{12}S_{21}^*). \tag{3.7.26}$$

Finally, the modified Stokes and circular-polarization phase matrices are given by

$$\mathbf{Z}^{\text{MS}}(\hat{\mathbf{n}}^{\text{sca}}, \hat{\mathbf{n}}^{\text{inc}}) = \mathbf{B}\mathbf{Z}(\hat{\mathbf{n}}^{\text{sca}}, \hat{\mathbf{n}}^{\text{inc}})\mathbf{B}^{-1} \tag{3.7.27}$$

and

$$\mathbf{Z}^{\text{CP}}(\hat{\mathbf{n}}^{\text{sca}}, \hat{\mathbf{n}}^{\text{inc}}) = \mathbf{A}\mathbf{Z}(\hat{\mathbf{n}}^{\text{sca}}, \hat{\mathbf{n}}^{\text{inc}})\mathbf{A}^{-1}, \tag{3.7.28}$$

respectively (see Eqs. (2.6.9)–(2.6.16)). The elements of the matrices $\mathbf{Z}$ and $\mathbf{Z}^{\text{MS}}$ are real-valued. Like the amplitude scattering matrix, the phase matrices explicitly depend on φ^{inc} and φ^{sca} even when the incident and/or scattered light propagates along the z-axis.

The elements of all phase matrices have the dimension of area. It is easy to see that the dimensionless products of k_1^2 and the phase matrix elements satisfy the scale invariance rule (see Section 3.5). Another way to create scale-invariant quantities is to divide the phase matrix elements by a^2.

Up until now we have been considering only the scattering of monochromatic plane waves. However, we already pointed out in Section 2.9 that the formalism based on the solution of time-harmonic Maxwell equations must also be applicable to quasi-monochromatic light. Therefore, Eqs. (3.7.4) and (3.7.8) remain valid even when the incident radiation is a parallel quasi-monochromatic beam provided that the coherency and Stokes column vectors entering Eqs. (3.7.4) and (3.7.8) are averages taken over a sufficiently long time interval.

In general, all 16 elements of any of the phase matrices introduced above are non-zero. However, the phase matrix elements of a single particle are expressed in terms of only seven independent real numbers resulting from the four moduli $|S_{ij}|$ $(i, j = 1, 2)$ and three differences in phase between the S_{ij}. Therefore, only seven of the phase matrix elements are actually independent, and there must be nine unique relations among the sixteen phase matrix elements. Furthermore, the specific mathematical structure of the phase matrix can also be used to derive many useful linear and quadratic inequalities for the phase matrix elements. The most important of these inequalities are

$$Z_{11} \geq 0 \tag{3.7.29}$$

(this property follows directly from Eq. (3.7.11)) and

$$|Z_{ij}| \leq Z_{11} \quad (i, j = 1, \ldots, 4). \tag{3.7.30}$$

The reader is referred to Hovenier *et al.* (1986), Cloude and Pottier (1996), and Hovenier and van der Mee (1996, 2000) for a review of this subject and a discussion of how the general properties of the phase matrix can be used for testing the results of theoretical computations and laboratory measurements.

From Eqs. (3.7.11)–(3.7.26) and (3.4.21) we derive the reciprocity relation for the Stokes phase matrix:

$$\mathbf{Z}(-\hat{\mathbf{n}}^{\text{inc}}, -\hat{\mathbf{n}}^{\text{sca}}) = \mathbf{\Delta}_3[\mathbf{Z}(\hat{\mathbf{n}}^{\text{sca}}, \hat{\mathbf{n}}^{\text{inc}})]^{\text{T}}\mathbf{\Delta}_3, \tag{3.7.31}$$

where

$$\mathbf{\Delta}_3 = \mathbf{\Delta}_3^{\text{T}} = \mathbf{\Delta}_3^{-1} = \begin{bmatrix} 1 & 0 & 0 & 0 \\ 0 & 1 & 0 & 0 \\ 0 & 0 & -1 & 0 \\ 0 & 0 & 0 & 1 \end{bmatrix}. \tag{3.7.32}$$

The reciprocity relations for the other phase matrices can be easily obtained from Eqs. (3.7.10), (3.7.27), and (3.7.28):

$$\begin{aligned}\mathbf{Z}^J(-\hat{\mathbf{n}}^{\rm inc},-\hat{\mathbf{n}}^{\rm sca}) &= \mathbf{D}^{-1}\mathbf{Z}(-\hat{\mathbf{n}}^{\rm inc},-\hat{\mathbf{n}}^{\rm sca})\mathbf{D}\\ &= \mathbf{D}^{-1}\mathbf{\Delta}_3[\mathbf{Z}(\hat{\mathbf{n}}^{\rm sca},\hat{\mathbf{n}}^{\rm inc})]^{\rm T}\mathbf{\Delta}_3\mathbf{D}\\ &= \mathbf{D}^{-1}\mathbf{\Delta}_3[\mathbf{D}\mathbf{Z}^J(\hat{\mathbf{n}}^{\rm sca},\hat{\mathbf{n}}^{\rm inc})\mathbf{D}^{-1}]^{\rm T}\mathbf{\Delta}_3\mathbf{D}\\ &= \mathbf{D}^{-1}\mathbf{\Delta}_3[\mathbf{D}^{-1}]^{\rm T}[\mathbf{Z}^J(\hat{\mathbf{n}}^{\rm sca},\hat{\mathbf{n}}^{\rm inc})]^{\rm T}\mathbf{D}^{\rm T}\mathbf{\Delta}_3\mathbf{D}\\ &= \mathbf{\Delta}_{23}[\mathbf{Z}^J(\hat{\mathbf{n}}^{\rm sca},\hat{\mathbf{n}}^{\rm inc})]^{\rm T}\mathbf{\Delta}_{23},\end{aligned} \tag{3.7.33}$$

$$\begin{aligned}\mathbf{Z}^{\rm MS}(-\hat{\mathbf{n}}^{\rm inc},-\hat{\mathbf{n}}^{\rm sca}) &= \mathbf{B}\mathbf{Z}(-\hat{\mathbf{n}}^{\rm inc},-\hat{\mathbf{n}}^{\rm sca})\mathbf{B}^{-1}\\ &= \mathbf{B}\mathbf{\Delta}_3[\mathbf{Z}(\hat{\mathbf{n}}^{\rm sca},\hat{\mathbf{n}}^{\rm inc})]^{\rm T}\mathbf{\Delta}_3\mathbf{B}^{-1}\\ &= \mathbf{B}\mathbf{\Delta}_3[\mathbf{B}^{-1}\mathbf{Z}^{\rm MS}(\hat{\mathbf{n}}^{\rm sca},\hat{\mathbf{n}}^{\rm inc})\mathbf{B}]^{\rm T}\mathbf{\Delta}_3\mathbf{B}^{-1}\\ &= \mathbf{B}\mathbf{\Delta}_3\mathbf{B}^{\rm T}[\mathbf{Z}^{\rm MS}(\hat{\mathbf{n}}^{\rm sca},\hat{\mathbf{n}}^{\rm inc})]^{\rm T}[\mathbf{B}^{-1}]^{\rm T}\mathbf{\Delta}_3\mathbf{B}^{-1}\\ &= \mathbf{\Delta}^{\rm MS}[\mathbf{Z}^{\rm MS}(\hat{\mathbf{n}}^{\rm sca},\hat{\mathbf{n}}^{\rm inc})]^{\rm T}[\mathbf{\Delta}^{\rm MS}]^{-1},\end{aligned} \tag{3.7.34}$$

$$\begin{aligned}\mathbf{Z}^{\rm CP}(-\hat{\mathbf{n}}^{\rm inc},-\hat{\mathbf{n}}^{\rm sca}) &= \mathbf{A}\mathbf{\Delta}_3\mathbf{A}^{\rm T}[\mathbf{Z}^{\rm CP}(\hat{\mathbf{n}}^{\rm sca},\hat{\mathbf{n}}^{\rm inc})]^{\rm T}[\mathbf{A}^{-1}]^{\rm T}\mathbf{\Delta}_3\mathbf{A}^{-1}\\ &= [\mathbf{Z}^{\rm CP}(\hat{\mathbf{n}}^{\rm sca},\hat{\mathbf{n}}^{\rm inc})]^{\rm T},\end{aligned} \tag{3.7.35}$$

where

$$\mathbf{\Delta}_{23} = \mathbf{\Delta}_{23}^{\rm T} = \mathbf{\Delta}_{23}^{-1} = \begin{bmatrix}1 & 0 & 0 & 0\\ 0 & -1 & 0 & 0\\ 0 & 0 & -1 & 0\\ 0 & 0 & 0 & 1\end{bmatrix}, \tag{3.7.36}$$

$$\mathbf{\Delta}^{\rm MS} = [\mathbf{\Delta}^{\rm MS}]^{\rm T} = \begin{bmatrix}1/2 & 0 & 0 & 0\\ 0 & 1/2 & 0 & 0\\ 0 & 0 & -1 & 0\\ 0 & 0 & 0 & 1\end{bmatrix}, \quad [\mathbf{\Delta}^{\rm MS}]^{-1} = \begin{bmatrix}2 & 0 & 0 & 0\\ 0 & 2 & 0 & 0\\ 0 & 0 & -1 & 0\\ 0 & 0 & 0 & 1\end{bmatrix}. \tag{3.7.37}$$

The backscattering theorem, Eq. (3.4.22), along with Eqs. (3.7.11), (3.7.16), (3.7.21), and (3.7.26), leads to the following general property of the backscattering Stokes phase matrix (Mishchenko *et al.*, 2000b):

$$Z_{11}(-\hat{\mathbf{n}},\hat{\mathbf{n}}) - Z_{22}(-\hat{\mathbf{n}},\hat{\mathbf{n}}) + Z_{33}(-\hat{\mathbf{n}},\hat{\mathbf{n}}) - Z_{44}(-\hat{\mathbf{n}},\hat{\mathbf{n}}) = 0. \tag{3.7.38}$$

Electromagnetic scattering most typically produces light with polarization characteristics different from those of the incident beam. If the incident beam is unpolarized, i.e., $\mathbf{I}^{\rm inc} = [I^{\rm inc}\ \ 0\ \ 0\ \ 0]^{\rm T}$, the scattered light generally has at least one nonzero Stokes parameter other than intensity:

$$I^{\text{sca}} = Z_{11}I^{\text{inc}}, \quad Q^{\text{sca}} = Z_{21}I^{\text{inc}}, \quad U^{\text{sca}} = Z_{31}I^{\text{inc}}, \quad V^{\text{sca}} = Z_{41}I^{\text{inc}}. \tag{3.7.39}$$

This effect is traditionally called "polarization" and results in scattered light with non-zero degree of polarization:

$$P = \frac{\sqrt{Z_{21}^2 + Z_{31}^2 + Z_{41}^2}}{Z_{11}} \tag{3.7.40}$$

(see Eq. (2.9.20)). Obviously, if the incident light is unpolarized, then the element Z_{11} determines the angular distribution of the scattered intensity. When the incident beam is linearly polarized, i.e., $\mathbf{I}^{\text{inc}} = [I^{\text{inc}}\ Q^{\text{inc}}\ U^{\text{inc}}\ 0]^{\text{T}}$, the scattered light may become elliptically polarized ($V^{\text{sca}} \neq 0$). Conversely, when the incident light is circularly polarized, i.e., $\mathbf{I}^{\text{inc}} = [I^{\text{inc}}\ 0\ 0\ V^{\text{inc}}]^{\text{T}}$, the scattered light may become partially linearly polarized ($Q^{\text{sca}} \neq 0$ and/or $U^{\text{sca}} \neq 0$).

A general feature of scattering by a single particle is that if the incident beam is fully polarized ($P^{\text{inc}} = 1$), then the scattered light is also fully polarized. Hovenier *et al.* (1986) gave a proof of this property based on the general mathematical structure of the Stokes phase matrix. Thus, a single particle does not depolarize fully polarized incident light. However, single scattering by a collection of non-identical nonspherical particles (including particles of the same kind but with different orientations) can result in depolarization of incident polarized light, and this is another important property of electromagnetic scattering.

3.8 Extinction matrix

Let us now consider the special case of the *exact forward-scattering direction* ($\hat{\mathbf{r}} = \hat{\mathbf{n}}^{\text{inc}}$). Because now both the incident plane wave and the scattered outgoing spherical wave propagate in the same direction and are transverse, their superposition is also a transverse wave propagating in the forward direction. Therefore, we can define the coherency column vector of the total field for propagation directions $\hat{\mathbf{r}}$ very close to $\hat{\mathbf{n}}^{\text{inc}}$ as follows:

$$\mathbf{J}(r\hat{\mathbf{r}}) = \frac{1}{2}\sqrt{\frac{\epsilon_1}{\mu_0}} \begin{bmatrix} E_\theta(r\hat{\mathbf{r}})[E_\theta(r\hat{\mathbf{r}})]^* \\ E_\theta(r\hat{\mathbf{r}})[E_\varphi(r\hat{\mathbf{r}})]^* \\ E_\varphi(r\hat{\mathbf{r}})[E_\theta(r\hat{\mathbf{r}})]^* \\ E_\varphi(r\hat{\mathbf{r}})[E_\varphi(r\hat{\mathbf{r}})]^* \end{bmatrix}, \tag{3.8.1}$$

where the total electric field is given by

$$\mathbf{E}(r\hat{\mathbf{r}}) = \mathbf{E}^{\text{inc}}(r\hat{\mathbf{r}}) + \mathbf{E}^{\text{sca}}(r\hat{\mathbf{r}}). \tag{3.8.2}$$

Integrating the elements of $\mathbf{J}(r\hat{\mathbf{r}})$ over the surface of the collimated detector aligned normal to $\hat{\mathbf{n}}^{\text{inc}}$, one can derive for the coherency-vector representation of the polarized signal recorded by detector 1 in Fig. 3.6.1:

$$(\textbf{Signal 1})^J = \int_{\Delta S} \mathrm{d}S\, \mathbf{J}(r\hat{\mathbf{r}})$$

$$= \Delta S\, \mathbf{J}^{\mathrm{inc}} - \mathbf{K}^J(\hat{\mathbf{n}}^{\mathrm{inc}})\mathbf{J}^{\mathrm{inc}} + \frac{\Delta S}{r^2}\mathbf{Z}^J(\hat{\mathbf{n}}^{\mathrm{inc}}, \hat{\mathbf{n}}^{\mathrm{inc}})\mathbf{J}^{\mathrm{inc}} \tag{3.8.3a}$$

$$= \Delta S\, \mathbf{J}^{\mathrm{inc}} - \mathbf{K}^J(\hat{\mathbf{n}}^{\mathrm{inc}})\mathbf{J}^{\mathrm{inc}} + \mathbf{O}(r^{-2}), \tag{3.8.3b}$$

where $\mathbf{Z}^J(\hat{\mathbf{n}}^{\mathrm{inc}}, \hat{\mathbf{n}}^{\mathrm{inc}})$ is the forward-scattering coherency phase matrix, $\mathbf{O}(r^{-2})$ is a 4×4 matrix with elements vanishing at infinity as r^{-2}, and the elements of the 4×4 coherency extinction matrix $\mathbf{K}^J(\theta^{\mathrm{inc}}, \varphi^{\mathrm{inc}})$ are expressed in terms of the elements of the forward-scattering amplitude matrix $\mathbf{S}(\theta^{\mathrm{inc}}, \varphi^{\mathrm{inc}};\, \theta^{\mathrm{inc}}, \varphi^{\mathrm{inc}})$ as follows:

$$\mathbf{K}^J = \frac{\mathrm{i}2\pi}{k_1}\begin{bmatrix} S_{11}^* - S_{11} & S_{12}^* & -S_{12} & 0 \\ S_{21}^* & S_{22}^* - S_{11} & 0 & -S_{12} \\ -S_{21} & 0 & S_{11}^* - S_{22} & S_{12}^* \\ 0 & -S_{21} & S_{21}^* & S_{22}^* - S_{22} \end{bmatrix}. \tag{3.8.4}$$

In the Stokes-vector representation,

$$\textbf{Signal 1} = \int_{\Delta S} \mathrm{d}S\, \mathbf{I}(r\hat{\mathbf{r}})$$

$$= \Delta S\, \mathbf{I}^{\mathrm{inc}} - \mathbf{K}(\hat{\mathbf{n}}^{\mathrm{inc}})\mathbf{I}^{\mathrm{inc}} + \frac{\Delta S}{r^2}\mathbf{Z}(\hat{\mathbf{n}}^{\mathrm{inc}}, \hat{\mathbf{n}}^{\mathrm{inc}})\mathbf{I}^{\mathrm{inc}} \tag{3.8.5a}$$

$$= \Delta S\, \mathbf{I}^{\mathrm{inc}} - \mathbf{K}(\hat{\mathbf{n}}^{\mathrm{inc}})\mathbf{I}^{\mathrm{inc}} + \mathbf{O}(r^{-2}), \tag{3.8.5b}$$

where

$$\mathbf{I}(r\hat{\mathbf{n}}^{\mathrm{inc}}) = \mathbf{D}\mathbf{J}(r\hat{\mathbf{n}}^{\mathrm{inc}}). \tag{3.8.6}$$

The 4×4 Stokes extinction matrix $\mathbf{K}(\hat{\mathbf{n}}^{\mathrm{inc}})$ is given by

$$\mathbf{K}(\hat{\mathbf{n}}^{\mathrm{inc}}) = \mathbf{D}\mathbf{K}^J(\hat{\mathbf{n}}^{\mathrm{inc}})\mathbf{D}^{-1}. \tag{3.8.7}$$

The explicit formulas for the elements of this matrix in terms of the elements of the forward-scattering amplitude matrix $\mathbf{S}(\theta^{\mathrm{inc}}, \varphi^{\mathrm{inc}};\, \theta^{\mathrm{inc}}, \varphi^{\mathrm{inc}})$ are as follows:

$$K_{jj} = \frac{2\pi}{k_1}\mathrm{Im}(S_{11} + S_{22}), \qquad j = 1, \ldots, 4, \tag{3.8.8}$$

$$K_{12} = K_{21} = \frac{2\pi}{k_1}\mathrm{Im}(S_{11} - S_{22}), \tag{3.8.9}$$

$$K_{13} = K_{31} = -\frac{2\pi}{k_1}\mathrm{Im}(S_{12} + S_{21}), \tag{3.8.10}$$

$$K_{14} = K_{41} = \frac{2\pi}{k_1}\mathrm{Re}(S_{21} - S_{12}), \tag{3.8.11}$$

$$K_{23} = -K_{32} = \frac{2\pi}{k_1}\,\mathrm{Im}(S_{21} - S_{12}), \tag{3.8.12}$$

$$K_{24} = -K_{42} = -\frac{2\pi}{k_1}\,\mathrm{Re}(S_{12} + S_{21}), \tag{3.8.13}$$

$$K_{34} = -K_{43} = \frac{2\pi}{k_1}\,\mathrm{Re}(S_{22} - S_{11}). \tag{3.8.14}$$

The elements of the coherency and Stokes extinction matrices have the dimension of area. The dimensionless products of k_1^2 and the extinction matrix elements as well as the dimensionless ratios of the extinction matrix elements and a^2 satisfy the scale invariance rule (see Section 3.5).

Equations (3.8.3) and (3.8.5) represent the most general form of the optical theorem. They show that the presence of the scattering object changes not only the total power of the electromagnetic radiation received by the detector facing the incident wave (detector 1 in Fig. 3.6.1) but also, perhaps, its state of polarization. The latter phenomenon is called dichroism and results from different attenuation rates for different polarization components of the incident wave. Equations (3.8.3) and (3.8.5) remain valid if the incident radiation is a parallel quasi-monochromatic beam of light rather than a plane electromagnetic wave.

By placing detector 1 sufficiently far from the scatterer, one can make the contribution of the third term on the right-hand side of Eqs. (3.8.5a) and (3.8.5b) negligibly small:

$$\textbf{Signal 1} \underset{r\to\infty}{=} \Delta S\,\mathbf{I}^{\mathrm{inc}} - \mathbf{K}(\hat{\mathbf{n}}^{\mathrm{inc}})\mathbf{I}^{\mathrm{inc}}. \tag{3.8.15}$$

As a consequence, the extinction matrix becomes a directly observable quantity.

It is clear from Eqs. (3.8.8)–(3.8.14) that only seven of the sixteen elements of the Stokes extinction matrix are independent. It is easy to verify that this is also true of the coherency extinction matrix. The elements of both matrices explicitly depend on φ^{inc} even when the incident wave propagates along the z-axis.

From Eqs. (3.4.21) and (3.8.8)–(3.8.14) we obtain the reciprocity relation for the Stokes extinction matrix:

$$\mathbf{K}(-\hat{\mathbf{n}}^{\mathrm{inc}}) = \mathbf{\Delta}_3[\mathbf{K}(\hat{\mathbf{n}}^{\mathrm{inc}})]^{\mathrm{T}}\mathbf{\Delta}_3. \tag{3.8.16}$$

It is also straightforward to derive a related symmetry property:

$$\mathbf{K}(-\hat{\mathbf{n}}^{\mathrm{inc}}) = \begin{bmatrix} K_{11}(\hat{\mathbf{n}}^{\mathrm{inc}}) & K_{12}(\hat{\mathbf{n}}^{\mathrm{inc}}) & -K_{13}(\hat{\mathbf{n}}^{\mathrm{inc}}) & K_{14}(\hat{\mathbf{n}}^{\mathrm{inc}}) \\ K_{21}(\hat{\mathbf{n}}^{\mathrm{inc}}) & K_{22}(\hat{\mathbf{n}}^{\mathrm{inc}}) & K_{23}(\hat{\mathbf{n}}^{\mathrm{inc}}) & -K_{24}(\hat{\mathbf{n}}^{\mathrm{inc}}) \\ -K_{31}(\hat{\mathbf{n}}^{\mathrm{inc}}) & K_{32}(\hat{\mathbf{n}}^{\mathrm{inc}}) & K_{33}(\hat{\mathbf{n}}^{\mathrm{inc}}) & K_{34}(\hat{\mathbf{n}}^{\mathrm{inc}}) \\ K_{41}(\hat{\mathbf{n}}^{\mathrm{inc}}) & -K_{42}(\hat{\mathbf{n}}^{\mathrm{inc}}) & K_{43}(\hat{\mathbf{n}}^{\mathrm{inc}}) & K_{44}(\hat{\mathbf{n}}^{\mathrm{inc}}) \end{bmatrix}. \tag{3.8.17}$$

Thus, the only effect of reversing the direction of propagation is to change the sign of four elements of the Stokes extinction matrix.

The modified Stokes and circular-polarization extinction matrices are given by

$$\mathbf{K}^{\mathrm{MS}}(\hat{\mathbf{n}}^{\mathrm{inc}}) = \mathbf{B}\mathbf{K}(\hat{\mathbf{n}}^{\mathrm{inc}})\mathbf{B}^{-1}, \tag{3.8.18}$$

$$\mathbf{K}^{\mathrm{CP}}(\hat{\mathbf{n}}^{\mathrm{inc}}) = \mathbf{A}\mathbf{K}(\hat{\mathbf{n}}^{\mathrm{inc}})\mathbf{A}^{-1}. \tag{3.8.19}$$

Reciprocity relations for the matrices $\mathbf{K}^{J}(\hat{\mathbf{n}}^{\mathrm{inc}})$, $\mathbf{K}^{\mathrm{MS}}(\hat{\mathbf{n}}^{\mathrm{inc}})$, and $\mathbf{K}^{\mathrm{CP}}(\hat{\mathbf{n}}^{\mathrm{inc}})$ can be derived from Eq. (3.8.16) by analogy with Eqs. (3.7.33)–(3.7.35):

$$\mathbf{K}^{J}(-\hat{\mathbf{n}}^{\mathrm{inc}}) = \mathbf{\Delta}_{23}[\mathbf{K}^{J}(\hat{\mathbf{n}}^{\mathrm{inc}})]^{\mathrm{T}}\mathbf{\Delta}_{23}, \tag{3.8.20}$$

$$\mathbf{K}^{\mathrm{MS}}(-\hat{\mathbf{n}}^{\mathrm{inc}}) = \mathbf{\Delta}^{\mathrm{MS}}[\mathbf{K}^{\mathrm{MS}}(\hat{\mathbf{n}}^{\mathrm{inc}})]^{\mathrm{T}}[\mathbf{\Delta}^{\mathrm{MS}}]^{-1}, \tag{3.8.21}$$

$$\mathbf{K}^{\mathrm{CP}}(-\hat{\mathbf{n}}^{\mathrm{inc}}) = [\mathbf{K}^{\mathrm{CP}}(\hat{\mathbf{n}}^{\mathrm{inc}})]^{\mathrm{T}}. \tag{3.8.22}$$

3.9 Extinction, scattering, and absorption cross sections

The knowledge of the total electromagnetic field in the far-field zone also allows us to calculate such important optical characteristics of the scattering object as the total scattering, absorption, and extinction cross sections. These optical cross sections are defined as follows. The product of the scattering cross section C_{sca} and the incident monochromatic energy flux gives the total monochromatic power removed from the incident wave resulting solely from scattering of the incident radiation in all directions. Analogously, the product of the absorption cross section C_{abs} and the incident monochromatic energy flux gives the total monochromatic power removed from the incident wave as a result of absorption of light by the object. Of course, the absorbed electromagnetic energy does not disappear, but rather is converted into other forms of energy. Finally, the extinction cross section C_{ext} is the sum of the scattering and absorption cross sections and, when multiplied by the incident monochromatic energy flux, gives the total monochromatic power removed from the incident light due to the combined effect of scattering and absorption.

To determine the total optical cross sections, we surround the object by an imaginary sphere S of radius r large enough to be in the far-field zone. Since the surrounding medium is assumed to be nonabsorbing, the net rate at which the electromagnetic energy crosses the surface S of the sphere is always nonnegative and is equal to the power absorbed by the particle:

$$W^{\mathrm{abs}} = -\oint_{S} \mathrm{d}S\,\langle \boldsymbol{\mathcal{S}}(\mathbf{r},t)\rangle_t \cdot \hat{\mathbf{r}} = -r^2 \int_{4\pi} \mathrm{d}\hat{\mathbf{r}}\,\langle \boldsymbol{\mathcal{S}}(\mathbf{r},t)\rangle_t \cdot \hat{\mathbf{r}} \tag{3.9.1}$$

(see Eq. (2.4.15)). According to Eq. (3.6.3), W^{abs} can be written as a combination of three terms:

$$W^{\mathrm{abs}} = W^{\mathrm{inc}} - W^{\mathrm{sca}} + W^{\mathrm{ext}}, \tag{3.9.2}$$

where

$$W^{\rm inc} = -r^2 \int_{4\pi} d\hat{\mathbf{r}} \langle \boldsymbol{\mathcal{S}}^{\rm inc}(\mathbf{r},t)\rangle_t \cdot \hat{\mathbf{r}}, \tag{3.9.3}$$

$$W^{\rm sca} = r^2 \int_{4\pi} d\hat{\mathbf{r}} \langle \boldsymbol{\mathcal{S}}^{\rm sca}(\mathbf{r},t)\rangle_t \cdot \hat{\mathbf{r}}, \tag{3.9.4}$$

$$W^{\rm ext} = -r^2 \int_{4\pi} d\hat{\mathbf{r}} \langle \boldsymbol{\mathcal{S}}^{\rm ext}(\mathbf{r},t)\rangle_t \cdot \hat{\mathbf{r}}. \tag{3.9.5}$$

$W^{\rm inc}$ vanishes identically because the surrounding medium is nonabsorbing and $\langle \boldsymbol{\mathcal{S}}^{\rm inc}(\mathbf{r},t)\rangle_t$ is a constant vector independent of $\mathbf{r}$, whereas $W^{\rm sca}$ is the rate at which the scattered energy crosses the surface S in the outward direction. Therefore, $W^{\rm ext}$ is equal to the sum of the energy scattering rate and the energy absorption rate:

$$W^{\rm ext} = W^{\rm sca} + W^{\rm abs}. \tag{3.9.6}$$

Inserting Eqs. (3.6.5)–(3.6.11) in Eqs. (3.9.4) and (3.9.5) and recalling the definitions of the extinction and scattering cross sections, we derive after some algebra

$$C_{\rm ext} = \frac{W^{\rm ext}}{\frac{1}{2}\sqrt{\frac{\epsilon_1}{\mu_0}}\,|\mathbf{E}_0^{\rm inc}|^2} = \frac{4\pi}{k_1|\mathbf{E}_0^{\rm inc}|^2}\,{\rm Im}[\mathbf{E}_1^{\rm sca}(\hat{\mathbf{n}}^{\rm inc})\cdot(\mathbf{E}_0^{\rm inc})^*], \tag{3.9.7}$$

$$C_{\rm sca} = \frac{W^{\rm sca}}{\frac{1}{2}\sqrt{\frac{\epsilon_1}{\mu_0}}\,|\mathbf{E}_0^{\rm inc}|^2} = \frac{1}{|\mathbf{E}_0^{\rm inc}|^2}\int_{4\pi} d\hat{\mathbf{r}}\,|\mathbf{E}_1^{\rm sca}(\hat{\mathbf{r}})|^2. \tag{3.9.8}$$

In view of Eqs. (3.2.16), (3.3.6), (3.7.7)–(3.7.9), and (3.8.8)–(3.8.11), Eqs. (3.9.7) and (3.9.8) can be rewritten as follows:

$$C_{\rm ext} = \frac{1}{I^{\rm inc}}[K_{11}(\hat{\mathbf{n}}^{\rm inc})I^{\rm inc} + K_{12}(\hat{\mathbf{n}}^{\rm inc})Q^{\rm inc} + K_{13}(\hat{\mathbf{n}}^{\rm inc})U^{\rm inc} + K_{14}(\hat{\mathbf{n}}^{\rm inc})V^{\rm inc}], \tag{3.9.9}$$

$$\begin{aligned} C_{\rm sca} &= \frac{r^2}{I^{\rm inc}}\int_{4\pi} d\hat{\mathbf{r}}\, I^{\rm sca}(r\hat{\mathbf{r}}) \\ &= \frac{1}{I^{\rm inc}}\int_{4\pi} d\hat{\mathbf{r}}[Z_{11}(\hat{\mathbf{r}},\hat{\mathbf{n}}^{\rm inc})I^{\rm inc} + Z_{12}(\hat{\mathbf{r}},\hat{\mathbf{n}}^{\rm inc})Q^{\rm inc} \\ &\quad + Z_{13}(\hat{\mathbf{r}},\hat{\mathbf{n}}^{\rm inc})U^{\rm inc} + Z_{14}(\hat{\mathbf{r}},\hat{\mathbf{n}}^{\rm inc})V^{\rm inc}]. \end{aligned} \tag{3.9.10}$$

The absorption cross section is equal to the difference of the extinction and scattering cross sections:

$$C_{\rm abs} = C_{\rm ext} - C_{\rm sca} \geq 0. \tag{3.9.11}$$

The single-scattering albedo is defined as the ratio of the scattering and extinction cross sections:

$$\varpi = \frac{C_{\text{sca}}}{C_{\text{ext}}} \leq 1. \tag{3.9.12}$$

Obviously, $\varpi = 1$ for nonabsorbing particles.

Equations (3.9.9) and (3.9.10) (and thus Eqs. (3.9.11) and (3.9.12)) also hold for quasi-monochromatic incident light provided that the elements of the Stokes column vector entering these equations are averages over a time interval long compared with the period of fluctuations. All cross sections are inherently positive real quantities and have the dimension of area. They depend on the direction, polarization state, and wavelength of the incident light as well as on the particle size, morphology, relative refractive index, and orientation with respect to the reference frame. The products of the cross sections and k_1^2 obey the scale invariance rule.

Equation (3.9.9) is another representation of the optical theorem and, along with Eqs. (3.8.8)–(3.8.11), shows that although extinction is the combined effect of absorption and scattering in all directions by the object, it is determined only by the amplitude scattering matrix in the exact forward direction. This is a direct consequence of the fact that extinction results from the interference between the incident and scattered light (Eq. (3.6.6)) and the presence of delta-function terms in Eqs. (3.6.7) and (3.6.9).

Equation (3.6.18) shows that the extinction cross section is a well-defined, observable quantity and can be determined by measuring $W_{\Delta S}(\hat{\mathbf{n}}^{\text{inc}})$ without and with the scattering object interposed between the source of light and the detector. The net effect of the object is to reduce the detector area by "casting a shadow" of area C_{ext}. Of course, this does not mean that C_{ext} is merely given by the area G of the object's geometrical projection on the detector surface. However, this geometrical interpretation of the extinction cross section illustrates the rationale for introducing the dimensionless efficiency factor for extinction by dividing the extinction cross section by the geometrical cross section:

$$Q_{\text{ext}} = \frac{C_{\text{ext}}}{G}. \tag{3.9.13}$$

As demonstrated in Chapters 9 and 10 of MTL, Q_{ext} can be considerably greater or much less than unity. The efficiency factors for scattering and absorption are defined analogously:

$$Q_{\text{sca}} = \frac{C_{\text{sca}}}{G}, \qquad Q_{\text{abs}} = \frac{C_{\text{abs}}}{G}. \tag{3.9.14}$$

It is easy to see that the efficiency factors obey the scale invariance rule.

The quantity

$$\frac{\mathrm{d}C_{\text{sca}}}{\mathrm{d}\Omega} = \frac{I^{\text{sca}}(r\hat{\mathbf{r}})r^2}{I^{\text{inc}}}$$

$$= \frac{1}{I^{\text{inc}}}[Z_{11}(\hat{\mathbf{r}}, \hat{\mathbf{n}}^{\text{inc}})I^{\text{inc}} + Z_{12}(\hat{\mathbf{r}}, \hat{\mathbf{n}}^{\text{inc}})Q^{\text{inc}}$$

$$+ Z_{13}(\hat{\mathbf{r}}, \hat{\mathbf{n}}^{\text{inc}})U^{\text{inc}} + Z_{14}(\hat{\mathbf{r}}, \hat{\mathbf{n}}^{\text{inc}})V^{\text{inc}}] \tag{3.9.15}$$

has the dimension of area and is called the differential scattering cross section. It describes the angular distribution of the scattered light and specifies the electromagnetic power scattered into a unit solid angle about a given direction per unit incident intensity.[1] The differential scattering cross section depends on the polarization state of the incident light as well as on the incidence and scattering directions. Comparison of Eqs. (3.9.10) and (3.9.15) shows that

$$C_{\text{sca}} = \int_{4\pi} \mathrm{d}\hat{\mathbf{r}} \frac{\mathrm{d}C_{\text{sca}}}{\mathrm{d}\Omega}. \tag{3.9.16}$$

A quantity related to the differential scattering cross section is the phase function $p(\hat{\mathbf{r}}, \hat{\mathbf{n}}^{\text{inc}})$ defined as

$$p(\hat{\mathbf{r}}, \hat{\mathbf{n}}^{\text{inc}}) = \frac{4\pi}{C_{\text{sca}}} \frac{\mathrm{d}C_{\text{sca}}}{\mathrm{d}\Omega}. \tag{3.9.17}$$

The convenience of the phase function is that it is dimensionless and normalized:

$$\frac{1}{4\pi} \int_{4\pi} \mathrm{d}\hat{\mathbf{r}} p(\hat{\mathbf{r}}, \hat{\mathbf{n}}^{\text{inc}}) = 1. \tag{3.9.18}$$

The asymmetry parameter $\langle \cos\Theta \rangle$ is defined as the average cosine of the scattering angle $\Theta = \arccos(\hat{\mathbf{r}} \cdot \hat{\mathbf{n}}^{\text{inc}})$ (i.e., the angle between the incidence and scattering directions):

$$\begin{aligned} \langle \cos\Theta \rangle &= \frac{1}{4\pi} \int_{4\pi} \mathrm{d}\hat{\mathbf{r}}\, p(\hat{\mathbf{r}}, \hat{\mathbf{n}}^{\text{inc}}) \hat{\mathbf{r}} \cdot \hat{\mathbf{n}}^{\text{inc}} \\ &= \frac{1}{C_{\text{sca}}} \int_{4\pi} \mathrm{d}\hat{\mathbf{r}} \frac{\mathrm{d}C_{\text{sca}}}{\mathrm{d}\Omega} \hat{\mathbf{r}} \cdot \hat{\mathbf{n}}^{\text{inc}}. \end{aligned} \tag{3.9.19}$$

The asymmetry parameter is positive if the particle scatters more light toward the forward direction ($\Theta = 0$), is negative if more light is scattered toward the backscattering direction ($\Theta = \pi$), and vanishes if the scattering is symmetric with respect to the plane perpendicular to the incidence direction. Obviously, $\langle \cos\Theta \rangle \in [-1, +1]$. The limiting values correspond to the phase functions $4\pi\delta(\hat{\mathbf{r}} + \hat{\mathbf{n}}^{\text{inc}})$ and $4\pi\delta(\hat{\mathbf{r}} - \hat{\mathbf{n}}^{\text{inc}})$, respectively.

3.10 Coherency dyad of the total electric field

We explained in Section 3.6 that in order to characterize the directional flow and spa-

[1] Note that the symbol $\mathrm{d}C_{\text{sca}}/\mathrm{d}\Omega$ should not be interpreted as the derivative of a function of Ω.

tial distribution of electromagnetic energy that results from a scattering process, one must calculate the Poynting vector and the energy density of the total electromagnetic field. This, in turn, requires the knowledge of both the electric and the magnetic components of the field. It would be attractive, however, to develop a simplified formalism that would involve the electric field only and would make feasible the solution of more involved problems such as the development of a unified microphysical theory of radiative transfer and coherent backscattering (Chapters 8 and 14). Therefore, the aim of this section is to analyze whether the scattering process can be described adequately in terms of the coherency dyad introduced in Section 2.12.

As in Section 3.6, we begin by representing the coherency dyad of the total field in the far-field zone as the sum of three components:

$$\begin{aligned}\ddot{\rho}(\mathbf{r}) &= \mathbf{E}(\mathbf{r},t)\otimes[\mathbf{E}(\mathbf{r},t)]^* \\ &= \ddot{\rho}^{\text{inc}} + \ddot{\rho}^{\text{sca}}(\mathbf{r}) + \rho^{\text{int}}(\mathbf{r}),\end{aligned} \tag{3.10.1}$$

where

$$\ddot{\rho}^{\text{inc}} = \mathbf{E}^{\text{inc}}(\mathbf{r},t)\otimes[\mathbf{E}^{\text{inc}}(\mathbf{r},t)]^* = \mathbf{E}_0^{\text{inc}}\otimes(\mathbf{E}_0^{\text{inc}})^* \tag{3.10.2}$$

is the coherency dyad of the incident field,

$$\ddot{\rho}^{\text{sca}}(\mathbf{r}) = \mathbf{E}^{\text{sca}}(\mathbf{r},t)\otimes[\mathbf{E}^{\text{sca}}(\mathbf{r},t)]^* = \frac{1}{r^2}\mathbf{E}_1^{\text{sca}}(\hat{\mathbf{r}})\otimes[\mathbf{E}_1^{\text{sca}}(\hat{\mathbf{r}})]^* \tag{3.10.3}$$

is the coherency dyad of the scattered field, and the component

$$\begin{aligned}\ddot{\rho}^{\text{int}}(\mathbf{r}) &= \mathbf{E}^{\text{inc}}(\mathbf{r},t)\otimes[\mathbf{E}^{\text{sca}}(\mathbf{r},t)]^* + \mathbf{E}^{\text{sca}}(\mathbf{r},t)\otimes[\mathbf{E}^{\text{inc}}(\mathbf{r},t)]^* \\ &= \frac{\mathrm{i}2\pi}{k_1 r^2}\{[\delta(\hat{\mathbf{n}}^{\text{inc}}+\hat{\mathbf{r}})\exp(-\mathrm{i}2k_1 r) - \delta(\hat{\mathbf{n}}^{\text{inc}}-\hat{\mathbf{r}})]\mathbf{E}_0^{\text{inc}}\otimes[\mathbf{E}_1^{\text{sca}}(\hat{\mathbf{r}})]^* \\ &\qquad + [-\delta(\hat{\mathbf{n}}^{\text{inc}}+\hat{\mathbf{r}})\exp(\mathrm{i}2k_1 r) + \delta(\hat{\mathbf{n}}^{\text{inc}}-\hat{\mathbf{r}})]\mathbf{E}_1^{\text{sca}}(\hat{\mathbf{r}})\otimes(\mathbf{E}_0^{\text{inc}})^*\}\end{aligned} \tag{3.10.4}$$

can be interpreted as the result of interaction of the incident and scattered fields. The coherency dyad of the incident field yields directly the coherency and Stokes column vectors of the incident field via

$$\mathbf{J}^{\text{inc}} = \frac{1}{2}\sqrt{\frac{\epsilon_1}{\mu_0}}\begin{bmatrix}\hat{\boldsymbol{\theta}}^{\text{inc}}\cdot\ddot{\rho}^{\text{inc}}\cdot\hat{\boldsymbol{\theta}}^{\text{inc}} \\ \hat{\boldsymbol{\theta}}^{\text{inc}}\cdot\ddot{\rho}^{\text{inc}}\cdot\hat{\boldsymbol{\varphi}}^{\text{inc}} \\ \hat{\boldsymbol{\varphi}}^{\text{inc}}\cdot\ddot{\rho}^{\text{inc}}\cdot\hat{\boldsymbol{\theta}}^{\text{inc}} \\ \hat{\boldsymbol{\varphi}}^{\text{inc}}\cdot\ddot{\rho}^{\text{inc}}\cdot\hat{\boldsymbol{\varphi}}^{\text{inc}}\end{bmatrix} \tag{3.10.5}$$

and $\mathbf{I}^{\text{inc}} = \mathbf{D}\mathbf{J}^{\text{inc}}$. Furthermore, we can rewrite Eq. (3.10.3) in the form

$$\ddot{\rho}^{\text{sca}}(\mathbf{r}) = \frac{1}{r^2}[\ddot{A}(\hat{\mathbf{r}},\hat{\mathbf{n}}^{\text{inc}})\cdot\mathbf{E}_0^{\text{inc}}]\otimes[\ddot{A}(\hat{\mathbf{r}},\hat{\mathbf{n}}^{\text{inc}})\cdot\mathbf{E}_0^{\text{inc}}]^*$$

$$= \frac{1}{r^2} \ddot{A}(\hat{\mathbf{r}}, \hat{\mathbf{n}}^{\text{inc}}) \cdot \ddot{\rho}^{\text{inc}} \cdot [\ddot{A}(\hat{\mathbf{r}}, \hat{\mathbf{n}}^{\text{inc}})]^{\text{T}*}, \tag{3.10.6}$$

where we have used the dyadic identity (A.8). Recalling then that

$$\mathbf{E}_0^{\text{inc}} = E_{0\theta}^{\text{inc}} \hat{\boldsymbol{\theta}}^{\text{inc}} + E_{0\varphi}^{\text{inc}} \hat{\boldsymbol{\varphi}}^{\text{inc}}$$

and

$$\mathbf{E}_1^{\text{sca}}(\hat{\mathbf{r}}) = E_{1\theta}^{\text{sca}}(\hat{\mathbf{r}}) \hat{\boldsymbol{\theta}}^{\text{sca}} + E_{1\varphi}^{\text{sca}}(\hat{\mathbf{r}}) \hat{\boldsymbol{\varphi}}^{\text{sca}}$$

and using Eqs. (3.3.6)–(3.3.11) as well as the identity (A.12) and the transversality conditions (3.3.3) and (3.3.4), we easily recover Eqs. (3.7.4) and (3.7.8), in which

$$\mathsf{J}^{\text{sca}}(r\hat{\mathbf{r}}) = \frac{1}{2} \sqrt{\frac{\epsilon_1}{\mu_0}} \begin{bmatrix} \hat{\boldsymbol{\theta}}^{\text{sca}} \cdot \ddot{\rho}^{\text{sca}}(r\hat{\mathbf{r}}) \cdot \hat{\boldsymbol{\theta}}^{\text{sca}} \\ \hat{\boldsymbol{\theta}}^{\text{sca}} \cdot \ddot{\rho}^{\text{sca}}(r\hat{\mathbf{r}}) \cdot \hat{\boldsymbol{\varphi}}^{\text{sca}} \\ \hat{\boldsymbol{\varphi}}^{\text{sca}} \cdot \ddot{\rho}^{\text{sca}}(r\hat{\mathbf{r}}) \cdot \hat{\boldsymbol{\theta}}^{\text{sca}} \\ \hat{\boldsymbol{\varphi}}^{\text{sca}} \cdot \ddot{\rho}^{\text{sca}}(r\hat{\mathbf{r}}) \cdot \hat{\boldsymbol{\varphi}}^{\text{sca}} \end{bmatrix} \tag{3.10.7}$$

and $\mathsf{I}^{\text{sca}}(r\hat{\mathbf{r}}) = \mathbf{D}\mathsf{J}^{\text{sca}}(r\hat{\mathbf{r}})$. Finally, by integrating the coherency dyad of the total field over the surface of detector 1 in Fig. 3.6.1 and applying similar algebra, we recover Eqs. (3.8.3) and (3.8.5).

Thus, the use of the coherency dyad to describe the electromagnetic scattering process appears to be consistent with the main results of Sections 3.7 and 3.8. However, one encounters a problem when the response of the detector facing the exact backscattering direction is being considered. When the right-hand side of Eq. (3.10.4) is integrated over the surface of this detector, the term proportional to $\delta(\hat{\mathbf{n}}^{\text{inc}} + \hat{\mathbf{r}})$ gives a nonzero contribution due to apparent interference of the incident and the backscattered field. However, in view of the discussion in Section 3.6 this contribution is unphysical. Indeed, the effect of interference of the incident and backscattered fields is annihilated by the real filter Re on the right-hand side of Eq. (3.6.6) and by the fact that the corresponding electric and magnetic contributions to $\langle \mathcal{U}^{\text{ext}}(\mathbf{r}, t) \rangle_t$ in Eq. (3.6.17) cancel each other. It is thus clear that one must exercise caution if the coherency dyad is used as a basic characteristic of the electromagnetic scattering process.

As we pointed out in Section 2.12, one of the main advantages of the formalism based on the concept of the coherency dyad is that it does not require the electric field to be transverse. Thus it can potentially be used to:

- Characterize the total field everywhere in space rather than in the far-field zone only.
- Describe situations in which an object is illuminated by two or more sources of radiation.
- Analyze electromagnetic scattering by time-variable objects.

Several examples of this versatility will be given in the remainder of this and in the following two sections.

Our first step is to include explicitly the time-harmonic factor and rewrite Eq. (3.1.23) as follows:

$$\mathbf{E}^{\text{sca}}(\mathbf{r},t) = \int_{V_{\text{INT}}} \mathrm{d}\mathbf{r}'\,\ddot{G}(\mathbf{r},\mathbf{r}')\cdot \int_{V_{\text{INT}}} \mathrm{d}\mathbf{r}''\exp(\mathrm{i}k_1\hat{\mathbf{n}}^{\text{inc}}\cdot\mathbf{r}'')\,\ddot{T}(\mathbf{r}',\mathbf{r}'')\cdot\mathbf{E}_0^{\text{inc}} \times\exp(-\mathrm{i}\omega t), \qquad \mathbf{r}\in\Re^3, \tag{3.10.8}$$

where it is assumed that the incident field is a plane electromagnetic wave incident in the direction $\hat{\mathbf{n}}^{\text{inc}}$. Hence, the total field can be expressed as

$$\mathbf{E}(\mathbf{r},t) = \ddot{\mathcal{T}}(\omega,\mathbf{r},\hat{\mathbf{n}}^{\text{inc}})\cdot\mathbf{E}_0^{\text{inc}}\exp(-\mathrm{i}\omega t), \qquad \mathbf{r}\in\Re^3, \tag{3.10.9}$$

where $\ddot{\mathcal{T}}$ is a transformation dyadic given by

$$\ddot{\mathcal{T}}(\omega,\mathbf{r},\hat{\mathbf{n}}^{\text{inc}}) = \exp(\mathrm{i}k_1\hat{\mathbf{n}}^{\text{inc}}\cdot\mathbf{r})\ddot{I} + \int_{V_{\text{INT}}} \mathrm{d}\mathbf{r}'\,\ddot{G}(\mathbf{r},\mathbf{r}')\cdot \int_{V_{\text{INT}}} \mathrm{d}\mathbf{r}''\exp(\mathrm{i}k_1\hat{\mathbf{n}}^{\text{inc}}\cdot\mathbf{r}'')\,\ddot{T}(\mathbf{r}',\mathbf{r}''). \tag{3.10.10}$$

The coherency dyad of the total field now takes the following form:

$$\begin{aligned}\ddot{\rho}(\mathbf{r}) &= \mathbf{E}(\mathbf{r},t)\otimes\mathbf{E}^*(\mathbf{r},t)\\ &= [\ddot{\mathcal{T}}(\omega,\mathbf{r},\hat{\mathbf{n}}^{\text{inc}})\cdot\mathbf{E}_0^{\text{inc}}]\otimes[\ddot{\mathcal{T}}(\omega,\mathbf{r},\hat{\mathbf{n}}^{\text{inc}})\cdot\mathbf{E}_0^{\text{inc}}]^*\\ &= \ddot{\mathcal{T}}(\omega,\mathbf{r},\hat{\mathbf{n}}^{\text{inc}})\cdot\ddot{\rho}^{\text{inc}}\cdot[\ddot{\mathcal{T}}(\omega,\mathbf{r},\hat{\mathbf{n}}^{\text{inc}})]^{\text{T}*}, \qquad \mathbf{r}\in\Re^3,\end{aligned} \tag{3.10.11}$$

where $\ddot{\rho}^{\text{inc}}$ is the coherency dyad of the incident field given by Eq. (3.10.2). Equation (3.10.11) generalizes Eqs. (3.10.1)–(3.10.4) and (3.10.6) and shows that the coherency dyad of the total electric field everywhere in space is linearly expressed in the coherency dyad of the incident electric field.

If the incident light is a parallel quasi-monochromatic beam then

$$\mathbf{E}(\mathbf{r},t) = \ddot{\mathcal{T}}(\omega,\mathbf{r},\hat{\mathbf{n}}^{\text{inc}})\cdot\mathbf{E}_0^{\text{inc}}(t)\exp(-\mathrm{i}\omega t), \qquad \mathbf{r}\in\Re^3, \tag{3.10.12}$$

where the fluctuating amplitude $\mathbf{E}_0^{\text{inc}}(t)$ changes in time much more slowly than the time-harmonic factor $\exp(-\mathrm{i}\omega t)$. The average of the coherency dyad of the total electric field over a time interval long compared with the typical period of fluctuation is now given by

$$\begin{aligned}\langle\ddot{\rho}(\mathbf{r})\rangle_t &= \langle\mathbf{E}(\mathbf{r},t)\otimes\mathbf{E}^*(\mathbf{r},t)\rangle_t\\ &= \langle[\ddot{\mathcal{T}}(\omega,\mathbf{r},\hat{\mathbf{n}}^{\text{inc}})\cdot\mathbf{E}_0^{\text{inc}}(t)]\otimes[\ddot{\mathcal{T}}(\omega,\mathbf{r},\hat{\mathbf{n}}^{\text{inc}})\cdot\mathbf{E}_0^{\text{inc}}(t)]^*\rangle_t\\ &= \ddot{\mathcal{T}}(\omega,\mathbf{r},\hat{\mathbf{n}}^{\text{inc}})\cdot\langle\ddot{\rho}^{\text{inc}}\rangle_t\cdot[\ddot{\mathcal{T}}(\omega,\mathbf{r},\hat{\mathbf{n}}^{\text{inc}})]^{\text{T}*},\end{aligned} \tag{3.10.13}$$

where

$$\langle\ddot{\rho}^{\text{inc}}\rangle_t = \langle\mathbf{E}_0^{\text{inc}}(t)\otimes[\mathbf{E}_0^{\text{inc}}(t)]^*\rangle_t. \tag{3.10.14}$$

Equation (3.10.13) demonstrates that the time average of the coherency dyad of the total field everywhere in space is linearly expressed in the time average of the coherency dyad of the incident quasi-monochromatic beam.

Comparison of Eqs. (3.10.11) and (3.10.13) reinforces the point made previously: the scattering formalism based on the introduction of actual observables having the dimension of electromagnetic energy flux applies equally to the situations when the incident light is a plane electromagnetic wave and when it is a parallel quasi-monochromatic beam.

3.11 Other types of illumination

Consider now a more complex case of illumination of an object by two monochromatic plane electromagnetic waves with angular frequencies ω_1 and $\omega_2 \neq \omega_1$, propagation directions $\hat{\mathbf{n}}_1^{\text{inc}}$ and $\hat{\mathbf{n}}_2^{\text{inc}}$, and amplitudes $\mathbf{E}_{01}^{\text{inc}}$ and $\mathbf{E}_{02}^{\text{inc}}$, respectively. Note that $\hat{\mathbf{n}}_2^{\text{inc}}$ may or may not coincide with $\hat{\mathbf{n}}_1^{\text{inc}}$. The total electric field is now the vector superposition of two partial fields:

$$\mathbf{E}(\mathbf{r},t) = \overleftrightarrow{T}(\omega_1,\mathbf{r},\hat{\mathbf{n}}_1^{\text{inc}})\cdot\mathbf{E}_{01}^{\text{inc}}\exp(-\mathrm{i}\omega_1 t) + \overleftrightarrow{T}(\omega_2,\mathbf{r},\hat{\mathbf{n}}_2^{\text{inc}})\cdot\mathbf{E}_{02}^{\text{inc}}\exp(-\mathrm{i}\omega_2 t),$$

$$\mathbf{r}\in\Re^3. \tag{3.11.1}$$

Since $\omega_2 \neq \omega_1$, the average of the product $\exp(-\mathrm{i}\omega_1 t)[\exp(-\mathrm{i}\omega_2 t)]^*$ over a time interval long compared with the period of time-harmonic oscillations vanishes:

$$\frac{1}{T}\int_t^{t+T} \mathrm{d}t' \exp[-\mathrm{i}(\omega_1 - \omega_2)t'] \underset{T\gg 2\pi/\omega}{=} 0. \tag{3.11.2}$$

Therefore, the time average of the coherency dyad of the total field is equal to the sum of the respective partial coherency dyads:

$$\langle\overleftrightarrow{\rho}(\mathbf{r})\rangle_t = \overleftrightarrow{\rho}_1(\mathbf{r}) + \overleftrightarrow{\rho}_2(\mathbf{r}), \tag{3.11.3}$$

where

$$\overleftrightarrow{\rho}_i(\mathbf{r}) = \overleftrightarrow{T}(\omega_i,\mathbf{r},\hat{\mathbf{n}}_i^{\text{inc}})\cdot\overleftrightarrow{\rho}_i^{\text{inc}}\cdot[\overleftrightarrow{T}(\omega_i,\mathbf{r},\hat{\mathbf{n}}_i^{\text{inc}})]^{\mathrm{T}*}, \quad i = 1,2, \tag{3.11.4}$$

$$\overleftrightarrow{\rho}_i^{\text{inc}} = \mathbf{E}_{0i}^{\text{inc}}\otimes(\mathbf{E}_{0i}^{\text{inc}})^*, \quad i = 1,2. \tag{3.11.5}$$

Equations (3.11.3)–(3.11.5) can be generalized to any number N of incident plane waves provided that all of them have different angular frequencies:

$$\langle\overleftrightarrow{\rho}(\mathbf{r})\rangle_t = \sum_{i=1}^{N}\overleftrightarrow{\rho}_i(\mathbf{r}). \tag{3.11.6}$$

Equation (3.11.6) has the following practical interpretation. Imagine N scattering experiments in which a fixed particle is illuminated sequentially by each of N sources of monochromatic light. The corresponding illumination directions are given by the unit vectors $\hat{\mathbf{n}}_i^{\text{inc}}$, $i = 1, ..., N$, and the angular frequencies of all the sources are different. One or several observable characteristics of the total radiation field are measured by a fixed detector. Now let us imagine that all the light sources are turned on simultaneously, and the total radiation is measured by the same detector. Then, according to Eq. (3.11.6), the reading of the detector will be equal to the sum of the N readings recorded during the N individual experiments.

The requirement that the angular frequencies of all the N sources of light be different becomes unnecessary if the light is quasi-monochromatic so that the amplitudes $\mathbf{E}_{0i}^{\text{inc}}(t)$ fluctuate in time. Indeed, now we have instead of Eq. (3.11.6):

$$\langle \vec{\rho}(\mathbf{r}) \rangle_t = \sum_{i=1}^{N} \langle \vec{\rho}_i(\mathbf{r}) \rangle_t, \tag{3.11.7}$$

where

$$\langle \vec{\rho}_i(\mathbf{r}) \rangle_t = \ddot{\mathcal{T}}(\omega_i, \mathbf{r}, \hat{\mathbf{n}}_i^{\text{inc}}) \cdot \langle \vec{\rho}_i^{\text{inc}} \rangle_t \cdot [\ddot{\mathcal{T}}(\omega_i, \mathbf{r}, \hat{\mathbf{n}}_i^{\text{inc}})]^{\text{T}*} \tag{3.11.8}$$

and $\langle \cdots \rangle_t$ denotes an average over a time period long compared with the typical period of fluctuation. In deriving Eq. (3.11.7), we have taken into account that the fluctuations of each amplitude $\mathbf{E}_{0i}^{\text{inc}}(t)$ occur randomly and independently of those of all the other amplitudes so that

$$\langle \mathbf{E}_{0i}^{\text{inc}}(t) \otimes [\mathbf{E}_{0j}^{\text{inc}}(t)]^* \rangle_t = \ddot{0} \tag{3.11.9}$$

for any $i \neq j$, where, as before, $\ddot{0}$ is a zero dyad.

3.12 Variable scatterers

Most scatterers encountered in practice change during the time necessary to take a measurement rather than remain fixed. However, the results of the two preceding sections remain valid provided that significant changes of the transformation dyadic require time intervals that are much longer than the period of time-harmonic oscillations and/or much longer than the typical period of fluctuation.

Specifically, let us first assume that a scatterer is illuminated by N monochromatic plane waves with different angular frequencies and arbitrary propagation directions. Let T_ω be the shortest time interval such that

$$\frac{1}{T_\omega} \int_t^{t+T_\omega} \mathrm{d}t' \exp[-\mathrm{i}(\omega_i - \omega_j)t'] = 0 \tag{3.12.1}$$

for any $i \neq j$. Let us also assume that the measurement is taken over a time interval T_{m} such that $T_\omega \ll T_{\text{m}}$. Then, recalling Eqs. (3.11.4)–(3.11.6), we may conclude that the reading of a detector of electromagnetic energy per unit time will be described by the following expression:

$$\frac{1}{T_{\rm m}}\sum_{i=1}^{N}\int_{t}^{t+T_{\rm m}} {\rm d}t'\, \ddot{\mathcal{T}}(t', \omega_i, \mathbf{r}, \hat{\mathbf{n}}_i^{\rm inc})\cdot \ddot{\rho}_i^{\rm inc}\cdot [\ddot{\mathcal{T}}(t', \omega_i, \mathbf{r}, \hat{\mathbf{n}}_i^{\rm inc})]^{\rm T*}, \tag{3.12.2}$$

in which the transformation dyadic depends on time explicitly. It is clear that Eq. (3.12.2) represents the sum of the N individual readings of the detector corresponding to illumination of the object by each monochromatic plane wave separately.

Let us now assume that a scatterer is illuminated by a single quasi-monochromatic beam and denote by $T_{\rm f}$ the shortest time interval such that averaging the coherency dyad of the incident field over $T_{\rm f}$ gives the result indistinguishable from that obtained by averaging over an infinite time interval:

$$\frac{1}{T_{\rm f}}\int_{t}^{t+T_{\rm f}} {\rm d}t'\, \ddot{\rho}^{\rm inc}(t') \approx \lim_{T\to\infty}\frac{1}{T}\int_{t}^{t+T} {\rm d}t'\, \ddot{\rho}^{\rm inc}(t') = \langle \ddot{\rho}^{\rm inc}\rangle_t \tag{3.12.3}$$

for any t. Let us also assume that the typical time during which the transformation dyadic changes appreciably, $T_{\mathcal{T}}$, is such that $T_{\rm f} \ll T_{\mathcal{T}}$ and that the measurement is taken over a time interval $T_{\rm m}$ such that $T_{\rm f} \ll T_{\rm m}$. Then, in view of Eq. (3.10.13), the reading of the detector of electromagnetic energy per unit time will be described by the expression

$$\frac{1}{T_{\rm m}}\int_{t}^{t+T_{\rm m}} {\rm d}t'\, \ddot{\mathcal{T}}(t', \omega, \mathbf{r}, \hat{\mathbf{n}}^{\rm inc})\cdot \langle\ddot{\rho}^{\rm inc}\rangle_t\cdot [\ddot{\mathcal{T}}(t', \omega, \mathbf{r}, \hat{\mathbf{n}}^{\rm inc})]^{\rm T*}. \tag{3.12.4}$$

Similarly, let us assume that the object is illuminated by N quasi-monochromatic beams with arbitrary frequencies and propagation directions and that

$$\frac{1}{T_{\rm f}}\int_{t}^{t+T_{\rm f}} {\rm d}t'\, \mathbf{E}_{0i}^{\rm inc}(t')\otimes[\mathbf{E}_{0j}^{\rm inc}(t')]^* = \ddot{0} \tag{3.12.5}$$

for any t and any $i \neq j$. Then the reading of the detector per unit time is given by

$$\frac{1}{T_{\rm m}}\sum_{i=1}^{N}\int_{t}^{t+T_{\rm m}} {\rm d}t'\, \ddot{\mathcal{T}}(t', \omega_i, \mathbf{r}, \hat{\mathbf{n}}_i^{\rm inc})\cdot \langle\ddot{\rho}_i^{\rm inc}\rangle_t\cdot [\ddot{\mathcal{T}}(t', \omega_i, \mathbf{r}, \hat{\mathbf{n}}_i^{\rm inc})]^{\rm T*}. \tag{3.12.6}$$

Again, Eq. (3.12.6) represents the sum of the N individual readings of the detector corresponding to illumination of the object by each quasi-monochromatic beam separately.

The results of this section are quite general and apply to all scattering objects that change in time not too rapidly. As such, they expand significantly the range of applicability of formulas derived in Chapters 6–8.

3.13 Thermal emission

If the particle's absolute temperature T is above zero, it can emit as well as scatter and absorb electromagnetic radiation. The emitted radiation in the far-field zone of the particle propagates in the radial direction, i.e., along the unit vector $\hat{\mathbf{r}} = \mathbf{r}/r$, where $\mathbf{r}$ is the position vector of the observation point with origin inside the particle. The energetic and polarization characteristics of the emitted radiation are described by a four-component emission Stokes column vector $\mathbf{K}_\mathrm{e}(\hat{\mathbf{r}}, T, \omega)$ defined in such a way that the net rate at which the emitted energy crosses a surface element ΔS normal to $\hat{\mathbf{r}}$ at a distance r from the particle at angular frequencies from ω to $\omega + \Delta\omega$ is

$$W^\mathrm{e} = \frac{1}{r^2}\Delta S \Delta\omega K_{\mathrm{e}1}(\hat{\mathbf{r}}, T, \omega). \tag{3.13.1}$$

The $K_{\mathrm{e}1}(\hat{\mathbf{r}}, T, \omega)$, the first component of the column vector, can also be interpreted as the amount of electromagnetic energy emitted by the particle in the direction $\hat{\mathbf{r}}$ per unit solid angle per unit frequency interval per unit time.

In order to calculate $\mathbf{K}_\mathrm{e}(\hat{\mathbf{r}}, T, \omega)$, let us assume that the particle is placed inside an opaque cavity of dimensions large compared with the particle and with any wavelength under consideration (Fig. 3.13.1(a)). If the cavity and the particle are maintained at the constant absolute temperature T, then the equilibrium electromagnetic radiation inside the cavity is isotropic, homogeneous, and unpolarized (Mandel and Wolf, 1995). This radiation can be represented as a collection of quasi-monochromatic, unpolarized, incoherent beams propagating in all directions and characterized by the Planck blackbody energy distribution $I_\mathrm{b}(T, \omega)$. Specifically, at any point inside the cavity the amount of radiant energy per unit frequency interval, confined to a small solid angle $\Delta\Omega$ about any direction, which crosses an area ΔS normal to this direction in unit time is given by

$$\Delta\Omega \Delta S I_\mathrm{b}(T, \omega) = \Delta\Omega \Delta S \frac{\hbar\omega^3}{4\pi^3 c^2 \left[\exp\left(\dfrac{\hbar\omega}{k_\mathrm{B} T}\right) - 1\right]}, \tag{3.13.2}$$

where $\hbar = h/2\pi$, h is Planck's constant, c is the speed of light in a vacuum, and k_B is Boltzmann's constant.

Consider an imaginary collimated, polarization-sensitive detector of electromagnetic radiation with surface ΔS and small solid-angle field of view $\Delta\Omega$, placed at a distance r from the particle (Fig. 3.13.1(a)). The dimension of the detector surface is much greater than any dimension of the particle and r is large enough to be in the far-field zone of the particle but smaller than $(\Delta S/\Delta\Omega)^{1/2}$. The latter condition ensures that all plane wave fronts incident on the detector in directions falling into its solid-angle field of view $\Delta\Omega$ are equally attenuated by the particle (Fig. 3.13.1(b)). The surface ΔS is aligned normal to and centered on $\hat{\mathbf{r}}$, where $\hat{\mathbf{r}}$ is the unit vector originating inside the particle and pointing toward the detector.

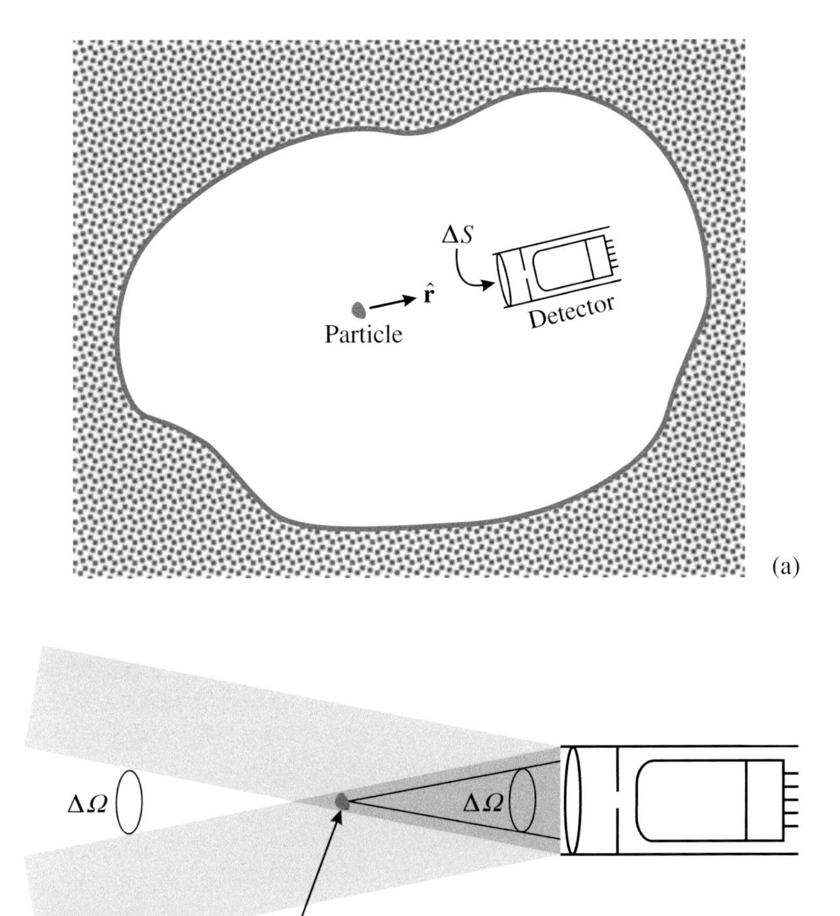

Figure 3.13.1. (a) Cavity, particle, and electromagnetic radiation field in thermal equilibrium. (b) Illumination geometry.

In the absence of the particle, the polarized signal per unit frequency interval measured by the detector would be given by

$$\Delta\Omega\Delta S\,\mathbf{I}_{\mathrm{b}}(T,\omega), \tag{3.13.3}$$

where

$$\mathbf{I}_{\mathrm{b}}(T,\omega)=\begin{bmatrix} I_{\mathrm{b}}(T,\omega) \\ 0 \\ 0 \\ 0 \end{bmatrix} \tag{3.13.4}$$

is the blackbody Stokes column vector. The particle attenuates the incident blackbody radiation, emits radiation, and scatters the blackbody radiation coming from all directions in the direction of the detector. Taking into account that only the radiation emitted and scattered by the particle within the solid-angle field of view $\Delta\Omega$ is detected (Fig. 3.13.1(b)), we conclude that the polarized signal measured by the detector in the presence of the particle is

$$\Delta\Omega\Delta S\mathbf{I}_\mathrm{b}(T,\omega) - \Delta\Omega\mathbf{K}(\hat{\mathbf{r}},\omega)\mathbf{I}_\mathrm{b}(T,\omega) + \Delta\Omega\mathbf{K}_\mathrm{e}(\hat{\mathbf{r}},T,\omega) + \Delta\Omega\int_{4\pi}\mathrm{d}\hat{\mathbf{r}}'\,\mathbf{Z}(\hat{\mathbf{r}},\hat{\mathbf{r}}',\omega)\mathbf{I}_\mathrm{b}(T,\omega) \tag{3.13.5}$$

(see Eqs. (3.8.5b) and (3.7.8)). However, in thermal equilibrium the presence of the particle does not change the distribution of radiation. Therefore, we can equate expressions (3.13.3) and (3.13.5) and finally derive for the ith component of $\mathbf{K}_\mathrm{e}$

$$K_{\mathrm{e}i}(\hat{\mathbf{r}},T,\omega) = I_\mathrm{b}(T,\omega)K_{i1}(\hat{\mathbf{r}},\omega) - I_\mathrm{b}(T,\omega)\int_{4\pi}\mathrm{d}\hat{\mathbf{r}}'\,Z_{i1}(\hat{\mathbf{r}},\hat{\mathbf{r}}',\omega), \qquad i=1,\ldots,4. \tag{3.13.6}$$

This important relation expresses the emission Stokes column vector in terms of the leftmost columns of the extinction and phase matrices and the Planck energy distribution.

Although our derivation assumed that the particle was in thermal equilibrium with the surrounding radiation field, emissivity is a property of the particle only. Therefore, Eq. (3.13.6) is valid for any particle, in equilibrium or in nonequilibrium.

3.14 Historical notes and further reading

Important early contributions to the subject of far-field electromagnetic scattering were made by Silver (1949) and Müller (1969). Formal mathematical aspects of the electromagnetic scattering theory, including basic existence and uniqueness theorems, are discussed in Müller (1969), Colton and Kress (1998), Doicu *et al.* (2000), and Pike and Sabatier (2001).

Chapter 4

Scattering by a fixed multi-particle group

The formalism described in the preceding section equally applies to a scatterer in the form of a single body and to a fixed multi-particle group. However, when the scattering object is a cluster consisting of touching and/or separated distinct components then it is often convenient to use an alternative formalism in which the total scattered electric field is explicitly represented as a vector superposition of the partial fields scattered by the cluster components. This approach is based on the system of integral so-called Foldy–Lax equations which follow directly from the macroscopic Maxwell equations and rigorously describe the scattered electric field at any point in space. In this chapter, we will derive both the exact form of the Foldy–Lax equations and an approximate far-field version. The latter applies to a group of widely separated particles and offers significant simplifications essential for a microphysical derivation of the radiative transfer equation.

4.1 Vector form of the Foldy–Lax equations

Consider electromagnetic scattering by a fixed group of N finite particles collectively occupying the interior region

$$V_{\mathrm{INT}} = \bigcup_{i=1}^{N} V_i, \tag{4.1.1}$$

where V_i is the (bounded) volume occupied by the ith particle (Fig. 4.1.1). As before, we assume that the particles are imbedded in an infinite, homogeneous, linear, isotropic, and nonabsorbing medium. Repeating the derivation of Section 3.1, we arrive at a similar volume integral equation describing the electric field everywhere in space:

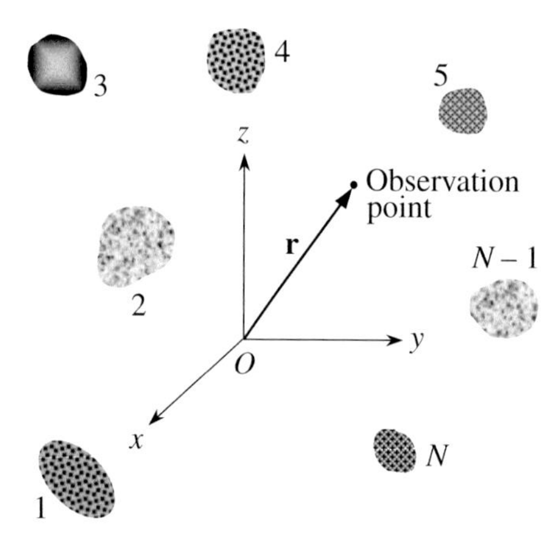

Figure 4.1.1. Scattering by a fixed group of N finite particles.

$$\mathbf{E}(\mathbf{r}) = \mathbf{E}^{\text{inc}}(\mathbf{r}) + \int_{\Re^3} d\mathbf{r}' U(\mathbf{r}') \overleftrightarrow{G}(\mathbf{r},\mathbf{r}') \cdot \mathbf{E}(\mathbf{r}'), \qquad \mathbf{r} \in \Re^3, \tag{4.1.2}$$

where the integration is performed over the entire space, the potential function $U(\mathbf{r})$ is given by

$$U(\mathbf{r}) = \sum_{i=1}^{N} U_i(\mathbf{r}), \qquad \mathbf{r} \in \Re^3, \tag{4.1.3}$$

and $U_i(\mathbf{r})$ is the ith-particle potential function. The latter is given by

$$U_i(\mathbf{r}) = \begin{cases} 0, & \mathbf{r} \notin V_i, \\ k_1^2[m_i^2(\mathbf{r}) - 1], & \mathbf{r} \in V_i, \end{cases} \tag{4.1.4}$$

where

$$m_i(\mathbf{r}) = k_{2i}(\mathbf{r})/k_1 \tag{4.1.5}$$

is the refractive index of particle i relative to that of the host medium. All position vectors originate at the origin O of an arbitrarily chosen laboratory coordinate system.

We will now show that the solution of Eq. (4.1.2) everywhere in space can be expressed as

$$\mathbf{E}(\mathbf{r}) = \mathbf{E}^{\text{inc}}(\mathbf{r}) + \sum_{i=1}^{N} \int_{V_i} d\mathbf{r}' \overleftrightarrow{G}(\mathbf{r},\mathbf{r}') \cdot \int_{V_i} d\mathbf{r}'' \overleftrightarrow{T}_i(\mathbf{r}',\mathbf{r}'') \cdot \mathbf{E}_i(\mathbf{r}''), \qquad \mathbf{r} \in \Re^3, \tag{4.1.6}$$

where the electric field $\mathbf{E}_i(\mathbf{r})$ "exciting" particle i is given by

$$\mathbf{E}_i(\mathbf{r}) = \mathbf{E}^{\text{inc}}(\mathbf{r}) + \sum_{j(\neq i)=1}^{N} \mathbf{E}_{ij}^{\text{exc}}(\mathbf{r}), \tag{4.1.7}$$

the $\mathbf{E}_{ij}^{\text{exc}}(\mathbf{r})$ are partial exciting fields given by

$$\mathbf{E}_{ij}^{\text{exc}}(\mathbf{r}) = \int_{V_j} d\mathbf{r}' \ddot{G}(\mathbf{r}, \mathbf{r}') \cdot \int_{V_j} d\mathbf{r}'' \ddot{T}_j(\mathbf{r}', \mathbf{r}'') \cdot \mathbf{E}_j(\mathbf{r}''), \qquad \mathbf{r} \in V_i, \tag{4.1.8}$$

and $\ddot{T}_i$ is the ith-particle dyadic transition operator with respect to the laboratory coordinate system and satisfies the following Lippmann–Schwinger equation:

$$\ddot{T}_i(\mathbf{r}, \mathbf{r}') = U_i(\mathbf{r})\delta(\mathbf{r}-\mathbf{r}')\ddot{I} + U_i(\mathbf{r}) \int_{V_i} d\mathbf{r}'' \ddot{G}(\mathbf{r}, \mathbf{r}'') \cdot \ddot{T}_i(\mathbf{r}'', \mathbf{r}'), \qquad \mathbf{r}, \mathbf{r}' \in V_i. \tag{4.1.9}$$

We first introduce the ith potential dyadic centered at the origin of the laboratory reference frame,

$$\ddot{U}_i(\mathbf{r}, \mathbf{r}') = U_i(\mathbf{r})\delta(\mathbf{r}-\mathbf{r}')\ddot{I}, \tag{4.1.10}$$

and rewrite Eqs. (4.1.2) and (4.1.6)–(4.1.9) in the following operator form:

$$E = E^{\text{inc}} + \hat{G}\hat{U}E, \tag{4.1.11}$$

$$E = E^{\text{inc}} + \sum_{i=1}^{N} \hat{G}\hat{T}_i E_i, \tag{4.1.12}$$

$$E_i = E^{\text{inc}} + \sum_{j(\neq i)=1}^{N} \hat{G}\hat{T}_j E_j, \tag{4.1.13}$$

$$\hat{T}_i = \hat{U}_i + \hat{U}_i \hat{G}\hat{T}_i, \tag{4.1.14}$$

where

$$\hat{U} = \sum_{i=1}^{N} \hat{U}_i \tag{4.1.15}$$

and

$$\hat{B}E = \int d\mathbf{r}' \ddot{B}(\mathbf{r}, \mathbf{r}') \cdot \mathbf{E}(\mathbf{r}'). \tag{4.1.16}$$

Note that the ordering of operators in Eqs. (4.1.11)–(4.1.14) is important and cannot be changed at will. Equations (4.1.15) and (4.1.14) yield

$$\begin{aligned} \hat{U}\hat{G}\hat{T}_i &= \hat{U}_i\hat{G}\hat{T}_i + \sum_{j(\neq i)=1}^{N} \hat{U}_j\hat{G}\hat{T}_i \\ &= \hat{T}_i - \hat{U}_i + \sum_{j(\neq i)=1}^{N} \hat{U}_j\hat{G}\hat{T}_i. \end{aligned} \tag{4.1.17}$$

Let us now evaluate the right-hand side of Eq. (4.1.11). Substituting sequentially Eqs.

(4.1.12), (4.1.17), and (4.1.13) and then again Eq. (4.1.12) gives

$$\begin{aligned}
E^{\mathrm{inc}} + \hat{G}\hat{U}E &= E^{\mathrm{inc}} + \hat{G}\hat{U}\left(E^{\mathrm{inc}} + \sum_{i=1}^{N} \hat{G}\hat{T}_i E_i\right) \\
&= E^{\mathrm{inc}} + \hat{G}\hat{U}E^{\mathrm{inc}} + \hat{G}\sum_{i=1}^{N}\left(\hat{T}_i - \hat{U}_i + \sum_{j(\neq i)=1}^{N} \hat{U}_j \hat{G}\hat{T}_i\right) E_i \\
&= E^{\mathrm{inc}} + \sum_{i=1}^{N} \hat{G}\hat{T}_i E_i + \hat{G}\hat{U}E^{\mathrm{inc}} + \hat{G}\sum_{i=1}^{N} \hat{U}_i \sum_{j(\neq i)=1}^{N} \hat{G}\hat{T}_j E_j \\
&\quad - \hat{G}\sum_{i=1}^{N} \hat{U}_i E_i \\
&= E^{\mathrm{inc}} + \sum_{i=1}^{N} \hat{G}\hat{T}_i E_i + \hat{G}\hat{U}E^{\mathrm{inc}} \\
&\quad + \hat{G}\sum_{i=1}^{N} \hat{U}_i \left(\sum_{j(\neq i)=1}^{N} \hat{G}\hat{T}_j E_j - E_i\right) \\
&= E^{\mathrm{inc}} + \sum_{i=1}^{N} \hat{G}\hat{T}_i E_i + \hat{G}\hat{U}E^{\mathrm{inc}} - \hat{G}\sum_{i=1}^{N} \hat{U}_i E^{\mathrm{inc}} \\
&= E.
\end{aligned} \tag{4.1.18}$$

Thus, the substitution of Eqs. (4.1.12)–(4.1.14) into the right-hand side of Eq. (4.1.11) yields the left-hand side, which proves that Eqs. (4.1.6)–(4.1.8) indeed give the solution of the volume integral equation (4.1.2) (Prishivalko *et al.*, 1984).

Equations (4.1.6)–(4.1.8) represent the vector form of the so-called Foldy–Lax equations (Foldy, 1945; Lax, 1951). They follow directly from the Maxwell equations and rigorously describe the process of multiple scattering by a fixed group of N particles. Indeed, Eq. (4.1.6) expresses the total field everywhere in space in terms of the vector sum of the incident field and the partial fields generated by each particle in response to the corresponding exciting fields, whereas Eqs. (4.1.7) and (4.1.8) show that the field exciting each particle consists of the incident field and the fields generated by all other particles. Importantly, $\vec{T}_i$ is the dyadic transition operator of particle i in the absence of all other particles (cf. Eqs. (3.1.24) and (4.1.9)).

4.2 Far-field version of the vector Foldy–Lax equations

Although the Foldy–Lax equations can be solved numerically in order to compute the electric field scattered by a finite cluster consisting of arbitrarily positioned components (Tsang *et al.*, 2001), the solution becomes increasingly problematic and eventually impracticable with increasing number of cluster components and/or their sizes relative to the wavelength. To make the problem more manageable, we will often assume that:

- The particles forming the group are separated widely enough that each of them is located in the far-field zones of all the other particles.
- The observation point is located in the far-field zone of any particle forming the group.

These approximations lead to a considerable simplification of the Foldy–Lax equations and will eventually enable us to derive the RTE.

Indeed, according to Eqs. (3.1.23), (3.2.16), and (4.1.8), the contribution of the jth particle to the field exciting the ith particle in Eq. (4.1.7) can now be represented as a simple outgoing spherical wave centered at the origin of particle j:

$$\mathbf{E}_{ij}^{\text{exc}}(\mathbf{r}) \approx G(r_j)\mathbf{E}_{1ij}(\hat{\mathbf{r}}_j) \tag{4.2.1a}$$

$$\approx \exp(-\mathrm{i}k_1\hat{\mathbf{R}}_{ij}\cdot\mathbf{R}_i)\mathbf{E}_{ij}\exp(\mathrm{i}k_1\hat{\mathbf{R}}_{ij}\cdot\mathbf{r}), \qquad \mathbf{r}\in V_i. \tag{4.2.1b}$$

Here,

$$G(r) = \frac{\exp(\mathrm{i}k_1 r)}{r}, \tag{4.2.2}$$

$$\mathbf{E}_{ij} = G(R_{ij})\mathbf{E}_{1ij}(\hat{\mathbf{R}}_{ij}), \qquad \mathbf{E}_{ij}\cdot\hat{\mathbf{R}}_{ij} = 0, \tag{4.2.3}$$

$$\hat{\mathbf{r}}_j = \frac{\mathbf{r}_j}{r_j}, \qquad \hat{\mathbf{R}}_{ij} = \frac{\mathbf{R}_{ij}}{R_{ij}}, \tag{4.2.4}$$

$$r_j = |\mathbf{R}_{ij}+\mathbf{r}-\mathbf{R}_i| \approx R_{ij} + \hat{\mathbf{R}}_{ij}\cdot(\mathbf{r}-\mathbf{R}_i) + \frac{|\mathbf{r}-\mathbf{R}_i|^2}{2R_{ij}}, \tag{4.2.5}$$

and the vectors $\mathbf{r}$, $\mathbf{r}_j$, $\mathbf{R}_i$, $\mathbf{R}_j$, and $\mathbf{R}_{ij}$ are shown in Fig. 4.2.1(a). According to the results of Section 3.2, Eq. (4.2.1a) is valid provided that any point inside particle i is located in the far-field zone of particle j:

$$k_1(R_{ij} - a_i - a_j) \gg 1, \qquad R_{ij} - a_i \gg a_j, \qquad R_{ij} - a_i \gg \frac{k_1 a_j^2}{2},$$

where a_i and a_j are the radii of the smallest circumscribing spheres of particles i and j, respectively. Equation (4.2.1b) follows from similar criteria:

$$k_1 R_{ij} \gg 1, \qquad R_{ij} \gg |\mathbf{r}-\mathbf{R}_i|, \qquad R_{ij} \gg \frac{k_1|\mathbf{r}-\mathbf{R}_i|^2}{2}. \tag{4.2.6}$$

Note that we use a lower case bold letter to denote a vector ending at an observation point, a capital bold letter to denote a vector ending at a particle origin, and a caret above a vector to denote a unit vector in the corresponding direction.

Obviously, $\mathbf{E}_{ij}$ is the partial exciting field at the origin of the ith particle (i.e., at

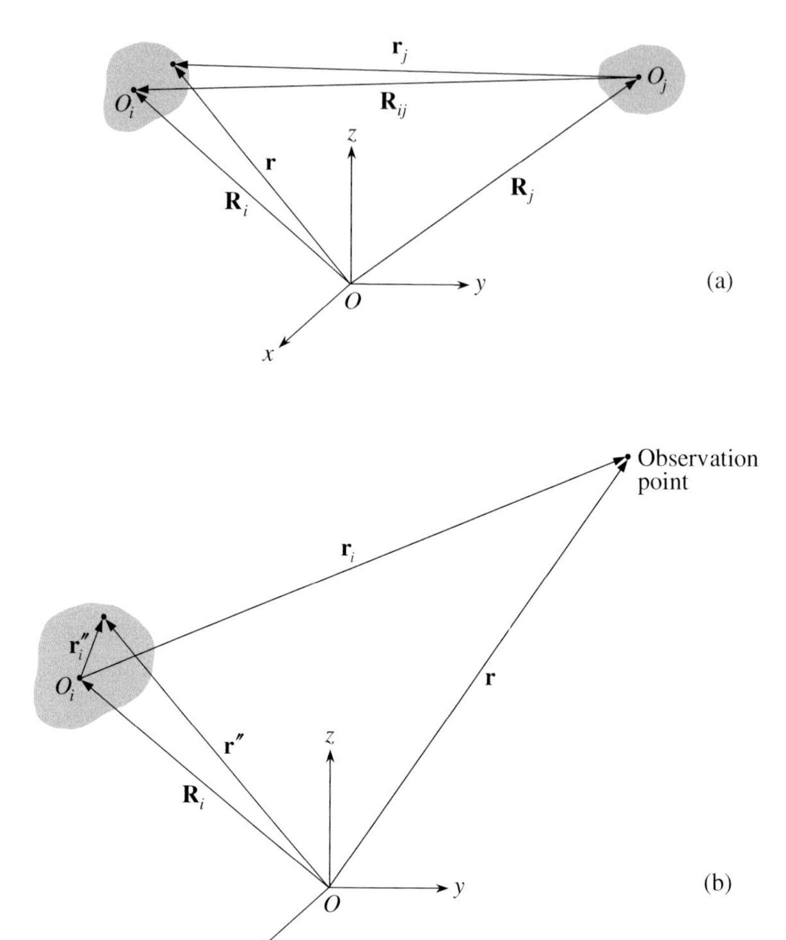

Figure 4.2.1. Scattering by widely separated particles. The local origins O_i and O_j are chosen arbitrarily inside particles i and j, respectively.

$\mathbf{r} = \mathbf{R}_i$) generated by the jth particle. Thus, Eqs. (4.1.7) and (4.2.1b) show that each particle is excited by the external field and the superposition of *locally* plane waves with amplitudes $\exp(-\mathrm{i}k_1\hat{\mathbf{R}}_{ij}\cdot\mathbf{R}_i)\mathbf{E}_{ij}$ and propagation directions $\hat{\mathbf{R}}_{ij}$:

$$\mathbf{E}_i(\mathbf{r}) \approx \mathbf{E}_0^{\mathrm{inc}}\exp(\mathrm{i}k_1\hat{\mathbf{s}}\cdot\mathbf{r}) + \sum_{j(\neq i)=1}^{N}\exp(-\mathrm{i}k_1\hat{\mathbf{R}}_{ij}\cdot\mathbf{R}_i)\mathbf{E}_{ij}\exp(\mathrm{i}k_1\hat{\mathbf{R}}_{ij}\cdot\mathbf{r}), \qquad \mathbf{r}\in V_i, \tag{4.2.7}$$

where we have assumed that the external incident field is a plane electromagnetic wave:

$$\mathbf{E}^{\mathrm{inc}}(\mathbf{r}) = \mathbf{E}_0^{\mathrm{inc}}\exp(\mathrm{i}k_1\hat{\mathbf{s}}\cdot\mathbf{r}), \qquad \mathbf{E}_0^{\mathrm{inc}}\cdot\hat{\mathbf{s}} = 0. \tag{4.2.8}$$

According to Eq. (3.3.2), the outgoing spherical wave generated by the jth particle in response to a plane-wave excitation of the form $\mathbf{E}_0^{\text{inc}}\exp(\mathrm{i}k_1\hat{\mathbf{s}}\cdot\mathbf{r}_j)$ is given by

$$G(r_j)\,\overset{\leftrightarrow}{A}_j(\hat{\mathbf{r}}_j,\hat{\mathbf{s}})\cdot\mathbf{E}_0^{\text{inc}},$$

where $\mathbf{r}_j$ originates at O_j and $\overset{\leftrightarrow}{A}_j(\hat{\mathbf{r}}_j,\hat{\mathbf{s}})$ is the jth particle scattering dyad centered at O_j. To exploit this fact, we must rewrite Eq. (4.2.7) for particle j with respect to the jth-particle coordinate system centered at O_j, Fig. 4.2.1(a). Taking into account that $\mathbf{r} = \mathbf{r}_j + \mathbf{R}_j$ yields

$$\mathbf{E}_j(\mathbf{r}) \approx \mathbf{E}^{\text{inc}}(\mathbf{R}_j)\exp(\mathrm{i}k_1\hat{\mathbf{s}}\cdot\mathbf{r}_j) + \sum_{l(\neq j)=1}^{N}\mathbf{E}_{jl}\exp(\mathrm{i}k_1\hat{\mathbf{R}}_{jl}\cdot\mathbf{r}_j), \qquad \mathbf{r}\in V_j. \tag{4.2.9}$$

The electric field at O_i generated in response to this excitation is simply

$$G(R_{ij})\left[\overset{\leftrightarrow}{A}_j(\hat{\mathbf{R}}_{ij},\hat{\mathbf{s}})\cdot\mathbf{E}^{\text{inc}}(\mathbf{R}_j) + \sum_{l(\neq j)=1}^{N}\overset{\leftrightarrow}{A}_j(\hat{\mathbf{R}}_{ij},\hat{\mathbf{R}}_{jl})\cdot\mathbf{E}_{jl}\right]. \tag{4.2.10}$$

Equating Eq. (4.2.10) with the right-hand side of Eq. (4.2.1b) evaluated for $\mathbf{r} = \mathbf{R}_i$ finally yields a system of linear algebraic equations for determining the partial exciting fields $\mathbf{E}_{ij}$:

$$\mathbf{E}_{ij} = G(R_{ij})\overset{\leftrightarrow}{A}_j(\hat{\mathbf{R}}_{ij},\hat{\mathbf{s}})\cdot\mathbf{E}^{\text{inc}}(\mathbf{R}_j) + G(R_{ij})\sum_{l(\neq j)=1}^{N}\overset{\leftrightarrow}{A}_j(\hat{\mathbf{R}}_{ij},\hat{\mathbf{R}}_{jl})\cdot\mathbf{E}_{jl},$$
$$i,\ j = 1,\ldots,N, \qquad j\neq i. \tag{4.2.11}$$

This system is much simpler than the original system of integral equations (4.1.7)–(4.1.8) and can be readily solved on a computer provided that N is not too large.

After the system (4.2.11) has been solved, one can find the electric field exciting each particle and the total field. Indeed, Eq. (4.2.7) gives for a point $\mathbf{r}''\in V_i$:

$$\mathbf{E}_i(\mathbf{r}'') \approx \mathbf{E}^{\text{inc}}(\mathbf{R}_i)\exp(\mathrm{i}k_1\hat{\mathbf{s}}\cdot\mathbf{r}_i'') + \sum_{j(\neq i)=1}^{N}\mathbf{E}_{ij}\exp(\mathrm{i}k_1\hat{\mathbf{R}}_{ij}\cdot\mathbf{r}_i''), \qquad \mathbf{r}''\in V_i \tag{4.2.12}$$

(see Fig. 4.2.1(b)), which is a vector superposition of plane waves. Substituting $\mathbf{r}_i'' = \mathbf{0}$ in Eq. (4.2.12) gives a simple formula for the exiting field at the origin of particle i:

$$\mathbf{E}_i(\mathbf{R}_i) = \mathbf{E}^{\text{inc}}(\mathbf{R}_i) + \sum_{j(\neq i)=1}^{N}\mathbf{E}_{ij}. \tag{4.2.13}$$

Finally, substituting Eq. (4.2.12) in Eq. (4.1.6) and recalling the mathematical form of the far-field response of a particle to a plane-wave excitation, we derive for the total electric field:

$$\mathbf{E}(\mathbf{r}) = \mathbf{E}^{\text{inc}}(\mathbf{r}) + \sum_{i=1}^{N} G(r_i) \overleftrightarrow{A}_i(\hat{\mathbf{r}}_i, \hat{\mathbf{s}}) \cdot \mathbf{E}^{\text{inc}}(\mathbf{R}_i) + \sum_{i=1}^{N} G(r_i) \sum_{j(\neq i)=1}^{N} \overleftrightarrow{A}_i(\hat{\mathbf{r}}_i, \hat{\mathbf{R}}_{ij}) \cdot \mathbf{E}_{ij}, \tag{4.2.14}$$

where the observation point $\mathbf{r}$, Fig. 4.2.1(b), is assumed to be in the far-field zone of any particle forming the group:

$$k_1(r_i - a_i) \gg 1, \qquad r_i \gg a_i, \qquad r_i \gg \frac{k_1 a_i^2}{2} \tag{4.2.15}$$

for any i.

Equation (4.2.14) can also be re-written as follows:

$$\mathbf{E}(\mathbf{r}) = \mathbf{E}^{\text{inc}}(\mathbf{r}) + \sum_{i=1}^{N} \mathbf{E}_i^{\text{sca}}(\mathbf{r}), \tag{4.2.16}$$

where

$$\mathbf{E}_i^{\text{sca}}(\mathbf{r}) = G(r_i) \left[\overleftrightarrow{A}_i(\hat{\mathbf{r}}_i, \hat{\mathbf{s}}) \cdot \mathbf{E}^{\text{inc}}(\mathbf{R}_i) + \sum_{j(\neq i)=1}^{N} \overleftrightarrow{A}_i(\hat{\mathbf{r}}_i, \hat{\mathbf{R}}_{ij}) \cdot \mathbf{E}_{ij} \right]. \tag{4.2.17}$$

These formulas show that the total field at any point located sufficiently far from any particle in the group is the superposition of the incident plane wave and N spherical waves generated by the N particles.

Chapter 5

Statistical averaging

Most of our discussion of electromagnetic scattering in Chapters 3 and 4 was based on the assumption that the position and orientation of the scattering object are fixed. Although there are practical circumstances in which this assumption is true, more often than not the scatterer changes its position and orientation randomly during the time necessary to take a measurement. Moreover, one usually has to deal with a scattering object in the form of a group of many discrete particles randomly rotating and moving relative to each other. Important examples of such "stochastic" scattering objects are clouds consisting of water droplets and/or ice crystals and plumes of aerosol particles.

At any given moment in time, a cloud can be represented by a fixed static group of discrete particles. However, any measurement takes a finite amount of time during which the cloud goes through an infinite succession of varying discrete states. Although the result of the measurement can be modeled numerically by computing the scattered signal for many different discrete cloud states and then taking the average, a more efficient approach is to use methods of mathematical statistics and attempt to derive the average analytically. Specifically, all further discussion in this book will be based on the following two fundamental premises:

- The scattering object can be adequately characterized at any moment in time by a finite set of physical parameters.
- The scattering object is sufficiently variable in time and the time interval necessary to take a measurement is sufficiently long that averaging the scattering signal over this interval is essentially equivalent to averaging the signal over an appropriate analytical probability distribution of the physical parameters characterizing the scattering object.

According to the discussion in Section 1.5, the second premise is equivalent to the ergodic hypothesis and means that averaging over time for one specific realization of a random scattering process is equivalent to ensemble averaging, Eq. (1.5.3).

The specific aim of this chapter is to introduce the concepts of mathematical statistics necessary for the following discussion of electromagnetic scattering by a group of randomly distributed particles. We will also give examples of analytical distribution functions frequently used to describe statistical characteristics of particles encountered in natural and artificial environments.

5.1 Statistical averages

It is convenient to describe a large group of N arbitrarily oriented and randomly distributed particles using the probability density function $p(\mathbf{R}_1, \xi_1; ...; \mathbf{R}_i, \xi_i; ...; \mathbf{R}_N, \xi_N)$ defined with respect to a common laboratory coordinate system. The probability of finding the first particle in the volume element $\mathrm{d}\mathbf{R}_1$ centered at $\mathbf{R}_1$ and with its state in the region $\mathrm{d}\xi_1$ centered at ξ_1, …, the ith particle in the volume element $\mathrm{d}\mathbf{R}_i$ centered at $\mathbf{R}_i$ and with its state in the region $\mathrm{d}\xi_i$ centered at ξ_i, …, and the Nth particle in the volume element $\mathrm{d}\mathbf{R}_N$ centered at $\mathbf{R}_N$ and with its state in the region $\mathrm{d}\xi_N$ centered at ξ_N is given by

$$p(\mathbf{R}_1, \xi_1; ...; \mathbf{R}_N, \xi_N) \prod_{i=1}^{N} \mathrm{d}\mathbf{R}_i \, \mathrm{d}\xi_i. \tag{5.1.1}$$

The state of a particle can collectively indicate its size, refractive index, shape, orientation, etc. Hence,

$$\begin{aligned} \mathrm{d}\xi_i &= \mathrm{d}(\text{size of particle } i) \times \mathrm{d}(\text{refractive index of particle } i) \\ &\quad \times \mathrm{d}(\text{shape of particle } i) \times \mathrm{d}(\text{orientation of particle } i) \times \cdots. \end{aligned} \tag{5.1.2}$$

The probability density function is normalized to unity:

$$\int \prod_{i=1}^{N} \mathrm{d}\mathbf{R}_i \, \mathrm{d}\xi_i \, p(\mathbf{R}_1, \xi_1; ...; \mathbf{R}_N, \xi_N) = 1, \tag{5.1.3}$$

where the integration is performed over the entire range of particle positions and states. The statistical average of a random function f depending on all N particles is given by

$$\langle f \rangle_{\mathbf{R},\xi} = \int \prod_{i=1}^{N} \mathrm{d}\mathbf{R}_i \, \mathrm{d}\xi_i \, f(\mathbf{R}_1, \xi_1; ...; \mathbf{R}_N, \xi_N) p(\mathbf{R}_1, \xi_1; ...; \mathbf{R}_N, \xi_N). \tag{5.1.4}$$

If the position and state of each particle are independent of those of all other particles then

$$p(\mathbf{R}_1, \xi_1; ...; \mathbf{R}_N, \xi_N) = \prod_{i=1}^{N} p_i(\mathbf{R}_i, \xi_i). \tag{5.1.5}$$

This is a good approximation when particles are sparsely distributed so that the finite size of the particles can be neglected. In this case the effect of size appears only in the single-particle scattering and absorption characteristics. Obviously,

$$\int \mathrm{d}\mathbf{R}_i \, \mathrm{d}\xi_i \, p_i(\mathbf{R}_i, \xi_i) = 1. \tag{5.1.6}$$

If, furthermore, the state of each particle is independent of its position, then

$$p_i(\mathbf{R}_i, \xi_i) = p_{\mathbf{R}i}(\mathbf{R}_i) p_{\xi i}(\xi_i) \tag{5.1.7}$$

with

$$\int \mathrm{d}\mathbf{R}_i \, p_{\mathbf{R}i}(\mathbf{R}_i) = 1, \tag{5.1.8}$$

$$\int \mathrm{d}\xi_i \, p_{\xi i}(\xi_i) = 1. \tag{5.1.9}$$

Equation (5.1.7) allows one to separate the configurational averaging (i.e., averaging over the particle positions) from the averaging over the particle states.

Finally, assuming that all particles have the same statistical characteristics, we have

$$p_i(\mathbf{R}_i, \xi_i) \equiv p(\mathbf{R}_i, \xi_i) = p_{\mathbf{R}}(\mathbf{R}_i) p_{\xi}(\xi_i), \tag{5.1.10}$$

$$p(\mathbf{R}_1, \xi_1; \ldots; \mathbf{R}_N, \xi_N) = \prod_{i=1}^{N} p_{\mathbf{R}}(\mathbf{R}_i) p_{\xi}(\xi_i), \tag{5.1.11}$$

$$\int \mathrm{d}\mathbf{R} \, p_{\mathbf{R}}(\mathbf{R}) = 1, \tag{5.1.12}$$

$$\int \mathrm{d}\xi \, p_{\xi}(\xi) = 1. \tag{5.1.13}$$

The interpretation of the probability density function $p_{\mathbf{R}}(\mathbf{R})$ is simple:

$$\begin{aligned} p_{\mathbf{R}}(\mathbf{R})\mathrm{d}\mathbf{R} &= \text{probability of finding a particle within volume } \mathrm{d}\mathbf{R} \text{ centered at } \mathbf{R} \\ &= \frac{\text{number of particles within } \mathrm{d}\mathbf{R}}{\text{total number of particles}} \\ &= \frac{n_0(\mathbf{R})\mathrm{d}\mathbf{R}}{N}, \end{aligned} \tag{5.1.14}$$

where $n_0(\mathbf{R})$ is the local particle number density defined as the number of particles per unit volume in the vicinity of $\mathbf{R}$. Thus,

$$p_{\mathbf{R}}(\mathbf{R}) = \frac{n_0(\mathbf{R})}{N}. \tag{5.1.15}$$

5.2 Configurational averaging

In what follows, we will often assume that particles forming a multi-particle group are confined to a finite bounded volume of space V. Furthermore, we will always assume that the spatial distribution of the N particles throughout the volume V is statistically uniform. Then

$$n_0(\mathbf{R}) = \begin{cases} n_0 = \dfrac{N}{V} & \text{if } \mathbf{R} \in V, \\ 0 & \text{if } \mathbf{R} \notin V \end{cases} \tag{5.2.1}$$

and

$$p_{\mathbf{R}}(\mathbf{R}) = \begin{cases} \dfrac{1}{V} & \text{if } \mathbf{R} \in V, \\ 0 & \text{if } \mathbf{R} \notin V. \end{cases} \tag{5.2.2}$$

5.3 Averaging over particle states

The computation of averages over the particle states is, in principle, rather straightforward. The orientation of a particle with respect to the laboratory coordinate system can be specified by affixing a Cartesian coordinate system to the particle and specifying the Euler angles α, β, and γ that transform the laboratory coordinate system into the particle coordinate system (see Appendix C). If a multi-particle group consists, for example, of homogeneous ellipsoids with semi-axes $a \in [a_{\min}, a_{\max}]$, $b \in [b_{\min}, b_{\max}]$, and $c \in [c_{\min}, c_{\max}]$ and the same refractive index then the ensemble average of a scattering or absorption characteristic ς per particle is given by

$$\langle \varsigma \rangle_\xi = \int_0^{2\pi} \mathrm{d}\alpha \int_0^{\pi} \mathrm{d}\beta \sin\beta \int_0^{2\pi} \mathrm{d}\gamma \int_{a_{\min}}^{a_{\max}} \mathrm{d}a \int_{b_{\min}}^{b_{\max}} \mathrm{d}b \int_{c_{\min}}^{c_{\max}} \mathrm{d}c \times p_\xi(\alpha, \beta, \gamma; a, b, c) \varsigma(\alpha, \beta, \gamma; a, b, c), \tag{5.3.1}$$

where the probability density function $p_\xi(\alpha, \beta, \gamma; a, b, c)$ satisfies the following normalization condition:

$$\int_0^{2\pi} \mathrm{d}\alpha \int_0^{\pi} \mathrm{d}\beta \sin\beta \int_0^{2\pi} \mathrm{d}\gamma \int_{a_{\min}}^{a_{\max}} \mathrm{d}a \int_{b_{\min}}^{b_{\max}} \mathrm{d}b \int_{c_{\min}}^{c_{\max}} \mathrm{d}c \, p_\xi(\alpha, \beta, \gamma; a, b, c) = 1. \tag{5.3.2}$$

The integrals in Eq. (5.3.1) are usually evaluated numerically by using appropriate quadrature formulas (see Appendix D). Some theoretical techniques (e.g., the T-matrix method described in Chapter 5 of MTL) allow analytical averaging over particle orientations, thereby bypassing time-consuming numerical integration over the Euler angles.

It is often assumed that the shape/size and orientation distributions are statistically independent. The total probability density function can then be simplified by representing it as a product of two functions, one of which, $p_{\mathrm{o}}(\alpha, \beta, \gamma)$, describes the distribution of particle orientations, and the other one, $p_{\mathrm{s}}(a, b, c)$, describes the particle shape/size distribution:

$$p_{\xi}(\alpha, \beta, \gamma; a, b, c) = p_{\mathrm{o}}(\alpha, \beta, \gamma) p_{\mathrm{s}}(a, b, c), \tag{5.3.3}$$

each normalized to unity:

$$\int_0^{2\pi} \mathrm{d}\alpha \int_0^{\pi} \mathrm{d}\beta \sin\beta \int_0^{2\pi} \mathrm{d}\gamma \, p_{\mathrm{o}}(\alpha, \beta, \gamma) = 1, \tag{5.3.4}$$

$$\int_{a_{\min}}^{a_{\max}} \mathrm{d}a \int_{b_{\min}}^{b_{\max}} \mathrm{d}b \int_{c_{\min}}^{c_{\max}} \mathrm{d}c \, p_{\mathrm{s}}(a, b, c) = 1. \tag{5.3.5}$$

As a consequence, the problems of computing shape/size and orientation averages are separated.

Similarly, it is often convenient to separate averaging over shapes and sizes by assuming that particle shapes and sizes are statistically independent. For example, the shape of a spheroidal particle can be specified by its aspect ratio ε (ratio of the largest to the smallest axes) along with the designation of either prolate or oblate, whereas the particle size can be specified by an equivalent-sphere radius r. Then the shape/size probability density function $p_{\mathrm{s}}(\varepsilon, r)$ can be represented as a product

$$p_{\mathrm{s}}(\varepsilon, r) = p(\varepsilon) n(r), \tag{5.3.6}$$

where $p(\varepsilon)$ describes the distribution of spheroid aspect ratios and $n(r)$ is the distribution of equivalent-sphere radii. Again, both $p(\varepsilon)$ and $n(r)$ are normalized to unity:

$$\int_{\varepsilon_{\min}}^{\varepsilon_{\max}} \mathrm{d}\varepsilon \, p(\varepsilon) = 1, \tag{5.3.7}$$

$$\int_{r_{\min}}^{r_{\max}} \mathrm{d}r \, n(r) = 1. \tag{5.3.8}$$

In the absence of external forces such as magnetic, electrostatic, or aerodynamical forces, all orientations of a nonspherical particle are equiprobable. In this practically important case of randomly oriented particles, the orientation distribution function is uniform with respect to the Euler angles of rotation, and we have

$$p_{\mathrm{o,random}}(\alpha, \beta, \gamma) = \frac{1}{8\pi^2}. \tag{5.3.9}$$

An external force can make the orientation distribution axially symmetric with the axis of symmetry given by the direction of the force. In this case it is convenient to choose the laboratory reference frame with the z-axis along the external force direction so that the orientation distribution is uniform with respect to the Euler angles α and γ:

$$p_{\mathrm{o,axial}}(\alpha, \beta, \gamma) = \frac{1}{4\pi^2} p_{\mathrm{o}}(\beta). \tag{5.3.10}$$

Particular details of the particle shape can also simplify the orientation distribution function. For example, for rotationally symmetric bodies it is convenient to direct the z-axis of the particle reference frame along the axis of rotation, in which case the orientation distribution function in the laboratory reference frame becomes independent of the Euler angle γ:

$$p_{\mathrm{o}}(\alpha, \beta, \gamma) = \frac{1}{2\pi} p_{\mathrm{o}}(\alpha, \beta). \tag{5.3.11}$$

Natural size distributions are often approximated using convenient analytical functions. The analytical size distribution functions used most typically are the following:

- The modified gamma distribution

$$n(r) = \text{constant} \times r^{\alpha} \exp\left(-\frac{\alpha r^{\gamma}}{\gamma r_c^{\gamma}}\right). \tag{5.3.12}$$

- The log normal distribution

$$n(r) = \text{constant} \times r^{-1} \exp\left[-\frac{(\ln r - \ln r_g)^2}{2\ln^2 \sigma_g}\right]. \tag{5.3.13}$$

- The power law distribution

$$n(r) = \begin{cases} \text{constant} \times r^{-3}, & r_1 \le r \le r_2, \\ 0, & \text{otherwise}. \end{cases} \tag{5.3.14}$$

- The gamma distribution

$$n(r) = \text{constant} \times r^{(1-3b)/b} \exp\left(-\frac{r}{ab}\right), \quad b \in (0, 0.5). \tag{5.3.15}$$

- The modified power law distribution

$$n(r) = \begin{cases} \text{constant}, & 0 \le r \le r_1, \\ \text{constant} \times (r/r_1)^{\alpha}, & r_1 \le r \le r_2, \\ 0, & r_2 < r. \end{cases} \tag{5.3.16}$$

- The modified bimodal log normal distribution

$$n(r) = \text{constant} \times r^{-4}\left\{\exp\left[-\frac{(\ln r - \ln r_{g1})^2}{2\ln^2\sigma_{g1}}\right] + \gamma\exp\left[-\frac{(\ln r - \ln r_{g2})^2}{2\ln^2\sigma_{g2}}\right]\right\}. \quad (5.3.17)$$

The constant for each size distribution is chosen such that the size distribution satisfies the standard normalization of Eq. (5.3.8).

Implicitly, particle radii in the modified gamma, log normal, gamma, and modified bimodal log normal distributions extend to infinity. However, a finite $r_{\max}$ must be chosen in actual computer calculations. There are two different practical interpretations of a truncated size distribution. The first one assumes that $r_{\max}$ is increased iteratively until the scattering and absorption characteristics of the size distribution converge within a prescribed numerical accuracy. In this case the converged truncated size distribution is numerically equivalent to the distribution with $r_{\max} = \infty$. In the second interpretation, the truncated distribution with a specified $r_{\max}$ can be considered as a specific size distribution with scattering and absorption characteristics distinctly different from those for the distribution with $r_{\max} = \infty$. Similar considerations apply to the parameter $r_{\min}$, whose implicit value for the modified gamma, log normal, gamma, and modified bimodal log normal distributions is zero, but in practice can be any number smaller than $r_{\max}$. In what follows, we always adopt the first interpretation of a truncated size distribution.

Two important integral characteristics of a size distribution are the effective radius r_{eff} and effective variance v_{eff} defined by

$$r_{\text{eff}} = \frac{1}{\langle G\rangle}\int_{r_{\min}}^{r_{\max}} \mathrm{d}r\, n(r) r \pi r^2, \quad (5.3.18)$$

$$v_{\text{eff}} = \frac{1}{\langle G\rangle r_{\text{eff}}^2}\int_{r_{\min}}^{r_{\max}} \mathrm{d}r\, n(r)(r - r_{\text{eff}})^2 \pi r^2, \quad (5.3.19)$$

where

$$\langle G\rangle = \int_{r_{\min}}^{r_{\max}} \mathrm{d}r\, n(r)\pi r^2 \quad (5.3.20)$$

is the average area of the geometric projection per particle. The r_{eff} is simply the projected-area-weighted mean radius, whereas the dimensionless effective variance provides a measure of the width of the size distribution. Hansen and Travis (1974) and Mishchenko and Travis (1994a) have shown that different moderately broad size distributions that have the same values of r_{eff} and v_{eff} can be expected to have similar *dimensionless* scattering and absorption characteristics.

Note that for the gamma distribution with $r_{\min} = 0$ and $r_{\max} = \infty$, a and b coincide with r_{eff} and v_{eff}, respectively. For the other size distributions with specific values of

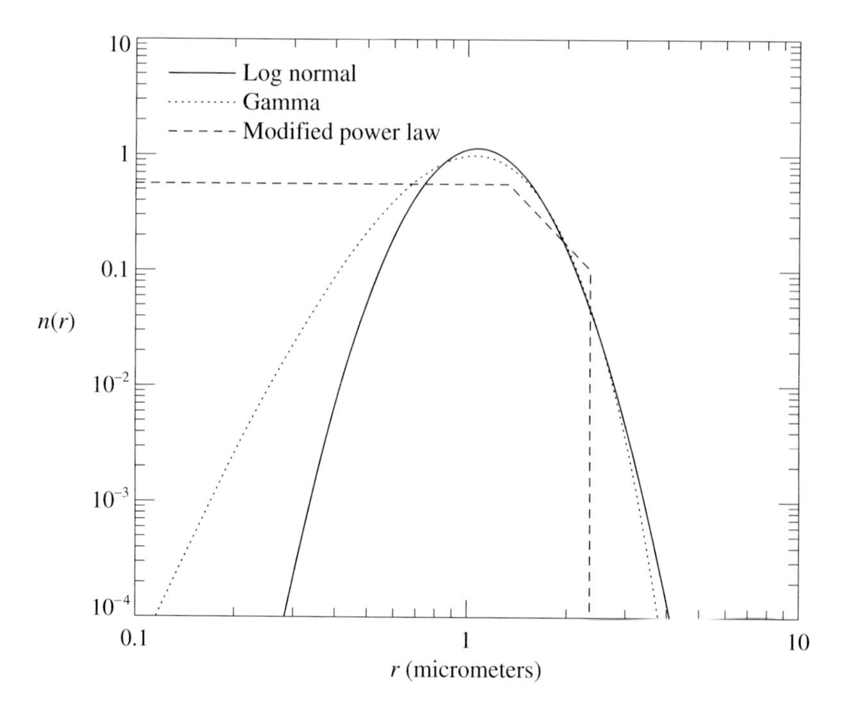

Figure 5.3.1. Log normal, gamma, and modified power law size distributions with $r_{\text{eff}} = 1.5\,\mu\text{m}$ and $v_{\text{eff}} = 0.1$. The power exponent of the modified power law size distribution is $\alpha = -3$.

$r_{\min}$ and $r_{\max}$ the effective radius and effective variance must be determined either analytically or numerically.

As an example, Fig. 5.3.1 shows three kinds of size distribution with the same values of the effective radius and the effective variance. It is rather obvious that the modified power law distribution has an important practical advantage in that its maximal radius $r_{\max}$ is finite by definition and can be significantly smaller than the corresponding convergent maximal radii for the "equivalent" log normal and gamma distributions.

Chapter 6

Scattering by a single random particle

The simplest kind of a stochastic scattering object is a single particle that moves, rotates, and perhaps changes its size and/or shape during the measurement. A typical example is the scattering by a single particle suspended in air or vacuum with one of the existing levitation techniques (Davis and Schweiger, 2002). The particle position within the trap volume of the levitator as well as the particle orientation are never perfectly fixed, and the particle can undergo random or periodic movements and can spin during the time interval necessary to take a measurement. The particle may also change its size and shape owing to evaporation, sublimation, condensation, or melting. The shape of a liquid particle can also change owing to surface oscillations.

The results of Sections 3.2 and 3.6–3.9 are not applicable directly to electromagnetic scattering by such a "random" particle. However, we will show in this chapter that under certain assumptions one can still use most of those results in combination with the statistical averaging concepts introduced in Chapter 5.

The discussion in this chapter is explicitly based on the assumption that the scattering object is illuminated by a plane electromagnetic wave. However, the results can be generalized easily to cover the more general cases of illumination considered in Sections 3.10–3.12.

6.1 Scattering in the far-field zone of the trap volume

To model electromagnetic scattering by a random particle trapped in a finite volume, let us assume that at any moment during the measurement the particle can be anywhere inside a small volume V with radius $R_V \geq a$, where a is the radius of the smallest circumscribing sphere of the scatterer (see Fig. 6.1.1). The geometrical center of the volume serves as the origin O of the laboratory coordinate system. Let the

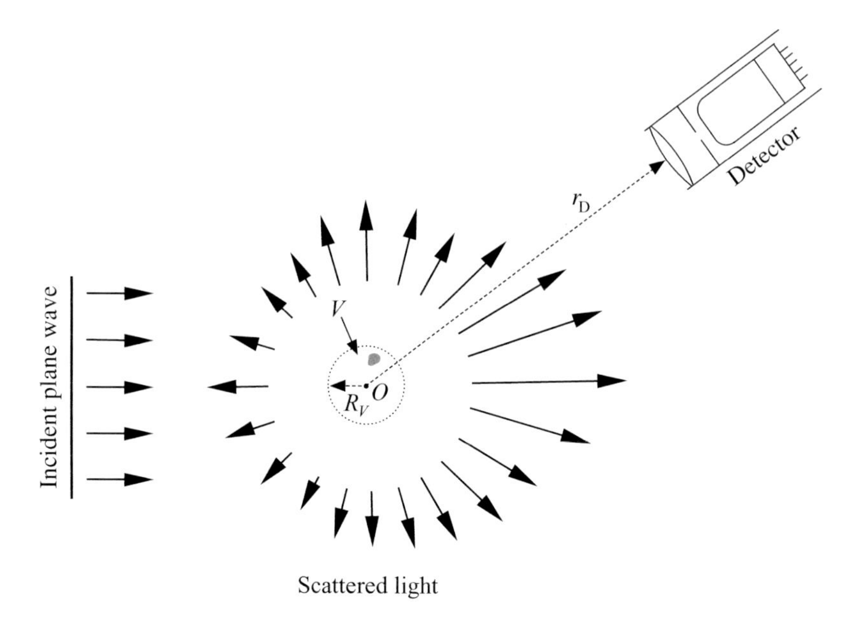

Figure 6.1.1. Scattering by a single random particle.

electric field of the incident plane wave be given by

$$\mathbf{E}^{\text{inc}}(\mathbf{r}) = \mathbf{E}_0^{\text{inc}} \exp(\mathrm{i}k_1 \hat{\mathbf{n}}^{\text{inc}} \cdot \mathbf{r}), \tag{6.1.1}$$

where $\mathbf{r}$ is the position vector originating at O, and let $\mathbf{r}'$ be the position vector of the same observation point but originating at the particle origin O' (Fig. 6.1.2). Since

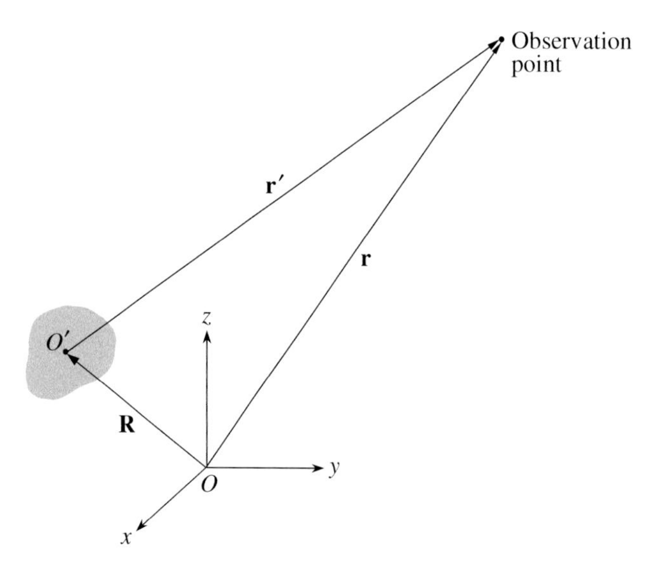

Figure 6.1.2. The origin of the particle reference frame does not coincide with that of the laboratory reference frame.

$\mathbf{r} = \mathbf{r}' + \mathbf{R}$, where $\mathbf{R}$ connects the origin of the laboratory coordinate system with the particle origin, the incident electric field at the observation point can also be written as follows:

$$\mathbf{E}^{\text{inc}}(\mathbf{r}) = \mathbf{E}_0^{\text{inc}} \exp(\mathrm{i}k_1 \hat{\mathbf{n}}^{\text{inc}} \cdot \mathbf{r}') \exp(\mathrm{i}k_1 \hat{\mathbf{n}}^{\text{inc}} \cdot \mathbf{R}). \tag{6.1.2}$$

We know that the outgoing spherical wave generated by the particle in response to a plane-wave excitation of the form $\mathbf{E}_0^{\text{inc}} \exp(\mathrm{i}k_1 \hat{\mathbf{n}}^{\text{inc}} \cdot \mathbf{r}')$ in the far-field zone is given by

$$\frac{\exp(\mathrm{i}k_1 r')}{r'} \overset{\leftrightarrow}{A}(\hat{\mathbf{r}}', \hat{\mathbf{n}}^{\text{inc}}) \cdot \mathbf{E}_0^{\text{inc}},$$

where $\hat{\mathbf{r}}' = \mathbf{r}'/r'$ is the scattering direction centered at the particle and $\overset{\leftrightarrow}{A}(\hat{\mathbf{r}}', \hat{\mathbf{n}}^{\text{inc}})$ is the scattering dyadic with respect to the particle reference frame. Therefore, the incident field (6.1.2) results in the following scattered field:

$$\mathbf{E}^{\text{sca}}(\mathbf{r}) = \frac{\exp(\mathrm{i}k_1 r')}{r'} \exp(\mathrm{i}k_1 \hat{\mathbf{n}}^{\text{inc}} \cdot \mathbf{R}) \overset{\leftrightarrow}{A}(\hat{\mathbf{r}}', \hat{\mathbf{n}}^{\text{inc}}) \cdot \mathbf{E}_0^{\text{inc}}. \tag{6.1.3}$$

This formula is valid provided that the following far-field criteria are satisfied (see Eqs. (3.2.17)–(3.2.19)):

$$k_1(r' - a) \gg 1, \tag{6.1.4}$$

$$r' \gg a, \tag{6.1.5}$$

$$r' \gg \frac{k_1 a^2}{2}. \tag{6.1.6}$$

Using the law of cosines,

$$r'^2 = r^2 + R^2 - 2\mathbf{r} \cdot \mathbf{R}, \tag{6.1.7}$$

we finally obtain

$$\mathbf{E}^{\text{sca}}(\mathbf{r}) = \frac{\exp(\mathrm{i}k_1 r)}{r} \overset{\leftrightarrow}{A}(\hat{\mathbf{r}}, \hat{\mathbf{n}}^{\text{inc}}; \mathbf{R}) \cdot \mathbf{E}_0^{\text{inc}}, \tag{6.1.8}$$

where $\hat{\mathbf{r}} = \mathbf{r}/r$ is the scattering direction centered at the origin of the laboratory coordinate system,

$$\overset{\leftrightarrow}{A}(\hat{\mathbf{r}}, \hat{\mathbf{n}}^{\text{inc}}; \mathbf{R}) = \exp(\mathrm{i}\Delta) \overset{\leftrightarrow}{A}(\hat{\mathbf{r}}, \hat{\mathbf{n}}^{\text{inc}}) \tag{6.1.9}$$

is the scattering dyadic of the particle with respect to the laboratory coordinate system,

$$\Delta = k_1(\hat{\mathbf{n}}^{\text{inc}} - \hat{\mathbf{r}}) \cdot \mathbf{R}, \tag{6.1.10}$$

and we have further assumed that

$$r \gg R, \tag{6.1.11}$$

$$\frac{k_1 R^2}{2r} \ll 1, \tag{6.1.12}$$

and

$$\ddot{A}(\hat{\mathbf{r}}', \hat{\mathbf{n}}^{\text{inc}}) \approx \ddot{A}(\hat{\mathbf{r}}, \hat{\mathbf{n}}^{\text{inc}}). \tag{6.1.13}$$

As should have been expected,

$$\ddot{A}(\hat{\mathbf{r}}, \hat{\mathbf{n}}^{\text{inc}}; \mathbf{0}) = \ddot{A}(\hat{\mathbf{r}}, \hat{\mathbf{n}}^{\text{inc}}). \tag{6.1.14}$$

Equation (6.1.8) describes a transverse outgoing spherical wave centered at the origin of the laboratory reference frame. This allows us to proceed in exactly the same way as we did in Chapter 3. Specifically, exploiting the transverse character of the wave yields

$$\mathbf{E}^{\text{sca}}(\mathbf{r}) = \frac{\exp(\mathrm{i}k_1 r)}{r} \mathbf{S}(\hat{\mathbf{r}}, \hat{\mathbf{n}}^{\text{inc}}; \mathbf{R}) \cdot \mathbf{E}_0^{\text{inc}}, \tag{6.1.15}$$

where we have used the notation of Eq. (3.3.7), and the amplitude scattering matrix of the particle with respect to the laboratory coordinate system, $\mathbf{S}(\hat{\mathbf{r}}, \hat{\mathbf{n}}^{\text{inc}}; \mathbf{R})$, is expressed in terms of that with respect to the particle coordinate system, $\mathbf{S}(\hat{\mathbf{r}}, \hat{\mathbf{n}}^{\text{inc}})$, as follows:

$$\mathbf{S}(\hat{\mathbf{r}}, \hat{\mathbf{n}}^{\text{inc}}; \mathbf{R}) = \exp(\mathrm{i}\Delta) \mathbf{S}(\hat{\mathbf{r}}, \hat{\mathbf{n}}^{\text{inc}}). \tag{6.1.16}$$

Of course, Eq. (6.1.16) implies that the spatial orientations of the two coordinate systems are the same. As before,

$$\mathbf{S}(\hat{\mathbf{r}}, \hat{\mathbf{n}}^{\text{inc}}; \mathbf{0}) = \mathbf{S}(\hat{\mathbf{r}}, \hat{\mathbf{n}}^{\text{inc}}). \tag{6.1.17}$$

Substituting Eq. (6.1.16) in Eqs. (3.7.11)–(3.7.26) and Eqs. (3.8.8)–(3.8.14) shows that irrespective of the particle position within the trap volume, the phase and extinction matrices of the particle with respect to the laboratory reference frame remain the same and are equal to those with respect to the particle reference frame:

$$\mathbf{Z}(\hat{\mathbf{r}}, \hat{\mathbf{n}}^{\text{inc}}; \mathbf{R}) \equiv \mathbf{Z}(\hat{\mathbf{r}}, \hat{\mathbf{n}}^{\text{inc}}; \mathbf{0}) = \mathbf{Z}(\hat{\mathbf{r}}, \hat{\mathbf{n}}^{\text{inc}}), \tag{6.1.18}$$

$$\mathbf{K}(\hat{\mathbf{n}}^{\text{inc}}; \mathbf{R}) \equiv \mathbf{K}(\hat{\mathbf{n}}^{\text{inc}}; \mathbf{0}) = \mathbf{K}(\hat{\mathbf{n}}^{\text{inc}}). \tag{6.1.19}$$

Indeed, the factor $\exp(\mathrm{i}\Delta)$ is common to all elements of the amplitude scattering matrix centered at the origin of the laboratory reference frame and disappears when multiplied by its complex-conjugate counterpart, whereas the phase Δ vanishes identically in the exact forward-scattering direction. It is straightforward to verify that all the optical cross sections and efficiency factors, the single-scattering albedo, the phase function, and the asymmetry parameter are also invariant with respect to changing **R**.

This important result indicates that to model the cumulative signal measured by a distant detector over a finite time interval, one may use Eqs. (3.7.6), (3.7.8), and

(3.8.5b), in which $r = r_D$ is the distance from the origin of the laboratory reference frame to the detector (Fig. 6.1.1), and the phase and extinction matrices are obtained by averaging the matrices $\mathbf{Z}(\hat{\mathbf{r}}, \hat{\mathbf{n}}^{\text{inc}})$ and $\mathbf{K}(\hat{\mathbf{n}}^{\text{inc}})$ over particle states:

$$\langle \mathbf{Z}(\hat{\mathbf{r}}, \hat{\mathbf{n}}^{\text{inc}}; \mathbf{R}) \rangle_{\mathbf{R},\xi} = \langle \mathbf{Z}(\hat{\mathbf{r}}, \hat{\mathbf{n}}^{\text{inc}}) \rangle_{\xi}, \tag{6.1.20}$$

$$\langle \mathbf{K}(\hat{\mathbf{n}}^{\text{inc}}; \mathbf{R}) \rangle_{\mathbf{R},\xi} = \langle \mathbf{K}(\hat{\mathbf{n}}^{\text{inc}}) \rangle_{\xi}. \tag{6.1.21}$$

The averaging over particle states incorporates the possible effects of variable particle orientation, size, and/or shape during the measurement. Thus, a moving particle can be effectively replaced by a particle fixed at the origin of the laboratory coordinate system. The latter is still partially random in that it may change its orientation, size, and/or shape.

Let us now analyze the conditions of applicability of Eqs. (6.1.20) and (6.1.21). First, the very concept of using a detector of electromagnetic radiation implies that the following criteria, adapted from Eqs. (3.6.1), (3.6.19), and (3.6.21), must be satisfied:

$$\frac{\Delta S}{r_D^2} = \frac{\pi D^2}{4 r_D^2} < \Delta\Omega, \tag{6.1.22}$$

$$D \gg 2R_V, \tag{6.1.23}$$

$$r_D \gg \frac{D k_1 a}{\pi}, \tag{6.1.24}$$

where D is the diameter of the sensitive surface of the detector, ΔS is its area, and $\Delta\Omega$ is the detector angular aperture. As in Section 3.6, the criterion (6.1.23) applies only to the detector facing the incident light. It reflects the fact that the interaction of the incident plane wave and the scattered spherical wave occurs along the line drawn through the particle origin in the direction of the unit vector $\hat{\mathbf{n}}^{\text{inc}}$ (Fig. 6.1.3(a)). In order to capture this interaction irrespective of the particle position within the trap volume, the sensitive area of the detector centered at O must be sufficiently large to always contain the geometrical shadow cast by the particle (Fig. 6.1.3(b)):

$$\text{Particle shadow} \in \Delta S. \tag{6.1.25}$$

Second, the inequalities (6.1.4)–(6.1.6), (6.1.11), and (6.1.12) must be valid for any position of the particle within the volume element V. This yields

$$k_1 (r_D - R_V) \gg 1, \tag{6.1.26}$$

$$r_D - R_V + a \gg a, \tag{6.1.27}$$

$$r_D - R_V + a \gg \frac{k_1 a^2}{2}, \tag{6.1.28}$$

$$r_D \gg R_V - a, \tag{6.1.29}$$

$$r_D \gg \frac{k_1 (R_V - a)^2}{2}. \tag{6.1.30}$$

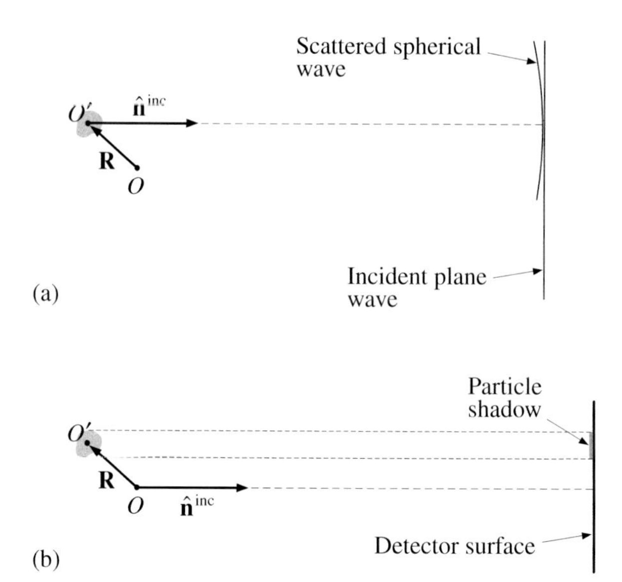

Figure 6.1.3. The geometrical shadow cast by the particle must be within the sensitive area of the detector facing the incident light.

It is obvious that if $R_V \gg a$ then the criteria (6.1.29) and (6.1.30) are stronger than the criteria (6.1.27) and (6.1.28), respectively, so that the latter can be neglected. Furthermore, comparison with Eqs. (3.2.17)–(3.2.19) shows that the criteria (6.1.26), (6.1.29), and (6.1.30) require the detector to be located in the far-field zone of the entire trap volume.

Finally, the approximate equality (6.1.13) used to derive Eq. (6.1.8) means that the angular pattern of light scattering by the particle is assumed to change insignificantly over the range of scattering directions equal to the angular size of the trap volume as viewed from the detector. Therefore, it follows from the discussion preceding Eq. (3.6.21) that the distance $r_{\rm D}$ from the volume element to the detector must obey the following additional inequality:

$$r_{\rm D} \gg \frac{2(R_V - a)k_1 a}{\pi}. \tag{6.1.31}$$

For a particle with size parameter $k_1 a$ significantly exceeding unity, the condition (6.1.31) becomes much more restrictive than the condition (6.1.29).

6.2 "Near-field" scattering

The approach described in the preceding section is based on the assumption that the detector is positioned so far from the origin of the laboratory reference frame O that for any position of the particle within the volume V the scattered wavefront at the detector is indistinguishable from that created by the particle centered at the origin of the laboratory reference frame. The advantage of this far-field approach is that it allows one to conveniently use the formalism of Chapter 3 without any change by invoking the concepts of O-centered phase and extinction matrices. The price that one has to pay for this simplicity is the condition (6.1.30), which can become very onerous if the radius of the volume element R_V significantly exceeds the wavelength. Therefore, it is desirable to develop an alternative "near-field" approach that would not require the far-field condition (6.1.30).

Let us first rewrite Eq. (6.1.3) in the form

$$\mathbf{E}^{\mathrm{sca}}(\mathbf{r}') = \frac{\exp(\mathrm{i}k_1 r')}{r'} \tensor{A}(\hat{\mathbf{r}}', \hat{\mathbf{n}}^{\mathrm{inc}}) \cdot \mathbf{E}_{0\mathbf{R}}^{\mathrm{inc}}, \tag{6.2.1}$$

where

$$\mathbf{E}_{0\mathbf{R}}^{\mathrm{inc}} = \exp(\mathrm{i}k_1 \hat{\mathbf{n}}^{\mathrm{inc}} \cdot \mathbf{R}) \mathbf{E}_0^{\mathrm{inc}}. \tag{6.2.2}$$

It is clear that Eq. (6.2.1) describes far-field scattering with respect to the particle reference frame. The only difference from the case studied in Chapter 3 is that the original amplitude of the electric (and thus the magnetic) field is multiplied by the exponential factor $\exp(\mathrm{i}k_1 \hat{\mathbf{n}}^{\mathrm{inc}} \cdot \mathbf{R})$. It is easy to see, however, that this factor has no effect on the final formulas of Sections 3.6–3.9 because it always gets multiplied by its own complex conjugate value and thereby disappears.

This is a very important result which shows that one can use the formulas of Sections 3.6–3.9 without any modification to describe the response of a detector with its sensitive surface centered at and normal to the position vector $\mathbf{r}'$ for any value of the particle position vector $\mathbf{R}$ (Fig. 6.2.1).

We will now use this result to quantify the response of the original detector with its sensitive surface ΔS centered on and normal to the unit vector $\hat{\mathbf{r}}$ originating at O (Figs. 6.1.1 and 6.2.2). The unit vector $\hat{\mathbf{r}}'$ originating at O' still points towards the center of the sensitive surface but is not normal to it. As a consequence, the particle sees an "effective" detector with a "sensitive surface" $\Delta S'$ centered at and normal to $\hat{\mathbf{r}}'$ such that $\Delta S' < \Delta S$ (Fig. 6.2.2). Let us, however, assume that $\hat{\mathbf{r}}'$ and $\hat{\mathbf{r}}$ are close enough that the approximate equality (6.1.13) holds and that

$$\Delta S' \approx \Delta S. \tag{6.2.3}$$

Furthermore, we assume that

$$r'_{\mathrm{D}} \approx r_{\mathrm{D}}, \tag{6.2.4}$$

where r'_{D} is the distance from O' to the center of the detector sensitive surface. It then becomes clear from the discussion in Sections 3.3 and 3.6 that if the detector is not facing the incident wave then its instantaneous response can be accurately described by Eqs. (3.7.6) and (3.7.8) in which $r = r_{\mathrm{D}}$ and the phase matrix is that with respect to the particle reference frame.

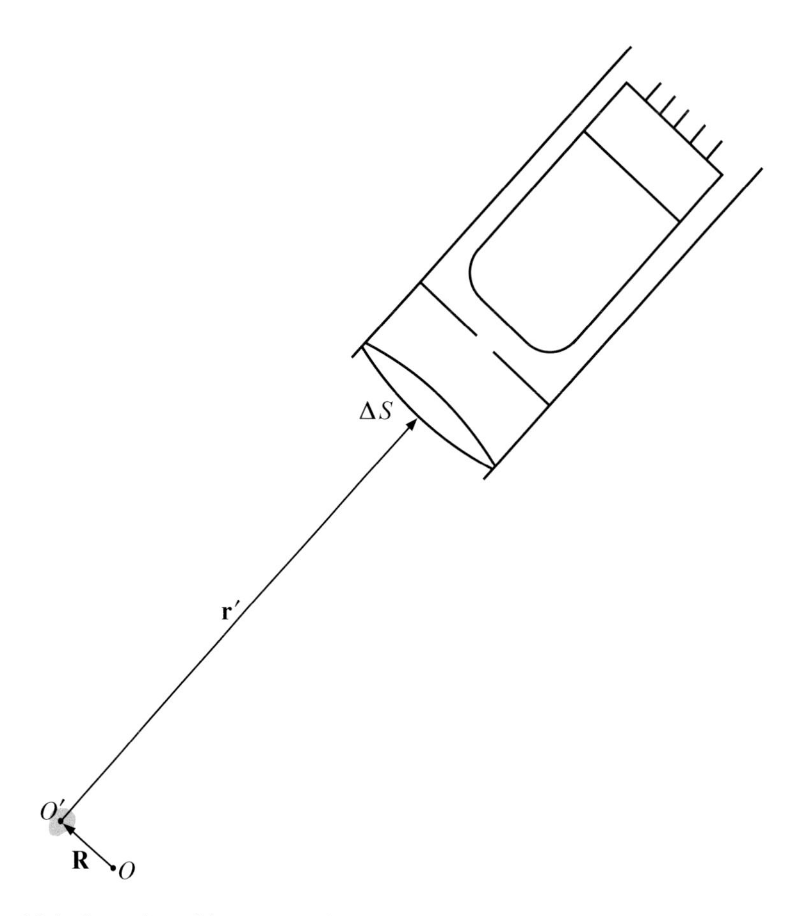

Figure 6.2.1. Scattering with respect to the particle origin.

To describe the response of the O-centered detector normal to the incidence direction ($\hat{\mathbf{r}} = \hat{\mathbf{n}}^{\mathrm{inc}}$) and provide for the possibility of a meaningful measurement of extinction, we further require that for any location of the particle within the volume element V, the geometrical shadow cast by the particle be within the sensitive surface of the detector (Fig. 6.1.3(b) and Eq. (6.1.25)). Then it follows from Sections 3.3 and 3.6 that the instantaneous response of the detector is accurately described by Eq. (3.8.5b), in which the extinction matrix is that with respect to the particle reference frame. The corresponding time-averaged detector responses are described by the ensemble averaged phase and extinction matrices $\langle \mathbf{Z}(\hat{\mathbf{r}}, \hat{\mathbf{n}}^{\mathrm{inc}}) \rangle_\xi$ and $\langle \mathbf{K}(\hat{\mathbf{n}}^{\mathrm{inc}}) \rangle_\xi$, respectively.

This is substantially the same result as that obtained in the preceding section. However, the conditions of applicability are now somewhat different. Indeed, it is easy to see that the inequalities (6.1.22)–(6.1.29) and (6.1.31) must still apply, but they must be supplemented by a new condition reflecting the fact that the observation point may now be so close to the origin of the laboratory reference frame that the light scattered by a particle located at the boundary of the trap volume can come to the

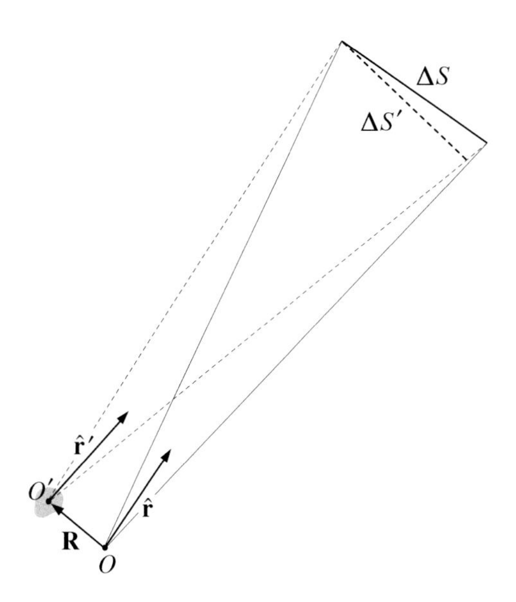

Figure 6.2.2. Scattering with respect to the origin of the laboratory reference frame.

observation point from a direction distinctly different from that originating at O. We must still require that all the light scattered by this off-centered particle and impinging on the detector surface be detected, which means that the entire trap volume as viewed from the detector must be within the detector angular aperture:

$$\frac{\pi(\frac{1}{2}D + R_V - a)^2}{r_\mathrm{D}^2} < \Delta\Omega. \tag{6.2.5}$$

On the other hand, the potentially most demanding condition of the far-field approach, Eq. (6.1.30), is now excluded. This means that the detector is allowed to be in the near-field zone of the volume element V, which justifies the title of this section. The detector must still be sufficiently distant in order to be in the far-field zone of the particle irrespective of its location within V (see Eqs. (6.1.26)–(6.1.28)).

Thus the net difference between the far-field and near-field approaches is that the conditions (6.1.22) and (6.1.30) are replaced by the condition (6.2.5). The relative importance of these conditions may vary depending on the specific measurement situation. Obviously, to apply Eqs. (6.1.20) and (6.1.21) one should verify whether:

- The conditions (6.1.23)–(6.1.29) and (6.1.31) are met *and*
- Either the combination of the inequalities (6.1.22) and (6.1.30) or the inequality (6.2.5) is satisfied.

Chapter 7

Single scattering by a small random particle group

The next problem in order of increasing complexity is electromagnetic scattering by a sparse group of particles randomly distributed through a small element of space. The concept of single scattering of light by a "differential" volume element has been central to the phenomenological theory of radiative transfer. With the development of the microphysical approach to radiative transfer (Chapter 8), the differential volume element has lost its long-cherished role as an elementary scattering unit in a macroscopic medium composed of a very large number of randomly positioned discrete particles. However, the concept of a small volume element filled with sparsely and randomly positioned particles remains a useful modeling tool in practical applications in which:

- The scattering medium is observed from a distance much greater than its maximal linear dimension.
- The total number of particles in the medium is insufficiently large to cause a significant multiple-scattering contribution to the total radiation leaving the medium in all directions.

A prime example of such applications is the analysis and interpretation of laboratory measurements of light scattering by tenuous collections of natural and artificial small particles (Section 9.3). Hence, the objective of this chapter is to discuss how one can model theoretically the response of a polarization-sensitive well-collimated detector placed at a large distance from a small volume element filled with randomly and sparsely distributed discrete scatterers (Mishchenko *et al.*, 2004b).

By analogy with the previous chapter, we will assume that the incident light is a plane electromagnetic wave. However, the results can be generalized easily to encompass the more general cases of illumination considered in Sections 3.10–3.12.

7.1 Single-scattering approximation for a fixed group of particles

We have seen in Section 4.1 that electromagnetic scattering by an arbitrary fixed group of N finite particles (Fig. 4.1.1) is rigorously described by the vector form of the Foldy–Lax equations (4.1.6)–(4.1.8). Let us now assume that the second term on the right-hand side of Eq. (4.1.7) is small in comparison with the first term (specific conditions under which this assumption holds will be discussed in Section 7.6). This means that each particle is excited only by the external incident field, which is the gist of the *single-scattering approximation* (SSA) for the fixed N-particle aggregate. We then have instead of Eq. (4.1.6):

$$\mathbf{E}(\mathbf{r}) = \mathbf{E}^{\mathrm{inc}}(\mathbf{r}) + \mathbf{E}^{\mathrm{sca}}(\mathbf{r}), \qquad \mathbf{r} \in \Re^3, \tag{7.1.1}$$

where the total scattered field is a vector sum of the partial scattered fields:

$$\mathbf{E}^{\mathrm{sca}}(\mathbf{r}) = \sum_{i=1}^{N} \mathbf{E}_i^{\mathrm{sca}}(\mathbf{r}), \tag{7.1.2}$$

$$\mathbf{E}_i^{\mathrm{sca}}(\mathbf{r}) = \int_{V_i} \mathrm{d}\mathbf{r}'\, \overleftrightarrow{G}(\mathbf{r}, \mathbf{r}') \cdot \int_{V_i} \mathrm{d}\mathbf{r}''\, \overleftrightarrow{T}_i(\mathbf{r}', \mathbf{r}'') \cdot \mathbf{E}^{\mathrm{inc}}(\mathbf{r}''). \tag{7.1.3}$$

It is clear that each partial field is independent of the partial fields scattered by all other particles forming the group (see Eq. (3.1.23)).

Let us choose the origin O of the laboratory coordinate system close to the geometrical center of the group, illuminate the fixed N-particle group by a plane electromagnetic wave incident in the direction of the unit vector $\hat{\mathbf{s}}$,

$$\mathbf{E}^{\mathrm{inc}}(\mathbf{r}) = \mathbf{E}_0^{\mathrm{inc}} \exp(\mathrm{i}k_1 \hat{\mathbf{s}} \cdot \mathbf{r}), \qquad \mathbf{E}_0^{\mathrm{inc}} \cdot \hat{\mathbf{s}} = 0, \tag{7.1.4}$$

assume that the observation point is located in the far-field zone of any particle forming the group (Fig. 7.1.1), and recall Eqs. (3.3.1) and (3.3.2). The latter indicate that the outgoing spherical wave generated by particle i in response to a plane-wave excitation of the form $\mathbf{E}_0^{\mathrm{inc}} \exp(\mathrm{i}k_1 \hat{\mathbf{s}} \cdot \mathbf{r}_i)$ in the far-field zone of this particle is given by

$$\frac{\exp(\mathrm{i}k_1 r_i)}{r_i} \overleftrightarrow{A}_i(\hat{\mathbf{r}}_i, \hat{\mathbf{s}}) \cdot \mathbf{E}_0^{\mathrm{inc}},$$

where $\mathbf{r}_i$ originates inside particle i (Fig. 7.1.1), $\overleftrightarrow{A}_i(\hat{\mathbf{r}}_i, \hat{\mathbf{s}})$ is the ith particle scattering dyadic centered at the particle origin, and $\hat{\mathbf{r}}_i = \mathbf{r}_i / r_i$ is the unit vector in the scattering direction. To make use of this fact, we must rewrite Eq. (7.1.4) in the following form:

$$\mathbf{E}^{\mathrm{inc}}(\mathbf{r}) = \mathbf{E}_0^{\mathrm{inc}} \exp(\mathrm{i}k_1 \hat{\mathbf{s}} \cdot \mathbf{r}_i) \exp(\mathrm{i}k_1 \hat{\mathbf{s}} \cdot \mathbf{R}_i), \tag{7.1.5}$$

where $\mathbf{R}_i$ connects the origin of the laboratory coordinate system with the origin of particle i (Fig. 7.1.1). This yields

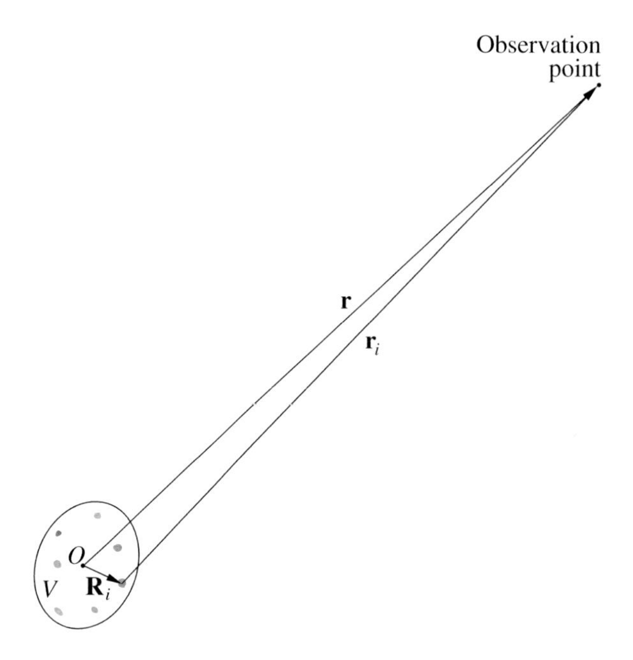

Figure 7.1.1. Far-field scattering by a group of particles occupying collectively a small volume element V.

$$\mathbf{E}_i^{\text{sca}}(\mathbf{r}) = \exp(\mathrm{i}k_1\hat{\mathbf{s}} \cdot \mathbf{R}_i)\frac{\exp(\mathrm{i}k_1 r_i)}{r_i}\overleftrightarrow{A}_i(\hat{\mathbf{r}}_i, \hat{\mathbf{s}}) \cdot \mathbf{E}_0^{\text{inc}}. \tag{7.1.6}$$

This formula is valid provided that the following inequalities hold for each particle of the group:

$$k_1(r_i - a_i) \gg 1, \tag{7.1.7}$$

$$r_i \gg a_i, \tag{7.1.8}$$

$$r_i \gg \frac{k_1 a_i^2}{2}, \tag{7.1.9}$$

where a_i is the smallest circumscribing sphere of particle i (cf. Eqs. (3.2.17)–(3.2.19)).

7.2 Far-field single-scattering approximation for a fixed particle group

Assuming that the observation point is located so far from the center of the particle group that $r \gg R_i$ for any i yields

$$r_i = |\mathbf{r} - \mathbf{R}_i|$$

$$= r\sqrt{1 - \frac{2\hat{\mathbf{r}} \cdot \mathbf{R}_i}{r} + \frac{R_i^2}{r^2}}$$

$$\approx r - \hat{\mathbf{r}} \cdot \mathbf{R}_i + \frac{R_i^2}{2r}. \tag{7.2.1}$$

Therefore,

$$\mathbf{E}_i^{\text{sca}}(\mathbf{r}) = \frac{\exp(\mathrm{i}k_1 r)}{r} \exp(\mathrm{i}\Delta_i) \overleftrightarrow{A}_i(\hat{\mathbf{r}}, \hat{\mathbf{s}}) \cdot \mathbf{E}_0^{\text{inc}}, \tag{7.2.2}$$

where

$$\Delta_i = k_1(\hat{\mathbf{s}} - \hat{\mathbf{r}}) \cdot \mathbf{R}_i, \tag{7.2.3}$$

and it is further assumed that

$$r \gg \frac{k_1 R_i^2}{2} \tag{7.2.4}$$

and

$$\overleftrightarrow{A}_i(\hat{\mathbf{r}}_i, \hat{\mathbf{s}}) \approx \overleftrightarrow{A}_i(\hat{\mathbf{r}}, \hat{\mathbf{s}}) \tag{7.2.5}$$

for any i. We can now rewrite Eq. (7.1.2) as

$$\mathbf{E}^{\text{sca}}(\mathbf{r}) = \frac{\exp(\mathrm{i}k_1 r)}{r} \overleftrightarrow{A}(\hat{\mathbf{r}}, \hat{\mathbf{s}}) \cdot \mathbf{E}_0^{\text{inc}}, \tag{7.2.6}$$

where the scattering dyadic of the entire group is given by

$$\overleftrightarrow{A}(\hat{\mathbf{r}}, \hat{\mathbf{s}}) = \sum_{i=1}^{N} \exp(\mathrm{i}\Delta_i) \overleftrightarrow{A}_i(\hat{\mathbf{r}}, \hat{\mathbf{s}}). \tag{7.2.7}$$

It is clear that Eq. (7.2.6) describes a transverse outgoing spherical wave centered at O. Exploiting the transverse character of the wave yields

$$\mathbf{E}^{\text{sca}}(\mathbf{r}) = \frac{\exp(\mathrm{i}k_1 r)}{r} \mathbf{S}(\hat{\mathbf{r}}, \hat{\mathbf{s}}) \cdot \mathbf{E}_0^{\text{inc}}, \tag{7.2.8}$$

where we have used the notation of Eq. (3.3.7) and expressed the total amplitude matrix of the group $\mathbf{S}(\hat{\mathbf{r}}, \hat{\mathbf{s}})$ in terms of the partial amplitude matrices $\mathbf{S}_i(\hat{\mathbf{r}}, \hat{\mathbf{s}})$ centered at the respective particle origins as follows:

$$\mathbf{S}(\hat{\mathbf{r}}, \hat{\mathbf{s}}) = \sum_{i=1}^{N} \exp(\mathrm{i}\Delta_i) \mathbf{S}_i(\hat{\mathbf{r}}, \hat{\mathbf{s}}). \tag{7.2.9}$$

This formula is based on the assumption that the orientations of the laboratory and particle-centered reference frames are the same.

The approximate equality (7.2.5) used to derive Eq. (7.2.6) means that the distance r from the center of the particle group to the observation point must satisfy the inequality $\pi/(2k_1 a_i) \gg L/2r$ (cf. Eq. (6.1.31)), where L is the maximal linear dimension of the volume element V occupied collectively by the group; we assume, for simplicity, that $L/2$ is much greater than any a_i. Furthermore, the assumption $r \gg R_i$ leads to the inequality $r \gg L/2$. Thus, our derivation and discussion show that the criteria of applicability of Eqs. (7.2.6) and (7.2.9) can be summarized as follows:

$$k_1 r \gg 1, \tag{7.2.10}$$

$$r \gg \frac{L}{2}, \tag{7.2.11}$$

$$r \gg \frac{k_1 L^2}{8}, \tag{7.2.12}$$

$$r \gg \frac{L k_1 a_i}{\pi} \quad \text{or} \quad r \gg \frac{2 L a_i}{\lambda_1} \quad \text{for } i = 1, \ldots, N, \tag{7.2.13}$$

where, as before, $\lambda_1 = 2\pi/k_1$ is the wavelength in the surrounding medium.

Equations (7.2.6)–(7.2.9) imply that the entire particle group behaves like an effective point-like scatterer generating a unified outgoing spherical wave and characterized by a cumulative scattering dyadic and a cumulative amplitude scattering matrix. It is, therefore, not surprising that the inequalities (7.2.10)–(7.2.12) are essentially equivalent to the criteria (3.2.17)–(3.2.19) of far-field scattering as applied to the particle group as a whole. Hence, Eqs. (7.2.6)–(7.2.9) summarize what can be called the *far-field single-scattering approximation* for the multi-particle group.

The critical advantage of the approximate formula (7.2.7) over the exact formula (3.3.5) is that the former provides a much simpler way to compute the scattering dyadic of the multi-particle group provided that the individual scattering dyadics of the component particles are known. Equation (7.2.9) can then be used to compute all observable scattering and absorption characteristics of the group introduced in Sections 3.6–3.9.

In particular, since the Δ_i vanish in the exact forward-scattering direction ($\hat{\mathbf{r}} = \hat{\mathbf{s}}$), substituting Eq. (7.2.9) in Eqs. (3.8.8)–(3.8.14) and Eq. (3.9.9) yields

$$\mathbf{K} = \sum_{i=1}^{N} \mathbf{K}_i, \tag{7.2.14}$$

$$C_{\text{ext}} = \sum_{i=1}^{N} C_{\text{ext},i}. \tag{7.2.15}$$

In other words, the extinction matrix and the extinction cross section of the fixed N-particle group in the framework of the far-field SSA are obtained by adding the respective optical characteristics of all the individual particles forming the group. One can also substitute Eq. (7.2.9) in Eqs. (3.7.11)–(3.7.26) and derive the corresponding formulas for the elements of the total Stokes phase matrix. However, we will not do that explicitly but rather will derive, in the following section, a formula for the total phase matrix under additional simplifying assumptions.

7.3 Far-field uncorrelated single-scattering approximation and modified uncorrelated single-scattering approximation

Let us now make two further assumptions:

- The N particles filling the volume element V (Fig. 7.1.1) move during the time necessary to take a measurement in such a way that their positions are random and uncorrelated with each other.
- The criteria of validity of Eqs. (7.2.8) and (7.2.9) are satisfied at each moment during the measurement.

Collectively, these assumptions define what can be called the far-field *uncorrelated single-scattering approximation* (USSA) for a small volume element. Obviously, these assumptions do not change Eqs. (7.2.14) and (7.2.15) since the latter are independent of the specific particle positions at any moment during the measurement. Therefore, Eqs. (7.2.14) and (7.2.15) are also the formulas for the time-averaged or, equivalently, configuration-averaged total extinction matrix and extinction cross section of the volume element:

$$\langle \mathbf{K} \rangle_{\mathbf{R}} = \sum_{i=1}^{N} \mathbf{K}_i, \tag{7.3.1}$$

$$\langle C_{\text{ext}} \rangle_{\mathbf{R}} = \sum_{i=1}^{N} C_{\text{ext},i}. \tag{7.3.2}$$

Our next step is to substitute Eq. (7.2.9) in Eqs. (3.7.11)–(3.7.26) and assume that the randomness of particle positions during the measurement leads to the following inequalities:

$$\left| \operatorname{Re} \sum_{i=1}^{N} \sum_{i'(\neq i)=1}^{N} [\mathbf{S}_i(\hat{\mathbf{r}}, \hat{\mathbf{s}})]_{kl} [\mathbf{S}_{i'}(\hat{\mathbf{r}}, \hat{\mathbf{s}})]^*_{pq} \langle \exp[\mathrm{i}(\Delta_i - \Delta_{i'})] \rangle_{\mathbf{R}} \right|$$

$$\ll \left| \operatorname{Re} \sum_{i=1}^{N} [\mathbf{S}_i(\hat{\mathbf{r}}, \hat{\mathbf{s}})]_{kl} [\mathbf{S}_i(\hat{\mathbf{r}}, \hat{\mathbf{s}})]^*_{pq} \right|, \tag{7.3.3}$$

and, if $k \neq p$ or $l \neq q$,

$$\left| \operatorname{Im} \sum_{i=1}^{N} \sum_{i'(\neq i)=1}^{N} [\mathbf{S}_i(\hat{\mathbf{r}}, \hat{\mathbf{s}})]_{kl} [\mathbf{S}_{i'}(\hat{\mathbf{r}}, \hat{\mathbf{s}})]^*_{pq} \langle \exp[\mathrm{i}(\Delta_i - \Delta_{i'})] \rangle_{\mathbf{R}} \right|$$

$$\ll \left| \operatorname{Im} \sum_{i=1}^{N} [\mathbf{S}_i(\hat{\mathbf{r}}, \hat{\mathbf{s}})]_{kl} [\mathbf{S}_i(\hat{\mathbf{r}}, \hat{\mathbf{s}})]^*_{pq} \right|, \qquad k, l, p, q = 1, 2. \tag{7.3.4}$$

Equation (7.2.3) suggests that for the left-hand sides of the inequalities (7.3.3) and (7.3.4) to vanish, the positions of particles i and i' must change randomly by a few wavelengths or more, thereby causing the real and imaginary parts of the factor $\exp[\mathrm{i}(\Delta_i - \Delta_{i'})]$ to vary randomly between -1 and $+1$. It is then straightforward to show that the configuration average of the total phase matrix of the volume element is also given by the "incoherent" sum of the partial phase matrices:

$$\langle \mathbf{Z} \rangle_{\mathbf{R}} = \sum_{i=1}^{N} \mathbf{Z}_i. \tag{7.3.5}$$

Finally, Eqs. (3.9.10), (3.9.11), and (7.2.15) yield the configuration-averaged total scattering and absorption cross sections of the volume element as sums of the respective partial optical characteristics:

$$\langle C_{\mathrm{sca}} \rangle_{\mathbf{R}} = \sum_{i=1}^{N} C_{\mathrm{sca},i}, \tag{7.3.6}$$

$$\langle C_{\mathrm{abs}} \rangle_{\mathbf{R}} = \sum_{i=1}^{N} C_{\mathrm{abs},i}. \tag{7.3.7}$$

Although the presence of the rapidly oscillating complex exponential factors indeed causes the left-hand sides of the inequalities (7.3.3) and (7.3.4) to vanish upon configurational averaging in most cases, it is clear that both inequalities are violated in the vicinity of the exact forward-scattering direction ($\hat{\mathbf{r}} \approx \hat{\mathbf{s}}$), when all the Δ_i vanish or become very small (see Eq. (7.2.3)) and all the factors $\exp[\mathrm{i}(\Delta_i - \Delta_{i'})]$ reduce to unity. This means that single scattering by constituent particles in directions close to the exact forward direction is always coherent or almost coherent irrespective of specific particle positions and must result in an additional enhancement of intensity due to constructive interference. Therefore, Eqs. (7.3.5)–(7.3.7) are *not* a direct consequence of the USSA, but rather are based on the USSA and the additional assumption that the forward-scattering interference can be neglected. The latter assumption, along with the USSA, defines the far-field *modified uncorrelated single-scattering approximation* (MUSSA) for a small volume element.

Equations (7.2.14), (7.2.15), and (7.3.5)–(7.3.7) are usually adopted without rigorous proof and form the basis for treating single scattering by random particle ensembles in virtually every book on light scattering and radiative transfer. It is clear from our detailed derivation that Eqs. (7.2.14) and (7.2.15) are a consequence of the simple far-field SSA as applied to any particle group, either fixed or random, whereas Eqs. (7.3.5)–(7.3.7) are strictly valid only in the framework of the far-field MUSSA.

Spatial coordinates are not the only particle characteristics that can vary with time.

In principle, the particles may also change their sizes, shapes, and/or orientations. A traditional approach in such cases is to assume that temporal changes of particle states are totally uncorrelated with temporal changes of their coordinates (Section 5.1). As a consequence, one may average the right-hand sides of Eqs. (7.2.14), (7.2.15), and (7.3.5)–(7.3.7) over the varying particle states and obtain the following formulas for the cumulative ensemble-averaged optical characteristics of the entire volume element:

$$\langle \mathbf{K} \rangle_{\mathbf{R},\xi} = N\langle \mathbf{K}_1 \rangle_\xi, \tag{7.3.8}$$

$$\langle \mathbf{Z} \rangle_{\mathbf{R},\xi} = N\langle \mathbf{Z}_1 \rangle_\xi, \tag{7.3.9}$$

$$\langle C_{\text{ext}} \rangle_{\mathbf{R},\xi} = N\langle C_{\text{ext},1} \rangle_\xi, \tag{7.3.10}$$

$$\langle C_{\text{sca}} \rangle_{\mathbf{R},\xi} = N\langle C_{\text{sca},1} \rangle_\xi, \tag{7.3.11}$$

$$\langle C_{\text{abs}} \rangle_{\mathbf{R},\xi} = N\langle C_{\text{abs},1} \rangle_\xi, \tag{7.3.12}$$

where the angular brackets on the right-hand side denote averages of the respective *single-particle* characteristics over the particle states.

7.4 Forward-scattering interference

To demonstrate the forward-scattering interference effect, Fig. 7.4.1 shows the element $\langle F_{11} \rangle_\text{o}$ of the scattering matrix for a simple two-sphere system in random orientation computed using the exact superposition T-matrix method (Mishchenko and Mackowski 1994). As will be discussed in greater detail in Chapter 11, the orientation-averaged scattering matrix is defined as

$$\langle \mathbf{F}(\Theta) \rangle_\text{o} = \langle \mathbf{Z}(\theta^{\text{sca}} = \Theta, \varphi^{\text{sca}} = 0; \theta^{\text{inc}} = 0, \varphi^{\text{inc}} = 0) \rangle_\text{o}, \tag{7.4.1}$$

which means that the plane through the incidence and scattering directions is used as a reference for defining the Stokes parameters of both the incident and the scattered light. Averaging over the uniform orientation distribution of a two-sphere cluster with a fixed distance between the components is intended to approximately model the randomness of the component-sphere positions. Also shown are the results for two equivalent spheres that scatter light in total isolation from each other.

It is clearly seen indeed that the main difference in the curves for two interacting spheres from those for two non-interacting spheres is the presence of a pronounced oscillating pattern at forward-scattering angles. To demonstrate unequivocally that the latter is caused by the interference, we note that, as follows from Eqs. (3.7.11)–(3.7.26), (7.2.3), and (7.2.9), the interference contributions ($i \neq i'$) to the total phase matrix of a two-sphere cluster differ from the incoherent contributions ($i = i'$) in that each of them includes an additional factor $\exp[\mathrm{i}k_1(\hat{\mathbf{s}} - \hat{\mathbf{r}}) \cdot (\mathbf{R}_1 - \mathbf{R}_2)]$ or $\exp[-\mathrm{i}k_1(\hat{\mathbf{s}} - \hat{\mathbf{r}}) \cdot (\mathbf{R}_1 - \mathbf{R}_2)]$, where $\mathbf{R}_1$ and $\mathbf{R}_2$ connect the origin of the laboratory coordinate

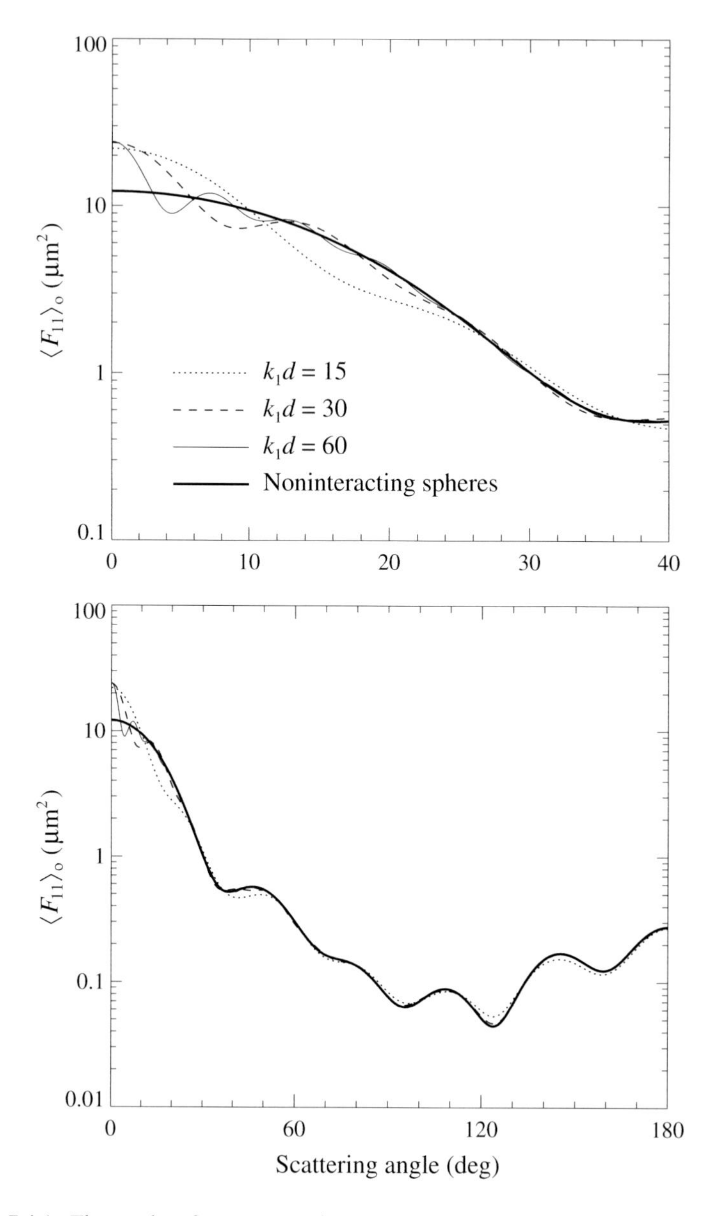

Figure 7.4.1. The results of exact T-matrix computations of the $\langle F_{11} \rangle_{\text{O}}$ element of the scattering matrix versus the scattering angle Θ for a two-sphere cluster in random orientation. The d is the distance between the centers of the component spheres and, for the three cases studied, increases such that the product $k_1 d$ grows from 15 to 60. The radius a of each sphere is 0.5 μm, their relative refractive index is 1.5, and the wavelength in the surrounding medium is 0.6283 μm. For comparison, the thick curves show $\langle F_{11} \rangle_{\text{O}}$ for two noninteracting spheres of the same size and relative refractive index.

system with the centers of spheres 1 and 2, respectively. By writing $\mathbf{R}_2 - \mathbf{R}_1 = \mathbf{d} = d\hat{\mathbf{d}}$, where d is the distance between the component sphere centers and the unit

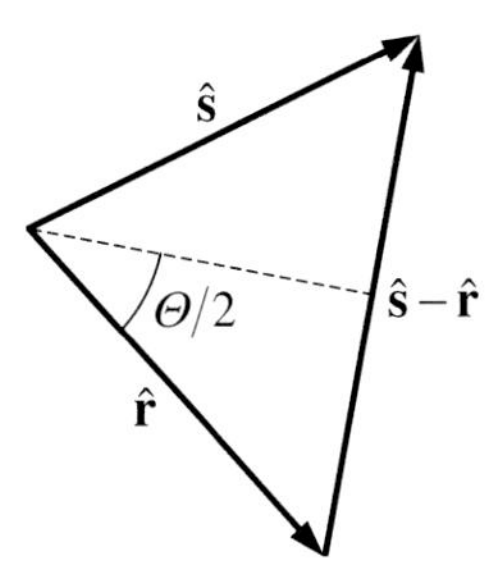

Figure 7.4.2. Illustration of the equality $|\hat{\mathbf{s}}-\hat{\mathbf{r}}| = 2\sin(\Theta/2)$.

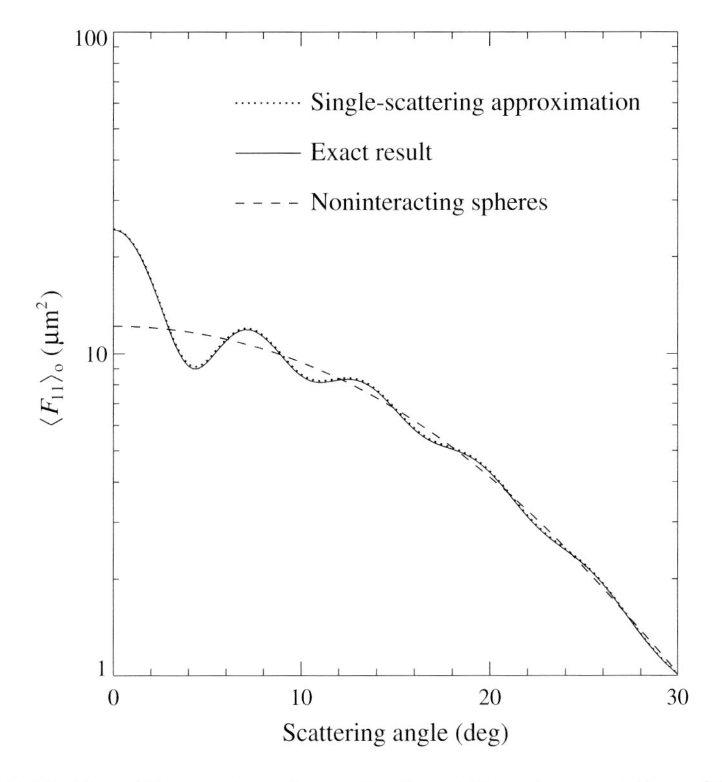

Figure 7.4.3. The solid curve shows the results of exact T-matrix computations of the $\langle F_{11}\rangle_{\mathrm{o}}$ element of the scattering matrix versus the scattering angle Θ for a two-sphere cluster in random orientation with $k_1 d = 60$, $a = 0.5\ \mu\mathrm{m}$, $m = 1.5$, and $\lambda_1 = 0.6283\ \mu\mathrm{m}$. For comparison, the dotted curve shows the result of using Eq. (7.4.5), whereas the dashed curve depicts the $\langle F_{11}\rangle_{\mathrm{o}}$ for two noninteracting spheres of the same size and relative refractive index.

vector $\hat{\mathbf{d}}$ specifies the cluster orientation, and averaging over all $\hat{\mathbf{d}}$, we derive

$$\frac{1}{4\pi}\int_{4\pi} \mathrm{d}\hat{\mathbf{d}} \exp[\mathrm{i}k_1 d(\hat{\mathbf{s}}-\hat{\mathbf{r}})\cdot\hat{\mathbf{d}}] = \frac{1}{4\pi}\int_{4\pi} \mathrm{d}\hat{\mathbf{d}} \exp[-\mathrm{i}k_1 d(\hat{\mathbf{s}}-\hat{\mathbf{r}})\cdot\hat{\mathbf{d}}]$$

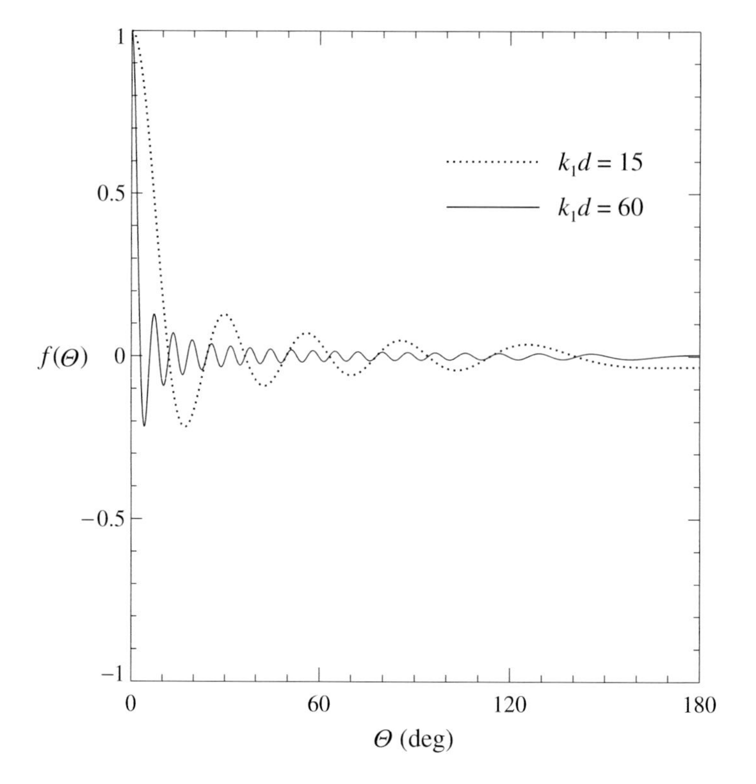

Figure 7.4.4. $f(\Theta)$ versus Θ for $k_1d = 15$ and 60.

$$= \frac{\sin(k_1 d\,|\hat{\mathbf{s}} - \hat{\mathbf{r}}|)}{k_1 d\,|\hat{\mathbf{s}} - \hat{\mathbf{r}}|}$$

$$= f(\Theta), \tag{7.4.2}$$

where $\Theta = \arccos(\hat{\mathbf{r}} \cdot \hat{\mathbf{s}})$ is the scattering angle and

$$f(\Theta) = \frac{\sin[2k_1 d \sin(\Theta/2)]}{2k_1 d \sin(\Theta/2)}, \tag{7.4.3}$$

since $|\hat{\mathbf{s}} - \hat{\mathbf{r}}| = 2\sin(\Theta/2)$ as shown in Fig. 7.4.2. Thus, the orientation-averaged total two-sphere phase and scattering matrices in the single-scattering approximation are given by

$$\langle \mathbf{Z}(\hat{\mathbf{r}}, \hat{\mathbf{s}}) \rangle_{\mathrm{o}} = 2\mathbf{Z}_1(\hat{\mathbf{r}}, \hat{\mathbf{s}})[1 + f(\Theta)], \tag{7.4.4}$$

$$\langle \mathbf{F}(\Theta) \rangle_{\mathrm{o}} = 2\mathbf{F}_1(\Theta)[1 + f(\Theta)], \tag{7.4.5}$$

where $\mathbf{Z}_1(\hat{\mathbf{r}}, \hat{\mathbf{s}})$ and $\mathbf{F}_1(\Theta)$ are the single-sphere phase and scattering matrices, respectively.

Figure 7.4.3 demonstrates that for a sufficiently large value of $k_1 d$, these simple formulas provide a nearly perfect fit to the exact *T*-matrix result. The $f(\Theta)$ has a sharp and narrow maximum at $\Theta = 0$ followed by a succession of maxima and minima with decreasing frequency and magnitude (see Fig. 7.4.4). The magnitude of all maxima and minima is inversely proportional to $k_1 d$ with the exception of the first interference maximum at $\Theta = 0$ whose magnitude is always equal to unity owing to the well-known limit

$$\frac{\sin x}{x} \underset{x \to 0}{\to} 1.$$

This explains the diminishing effect of the interference with increasing $k_1 d$ and Θ at side- and backscattering angles in Fig. 7.4.1.

7.5 Energy conservation

As we have already mentioned, the presence of the interference pattern at forward-scattering angles means that Eqs. (7.3.5)–(7.3.7) for the configuration-averaged total phase matrix and total scattering and absorption cross sections are only approximate consequences of the far-field USSA. Unfortunately, this also implies that the USSA violates the energy conservation law. Indeed, energy conservation requires that the total scattering cross section of the particle collection $\langle C_{\text{sca}} \rangle_{\mathbf{R}}$ be equal to the total extinction cross section $\langle C_{\text{ext}} \rangle_{\mathbf{R}}$ if all the constituent particles are nonabsorbing so that $C_{\text{sca},i} = C_{\text{ext},i}$ for each *i*. One can see that Eqs. (7.2.15) and (7.3.6) already lead to $\langle C_{\text{sca}} \rangle_{\mathbf{R}} = \langle C_{\text{ext}} \rangle_{\mathbf{R}}$ even though Eq. (7.3.6) does *not* include the contribution of the forward-scattering interference. Adding this contribution breaks the energy balance and leads to the unphysical result $\langle C_{\text{sca}} \rangle_{\mathbf{R}} \neq \langle C_{\text{ext}} \rangle_{\mathbf{R}}$.

The fact that the MUSSA satisfies the energy conservation law precisely whereas the presumably more accurate USSA does not seems to be rather strange. The explanation of this paradox is that the USSA includes two-particle electromagnetic interactions in the calculation of the total phase matrix and the total scattering cross section, but not in the calculation of the total extinction matrix and the total extinction cross section. It can in fact be shown that energy conservation would be restored if one were to take into account two-particle interactions in the calculation of $\langle \mathbf{K} \rangle_{\mathbf{R}}$ and $\langle C_{\text{ext}} \rangle_{\mathbf{R}}$ by including the contribution of light scattered twice, but this would go beyond the framework of the SSA. Therefore, the implicit (and not the best) way in which energy conservation is restored in the MUSSA is by neglecting artificially the forward-scattering interference.

7.6 Conditions of validity of the far-field modified uncorrelated single-scattering approximation

Let us now consider what happens with increasing average distance $\langle d \rangle$ between particles in a random group. Figure 7.4.1 shows that increasing the distance between two interacting spheres makes the main interference maximum narrower, whereas the $\langle F_{11} \rangle_o$ values at other scattering angles approach those obtained by doubling the corresponding single-sphere values. Also it is seen that the $\langle F_{11}(0) \rangle_o$ value for two interacting spheres remains approximately constant with varying distance between the sphere centers and is close to twice that computed for two non-interacting spheres, as it should be (the square of the sum of two equal electric fields is equal to twice the sum of the squares of the fields: $|\mathbf{E} + \mathbf{E}|^2 = 2(|\mathbf{E}|^2 + |\mathbf{E}|^2)$). Thus we can conclude that the expected consequences of taking the limit $k_1 \langle d \rangle \to \infty$ are the following:

- The total amount of energy contained in the interference pattern decreases with increasing interparticle distance and eventually becomes negligible compared to the total energy scattered by the particles.
- The angular width of the main interference peak becomes so small that the peak becomes hardly distinguishable from the incident beam.

Therefore, the MUSSA can be expected to give essentially the same results as the USSA provided that the particles are separated widely enough. This is a welcome conclusion since the MUSSA is significantly simpler than the USSA.

The first zero of the function $f(\Theta)$ occurs at $\Theta = \Theta_0 = 2\arcsin[\pi/(2k_1 d)]$. Therefore, to make the amount of energy contained in the interference pattern for a two-particle system negligibly small, this angle must be much smaller than π, which means that $k_1 d$ must be much greater than unity. Furthermore, it is well known that at least half of the energy scattered by large particles ($k_1 a \gg 1$) is contained in the narrow diffraction peak and mostly at scattering angles $\Theta < 4/(k_1 a)$ (see Section 7.4 of MTL). Therefore, we must also require that $\Theta_0 \ll 4/(k_1 a)$, which leads to $d \gg a$.

Although the forward-scattering interference pattern for a many-particle system can be significantly more complex than that shown in Figs. 7.4.1 and 7.4.3, it is clear that the conditions of validity of Eqs. (7.3.5)–(7.3.7) imposed by the presence of the interference pattern should be as follows:

$$\frac{k_1 L}{2} \gg 1, \tag{7.6.1}$$

$$\frac{L}{2} \gg a_i, \qquad i = 1, \ldots, N. \tag{7.6.2}$$

These inequalities reflect the obvious fact that the angular width of the forward-scattering interference peak generated by a many-particle group is controlled by the average distance between any two particles from the group rather than that between two neighboring particles. To ensure that particle positions are uncorrelated during the measurement (the position of each particle is not affected by the presence of the other particles), we must also require that the average distance $\langle d_n \rangle$ between neighboring particles be much greater than their sizes:

$$\langle d_n \rangle \gg a_i, \qquad i = 1, \ldots, N. \tag{7.6.3}$$

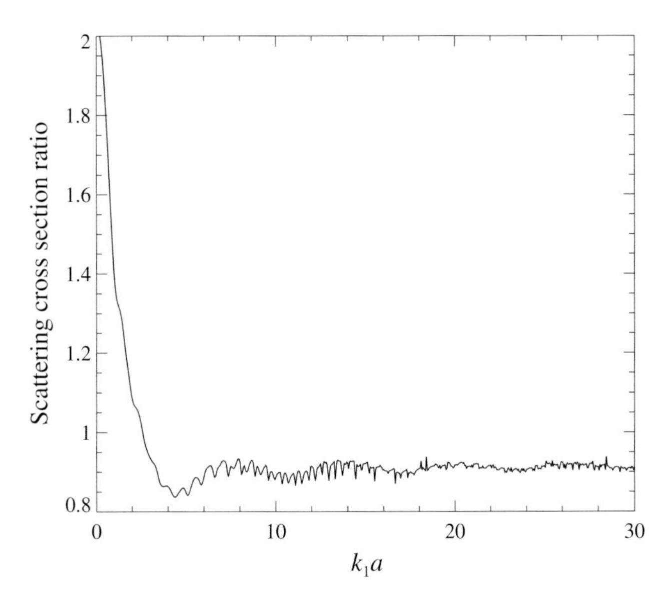

Figure 7.6.1. The ratio of the total scattering cross section for a two-particle cluster with identical touching components and in random orientation to the sum of the scattering cross sections of two noninteracting spheres of the same radius as a function of the sphere size parameter. The relative refractive index of the spheres is 1.5.

Let us now discuss the conditions of validity of the main assumption of the SSA, viz., that each particle is excited only by the incident field. First of all, it is obvious that the total amount of energy scattered by the particles filling a volume element must be much smaller than the amount of incident energy passing through the volume-element's geometrical cross section:

$$\sum_{i=1}^{N} C_{\text{sca},i} \ll L^2. \tag{7.6.4}$$

Besides this generic constraint, one may look at specific manifestations of close-proximity effects and how they behave with increasing interparticle separation. For example, if the line connecting the centers of two particles is nearly parallel to the incidence direction, then the field scattered by the particle located closer to the source of illumination can attenuate the incident field when it reaches the other particle. For particles much larger than the wavelength, this effect can be qualitatively interpreted as a "shadow" cast by the first particle upon the second particle.

To illustrate this phenomenon, Fig. 7.6.1 shows the results of *T*-matrix computations of the ratio ρ of the total scattering cross section for a two-particle cluster with identical touching components and in random orientation to the sum of the scattering cross sections of two non-interacting spheres as a function of the sphere size parameter $x = k_1 a$. In the geometrical optics limit, the scattering cross section of a nonabsorbing particle is equal to twice the area of the particle projection on the plane per-

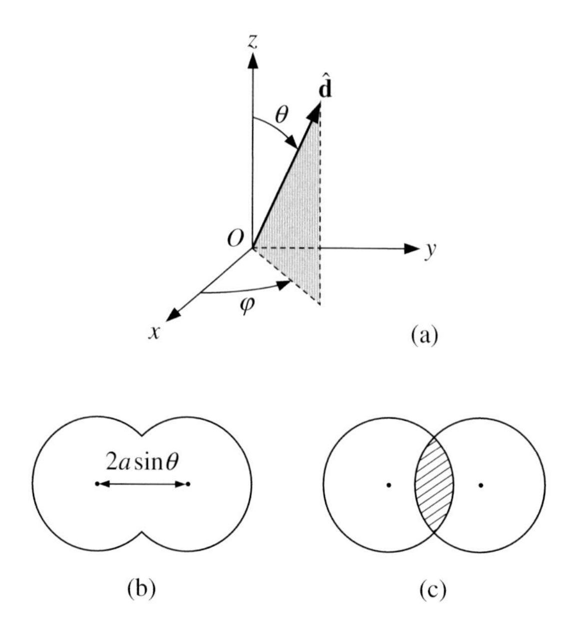

Figure 7.6.2. Computation of $\langle G \rangle_{\rm o}$ for a randomly oriented cluster consisting of two identical touching spheres.

pendicular to the incidence direction (see Section 7.4 of MTL). Therefore, in the limit $k_1 a \to \infty$ the ratio ρ should approach the value $\langle G \rangle_{\rm o}/(2\pi a^2)$, where $\langle G \rangle_{\rm o}$ is the orientation average of the projected area of the two-sphere cluster. Obviously, ρ would be very close to unity if the distance between the sphere centers were much greater than their radii, but should be significantly smaller than unity for touching spheres.

Figure 7.6.2 illustrates the computation of $\langle G \rangle_{\rm o}$ for the case of a randomly oriented two-sphere cluster with identical touching components. As before, the cluster orientation is specified by the direction of the unit vector $\hat{\mathbf{d}}$ or, equivalently, by its polar angle θ and azimuth angle φ (Fig. 7.6.2(a)). Let us assume for the sake of simplicity that the incident light propagates in the direction of the positive *z*-axis. Then the area of the bisphere projection onto the *xy*-plane is independent of the azimuth angle, so that

$$\begin{aligned}
\langle G \rangle_{\rm o} &= \frac{1}{4\pi} \int_{4\pi} \mathrm{d}\hat{\mathbf{d}}\, G(\hat{\mathbf{d}}) \\
&= \frac{1}{4\pi} \int_0^{2\pi} \mathrm{d}\varphi \int_0^{\pi} \mathrm{d}\theta \sin\theta\, G(\theta) \\
&= \int_0^{\pi/2} \mathrm{d}\theta \sin\theta\, G(\theta),
\end{aligned} \tag{7.6.5}$$

where $G(\theta)$ is the area of the shape shown in Fig. 7.6.2(b). Obviously, the latter is

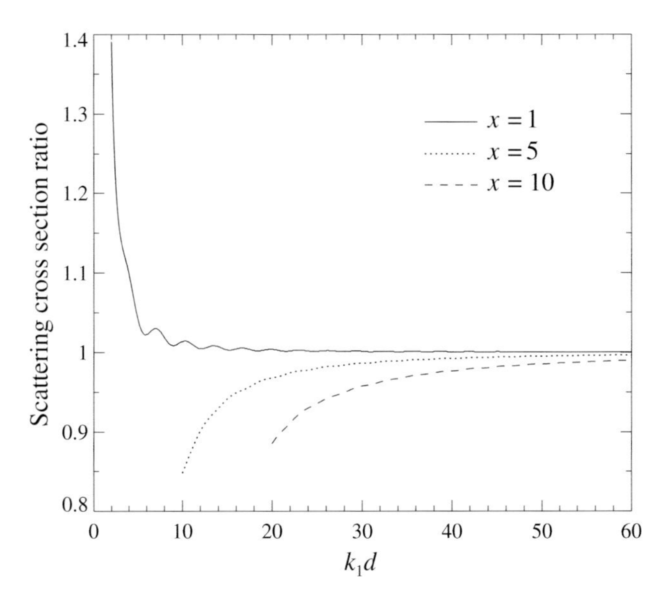

Figure 7.6.3. The ratio of the scattering cross section of a two-sphere cluster with equal components and in random orientation to the sum of the scattering cross sections of two noninteracting spheres of the same radius as a function of k_1d. The relative refractive index of the spheres is 1.5 and their size parameter $x = k_1a$ varies from 1 to 10.

equal to $2\pi a^2$ minus twice the common area of the two overlapping circles in Fig. 7.6.2(c), or $G(\theta) = a^2(\pi/2 - \theta - \sin\theta\cos\theta)$, thereby yielding $\langle G\rangle_{\mathrm{o}}/(2\pi a^2) = 1/2 + 4/(3\pi) \approx 0.9244$. The actual scattering cross section ratio in Fig. 7.6.1 indeed tends to this asymptotic value as the size parameter increases, thereby corroborating the presence and the importance of the shadowing effect.

Of course, the shadowing effect and the forward-scattering interference are not the only manifestations of the electromagnetic interaction between the particles forming a group and not the only factors that limit the accuracy of the far-field MUSSA and its range of applicability in terms of the smallest allowable interparticle separation. Unfortunately, it is difficult to perform a detailed theoretical analysis of this problem for many-particle groups consisting of arbitrary components. We hope, therefore, that exact numerical results for a few simple cases can provide at least qualitative guidance.

Figure 7.6.3 depicts the ratio of the total scattering cross section of a two-sphere cluster in random orientation to that of two noninteracting spheres of the same size. It is clear that in order for this ratio to be sufficiently close to unity, the distance between the centers of the interacting spheres must be at least several times greater than the sphere radii. The corresponding asymmetry parameter ratio (Fig. 7.6.4) is much closer to unity and is essentially independent of k_1d for the larger spheres with $x = 5$ and 10, but still requires interparticle distances $d \gg a$ for the spheres with $x = 1$ in order to reach the asymptotic value unity.

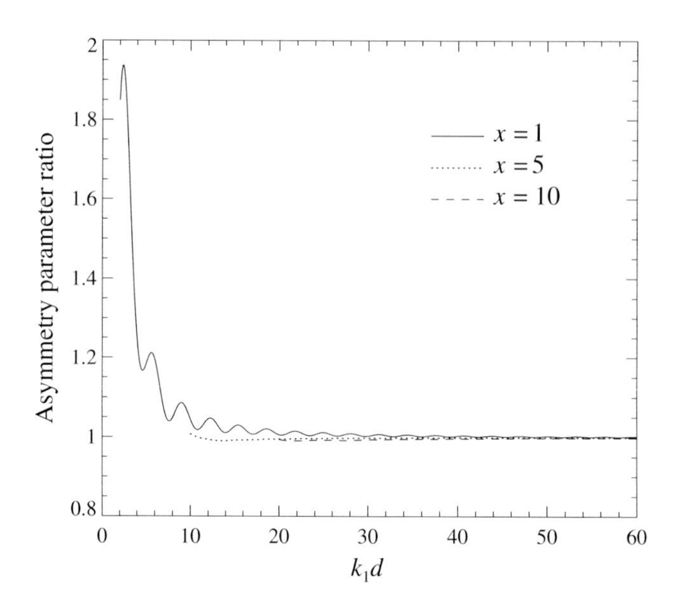

Figure 7.6.4. As in Fig. 7.6.3, but for the ratio of the asymmetry parameter of a two-sphere cluster with equal components and in random orientation to the asymmetry parameter of two noninteracting spheres of the same radius.

Figures 7.4.1 and 7.6.5 demonstrate how increasing the interparticle separation affects the scattering matrix element $\langle F_{11}\rangle_{\mathrm{o}}$ and the ratios $\langle F_{22}\rangle_{\mathrm{o}}/\langle F_{11}\rangle_{\mathrm{o}}$ and $-\langle F_{21}\rangle_{\mathrm{o}}/\langle F_{11}\rangle_{\mathrm{o}}$ for two interacting wavelength-sized spheres with a size parameter $x = 5$. The behavior of the ratio $\langle F_{22}\rangle_{\mathrm{o}}/\langle F_{11}\rangle_{\mathrm{o}}$ is especially revealing since it must be identically equal to unity for noninteracting spheres. Obviously, this asymptotic regime is approximately reached when the distance between the sphere centers exceeds several times their radii. We have seen before that no distance between the interacting spheres can eliminate the forward-scattering interference pattern (Fig. 7.4.1). However, this pattern becomes very narrow when d exceeds several times the sphere radii (or several times the wavelength for subwavelength-sized particles) and eventually becomes indistinguishable from the incident light. Although the data depicted in Figs. 7.4.1 and 7.6.3–7.6.5 were computed for two-sphere clusters with equal components, analogous T-matrix results for bispheres with different components (not shown here) exhibit the same basic features and lead to the same conclusions.

Our final note concerns the relative importance of the far-field-zone criteria (3.2.19) and (7.2.12) for a single component particle and for the entire particle group, respectively. For a single particle with a size parameter $k_1 a = 10$ the inequality (3.2.19) implies that the far-field zone begins at a distance from the particle much greater than five particle radii, which is not much stricter than the inequality (3.2.18). However, for a volume element with a size parameter $k_1 L/2 = 10^4$ the inequality (7.2.12) yields $r \gg 0.25 \times 10^4 L$, which moves the far-field zone much farther from the volume element than the inequalities (7.2.10) and (7.2.11) would require. This

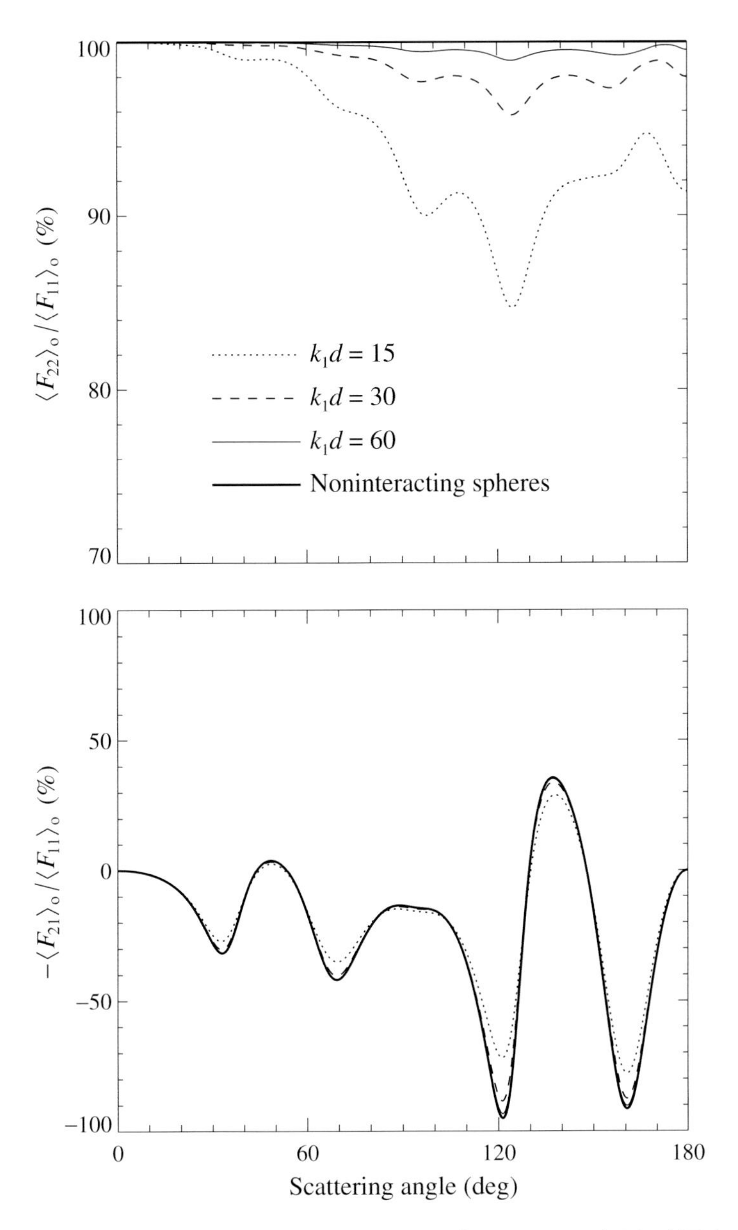

Figure 7.6.5. As in Fig. 7.4.1, but for the ratios $\langle F_{22}\rangle_o/\langle F_{11}\rangle_o$ and $-\langle F_{21}\rangle_o/\langle F_{11}\rangle_o$.

implies that if one wants to apply the MUSSA to a volume element with $L = 2$ mm assuming a source of illumination with a wavelength of $0.6283\,\mu\text{m}$ (thereby yielding $k_1L/2 = 10^4$), then the observation point must be moved from the volume element by many meters. However, the following section will demonstrate that in many circumstances, one can theoretically model the response of a detector located at a distance much greater than the volume element size but perhaps not as far as the inequality Eq. (7.2.12) would necessitate.

7.7 First-order-scattering approximation

In this section we will take another look at single scattering of light by a small volume element by assuming that detectors of the scattered light are located sufficiently far from the volume element that the inequalities (7.1.9), (7.2.10), (7.2.11), and (7.2.13) are satisfied, whereas the condition (7.2.12) will not be enforced. As a consequence, the volume element can no longer be considered at each moment in time as an effective point-like scatterer and characterized by a cumulative amplitude scattering matrix. Instead, it must be explicitly treated as a macroscopic random cloud of particles. The alternative approach described below will be based on the concept of the coherency dyad outlined in Section 3.10.

As before, we start with the SSA equations (7.1.1), (7.1.2), and (7.1.6). We then define the coherency dyad of the total electric field at the observation point as

$$\begin{aligned}\ddot{\rho}(\mathbf{r}) &= \mathbf{E}(\mathbf{r}) \otimes [\mathbf{E}(\mathbf{r})]^* \\ &= \mathbf{E}^{\text{inc}}(\mathbf{r}) \otimes [\mathbf{E}^{\text{inc}}(\mathbf{r})]^* + \mathbf{E}^{\text{inc}}(\mathbf{r}) \otimes \sum_{i=1}^{N} [\mathbf{E}_i^{\text{sca}}(\mathbf{r})]^* \\ &\quad + \sum_{i=1}^{N} \mathbf{E}_i^{\text{sca}}(\mathbf{r}) \otimes [\mathbf{E}^{\text{inc}}(\mathbf{r})]^* + \sum_{i=1}^{N} \mathbf{E}_i^{\text{sca}}(\mathbf{r}) \otimes \sum_{j(\neq i)=1}^{N} [\mathbf{E}_j^{\text{sca}}(\mathbf{r})]^* \\ &\quad + \sum_{i=1}^{N} \mathbf{E}_i^{\text{sca}}(\mathbf{r}) \otimes [\mathbf{E}_i^{\text{sca}}(\mathbf{r})]^* \end{aligned} \tag{7.7.1}$$

and assume that during the time necessary to take a measurement, the positions of all particles inside the volume V are totally random (Section 5.2). The latter assumption implies that the average distance between neighboring particles is much greater than the particle sizes, Eq. (7.6.3). Thus, the configuration-averaged coherency dyad is given by

$$\begin{aligned}\langle \ddot{\rho}(\mathbf{r}) \rangle_{\mathbf{R}} &= \mathbf{E}^{\text{inc}}(\mathbf{r}) \otimes [\mathbf{E}^{\text{inc}}(\mathbf{r})]^* + \mathbf{E}^{\text{inc}}(\mathbf{r}) \otimes \sum_{i=1}^{N} \langle [\mathbf{E}_i^{\text{sca}}(\mathbf{r})]^* \rangle_{\mathbf{R}} \\ &\quad + \sum_{i=1}^{N} \langle \mathbf{E}_i^{\text{sca}}(\mathbf{r}) \rangle_{\mathbf{R}} \otimes [\mathbf{E}^{\text{inc}}(\mathbf{r})]^* \\ &\quad + \sum_{i=1}^{N} \langle \mathbf{E}_i^{\text{sca}}(\mathbf{r}) \rangle_{\mathbf{R}} \otimes \sum_{j(\neq i)=1}^{N} \langle [\mathbf{E}_j^{\text{sca}}(\mathbf{r})]^* \rangle_{\mathbf{R}} \\ &\quad + \sum_{i=1}^{N} \langle \mathbf{E}_i^{\text{sca}}(\mathbf{r}) \otimes [\mathbf{E}_i^{\text{sca}}(\mathbf{r})]^* \rangle_{\mathbf{R}}, \end{aligned} \tag{7.7.2}$$

where the configuration averaging is performed assuming the probability distribution function (5.2.2). The first term on the right-hand side of this formula is the coherency dyad of the incident field, the second and third terms describe the interference of the incident and scattered fields, the fourth term describes the interference of the partial fields singly scattered by different particles, and the fifth term is the sum of the coherency dyads of the partial scattered fields.

Averaging the interference terms over particle positions involves the evaluation of the integrals

$$\begin{aligned}\langle \mathbf{E}_i^{\mathrm{sca}}(\mathbf{r})\rangle_{\mathbf{R}} &= \int_{\Re^3} \mathrm{d}\mathbf{R}_i p_{\mathbf{R}}(\mathbf{R}_i)\mathbf{E}_i^{\mathrm{sca}}(\mathbf{r}) \\ &= \int_{\Re^3} \mathrm{d}\mathbf{R}_i p_{\mathbf{R}}(\mathbf{R}_i)\exp(\mathrm{i}k_1\hat{\mathbf{s}}\cdot\mathbf{R}_i)\frac{\exp(\mathrm{i}k_1 r_i)}{r_i}\overleftrightarrow{A}_i(\hat{\mathbf{r}}_i,\hat{\mathbf{s}})\cdot\mathbf{E}_0^{\mathrm{inc}} \qquad (7.7.3)\end{aligned}$$

(cf. Eq. (7.1.6)), which give the average partial scattered fields at the observation point. It is convenient to perform the integration in the spherical coordinate system originating at the observation point (Fig. 7.7.1(a)). Taking into account that $\mathbf{R}_i = \mathbf{r} + \mathbf{R}'_i$, where the vector $\mathbf{R}'_i$ connects the observation point and particle i, and using the Saxon asymptotic expression (B.10), we obtain

$$\begin{aligned}\langle \mathbf{E}_i^{\mathrm{sca}}(\mathbf{r})\rangle_{\mathbf{R}} = \frac{\mathrm{i}2\pi}{k_1}\int_{4\pi}\mathrm{d}\hat{\mathbf{R}}'_i\int_0^{\infty}\mathrm{d}R'_i\,[\delta(\hat{\mathbf{s}}+\hat{\mathbf{R}}'_i) - \delta(\hat{\mathbf{s}}-\hat{\mathbf{R}}'_i)\exp(2\mathrm{i}k_1R'_i)] \\ \times p_{\mathbf{R}}(\mathbf{R}_i)\overleftrightarrow{A}_i(-\hat{\mathbf{R}}'_i,\hat{\mathbf{s}})\cdot\mathbf{E}^{\mathrm{inc}}(\mathbf{r}). \qquad (7.7.4)\end{aligned}$$

This formula shows that each average partial scattered field is contributed to only by those points of the volume element that belong to the segment $\Delta s(\mathbf{r})$ of the infinite straight line through the observation point and the source of illumination (Figs. 7.7.1(a) and 7.7.1(b)). Hence the following three situations must be considered: the observation point can either be behind the scattering volume as viewed from the source of illumination (e.g., point 1 in Fig. 7.7.1(b)), or between the source of illumination and the scattering volume (e.g., point 2), or lie on a line which is parallel to the incidence direction and does not go through the scattering volume (e.g., point 3).

It is obvious that $\langle \mathbf{E}_i^{\mathrm{sca}}(\mathbf{r}_3)\rangle_{\mathbf{R}}$ at point 3 is equal to zero and that the average field at point 1 is given by

$$\langle \mathbf{E}_i^{\mathrm{sca}}(\mathbf{r}_1)\rangle_{\mathbf{R}} = \frac{\mathrm{i}2\pi}{k_1 V}\Delta s(\mathbf{r}_1)\overleftrightarrow{A}_i(\hat{\mathbf{s}},\hat{\mathbf{s}})\cdot\mathbf{E}^{\mathrm{inc}}(\mathbf{r}_1). \qquad (7.7.5)$$

The radial integral for point 2 contains a rapidly oscillating factor $\exp(2\mathrm{i}k_1R'_i)$, which makes $\langle \mathbf{E}_i^{\mathrm{sca}}(\mathbf{r}_2)\rangle_{\mathbf{R}}$ much smaller than $\langle \mathbf{E}_i^{\mathrm{sca}}(\mathbf{r}_1)\rangle_{\mathbf{R}}$ provided that $k_1\Delta s(\mathbf{r}_2) \gg 1$. The latter condition is equivalent to the inequality (7.6.1). More fundamentally, the presence of the delta function $\delta(\hat{\mathbf{s}}-\hat{\mathbf{R}}'_i)$ implies the existence of the interference of the incident and backscattered fields, which is unphysical (recall the warning issued on p. 107). Thus we can conclude that $\langle \mathbf{E}_i^{\mathrm{sca}}(\mathbf{r})\rangle_{\mathbf{R}}$ is given by Eq. (7.7.5) if the observation point is "shadowed" by the volume element and vanishes otherwise.

It is clear from Eqs. (7.7.5) and (3.3.3) that the average partial field created by

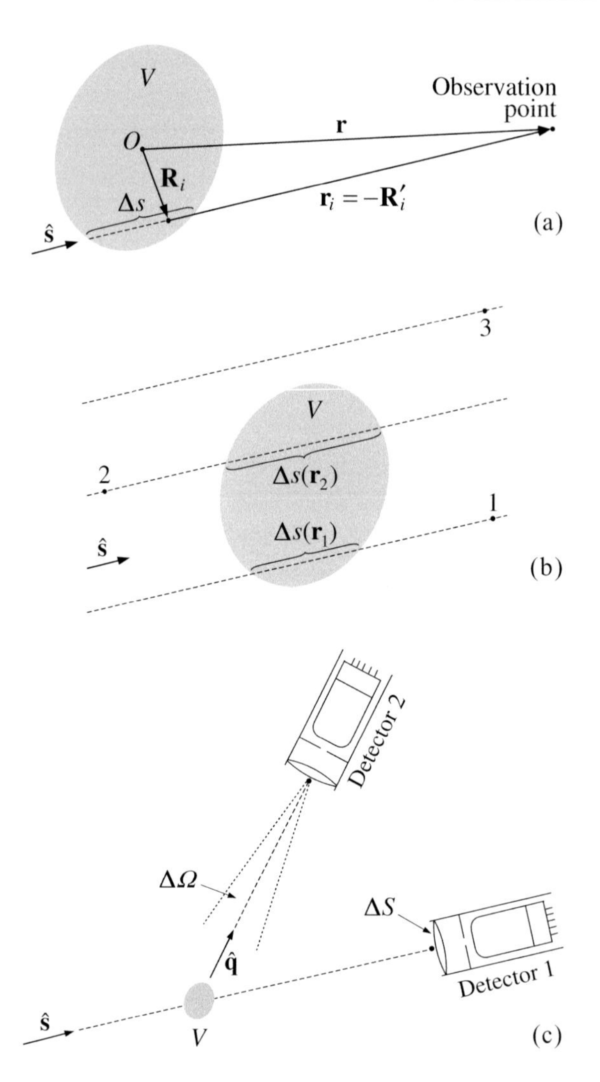

Figure 7.7.1. First-order scattering by a small volume element.

particle i at a "shadowed" distant observation point is a transverse plane wave propagating in the direction of the incident plane wave. Therefore, the second and third terms on the right-hand side of Eq. (7.7.2) describe the interference of pairs of transverse plane waves propagating in the same direction.

It follows from Eqs. (3.3.8)–(3.3.11), (3.8.8)–(3.8.14), and (3.9.9) that the factor

$$\frac{2\pi}{k_1 V}\Delta s(\mathbf{r}_1)\sum_{i=1}^{N}\vec{A}_i(\hat{\mathbf{s}},\hat{\mathbf{s}})$$

is of the same order of magnitude as the sum of the extinction cross sections of all the particles filling the volume element divided by the volume element's geometrical cross section:

$$\frac{1}{L^2}\sum_{i=1}^{N} C_{\text{ext},i}.$$

Assuming that this ratio is much smaller than unity,

$$\sum_{i=1}^{N} C_{\text{ext},i} \ll L^2, \tag{7.7.6}$$

we can neglect the fourth term on the right-hand side of Eq. (7.7.2) in comparison with the second and third terms.

Integrating the last term on the right-hand side of Eq. (7.7.2) over all particle positions and recalling the inequalities (7.2.11) and (7.2.13) yields

$$\frac{1}{r^2}\sum_{i=1}^{N}[\vec{A}_i(\hat{\mathbf{r}},\hat{\mathbf{s}})\cdot\mathbf{E}_0^{\text{inc}}]\otimes[\vec{A}_i(\hat{\mathbf{r}},\hat{\mathbf{s}})\cdot\mathbf{E}_0^{\text{inc}}]^*$$

$$= \frac{1}{r^2}\sum_{i=1}^{N}\vec{A}_i(\hat{\mathbf{r}},\hat{\mathbf{s}})\cdot\vec{\rho}^{\,\text{inc}}\cdot[\vec{A}_i(\hat{\mathbf{r}},\hat{\mathbf{s}})]^{\text{T}*},$$

where $\vec{\rho}^{\,\text{inc}}$ is the coherency dyad of the incident field. This is simply an "incoherent" sum of partial coherency dyads at the observation point, each partial dyad being due to a transverse spherical wave propagating in the same direction given by the unit vector $\hat{\mathbf{r}}$.

We can now make use of the transverse character of the plane and spherical waves involved in the first, second, third, and fifth terms on the right-hand side of Eq. (7.7.2) and rewrite this equation in terms of the Stokes vector using Eq. (2.12.3). After tedious but simple manipulations, we derive

$$\mathbf{I}(\mathbf{r}) = \mathbf{I}^{\text{inc}} - \frac{\Delta s(\mathbf{r})}{V}\sum_{i=1}^{N}\mathbf{K}_i(\hat{\mathbf{s}})\mathbf{I}^{\text{inc}} + \frac{1}{r^2}\sum_{i=1}^{N}\mathbf{Z}_i(\hat{\mathbf{s}},\hat{\mathbf{s}})\mathbf{I}^{\text{inc}} \tag{7.7.7}$$

if the observation point is shadowed by the volume element and

$$\mathbf{I}(\mathbf{r}) = \frac{1}{r^2}\sum_{i=1}^{N}\mathbf{Z}_i(\hat{\mathbf{r}},\hat{\mathbf{s}})\mathbf{I}^{\text{inc}} \tag{7.7.8}$$

otherwise. We will refer to the totality of approximations made in the derivation of Eqs. (7.7.7) and (7.7.8) as the first-order-scattering approximation (FOSA).

Let us now consider the measurement situation shown schematically in Fig. 7.7.1(c). The integration of Eqs. (7.7.7) and (7.7.8) over the acceptance area of the detectors shows that the polarized signal measured by detector 1 per unit time is given by

$$\textbf{Signal 1} = \Delta S\,\mathbf{I}^{\text{inc}} - \sum_{i=1}^{N}\mathbf{K}_i(\hat{\mathbf{s}})\mathbf{I}^{\text{inc}} + \frac{\Delta S}{r^2}\sum_{i=1}^{N}\mathbf{Z}_i(\hat{\mathbf{s}},\hat{\mathbf{s}})\mathbf{I}^{\text{inc}}, \tag{7.7.9}$$

whereas that measured by detector 2 per unit time is given by

$$\textbf{Signal 2} = \frac{\Delta S}{r^2} \sum_{i=1}^{N} \mathbf{Z}_i(\hat{\mathbf{q}}, \hat{\mathbf{s}}) \mathbf{I}^{\text{inc}}. \tag{7.7.10}$$

By choosing r to be sufficiently large, one can minimize the third term on the right-hand side of Eq. (7.7.19) relative to the second term. As a consequence, the response of detector 1 becomes

$$\textbf{Signal 1} \underset{r\to\infty}{=} \Delta S \mathbf{I}^{\text{inc}} - \sum_{i=1}^{N} \mathbf{K}_i(\hat{\mathbf{s}}) \mathbf{I}^{\text{inc}}. \tag{7.7.11}$$

Equations (7.7.10) and (7.7.11) represent the main result of the FOSA. Comparison with Eqs. (3.7.6), (3.7.8), (3.8.5b), (7.2.14), and (7.3.5) shows that the FOSA predicts essentially the same electromagnetic response of the distant detectors as the far-field MUSSA but without requiring that the detectors be placed as far from the volume element as to satisfy the inequality (7.2.12). However, since the volume element is now treated explicitly as a macroscopic object subtending a nonzero solid angle when viewed from the observation point, we must require that it be fully within the detector angular aperture $\Delta\Omega$. This implies that the distance r must be large enough to satisfy the inequality

$$\frac{L^2}{r^2} < \Delta\Omega. \tag{7.7.12}$$

This condition can be rather onerous in the case of a well-collimated detector. Like the MUSSA, the FOSA is based on ignoring the interference of light singly scattered by different particles in the forward direction (i.e., the fourth term on the right-hand side of Eq. (7.7.2)) and, as a consequence, satisfies the energy conservation law.

Assuming, as before, that the temporal changes of the particle states are uncorrelated with temporal changes of their coordinates, we obtain

$$\textbf{Signal 1} = \Delta S \mathbf{I}^{\text{inc}} - N\langle \mathbf{K}_1(\hat{\mathbf{s}})\rangle_\xi \mathbf{I}^{\text{inc}} + \frac{\Delta S}{r^2} N\langle \mathbf{Z}(\hat{\mathbf{s}}, \hat{\mathbf{s}})\rangle_\xi \mathbf{I}^{\text{inc}}$$

$$\underset{r\to\infty}{=} \Delta S \mathbf{I}^{\text{inc}} - N\langle \mathbf{K}_1(\hat{\mathbf{s}})\rangle_\xi \mathbf{I}^{\text{inc}}, \tag{7.7.13}$$

$$\textbf{Signal 2} = \frac{\Delta S}{r^2} N\langle \mathbf{Z}_1(\hat{\mathbf{q}}, \hat{\mathbf{s}})\rangle_\xi \mathbf{I}^{\text{inc}}, \tag{7.7.14}$$

where the angular brackets denote averages of the single-particle extinction and phase matrices over the particle states. Again, this is the same result as that predicted by the far-field MUSSA (cf. Eqs. (7.3.8) and (7.3.9)). It is also straightforward to verify that all formulas of this section remain unchanged if the volume element is illuminated by a parallel quasi-monochromatic beam rather than a plane electromagnetic wave.

7.8 Discussion

The traditional way to define the Stokes parameters applies only to transverse electromagnetic waves such as plane and spherical waves. It was, therefore, logical to start the analysis of single scattering by a small volume element using the far-field SSA, which treats the volume element at each moment in time as a unified scatterer generating a single outgoing spherical wave and makes possible the introduction of the cumulative amplitude scattering matrix.

An important result of our analysis of the far-field SSA applied to a random group of particles is that one must distinguish between the simple USSA and the MUSSA. The MUSSA satisfies the energy conservation law, is widely used in practice, and is a cornerstone of the phenomenological theory of radiative transfer. However, one should be aware of the fact that the MUSSA goes beyond the USSA by neglecting the interference of light scattered by various particles in the vicinity of the exact forward direction and thus may be inapplicable in circumstances involving precise computations or measurements at scattering angles approaching zero (e.g., Ivanov *et al.*, 1970). Otherwise, the MUSSA can be expected to give satisfactory results provided that the following conditions are met:

- The observation point is located far enough to satisfy the inequalities (7.2.10)–(7.2.13).
- The inequalities (7.6.1) and (7.6.2) are satisfied.
- Particle positions are uncorrelated, Eq. (7.6.3), and change by approximately a few wavelengths or more during the time interval necessary to take a measurement.
- The geometrical cross section of the volume element is much greater than its total scattering cross section, the inequality (7.6.4).
- The following inequalities analogous to the inequalities (6.1.22)–(6.1.24) are satisfied:

$$\frac{\Delta S}{r^2} < \Delta \Omega, \tag{7.8.1}$$

$$D \gg L, \tag{7.8.2}$$

$$r \gg \frac{D k_1 a_i}{\pi}, \tag{7.8.3}$$

where, as before, D is the diameter of the sensitive surface of the detector, ΔS is its area, and $\Delta \Omega$ is the detector angular aperture. As in Sections 3.6 and 6.1, the inequality (7.8.2) applies only to the detector facing the incident light. The inequality (7.8.3) must be valid for each particle of the group.

Since for large nonabsorbing particles the scattering cross section is approximately equal to twice the area of the particle geometrical cross section (e.g., Section 7.4 of MTL), the inequality (7.6.4) can be rewritten in the form

$$L \gg \langle a \rangle \sqrt{2\pi N}. \tag{7.8.4}$$

If the distance from the center of the volume element to the observation point does not satisfy the inequality (7.2.12), then the total field scattered by the volume element at a moment in time cannot be approximated by a single spherical wave. In this case, it is impossible to define the amplitude scattering matrix of the volume element as a whole, and a different approximate way to model the response of a detector measuring electromagnetic scattering by the small volume element is called for. One such approach is to apply the FOSA, which is based on the following assumptions:

- The observation point is located far enough to satisfy the inequalities (7.1.9), (7.2.10), (7.2.11), and (7.2.13).
- Particle positions within the volume element are completely random during the time interval necessary to take a measurement, Eqs. (5.2.2) and (7.6.3).
- The geometrical cross section of the volume element is much greater than its total scattering cross section, the inequality (7.6.4).
- The sum of the extinction cross sections of the particles filling the volume element is much smaller than the volume element geometrical cross section, Eq. (7.7.6). For particles larger than the wavelength, this assumption is roughly equivalent to the inequality (7.6.4).
- The inequalities (7.7.12) and (7.8.1)–(7.8.3) are satisfied.

We have demonstrated that if these conditions are met, then the FOSA leads to essentially the same result as the far-field MUSSA in terms of the response of a distant polarization-sensitive detector.

In summary, the far-field MUSSA and the FOSA can be viewed as alternative ways to model electromagnetic scattering by a small volume element filled with randomly distributed particles. The far-field MUSSA treats the entire volume element at each moment in time as an effective point-like scatterer, whereas the FOSA explicitly considers the volume element as a macroscopic random cloud of particles. However, both approximations give substantially the same result in terms of the polarization response of a sufficiently distant detector. This allows one to use Eqs. (7.7.10) and (7.7.11) whenever a specific scattering situation satisfies the conditions of applicability of either approximation.

Chapter 8

Radiative transfer equation

The radiative transfer theory originated more than a century ago in the papers by Lommel (1887) and Chwolson (1889). Since then analytical studies of the radiative transfer equation have become an independent discipline of mathematical physics and have resulted in numerous new techniques for solving integral and integro-differential equations. The RTE has also found remarkably diverse applications in a variety of science and engineering disciplines dealing with multiple scattering of light by randomly and sparsely distributed discrete particles. However, the usual way to introduce the RTE has been based on deceptively simple principles of phenomenological radiometry. This has led to the widespread ignorance of the fact that the real derivation of the RTE and the clarification of the physical meaning of all participating quantities must be based on fundamental principles of classical electromagnetics as applied to discrete random media.

During the past three decades, there has been significant progress in reconsideration of the RTT in terms of the statistical wave formalism (e.g., Barabanenkov, 1975; Ishimaru, 1978; Apresyan and Kravtsov, 1996; Tsang and Kong, 2001). This research has ultimately led to the RTE becoming a corollary of the electromagnetic theory (Mishchenko, 2002, 2003). Hence the aim of this chapter is to provide a detailed and systematic microphysical derivation of the RTE from first principles.

Our point of departure is the far-field version of the vector Foldy–Lax equations for a fixed N-particle system which allows one to represent the total electric field at an observation point as a superposition of the incident plane wave and N spherical waves centered at the particles. We will then assume that particle positions are completely random and will apply the so-called Twersky and ladder approximations to the average coherency dyad of the total electric field in the limit $N \to \infty$. Separate sections will provide a summary of approximations necessary to derive the RTE, a discussion of the physical meaning of all participating quantities, and a detailed comparison of the microphysical and phenomenological approaches to radiative transfer.

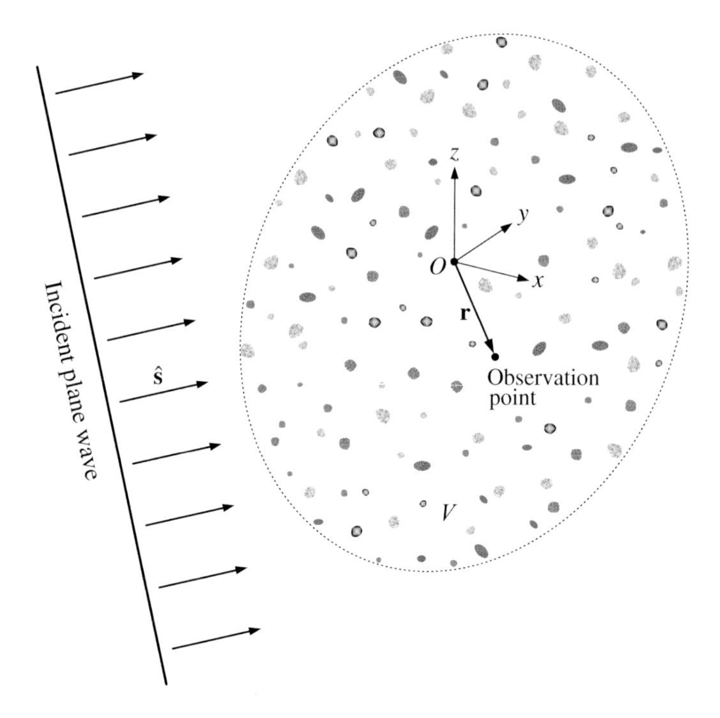

Figure 8.1.1. Electromagnetic scattering by a large number of discrete particles sparsely distributed throughout a macroscopic volume V.

8.1 The Twersky approximation

Let us consider electromagnetic scattering by a large group of particles imbedded in an infinite, homogeneous, isotropic, and nonabsorbing medium. The particles are sparsely distributed throughout a macroscopic volume V and are illuminated by a plane electromagnetic wave propagating in the direction of the unit vector $\hat{\mathbf{s}}$, Eq. (4.2.8) (see Fig. 8.1.1). We will assume that:

- The particles are separated widely enough that each of them is located in the far-field zones of all the other particles.
- The observation point is also located in the far-field zones of all the particles.

These assumptions make applicable the far-field version of the vector Foldy–Lax equations, Section 4.2.

Let us first rewrite Eqs. (4.2.14) and (4.2.11) in a compact symbolic form:

$$\mathbf{E} = \mathbf{E}^{\text{inc}} + \sum_{i=1}^{N} \vec{\vec{B}}_{ri0} \cdot \mathbf{E}_i^{\text{inc}} + \sum_{i=1}^{N} \sum_{j(\neq i)=1}^{N} \vec{\vec{B}}_{rij} \cdot \mathbf{E}_{ij}, \tag{8.1.1}$$

$$\mathbf{E}_{ij} = \vec{\vec{B}}_{ij0} \cdot \mathbf{E}_j^{\text{inc}} + \sum_{l(\neq j)=1}^{N} \vec{\vec{B}}_{ijl} \cdot \mathbf{E}_{jl}, \tag{8.1.2}$$

where N is the total number of the particles,

$$\mathbf{E} = \mathbf{E}(\mathbf{r}), \tag{8.1.3}$$

$$\mathbf{E}^{\text{inc}} = \mathbf{E}^{\text{inc}}(\mathbf{r}), \tag{8.1.4}$$

$$\mathbf{E}_i^{\text{inc}} = \mathbf{E}^{\text{inc}}(\mathbf{R}_i), \tag{8.1.5}$$

$$\overleftrightarrow{B}_{ri0} = G(r_i)\overleftrightarrow{A}_i(\hat{\mathbf{r}}_i, \hat{\mathbf{s}}), \tag{8.1.6}$$

$$\overleftrightarrow{B}_{rij} = G(r_i)\overleftrightarrow{A}_i(\hat{\mathbf{r}}_i, \hat{\mathbf{R}}_{ij}), \tag{8.1.7}$$

$$\overleftrightarrow{B}_{ij0} = G(R_{ij})\overleftrightarrow{A}_j(\hat{\mathbf{R}}_{ij}, \hat{\mathbf{s}}), \tag{8.1.8}$$

$$\overleftrightarrow{B}_{ijl} = G(R_{ij})\overleftrightarrow{A}_j(\hat{\mathbf{R}}_{ij}, \hat{\mathbf{R}}_{jl}). \tag{8.1.9}$$

The notation on the right-hand sides of Eqs. (8.1.5)–(8.1.9) follows that introduced in Section 4.2. Iterating Eq. (8.1.2) yields

$$\begin{aligned}\mathbf{E}_{ij} = {} & \overleftrightarrow{B}_{ij0}\cdot\mathbf{E}_j^{\text{inc}} + \sum_{\substack{l=1\\ l\neq j}}^{N}\overleftrightarrow{B}_{ijl}\cdot\overleftrightarrow{B}_{jl0}\cdot\mathbf{E}_l^{\text{inc}} + \sum_{\substack{l=1\\ l\neq j}}^{N}\sum_{\substack{m=1\\ m\neq l}}^{N}\overleftrightarrow{B}_{ijl}\cdot\overleftrightarrow{B}_{jlm}\cdot\overleftrightarrow{B}_{lm0}\cdot\mathbf{E}_m^{\text{inc}}\\ & + \sum_{\substack{l=1\\ l\neq j}}^{N}\sum_{\substack{m=1\\ m\neq l}}^{N}\sum_{\substack{n=1\\ n\neq m}}^{N}\overleftrightarrow{B}_{ijl}\cdot\overleftrightarrow{B}_{jlm}\cdot\overleftrightarrow{B}_{lmn}\cdot\overleftrightarrow{B}_{mn0}\cdot\mathbf{E}_n^{\text{inc}} + \cdots,\end{aligned} \tag{8.1.10}$$

whereas substituting Eq. (8.1.10) in Eq. (8.1.1) gives an order-of-scattering expansion of the total electric field:

$$\begin{aligned}\mathbf{E} = {} & \mathbf{E}^{\text{inc}} + \sum_{i=1}^{N}\overleftrightarrow{B}_{ri0}\cdot\mathbf{E}_i^{\text{inc}} + \sum_{i=1}^{N}\sum_{\substack{j=1\\ j\neq i}}^{N}\overleftrightarrow{B}_{rij}\cdot\overleftrightarrow{B}_{ij0}\cdot\mathbf{E}_j^{\text{inc}}\\ & + \sum_{i=1}^{N}\sum_{\substack{j=1\\ j\neq i}}^{N}\sum_{\substack{l=1\\ l\neq j}}^{N}\overleftrightarrow{B}_{rij}\cdot\overleftrightarrow{B}_{ijl}\cdot\overleftrightarrow{B}_{jl0}\cdot\mathbf{E}_l^{\text{inc}}\\ & + \sum_{i=1}^{N}\sum_{\substack{j=1\\ j\neq i}}^{N}\sum_{\substack{l=1\\ l\neq j}}^{N}\sum_{\substack{m=1\\ m\neq l}}^{N}\overleftrightarrow{B}_{rij}\cdot\overleftrightarrow{B}_{ijl}\cdot\overleftrightarrow{B}_{jlm}\cdot\overleftrightarrow{B}_{lm0}\cdot\mathbf{E}_m^{\text{inc}} + \cdots\end{aligned} \tag{8.1.11}$$

(cf. Twersky 1964). Indeed, the first term on the right-hand side of Eq. (8.1.11) is the incident field, the second term is the sum of all single-scattering contributions, the third term is the sum of all double-scattering contributions, etc., as shown schematically in Fig. 8.1.2.

The terms with $j=i$ and $l=j$ in the triple summation on the right-hand side of Eq. (8.1.11) are excluded, but the terms with $l=i$ are not. Therefore, we can decompose this summation as follows:

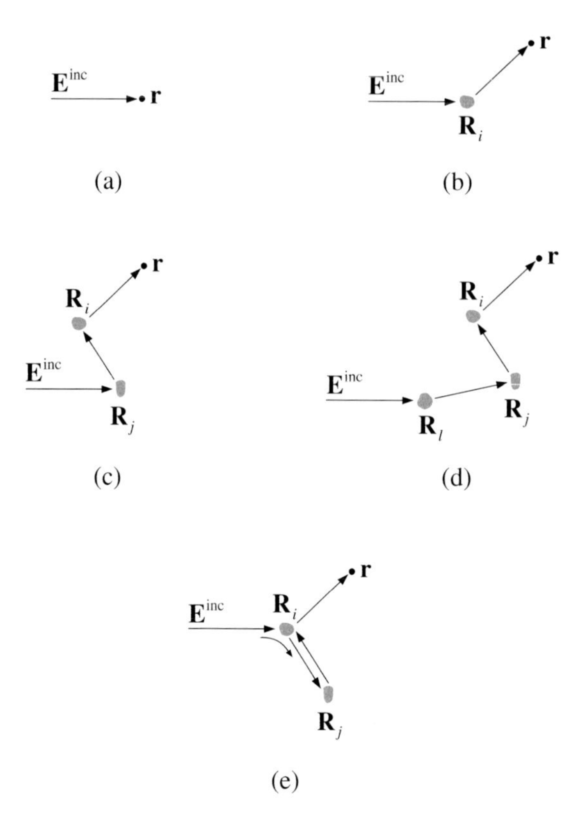

Figure 8.1.2. (a) Incident field, (b) single scattering, (c) double scattering, (d) triple scattering through a self-avoiding path, and (e) triple scattering through a path that goes through particle i twice.

$$\sum_{i=1}^{N}\sum_{\substack{j=1\\ j\neq i}}^{N}\sum_{\substack{l=1\\ l\neq j}}^{N} \overleftrightarrow{B}_{rij}\cdot\overleftrightarrow{B}_{ijl}\cdot\overleftrightarrow{B}_{jl0}\cdot\mathbf{E}_l^{\text{inc}} = \sum_{i=1}^{N}\sum_{\substack{j=1\\ j\neq i}}^{N}\sum_{\substack{l=1\\ l\neq i\\ l\neq j}}^{N} \overleftrightarrow{B}_{rij}\cdot\overleftrightarrow{B}_{ijl}\cdot\overleftrightarrow{B}_{jl0}\cdot\mathbf{E}_l^{\text{inc}} + \sum_{i=1}^{N}\sum_{\substack{j=1\\ j\neq i}}^{N} \overleftrightarrow{B}_{rij}\cdot\overleftrightarrow{B}_{iji}\cdot\overleftrightarrow{B}_{ji0}\cdot\mathbf{E}_i^{\text{inc}}. \quad (8.1.12)$$

The triple summation on the right-hand side of Eq. (8.1.12) is illustrated in Fig. 8.1.2(d) and includes scattering paths going through a particle only once (so-called self-avoiding paths), whereas the double summation involves the paths that go through the same particle more than once, as shown schematically in Fig. 8.1.2(e). Higher-order summations in Eq. (8.1.11) can be decomposed similarly.

Hence, the total field at an observation point $\mathbf{r}$ is composed of the incident field and single- and multiple-scattering contributions that can be divided into two groups. The first one includes all the terms that correspond to self-avoiding scattering paths,

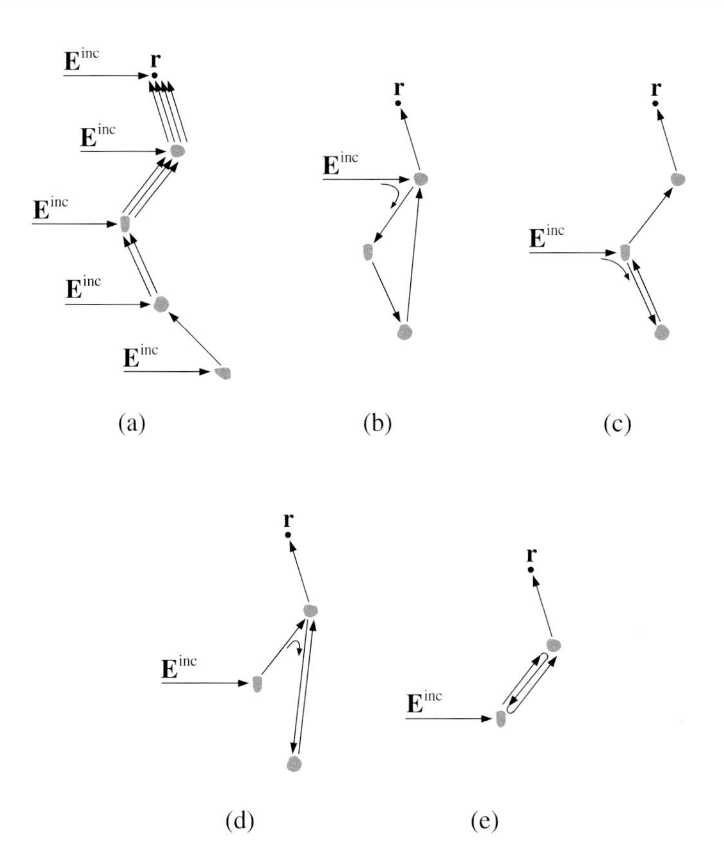

Figure 8.1.3. (a) Self-avoiding scattering paths and (b)–(e) paths involving four scattering events and going through a particle more than once.

Fig. 8.1.3(a), whereas the second group includes all the terms corresponding to the paths that go through a scatterer more than once, Fig. 8.1.3(b)–8.1.3(e).

The so-called Twersky approximation neglects the terms belonging to the second group and retains only the terms from the first group:

$$
\begin{aligned}
\mathbf{E} \approx\ & \mathbf{E}^{\mathrm{inc}} + \sum_{i=1}^{N} \ddot{B}_{ri0} \cdot \mathbf{E}_i^{\mathrm{inc}} + \sum_{i=1}^{N} \sum_{\substack{j=1 \\ j \neq i}}^{N} \ddot{B}_{rij} \cdot \ddot{B}_{ij0} \cdot \mathbf{E}_j^{\mathrm{inc}} \\
& + \sum_{i=1}^{N} \sum_{\substack{j=1 \\ j \neq i}}^{N} \sum_{\substack{l=1 \\ l \neq i \\ l \neq j}}^{N} \ddot{B}_{rij} \cdot \ddot{B}_{ijl} \cdot \ddot{B}_{jl0} \cdot \mathbf{E}_l^{\mathrm{inc}} \\
& + \sum_{i=1}^{N} \sum_{\substack{j=1 \\ j \neq i}}^{N} \sum_{\substack{l=1 \\ l \neq i \\ l \neq j}}^{N} \sum_{\substack{m=1 \\ m \neq i \\ m \neq j \\ m \neq l}}^{N} \ddot{B}_{rij} \cdot \ddot{B}_{ijl} \cdot \ddot{B}_{jlm} \cdot \ddot{B}_{lm0} \cdot \mathbf{E}_m^{\mathrm{inc}} + \cdots
\end{aligned}
\tag{8.1.13}
$$

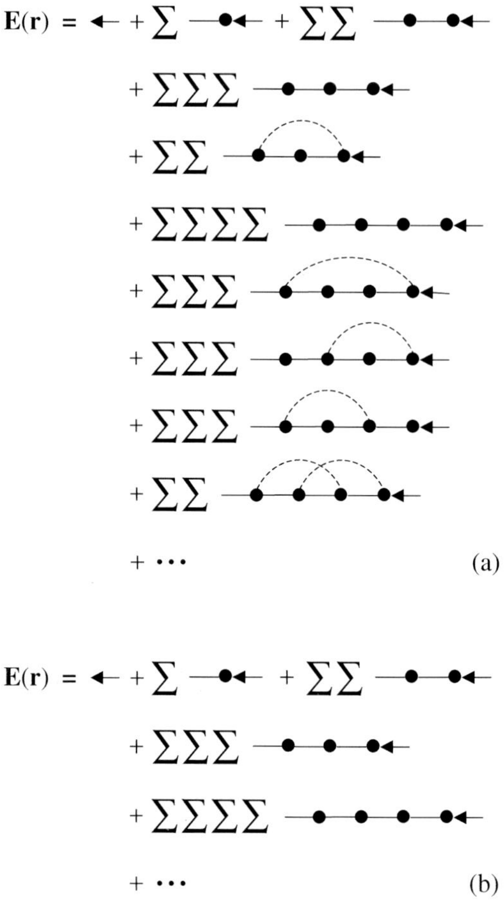

Figure 8.1.4. Diagrammatic representations of (a) Eq. (8.1.11) and (b) Eq. (8.1.13).

(Twersky, 1964). The physical meaning of the Twersky approximation is rather transparent. Indeed, a close look at Eq. (8.1.13) shows that the electric field exciting each particle i ($i = 1, \ldots, N$) is now replaced by the total electric field that would exist at the origin of particle i if this particle were removed from the group. We will see in the following sections that switching from the full order-of-scattering expansion (8.1.11) to the partial Twersky expansion (8.1.13) is a crucial step in the derivation of the RTE.

It is straightforward to show that for a large N, the Twersky approximation includes the majority of multiple-scattering paths. Specifically, an L-fold summation with $L > 2$ on the right-hand side of the exact expansion (8.1.11) contains $N(N-1)^{L-1}$ terms, whereas that in the approximate expansion (8.1.13) contains $N!/(N-L)!$ terms. The ratio of these two numbers indeed tends to unity as $N \to \infty$, which suggests that one can expect the Twersky approximation to yield rather accurate results provided that the number of particles is sufficiently large.

It is convenient to represent order-of-scattering expansions of the electric field using the diagram method. Panel (a) of Fig. 8.1.4 visualizes the full expansion (8.1.11), whereas panel (b) illustrates the Twersky approximation (8.1.13). The arrows in these diagrams represent the incident field, the symbol —● denotes multiplying a field by a $\vec{B}$ dyadic, and the dashed curve indicates that two scattering events involve the same particle.

8.2 The Twersky expansion of the coherent field

Let us now assume that the particles filling the volume V are randomly moving and consider the field $\mathbf{E}(\mathbf{r})$ at an internal point $\mathbf{r} \in V$. In general, $\mathbf{E}(\mathbf{r})$ varies (fluctuates) in time because of the random temporal variations of particle coordinates and states, albeit at a much slower rate than the time-harmonic factor $\exp(-\mathrm{i}\omega t)$. A typical measurement takes a significant amount of time during which the electromagnetic signal is averaged over a representative set of particle positions and states. Therefore, it is often convenient to decompose $\mathbf{E}(\mathbf{r})$ into the average (or coherent) part $\mathbf{E}_\mathrm{c}(\mathbf{r})$ and the fluctuating part $\mathbf{E}_\mathrm{f}(\mathbf{r})$:

$$\mathbf{E}(\mathbf{r}) = \mathbf{E}_\mathrm{c}(\mathbf{r}) + \mathbf{E}_\mathrm{f}(\mathbf{r}), \tag{8.2.1}$$

where, upon assuming that the particle ensemble is fully ergodic,

$$\mathbf{E}_\mathrm{c}(\mathbf{r}) = \langle \mathbf{E}(\mathbf{r}) \rangle_t = \langle \mathbf{E}(\mathbf{r}) \rangle_{\mathbf{R},\xi}, \tag{8.2.2}$$

$$\langle \mathbf{E}_\mathrm{f}(\mathbf{r}) \rangle_t = \langle \mathbf{E}_\mathrm{f}(\mathbf{r}) \rangle_{\mathbf{R},\xi} = \mathbf{0}. \tag{8.2.3}$$

The statistical averaging is performed over those coordinates and states of all the particles that are physically realizable during the time of the measurement.

It is very important to recognize that the coherent field $\mathbf{E}_\mathrm{c}(\mathbf{r})$ thus defined is not a real physical field but rather is a purely mathematical construction. Indeed, if we restore the time-harmonic factor $\exp(-\mathrm{i}\omega t)$, which we have been omitting so far for the sake of brevity, then we must conclude that the time average of the actual electric field is equal to zero, Eq. (1.2.1). In contrast, the coherent field does not vanish because it is defined as the time average of the part of the electric field that does not include the factor $\exp(-\mathrm{i}\omega t)$. The only reason to introduce the coherent field in the first place is that it will eventually appear in formulas for quantities that describe the multiply scattered radiation and can be actually measured with a suitable optical device. These quantities are defined in such a way that the factor $\exp(-\mathrm{i}\omega t)$ naturally disappears upon multiplication by its complex-conjugate counterpart.

Assuming that all particles have the same statistical characteristics and that the state of each particle is independent of its coordinates, we have from Eqs. (8.1.13) and (5.1.11):

$$\begin{aligned}
\mathbf{E}_{\mathrm{c}} &= \mathbf{E}^{\mathrm{inc}} + \sum_{i=1}^{N} \langle \ddot{B}_{ri0} \cdot \mathbf{E}_i^{\mathrm{inc}} \rangle_{\mathbf{R},\xi} + \sum_{i=1}^{N} \sum_{\substack{j=1 \\ j \neq i}}^{N} \langle \ddot{B}_{rij} \cdot \ddot{B}_{ij0} \cdot \mathbf{E}_j^{\mathrm{inc}} \rangle_{\mathbf{R},\xi} \\
&+ \sum_{i=1}^{N} \sum_{\substack{j=1 \\ j \neq i}}^{N} \sum_{\substack{l=1 \\ l \neq i \\ l \neq j}}^{N} \langle \ddot{B}_{rij} \cdot \ddot{B}_{ijl} \cdot \ddot{B}_{jl0} \cdot \mathbf{E}_l^{\mathrm{inc}} \rangle_{\mathbf{R},\xi} + \cdots \\
&= \mathbf{E}^{\mathrm{inc}} + \sum_{i=1}^{N} \int \mathrm{d}\mathbf{R}_i \, \mathrm{d}\xi_i \, p_{\mathbf{R}}(\mathbf{R}_i) p_{\xi}(\xi_i) \, \ddot{B}_{ri0} \cdot \mathbf{E}_i^{\mathrm{inc}} \\
&+ \sum_{i=1}^{N} \sum_{\substack{j=1 \\ j \neq i}}^{N} \int \mathrm{d}\mathbf{R}_i \, \mathrm{d}\xi_i \, \mathrm{d}\mathbf{R}_j \, \mathrm{d}\xi_j \, p_{\mathbf{R}}(\mathbf{R}_i) p_{\xi}(\xi_i) p_{\mathbf{R}}(\mathbf{R}_j) p_{\xi}(\xi_j) \\
&\qquad \times \ddot{B}_{rij} \cdot \ddot{B}_{ij0} \cdot \mathbf{E}_j^{\mathrm{inc}} \\
&+ \sum_{i=1}^{N} \sum_{\substack{j=1 \\ j \neq i}}^{N} \sum_{\substack{l=1 \\ l \neq i \\ l \neq j}}^{N} \int \mathrm{d}\mathbf{R}_i \, \mathrm{d}\xi_i \, \mathrm{d}\mathbf{R}_j \, \mathrm{d}\xi_j \, \mathrm{d}\mathbf{R}_l \, \mathrm{d}\xi_l \, p_{\mathbf{R}}(\mathbf{R}_i) p_{\xi}(\xi_i) p_{\mathbf{R}}(\mathbf{R}_j) \\
&\qquad \times p_{\xi}(\xi_j) p_{\mathbf{R}}(\mathbf{R}_l) p_{\xi}(\xi_l) \, \ddot{B}_{rij} \cdot \ddot{B}_{ijl} \cdot \ddot{B}_{jl0} \cdot \mathbf{E}_l^{\mathrm{inc}} \\
&+ \cdots. \qquad (8.2.4)
\end{aligned}$$

The spatial integrations are performed over the entire volume V.

Note that since Eqs. (8.1.1) and (8.1.2) are valid only in the far-field zones of all the particles filling the scattering volume, each integral on the right-hand side of Eq. (8.2.4) should, in principle, exclude a spherical volume element centered at the observation point $\mathbf{r}$ or at a particle origin $\mathbf{R}_i$ $(i = 1, \ldots, N)$ and having a radius satisfying the inequalities (3.2.17)–(3.2.19). However, usually this volume element is much smaller than V, and its relative contribution to the integrals can be expected to be negligible.

Equations (8.1.6)–(8.1.9) yield

$$\begin{aligned}
\mathbf{E}_{\mathrm{c}} &= \mathbf{E}^{\mathrm{inc}} + \sum_{i=1}^{N} \int_V \mathrm{d}\mathbf{R}_i \, p_{\mathbf{R}}(\mathbf{R}_i) G(r_i) \langle \ddot{A}(\hat{\mathbf{r}}_i, \hat{\mathbf{s}}) \rangle_{\xi} \cdot \mathbf{E}_i^{\mathrm{inc}} \\
&+ \sum_{i=1}^{N} \sum_{\substack{j=1 \\ j \neq i}}^{N} \int_V \mathrm{d}\mathbf{R}_i \, \mathrm{d}\mathbf{R}_j \, p_{\mathbf{R}}(\mathbf{R}_i) p_{\mathbf{R}}(\mathbf{R}_j) G(r_i) G(R_{ij}) \\
&\qquad \times \langle \ddot{A}(\hat{\mathbf{r}}_i, \hat{\mathbf{R}}_{ij}) \rangle_{\xi} \cdot \langle \ddot{A}(\hat{\mathbf{R}}_{ij}, \hat{\mathbf{s}}) \rangle_{\xi} \cdot \mathbf{E}_j^{\mathrm{inc}} \\
&+ \sum_{i=1}^{N} \sum_{\substack{j=1 \\ j \neq i}}^{N} \sum_{\substack{l=1 \\ l \neq i \\ l \neq j}}^{N} \int_V \mathrm{d}\mathbf{R}_i \, \mathrm{d}\mathbf{R}_j \, \mathrm{d}\mathbf{R}_l \, p_{\mathbf{R}}(\mathbf{R}_i) p_{\mathbf{R}}(\mathbf{R}_j) p_{\mathbf{R}}(\mathbf{R}_l)
\end{aligned}$$

$$
\begin{aligned}
&\qquad\qquad \times G(r_i)G(R_{ij})G(R_{jl})\langle \overleftrightarrow{A}(\hat{\mathbf{r}}_i, \hat{\mathbf{R}}_{ij})\rangle_\xi \\
&\qquad\qquad \cdot \langle \overleftrightarrow{A}(\hat{\mathbf{R}}_{ij}, \hat{\mathbf{R}}_{jl})\rangle_\xi \cdot \langle \overleftrightarrow{A}(\hat{\mathbf{R}}_{jl}, \hat{\mathbf{s}})\rangle_\xi \cdot \mathbf{E}_l^{\text{inc}} \\
&+ \cdots, \qquad (8.2.5)
\end{aligned}
$$

where $\langle \overleftrightarrow{A}(\hat{\mathbf{m}}, \hat{\mathbf{n}})\rangle_\xi$ is the average of the single-particle scattering dyadic over the particle states. Finally, recalling Eq. (5.1.15), we obtain

$$
\begin{aligned}
\mathbf{E}_\text{c} &= \mathbf{E}^{\text{inc}} + \int_V \mathrm{d}\mathbf{R}_i\, n_0(\mathbf{R}_i) G(r_i)\langle \overleftrightarrow{A}(\hat{\mathbf{r}}_i, \hat{\mathbf{s}})\rangle_\xi \cdot \mathbf{E}_i^{\text{inc}} \\
&+ \frac{N(N-1)}{N^2}\int_V \mathrm{d}\mathbf{R}_i\, \mathrm{d}\mathbf{R}_j\, n_0(\mathbf{R}_i)n_0(\mathbf{R}_j)G(r_i)G(R_{ij}) \\
&\qquad \times \langle \overleftrightarrow{A}(\hat{\mathbf{r}}_i, \hat{\mathbf{R}}_{ij})\rangle_\xi \cdot \langle \overleftrightarrow{A}(\hat{\mathbf{R}}_{ij}, \hat{\mathbf{s}})\rangle_\xi \cdot \mathbf{E}_j^{\text{inc}} \\
&+ \frac{N(N-1)(N-2)}{N^3}\int_V \mathrm{d}\mathbf{R}_i\, \mathrm{d}\mathbf{R}_j\, \mathrm{d}\mathbf{R}_l\, n_0(\mathbf{R}_i)n_0(\mathbf{R}_j)n_0(\mathbf{R}_l) \\
&\qquad \times G(r_i)G(R_{ij})G(R_{jl})\langle \overleftrightarrow{A}(\hat{\mathbf{r}}_i, \hat{\mathbf{R}}_{ij})\rangle_\xi \\
&\qquad \cdot \langle \overleftrightarrow{A}(\hat{\mathbf{R}}_{ij}, \hat{\mathbf{R}}_{jl})\rangle_\xi \cdot \langle \overleftrightarrow{A}(\hat{\mathbf{R}}_{jl}, \hat{\mathbf{s}})\rangle_\xi \cdot \mathbf{E}_l^{\text{inc}} \\
&+ \cdots \qquad (8.2.6)
\end{aligned}
$$

or, in the limit $N \to \infty$,

$$
\begin{aligned}
\mathbf{E}_\text{c} \underset{N\to\infty}{=} \mathbf{E}^{\text{inc}} &+ \int_V \mathrm{d}\mathbf{R}_i\, n_0(\mathbf{R}_i) G(r_i)\langle \overleftrightarrow{A}(\hat{\mathbf{r}}_i, \hat{\mathbf{s}})\rangle_\xi \cdot \mathbf{E}_i^{\text{inc}} \\
&+ \int_V \mathrm{d}\mathbf{R}_i\, \mathrm{d}\mathbf{R}_j\, n_0(\mathbf{R}_i)n_0(\mathbf{R}_j)G(r_i)G(R_{ij}) \\
&\qquad \times \langle \overleftrightarrow{A}(\hat{\mathbf{r}}_i, \hat{\mathbf{R}}_{ij})\rangle_\xi \cdot \langle \overleftrightarrow{A}(\hat{\mathbf{R}}_{ij}, \hat{\mathbf{s}})\rangle_\xi \cdot \mathbf{E}_j^{\text{inc}} \\
&+ \int_V \mathrm{d}\mathbf{R}_i\, \mathrm{d}\mathbf{R}_j\, \mathrm{d}\mathbf{R}_l\, n_0(\mathbf{R}_i)n_0(\mathbf{R}_j)n_0(\mathbf{R}_l)G(r_i)G(R_{ij})G(R_{jl}) \\
&\qquad \times \langle \overleftrightarrow{A}(\hat{\mathbf{r}}_i, \hat{\mathbf{R}}_{ij})\rangle_\xi \cdot \langle \overleftrightarrow{A}(\hat{\mathbf{R}}_{ij}, \hat{\mathbf{R}}_{jl})\rangle_\xi \cdot \langle \overleftrightarrow{A}(\hat{\mathbf{R}}_{jl}, \hat{\mathbf{s}})\rangle_\xi \cdot \mathbf{E}_l^{\text{inc}} \\
&+ \cdots. \qquad (8.2.7)
\end{aligned}
$$

Note that the subscripts $i, j, \ldots$ are no longer summation indices and are only used to label different integration variables. Equation (8.2.7) is the full vector version of the expansion derived by Twersky (1964) for scalar waves.

8.3 Coherent field

Let us now assume for the sake of simplicity that the distribution of the particles throughout the volume V is statistically uniform, Eq. (5.2.1), and that the volume has a concave boundary. The latter assumption ensures that all points of a straight line

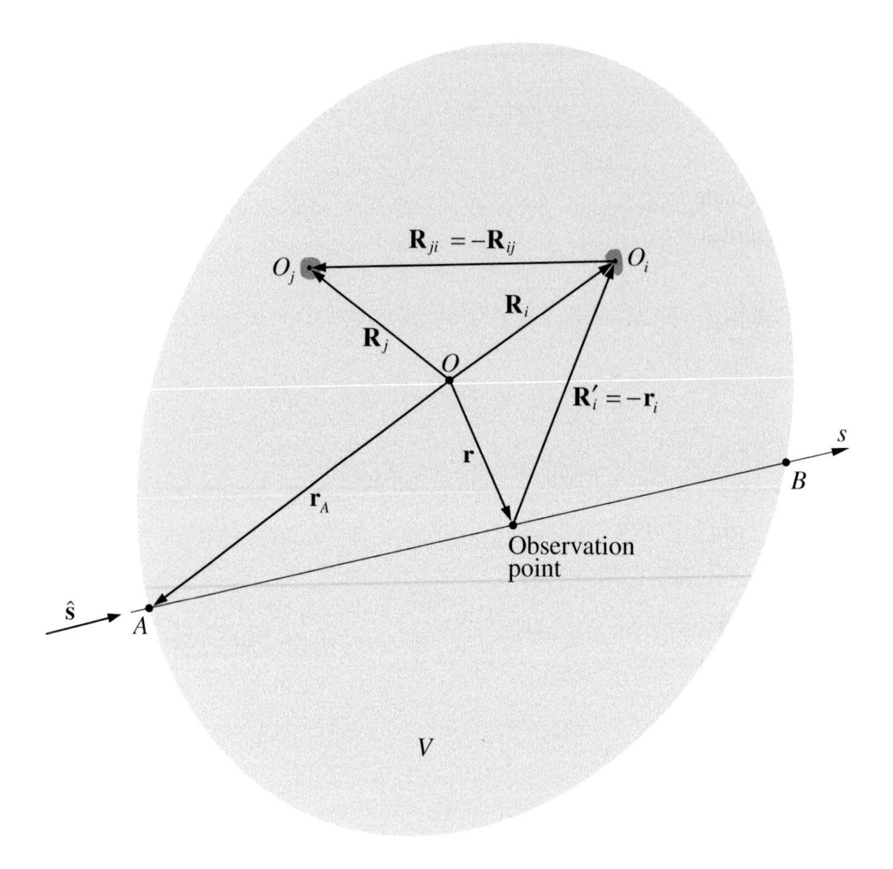

Figure 8.3.1. Geometry showing the quantities used in the derivation of Eq. (8.3.8).

connecting any two points of the medium are inside the medium.

It is convenient for our purposes to introduce an s-axis parallel to the incidence direction and going through the observation point. This axis enters the volume V at the point A such that $s(A) = 0$ and exits it at point B (Fig. 8.3.1). Let us consider the first integral on the right-hand side of Eq. (8.2.7) and denote it $\mathbf{I}_1$. From $\mathbf{R}_i = \mathbf{r} + \mathbf{R}'_i$, we have

$$\begin{aligned} \mathbf{I}_1 &= n_0 \int_V \mathrm{d}\mathbf{R}_i \, G(r_i) \langle \vec{\vec{A}}(\hat{\mathbf{r}}_i, \hat{\mathbf{s}}) \rangle_\xi \cdot \mathbf{E}_i^{\mathrm{inc}} \\ &= n_0 \int_V \mathrm{d}\mathbf{R}'_i \exp(\mathrm{i}k_1 \hat{\mathbf{s}} \cdot \mathbf{R}'_i) \frac{\exp(\mathrm{i}k_1 R'_i)}{R'_i} \langle \vec{\vec{A}}(-\hat{\mathbf{R}}'_i, \hat{\mathbf{s}}) \rangle_\xi \cdot \mathbf{E}^{\mathrm{inc}}(\mathbf{r}). \end{aligned} \tag{8.3.1}$$

The observation point is assumed to be in the far-field zone of any particle, which means that $k_1 R'_i \gg 1$. We may, therefore, use in Eq. (8.3.1) the Saxon asymptotic expansion of a plane wave in spherical waves (see Appendix B):

$$\exp(\mathrm{i}k_1\hat{\mathbf{s}}\cdot\mathbf{R}_i') \underset{k_1R_i'\to\infty}{=} \frac{\mathrm{i}2\pi}{k_1R_i'}[\delta(\hat{\mathbf{s}}+\hat{\mathbf{R}}_i')\exp(-\mathrm{i}k_1R_i') - \delta(\hat{\mathbf{s}}-\hat{\mathbf{R}}_i')\exp(\mathrm{i}k_1R_i')]. \tag{8.3.2}$$

In order to evaluate the integral (8.3.1), we use a spherical polar coordinate system with origin at the observation point and with the z-axis directed along the s-axis. We thus have

$$\begin{aligned}\mathbf{I}_1 &= \frac{\mathrm{i}2\pi n_0}{k_1}\int_{4\pi}\mathrm{d}\hat{\mathbf{R}}_i'\int \mathrm{d}R_i'\langle\vec{A}(-\hat{\mathbf{R}}_i',\hat{\mathbf{s}})\rangle_\xi\cdot\mathbf{E}^{\mathrm{inc}}(\mathbf{r})\\ &\qquad\times[\delta(\hat{\mathbf{s}}+\hat{\mathbf{R}}_i') - \delta(\hat{\mathbf{s}}-\hat{\mathbf{R}}_i')\exp(2\mathrm{i}k_1R_i')]\\ &= \frac{\mathrm{i}2\pi n_0}{k_1}s(\mathbf{r})\langle\vec{A}(\hat{\mathbf{s}},\hat{\mathbf{s}})\rangle_\xi\cdot\mathbf{E}^{\mathrm{inc}}(\mathbf{r})\\ &\quad - \frac{\pi n_0}{k_1^2}\exp\{\mathrm{i}2k_1[s(B)-s(\mathbf{r})]\}\langle\vec{A}(-\hat{\mathbf{s}},\hat{\mathbf{s}})\rangle_\xi\cdot\mathbf{E}^{\mathrm{inc}}(\mathbf{r}).\end{aligned} \tag{8.3.3}$$

Let us now recall that the number of particles filling the volume V is assumed to be very large, and the particles are assumed to be separated by distances greatly exceeding the wavelength (Eq. (4.2.6)). As a consequence, $s(\mathbf{r}) \gg 1/k_1$ (except for points in the immediate vicinity of the boundary), which suggests that the second term on the right-hand side of Eq. (8.3.3) must be much smaller than the first term. Hence,

$$\mathbf{I}_1 = \frac{\mathrm{i}2\pi n_0}{k_1}s(\mathbf{r})\langle\vec{A}(\hat{\mathbf{s}},\hat{\mathbf{s}})\rangle_\xi\cdot\mathbf{E}^{\mathrm{inc}}(\mathbf{r}). \tag{8.3.4}$$

Consider now the second integral on the right-hand side of Eq. (8.2.7) and denote it $\mathbf{I}_2$. Since $\mathbf{R}_j = \mathbf{r}+\mathbf{R}_i'+\mathbf{R}_{ji}$, we have

$$\begin{aligned}\mathbf{I}_2 &= n_0^2\int \mathrm{d}R_i' R_i'^2 G(R_i')\int_{4\pi}\mathrm{d}\hat{\mathbf{R}}_i'\int \mathrm{d}R_{ji}R_{ji}^2 G(R_{ji})\int_{4\pi}\mathrm{d}\hat{\mathbf{R}}_{ji}\\ &\qquad\times\langle\vec{A}(-\hat{\mathbf{R}}_i',-\hat{\mathbf{R}}_{ji})\rangle_\xi\cdot\langle\vec{A}(-\hat{\mathbf{R}}_{ji},\hat{\mathbf{s}})\rangle_\xi\cdot\mathbf{E}_j^{\mathrm{inc}},\end{aligned} \tag{8.3.5}$$

where

$$\begin{aligned}\mathbf{E}_j^{\mathrm{inc}} &= \exp(\mathrm{i}k_1\hat{\mathbf{s}}\cdot\mathbf{R}_j)\mathbf{E}_0^{\mathrm{inc}}\\ &= \exp(\mathrm{i}k_1\hat{\mathbf{s}}\cdot\mathbf{R}_i')\exp(\mathrm{i}k_1\hat{\mathbf{s}}\cdot\mathbf{R}_{ji})\mathbf{E}^{\mathrm{inc}}(\mathbf{r})\\ &= \left(\frac{\mathrm{i}2\pi}{k_1}\right)^2\frac{1}{R_i'}[\delta(\hat{\mathbf{s}}+\hat{\mathbf{R}}_i')\exp(-\mathrm{i}k_1R_i') - \delta(\hat{\mathbf{s}}-\hat{\mathbf{R}}_i')\exp(\mathrm{i}k_1R_i')]\\ &\quad\times\frac{1}{R_{ji}}[\delta(\hat{\mathbf{s}}+\hat{\mathbf{R}}_{ji})\exp(-\mathrm{i}k_1R_{ji}) - \delta(\hat{\mathbf{s}}-\hat{\mathbf{R}}_{ji})\exp(\mathrm{i}k_1R_{ji})]\mathbf{E}^{\mathrm{inc}}(\mathbf{r}).\end{aligned} \tag{8.3.6}$$

It is thus clear that only particles with origins on the s-axis contribute to $\mathbf{I}_2$. Substituting Eq. (8.3.6) in Eq. (8.3.5) yields

$$\mathbf{I}_2 = \frac{1}{2}\left[\frac{\mathrm{i}2\pi n_0}{k_1}s(\mathbf{r})\right]^2 \langle\ddot{A}(\hat{\mathbf{s}},\hat{\mathbf{s}})\rangle_\xi\cdot\langle\ddot{A}(\hat{\mathbf{s}},\hat{\mathbf{s}})\rangle_\xi\cdot\mathbf{E}^{\mathrm{inc}}(\mathbf{r}). \tag{8.3.7}$$

The remaining integrals in Eq. (8.2.7) are evaluated analogously. The final result is as follows:

$$\mathbf{E}_{\mathrm{c}}(\mathbf{r}) = \exp\left[\frac{\mathrm{i}2\pi n_0}{k_1}s(\mathbf{r})\langle\ddot{A}(\hat{\mathbf{s}},\hat{\mathbf{s}})\rangle_\xi\right]\cdot\mathbf{E}^{\mathrm{inc}}(\mathbf{r}), \tag{8.3.8}$$

where the dyadic exponential is defined as

$$\exp\ddot{B} = \ddot{I} + \ddot{B} + \frac{1}{2!}\ddot{B}\cdot\ddot{B} + \frac{1}{3!}\ddot{B}\cdot\ddot{B}\cdot\ddot{B} + \cdots. \tag{8.3.9}$$

It is clear from the derivation of Eq. (8.3.8) that the coherent field is a superposition of the incident field and the fields that are singly and multiply scattered in the *exact forward direction*. In other words, all single- and multiple-scattering paths that contribute to the coherent field at an internal observation point lie on the straight line parallel to the incidence direction and going through the observation point. Furthermore, all particles that do contribute to the coherent field lie between the source of illumination and the observation point. It is important to recognize that it was the inclusion of all orders of multiple forward scattering that led to the exponential s-dependence of the coherent field.

The fact that the coherent field is controlled by the forward-scattering dyadic is not surprising. Indeed, the fluctuating component of the total field is the vector sum of the partial fields generated by different particles. Random movements of the particles involve large phase shifts in the partial fields, thereby causing the fluctuating field to vanish when it is averaged over particle positions. The exact forward-scattering direction is different because the phase of the partial wave forward-scattered by a particle towards the observation point in response to the incident wave does not depend on the particle position along the line connecting the source of illumination and the observation point (see Fig. 8.3.2). Therefore, the interference of the incident wave and the forward-scattered partial wave is always the same irrespective of the precise position of the particle, and the result of the interference does not vanish upon statistical averaging over particle positions. The same is true of the interference of the incident field and a wave forward scattered along a multi-particle path of any order as well as of the mutual interference of different forward-scattered waves.

Since $\mathbf{r} = \mathbf{r}_A + s(\mathbf{r})\hat{\mathbf{s}}$ (Fig. 8.3.1), we have from Eq. (8.3.8):

$$\begin{aligned}\mathbf{E}_{\mathrm{c}}(\mathbf{r}) &= \exp(\mathrm{i}k_1\hat{\mathbf{s}}\cdot\mathbf{r}_A)\exp[\mathrm{i}k_1 s(\mathbf{r})]\exp\left[\frac{\mathrm{i}2\pi n_0}{k_1}s(\mathbf{r})\langle\ddot{A}(\hat{\mathbf{s}},\hat{\mathbf{s}})\rangle_\xi\right]\cdot\mathbf{E}_0^{\mathrm{inc}}\\ &= \exp[\mathrm{i}\ddot{\kappa}(\hat{\mathbf{s}})s(\mathbf{r})]\cdot\mathbf{E}^{\mathrm{inc}}(\mathbf{r}_A)\end{aligned} \tag{8.3.10}$$

or

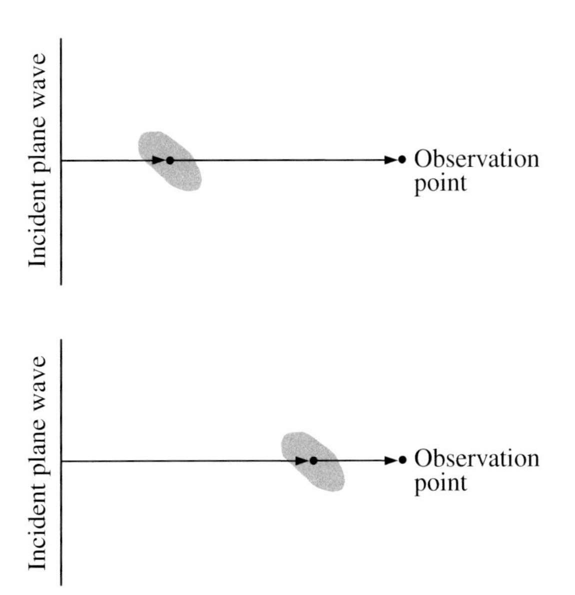

Figure 8.3.2. The phase of the wave forward scattered by a particle in response to the incident plane wave is the same irrespective of the exact position of the particle on the line connecting the source of illumination and the observation point.

$$\mathbf{E}_{\mathrm{c}}(s) = \vec{\eta}(\hat{\mathbf{s}}, s) \cdot \mathbf{E}_{\mathrm{c}}(s=0), \tag{8.3.11}$$

where

$$\vec{\kappa}(\hat{\mathbf{s}}) = k_1 \vec{I} + \frac{2\pi n_0}{k_1} \langle \vec{A}(\hat{\mathbf{s}}, \hat{\mathbf{s}}) \rangle_{\xi} \tag{8.3.12}$$

is the dyadic propagation constant for the propagation direction $\hat{\mathbf{s}}$,

$$\vec{\eta}(\hat{\mathbf{s}}, s) = \exp[\mathrm{i}\vec{\kappa}(\hat{\mathbf{s}})s] \tag{8.3.13}$$

is the coherent transmission dyadic, and

$$\mathbf{E}_{\mathrm{c}}(s=0) = \mathbf{E}^{\mathrm{inc}}(\mathbf{r}_A) \tag{8.3.14}$$

is the boundary value of the coherent field. This is the general vector form of the Foldy approximation for the coherent field (cf. Foldy, 1945). Another form of Eq. (8.3.10) is

$$\frac{\mathrm{d}\mathbf{E}_{\mathrm{c}}(\mathbf{r})}{\mathrm{d}s} = \mathrm{i}\vec{\kappa}(\hat{\mathbf{s}}) \cdot \mathbf{E}_{\mathrm{c}}(\mathbf{r}). \tag{8.3.15}$$

These results have several important implications. First, since the products

$$\langle \vec{A}(\hat{\mathbf{s}}, \hat{\mathbf{s}}) \rangle_{\xi} \cdot \mathbf{E}_0^{\mathrm{inc}}, \quad \langle \vec{A}(\hat{\mathbf{s}}, \hat{\mathbf{s}}) \rangle_{\xi} \cdot \langle \vec{A}(\hat{\mathbf{s}}, \hat{\mathbf{s}}) \rangle_{\xi} \cdot \mathbf{E}_0^{\mathrm{inc}}, \quad \text{etc.}$$

always yield electric vectors perpendicular to $\hat{\mathbf{s}}$, the coherent field satisfies the transversality condition,

$$\mathbf{E}_c(\mathbf{r}) \cdot \hat{\mathbf{s}} = 0. \tag{8.3.16}$$

Second, the coherent field describes a superposition of transverse waves propagating in the direction of $\hat{\mathbf{s}}$ and, therefore, may be associated with the transport of electromagnetic energy in the same direction. Third, Eq. (8.3.11) generalizes the optical theorem to the case of many scatterers by expressing the dyadic propagation constant in terms of the forward-scattering dyadic averaged over the particle states.

Although Eqs. (8.3.10) and (8.3.16) may appear to describe a transverse electromagnetic wave, the reader should not forget that $\mathbf{E}_c(\mathbf{r})$ is not a real physical field. Furthermore, the coherent field was computed by taking an average over a uniform distribution of particle positions as well as over all physically realizable particle states. Therefore, it is not defined at any given moment in time. The physical meaning of the coherent field will be further discussed in later sections.

We can exploit the transverse character of the coherent field to rewrite the above equations in a simpler matrix form. As in Section 2.6, we characterize the direction of propagation $\hat{\mathbf{s}}$ at the observation point $\mathbf{r}$ using the corresponding polar and azimuth angles in the local coordinate system which is centered at the observation point and has the same spatial orientation as the laboratory coordinate system (see Fig. 8.3.3). Then the electric vector of the coherent field can be written as the vector sum of the corresponding θ- and φ-components:

$$\mathbf{E}_c(\mathbf{r}) = E_{c\theta}(\mathbf{r})\hat{\boldsymbol{\theta}} + E_{c\varphi}(\mathbf{r})\hat{\boldsymbol{\varphi}}. \tag{8.3.17}$$

Defining the two-component electric column vector of the coherent field according to

$$\mathsf{E}_c(\mathbf{r}) = \begin{bmatrix} E_{c\theta}(\mathbf{r}) \\ E_{c\varphi}(\mathbf{r}) \end{bmatrix}, \tag{8.3.18}$$

we have instead of Eq. (8.3.15)

$$\frac{\mathrm{d}\mathsf{E}_c(\mathbf{r})}{\mathrm{d}s} = \mathrm{i}\mathsf{k}(\hat{\mathbf{s}})\mathsf{E}_c(\mathbf{r}), \tag{8.3.19}$$

where $\mathsf{k}(\hat{\mathbf{s}})$ is the 2×2 matrix propagation constant with elements

$$k_{11}(\hat{\mathbf{s}}) = \hat{\boldsymbol{\theta}}(\hat{\mathbf{s}}) \cdot \ddot{\kappa}(\hat{\mathbf{s}}) \cdot \hat{\boldsymbol{\theta}}(\hat{\mathbf{s}}), \tag{8.3.20}$$

$$k_{12}(\hat{\mathbf{s}}) = \hat{\boldsymbol{\theta}}(\hat{\mathbf{s}}) \cdot \ddot{\kappa}(\hat{\mathbf{s}}) \cdot \hat{\boldsymbol{\varphi}}(\hat{\mathbf{s}}), \tag{8.3.21}$$

$$k_{21}(\hat{\mathbf{s}}) = \hat{\boldsymbol{\varphi}}(\hat{\mathbf{s}}) \cdot \ddot{\kappa}(\hat{\mathbf{s}}) \cdot \hat{\boldsymbol{\theta}}(\hat{\mathbf{s}}), \tag{8.3.22}$$

$$k_{22}(\hat{\mathbf{s}}) = \hat{\boldsymbol{\varphi}}(\hat{\mathbf{s}}) \cdot \ddot{\kappa}(\hat{\mathbf{s}}) \cdot \hat{\boldsymbol{\varphi}}(\hat{\mathbf{s}}). \tag{8.3.23}$$

Obviously,

$$\mathsf{k}(\hat{\mathbf{s}}) = k_1\begin{bmatrix} 1 & 0 \\ 0 & 1 \end{bmatrix} + \frac{2\pi n_0}{k_1}\langle \mathsf{S}(\hat{\mathbf{s}}, \hat{\mathbf{s}})\rangle_\xi, \tag{8.3.24}$$

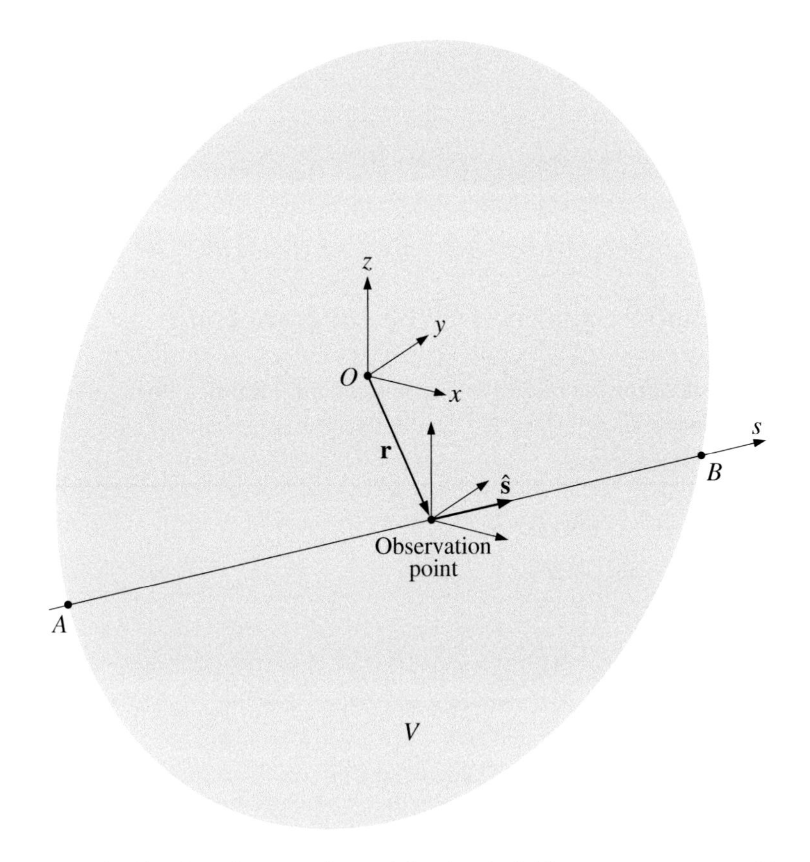

Figure 8.3.3. The direction of propagation and the electric field vector components of the coherent field at an observation point **r** are specified using a local coordinate system with the same orientation as the laboratory coordinate system centered at O.

where $\langle \mathbf{S}(\hat{\mathbf{s}}, \hat{\mathbf{s}}) \rangle_\xi$ is the forward-scattering amplitude matrix averaged over the particle states (cf. Eqs. (3.3.8)–(3.3.11)).

It is often convenient to rewrite Eq. (8.3.19) in the form

$$\mathbf{E}_c(s) = \mathbf{h}(\hat{\mathbf{s}}, s)\mathbf{E}_c(s=0), \tag{8.3.25}$$

where

$$\mathbf{h}(\hat{\mathbf{s}}, s) = \exp[\mathrm{i}s\mathbf{k}(\hat{\mathbf{s}})] \tag{8.3.26}$$

is the coherent transmission amplitude matrix and the 2×2 matrix exponential is defined as follows:

$$\exp\mathbf{B} = \begin{bmatrix} 1 & 0 \\ 0 & 1 \end{bmatrix} + \mathbf{B} + \frac{1}{2!}\mathbf{BB} + \frac{1}{3!}\mathbf{BBB} + \cdots. \tag{8.3.27}$$

From Eq. (A.7), the matrix identity $(\mathbf{AB})^{\mathrm{T}} = \mathbf{B}^{\mathrm{T}}\mathbf{A}^{\mathrm{T}}$, and the reciprocity relations (3.4.19) and (3.4.21), we easily derive the following reciprocity relations for the coherent transmission dyadic and the coherent transmission amplitude matrix:

$$\vec{\eta}(-\hat{\mathbf{s}}, s) = [\vec{\eta}(\hat{\mathbf{s}}, s)]^{\mathrm{T}}, \tag{8.3.28}$$

$$\mathbf{h}(-\hat{\mathbf{s}}, s) = \begin{bmatrix} 1 & 0 \\ 0 & -1 \end{bmatrix} [\mathbf{h}(\hat{\mathbf{s}}, s)]^{\mathrm{T}} \begin{bmatrix} 1 & 0 \\ 0 & -1 \end{bmatrix}. \tag{8.3.29}$$

8.4 Transfer equation for the coherent field

We will now describe the coherent field in terms of quantities having the dimension of monochromatic energy flux. We first define the coherency column vector of the coherent field according to

$$\mathbf{J}_{\mathrm{c}} = \frac{1}{2}\sqrt{\frac{\epsilon_1}{\mu_0}} \begin{bmatrix} E_{\mathrm{c}\theta} E_{\mathrm{c}\theta}^* \\ E_{\mathrm{c}\theta} E_{\mathrm{c}\varphi}^* \\ E_{\mathrm{c}\varphi} E_{\mathrm{c}\theta}^* \\ E_{\mathrm{c}\varphi} E_{\mathrm{c}\varphi}^* \end{bmatrix} \tag{8.4.1}$$

and easily derive from Eqs. (8.3.19) and (4.7.24) the following transfer equation:

$$\frac{\mathrm{d}\mathbf{J}_{\mathrm{c}}(\mathbf{r})}{\mathrm{d}s} = -n_0 \langle \mathbf{K}^J(\hat{\mathbf{s}}) \rangle_\xi \mathbf{J}_{\mathrm{c}}(\mathbf{r}), \tag{8.4.2}$$

where $\mathbf{K}^J$ is the coherency extinction matrix given by Eq. (3.8.4). The Stokes-vector representation of this equation is obtained by using the definition

$$\mathbf{I}_{\mathrm{c}} = \mathbf{D}\mathbf{J}_{\mathrm{c}} = \frac{1}{2}\sqrt{\frac{\epsilon_1}{\mu_0}} \begin{bmatrix} E_{\mathrm{c}\theta} E_{\mathrm{c}\theta}^* + E_{\mathrm{c}\varphi} E_{\mathrm{c}\varphi}^* \\ E_{\mathrm{c}\theta} E_{\mathrm{c}\theta}^* - E_{\mathrm{c}\varphi} E_{\mathrm{c}\varphi}^* \\ -2\,\mathrm{Re}(E_{\mathrm{c}\theta} E_{\mathrm{c}\varphi}^*) \\ 2\,\mathrm{Im}(E_{\mathrm{c}\theta} E_{\mathrm{c}\varphi}^*) \end{bmatrix} \tag{8.4.3}$$

and Eq. (3.8.7):

$$\frac{\mathrm{d}\mathbf{I}_{\mathrm{c}}(\mathbf{r})}{\mathrm{d}s} = -n_0 \langle \mathbf{K}(\hat{\mathbf{s}}) \rangle_\xi \mathbf{I}_{\mathrm{c}}(\mathbf{r}), \tag{8.4.4}$$

where $\mathbf{K}$ is the Stokes extinction matrix with elements given by Eqs. (3.8.8)–(3.8.14).

The formal solution of Eq. (8.4.4) can be written in the form

$$\begin{aligned} \mathbf{I}_{\mathrm{c}}(\mathbf{r}) &= \mathbf{H}[\hat{\mathbf{s}}, s(\mathbf{r})]\mathbf{I}_{\mathrm{c}}(\mathbf{r}_A) \\ &= \mathbf{H}[\hat{\mathbf{s}}, s(\mathbf{r})]\mathbf{I}^{\mathrm{inc}}, \end{aligned} \tag{8.4.5}$$

where $\mathbf{I}^{\mathrm{inc}}$ is the Stokes column vector of the incident plane wave and

$$\mathbf{H}(\hat{\mathbf{s}}, s) = \exp[-n_0 s \langle \mathbf{K}(\hat{\mathbf{s}}) \rangle_\xi] \tag{8.4.6}$$

is the coherent transmission Stokes matrix. As before, the 4×4 matrix exponential is defined by

$$\exp\mathbf{B} = \mathbf{\Delta} + \mathbf{B} + \frac{1}{2!}\mathbf{BB} + \frac{1}{3!}\mathbf{BBB} + \cdots, \tag{8.4.7}$$

where $\mathbf{\Delta}$ is the 4×4 unit matrix. In view of Eq. (3.8.16), the coherent transmission Stokes matrix obeys the following reciprocity relation:

$$\mathbf{H}(-\hat{\mathbf{s}}, s) = \mathbf{\Delta}_3[\mathbf{H}(\hat{\mathbf{s}}, s)]^{\mathrm{T}}\mathbf{\Delta}_3. \tag{8.4.8}$$

8.5 Dyadic correlation function in the ladder approximation

We are now well prepared to start the derivation of the RTE. Our first step is to introduce the so-called dyadic correlation function, which involves the total electric fields at two points inside the volume V and is defined as the following average dyadic product:

$$\langle \mathbf{E}(\mathbf{r}, t) \otimes [\mathbf{E}(\mathbf{r}', t)]^* \rangle_t = \langle \mathbf{E}(\mathbf{r}) \otimes [\mathbf{E}(\mathbf{r}')]^* \rangle_{\xi,\mathbf{R}}, \qquad \mathbf{r}, \mathbf{r}' \in V.$$

Note that the left-hand side of this equation involves the actual electric fields and that the time-harmonic factors $\exp(-\mathrm{i}\omega t)$ and $[\exp(-\mathrm{i}\omega t)]^*$ cancel each other without canceling the time average. Also, by equating the time average and the average over particle coordinates and states we assume the full ergodicity of the particle ensemble. Obviously, the product

$$\frac{1}{2}\sqrt{\frac{\epsilon_1}{\mu_0}} \langle \mathbf{E}(\mathbf{r}) \otimes [\mathbf{E}(\mathbf{r}')]^* \rangle_{\mathbf{R},\xi}$$

has the dimension of monochromatic energy flux and can potentially be used to define appropriate measurable quantities. That this is indeed the case will become clear in Section 8.12.

Recalling the Twersky approximation, Eq. (8.1.13), and Fig. 8.1.4(b), we conclude that the dyadic correlation function is given by the expression shown diagrammatically in Fig. 8.5.1. To classify the different terms entering the expanded expression inside the angular brackets on the right-hand side of this equation, we will use the notation illustrated in Fig. 8.5.2(a). In this particular case, the upper and the lower scattering paths go through different particles. However, the two paths can involve one or more common particles, as shown in panels (b)–(d) by using the dashed connectors. Furthermore, if the number of common particles is two or more, they can enter the upper and lower paths in the same order, as in panel (c), or in the reverse order, as in panel (d). Panel (e) shows a mixed diagram in which two common

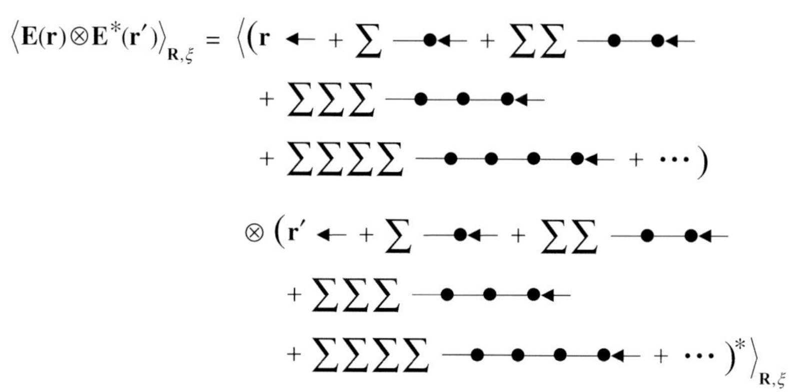

Figure 8.5.1. The Twersky expansion of the dyadic correlation function.

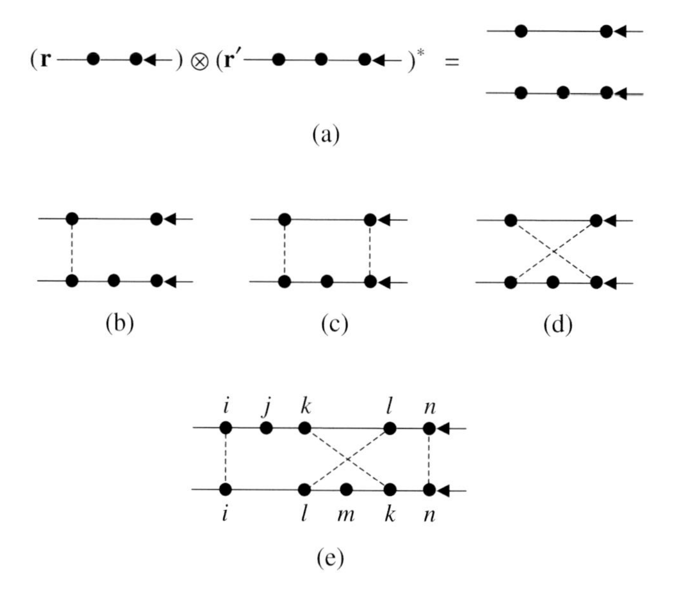

Figure 8.5.2. Classification of terms entering the Twersky expansion of the dyadic correlation function.

particles appear in the same order and two other common particles appear in the reverse order. The contribution of this diagram to the dyadic correlation function is simply

$$[\ddot{B}_{rij} \cdot \ddot{B}_{ijk} \cdot \ddot{B}_{jkl} \cdot \ddot{B}_{kln} \cdot \ddot{B}_{ln0} \cdot \mathbf{E}_n^{\mathrm{inc}}] \otimes [\ddot{B}_{r'il} \cdot \ddot{B}_{ilm} \cdot \ddot{B}_{lmk} \cdot \ddot{B}_{mkn} \cdot \ddot{B}_{kn0} \cdot \mathbf{E}_n^{\mathrm{inc}}]^*.$$

By the nature of the Twersky approximation, neither the upper path nor the lower path can go through a particle more than once. Therefore, no particle can be the origin of more than one connector.

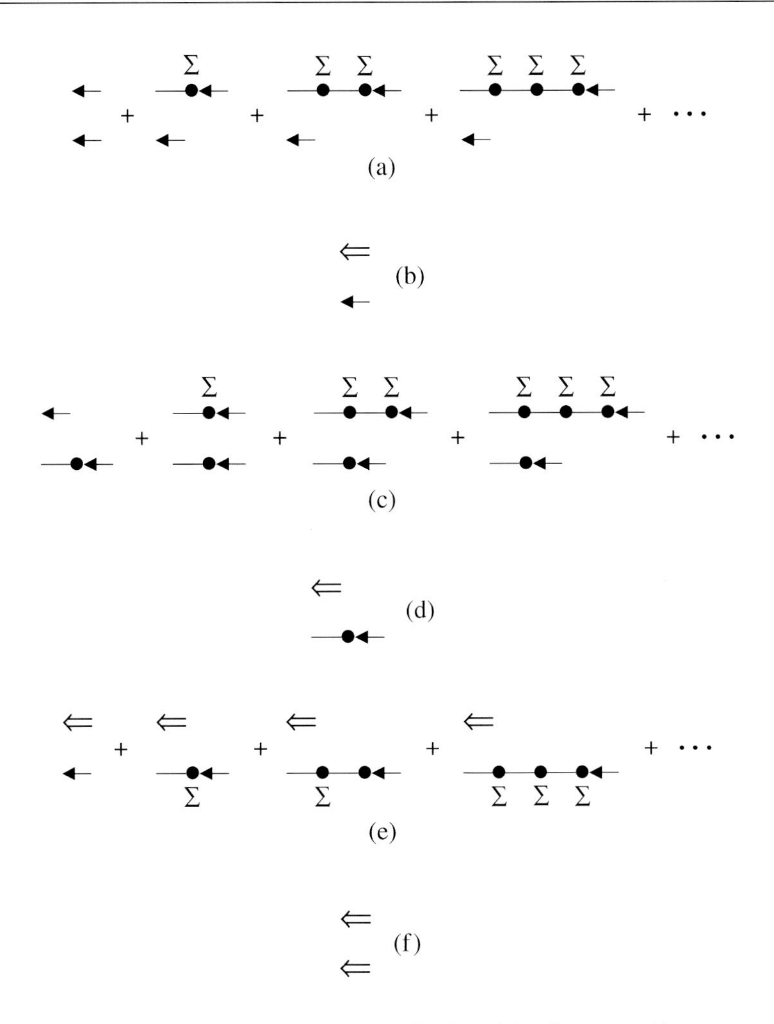

Figure 8.5.3. Calculation of the cumulative contribution of the diagrams with no connectors.

To sum and average all the diagrams entering the expanded expression for the dyadic correlation function in Fig. 8.5.1 is a very difficult problem which we will not try to address fully. Instead, we will neglect all diagrams with crossing connectors and will work with a truncated expansion that includes only the diagrams with vertical or no connectors. This approximation will allow us to sum and average large groups of diagrams independently and eventually derive the RTE. The consequences of neglecting the diagrams with crossing connectors will be discussed in Section 8.11.

Let us begin with diagrams that have no connectors. Since these diagrams do not involve common particles, the ensemble averaging of the upper and lower paths can be performed independently. Consider first the sum of the diagrams shown in Fig. 8.5.3(a), in which the symbol Σ indicates both the summation over all appropriate particles and the statistical averaging over the particle states and positions. According

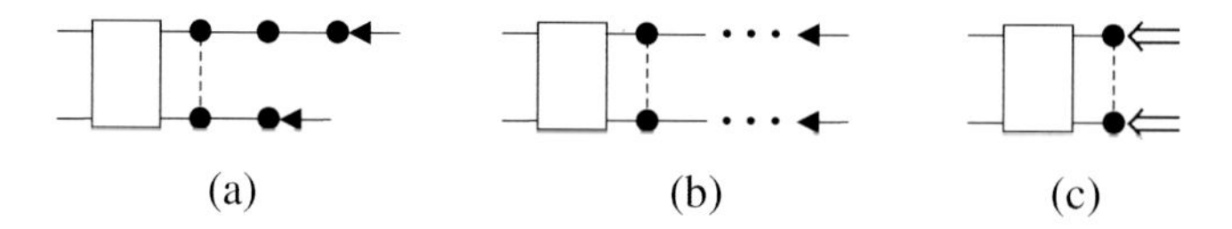

Figure 8.5.4. Diagrams with one or more vertical connectors.

to Section 8.3, summing the upper paths yields the coherent field at the point **r**. This result can be represented by the diagram shown in Fig. 8.5.3(b), in which the symbol $\Leftarrow$ denotes the coherent field.

Similarly, summing the upper paths of the diagram shown in panel (c) yields, in the limit $N \to \infty$, the diagram shown in panel (d). Indeed, since one particle is already "reserved" for the lower path, the number of particles contributing to the upper paths in panel (c) is $N-1$. However, the difference between the sum of the upper paths in panel (c) and the coherent field at **r** vanishes as N tends to infinity. We can continue this process and eventually conclude that the total contribution of the diagrams with no connectors is given by the sum of the diagrams shown in panel (e).

It is now clear that the final result can be represented by the diagram in panel (f), which means that the contribution of all the diagrams with no connectors to the dyadic correlation function is simply the dyadic product of the coherent fields at the points **r** and **r′**: $\mathbf{E}_c(\mathbf{r}) \otimes [\mathbf{E}_c(\mathbf{r}')]^*$. This result explains the usefulness of introducing the concept of the coherent field in Section 8.2 despite the fact that $\mathbf{E}_c(\mathbf{r})$ does not represent the actual time average of the electric field.

All other diagrams contributing to the dyadic correlation function have at least one vertical connector, as shown in Fig. 8.5.4(a). The part of the diagram on the right-hand side of the right-most connector will be called the tail, whereas the box represents collectively the part of the diagram on the left-hand side of the right-most connector and can, in principle, be empty. The right-most common particle and the box form the body of the diagram.

Let us first consider the group of diagrams with the same body but with different tails, as shown in Fig. 8.5.4(b). We can repeat the derivation of Section 8.3 and verify that in the limit $N \to \infty$, the sum of all diagrams in Fig. 8.5.5(a) gives the diagram shown in Fig. 8.5.5(c). Indeed, let particle q be the right-most connected particle and particle p be the right-most particle on the left-hand side of particle q in the upper scattering paths of the diagrams shown in Fig. 8.5.5(a). Consider the cumulative contribution of all the diagrams on the left-hand side of Fig. 8.5.5(b) to the total electric field created at the origin of particle p. Writing this contribution in the expanded form yields

$$
\begin{aligned}
&G(R_{pq})\ddot{A}(\hat{\mathbf{R}}_{pq}, \hat{\mathbf{s}}) \cdot \mathbf{E}_q^{\rm inc} \\
&+ \frac{N-n}{N} n_0 G(R_{pq}) \int_V \mathrm{d}\mathbf{R}_i\, G(R_{qi}) \ddot{A}(\hat{\mathbf{R}}_{pq}, \hat{\mathbf{R}}_{qi}) \cdot \langle \ddot{A}(\hat{\mathbf{R}}_{qi}, \hat{\mathbf{s}}) \rangle_\xi \cdot \mathbf{E}_i^{\rm inc}
\end{aligned}
$$

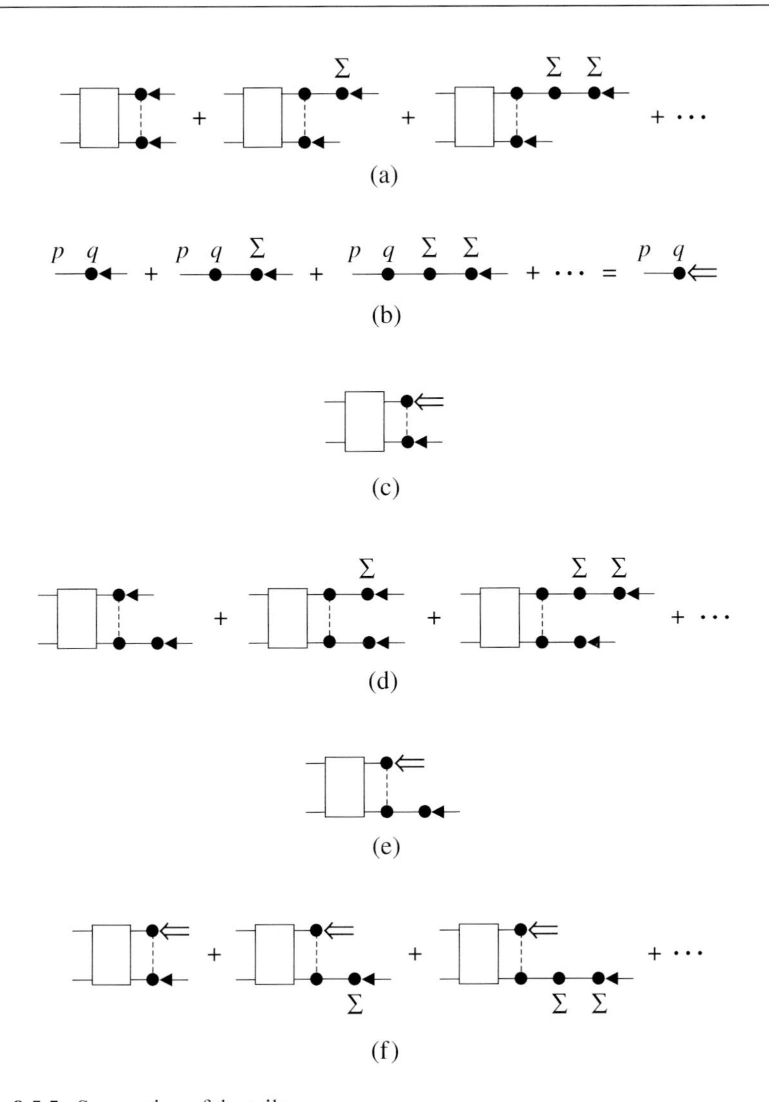

Figure 8.5.5. Summation of the tails.

$$+\ \frac{(N-n)(N-n-1)}{N^2} n_0^2 G(R_{pq}) \int_V \mathrm{d}\mathbf{R}_i\, \mathrm{d}\mathbf{R}_j\, G(R_{qi}) G(R_{ij})\, \ddot{A}(\hat{\mathbf{R}}_{pq}, \hat{\mathbf{R}}_{qi})$$

$$\cdot \langle \ddot{A}(\hat{\mathbf{R}}_{qi}, \hat{\mathbf{R}}_{ij}) \rangle_\xi \cdot \langle \ddot{A}(\hat{\mathbf{R}}_{ij}, \hat{\mathbf{s}}) \rangle_\xi \cdot \mathbf{E}_j^{\mathrm{inc}}$$

$$+ \cdots \underset{N \to \infty}{=} G(R_{pq})\, \ddot{A}(\hat{\mathbf{R}}_{pq}, \hat{\mathbf{s}}) \cdot \mathbf{E}_{\mathrm{c}}(\mathbf{R}_q), \tag{8.5.1}$$

where n is the number of particles in the common body of the diagrams. The right-hand side of Eq. (8.5.1) was derived under the assumption that N is so large that all factors of the type $(N-n)!/(N-n-k)!$ can be replaced by N^k. This result is summarized by the right-hand side of Fig. 8.5.5(b).

Analogously, the sum of the diagrams in Fig. 8.5.5(d) is given by the diagram in

Fig. 8.5.5(e), and so on. We can now sum up all diagrams in Fig. 8.5.5(f) and obtain the diagram shown in Fig. 8.5.4(c).

Thus the collective contribution to the dyadic correlation function of all the diagrams with the same body and all possible tails is equivalent to the contribution of a single diagram formed by the body alone, provided that the right-most common particle is excited by the coherent field rather than by the external incident field. This represents a radical difference from the initial expansion (8.1.13), in which the source of multiple scattering is the external field. This important result allows us to cut off all tails and consider only truncated diagrams of the type shown in Fig. 8.5.4(c).

Thus, the dyadic correlation function is equal to $\mathbf{E}_c(\mathbf{r}) \otimes [\mathbf{E}_c(\mathbf{r}')]^*$ plus the statistical average of the sum of all connected diagrams of the type illustrated by panels (a)–(c) of Fig. 8.5.6. The symbols $\cdots$ in these diagrams denote all possible combinations of unconnected particles. Let us, for example, consider the statistical average of the sum of all diagrams of the kind shown in panel (d) with the same fixed shaded part. We thus must evaluate the left-hand side of the equation shown in panel (f), where, as before, the symbol Σ indicates both the summation over all appropriate particles and the statistical averaging over the particle states and positions. Let particle w be the right-most particle on the left-hand side of particle p in the upper scattering paths of the diagrams on the left-hand side of panel (f) and u be the left-most particle on the right-hand side of particle q. The electric field created by particle p at the origin of particle w via the upper scattering paths of all the diagrams shown on the left-hand side of panel (f) is given by the left-hand side of the equation shown diagrammatically in panel (g) and can be written in expanded form as

$$\begin{aligned}
\mathbf{E}_w &= G(R_{wp})G(R_{pq})\vec{A}_p(\hat{\mathbf{R}}_{wp}, \hat{\mathbf{R}}_{pq}) \cdot \vec{A}_q(\hat{\mathbf{R}}_{pq}, \hat{\mathbf{R}}_{qu}) \cdot \mathbf{E}_q \\
&\quad + \sum_i G(R_{wp})\langle G(R_{pi})G(R_{iq})\vec{A}_p(\hat{\mathbf{R}}_{wp}, \hat{\mathbf{R}}_{pi}) \cdot \vec{A}_i(\hat{\mathbf{R}}_{pi}, \hat{\mathbf{R}}_{iq}) \\
&\qquad\qquad \cdot \vec{A}_q(\hat{\mathbf{R}}_{iq}, \hat{\mathbf{R}}_{qu})\rangle_{\mathbf{R},\xi} \cdot \mathbf{E}_q \\
&\quad + \sum_{ij} G(R_{wp})\langle G(R_{pi})G(R_{ij})G(R_{jq})\vec{A}_p(\hat{\mathbf{R}}_{wp}, \hat{\mathbf{R}}_{pi}) \cdot \vec{A}_i(\hat{\mathbf{R}}_{pi}, \hat{\mathbf{R}}_{ij}) \\
&\qquad\qquad \cdot \vec{A}_j(\hat{\mathbf{R}}_{ij}, \hat{\mathbf{R}}_{jq}) \cdot \vec{A}_q(\hat{\mathbf{R}}_{jq}, \hat{\mathbf{R}}_{qu})\rangle_{\mathbf{R},\xi} \cdot \mathbf{E}_q \\
&\quad + \cdots,
\end{aligned} \tag{8.5.2}$$

where $\mathbf{E}_q$ is the electric field coming to the origin of particle q via particle u and the summations and statistical averaging are performed over all appropriate unconnected particles (see Fig. 8.5.7). In the limit $N \to \infty$, Eq. (8.5.2) takes the form

$$\begin{aligned}
\mathbf{E}_w &= G(R_{wp})G(R_{pq})\vec{A}_p(\hat{\mathbf{R}}_{wp}, \hat{\mathbf{R}}_{pq}) \cdot \vec{A}_q(\hat{\mathbf{R}}_{pq}, \hat{\mathbf{R}}_{qu}) \cdot \mathbf{E}_q \\
&\quad + n_0 G(R_{wp}) \int_V \mathrm{d}\mathbf{R}_i\, G(R_{pi})G(R_{iq})\vec{A}_p(\hat{\mathbf{R}}_{wp}, \hat{\mathbf{R}}_{pi}) \cdot \langle \vec{A}(\hat{\mathbf{R}}_{pi}, \hat{\mathbf{R}}_{iq})\rangle_\xi
\end{aligned}$$

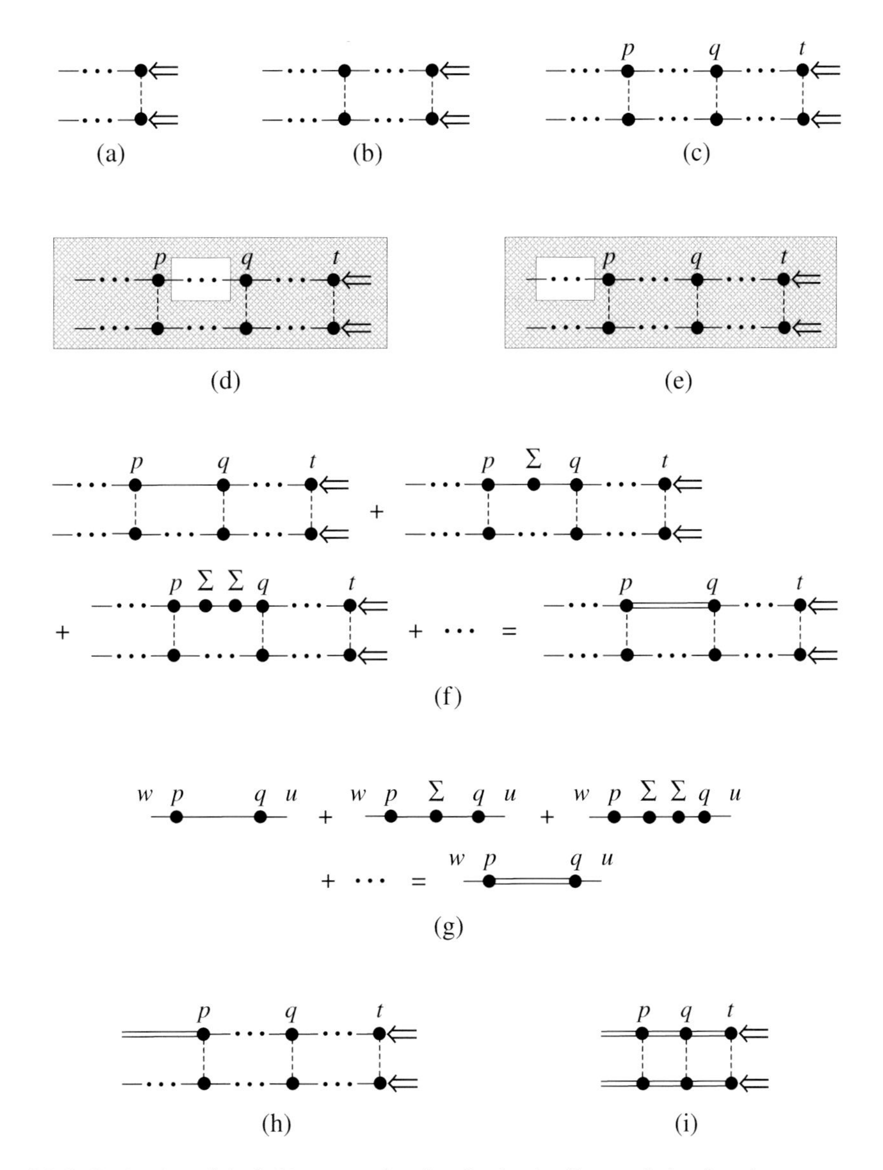

Figure 8.5.6. Derivation of the ladder approximation for the dyadic correlation function.

$$
\begin{aligned}
& \cdot \ddot{A}_q(\hat{\mathbf{R}}_{iq}, \hat{\mathbf{R}}_{qu}) \cdot \mathbf{E}_q \\
& + n_0^2 G(R_{wp}) \int_V \mathrm{d}\mathbf{R}_i \, \mathrm{d}\mathbf{R}_j \, G(R_{pi}) G(R_{ij}) G(R_{jq}) \ddot{A}_p(\hat{\mathbf{R}}_{wp}, \hat{\mathbf{R}}_{pi}) \\
& \qquad \cdot \langle \ddot{A}(\hat{\mathbf{R}}_{pi}, \hat{\mathbf{R}}_{ij}) \rangle_\xi \cdot \langle \ddot{A}(\hat{\mathbf{R}}_{ij}, \hat{\mathbf{R}}_{jq}) \rangle_\xi \cdot \ddot{A}_q(\hat{\mathbf{R}}_{jq}, \hat{\mathbf{R}}_{qu}) \cdot \mathbf{E}_q \\
& + \cdots,
\end{aligned}
\tag{8.5.3}
$$

where the angular brackets now denote amplitude matrices averaged over the particle states.

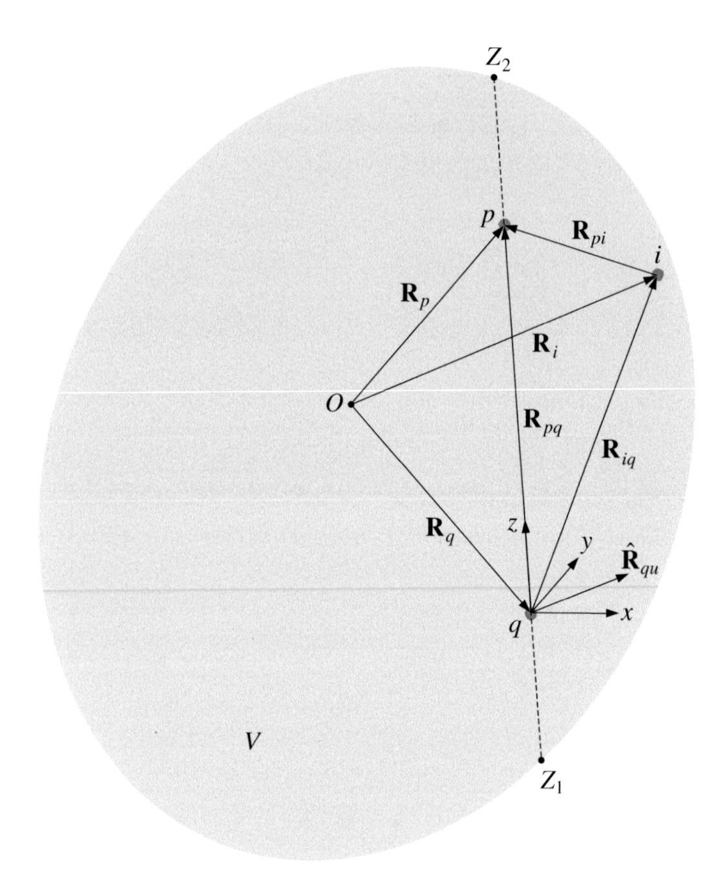

Figure 8.5.7. Calculation of the integrals entering Eq. (8.5.3).

Let us consider the first integral on the right-hand side of Eq. (8.5.3):

$$\mathbf{I}_1 = n_0 G(R_{wp}) \int_V \mathrm{d}\mathbf{R}_i \frac{\exp[\mathrm{i}k_1(R_{pi}+R_{iq})]}{R_{pi}R_{iq}} \ddot{A}_p(\hat{\mathbf{R}}_{wp}, \hat{\mathbf{R}}_{pi})$$
$$\cdot \langle \ddot{A}(\hat{\mathbf{R}}_{pi}, \hat{\mathbf{R}}_{iq}) \rangle_\xi \cdot \ddot{A}_q(\hat{\mathbf{R}}_{iq}, \hat{\mathbf{R}}_{qu}) \cdot \mathbf{E}_q. \tag{8.5.4}$$

Since the factor $\exp[\mathrm{i}k_1(R_{pi}+R_{iq})]$ is a rapidly oscillating function of $\mathbf{R}_i$, the contribution of a major part of V to $\mathbf{I}_1$ can be expected to zero out. The only exception is the small region around the straight line connecting particles q and p, where the phase $k_1(R_{pi}+R_{iq})$ is almost constant. Therefore, we can evaluate the integral (8.5.4) using the method of stationary phase (see Appendix E).

Using the Cartesian coordinate system with origin inside particle q and the z-axis along the vector $\mathbf{R}_{pq}$, as shown in Fig. 8.5.7, Eq. (8.5.4) can be rewritten in the form

$$\mathbf{I}_1 = n_0 G(R_{wp}) \int_{-\infty}^{+\infty} \mathrm{d}x_i \int_{-\infty}^{+\infty} \mathrm{d}y_i \int \mathrm{d}z_i \frac{\exp[\mathrm{i}k_1(R_{pi}+R_{iq})]}{R_{pi}R_{iq}}$$

$$\times \vec{A}_p(\hat{\mathbf{R}}_{wp}, \hat{\mathbf{R}}_{pi}) \cdot \langle \vec{A}(\hat{\mathbf{R}}_{pi}, \hat{\mathbf{R}}_{iq}) \rangle_{\xi} \cdot \vec{A}_q(\hat{\mathbf{R}}_{iq}, \hat{\mathbf{R}}_{qu}) \cdot \mathbf{E}_q, \quad (8.5.5)$$

where x_i, y_i, and z_i are the coordinates of particle i and the integration limits for x_i and y_i are set to infinity owing to the fact that only a small part of V along the z-axis contributes to $\mathbf{I}_1$. According to Eq. (E.10), the integral over z_i can be subdivided into three integrals covering the regions with $Z_1 < z_i < 0$, $0 < z_i < R_{pq}$, and $R_{pq} < z_i < Z_2$ (Fig. 8.5.7). The first and third integrals involve rapidly oscillating functions of z_i and vanish. Indeed,

$$\begin{aligned}
\int_{Z_1}^{0} \mathrm{d}z_i \frac{\exp[\mathrm{i}k_1(R_{pq} - 2z_i)]}{R_{pq} - 2z_i} &= \frac{1}{2}\int_{k_1 R_{pq}}^{\infty} \mathrm{d}t \frac{\exp(\mathrm{i}t)}{t} - \frac{1}{2}\int_{k_1(R_{pq}+2|Z_1|)}^{\infty} \mathrm{d}t \frac{\exp(\mathrm{i}t)}{t} \\
&= \tfrac{1}{2}\{-\mathrm{Ci}(k_1 R_{pq}) - \mathrm{i}\,\mathrm{si}(k_1 R_{pq}) \\
&\quad + \mathrm{Ci}[k_1(R_{pq} + 2|Z_1|)] + \mathrm{i}\,\mathrm{si}[k_1(R_{pq} + 2|Z_1|)]\} \\
&= O\left(\frac{1}{k_1 R_{pq}}\right),
\end{aligned} \quad (8.5.6)$$

where $t = k_1(R_{pq} - 2z_i)$,

$$\mathrm{si}(y) = -\int_{y}^{\infty} \mathrm{d}t \frac{\sin t}{t} \quad (8.5.7)$$

is the sine integral, and

$$\mathrm{Ci}(y) = -\int_{y}^{\infty} \mathrm{d}t \frac{\cos t}{t} \quad (8.5.8)$$

is the cosine integral (see Section 5.2 of Abramowitz and Stegun, 1964 or Section 5.10 of Arfken and Weber, 2001). Similarly,

$$\begin{aligned}
\int_{R_{pq}}^{Z_2} \mathrm{d}z_i \frac{\exp[\mathrm{i}k_1(R_{pq} + 2z_i)]}{R_{pq} + 2z_i} &= \frac{1}{2}\int_{3k_1 R_{pq}}^{\infty} \mathrm{d}t \frac{\exp(\mathrm{i}t)}{t} - \frac{1}{2}\int_{k_1(R_{pq}+2Z_2)}^{\infty} \mathrm{d}t \frac{\exp(\mathrm{i}t)}{t} \\
&= \tfrac{1}{2}\{-\mathrm{Ci}(3k_1 R_{pq}) - \mathrm{i}\,\mathrm{si}(3k_1 R_{pq}) \\
&\quad + \mathrm{Ci}[k_1(R_{pq} + 2Z_2)] + \mathrm{i}\,\mathrm{si}[k_1(R_{pq} + 2Z_2)]\} \\
&= O\left(\frac{1}{k_1 R_{pq}}\right),
\end{aligned} \quad (8.5.9)$$

where $t = k_1(R_{pq} + 2z_i)$. Thus only the interval $0 < z_i < R_{pq}$ gives a nonzero contribution. We can now use Eq. (E.10),

$$\int_{-\infty}^{+\infty} \mathrm{d}x_i \int_{-\infty}^{+\infty} \mathrm{d}y_i \frac{\exp[\mathrm{i}k_1(R_{pi} + R_{iq})]}{R_{pi}R_{iq}} \vec{A}_p(\hat{\mathbf{R}}_{wp}, \hat{\mathbf{R}}_{pi}) \cdot \langle \vec{A}(\hat{\mathbf{R}}_{pi}, \hat{\mathbf{R}}_{iq}) \rangle_{\xi}$$

$$\cdot \ddot{A}_q(\hat{\mathbf{R}}_{iq}, \hat{\mathbf{R}}_{qu}) \cdot \mathbf{E}_q$$

$$\approx \frac{\mathrm{i}2\pi}{k_1} \frac{\exp(\mathrm{i}k_1 R_{pq})}{R_{pq}} \ddot{A}_p(\hat{\mathbf{R}}_{wp}, \hat{\mathbf{R}}_{pq}) \cdot \langle \ddot{A}(\hat{\mathbf{R}}_{pq}, \hat{\mathbf{R}}_{pq}) \rangle_\xi \cdot \ddot{A}_q(\hat{\mathbf{R}}_{pq}, \hat{\mathbf{R}}_{qu}) \cdot \mathbf{E}_q \tag{8.5.10}$$

to derive

$$\mathbf{I}_1 = G(R_{wp}) \frac{\exp(\mathrm{i}k_1 R_{pq})}{R_{pq}} \frac{\mathrm{i}2\pi n_0 R_{pq}}{k_1} \ddot{A}_p(\hat{\mathbf{R}}_{wp}, \hat{\mathbf{R}}_{pq}) \cdot \langle \ddot{A}(\hat{\mathbf{R}}_{pq}, \hat{\mathbf{R}}_{pq}) \rangle_\xi \cdot \ddot{A}_q(\hat{\mathbf{R}}_{pq}, \hat{\mathbf{R}}_{qu}) \cdot \mathbf{E}_q. \tag{8.5.11}$$

The other integrals on the right-hand side of Eq. (8.5.3) are computed analogously. The final result is

$$\mathbf{E}_w = G(R_{wp}) \ddot{A}_p(\hat{\mathbf{R}}_{wp}, \hat{\mathbf{R}}_{pq}) \cdot \frac{\exp[\mathrm{i}\ddot{\kappa}(\hat{\mathbf{R}}_{pq}) R_{pq}]}{R_{pq}} \cdot \ddot{A}_q(\hat{\mathbf{R}}_{pq}, \hat{\mathbf{R}}_{qu}) \cdot \mathbf{E}_q \tag{8.5.12a}$$

$$= G(R_{wp}) \ddot{A}_p(\hat{\mathbf{R}}_{wp}, \hat{\mathbf{R}}_{pq}) \cdot \frac{\ddot{\eta}(\hat{\mathbf{R}}_{pq}, R_{pq})}{R_{pq}} \cdot \ddot{A}_q(\hat{\mathbf{R}}_{pq}, \hat{\mathbf{R}}_{qu}) \cdot \mathbf{E}_q, \tag{8.5.12b}$$

where the dyadic propagation constant $\ddot{\kappa}$ and the coherent transmission dyad $\ddot{\eta}$ are given by Eqs. (8.3.12) and (8.3.13), respectively. These equations are similar to Eqs. (8.3.10) and (8.3.11) for the coherent field and are yet another manifestation of the forward-scattering optical theorem. Obviously, they can be interpreted as describing the coherent propagation of the wave scattered by particle q towards particle p through the discrete scattering medium. The presence of other particles on the line of sight causes attenuation and, potentially, a change in polarization state of the wave. The exponential form of the coherent transmission dyad in Eq. (8.5.12a) is again the consequence of taking into account all orders of multiple forward scattering by unconnected particles. The notable difference from Eqs. (8.3.10) and (8.3.11) is the factor $1/R_{pq}$, which is a reminder that the wave scattered by a particle is spherical, whereas the coherent field is mathematically represented by a plane wave.

Equation (8.5.12a) can be summarized by the diagram on the right-hand side of Fig. 8.5.6(g), thereby yielding the right-hand side of the equation in Fig. 8.5.6(f). The double rather than a single line indicates that the scalar factor $\exp(\mathrm{i}k_1 R_{pq})/R_{pq}$ has been replaced by the dyadic factor $\exp[\mathrm{i}\ddot{\kappa}(\hat{\mathbf{R}}_{pq}) R_{pq}]/R_{pq}$.

In a quite similar way one can show that the sum of all diagrams of the kind shown in Fig. 8.5.6(e) with the same fixed shaded part is given by the diagram shown in Fig. 8.5.6(h).

It is now clear that the total contribution of all diagrams with three fixed common particles t, q, and p to the dyadic correlation function can be represented by the diagram in Fig. 8.5.6(i) or, in expanded form, by the statistical average of the following product:

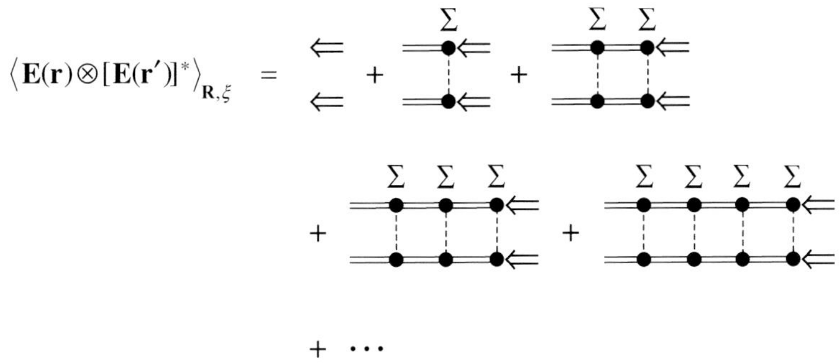

$$\langle \mathbf{E}(\mathbf{r}) \otimes [\mathbf{E}(\mathbf{r}')]^* \rangle_{\mathbf{R},\xi} = \quad + \quad \Sigma \quad + \quad \Sigma \; \Sigma \quad + \quad \Sigma \; \Sigma \; \Sigma \quad + \quad \Sigma \; \Sigma \; \Sigma \; \Sigma \quad + \cdots$$

Figure 8.5.8. Ladder approximation for the dyadic correlation function.

$$\left[\frac{\vec{\eta}(\hat{\mathbf{R}}_{rp}, R_{rp})}{R_{rp}} \cdot \vec{A}_p(\hat{\mathbf{R}}_{rp}, \hat{\mathbf{R}}_{pq}) \cdot \frac{\vec{\eta}(\hat{\mathbf{R}}_{pq}, R_{pq})}{R_{pq}} \cdot \vec{A}_q(\hat{\mathbf{R}}_{pq}, \hat{\mathbf{R}}_{qt}) \cdot \frac{\vec{\eta}(\hat{\mathbf{R}}_{qt}, R_{qt})}{R_{qt}} \cdot \vec{A}_t(\hat{\mathbf{R}}_{qt}, \hat{\mathbf{s}}) \cdot \mathbf{E}_c(\mathbf{R}_t) \right]$$
$$\otimes \left[\frac{\vec{\eta}(\hat{\mathbf{R}}_{r'p}, R_{r'p})}{R_{r'p}} \cdot \vec{A}_p(\hat{\mathbf{R}}_{r'p}, \hat{\mathbf{R}}_{pq}) \cdot \frac{\vec{\eta}(\hat{\mathbf{R}}_{pq}, R_{pq})}{R_{pq}} \cdot \vec{A}_q(\hat{\mathbf{R}}_{pq}, \hat{\mathbf{R}}_{qt}) \cdot \frac{\vec{\eta}(\hat{\mathbf{R}}_{qt}, R_{qt})}{R_{qt}} \cdot \vec{A}_t(\hat{\mathbf{R}}_{qt}, \hat{\mathbf{s}}) \cdot \mathbf{E}_c(\mathbf{R}_t) \right]^*, \tag{8.5.13}$$

where the subscripts r and r' refer to the observation points $\mathbf{r}$ and $\mathbf{r}'$, respectively.

After we have neglected all the diagrams with crossing connectors, computed the contribution of all the diagrams with no connectors, and figured out how to calculate the contributions from various diagrams with one or more vertical connectors, we are perfectly positioned to complete the derivation of the dyadic correlation function. The final result is shown in Fig. 8.5.8, in which the symbols Σ have the usual meaning. Owing to their appearance, the diagrams on the right-hand side of this equation are called ladder diagrams. Therefore, this entire diagrammatic formula can be called the ladder approximation for the dyadic correlation function.

8.6 Integral equation for the ladder specific coherency dyadic

Unlike the dyadic correlation function, which is defined in terms of the electric field vectors at two different observation points $\mathbf{r}$ and $\mathbf{r}'$, the coherency dyadic is a statistical characteristic of the random electric field at a single observation point and is defined as the time average of the coherency dyad of the total local electric field:

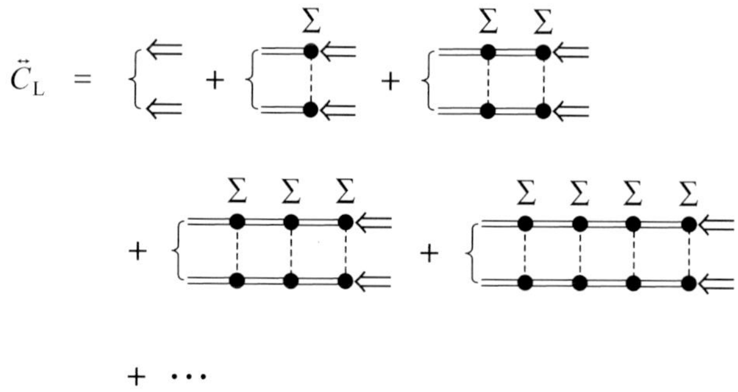

Figure 8.6.1. Ladder approximation for the coherency dyadic.

$$\ddot{C}(\mathbf{r}) = \langle \ddot{\rho}(\mathbf{r}, t) \rangle_t = \langle \mathbf{E}(\mathbf{r}, t) \otimes [\mathbf{E}(\mathbf{r}, t)]^* \rangle_t = \langle \mathbf{E}(\mathbf{r}) \otimes [\mathbf{E}(\mathbf{r})]^* \rangle_{\mathbf{R},\xi}. \tag{8.6.1}$$

The ladder approximation for $\ddot{C}(\mathbf{r})$ is shown in Fig. 8.6.1, in which the subscript L stands for "ladder" and the curly brackets serve to indicate that $\mathbf{r}' = \mathbf{r}$. The expanded form of this approximation follows from Fig. 8.6.2:

$$\begin{aligned}
\ddot{C}_\mathrm{L}(\mathbf{r}) = {} & \ddot{C}_\mathrm{c}(\mathbf{r}) + n_0 \int \mathrm{d}\mathbf{R}_1 \,\mathrm{d}\xi_1 \frac{\ddot{\eta}(\hat{\mathbf{r}}_1, r_1)}{r_1} \cdot \ddot{A}_1(\hat{\mathbf{r}}_1, \hat{\mathbf{s}}) \cdot \ddot{C}_\mathrm{c}(\mathbf{R}_1) \cdot [\ddot{A}_1(\hat{\mathbf{r}}_1, \hat{\mathbf{s}})]^{\mathrm{T}*} \\
& \cdot \frac{[\ddot{\eta}(\hat{\mathbf{r}}_1, r_1)]^{\mathrm{T}*}}{r_1} \\
& + n_0^2 \int \mathrm{d}\mathbf{R}_1 \,\mathrm{d}\xi_1 \int \mathrm{d}\mathbf{R}_2 \,\mathrm{d}\xi_2 \frac{\ddot{\eta}(\hat{\mathbf{r}}_1, r_1)}{r_1} \cdot \ddot{A}_1(\hat{\mathbf{r}}_1, \hat{\mathbf{R}}_{12}) \cdot \frac{\ddot{\eta}(\hat{\mathbf{R}}_{12}, R_{12})}{R_{12}} \\
& \cdot \ddot{A}_2(\hat{\mathbf{R}}_{12}, \hat{\mathbf{s}}) \cdot \ddot{C}_\mathrm{c}(\mathbf{R}_2) \cdot [\ddot{A}_2(\hat{\mathbf{R}}_{12}, \hat{\mathbf{s}})]^{\mathrm{T}*} \cdot \frac{[\ddot{\eta}(\hat{\mathbf{R}}_{12}, R_{12})]^{\mathrm{T}*}}{R_{12}} \\
& \cdot [\ddot{A}_1(\hat{\mathbf{r}}_1, \hat{\mathbf{R}}_{12})]^{\mathrm{T}*} \cdot \frac{[\ddot{\eta}(\hat{\mathbf{r}}_1, r_1)]^{\mathrm{T}*}}{r_1} \\
& + n_0^3 \int \mathrm{d}\mathbf{R}_1 \,\mathrm{d}\xi_1 \int \mathrm{d}\mathbf{R}_2 \,\mathrm{d}\xi_2 \int \mathrm{d}\mathbf{R}_3 \,\mathrm{d}\xi_3 \frac{\ddot{\eta}(\hat{\mathbf{r}}_1, r_1)}{r_1} \cdot \ddot{A}_1(\hat{\mathbf{r}}_1, \hat{\mathbf{R}}_{12}) \\
& \cdot \frac{\ddot{\eta}(\hat{\mathbf{R}}_{12}, R_{12})}{R_{12}} \cdot \ddot{A}_2(\hat{\mathbf{R}}_{12}, \hat{\mathbf{R}}_{23}) \cdot \frac{\ddot{\eta}(\hat{\mathbf{R}}_{23}, R_{23})}{R_{23}} \cdot \ddot{A}_3(\hat{\mathbf{R}}_{23}, \hat{\mathbf{s}}) \\
& \cdot \ddot{C}_\mathrm{c}(\mathbf{R}_3) \cdot [\ddot{A}_3(\hat{\mathbf{R}}_{23}, \hat{\mathbf{s}})]^{\mathrm{T}*} \cdot \frac{[\ddot{\eta}(\hat{\mathbf{R}}_{23}, R_{23})]^{\mathrm{T}*}}{R_{23}} \cdot [\ddot{A}_2(\hat{\mathbf{R}}_{12}, \hat{\mathbf{R}}_{23})]^{\mathrm{T}*} \\
& \cdot \frac{[\ddot{\eta}(\hat{\mathbf{R}}_{12}, R_{12})]^{\mathrm{T}*}}{R_{12}} \cdot [\ddot{A}_1(\hat{\mathbf{r}}_1, \hat{\mathbf{R}}_{12})]^{\mathrm{T}*} \cdot \frac{[\ddot{\eta}(\hat{\mathbf{r}}_1, r_1)]^{\mathrm{T}*}}{r_1}
\end{aligned}$$

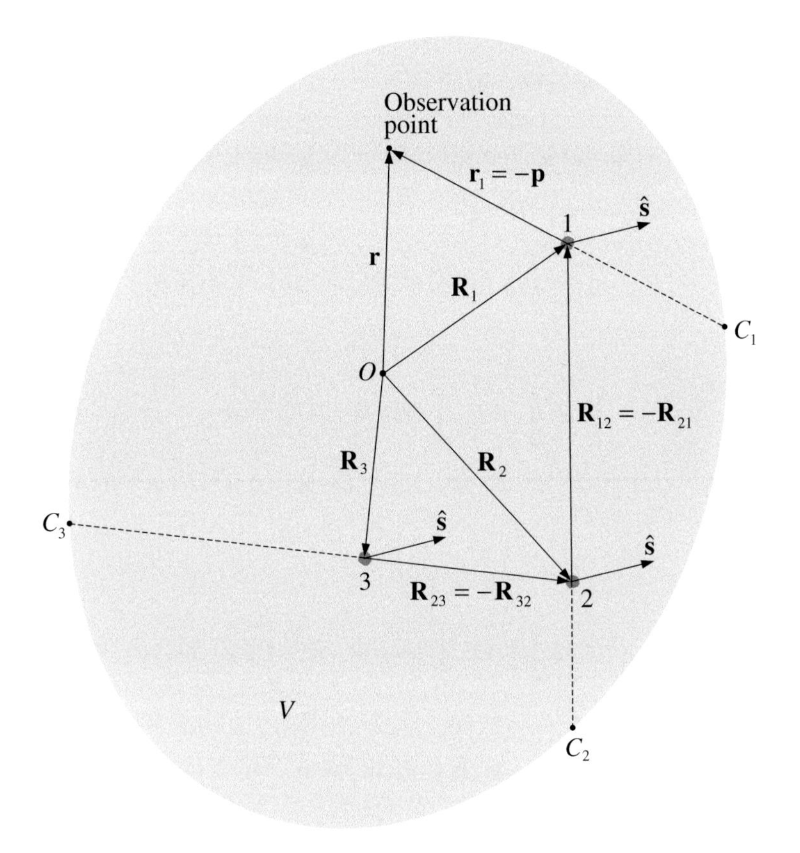

Figure 8.6.2. Geometry showing the quantities used in Eq. (8.6.2).

$$+ \cdots, \tag{8.6.2}$$

where

$$\tilde{C}_c(\mathbf{r}) = \mathbf{E}_c(\mathbf{r}) \otimes [\mathbf{E}_c(\mathbf{r})]^* \tag{8.6.3}$$

is the coherent part of the coherency dyadic, and we have taken into account Eqs. (A.7) and (A.8).

It is convenient to integrate over all positions of particle 1 using a local coordinate system with origin at the observation point, integrate over all positions of particle 2 using a local coordinate system with origin at the origin of particle 1, integrate over all positions of particle 3 using a local coordinate system with origin at the origin of particle 2, etc. Using the notation introduced in Fig. 8.6.2 and taking into account that

$$\mathrm{d}\mathbf{p} = r_1^2 \mathrm{d}p \mathrm{d}\hat{\mathbf{p}}, \tag{8.6.4}$$

$$\mathrm{d}\mathbf{R}_{21} = R_{12}^2 \mathrm{d}R_{21} \mathrm{d}\hat{\mathbf{R}}_{21}, \tag{8.6.5}$$

and so on, we get from Eq. (8.6.2)

$$\ddot{C}_{\mathrm{L}}(\mathbf{r}) = \int_{4\pi} \mathrm{d}\hat{\mathbf{p}}\, \ddot{\Sigma}_{\mathrm{L}}(\mathbf{r}, -\hat{\mathbf{p}}), \tag{8.6.6}$$

where $\ddot{\Sigma}_{\mathrm{L}}(\mathbf{r}, -\hat{\mathbf{p}})$ is the ladder specific coherency dyadic defined by

$$\begin{aligned}
\ddot{\Sigma}_{\mathrm{L}}(\mathbf{r}, -\hat{\mathbf{p}}) = {} & \delta(\hat{\mathbf{p}}+\hat{\mathbf{s}})\ddot{C}_{\mathrm{c}}(\mathbf{r}) \\
& + n_0 \int \mathrm{d}p\, \mathrm{d}\xi_1\, \ddot{\eta}(-\hat{\mathbf{p}}, p) \cdot \ddot{A}_1(-\hat{\mathbf{p}}, \hat{\mathbf{s}}) \cdot \ddot{C}_{\mathrm{c}}(\mathbf{r}+\mathbf{p}) \\
& \qquad \cdot [\ddot{A}_1(-\hat{\mathbf{p}}, \hat{\mathbf{s}})]^{\mathrm{T}*} \cdot [\ddot{\eta}(-\hat{\mathbf{p}}, p)]^{\mathrm{T}*} \\
& + n_0^2 \int \mathrm{d}p\, \mathrm{d}\xi_1 \int \mathrm{d}R_{21}\, \mathrm{d}\hat{\mathbf{R}}_{21}\, \mathrm{d}\xi_2\, \ddot{\eta}(-\hat{\mathbf{p}}, p) \cdot \ddot{A}_1(-\hat{\mathbf{p}}, -\hat{\mathbf{R}}_{21}) \\
& \qquad \cdot \ddot{\eta}(-\hat{\mathbf{R}}_{21}, R_{21}) \cdot \ddot{A}_2(-\hat{\mathbf{R}}_{21}, \hat{\mathbf{s}}) \cdot \ddot{C}_{\mathrm{c}}(\mathbf{r}+\mathbf{p}+\mathbf{R}_{21}) \\
& \qquad \cdot [\ddot{A}_2(-\hat{\mathbf{R}}_{21}, \hat{\mathbf{s}})]^{\mathrm{T}*} \cdot [\ddot{\eta}(-\hat{\mathbf{R}}_{21}, R_{21})]^{\mathrm{T}*} \cdot [\ddot{A}_1(-\hat{\mathbf{p}}, -\hat{\mathbf{R}}_{21})]^{\mathrm{T}*} \\
& \qquad \cdot [\ddot{\eta}(-\hat{\mathbf{p}}, p)]^{\mathrm{T}*} \\
& + n_0^3 \int \mathrm{d}p\, \mathrm{d}\xi_1 \int \mathrm{d}R_{21}\, \mathrm{d}\hat{\mathbf{R}}_{21}\, \mathrm{d}\xi_2 \int \mathrm{d}R_{32}\, \mathrm{d}\hat{\mathbf{R}}_{32}\, \mathrm{d}\xi_3\, \ddot{\eta}(-\hat{\mathbf{p}}, p) \\
& \qquad \cdot \ddot{A}_1(-\hat{\mathbf{p}}, -\hat{\mathbf{R}}_{21}) \cdot \ddot{\eta}(-\hat{\mathbf{R}}_{21}, R_{21}) \cdot \ddot{A}_2(-\hat{\mathbf{R}}_{21}, -\hat{\mathbf{R}}_{32}) \\
& \qquad \cdot \ddot{\eta}(-\hat{\mathbf{R}}_{32}, R_{32}) \cdot \ddot{A}_3(-\hat{\mathbf{R}}_{32}, \hat{\mathbf{s}}) \cdot \ddot{C}_{\mathrm{c}}(\mathbf{r}+\mathbf{p}+\mathbf{R}_{21}+\mathbf{R}_{32}) \\
& \qquad \cdot [\ddot{A}_3(-\hat{\mathbf{R}}_{32}, \hat{\mathbf{s}})]^{\mathrm{T}*} \cdot [\ddot{\eta}(-\hat{\mathbf{R}}_{32}, R_{32})]^{\mathrm{T}*} \cdot [\ddot{A}_2(-\hat{\mathbf{R}}_{21}, -\hat{\mathbf{R}}_{32})]^{\mathrm{T}*} \\
& \qquad \cdot [\ddot{\eta}(-\hat{\mathbf{R}}_{21}, R_{21})]^{\mathrm{T}*} \cdot [\ddot{A}_1(-\hat{\mathbf{p}}, -\hat{\mathbf{R}}_{21})]^{\mathrm{T}*} \cdot [\ddot{\eta}(-\hat{\mathbf{p}}, p)]^{\mathrm{T}*} \\
& + \cdots .
\end{aligned} \tag{8.6.7}$$

Note that p ranges from zero at the observation point to the corresponding value at the point where the straight line in the $\hat{\mathbf{p}}$ direction crosses the boundary of the medium (point C_1 in Fig. 8.6.2), R_{21} ranges from zero at the origin of particle 1 to the corresponding value at point C_2, etc. Importantly, the ladder specific coherency dyadic has the dimension of specific intensity or radiance ($\mathrm{W\,m^{-2}\,sr^{-1}}$) rather than that of intensity ($\mathrm{W\,m^{-2}}$).

It can be easily verified that the ladder specific coherency dyadic satisfies the following integral equation:

$$\begin{aligned}
\ddot{\Sigma}_{\mathrm{L}}(\mathbf{r}, -\hat{\mathbf{p}}) = {} & \delta(\hat{\mathbf{p}}+\hat{\mathbf{s}})\ddot{C}_{\mathrm{c}}(\mathbf{r}) \\
& + n_0 \int \mathrm{d}p\, \mathrm{d}\hat{\mathbf{p}}'\, \mathrm{d}\xi\, \ddot{\eta}(-\hat{\mathbf{p}}, p) \cdot \ddot{A}(\xi; -\hat{\mathbf{p}}, -\hat{\mathbf{p}}') \cdot \ddot{\Sigma}_{\mathrm{L}}(\mathbf{r}+\mathbf{p}, -\hat{\mathbf{p}}') \\
& \qquad \cdot [\ddot{A}(\xi; -\hat{\mathbf{p}}, -\hat{\mathbf{p}}')]^{\mathrm{T}*} \cdot [\ddot{\eta}(-\hat{\mathbf{p}}, p)]^{\mathrm{T}*}.
\end{aligned} \tag{8.6.8}$$

Indeed, using $\delta(\hat{\mathbf{p}}+\hat{\mathbf{s}})\ddot{C}_{\mathrm{c}}(\mathbf{r})$ as an initial approximation for $\ddot{\Sigma}_{\mathrm{L}}(\mathbf{r}, -\hat{\mathbf{p}})$, we can substitute it in the integral on the right-hand side of Eq. (8.6.8) and obtain an improved approximation. By continuing this iterative process, we arrive at Eq. (8.6.7), which is simply the Neumann order-of-scattering expansion of the ladder specific coherency dyadic with coherent field serving as the source of multiple scattering.

The interpretation of Eq. (8.6.8) is very transparent: the ladder specific coherency dyadic for a direction $-\hat{\mathbf{p}}$ at a point $\mathbf{r}$ consists of a coherent part and an incoherent part. The latter is a cumulative contribution of all particles located along the straight line in the $\hat{\mathbf{p}}$-direction and scattering radiation coming from all directions $-\hat{\mathbf{p}}'$ into the direction $-\hat{\mathbf{p}}$.

8.7 Integro-differential equation for the diffuse specific coherency dyadic

To derive the integro-differential form of Eq. (8.6.8), we introduce a q-axis as shown in Fig. 8.7.1. This axis originates at point C and goes through the observation point in the direction of the unit vector $\hat{\mathbf{q}} = -\hat{\mathbf{p}}$ (see Fig. 8.6.2). We can now rewrite Eq. (8.6.8) as

$$\begin{aligned}\ddot{\Sigma}_{\mathrm{L}}(Q, \hat{\mathbf{q}}) &= \delta(\hat{\mathbf{q}} - \hat{\mathbf{s}})\ddot{C}_{\mathrm{c}}(Q) \\ &\quad + n_0 \int_0^Q \mathrm{d}q \int \mathrm{d}\xi \int_{4\pi} \mathrm{d}\hat{\mathbf{q}}' \ddot{\eta}(\hat{\mathbf{q}}, Q-q) \cdot \ddot{A}(\xi; \hat{\mathbf{q}}, \hat{\mathbf{q}}') \cdot \ddot{\Sigma}_{\mathrm{L}}(q, \hat{\mathbf{q}}') \\ &\quad \cdot [\ddot{A}(\xi; \hat{\mathbf{q}}, \hat{\mathbf{q}}')]^{\mathrm{T}*} \cdot [\ddot{\eta}(\hat{\mathbf{q}}, Q-q)]^{\mathrm{T}*}. \end{aligned} \tag{8.7.1}$$

The diffuse specific coherency dyadic is defined as the difference between the full ladder specific coherency dyadic and its coherent component:

$$\ddot{\Sigma}_{\mathrm{d}}(Q, \hat{\mathbf{q}}) = \ddot{\Sigma}_{\mathrm{L}}(Q, \hat{\mathbf{q}}) - \delta(\hat{\mathbf{q}} - \hat{\mathbf{s}})\ddot{C}_{\mathrm{c}}(Q). \tag{8.7.2}$$

The integral equation for $\ddot{\Sigma}_{\mathrm{d}}(Q, \hat{\mathbf{q}})$ follows from Eqs. (8.7.1) and (8.7.2):

$$\begin{aligned}\ddot{\Sigma}_{\mathrm{d}}(Q, \hat{\mathbf{q}}) &= n_0 \int_0^Q \mathrm{d}q \int \mathrm{d}\xi\, \ddot{\eta}(\hat{\mathbf{q}}, Q-q) \cdot \ddot{A}(\xi; \hat{\mathbf{q}}, \hat{\mathbf{s}}) \cdot \ddot{C}_{\mathrm{c}}(q) \cdot [\ddot{A}(\xi; \hat{\mathbf{q}}, \hat{\mathbf{s}})]^{\mathrm{T}*} \\ &\quad \cdot [\ddot{\eta}(\hat{\mathbf{q}}, Q-q)]^{\mathrm{T}*} \\ &\quad + n_0 \int_0^Q \mathrm{d}q \int \mathrm{d}\xi \int_{4\pi} \mathrm{d}\hat{\mathbf{q}}' \ddot{\eta}(\hat{\mathbf{q}}, Q-q) \cdot \ddot{A}(\xi; \hat{\mathbf{q}}, \hat{\mathbf{q}}') \cdot \ddot{\Sigma}_{\mathrm{d}}(q, \hat{\mathbf{q}}') \\ &\quad \cdot [\ddot{A}(\xi; \hat{\mathbf{q}}, \hat{\mathbf{q}}')]^{\mathrm{T}*} \cdot [\ddot{\eta}(\hat{\mathbf{q}}, Q-q)]^{\mathrm{T}*}. \end{aligned} \tag{8.7.3}$$

Differentiating both sides of Eq. (8.7.3) and applying the Leibniz rule,

$$\frac{\partial}{\partial b} \int_a^b \mathrm{d}x\, f(x, b) = f(b, b) + \int_a^b \mathrm{d}x \frac{\partial f(x, b)}{\partial b},$$

finally yields

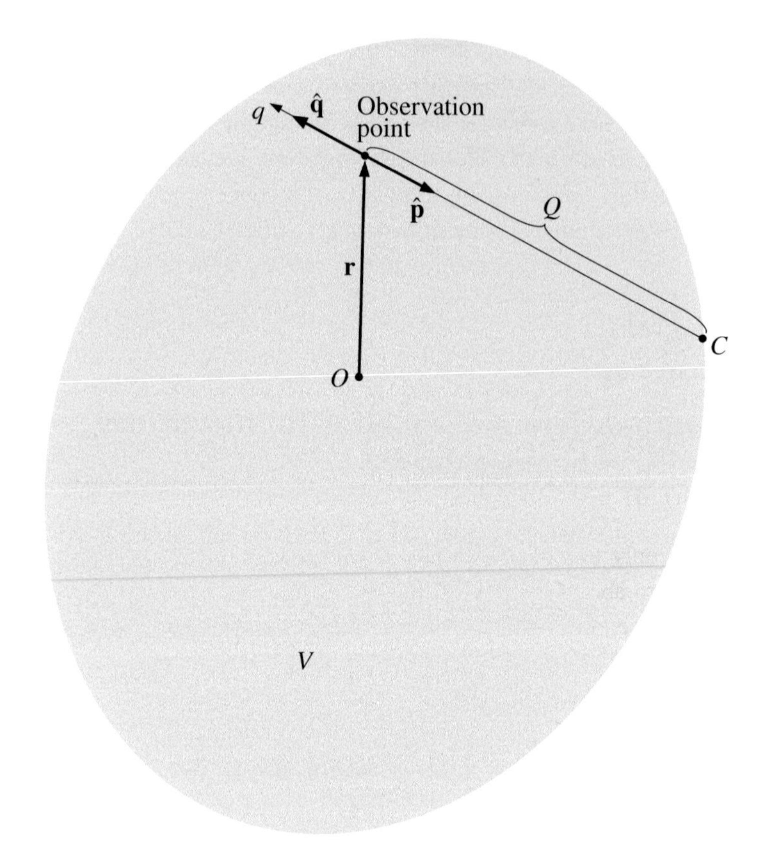

Figure 8.7.1. Geometry showing the quantities used in the derivation of the integro-differential form of the RTE.

$$
\begin{aligned}
\frac{\mathrm{d}\ddot{\Sigma}_{\mathrm{d}}(Q,\hat{\mathbf{q}})}{\mathrm{d}Q} &= \mathrm{i}\ddot{\kappa}(\hat{\mathbf{q}})\cdot\ddot{\Sigma}_{\mathrm{d}}(Q,\hat{\mathbf{q}}) - \mathrm{i}\ddot{\Sigma}_{\mathrm{d}}(Q,\hat{\mathbf{q}})\cdot[\ddot{\kappa}(\hat{\mathbf{q}})]^{\mathrm{T}*} \\
&+ n_0\int \mathrm{d}\xi \int_{4\pi}\mathrm{d}\hat{\mathbf{q}}'\,\ddot{A}(\xi;\hat{\mathbf{q}},\hat{\mathbf{q}}')\cdot\ddot{\Sigma}_{\mathrm{d}}(Q,\hat{\mathbf{q}}')\cdot[\ddot{A}(\xi;\hat{\mathbf{q}},\hat{\mathbf{q}}')]^{\mathrm{T}*} \\
&+ n_0\int \mathrm{d}\xi\,\ddot{A}(\xi;\hat{\mathbf{q}},\hat{\mathbf{s}})\cdot\ddot{C}_{\mathrm{c}}(Q)\cdot[\ddot{A}(\xi;\hat{\mathbf{q}},\hat{\mathbf{s}})]^{\mathrm{T}*}.
\end{aligned}
\tag{8.7.4}
$$

For further use, it is more convenient to rewrite Eqs. (8.7.1) and (8.7.4) in the following form:

$$
\ddot{\Sigma}_{\mathrm{d}}(\mathbf{r},\hat{\mathbf{q}}) = \ddot{\Sigma}_{\mathrm{L}}(\mathbf{r},\hat{\mathbf{q}}) - \delta(\hat{\mathbf{q}}-\hat{\mathbf{s}})\ddot{C}_{\mathrm{c}}(\mathbf{r}),
\tag{8.7.5}
$$

$$
\begin{aligned}
\frac{\mathrm{d}\ddot{\Sigma}_{\mathrm{d}}(\mathbf{r},\hat{\mathbf{q}})}{\mathrm{d}q} &= \mathrm{i}\ddot{\kappa}(\hat{\mathbf{q}})\cdot\ddot{\Sigma}_{\mathrm{d}}(\mathbf{r},\hat{\mathbf{q}}) - \mathrm{i}\ddot{\Sigma}_{\mathrm{d}}(\mathbf{r},\hat{\mathbf{q}})\cdot[\ddot{\kappa}(\hat{\mathbf{q}})]^{\mathrm{T}*} \\
&+ n_0\int \mathrm{d}\xi \int_{4\pi}\mathrm{d}\hat{\mathbf{q}}'\,\ddot{A}(\xi;\hat{\mathbf{q}},\hat{\mathbf{q}}')\cdot\ddot{\Sigma}_{\mathrm{d}}(\mathbf{r},\hat{\mathbf{q}}')\cdot[\ddot{A}(\xi;\hat{\mathbf{q}},\hat{\mathbf{q}}')]^{\mathrm{T}*}
\end{aligned}
$$

$$+ n_0 \int d\xi \, \ddot{A}(\xi; \hat{\mathbf{q}}, \hat{\mathbf{s}}) \cdot \ddot{C}_c(\mathbf{r}) \cdot [\ddot{A}(\xi; \hat{\mathbf{q}}, \hat{\mathbf{s}})]^{T*}, \tag{8.7.6}$$

where the path-length element dq is measured along the unit vector $\hat{\mathbf{q}}$. Equation (8.7.6) is the integro-differential radiative transfer equation for the diffuse specific coherency dyadic.

8.8 Integral and integro-differential equations for the diffuse specific coherency matrix

It follows from Eq. (8.7.3) that

$$\hat{\mathbf{q}} \cdot \ddot{\Sigma}_d(\mathbf{r}, \hat{\mathbf{q}}) = \ddot{\Sigma}_d(\mathbf{r}, \hat{\mathbf{q}}) \cdot \hat{\mathbf{q}} = \mathbf{0}. \tag{8.8.1}$$

This means that only four out of nine components of the diffuse specific coherency dyadic are nonzero and allows us to introduce the diffuse specific coherency matrix $\widetilde{\boldsymbol{\rho}}_d$ using the local coordinate system with origin at the observation point and orientation identical to that of the laboratory coordinate system:

$$\widetilde{\boldsymbol{\rho}}_d(\mathbf{r}, \hat{\mathbf{q}}) = \frac{1}{2}\sqrt{\frac{\epsilon_1}{\mu_0}} \begin{bmatrix} \hat{\boldsymbol{\theta}}(\hat{\mathbf{q}}) \cdot \ddot{\Sigma}_d(\mathbf{r}, \hat{\mathbf{q}}) \cdot \hat{\boldsymbol{\theta}}(\hat{\mathbf{q}}) & \hat{\boldsymbol{\theta}}(\hat{\mathbf{q}}) \cdot \ddot{\Sigma}_d(\mathbf{r}, \hat{\mathbf{q}}) \cdot \hat{\boldsymbol{\varphi}}(\hat{\mathbf{q}}) \\ \hat{\boldsymbol{\varphi}}(\hat{\mathbf{q}}) \cdot \ddot{\Sigma}_d(\mathbf{r}, \hat{\mathbf{q}}) \cdot \hat{\boldsymbol{\theta}}(\hat{\mathbf{q}}) & \hat{\boldsymbol{\varphi}}(\hat{\mathbf{q}}) \cdot \ddot{\Sigma}_d(\mathbf{r}, \hat{\mathbf{q}}) \cdot \hat{\boldsymbol{\varphi}}(\hat{\mathbf{q}}) \end{bmatrix}. \tag{8.8.2}$$

Note that we use a tilde in order to distinguish between the specific coherency matrix and the coherency matrix defined by Eq. (2.6.2). We can then rewrite Eqs. (8.7.3) and (8.7.6) in the form of the integral and integro-differential equations for the diffuse specific coherency matrix:

$$\begin{aligned} \widetilde{\boldsymbol{\rho}}_d(Q, \hat{\mathbf{q}}) = {} & n_0 \int_0^Q dq \int d\xi \, \mathbf{h}(\hat{\mathbf{q}}, Q-q) \mathbf{S}(\xi; \hat{\mathbf{q}}, \hat{\mathbf{s}}) \boldsymbol{\rho}_c(q) \\ & \times [\mathbf{S}(\xi; \hat{\mathbf{q}}, \hat{\mathbf{s}})]^{T*} [\mathbf{h}(\hat{\mathbf{q}}, Q-q)]^{T*} \\ & + n_0 \int_0^Q dq \int d\xi \int_{4\pi} d\hat{\mathbf{q}}' \, \mathbf{h}(\hat{\mathbf{q}}, Q-q) \mathbf{S}(\xi; \hat{\mathbf{q}}, \hat{\mathbf{q}}') \widetilde{\boldsymbol{\rho}}_d(q, \hat{\mathbf{q}}') \\ & \times [\mathbf{S}(\xi; \hat{\mathbf{q}}, \hat{\mathbf{q}}')]^{T*} [\mathbf{h}(\hat{\mathbf{q}}, Q-q)]^{T*}, \end{aligned} \tag{8.8.3}$$

$$\begin{aligned} \frac{d\widetilde{\boldsymbol{\rho}}_d(\mathbf{r}, \hat{\mathbf{q}})}{dq} = {} & i\mathbf{k}(\hat{\mathbf{q}}) \widetilde{\boldsymbol{\rho}}_d(\mathbf{r}, \hat{\mathbf{q}}) - i\widetilde{\boldsymbol{\rho}}_d(\mathbf{r}, \hat{\mathbf{q}}) [\mathbf{k}(\hat{\mathbf{q}})]^{T*} \\ & + n_0 \int d\xi \int_{4\pi} d\hat{\mathbf{q}}' \, \mathbf{S}(\xi; \hat{\mathbf{q}}, \hat{\mathbf{q}}') \widetilde{\boldsymbol{\rho}}_d(\mathbf{r}, \hat{\mathbf{q}}') [\mathbf{S}(\xi; \hat{\mathbf{q}}, \hat{\mathbf{q}}')]^{T*} \\ & + n_0 \int d\xi \, \mathbf{S}(\xi; \hat{\mathbf{q}}, \hat{\mathbf{s}}) \boldsymbol{\rho}_c(\mathbf{r}) [\mathbf{S}(\xi; \hat{\mathbf{q}}, \hat{\mathbf{s}})]^{T*}, \end{aligned} \tag{8.8.4}$$

where $\mathbf{S}$ is the amplitude scattering matrix, $\mathbf{h}$ is the coherent transmission amplitude matrix given by Eq. (8.3.26), $\mathbf{k}$ is the matrix propagation constant given by Eqs. (8.3.24), and

$$\boldsymbol{\rho}_{\rm c}(\mathbf{r}) = \frac{1}{2}\sqrt{\frac{\epsilon_1}{\mu_0}}\begin{bmatrix} \hat{\boldsymbol{\theta}}(\hat{\mathbf{s}})\cdot\ddot{C}_{\rm c}(\mathbf{r})\cdot\hat{\boldsymbol{\theta}}(\hat{\mathbf{s}}) & \hat{\boldsymbol{\theta}}(\hat{\mathbf{s}})\cdot\ddot{C}_{\rm c}(\mathbf{r})\cdot\hat{\boldsymbol{\varphi}}(\hat{\mathbf{s}}) \\ \hat{\boldsymbol{\varphi}}(\hat{\mathbf{s}})\cdot\ddot{C}_{\rm c}(\mathbf{r})\cdot\hat{\boldsymbol{\theta}}(\hat{\mathbf{s}}) & \hat{\boldsymbol{\varphi}}(\hat{\mathbf{s}})\cdot\ddot{C}_{\rm c}(\mathbf{r})\cdot\hat{\boldsymbol{\varphi}}(\hat{\mathbf{s}}) \end{bmatrix}. \tag{8.8.5}$$

8.9 Integral and integro-differential equations for the diffuse specific coherency column vector

The next obvious step is to introduce the corresponding coherency column vectors $\widetilde{\mathbf{J}}_{\rm d}$ and $\mathbf{J}_{\rm c}$ in terms of $\widetilde{\boldsymbol{\rho}}_{\rm d}$ and $\boldsymbol{\rho}_{\rm c}$:

$$\widetilde{\mathbf{J}}_{\rm d}(\mathbf{r},\hat{\mathbf{q}}) = \begin{bmatrix} \widetilde{\rho}_{\rm d11}(\mathbf{r},\hat{\mathbf{q}}) \\ \widetilde{\rho}_{\rm d12}(\mathbf{r},\hat{\mathbf{q}}) \\ \widetilde{\rho}_{\rm d21}(\mathbf{r},\hat{\mathbf{q}}) \\ \widetilde{\rho}_{\rm d22}(\mathbf{r},\hat{\mathbf{q}}) \end{bmatrix}, \tag{8.9.1}$$

$$\mathbf{J}_{\rm c}(\mathbf{r}) = \begin{bmatrix} \rho_{\rm c11}(\mathbf{r}) \\ \rho_{\rm c12}(\mathbf{r}) \\ \rho_{\rm c21}(\mathbf{r}) \\ \rho_{\rm c22}(\mathbf{r}) \end{bmatrix} \tag{8.9.2}$$

(cf. Eq. (2.6.3)). Again we use a tilde to emphasize that $\widetilde{\mathbf{J}}_{\rm d}$ has the dimension of specific intensity. After lengthy, but simple algebraic manipulations, we get

$$\widetilde{\mathbf{J}}_{\rm d}(Q,\hat{\mathbf{q}}) = n_0\int_0^Q {\rm d}q\,\mathbf{H}^J(\hat{\mathbf{q}},Q-q)\langle\mathbf{Z}^J(\hat{\mathbf{q}},\hat{\mathbf{s}})\rangle_\xi\,\mathbf{J}_{\rm c}(q)$$
$$+\, n_0\int_0^Q {\rm d}q\int_{4\pi}{\rm d}\hat{\mathbf{q}}'\,\mathbf{H}^J(\hat{\mathbf{q}},Q-q)\langle\mathbf{Z}^J(\hat{\mathbf{q}},\hat{\mathbf{q}}')\rangle_\xi\,\widetilde{\mathbf{J}}_{\rm d}(q,\hat{\mathbf{q}}'), \tag{8.9.3}$$

$$\frac{{\rm d}\widetilde{\mathbf{J}}_{\rm d}(\mathbf{r},\hat{\mathbf{q}})}{{\rm d}q} = -n_0\langle\mathbf{K}^J(\hat{\mathbf{q}})\rangle_\xi\,\widetilde{\mathbf{J}}_{\rm d}(\mathbf{r},\hat{\mathbf{q}}) + n_0\int_{4\pi}{\rm d}\hat{\mathbf{q}}'\langle\mathbf{Z}^J(\hat{\mathbf{q}},\hat{\mathbf{q}}')\rangle_\xi\,\widetilde{\mathbf{J}}_{\rm d}(\mathbf{r},\hat{\mathbf{q}}')$$
$$+\, n_0\langle\mathbf{Z}^J(\hat{\mathbf{q}},\hat{\mathbf{s}})\rangle_\xi\,\mathbf{J}_{\rm c}(\mathbf{r}), \tag{8.9.4}$$

where

$$\mathbf{H}^J(\hat{\mathbf{s}},s) = \exp\{-n_0 s\langle\mathbf{K}^J(\hat{\mathbf{s}})\rangle_\xi\}, \tag{8.9.5}$$

$\langle\mathbf{K}^J(\hat{\mathbf{q}})\rangle_\xi$ is the coherency extinction matrix averaged over the particle states, and $\langle\mathbf{Z}^J(\hat{\mathbf{q}},\hat{\mathbf{q}}')\rangle_\xi$ is the ensemble average of the coherency phase matrix given by Eq. (3.7.5). The column vector $\mathbf{J}_{\rm c}(\mathbf{r})$ satisfies the transfer equation (8.4.2).

8.10 Integral and integro-differential equations for the specific intensity column vector

Our final step is to define the diffuse specific intensity column vector and the coherent Stokes column vector,

$$\tilde{\mathbf{I}}_{\rm d}(\mathbf{r},\hat{\mathbf{q}}) = \begin{bmatrix} \tilde{I}_{\rm d}(\mathbf{r},\hat{\mathbf{q}}) \\ \tilde{Q}_{\rm d}(\mathbf{r},\hat{\mathbf{q}}) \\ \tilde{U}_{\rm d}(\mathbf{r},\hat{\mathbf{q}}) \\ \tilde{V}_{\rm d}(\mathbf{r},\hat{\mathbf{q}}) \end{bmatrix} = \mathbf{D}\tilde{\mathbf{J}}_{\rm d}(\mathbf{r},\hat{\mathbf{q}}), \tag{8.10.1}$$

$$\mathbf{I}_{\rm c}(\mathbf{r}) = \begin{bmatrix} I_{\rm c}(\mathbf{r}) \\ Q_{\rm c}(\mathbf{r}) \\ U_{\rm c}(\mathbf{r}) \\ V_{\rm c}(\mathbf{r}) \end{bmatrix} = \mathbf{D}\mathbf{J}_{\rm c}(\mathbf{r}) \tag{8.10.2}$$

(cf. Eqs. (2.6.4) and (2.6.5)), and rewrite Eqs. (8.9.3) and (8.9.4) in the form

$$\begin{aligned} \tilde{\mathbf{I}}_{\rm d}(Q,\hat{\mathbf{q}}) &= n_0 \int_0^Q {\rm d}q\, \mathbf{H}(\hat{\mathbf{q}}, Q-q)\langle \mathbf{Z}(\hat{\mathbf{q}},\hat{\mathbf{s}})\rangle_\xi \mathbf{I}_{\rm c}(q) \\ &\quad + n_0 \int_0^Q {\rm d}q \int_{4\pi} {\rm d}\hat{\mathbf{q}}'\, \mathbf{H}(\hat{\mathbf{q}}, Q-q)\langle \mathbf{Z}(\hat{\mathbf{q}},\hat{\mathbf{q}}')\rangle_\xi \tilde{\mathbf{I}}_{\rm d}(q,\hat{\mathbf{q}}'), \end{aligned} \tag{8.10.3}$$

$$\begin{aligned} \frac{{\rm d}\tilde{\mathbf{I}}_{\rm d}(\mathbf{r},\hat{\mathbf{q}})}{{\rm d}q} &= -n_0\langle \mathbf{K}(\hat{\mathbf{q}})\rangle_\xi \tilde{\mathbf{I}}_{\rm d}(\mathbf{r},\hat{\mathbf{q}}) + n_0 \int_{4\pi} {\rm d}\hat{\mathbf{q}}' \langle \mathbf{Z}(\hat{\mathbf{q}},\hat{\mathbf{q}}')\rangle_\xi \tilde{\mathbf{I}}_{\rm d}(\mathbf{r},\hat{\mathbf{q}}') \\ &\quad + n_0\langle \mathbf{Z}(\hat{\mathbf{q}},\hat{\mathbf{s}})\rangle_\xi \mathbf{I}_{\rm c}(\mathbf{r}), \end{aligned} \tag{8.10.4}$$

where $\langle \mathbf{K}(\hat{\mathbf{q}})\rangle_\xi$ is the ensemble average of the Stokes extinction matrix given by Eq. (3.8.7) and $\langle \mathbf{Z}(\hat{\mathbf{q}},\hat{\mathbf{q}}')\rangle_\xi$ is the ensemble average of the Stokes phase matrix given by Eq. (3.7.10). The coherent Stokes column vector $\mathbf{I}_{\rm c}(\mathbf{r})$ satisfies the transfer equation (8.4.4).

Equations (8.4.4) and (8.10.4) can also be written as

$$\hat{\mathbf{s}}\cdot\nabla\mathbf{I}_{\rm c}(\mathbf{r}) = -n_0\langle \mathbf{K}(\hat{\mathbf{s}})\rangle_\xi \mathbf{I}_{\rm c}(\mathbf{r}), \tag{8.10.5}$$

$$\begin{aligned} \hat{\mathbf{q}}\cdot\nabla\tilde{\mathbf{I}}_{\rm d}(\mathbf{r},\hat{\mathbf{q}}) &= -n_0\langle \mathbf{K}(\hat{\mathbf{q}})\rangle_\xi \tilde{\mathbf{I}}_{\rm d}(\mathbf{r},\hat{\mathbf{q}}) + n_0 \int_{4\pi} {\rm d}\hat{\mathbf{q}}' \langle \mathbf{Z}(\hat{\mathbf{q}},\hat{\mathbf{q}}')\rangle_\xi \tilde{\mathbf{I}}_{\rm d}(\mathbf{r},\hat{\mathbf{q}}') \\ &\quad + n_0\langle \mathbf{Z}(\hat{\mathbf{q}},\hat{\mathbf{s}})\rangle_\xi \mathbf{I}_{\rm c}(\mathbf{r}). \end{aligned} \tag{8.10.6}$$

These equations represent the classical integro-differential form of the vector RTE (VRTE) applicable to arbitrarily shaped and arbitrarily oriented particles. They were initially introduced by Rozenberg (1955) on the basis of heuristic, phenomenological considerations. In contrast, our detailed microphysical derivation is based on fundamental principles of statistical electromagnetics. It naturally replaces the original incident field as the source of multiple scattering in Eq. (8.1.13) by the decaying coherent field in Fig. 8.5.4(c) and leads to the introduction of the diffuse specific intensity column vector describing the photometric and polarimetric characteristics of the multiply scattered light (see the discussion in Section 8.12). Importantly, the microphysical derivation yields directly the integral form of the RTE, the integro-differential form being a corollary of the integral form. This is a striking contrast to the phenomenological approach, which starts with a confusing notion of an elementary volume element and the integro-differential form of the RTE (see Section 8.16).

It is often convenient to introduce the full specific intensity column vector,

$$\tilde{\mathbf{I}}(\mathbf{r},\hat{\mathbf{q}}) = \begin{bmatrix} \tilde{I}(\mathbf{r},\hat{\mathbf{q}}) \\ \tilde{Q}(\mathbf{r},\hat{\mathbf{q}}) \\ \tilde{U}(\mathbf{r},\hat{\mathbf{q}}) \\ \tilde{V}(\mathbf{r},\hat{\mathbf{q}}) \end{bmatrix} = \delta(\hat{\mathbf{q}}-\hat{\mathbf{s}})\mathbf{I}_{\mathrm{c}}(\mathbf{r}) + \tilde{\mathbf{I}}_{\mathrm{d}}(\mathbf{r},\hat{\mathbf{q}}), \tag{8.10.7}$$

and rewrite Eqs. (8.10.5) and (8.10.6) as a single integro-differential equation:

$$\hat{\mathbf{q}}\cdot\nabla\tilde{\mathbf{I}}(\mathbf{r},\hat{\mathbf{q}}) = -n_0\langle\mathbf{K}(\hat{\mathbf{q}})\rangle_\xi\tilde{\mathbf{I}}(\mathbf{r},\hat{\mathbf{q}}) + n_0\int_{4\pi} d\hat{\mathbf{q}}'\langle\mathbf{Z}(\hat{\mathbf{q}},\hat{\mathbf{q}}')\rangle_\xi\tilde{\mathbf{I}}(\mathbf{r},\hat{\mathbf{q}}'). \tag{8.10.8}$$

Accordingly, Eq. (8.10.3) takes the form

$$\tilde{\mathbf{I}}(Q,\hat{\mathbf{q}}) = \delta(\hat{\mathbf{q}}-\hat{\mathbf{s}})\mathbf{I}_{\mathrm{c}}(Q) + n_0\int_0^Q dq\int_{4\pi} d\hat{\mathbf{q}}'\mathbf{H}(\hat{\mathbf{q}},Q-q)\langle\mathbf{Z}(\hat{\mathbf{q}},\hat{\mathbf{q}}')\rangle_\xi\tilde{\mathbf{I}}(q,\hat{\mathbf{q}}'). \tag{8.10.9}$$

In the absence of particles, the coherent Stokes column vector, the diffuse specific intensity column vector, and the full specific intensity column vector become independent of the spatial coordinates. This property follows directly from Eqs. (8.10.5), (8.10.6), and (8.10.8) in the limit $n_0 \to 0$.

8.11 Summary of assumptions and approximations

Since the microphysical derivation of the VRTE is rather lengthy, it is useful to summarize what specific assumptions and approximations had to be made at various stages:

1. We assumed that the scattering medium is illuminated by a plane electromagnetic wave. However, as will be discussed in Section 8.15, the VRTE remains valid in the case of illumination by quasi-monochromatic light.
2. We assumed that each particle is located in the far-field zones of all the other particles and that the observation point is also located in the far-field zones of all the particles forming the scattering medium (Section 4.2).

3. We neglected all scattering paths going through a particle two or more times (the Twersky approximation). As we have seen in Section 8.1, doing this is justified when the total number of particles in the medium is very large.
4. We assumed that the scattering system is ergodic and that averaging over time can be replaced by averaging over particle positions and states.
5. We assumed that (i) the position and state of each particle are statistically independent of each other and of those of all the other particles, and (ii) the spatial distribution of the particles throughout the medium is random and statistically uniform (Section 8.3).
6. We assumed that the scattering medium is convex, which assured that a wave exiting the medium cannot re-enter it (Section 8.3).
7. We assumed that the number of particles N forming the scattering medium is very large and replaced all factors of the type $(N-n)!/(N-n-k)!$ by N^k (Sections 8.2 and 8.5).
8. We ignored all diagrams with crossing connectors in the diagrammatic expansion of the coherency dyadic (the ladder approximation, Sections 8.5 and 8.6).

Assumptions 2 and 7 imply that the overall size of the scattering medium must be much greater than the wavelength, average particle size, and average distance between two neighboring particles. They ensure, in particular, that the exponential factors of the type $\exp(ik_1 r)$ oscillate many times over the distances traveled by the particles during the measurement, thereby leading to Eqs. (8.3.4) and (8.5.11) and, ultimately, to Eqs. (8.3.8) and (8.5.12).

Randomly positioned particles located in the far-field zones of each other are called independent scatterers. Thus, assumptions 2 and 5 explicitly indicate that the requirement of independent scattering is a necessary condition of validity of the RTT. It is these assumptions that are largely responsible for the fact that the VRTE contains single-particle extinction and phase matrices rather than some "group" scattering properties. In other words, each particle is identified as an individual scatterer with scattering and absorption properties calculated under the implicit assumption that all other particles do not exist. Hence the term "independent scattering".

Another consequence of assumptions 2 and 5 is that the average particle number density in the scattering medium must be rather small. Therefore, the VRTE may not be expected to perform well for densely packed media (e.g., Tsang and Kong, 2001).

To justify approximation 8, let us consider, for example, the contributions of two simple two-particle diagrams, shown in panels (a) and (b) of Fig. 8.11.1, to the coherency dyadic of the total electric field. According to Eqs. (8.6.1) and (8.1.11), these contributions are given by

$$\langle (r_i^2 R_{ij}^2)^{-1} [\vec{A}_i(\hat{\mathbf{r}}_i, \hat{\mathbf{R}}_{ij}) \cdot \vec{A}_j(\hat{\mathbf{R}}_{ij}, \hat{\mathbf{s}}) \cdot \mathbf{E}_0^{\mathrm{inc}}] \otimes [\vec{A}_i(\hat{\mathbf{r}}_i, \hat{\mathbf{R}}_{ij}) \cdot \vec{A}_j(\hat{\mathbf{R}}_{ij}, \hat{\mathbf{s}}) \cdot \mathbf{E}_0^{\mathrm{inc}}]^* \rangle_{\mathbf{R},\xi} \tag{8.11.1}$$

and

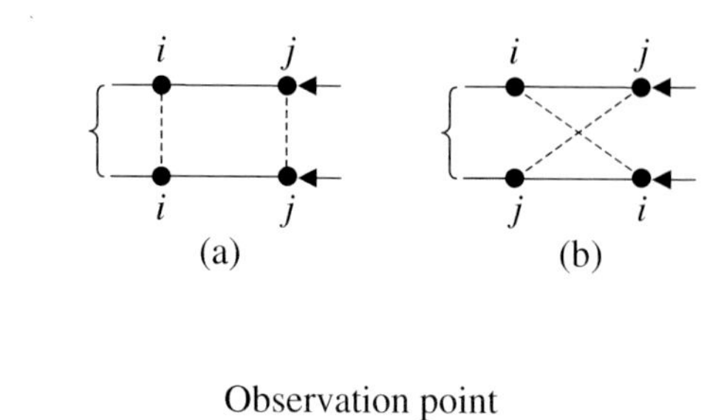

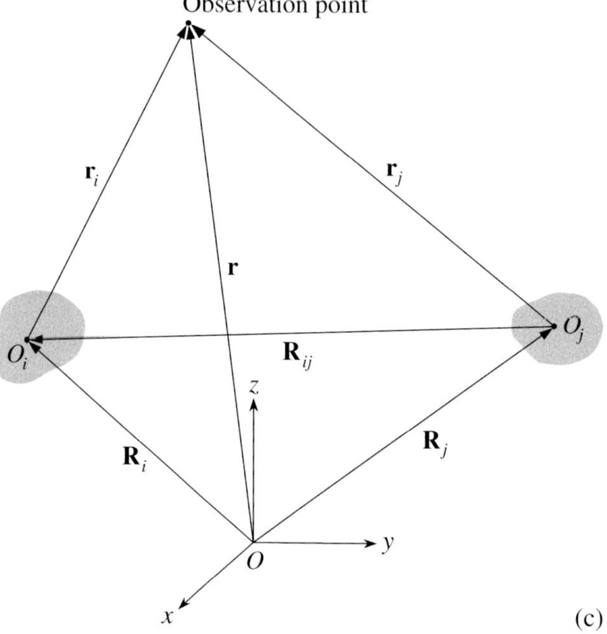

Figure 8.11.1. Two-particle diagrams with vertical and crossing connectors.

$$\langle (r_i r_j R_{ij}^2)^{-1} \exp[\mathrm{i}k_1(r_i - r_j - \hat{\mathbf{s}}\cdot\hat{\mathbf{R}}_{ij})][\vec{A}_i(\hat{\mathbf{r}}_i, \hat{\mathbf{R}}_{ij})\cdot\vec{A}_j(\hat{\mathbf{R}}_{ij}, \hat{\mathbf{s}})\cdot\mathbf{E}_0^{\mathrm{inc}}]$$
$$\otimes [\vec{A}_j(\hat{\mathbf{r}}_j, -\hat{\mathbf{R}}_{ij})\cdot\vec{A}_i(-\hat{\mathbf{R}}_{ij}, \hat{\mathbf{s}})\cdot\mathbf{E}_0^{\mathrm{inc}}]^*\rangle_{\mathbf{R},\xi}, \qquad (8.11.2)$$

respectively, where the notation follows that in Fig. 8.11.1(c). The main difference between the expressions inside the angular brackets in these formulas is that the latter contains a rapidly oscillating exponential factor, which changes with $\mathbf{r}_i$ and $\mathbf{r}_j$ much faster than all other participating factors. The presence of this exponential factor causes the contribution given by Eq. (8.11.2) to vanish upon the configurational averaging. The reader can verify that this is true of any diagram with crossing connectors and explains why their cumulative contribution to the specific coherency dyadic and thus to the diffuse specific intensity column vector is negligible relative to the contribution of the diagrams with vertical connectors.

An important exception is the situation where the observation point is located infinitely far from the scattering volume and is in the direction opposite to the direction of incidence. Then the phase difference

$$\Delta = k_1(r_i - r_j - \hat{\mathbf{s}} \cdot \hat{\mathbf{R}}_{ij})$$

vanishes, the exponential factor becomes identically equal to unity, and the contribution of the diagrams with crossing connectors becomes comparable to the contribution of the ladder diagrams. This remarkable effect, called coherent backscattering, will be discussed specifically in Chapter 14.

The assumption that the scattering medium is statistically uniform simplified greatly the derivation of the VRTE. It is reasonable to expect, however, that the VRTE remains valid for an inhomogeneous medium provided that the macrophysical properties of the medium change on spatial scales much greater than the average distance that light travels between two successive scattering events. In this case the extinction and phase matrices averaged over particle states and the particle number density become functions of spatial coordinates, and Eqs. (8.10.5), (8.10.6), (8.10.8) take the form

$$\hat{\mathbf{s}} \cdot \nabla \mathbf{I}_c(\mathbf{r}) = -n_0(\mathbf{r})\langle \mathbf{K}(\mathbf{r}, \hat{\mathbf{s}})\rangle_\xi \mathbf{I}_c(\mathbf{r}), \tag{8.11.3}$$

$$\begin{aligned} \hat{\mathbf{q}} \cdot \nabla \tilde{\mathbf{I}}_d(\mathbf{r}, \hat{\mathbf{q}}) = &-n_0(\mathbf{r})\langle \mathbf{K}(\mathbf{r}, \hat{\mathbf{q}})\rangle_\xi \tilde{\mathbf{I}}_d(\mathbf{r}, \hat{\mathbf{q}}) \\ &+ n_0(\mathbf{r}) \int_{4\pi} d\hat{\mathbf{q}}' \langle \mathbf{Z}(\mathbf{r}, \hat{\mathbf{q}}, \hat{\mathbf{q}}')\rangle_\xi \tilde{\mathbf{I}}_d(\mathbf{r}, \hat{\mathbf{q}}') \\ &+ n_0(\mathbf{r})\langle \mathbf{Z}(\mathbf{r}, \hat{\mathbf{q}}, \hat{\mathbf{s}})\rangle_\xi \mathbf{I}_c(\mathbf{r}), \end{aligned} \tag{8.11.4}$$

$$\hat{\mathbf{q}} \cdot \nabla \tilde{\mathbf{I}}(\mathbf{r}, \hat{\mathbf{q}}) = -n_0(\mathbf{r})\langle \mathbf{K}(\mathbf{r}, \hat{\mathbf{q}})\rangle_\xi \tilde{\mathbf{I}}(\mathbf{r}, \hat{\mathbf{q}}) + n_0(\mathbf{r}) \int_{4\pi} d\hat{\mathbf{q}}' \langle \mathbf{Z}(\mathbf{r}, \hat{\mathbf{q}}, \hat{\mathbf{q}}')\rangle_\xi \tilde{\mathbf{I}}(\mathbf{r}, \hat{\mathbf{q}}'). \tag{8.11.5}$$

8.12 Physical meaning of the diffuse specific intensity column vector and the coherent Stokes column vector

It might be fair to say that the way in which we derived the VRTE is rather mathematical. Indeed, what we have done so far was to introduce the coherency dyadic and a sequence of *ad hoc* derivative quantities and to see what equations these quantities satisfy. However, the final result can be meaningful and useful only if we can demonstrate the physical relevance of the quantities described by the VRTE.

It turns out that the physical interpretation of the diffuse specific intensity column vector is rather transparent and follows directly from the integral form of the VRTE. Indeed, imagine a well-collimated polarization-sensitive detector centered at the observation point and facing the direction $\hat{\mathbf{q}}$ ($\neq \hat{\mathbf{s}}$) (Figs. 8.7.1 and 8.12.1). Let ΔS be the area of the sensitive surface of the detector and $\Delta\Omega$ its (small) acceptance solid angle. Each infinitesimal element of the detector surface reacts to the radiant energy coming from the directions confined to a narrow cone with the small solid-angle

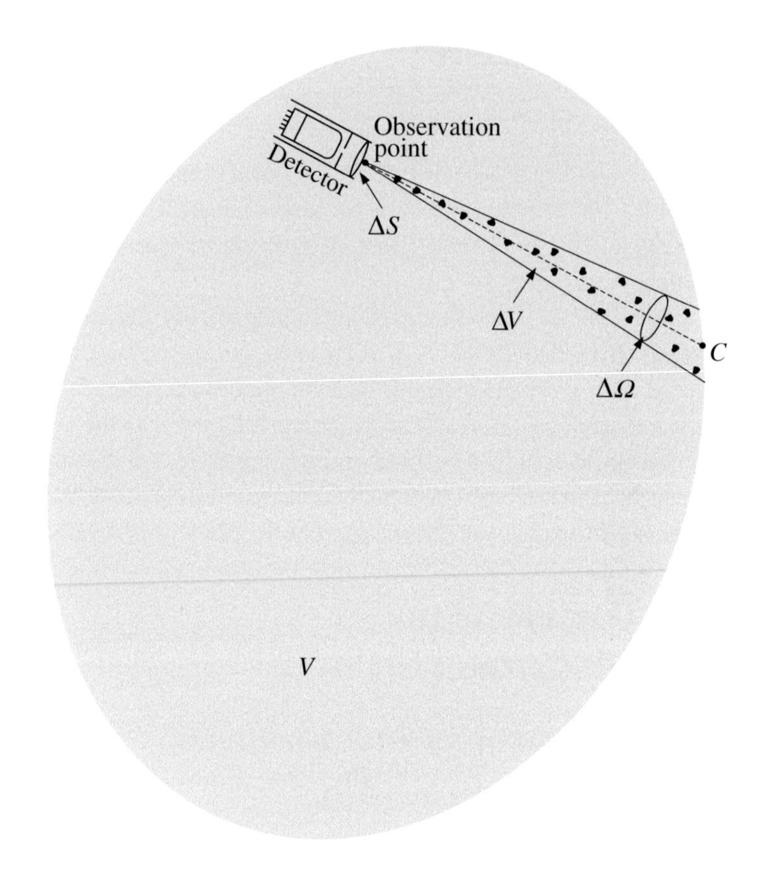

Figure 8.12.1. Physical meaning of the diffuse specific intensity column vector.

aperture $\Delta\Omega$ centered around $\hat{\mathbf{q}}$. On the other hand, we can use Eq. (8.10.3) to write

$$\Delta\Omega\,\tilde{\mathbf{I}}_{\mathrm{d}}(\mathbf{r},\hat{\mathbf{q}}) \approx n_0 \int_{\Delta V} \mathrm{d}\mathbf{p}\, \frac{\mathbf{H}(\hat{\mathbf{q}},p)}{p^2} \times \left[\langle \mathbf{Z}(\hat{\mathbf{q}},\hat{\mathbf{s}})\rangle_\xi \mathbf{I}_{\mathrm{c}}(\mathbf{r}+\mathbf{p}) + \int_{4\pi} \mathrm{d}\hat{\mathbf{q}}' \langle \mathbf{Z}(\hat{\mathbf{q}},\hat{\mathbf{q}}')\rangle_\xi \tilde{\mathbf{I}}_{\mathrm{d}}(\mathbf{r}+\mathbf{p},\hat{\mathbf{q}}') \right], \tag{8.12.1}$$

where $\mathbf{p}$ originates at the observation point $\mathbf{r}$ (Fig. 8.7.1) and the integration is performed over the conical volume element ΔV having the solid-angle aperture $\Delta\Omega$ and extending from the observation point to point C as shown in Fig. 8.12.1. The right-hand side of Eq. (8.12.1) is simply the integral of the scattering signal per unit surface area perpendicular to $\hat{\mathbf{q}}$ per unit time over all particles contained in the conical volume element. It is, thus, clear what quantity describes the total polarized signal measured by the detector per unit time: it is the product

$$\Delta S \Delta\Omega\,\tilde{\mathbf{I}}_{\mathrm{d}}(\mathbf{r},\hat{\mathbf{q}}),$$

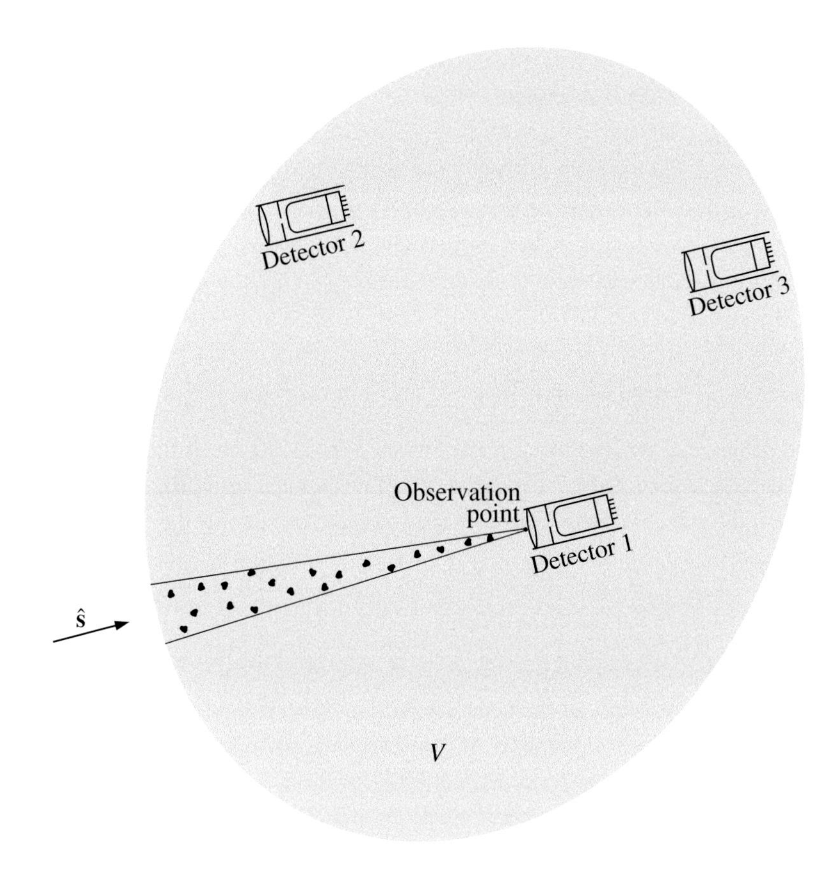

Figure 8.12.2. Detectors facing the external light.

which has the dimension of power (W). The fact that the diffuse specific intensity column vector can be measured by a polarization-sensitive optical device as well as computed theoretically by solving the VRTE explains the practical usefulness of this quantity in a great variety of applications.

Let us now consider a collimated detector aligned such that the direction $\hat{\mathbf{s}}$ of the external incident wave falls within its acceptance solid angle $\Delta\Omega$. Applying Eqs. (8.10.9) and (8.10.7) to the conical volume element shown in Fig. 8.12.2, we conclude that the polarized signal measured by detector 1 per unit time is given by

$$\Delta S \mathbf{I}_{\mathrm{c}}(\mathbf{r}) + \Delta S \Delta\Omega\, \tilde{\mathbf{I}}_{\mathrm{d}}(\mathbf{r}, \hat{\mathbf{s}}), \tag{8.12.2}$$

where $\mathbf{r}$ is the position vector of the observation point. This result explains the physical meaning of the coherent Stokes column vector. Indeed, if the acceptance solid angle of the detector were infinitely small and the axis of the detector were perfectly parallel to the incidence direction $(\hat{\mathbf{q}} = \hat{\mathbf{s}})$ then the detector response would be equal to $\Delta S \mathbf{I}_{\mathrm{c}}(\mathbf{r})$, which means that the detector would measure only the Stokes column vector of the coherent field.

The interpretation of Eqs. (8.4.5) and (8.12.2) is most transparent when the average extinction matrix is diagonal:

$$\langle \mathbf{K}(\hat{\mathbf{s}}) \rangle_\xi = \langle C_{\text{ext}} \rangle_\xi \boldsymbol{\Delta}, \tag{8.12.3}$$

where $\langle C_{\text{ext}} \rangle_\xi$ is the extinction cross section per particle averaged over particle states. This happens, for example, when the particles are spherically symmetric and are made of an optically isotropic material. In this case Eq. (8.4.5) becomes

$$\begin{aligned} \mathbf{I}_\text{c}(\mathbf{r}) &= \exp[-n_0 \langle C_{\text{ext}} \rangle_\xi s(\mathbf{r})] \mathbf{I}^{\text{inc}} \\ &= \exp[-\alpha_{\text{ext}} s(\mathbf{r})] \mathbf{I}^{\text{inc}}, \end{aligned} \tag{8.12.4}$$

which means that the elements of the Stokes column vector of the coherent field are exponentially attenuated with increasing s. The attenuation rates for all four components are the same, which means that the polarization state of the coherent field does not change with s. Equation (8.12.4) is the standard Bouguer–Beer law, in which

$$\alpha_{\text{ext}} = n_0 \langle C_{\text{ext}} \rangle_\xi = n_0 [\langle C_{\text{sca}} \rangle_\xi + \langle C_{\text{abs}} \rangle_\xi] \tag{8.12.5}$$

is the attenuation (or extinction) coefficient. We see that the exponential attenuation of the Stokes parameters of the coherent field is an inalienable property of all scattering media, even those composed of nonabsorbing particles with $\langle C_{\text{abs}} \rangle_\xi = 0$. The attenuation is a combined result of scattering of the coherent field by particles in all directions and, possibly, absorption inside the particles. It is also important to remember that it was the inclusion of all orders of multiple scattering in the coherent field that ultimately led to the exponential attenuation in the Bouguer–Beer law (8.12.4) as well as to the exponential s-dependence of the general coherent transmission matrix (8.4.6).

In general, the extinction matrix is not diagonal and can explicitly depend on the propagation direction. This occurs, for example, when the scattering medium is composed of nonrandomly oriented nonspherical particles. Then the coherent transmission matrix $\mathbf{H}$ in Eq. (8.4.5) may also have nonzero off-diagonal elements, thereby yielding different attenuation rates for different components of the Stokes column vector and causing a change in the polarization state of the coherent field with increasing s. This phenomenon is called dichroism. A typical example of dichroic scattering media are clouds of nonspherical interstellar grains preferentially oriented by galactic magnetic fields. Unpolarized light emitted by spherically symmetric stars becomes partially polarized after it passes one or several such dust clouds. Observations of this phenomenon, traditionally called interstellar polarization, can provide valuable information about sizes, shapes, and refractive indices of cosmic dust particles (Martin, 1978; Dolginov *et al.*, 1995).

In reality, $\Delta\Omega$ is never equal to zero, and detectors always pick up at least some of the diffuse light. Still if both $\Delta\Omega$ and $\tilde{\mathbf{I}}_\text{d}(\mathbf{r}, \hat{\mathbf{s}})$ are sufficiently small and $\mathbf{I}_\text{c}(\mathbf{r})$ is sufficiently large then the response of a detector facing the incident light is mostly

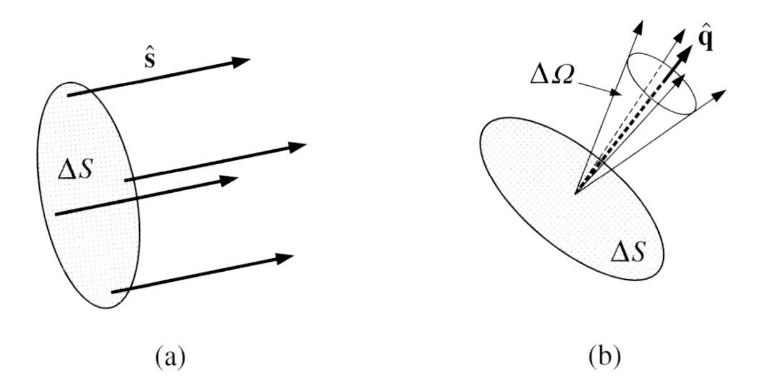

Figure 8.12.3. Physical meaning of (a) coherent intensity and (b) specific intensity.

determined by the Stokes column vector of the coherent field and is given by the first term of Eq. (8.12.2). It is reasonable to expect that this happens when the detector is located close to the volume boundary illuminated by the incident light (detector 2 in Fig. 8.12.2) so that the coherent field is still weakly attenuated and there is not much diffuse light propagating in directions close to the direction of incidence $\hat{\mathbf{s}}$.

As the detector moves farther from the boundary (detector 1), the coherent field is increasingly attenuated and more diffuse light propagates in directions close to $\hat{\mathbf{s}}$, thereby making the second term in Eq. (8.12.2) comparable to the first term. Ultimately, when the detector is placed deeply inside an optically thick medium (detector 3), the detector response is heavily dominated by the diffuse light and is given by the second term of Eq. (8.12.2).

We chose to discuss the physical meaning of $\tilde{\mathbf{I}}_{\mathrm{d}}(\mathbf{r}, \hat{\mathbf{q}})$ and $\mathbf{I}_{\mathrm{c}}(\mathbf{r})$ in terms of the concept of a detector of electromagnetic energy just to emphasize the polarization content of these quantities. It is clear, however, that they describe the directional flow of electromagnetic radiation through an arbitrary element of surface area and can be used to quantify the energy budget of objects such as cloud and aerosol layers in planetary atmospheres. This explains the usefulness of the RTE in radiation balance as well as remote sensing and particle characterization applications.

The fundamental difference between the coherent Stokes column vector and the diffuse specific intensity column vector is that the former describes a monodirectional whereas the latter describes an uncollimated flow of electromagnetic energy. In particular, the first element of the coherent Stokes column vector, i.e., the coherent intensity $I_{\mathrm{c}}(\mathbf{r})$, is the electromagnetic power per unit area of a small surface element perpendicular to $\hat{\mathbf{s}}$, whereas the first element of the diffuse specific intensity column vector, i.e., the diffuse specific intensity $\tilde{I}_{\mathrm{d}}(\mathbf{r}, \hat{\mathbf{q}})$, is the electromagnetic power per unit area of a small surface element perpendicular to $\hat{\mathbf{q}}$ per one steradian of a small solid angle centered around $\hat{\mathbf{q}}$ (Fig. 8.12.3).

The intensity can be considered to be the limit of a "highly collimated" specific intensity, which explains the presence of the solid-angle delta-function factor $\delta(\hat{\mathbf{q}} - \hat{\mathbf{s}})$ in the definition of the full specific intensity vector, Eq. (8.10.7). The dimension of a delta function is that of the inverse of its argument, which ensures that the dimension of the product $\delta(\hat{\mathbf{q}} - \hat{\mathbf{s}})\mathbf{I}_c(\mathbf{r})$ is that of specific intensity.

Perhaps the most important conclusion to remember is as follows. Since the microphysical derivation of the RTE involved statistical averaging over particle states and positions, neither the coherent Stokes column vector nor the diffuse specific intensity column vector characterize the instantaneous distribution of the radiation field inside the scattering medium. Instead, they characterize the directional flow of electromagnetic radiation averaged over a sufficiently long period of time. Although the minimal averaging time necessary to ensure ergodicity may be different for different scattering systems, it is safe to say that the longer the averaging time the more accurate should be the theoretical prediction based on the RTE.[1]

8.13 Energy conservation

A fundamental and practically important property of the RTE is that it satisfies precisely the energy conservation law. Indeed, using the vector identity

$$\mathbf{a} \cdot \nabla f = \nabla \cdot (\mathbf{a} f) - f \nabla \cdot \mathbf{a}, \tag{8.13.1}$$

where f is any scalar function of spatial coordinates, and taking into account that $\hat{\mathbf{q}}$ is a constant vector, we can rewrite Eq. (8.11.5) in the form

$$\nabla \cdot [\hat{\mathbf{q}}\, \tilde{\mathbf{I}}(\mathbf{r}, \hat{\mathbf{q}})] = -n_0(\mathbf{r}) \langle \mathbf{K}(\mathbf{r}, \hat{\mathbf{q}}) \rangle_\xi \tilde{\mathbf{I}}(\mathbf{r}, \hat{\mathbf{q}}) + n_0(\mathbf{r}) \int_{4\pi} \mathrm{d}\hat{\mathbf{q}}' \langle \mathbf{Z}(\mathbf{r}, \hat{\mathbf{q}}, \hat{\mathbf{q}}') \rangle_\xi \tilde{\mathbf{I}}(\mathbf{r}, \hat{\mathbf{q}}'). \tag{8.13.2}$$

Let us now introduce the flux density vector as

$$\mathbf{F}(\mathbf{r}) = \int_{4\pi} \mathrm{d}\hat{\mathbf{q}}\, \hat{\mathbf{q}}\, \tilde{I}(\mathbf{r}, \hat{\mathbf{q}}). \tag{8.13.3}$$

Obviously, the product $\hat{\mathbf{p}} \cdot \mathbf{F}(\mathbf{r}) \mathrm{d}S$ gives the amount and the direction of the net flow of power through a surface element dS normal to $\hat{\mathbf{p}}$ (see Fig. 8.13.1). Integrating both sides of Eq. (8.13.2) over all directions $\hat{\mathbf{q}}$ and recalling Eqs. (3.9.9)–(3.9.11), we obtain

$$-\nabla \cdot \mathbf{F}(\mathbf{r}) = n_0 \int_{4\pi} \mathrm{d}\hat{\mathbf{q}} \langle C_{\mathrm{abs}}(\mathbf{r}, \hat{\mathbf{q}}) \rangle_\xi \tilde{I}(\mathbf{r}, \mathbf{q}). \tag{8.13.4}$$

The physical meaning of this formula is very transparent: the net inflow of electro-

[1] Note that accumulating a signal over an extended period of time is often used to improve the accuracy of a measurement by reducing the effect of random noise. However, the situation with the RTT is fundamentally different in that averaging the signal over an extended period of time is necessary to ensure the very applicability of the RTE.

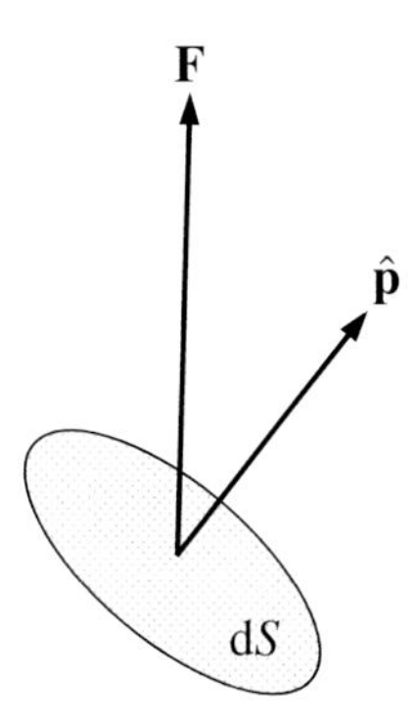

Figure 8.13.1. Electromagnetic power flow through an elementary surface element.

magnetic power per unit volume is equal to the total power absorbed per unit volume. If the particles forming the scattering medium are nonabsorbing so that $\langle C_{\text{abs}}(\mathbf{r}, \hat{\mathbf{q}}) \rangle_\xi = 0$, then the flux density vector is divergence-free:

$$\nabla \cdot \mathbf{F}(\mathbf{r}) = 0. \tag{8.13.5}$$

This is a manifestation of the conservation of the power flux, which means that the amount of electromagnetic energy entering a volume element per unit time is equal to the amount of electromagnetic energy leaving the volume element per unit time. This important result can be used for testing various numerical techniques for solving the RTE and is a particularly attractive feature of the RTT.

The previous discussion clearly shows that the VRTE follows from the Maxwell equations only after several simplifying assumptions are made. Still it is very rewarding to see that these approximations are sufficiently consistent with each other in that the final result fully complies with the energy conservation law.

8.14 External observation points

The derivation of the VRTE presented in the previous sections implied that the observation point was located *inside* the scattering volume. In this section we will explain how the solution of the VRTE can be used to calculate the response of a collimated detector placed *outside* the scattering volume. This problem is important in practice since scattering objects are often studied using external detectors of electromagnetic radiation. Typical examples are remote-sensing observations of the terrestrial atmosphere from earth-orbiting satellites, ground-based telescopic observations of other planets and various astrophysical objects, and bi-directional (polarized) reflectometry of particle suspensions and particulate surfaces.

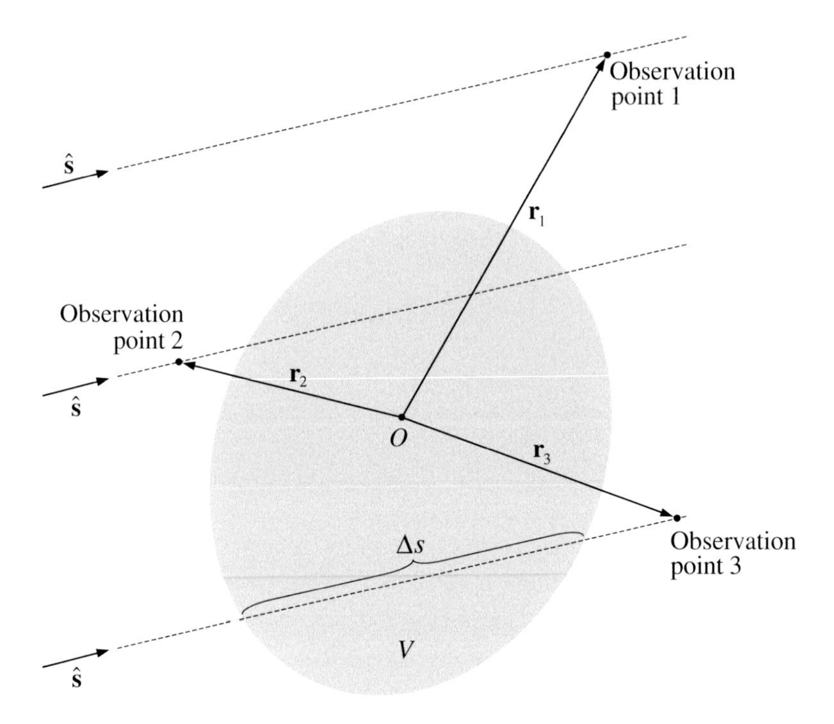

Figure 8.14.1. Coherency field at external observation points.

8.14.1 Coherent field

Let us first consider the computation of the coherent field at an external observation point $\mathbf{r} \notin V$. The analysis described in Sections 8.2 and 8.3 indicates that only forward-scattering particles that lie on the line connecting the source of illumination and the observation point can contribute to the coherent field. Hence, let us consider three possible types of location of the observation point with respect to the scattering volume as shown in Fig. 8.14.1. The line connecting the source of illumination and observation point 1 does not go through the scattering volume, whereas the lines through the source of illumination and observation points 2 and 3 do. However, only in the case of observation point 3 does the scattering volume lie between the source of illumination and the observation point. Therefore, repeating the derivation of Sections 8.2 and 8.3 yields

$$\mathbf{E}_c^{\text{ex}}(\mathbf{r}_1) = \mathbf{E}^{\text{inc}}(\mathbf{r}_1), \tag{8.14.1}$$

$$\mathbf{E}_c^{\text{ex}}(\mathbf{r}_2) = \mathbf{E}^{\text{inc}}(\mathbf{r}_2), \tag{8.14.2}$$

$$\mathbf{E}_c^{\text{ex}}(\mathbf{r}_3) = \exp\left[\frac{\mathrm{i}2\pi n_0}{k_1}\Delta s\,\langle \vec{A}(\hat{\mathbf{s}},\hat{\mathbf{s}})\rangle_\xi\right]\cdot \mathbf{E}^{\text{inc}}(\mathbf{r}_3), \tag{8.14.3}$$

where "ex" stands for external and Δs is the length of the light path inside the scattering volume as shown in Fig. 8.14.1. This result can be summarized by the following formula:

$$\mathbf{E}_{\mathrm{c}}^{\mathrm{ex}}(\mathbf{r}) = \begin{cases} \mathbf{E}^{\mathrm{inc}}(\mathbf{r}) & \text{if } \mathbf{r} \text{ is not "shadowed" by } V, \\ \exp\left[\dfrac{\mathrm{i}2\pi n_0}{k_1}\Delta s(\mathbf{r})\langle \ddot{A}(\hat{\mathbf{s}},\hat{\mathbf{s}})\rangle_\xi\right]\cdot \mathbf{E}^{\mathrm{inc}}(\mathbf{r}) & \text{if } \mathbf{r} \text{ is "shadowed" by } V, \end{cases} \tag{8.14.4}$$

where Δs is a function of $\mathbf{r}$.

By analogy with Section 8.4, Eq. (8.14.4) can be rewritten in terms of the Stokes column vector of the external coherent field $\mathbf{I}_{\mathrm{c}}^{\mathrm{ex}}(\mathbf{r})$:

$$\mathbf{I}_{\mathrm{c}}^{\mathrm{ex}}(\mathbf{r}) = \begin{cases} \mathbf{I}^{\mathrm{inc}} & \text{if } \mathbf{r} \text{ is not "shadowed" by } V, \\ \exp[-n_0\,\Delta s(\mathbf{r})\langle \mathbf{K}(\hat{\mathbf{s}})\rangle_\xi]\mathbf{I}^{\mathrm{inc}} & \text{if } \mathbf{r} \text{ is "shadowed" by } V, \end{cases} \tag{8.14.5}$$

where $\mathbf{I}^{\mathrm{inc}}$ is the Stokes column vector of the incident field. The physical interpretation of this formula is very simple: the intensity of the coherent wave is exponentially attenuated and its polarization state changes if and only if the wave travels through the scattering medium.

8.14.2 Ladder coherency dyadic

The derivation of the ladder approximation for the coherency dyadic

$$\ddot{C}^{\mathrm{ex}}(\mathbf{r}) = \langle \mathbf{E}(\mathbf{r}) \otimes [\mathbf{E}(\mathbf{r})]^*\rangle_{\mathbf{R},\xi}$$

defined in terms of the total electric field $\mathbf{E}(\mathbf{r})$ at an external observation point $\mathbf{r} \notin V$ is very similar to that for the coherency dyadic at an internal point, as described in sections 8.5 and 8.6. The only significant difference is that now only a part of the line connecting the observation point and particle 1 (see Fig. 8.14.2) lies inside the scattering volume (cf. Fig. 8.6.2). Therefore, the final result is as follows:

$$\ddot{C}_{\mathrm{L}}^{\mathrm{ex}}(\mathbf{r}) = \int_{4\pi} \mathrm{d}\hat{\mathbf{p}}\, \ddot{\Sigma}_{\mathrm{L}}^{\mathrm{ex}}(\mathbf{r},-\hat{\mathbf{p}}), \tag{8.14.6}$$

where the vector $\mathbf{p}$ originates at the observation point, $\hat{\mathbf{p}} = \mathbf{p}/p$ is the unit vector in the direction of $\mathbf{p}$, and the external ladder specific coherency dyadic is the sum of the coherent and diffuse parts:

$$\ddot{\Sigma}_{\mathrm{L}}^{\mathrm{ex}}(\mathbf{r},-\hat{\mathbf{p}}) = \ddot{\Sigma}_{\mathrm{c}}^{\mathrm{ex}}(\mathbf{r},-\hat{\mathbf{p}}) + \ddot{\Sigma}_{\mathrm{d}}^{\mathrm{ex}}(\mathbf{r},-\hat{\mathbf{p}}). \tag{8.14.7}$$

The coherent part is given by

$$\ddot{\Sigma}_{\mathrm{c}}^{\mathrm{ex}}(\mathbf{r},-\hat{\mathbf{p}}) = \delta(\hat{\mathbf{p}}+\hat{\mathbf{s}})\ddot{C}_{\mathrm{c}}^{\mathrm{ex}}(\mathbf{r}), \tag{8.14.8}$$

where

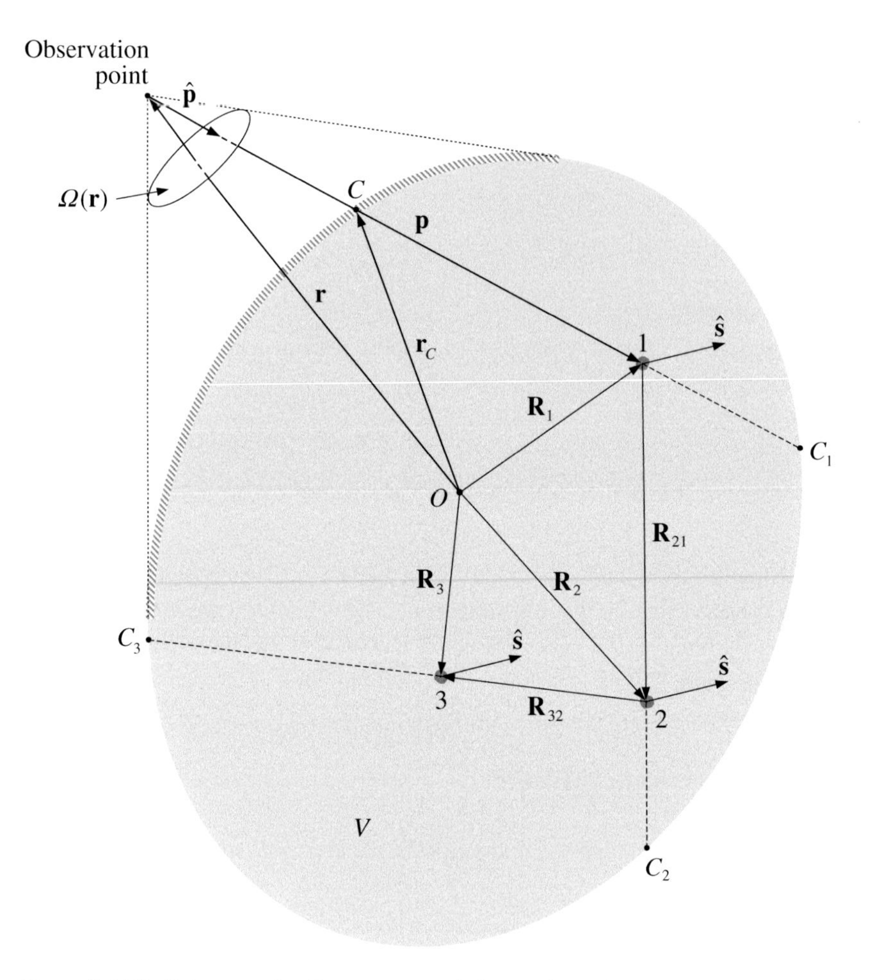

Figure 8.14.2. Coherency dyadic at an external observation point.

$$\ddot{C}_{\mathrm{c}}^{\mathrm{ex}}(\mathbf{r}) = \mathbf{E}_{\mathrm{c}}^{\mathrm{ex}}(\mathbf{r}) \otimes [\mathbf{E}_{\mathrm{c}}^{\mathrm{ex}}(\mathbf{r})]^{*} \tag{8.14.9}$$

is the coherent part of the "external" coherency dyadic $\ddot{C}^{\mathrm{ex}}(\mathbf{r})$ and the external coherent field $\mathbf{E}_{\mathrm{c}}^{\mathrm{ex}}(\mathbf{r})$ is given by Eq. (8.14.4). The diffuse part of the external ladder specific coherency dyadic vanishes if $\hat{\mathbf{p}} \notin \Omega(\mathbf{r})$, where $\Omega(\mathbf{r})$ is the solid angle subtended by the scattering volume when it is viewed from the external observation point $\mathbf{r}$ (see Fig. 8.14.2). Otherwise it is given by

$$\begin{aligned}
\ddot{\Sigma}_{\mathrm{d}}^{\mathrm{ex}}(\mathbf{r}, -\hat{\mathbf{p}}) = {} & n_0 \int_{p(C)}^{p(C_1)} \mathrm{d}p \int \mathrm{d}\xi_1\, \ddot{\eta}(-\hat{\mathbf{p}}, p) \cdot \ddot{A}_1(-\hat{\mathbf{p}}, \hat{\mathbf{s}}) \cdot \ddot{C}_{\mathrm{c}}(\mathbf{r} + \mathbf{p}) \\
& \cdot [\ddot{A}_1(-\hat{\mathbf{p}}, \hat{\mathbf{s}})]^{\mathrm{T}*} \cdot [\ddot{\eta}(-\hat{\mathbf{p}}, p)]^{\mathrm{T}*} \\
& + n_0^2 \int_{p(C)}^{p(C_1)} \mathrm{d}p \int \mathrm{d}\xi_1 \int \mathrm{d}R_{21}\, \mathrm{d}\hat{\mathbf{R}}_{21}\, \mathrm{d}\xi_2\, \ddot{\eta}(-\hat{\mathbf{p}}, p) \cdot \ddot{A}_1(-\hat{\mathbf{p}}, -\hat{\mathbf{R}}_{21})
\end{aligned}$$

$$
\begin{aligned}
& \cdot \ddot{\eta}(-\hat{\mathbf{R}}_{21}, R_{21}) \cdot \ddot{A}_2(-\hat{\mathbf{R}}_{21}, \hat{\mathbf{s}}) \cdot \ddot{C}_c(\mathbf{r} + \mathbf{p} + \mathbf{R}_{21}) \\
& \cdot [\ddot{A}_2(-\hat{\mathbf{R}}_{21}, \hat{\mathbf{s}})]^{T*} \cdot [\ddot{\eta}(-\hat{\mathbf{R}}_{21}, R_{21})]^{T*} \\
& \cdot [\ddot{A}_1(-\hat{\mathbf{p}}, -\hat{\mathbf{R}}_{21})]^{T*} \cdot [\ddot{\eta}(-\hat{\mathbf{p}}, p)]^{T*} \\
+\, n_0^3 \int_{p(C)}^{p(C_1)} \mathrm{d}p \int \mathrm{d}\xi_1 \int \mathrm{d}R_{21}\, \mathrm{d}\hat{\mathbf{R}}_{21}\, \mathrm{d}\xi_2 \int \mathrm{d}R_{32}\, \mathrm{d}\hat{\mathbf{R}}_{32}\, \mathrm{d}\xi_3\, \ddot{\eta}(-\hat{\mathbf{p}}, p) \\
& \cdot \ddot{A}_1(-\hat{\mathbf{p}}, -\hat{\mathbf{R}}_{21}) \cdot \ddot{\eta}(-\hat{\mathbf{R}}_{21}, R_{21}) \cdot \ddot{A}_2(-\hat{\mathbf{R}}_{21}, -\hat{\mathbf{R}}_{32}) \\
& \cdot \ddot{\eta}(-\hat{\mathbf{R}}_{32}, R_{32}) \cdot \ddot{A}_3(-\hat{\mathbf{R}}_{32}, \hat{\mathbf{s}}) \cdot \ddot{C}_c(\mathbf{r} + \mathbf{p} + \mathbf{R}_{21} + \mathbf{R}_{32}) \\
& \cdot [\ddot{A}_3(-\hat{\mathbf{R}}_{32}, \hat{\mathbf{s}})]^{T*} \cdot [\ddot{\eta}(-\hat{\mathbf{R}}_{32}, R_{32})]^{T*} \\
& \cdot [\ddot{A}_2(-\hat{\mathbf{R}}_{21}, -\hat{\mathbf{R}}_{32})]^{T*} \cdot [\ddot{\eta}(-\hat{\mathbf{R}}_{21}, R_{21})]^{T*} \\
& \cdot [\ddot{A}_1(-\hat{\mathbf{p}}, -\hat{\mathbf{R}}_{21})]^{T*} \cdot [\ddot{\eta}(-\hat{\mathbf{p}}, p)]^{T*} \\
+\, \cdots, \quad & \hat{\mathbf{p}} \in \Omega(\mathbf{r}). \qquad (8.14.10)
\end{aligned}
$$

The notation is clear from Fig. 8.14.2. Note that R_{21} ranges from zero at the origin of particle 1 to the corresponding value at point C_2, R_{32} ranges from zero at the origin of particle 2 to the corresponding value at point C_3, etc.

Direct comparison of Eq. (8.14.10) with Eq. (8.6.7) leads us to a fundamental conclusion: the external diffuse specific coherency dyadic for a direction $-\hat{\mathbf{p}}$ such that $\hat{\mathbf{p}} \in \Omega(\mathbf{r})$ is equal to the internal diffuse specific coherency dyadic at a boundary point C where the line drawn through the observation point in the direction $\hat{\mathbf{p}}$ enters the scattering volume (see Fig. 8.14.2). Thus,

$$
\ddot{\Sigma}_d^{ex}(\mathbf{r}, -\hat{\mathbf{p}}) = \begin{cases} \ddot{0} & \text{if } \hat{\mathbf{p}} \notin \Omega(\mathbf{r}), \\ \ddot{\Sigma}_d[\mathbf{r}_C(\mathbf{r}, \hat{\mathbf{p}}), -\hat{\mathbf{p}}] & \text{if } \hat{\mathbf{p}} \in \Omega(\mathbf{r}), \end{cases} \qquad (8.14.11)
$$

where $\mathbf{r}_C$ is the position vector of the point C (see Fig. 8.14.2). Obviously, $\mathbf{r}_C$ is a function of $\mathbf{r}$ and $\hat{\mathbf{p}}$. Equations (8.14.6) and (8.14.7) then demonstrate that the ladder coherency dyadic at the external observation point can be expressed in terms of the internal diffuse specific coherency dyadic at those boundary points of the scattering volume that are "visible" from the observation point (the part of the boundary visible from the observation point $\mathbf{r}$ is highlighted in Fig. 8.14.2).

8.14.3 Specific intensity column vector

It is straightforward to rewrite Eqs. (8.14.7)–(8.14.9) and Eq. (8.14.11) in terms of the full specific intensity column vector at the external observation point:

$$
\tilde{\mathbf{I}}^{ex}(\mathbf{r}, -\hat{\mathbf{p}}) = \tilde{\mathbf{I}}_c^{ex}(\mathbf{r}, -\hat{\mathbf{p}}) + \tilde{\mathbf{I}}_d^{ex}(\mathbf{r}, -\hat{\mathbf{p}}), \qquad (8.14.12)
$$

where

$$\tilde{\mathbf{I}}_{\mathrm{c}}^{\mathrm{ex}}(\mathbf{r},-\hat{\mathbf{p}}) = \delta(\hat{\mathbf{p}}+\hat{\mathbf{s}})\mathbf{I}_{\mathrm{c}}^{\mathrm{ex}}(\mathbf{r}) \tag{8.14.13}$$

is the external coherent specific intensity column vector,

$$\tilde{\mathbf{I}}_{\mathrm{d}}^{\mathrm{ex}}(\mathbf{r},-\hat{\mathbf{p}}) = \begin{cases} \mathbf{0} & \text{if } \hat{\mathbf{p}} \notin \Omega(\mathbf{r}), \\ \tilde{\mathbf{I}}_{\mathrm{d}}[\mathbf{r}_C(\mathbf{r},\hat{\mathbf{p}}),-\hat{\mathbf{p}}] & \text{if } \hat{\mathbf{p}} \in \Omega(\mathbf{r}) \end{cases} \tag{8.14.14}$$

is the external diffuse specific intensity column vector, $\mathbf{I}_{\mathrm{c}}^{\mathrm{ex}}(\mathbf{r})$ is given by Eq. (8.14.5), and $\mathbf{0}$ is a 4×1 zero column. As was the case with the external diffuse specific coherency dyadic, the external diffuse specific intensity column vector for a direction $-\hat{\mathbf{p}}$ such that $\hat{\mathbf{p}} \in \Omega(\mathbf{r})$ is equal to the internal diffuse specific intensity column vector at that boundary point where the line drawn through the observation point in the direction $\hat{\mathbf{p}}$ enters the scattering volume (Fig. 8.14.2). Furthermore, it vanishes for all directions $-\hat{\mathbf{p}}$ such that $\hat{\mathbf{p}} \notin \Omega(\mathbf{r})$.

8.14.4 Discussion

The physical significance of these results is illustrated in Fig. 8.14.3. All four external polarization-sensitive, well-collimated detectors have a small surface area ΔS and a small angular aperture. However, the orientations of the detectors and their positions are different. In order to emphasize the difference in the orientations of the four detector acceptance solid angles, we denote the latter as $\Delta\Omega_1$, $\Delta\Omega_2$, $\Delta\Omega_3$, and $\Delta\Omega_4$, whereas the position vectors of the respective observation points will be denoted as $\mathbf{r}_1$, $\mathbf{r}_2$, $\mathbf{r}_3$, and $\mathbf{r}_4$.

Detector 1 faces the incident wave, but its acceptance solid angle $\Delta\Omega_1$ captures no boundary points of the scattering volume. Therefore, the polarization signal measured by the first detector per unit time is given by

$$\textbf{Signal 1} = \Delta S\,\mathbf{I}^{\mathrm{inc}}. \tag{8.14.15}$$

Detector 2 is positioned and oriented such that its acceptance solid angle $\Delta\Omega_2$ does not capture the incidence direction, but captures all points of the part of the boundary denoted S_2. Therefore, the polarized signal measured by this detector per unit time is given by

$$\begin{aligned} \textbf{Signal 2} &= \Delta S \int_{\Delta\Omega_2} \mathrm{d}\hat{\mathbf{p}}\, \tilde{\mathbf{I}}_{\mathrm{d}}^{\mathrm{ex}}(\mathbf{r}_2,-\hat{\mathbf{p}}) \\ &= \Delta S \int_{\Delta\Omega_2} \mathrm{d}\hat{\mathbf{p}}\, \tilde{\mathbf{I}}_{\mathrm{d}}[\mathbf{r}_C(\mathbf{r}_2,\hat{\mathbf{p}}),-\hat{\mathbf{p}}], \end{aligned} \tag{8.14.16}$$

where, as before, the unit vector $\hat{\mathbf{p}}$ originates at observation point 2 and $\mathbf{r}_C \in V$ is the position vector of the point where the line drawn through the observation point in the direction $\hat{\mathbf{p}}$ crosses the boundary of the scattering volume (see Fig. 8.14.2).

The acceptance solid angle of detector 3 captures both the incidence direction and

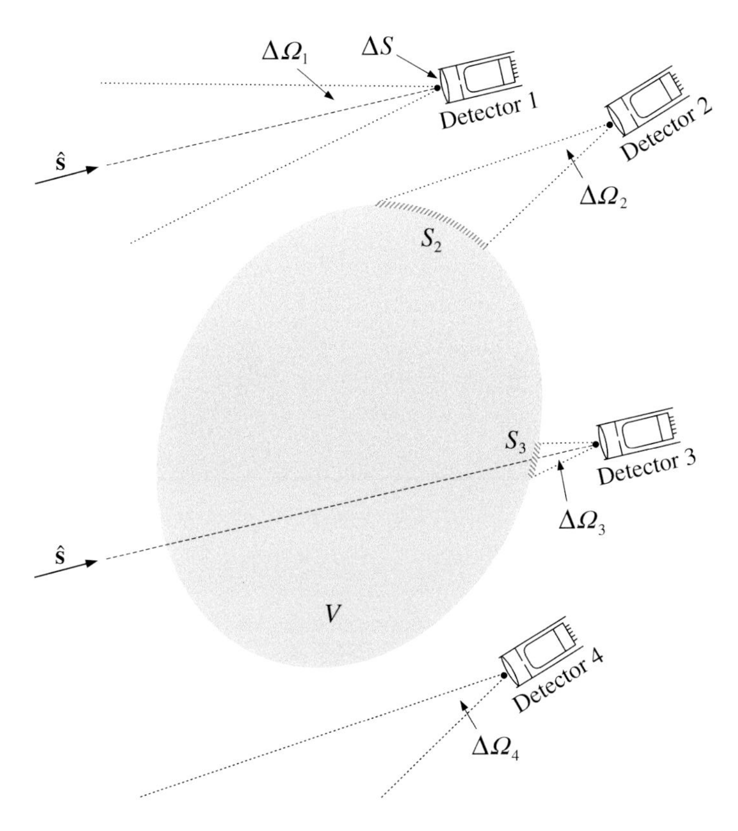

Figure 8.14.3. Polarized signal measured by an external collimated detector depends on the detector position and orientation with respect to the scattering volume.

all points of the part of the boundary denoted S_3. Therefore, the polarized signal measured by detector 3 per unit time is

$$\textbf{Signal 3} = \Delta S \exp[-n_0\, \Delta s(\mathbf{r}_3)\langle \mathbf{K}(\hat{\mathbf{s}})\rangle_\xi]\mathbf{I}^{\rm inc} + \Delta S \int_{\Delta\Omega_3} \mathrm{d}\hat{\mathbf{p}}\, \tilde{\mathbf{I}}_{\rm d}[\mathbf{r}_C(\mathbf{r}_3, \hat{\mathbf{p}}), -\hat{\mathbf{p}}], \tag{8.14.17}$$

where, as before, $\Delta s(\mathbf{r}_3)$ is the length of the path traveled by the coherent wave inside the scattering volume before it reaches observation point 3 (see Fig. 8.14.1).

Finally, neither the incidence direction nor any boundary point is captured by the acceptance solid angle of detector 4. Therefore, this detector measures no signal:

$$\textbf{Signal 4} = \mathbf{0}. \tag{8.14.18}$$

After the VRTE has been solved and, as a result, the diffuse specific intensity column vector is known at all points of the scattering volume, Eqs. (8.14.15)–(8.14.18) can be used to calculate the polarization response of an external collimated detector arbitrarily oriented and positioned with respect to the scattering volume V. Although heuristic equivalents of these formulas have been widely used in the framework of the

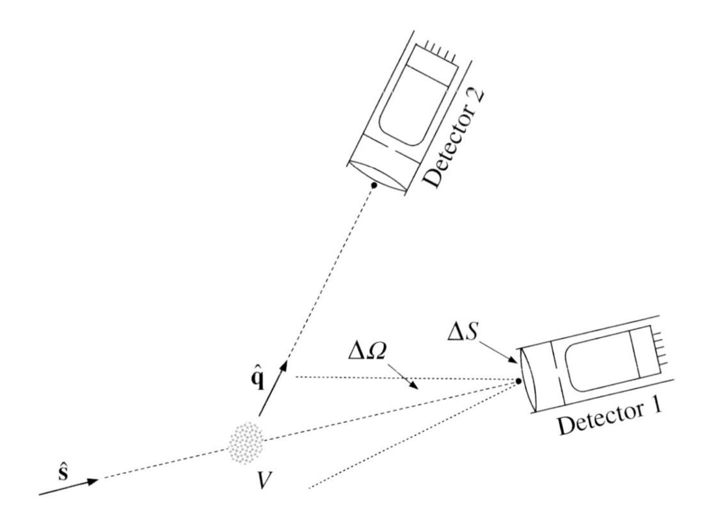

Figure 8.14.4. First-order scattering.

phenomenological RTT (see the discussion in Section 18.16), it is highly rewarding to see that they can be consistently derived using the microphysical approach.

8.14.5 Illustrative example: first-order scattering

To give an example of applying the above formulas, let us now assume that the number of particles in the scattering volume is sufficiently small that

$$|n_0 L[\langle \mathbf{K}(\hat{\mathbf{q}})\rangle_\xi]_{pq}| \ll 1 \tag{8.14.19}$$

and

$$|n_0 L[\langle \mathbf{Z}(\hat{\mathbf{q}}, \hat{\mathbf{q}}')\rangle_\xi]_{pq}| \ll 1 \tag{8.14.20}$$

for $p, q = 1, \ldots, 4$ and for any $\hat{\mathbf{q}}$ and $\hat{\mathbf{q}}'$, where L is the largest linear dimension of the volume element. As a consequence, one may neglect all terms proportional to powers of n_0 higher than the first and, thus, all orders of scattering higher than the first.

The scattering situation is shown schematically in Fig. 8.14.4, in which the diameter of the sensitive area of the detectors is assumed to be significantly greater than L and their angular aperture $\Delta\Omega$ is large enough to encompass the entire scattering volume. We will further assume that the distance r from the volume element to the detectors is much greater than L so that the waves scattered by different particles toward either detector propagate in essentially the same direction and the distance from the observation point to any particle inside the volume element is approximately the same. The electromagnetic response of either detector is calculated by integrating the full specific intensity column vector over the detector sensitive area and angular aperture. Let us recall the integral form of the VRTE, Eq. (8.10.3). We can now use Eqs. (8.14.5) and (8.14.14) to derive that the polarized signal measured by detector 1 per unit time is given by

$$\textbf{Signal 1} = \Delta S \mathbf{I}^{\text{inc}} - N\langle \mathbf{K}(\hat{\mathbf{s}})\rangle_{\xi} \mathbf{I}^{\text{inc}} + \frac{\Delta S}{r^2} N\langle \mathbf{Z}(\hat{\mathbf{s}}, \hat{\mathbf{s}})\rangle_{\xi} \mathbf{I}^{\text{inc}}, \tag{8.14.21}$$

whereas that measured by detector 2 per unit time is given by

$$\textbf{Signal 2} = \frac{\Delta S}{r^2} N\langle \mathbf{Z}(\hat{\mathbf{q}}, \hat{\mathbf{s}})\rangle_{\xi} \mathbf{I}^{\text{inc}}, \tag{8.14.22}$$

where $N = n_0 V$ is the total number of particles in the volume element. Not surprisingly, this is the same result as that obtained in the framework of the first-order-scattering approximation for a small volume element, Eqs. (7.7.13) and (7.7.14).

8.15 Other types of illumination

The above microphysical derivation of the VRTE was explicitly based on the assumption that the incident light is a plane electromagnetic wave. This was done primarily to make more natural the introduction of concepts such as the coherent field and to facilitate the comparison of the microphysical and phenomenological approaches to radiative transfer. However, we could have made the derivation of the VRTE more general by using the terminology introduced in Section 3.10. Specifically, one can express the total electric field everywhere in space and at any moment in terms of the transformation dyadic of the entire multi-particle group,

$$\mathbf{E}(\mathbf{r}, t) = \overset{\leftrightarrow}{\mathcal{T}}(t, \omega, \mathbf{r}, \hat{\mathbf{s}}) \cdot \mathbf{E}_0^{\text{inc}} \exp(-\mathrm{i}\omega t), \qquad \mathbf{r} \in \Re^3, \tag{8.15.1}$$

and then use the following Twersky approximation for the $\overset{\leftrightarrow}{\mathcal{T}}(t, \omega, \mathbf{r}, \hat{\mathbf{s}})$:

$$\begin{aligned}
\overset{\leftrightarrow}{\mathcal{T}}(t, \omega, \mathbf{r}, \hat{\mathbf{s}}) &= \exp(\mathrm{i}k_1 \hat{\mathbf{s}} \cdot \mathbf{r}) \overset{\leftrightarrow}{I} \\
&+ \sum_{i=1}^{N} \exp(\mathrm{i}k_1 \hat{\mathbf{s}} \cdot \mathbf{R}_i) \overset{\leftrightarrow}{B}_{ri0} + \sum_{i=1}^{N} \sum_{\substack{j=1 \\ j \neq i}}^{N} \exp(\mathrm{i}k_1 \hat{\mathbf{s}} \cdot \mathbf{R}_j) \overset{\leftrightarrow}{B}_{rij} \cdot \overset{\leftrightarrow}{B}_{ij0} \\
&+ \sum_{i=1}^{N} \sum_{\substack{j=1 \\ j \neq i}}^{N} \sum_{\substack{l=1 \\ l \neq i \\ l \neq j}}^{N} \exp(\mathrm{i}k_1 \hat{\mathbf{s}} \cdot \mathbf{R}_l) \overset{\leftrightarrow}{B}_{rij} \cdot \overset{\leftrightarrow}{B}_{ijl} \cdot \overset{\leftrightarrow}{B}_{jl0} \\
&+ \sum_{i=1}^{N} \sum_{\substack{j=1 \\ j \neq i}}^{N} \sum_{\substack{l=1 \\ l \neq i \\ l \neq j}}^{N} \sum_{\substack{m=1 \\ m \neq i \\ m \neq j \\ m \neq l}}^{N} \exp(\mathrm{i}k_1 \hat{\mathbf{s}} \cdot \mathbf{R}_m) \overset{\leftrightarrow}{B}_{rij} \cdot \overset{\leftrightarrow}{B}_{ijl} \cdot \overset{\leftrightarrow}{B}_{jlm} \cdot \overset{\leftrightarrow}{B}_{lm0} \\
&+ \cdots
\end{aligned} \tag{8.15.2}$$

(cf. Eq. (8.1.13)). The transformation dyadic depends on time explicitly in order to account for the temporal variability of the multi-particle configuration. Equation

(8.15.2) is valid at any point located in the far-field zones of all the particles forming the scattering medium. The next step is to consider one of the illumination scenarios discussed in Sections 3.10–3.12 and compute the time average of the coherency dyadic of the total field by using the diagrammatic technique introduced in the previous sections and the ladder approximation. This procedure also yields the VRTE.

Specifically, let us first assume that the discrete scattering medium is illuminated by a parallel quasi-monochromatic beam and that significant changes of the transformation dyadic occur much more slowly than the random oscillations of the electric field amplitude. It is then straightforward to show that Eqs. (8.4.5), (8.10.3) and (8.10.4) remain unchanged provided that the Stokes column vector of the incident plane wave in Eq. (8.4.5) is replaced by the time-averaged Stokes column vector of the quasi-monochromatic beam.

Second, if the medium is illuminated by N quasi-monochromatic beams with arbitrary propagation directions then it can be shown that the total specific intensity column vector is given by

$$\tilde{\mathbf{I}}(\mathbf{r}, \hat{\mathbf{q}}) = \sum_{i=1}^{N} \tilde{\mathbf{I}}_i(\mathbf{r}, \hat{\mathbf{q}}), \tag{8.15.3}$$

where

$$\tilde{\mathbf{I}}_i(\mathbf{r}, \hat{\mathbf{q}}) = \delta(\hat{\mathbf{q}} - \hat{\mathbf{s}}_i)\mathbf{I}_{ci}(\mathbf{r}) + \tilde{\mathbf{I}}_{di}(\mathbf{r}, \hat{\mathbf{q}}) \tag{8.15.4}$$

is the ith "partial" specific intensity column vector obtained by solving the VRTE under the assumption that the scattering medium is illuminated only by the ith quasi-monochromatic beam propagating in the direction $\hat{\mathbf{s}}_i$.

Finally, Eq. (8.15.3) remains valid if the medium is illuminated by N plane electromagnetic waves provided that all of them have different angular frequencies.

These important properties of the VRTE can be used to extend significantly its range of applications. In particular, the VRTE can be applied to situations in which the external source of light is multispectral, such as the sun.

8.16 Phenomenological approach to radiative transfer

After we have presented the detailed derivation of the VRTE from the Maxwell equations, it is interesting to compare the self-consistent microphysical methodology with the traditional phenomenological approach to radiative transfer (e.g., Chandrasekhar, 1950; Rozenberg, 1955; Preisendorfer, 1965). Since the latter cannot be used to *derive* many facts that appear as corollaries of classical electromagnetics in the framework of the microphysical approach, one has to *postulate* them. For example, it naturally follows from the microphysical derivation that the average (coherent) field inside the discrete random medium is exponentially attenuated and serves to replace the constant-amplitude incident field as the *de facto* source of multiple scattering (cf. the original order-of-scattering expansion (8.1.11), in which multiple scattering is initiated by the incident field, and Diagram 8.5.4(c), in which the source of multiple scattering is the coherent field). In contrast, the phenomenological approach begins with a *postulate* that the incident parallel beam of light is exponentially attenuated as it propagates through the medium and serves as the initial source of multiple scattering.

Another postulate of the phenomenological approach is that the diffuse radiation field at each point $\mathbf{r}$ inside the scattering medium and at each moment in time can be represented by a collection of elementary "rays" with a continuous distribution of propagation directions $\hat{\mathbf{q}}$ and can be characterized by the local four-component diffuse specific intensity column vector:

$$\tilde{\mathbf{I}}_{\mathrm{d}}(\mathbf{r}, \hat{\mathbf{q}}) = \begin{bmatrix} \tilde{I}_{\mathrm{d}}(\mathbf{r}, \hat{\mathbf{q}}) \\ \tilde{Q}_{\mathrm{d}}(\mathbf{r}, \hat{\mathbf{q}}) \\ \tilde{U}_{\mathrm{d}}(\mathbf{r}, \hat{\mathbf{q}}) \\ \tilde{V}_{\mathrm{d}}(\mathbf{r}, \hat{\mathbf{q}}) \end{bmatrix}. \tag{8.16.1}$$

The elementary rays are postulated to be mutually incoherent and make independent contributions to $\tilde{\mathbf{I}}_{\mathrm{d}}(\mathbf{r}, \hat{\mathbf{q}})$. The elements $\tilde{Q}_{\mathrm{d}}(\mathbf{r}, \hat{\mathbf{q}})$, $\tilde{U}_{\mathrm{d}}(\mathbf{r}, \hat{\mathbf{q}})$, and $\tilde{V}_{\mathrm{d}}(\mathbf{r}, \hat{\mathbf{q}})$ describe the polarization state of the ray propagating in the direction $\hat{\mathbf{q}}$ through the observation point specified by the position vector $\mathbf{r}$, whereas the product

$$\mathrm{d}S\, \mathrm{d}t\, \mathrm{d}\Omega\, \tilde{I}_{\mathrm{d}}(\mathbf{r}, \hat{\mathbf{q}})$$

gives the amount of electromagnetic energy transported through a surface element $\mathrm{d}S$ normal to $\hat{\mathbf{q}}$ and centered at $\mathbf{r}$ in a time interval $\mathrm{d}t$ in all directions confined to a solid angle element $\mathrm{d}\Omega$ centered at the direction of propagation $\hat{\mathbf{q}}$. All elements of the specific intensity column vector have the dimension of radiance. The direct propagation of the incident parallel beam of light through the medium is described by the "monodirectional" four-component Stokes column vector $\mathbf{I}_{\mathrm{c}}(\mathbf{r})$ having the dimension of intensity.

Thus, there is a fundamental difference between how the phenomenological and microphysical approaches treat the random radiation field. The phenomenological approach begins with a postulate of existence of the diffuse specific intensity column vector and the Stokes column vector of the direct light at each moment in time. In the framework of the microphysical approach these quantities are derived from more fundamental ones and are shown to describe the directional flow of electromagnetic radiation averaged over a sufficiently long period of time (Section 8.12).

It is easy to understand why the phenomenological way to introduce the specific intensity column vector is wrong. At a moment of time $t = t_1$, the particles constituting a turbid medium form a distinct configuration, as illustrated in Fig. 8.16.1. The straight line extending through the observation point $\mathbf{r}$ in the $-\hat{\mathbf{q}}_1$ direction does not go through any particle. Therefore, one has to conclude that

$$\left. \tilde{\mathbf{I}}_{\mathrm{d}}(\mathbf{r}, \hat{\mathbf{q}}_1) \right|_{t = t_1} = \mathbf{0}.$$

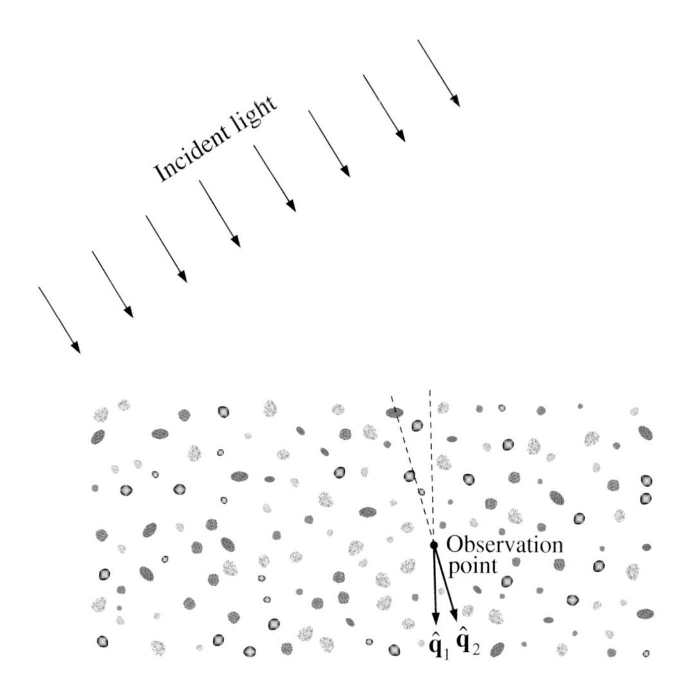

Figure 8.16.1. Specific intensity column vector cannot be defined at a moment in time.

The straight line extending through the observation point in the $-\hat{\mathbf{q}}_2$ direction goes through three particles, which suggests that

$$\tilde{\mathbf{I}}_{\mathrm{d}}(\mathbf{r}, \hat{\mathbf{q}}_2)\Big|_{t=t_1} \neq \mathbf{0}.$$

Thus, the phenomenological specific intensity column vector is not a continuous function of propagation direction at any moment in time. Rather, it is the explicit averaging of the coherency dyadic over particle positions in the microphysical approach that ensures that the specific intensity column vector depends on propagation direction continuously.

Another reason why it is impossible to define an instantaneous diffuse specific intensity column vector of the radiation field is as follows. The Stokes parameters can only be defined for a transverse electromagnetic wave or a superposition of transverse waves propagating in the same direction, whereas the instantaneous total field created by an N-particle group at an observation point is a superposition of the incident plane wave and N spherical waves coming from the N individual particles. The resulting field is neither a plane nor a spherical wave with a specific propagation direction and its electric and magnetic vectors are not always orthogonal, contrary to the assumption made on p. 393 of Preisendorfer (1965). Again, it is the explicit averaging of the coherency dyadic over particle positions in the microphysical approach that ultimately allows one to quantify the polarization response of a detector of electromagnetic energy in terms of an incoherent sum of the Stokes parameters of the spherical waves generated by the N individual particles.

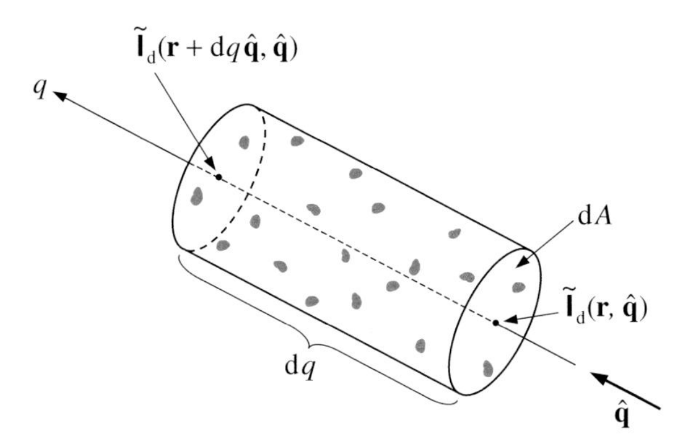

Figure 8.16.2. Phenomenological interpretation of the RTE.

The phenomenological RTT treats the medium filled with a large number of discrete, sparsely and randomly distributed particles as continuous and locally homogeneous and is fundamentally based on the concept of an elementary (or differential) volume element of the scattering medium. Specifically, it replaces the concept of single scattering and absorption by an individual particle with the concept of single scattering and absorption by an elementary volume element. It is assumed that the result of scattering is not the electromagnetic transformation of a plane incident wave into a spherical scattered wave in the far-field zone of the volume element, but rather the transformation of the diffuse specific intensity column vector of the incident light into the diffuse specific intensity column vector of the scattered light. This assumption appears to be especially artificial because the scattering transformation law is then written in the form

$$\tilde{\mathbf{I}}_d(\mathbf{r},\hat{\mathbf{q}}) \propto \mathbf{Z}_{dV}(\hat{\mathbf{q}},\hat{\mathbf{q}}')\,\tilde{\mathbf{I}}_d(\mathbf{r},\hat{\mathbf{q}}')$$

and $\mathbf{Z}_{dV}(\hat{\mathbf{q}},\hat{\mathbf{q}}')$, called the phase matrix of the elementary volume element, is computed from electromagnetics. Specifically, it is postulated that

$$\mathbf{Z}_{dV}(\hat{\mathbf{q}},\hat{\mathbf{q}}') = n_0\, dV \langle \mathbf{Z}(\hat{\mathbf{q}},\hat{\mathbf{q}}')\rangle_\xi,$$

where n_0 is the particle number density, dV is the size of the elementary volume element, $\mathbf{Z}(\hat{\mathbf{q}},\hat{\mathbf{q}}')$ is the single-particle phase matrix describing the transformation of an incident plane electromagnetic wave into the scattered spherical wave, and $\langle\cdots\rangle_\xi$ denotes an average over all physically realizable particle states.

It is further postulated that the change of the Stokes column vector of direct light $\mathbf{I}_c(\mathbf{r})$ over a differential length ds parallel to the incidence direction is caused by extinction and dichroism and can be described by Eq. (8.4.4) in which, again, the single-particle extinction matrix is computed from Maxwell's electromagnetics.

In addition, it is postulated that the cumulative change of the diffuse specific intensity column vector over the length dq of an elementary cylindrical volume element having bases of an area dA perpendicular to $\hat{\mathbf{q}}$ (see Fig. 8.16.2) is caused by:

- The effect of extinction and dichroism.
- The contribution of the diffuse light illuminating the volume element from all directions $\hat{\mathbf{q}}'$ and scattered into the direction $\hat{\mathbf{q}}$.
- The contribution of the attenuated external beam scattered into the direction $\hat{\mathbf{q}}$.

These three components are described by the first, second, and third terms, respectively, on the right-hand side of Eq. (8.10.4).

It is thus clear that the phenomenological approach is based on a rather eclectic combination of concepts borrowed from pure radiometry (light rays as geometrical trajectories along which radiant energy is assumed to be propagated, the concept of incoherent radiance) and pure electromagnetism (electromagnetic scattering of plane waves, Stokes parameters, phase and extinction matrices).

The concept of an elementary volume element is implicitly based on the modified uncorrelated single-scattering approximation discussed in Section 7.3. A fundamental problem here is that the MUSSA is only valid in the far-field zone of the elementary volume element as a whole and cannot be applied to adjacent volume elements having common boundaries. In particular, we have seen in Section 7.6 that the far-field zone of a volume element may begin at a distance exceeding the volume element's size by several orders of magnitude.

Another problem is caused by the assumption that the RTE describes the instantaneous state of the radiation field. Indeed, in order to justify the use of the phase and extinction matrices averaged over all particle states in Eqs. (8.4.4) and (8.10.4), one has to require that all physically realizable particle states (sizes, shapes, orientations, refractive indices, etc.) be well represented in each elementary volume element at any moment in time (e.g., West *et al.*, 1994). Since this requirement may imply an unrealistically large size of an elementary volume element, it has been concluded that the RTE may need a substantial modification when it is applied to scattering media such as terrestrial water clouds (e.g., Knyazikhin *et al.*, 2002). However, this conclusion does not take into account the following important consequences of the microphysical derivation of the RTE: (i) the concept of an elementary volume element has no actual relevance to the RTT, and (ii) the RTE describes a time average of the directional flow of electromagnetic radiation rather than its instantaneous pattern. Therefore, the range of applicability of the RTE is significantly wider than what the phenomenological approach may imply.

Of course, there is nothing wrong with the conception of postulating certain basic physical laws. In fact, any advanced physical theory must ultimately be based on a self-consistent set of well-defined axioms and have the formal structure of a mathematical theory (e.g., Sappes, 2002). The seemingly self-evident phenomenological concepts of radiative transfer have been taken for granted for more than a century and, with a few exceptions, have been traditionally presented as something that does not need proof. However, postulating phenomenological concepts such as the notion of the diffuse specific intensity or the Bouguer–Beer extinction law has the adverse effect of implying that the transfer of electromagnetic energy in discrete random media is controlled by fundamental physical laws other than the Maxwell equations.

One might argue that the microphysical derivation of the RTE from the Maxwell equations is too complicated and, unlike a half-a-page phenomenological "derivation", requires many pages of formulas and graphs. However, the microphysical approach has several decisive advantages. First, one can make certain that the RTT does not need any basic physical postulates other than the Maxwell equations. Second, the exact physical meaning of all participating quantities and their relation to more fundamental physical quantities become clear and unambiguous. Third, the range of applicability of the RTE becomes well characterized. Fourth, it becomes possible to establish the relation of the RTT to the effect of coherent backscattering (Chapter 14).

Another phenomenological way to introduce the RTE is to invoke Einstein's concept of photons (e.g., Ivanov, 1973), describe the radiation field in terms of a "photon gas", and postulate that the photon gas satisfies the Boltzmann kinetic equation (see, for example, Pomraning, 1991; Fernández *et al.*, 1993; Thomas and Stamnes, 1999; Mobley and Vo-Dinh, 2003). This approach is based on associating energy transport with the directional flow of localized particles of light, photons, each carrying energy of amount $h\nu$, where h is Planck's constant and ν is frequency. The diffuse specific intensity is then given by

$$\tilde{I}_{\mathrm{d}}(\mathbf{r}, \hat{\mathbf{q}}) = h\nu c f(\mathbf{r}, \hat{\mathbf{q}}),$$

where c is the speed of light and $f(\mathbf{r}, \hat{\mathbf{q}})$ is the photon distribution function such that $\mathrm{d}S\mathrm{d}\Omega c f(\mathbf{r}, \hat{\mathbf{q}})$ is the number of photons crossing an element of surface area $\mathrm{d}S$ normal to $\hat{\mathbf{q}}$ and centered at $\mathbf{r}$ in directions confined to an element of solid angle $\mathrm{d}\Omega$ centered around $\hat{\mathbf{q}}$ per unit time.

The concept of a photon as a localized particle of light was proposed by Albert Einstein in his 1905 paper on the photoelectric effect. Specifically, he suggested that the energy of a light ray spreading out from a point source is not continuously distributed over an increasing space but consists of a finite number of energy quanta which are localized at points in space (see Arons and Peppard, 1965).

However, it is known from quantum electrodynamics that there is no position operator for a photon and that it is impossible to introduce a photon wave function in the coordinate representation (e.g., Section 2.2 of Akhiezer and Berestetskii, 1965). In fact, photons are quantum excitations of the normal modes of the electromagnetic field and as such are associated with plane waves of definite wave vector and definite polarization but infinite lateral extent (Mandel and Wolf, 1995). This means that photons are not localized particles (Lamb, 1995; see also Section 8.18 below). Thus, the quantum theory of radiation does not allow one to associate the position variable $\mathbf{r}$ with a photon and even to speak about the probability of finding a photon at a particular point in space (Wolf, 1978). It is, therefore, impossible to define $f(\mathbf{r}, \hat{\mathbf{q}})$ as a function of photon coordinates and claim that it satisfies the Boltzmann transport equation.

Another fundamental problem with the "photonic" approach is that it is unclear why the phase and extinction matrices entering the VRTE are still defined and computed in the framework of classical macroscopic electromagnetics. Furthermore, it remains unknown whether there is a relation between the RTT and the effect of coherent backscattering.

It is worth emphasizing again that the detailed microphysical derivation of the VRTE described in this chapter leads quite naturally to the definition of the coherent and diffuse Stokes column vectors, clarifies the physical meaning of all quantities entering Eqs. (8.10.5) and (8.10.6), and makes unnecessary the multiple controversial assumptions of the phenomenological approach. In particular, it eliminates the need to introduce the troublesome and vague notion of an elementary volume element and avoids completely the use of the misleading "photonic" language.

8.17 Scattering media with thermal emission

If the absolute temperature of the particles forming the scattering medium becomes sufficiently high, the emitted component of the total radiation field can become comparable to the multiply scattered component, thereby making necessary a modification of the VRTE. This is usually accomplished by assuming that the emission process is not related directly to the scattering process. This implies that the light emitted by a particle at an angular frequency ω is optically independent of the light incident on and scattered by the particle at this frequency and depends only on the particle temperature, size, shape, orientation, and refractive index.[2] As a result, the radiation emitted by the particle is added incoherently to the radiation scattered by the particle, thereby contributing another term to the right-hand side of Eq. (8.11.5):

$$\begin{aligned}\hat{\mathbf{q}}\cdot\nabla\tilde{\mathbf{I}}(\mathbf{r},\hat{\mathbf{q}},\omega) &= -n_0(\mathbf{r})\langle\mathbf{K}(\mathbf{r},\hat{\mathbf{q}},\omega)\rangle_\xi\,\tilde{\mathbf{I}}(\mathbf{r},\hat{\mathbf{q}},\omega)\\ &\quad + n_0(\mathbf{r})\int_{4\pi} d\hat{\mathbf{q}}'\langle\mathbf{Z}(\mathbf{r},\hat{\mathbf{q}},\hat{\mathbf{q}}',\omega)\rangle_\xi\,\tilde{\mathbf{I}}(\mathbf{r},\hat{\mathbf{q}}',\omega)\\ &\quad + n_0(\mathbf{r})\langle\mathbf{K}_e[\mathbf{r},\hat{\mathbf{q}},T(\mathbf{r}),\omega]\rangle_\xi,\end{aligned} \tag{8.17.1}$$

where $\langle\mathbf{K}_e[\mathbf{r},\hat{\mathbf{q}},T(\mathbf{r}),\omega]\rangle_\xi$ is the single-particle emission Stokes column vector (Section 3.13) averaged over particle states. Note that we have allowed the particle temperature to vary with $\mathbf{r}$ and added the argument ω to explicitly indicate the dependence of the radiation field on angular frequency. The dimension of the elements of the frequency-dependent specific intensity column vector is $\mathrm{W\,m^{-2}\,sr^{-1}\,rad^{-1}\,s}$ rather than $\mathrm{W\,m^{-2}\,sr^{-1}}$.

In the following chapters, we will study only the emission-free VRTE (8.11.5), thereby assuming that the particle temperature is not high enough to make the emis-

[2] The particle temperature may itself depend on the characteristics of the incident radiation, most of all on its intensity and spectral distribution.

sion component of the radiation field comparable to the multiply scattered one.

8.18 Historical notes and further reading

The derivation of the VRTE in this chapter largely follows that given in Mishchenko (2002, 2003). Important early contributions to the microphysical derivation of the VRTE for discrete random media were made by Borovoy (1966) and Dolginov *et al.* (1970). Many aspects of the multiple-scattering theory for discrete random media are discussed by Tsang and Kong (2001).

The phenomenological RTT is outlined in the classical texts by Kourganoff (1952, 1969), Chandrasekhar (1950), Preisendorfer (1965), and Ishimaru (1978) (see also Mobley, 1994; Thomas and Stamnes, 1999). The early history of the phenomenological theory of radiative transfer is described by Ivanov (1994). He traces the origin of the simplest form of the RTE (no account of polarization, isotropically scattering particles) to papers by Lommel (1887) and Chwolson (1889). Unfortunately, those early publications have remained largely unnoticed, and the first introduction of the RTE has traditionally been attributed to the paper by Schuster (1905).

Gans (1924) was the first to consider the transfer of polarized light in a plane-parallel Rayleigh-scattering atmosphere; however, he analyzed only the special case of perpendicularly incident light and considered only the first two components of the Stokes column vector. The case of arbitrary illumination and arbitrary polarization was first studied by Chandrasekhar (1947a). Rozenberg (1955) introduced the most general form of the VRTE for scattering media composed of arbitrarily shaped and arbitrarily oriented particles.

The concept of photons has been thoroughly misused in the phenomenological treatment of radiative transfer. It is important to remember that photons are not localized particles (e.g., Section 4.10 of Bohm, 1951; §88 of Kramers, 1957; Section 5.1 of Power, 1964), which makes the words like "photon position", "photon path", "photon trajectory", or "local flow of photons" physically meaningless. Although the term "photon" is ubiquitous in quantum electrodynamics and quantum optics, there it means nothing more than a quantum of a single normal mode of the electromagnetic field (Mandel and Wolf, 1995). Since the normal modes have an infinite lateral extent, they cannot be interpreted as "particles". If the solution of a specific problem does require quantization of the electromagnetic field then the most one can say is that the photons represent a discrete character of light in that specific application but not a "particle" character.

If one is tempted to use the word "photon" to describe a relatively localized "packet" of radiation, it should be remembered that the Fourier analysis requires a wavepacket to consist of a superposition of normal modes. The drawback of this "particle" interpretation is that each source emits its own kind of wavepackets, which leaves one with a wide variety of analytical representations of a wavepacket or worse, no analytical representation at all (Meystre and Sargent, 1999).

The quest for a photon as a universal localized quantum of light appears to be as hopeless now as it has ever been, as revealed by the October 2003 supplement to *Optics and Photonics News* titled "The Nature of Light: What is a Photon?" (Roychoudhuri and Roy, 2003). Unfortunately, most undergraduate textbooks on modern physics and even many graduate texts remain profoundly confusing and often misleading on this issue.[3] Their authors keep relishing the so-called "wave-particle duality" of light which was discarded following the development of quantum electrodynamics seven decades ago. Furthermore, they appear not to realize that one does not need the concept of a photon as a particle of light to explain the photoelectric and Compton effects and that this concept is inconsistent with the Planck energy distribution law, the facts established in the 1910s and 1920s (see, for example, Kidd *et al.*, 1989 and references therein).[4] An excellent remedy to these textbooks are the thorough discussions of the concept of a photon and its history in Kidd *et al.* (1989) and Lamb (1995). In the former, the authors boldly assert that elementary texts would do well to drop the corpuscular photon (except, perhaps, as a historical topic) and switch to the semi-classical treatment as the first approximation to the modern quantum electrodynamics approach.

Unfortunately, the word "photon" is invoked most commonly in circumstances in which the electromagnetic field is classical and has no quantum character whatsoever. The word "photon" then serves as nothing more than a catchy synonym for "light". This usage of the word "photon" is especially misleading and should be avoided.

An interesting theoretical study of the range of applicability of the RTE was performed by Roux *et al.* (2001). They used an exact Monte-Carlo solution of the electromagnetic scattering problem for a slab containing randomly located parallel infinite cylinders and compared the results with those obtained using the corresponding two-dimensional radiative transfer theory (Mishchenko *et al.*, 1992).

There have been a few successful attempts to solve numerically the general VRTE without making overly restrictive assumptions about the morphology of the scattering medium. They have been documented in Haferman *et al.* (1997), Emde *et al.* (2004), and Battaglia and Mantovani (2005).

In this chapter, we derived the RTE for a medium composed of discrete scattering particles. However, a similar equation describes multiple scattering of light in a continuous medium with random fluctuations of the refractive index. The reader is referred to Barabanenkov *et al.* (1972), Papanicolaou and Burridge (1975), and Fante (1981) for discussions of this other branch of the RTT.

---

[3] The textbook by Lipson *et al.* (2001) is a rare exception.

[4] For modern semi-classical treatments of the photoelectric effect, see Chapter 11 of Schiff (1968) and the paper by Fearn and Lamb (1991).

Chapter 9

Calculations and measurements of single-particle characteristics

It follows from the structure of the VRTE that the first step in solving this equation for a specific scattering medium must be the determination of the single-particle extinction and phase matrices averaged over the relevant range of particle states. These quantities are also necessary to describe single scattering by an individual random particle as well as a small random particle group (Chapters 6 and 7). Given the unlimited variability of particles in natural and anthropogenic environments, as illustrated by Fig. 9.0.1, the computation of $\langle \mathbf{K}(\hat{\mathbf{s}})\rangle_\xi$ and $\langle \mathbf{Z}(\hat{\mathbf{q}}, \hat{\mathbf{q}}')\rangle_\xi$ can be a rather nontrivial problem. The case of spherically symmetric particles is an exception since it can be handled easily using the classical Lorenz–Mie theory or one of its extensions. However, the optical properties of nonspherical and heterogeneous particles must be either computed using a sophisticated theory or measured experimentally, both approaches having their strengths, weaknesses, and limitations.

The aim of this chapter is to provide a brief summary of the existing theoretical and experimental techniques for determination of the single-particle characteristics. More detailed information and further references can be found in MTL, in the book edited by Mishchenko *et al.* (2000a), and in a recent review by Kahnert (2003).

9.1 Exact theoretical techniques

Most of the existing exact theoretical approaches belong to one of two broad categories. Differential equation methods yield the scattered field via the solution to the Maxwell equations or the vector wave equation in the frequency or in the time domain, whereas integral equation methods are based on the volume or surface integral counterparts of the Maxwell equations.

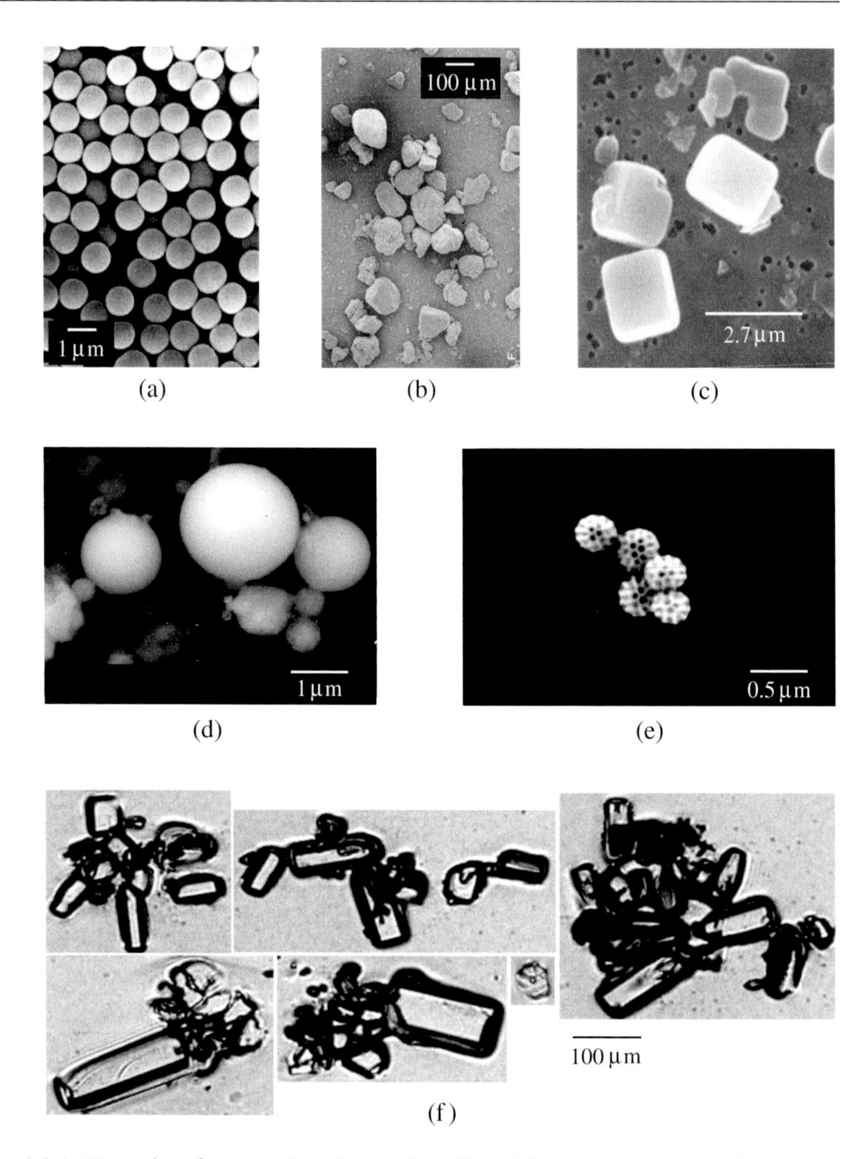

Figure 9.0.1. Examples of man-made and natural small particles. (a) Commercial glass spheres (after Bangs and Meza, 1995). (b) Sahara desert sand (after Volten *et al.*, 2001). (c) Dried sea-salt particles (after Chamaillard *et al.*, 2003). (d) Fly ash particles (after Ebert *et al.*, 2002). (e) Biological microparticles (after Ebert *et al.*, 2002). (f) Cirrus cloud crystals (after Arnott *et al.*, 1994).

The prime example of a differential equation method is the Lorenz–Mie theory (van de Hulst, 1957; Bohren and Huffman, 1983; MTL). The incident, internal, and scattered fields are expanded in suitable sets of vector spherical wave functions (VSWFs). The expansion coefficients of the incident plane wave can be computed analytically, whereas those of the incident and scattered fields are determined by satisfying the boundary conditions, Section 2.2, on the sphere surface. Owing to the orthogonality of the VSWFs, each expansion coefficient of either the internal or the scattered field is determined separately. This makes the Lorenz–Mie theory extremely efficient and numerically exact. Several computer implementations of the Lorenz–Mie solution are available on the World Wide Web. Section 5.10 of MTL provides a detailed user guide to the Lorenz–Mie code posted at http://www.giss.nasa.gov/~crmim. By implementing a recursive procedure, one can generalize the Lorenz–Mie solution and treat concentric multi-layer spheres (e.g., Babenko *et al.*, 2003 and references therein).

The separation of variables method (SVM) for spheroids provides the solution of the electromagnetic scattering problem in the spheroidal coordinate system by means of expanding the incident, internal, and scattered fields in vector spheroidal wave functions (Oguchi, 1973; Asano and Yamamoto, 1975). The expansion coefficients of the incident field are computed analytically, whereas the unknown expansion coefficients of the internal and scattered fields are determined by applying the boundary conditions, Section 2.2. Because the vector spheroidal wave functions are not orthogonal on the spheroid surface, this procedure yields an infinite set of linear algebraic equations for the unknown coefficients which must be truncated and solved numerically. The obvious limitation of the SVM is that it applies only to spheroidal scatterers, whereas its main advantages are the ability to produce very accurate results and the applicability to spheroids with extreme aspect ratios. This technique was significantly improved by Voshchinnikov and Farafonov (1993) and was extended to core-mantle spheroids by Onaka (1980), Cooray and Ciric (1992), and Farafonov *et al.* (1996). Further references can be found in the review by Ciric and Cooray (2000) and the book by Li *et al.* (2002).

The finite element method (FEM) is a differential equation technique that yields the scattered field by means of solving numerically the vector Helmholtz equation subject to the standard boundary conditions. The particle is imbedded in a finite computational domain discretized into many cells with about 10 to 20 cells per wavelength. The electric field values are specified at the nodes of the cells and are initially unknown. Using the boundary conditions, the differential equation is converted into a matrix equation for the unknown node field values. The latter is solved using the standard Gaussian elimination or preconditioned iterative techniques such as the conjugate gradient method. Although scattering in the far-field zone is an open-space problem, the FEM is always implemented in a finite computational domain in order to limit the number of unknowns. Therefore, approximate absorbing boundary conditions must be imposed at the outer boundary of the computational domain in order to suppress wave reflections back into the domain and permit the numerical analogs of the outward-propagating wave to exit the domain almost as if it were infinite. The FEM can be applied to arbitrarily shaped and inhomogeneous particles and is simple in concept and implementation. However, FEM computations are spread over the entire computational domain rather than confined to the scatterer itself, thereby making the technique slow and limited to size parameters less than about 10. The finite spatial discretization and the approximate absorbing boundary condition limit the accuracy of the method. Further information about the FEM can be found in the books by Silvester and Ferrari (1996), Volakis *et al.* (1998), and Jin (2002).

Unlike the FEM, the finite difference time domain method (FDTDM) yields the solution of the electromagnetic scattering problem in the time domain by directly solving the Maxwell time-dependent curl equations (2.1.2) and (2.1.4) (Yee, 1966). The space and time derivatives of the electric and magnetic fields are approximated using a finite difference scheme with space and time discretizations selected so that they constrain computational errors and ensure numerical stability of the algorithm. Since the scattering object is imbedded in a finite computational domain, absorbing boundary conditions are employed to model scattering in the open space. Modeling scattering objects with curved boundaries using rectangular grid cells causes a staircasing effect and increases numerical errors, especially for particles with large relative refractive indices. Since FDTDM yields the near field in the time domain, a special near-zone to far-zone transformation must be invoked in order to compute the scattered far field in the frequency domain. The FDTDM shares the advantages of the FEM as well as its limitations in terms of accuracy and size parameter range. Additional information on the FDTDM and its applications can be found in the books by Kunz and Luebbers (1993) and Taflove and Hagness (2000) as well as in the review by Yang and Liou (2000).

The point-matching method (PMM) is a differential equation technique based on expanding the incident and internal fields in VSWFs regular at the origin and expanding the scattered field outside the scatterer in outgoing VSWFs. The expansion coefficients of the incident field are computed analytically, whereas the coefficients of the internal and scattered fields are found by truncating the expansions to a finite size and matching the fields at the surface of the scatterer via the application of the boundary conditions. In the simple PMM, the fields are matched at as many points on the surface as there exist unknown expansion coefficients (Oguchi, 1973). The simple PMM often produces poorly converging and unstable results, which may be attributed to the fact that it relies on the so-called Rayleigh hypothesis. The convergence problem of the simple PMM appears to be partly ameliorated in the generalized PMM (GPMM) by creating an overdetermined system of equations for the unknown coefficients by means of matching the fields in the least squares sense at a number of surface points significantly greater than the number of unknowns (Morrison and Cross, 1974). The performance of the GPMM is further improved by employing multiple spherical expansions to describe the fields both inside and outside the scattering object. This multiple-expansion GPMM (ME-GPMM) does not rely on the Rayleigh hypothesis and is otherwise known as the generalized multipole technique, discrete sources method, and Yasuura method (Wriedt, 1999; Doicu *et al.*, 2000).

As we have seen in Section 3.1, the interaction of an incident electromagnetic wave with an object of volume $V_{\rm INT}$ is fully described by the volume integral equation (3.1.21). The calculation of the scattered field using Eq. (3.1.22) would be straightforward except that the internal electric field is unknown. Therefore, this equation must first be solved for the internal field. The integral in Eq. (3.1.21) is approximated by discretizing the interior region into N cubic cells of a volume ΔV with about 10 to 20 cells per wavelength and assuming that the electric field and the refractive index within each cell are constant:

$$\mathbf{E}(\mathbf{r}_i) = \mathbf{E}^{\mathrm{inc}}(\mathbf{r}_i) + k_1^2 \Delta V \sum_{j=1}^{N} \ddot{G}(\mathbf{r}_i, \mathbf{r}_j) \cdot \mathbf{E}(\mathbf{r}_j)[m^2(\mathbf{r}_j) - 1], \qquad i = 1, \dots, N, \tag{9.1.1}$$

where $\mathbf{r}_i \in V_{\mathrm{INT}}$ is the central point of the ith cell. Equations (9.1.1) form a system of N linear algebraic equations for the N unknown internal fields $\mathbf{E}(\mathbf{r}_i)$ and are solved numerically. Once the internal fields are found, the scattered field is determined from

$$\mathbf{E}^{\mathrm{sca}}(\mathbf{r}) = k_1^2 \Delta V \sum_{j=1}^{N} \ddot{G}(\mathbf{r}, \mathbf{r}_j) \cdot \mathbf{E}(\mathbf{r}_j)[m^2(\mathbf{r}_j) - 1], \qquad \mathbf{r} \in V_{\mathrm{EXT}}. \tag{9.1.2}$$

This version of the volume integral equation method (VIEM) is known as the method of moments (MOM). Since the free space dyadic Green's function becomes singular as $|\mathbf{r} - \mathbf{r}'| \to 0$, special techniques must be used to handle the self-interaction term $(j = i)$ in the sum on the right-hand side of Eq. (9.1.1). The straightforward approach to solving the MOM matrix equation using the standard Gaussian elimination is not practical for size parameters exceeding unity. The conjugate gradient method together with the fast Fourier transform (Peterson *et al.*, 1998) can be applied to significantly larger size parameters and substantially reduces computer memory requirements. The standard drawback of using a preconditioned iterative technique is that computations must be fully repeated for each new illumination direction.

Another version of the VIEM is the so-called discrete dipole approximation (DDA). Whereas the MOM deals with the *actual* electric fields in the central points of the cells, the DDA exploits the concept of *exciting* fields and is based on partitioning the particle into a number N of elementary polarizable units called dipoles. The electromagnetic response of the dipoles to the local electric field is assumed to be known. The field exciting a dipole is a superposition of the external field and the fields scattered by all other dipoles. This allows one to write a system of N linear equations for N fields exciting the N dipoles. The numerical solution of the DDA matrix equation is then used to compute the N partial fields scattered by the dipoles and thus the total scattered field. Although the original derivation of the DDA by Purcell and Pennypacker (1973) was heuristic, Lakhtakia and Mulholland (1993) showed that the DDA can be derived from Eq. (3.1.21) and is closely related to the MOM.

The major advantages of the MOM and the DDA are that they automatically satisfy the radiation condition at infinity, Eq. (3.2.16), are confined to the scatterer itself, thereby resulting in fewer unknowns than the differential equation methods, and can be applied to inhomogeneous, anisotropic, and optically active scatterers. However, the numerical accuracy of the methods is relatively low and improves slowly with increasing N, whereas the computer time grows rapidly with increasing size parameter. Another disadvantage of these techniques is the need to repeat the entire calculation for each new direction of incidence. Further information on the MOM and the DDA and their applications can be found in Miller *et al.* (1991) and Draine (2000).

Equation (3.1.21) is a Fredholm-type integral equation with a singular kernel at $\mathbf{r}' = \mathbf{r}$. Holt *et al.* (1978) removed the singularity by applying the Fourier transform to the internal field and converting the volume integral into an integral in the wave number coordinate space. Discretization of the latter integral results in a matrix equation which is solved numerically and gives the scattered field. A limitation of this Fredholm integral equation method (FIEM) is that the matrix elements must be evaluated analytically, thereby requiring different programs for each shape and restricting computations to only a few models such as spheroids, triaxial ellipsoids, and finite circular cylinders. The majority of reported FIEM computations pertain to size parameters smaller than five and tend to be rather time consuming (Holt, 1982).

The Lorenz–Mie theory can be extended to clusters of spheres by using the translation addition theorem for the VSWFs (Bruning and Lo, 1971a,b). The total field scattered by a multi-sphere cluster can be expressed as a superposition of individual fields scattered from each sphere. The external electric field illuminating the cluster and the individual fields scattered by the component spheres are expanded in VSWFs with origins at the individual sphere centers. The orthogonality of the VSWFs in the sphere boundary conditions is exploited by applying the translation addition theorem in which a VSWF centered at one sphere origin is re-expanded about another sphere origin. This procedure ultimately results in a matrix equation for the scattered-field expansion coefficients of each sphere. Numerical solution of this equation for the specific incident wave gives the individual scattered fields and thereby the total scattered field. Alternatively, inversion of the cluster matrix equation gives sphere-centered transition matrices (or T matrices) that transform the expansion coefficients of the incident wave into the expansion coefficients of the individual scattered fields. In the far-field region, the individual scattered-field expansions can be transformed into a single expansion centered at a single origin inside the cluster. This procedure gives the T matrix that transforms the incident-wave expansion coefficients into the single-origin expansion coefficients of the total scattered field (Mackowski, 1994) and can be used in the analytical averaging of scattering characteristics over cluster orientations (Mackowski and Mishchenko, 1996; Borghese *et al.*, 2003). The superposition method (SM) has been extended to spheres with one or more eccentrically positioned spherical inclusions (Fuller, 1995; Videen *et al.*, 1995; Borghese *et al.*, 2003) and to clusters of dielectric spheroids in an arbitrary configuration (Ciric and Cooray, 2000). Because of the analyticity of its mathematical formulation, the SM is capable of producing very accurate results. Fuller and Mackowski (2000) gave a detailed review of the SM for compounded spheres.

The T-matrix method (TMM) is based on expanding the incident field in VSWFs regular at the origin and expanding the scattered field outside a circumscribing sphere of the scatterer in VSWFs regular at infinity. The T matrix transforms the expansion coefficients of the incident field into those of the scattered field and, if known, can be used to compute any scattering characteristic of the particle. The TMM was initially developed by Waterman (1971) for single homogeneous objects and was generalized to multi-layered scatterers and arbitrary clusters of nonspherical particles by Peterson and Ström (1973, 1974). For spheres, all TMM formulas reduce to those of the Lorenz–Mie theory. In the case of clusters composed of spherical components, the *T*-matrix method reduces to the multi-sphere SM.

The *T* matrix for single homogeneous and multilayered scatterers is usually computed using the extended boundary condition method (EBCM; Waterman, 1971), which explicitly avoids the use of the Rayleigh hypothesis. The EBCM can be applied to any particle shape, although computations become much simpler and more efficient for bodies of revolution. Special procedures have been developed to improve the numerical stability of EBCM computations for large size parameters and/or extreme aspect ratios. Recent work has demonstrated the practical applicability of the EBCM to particles without axial symmetry, e.g., ellipsoids, cubes, and finite polyhedral cylinders. The computation of the *T* matrix for a cluster assumes that the *T* matrices of all components are known and is based on the use of the translation addition theorem for the VSWFs (Peterson and Ström, 1973). The loss of efficiency for particles with large aspect ratios or with shapes lacking axial symmetry is the main drawback of the TMM. The main advantages of the TMM are high accuracy and speed coupled with applicability to particles with equivalent-sphere size parameters exceeding 180. Mishchenko (1991a), Khlebtsov (1992), and Mackowski and Mishchenko (1996) have developed analytical orientation averaging procedures which make TMM computations for randomly oriented particles as fast as those for a particle in a fixed orientation.

Figure 9.1.1 gives examples of particles that can be treated using various implementations of the TMM. Further information on this technique can be found in Chapter 5 of MTL and in Kahnert (2003). A representative collection of public-domain *T*-matrix codes is posted on the World Wide Web at http://www.giss.nasa.gov/~crmim.

The only methods yielding very accurate results for particles comparable to and larger than a wavelength are the SVM, SM, and TMM. The SVM, SM, TMM, and ME-GPMM have been used in computations for particles significantly larger than a wavelength. The first three techniques appear to be the most efficient in application to bodies of revolution. The analytical orientation averaging procedure makes the TMM the most efficient technique for randomly oriented particles with moderate aspect ratios. Particles with larger aspect ratios can be treated with the SVM, an iterative EBCM, and the ME-GPMM. Computations for anisotropic objects and homogeneous and inhomogeneous particles lacking rotational symmetry often have to rely on more flexible techniques such as the FEM, FDTDM, MOM, and DDA. These techniques are simple in concept and computer implementation and have comparable performance characteristics, although their simplicity and flexibility are often accompanied by lower efficiency and accuracy and by stronger practical limitations on the maximal size parameter. A comprehensive collection of computer programs based on a variety

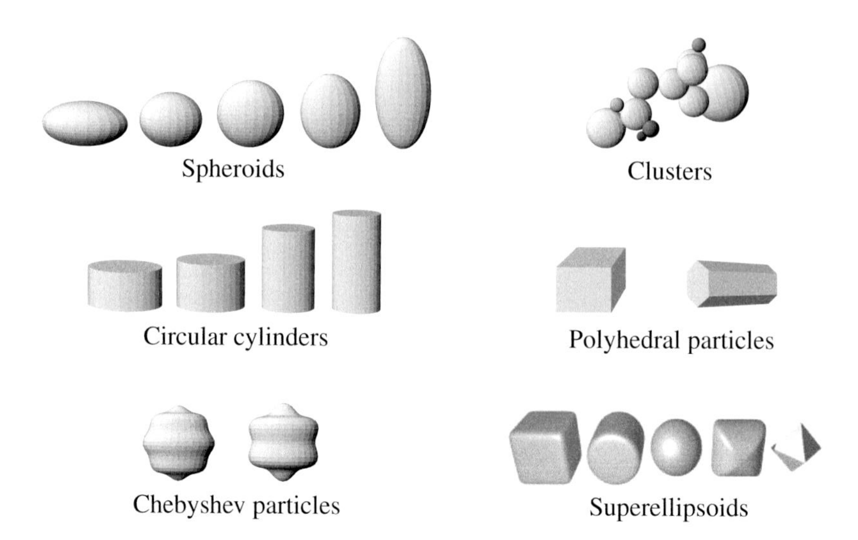

Figure 9.1.1. Types of particles that can be treated with the T-matrix method (after Wriedt, 2002 and Penttilä and Lumme, 2004).

of exact numerical techniques is posted at http://www.iwt-bremen.de/vt/laser/wriedt/index_ns.html.

9.2 Approximations

Any approximate theory of light scattering is based on a simplifying assumption that substantially limits its range of applicability. For example, Rayleigh (1897) derived an approximation for scattering in the small-particle limit ($x \ll 1$) by assuming that the incident field inside and near the particle behaves almost as an electrostatic field and the internal field is homogeneous. The conditions of validity of the Rayleigh–Gans approximation (otherwise known as the Rayleigh–Debye or Born approximation) are $x|m-1| \ll 1$ and $|m-1| \ll 1$. Hence particles are assumed to be not too large (although they may be larger than in the case of Rayleigh scattering) and optically "soft". The fundamental RGA assumption is that each small-volume element of the scattering object is excited only by the incident field. The scattered field is then computed from Eq. (3.2.14) after substituting $\mathbf{E}(\mathbf{r}') = \mathbf{E}^{\text{inc}}(\mathbf{r}')$. The anomalous diffraction approximation (ADA) was introduced by van de Hulst (1957) as a means of computing the extinction cross section for large, optically soft spheres with $x \gg 1$ and $|m-1| \ll 1$. Since the second condition means that rays are weakly deviated as they cross the particle boundary and are negligibly reflected, the ADA assumes that extinction is caused by absorption of light passing through the particle and by the interference of light passing through and around the particle.

The practical importance of approximate theories diminishes as various exact

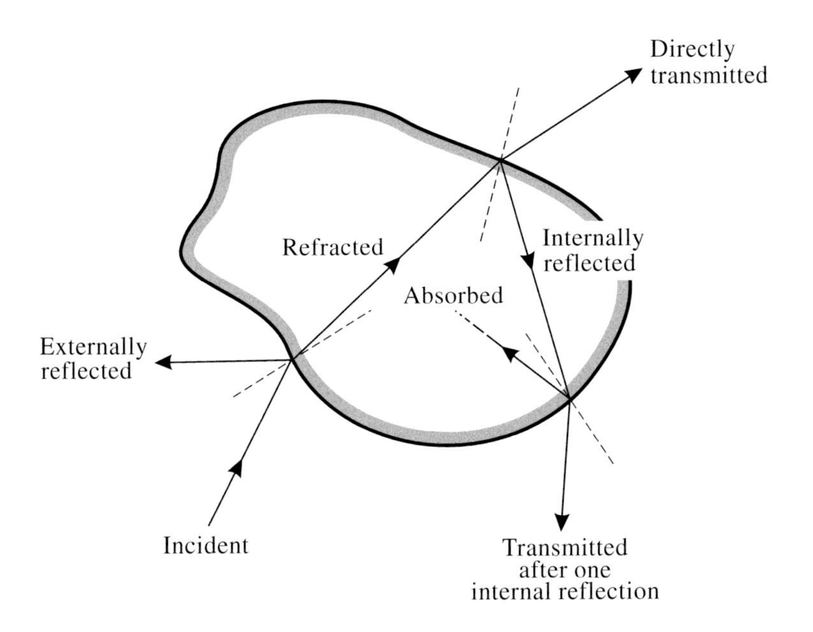

Figure 9.2.1. Ray-tracing diagram.

techniques mature and become applicable to a wider range of problems, while computers become ever more powerful. However, approximate theories still remain a valuable source of physical insight into the process of scattering and absorption by nonspherical particles. Furthermore, it is likely that at least one approximation, the geometrical optics method, will never become obsolete because its accuracy only improves as the particle size parameter grows, whereas all exact theoretical techniques for nonspherical particles cease to be practical when the size parameter exceeds a certain threshold.

The geometrical optics approximation (GOA) is a universal method for computing electromagnetic scattering by arbitrarily shaped particles with sizes much larger than the wavelength of the incident light. The GOA assumes that the incident plane wave can be represented as a collection of independent parallel rays. The history of each ray impinging on the particle surface is traced using Snell's law and Fresnel's formulas (see Fig. 9.2.1). Each incident ray is partially reflected and partially refracted into the particle. The refracted ray may emerge after another refraction, possibly after one or more internal reflections, and may be attenuated by absorption inside the particle. Each internal ray is traced until its intensity decreases below a prescribed cutoff value. Varying the polarization state of the incident rays, sampling all escaping rays into predefined narrow angular bins, and adding incoherently the respective Stokes parameters yields a quantitative representation of the particle's scattering properties in terms of the ray-tracing phase matrix $\mathbf{Z}^{\mathrm{RT}}$. Because all rays impinging on the particle surface are either scattered or absorbed irrespective of their polarization state, the ray-tracing extinction matrix is always diagonal and is given by $\mathbf{K}^{\mathrm{RT}} = C_{\mathrm{ext}}^{\mathrm{RT}}\mathbf{\Delta}$. The ray-tracing extinction cross section $C_{\rm ext}^{\rm RT}$ does not depend on the polarization state of the incident light and is equal to the geometrical area G of the particle projection on the plane perpendicular to the incidence direction. Since the presence of the particle modifies the incident plane wave front by eliminating a part that has the shape and size of the geometrical projection of the particle, the ray-tracing scattering pattern must be supplemented by the computation of Fraunhofer diffraction of the incident wave on the particle projection. The diffraction component of the phase matrix $\mathbf{Z}^{\rm D}$ is confined to a narrow angular cone centered at the exact forward-scattering direction and is usually computed in the Kirchhoff approximation (Jackson, 1998), thereby contributing only to the diagonal elements of the total phase matrix. The diffraction component $\mathbf{K}^{\rm D}$ of the total extinction matrix is equal to $\mathbf{K}^{\rm RT}$. We thus have

$$\mathbf{Z}^{\rm GO} = \mathbf{Z}^{\rm RT} + \mathbf{Z}^{\rm D} = \mathbf{Z}^{\rm RT} + Z_{11}^{\rm D}\boldsymbol{\Delta}, \tag{9.2.1}$$

$$\mathbf{K}^{\rm GO} = \mathbf{K}^{\rm RT} + \mathbf{K}^{\rm D} = C_{\rm ext}^{\rm GO}\boldsymbol{\Delta}, \tag{9.2.2}$$

where

$$C_{\rm ext}^{\rm GO} = C_{\rm ext}^{\rm RT} + C_{\rm ext}^{\rm D} = 2G. \tag{9.2.3}$$

The total scattering cross section is the sum of the ray-tracing and diffraction components:

$$C_{\rm sca}^{\rm GO} = C_{\rm sca}^{\rm RT} + C_{\rm sca}^{\rm D}. \tag{9.2.4}$$

Since the diffracted energy is not absorbed, the diffraction scattering cross section is equal to the diffraction extinction cross section:

$$C_{\rm sca}^{\rm D} = C_{\rm ext}^{\rm D} = G. \tag{9.2.5}$$

The ray-tracing scattering cross section $C_{\rm sca}^{\rm RT}$ is found from $\mathbf{Z}^{\rm RT}$ and Eq. (3.9.10).

The main advantage of the GOA is that it can be applied to essentially any shape. However, this technique is approximate by definition, and its range of applicability in terms of the smallest size parameter must be checked by comparing GOA results with exact numerical solutions of the Maxwell equations. It appears that although the main geometrical optics features can be qualitatively reproduced by particles with size parameters less than 100, obtaining good quantitative accuracy in GOA computations of the phase matrix still requires size parameters exceeding a few hundred. Even then the GOA fails to reproduce scattering features caused by interference and diffraction effects.

To improve the GOA, Ravey and Mazeron (1982) (see also Muinonen, 1989; Liou *et al.*, 2000) developed the so-called physical optics or Kirchhoff approximation (KA). This approach is based on expressing the scattered field in terms of the electric and magnetic fields on the exterior side of the particle surface. The latter are computed approximately using Fresnel formulas and the standard ray-tracing procedure. The KA partially preserves the phase information and reproduces some physical optics effects completely ignored by the standard GOA.

9.3 Measurement techniques

Existing laboratory measurement techniques fall into two categories:

- Scattering of visible or infrared light by particles with sizes from several hundredths of a micron to several hundred microns.
- Microwave scattering by millimeter- and centimeter-sized objects.

Measurements in the visible and infrared benefit from the availability of sensitive detectors, intense sources of radiation, and high-quality optical elements. They involve cheaper and more portable instrumentation and can be performed in the field as well as in the laboratory. However, they become problematic when experimental data for a fixed scattering object are needed and may be difficult to interpret because of lack of independent information on sample microphysics and composition. Microwave scattering experiments require more cumbersome and expensive instrumentation and large measurement facilities, but allow a much greater control over the scattering object.

Many detectors of electromagnetic energy in the visible and infrared spectral regions are polarization-insensitive, which means that the detector response is determined only by the first Stokes parameter of the beam impinging on the detector. Therefore, in order to measure all elements of the scattering matrix one must use various optical elements that can vary the polarization state of light before and after scattering in a controllable way (see Sections 2.10 and 3.7). Figure 9.3.1 (adapted from Hovenier, 2000) depicts the scheme of an advanced laboratory setup used to measure scattering matrix elements for random groups of natural and artificial particles. The light beam generated by a laser passes through a linear polarizer and a polarization modulator and then illuminates particles contained in the scattering chamber. Light scattered by the particles at an angle Θ relative to the incidence direction passes a quarter-wave plate and a polarization analyzer, after which its intensity is measured by a detector. Assuming that the scattering volume satisfies the criteria of applicability of the MUSSA (see Chapter 7), we can write for the Stokes column vector of the beam reaching the detector, $\mathbf{I}'$, the following expression:

$$\mathbf{I}' \propto \mathbf{AQZ}(\Theta)\mathbf{MPI} = \mathbf{AQ}N\langle\mathbf{Z}(\Theta)\rangle_{\xi}\mathbf{MPI}, \tag{9.3.1}$$

where $\mathbf{I}$ is the Stokes column vector of the beam leaving the light source, $\mathbf{A}$, $\mathbf{Q}$, $\mathbf{M}$, and $\mathbf{P}$ are 4×4 Mueller transformation matrices of the analyzer, quarter-wave plate, modulator, and polarizer, respectively, $\mathbf{Z}(\Theta)$ is the total phase matrix of the particles contributing to the scattered beam, N is the number of the particles, and $\langle\mathbf{Z}(\Theta)\rangle_{\xi}$ is the ensemble-averaged phase matrix per particle. It is assumed that the plane through the incidence and scattering directions serves as the azimuthal plane for defining the Stokes parameters. The Mueller matrices of the polarizer, modulator, quarter-wave plate, and analyzer depend on their orientation with respect to the scattering plane and can be precisely varied. Because the detector measures only the first element of the Stokes column vector $\mathbf{I}'$, several measurements with different orientations of the opti-

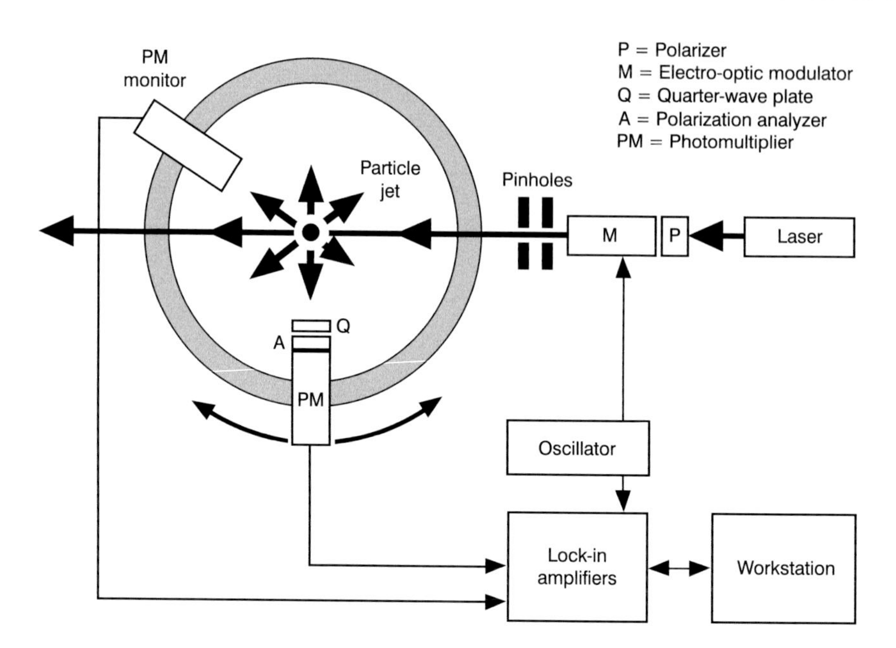

Figure 9.3.1. Schematic view of an experimental scattering setup using visible or infrared light.

cal components with respect to the scattering plane are required for the full determination of the phase matrix. This procedure is repeated at different scattering angles in order to determine the angular profile of the phase matrix.

Hunt and Huffman (1973) developed the technique of a high-frequency sinusoidal modulation in time of the polarization of light before scattering (Fig. 9.3.1) combined with intensity normalization. Followed by lock-in detection, this technique increases the experimental accuracy by enabling direct measurements of the phase matrix elements normalized by the (1, 1) element and yields the capability to determine several elements from only one detected signal.

In accordance with the scale invariance rule (Section 3.5), the main idea of the microwave analog technique is to manufacture a centimeter-sized scattering object with desired shape and refractive index, measure the scattering of a microwave beam by this object, and finally extrapolate the result to other wavelengths (e.g., visible or infrared) by keeping the ratio size/wavelength fixed. In a modern microwave scattering setup (see Fig. 9.3.2), radiation from a transmitting conical horn antenna passes through a collimating lens and a polarizer. The lens produces a nearly flat wave front which is scattered by an analog particle model target. The scattered wave passes through another polarizer and lens and is measured by a receiving horn antenna. The receiver end of the setup can be positioned at any scattering angle from 0° to $\Theta_{\max} \simeq 170°$, thereby providing measurements of the angular distribution of the scattered radiation. By varying the orientations of the two polarizers, one can measure all elements of the phase matrix.

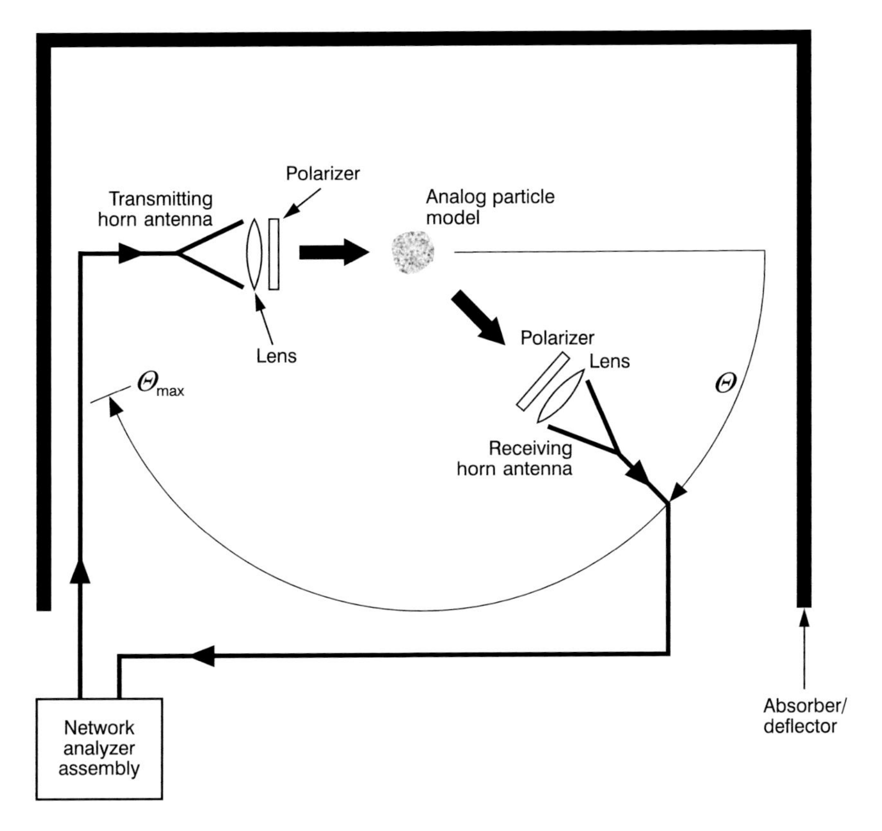

Figure 9.3.2. Layout of a modern microwave analog facility. (After Gustafson, 2000.)

9.4 Further reading

In addition to MTL, the collective monograph edited by Mishchenko *et al.* (2000a), and the recent review by Kahnert (2003), a plentiful source of information on electromagnetic scattering by nonspherical particles is the collection of special issues of the *Journal of Quantitative Spectroscopy and Radiative Transfer* edited by Hovenier (1996), Lumme (1998), Mishchenko *et al.* (1999a), Videen *et al.* (2001), Kolokolova *et al.* (2003), and Wriedt (2004). The book by Babenko *et al.* (2003) gives a detailed account of electromagnetic scattering by radially inhomogeneous and anisotropic spherical particles. A useful compendium of approximate formulas was provided by Kokhanovsky (2004).

Chapter 10

Radiative transfer in plane-parallel scattering media

In order to use the radiative transfer theory in analyses of laboratory measurements or remote-sensing observations, one needs efficient theoretical techniques for solving the VRTE in either the integral or the integro-differential form. Unfortunately, like many other integral and integro-differential equations, the VRTE is very difficult to solve analytically or numerically. In order to facilitate the solution, we will have to make several simplifying assumptions. The most important of them, which will be used throughout the remainder of the book, are the assumptions that the scattering medium:

- Is plane parallel.
- Has an infinite horizontal extent.
- Is illuminated from above by a plane electromagnetic wave or a parallel quasi-monochromatic beam of light of infinite lateral extent.

These assumptions mean that all properties of the medium and of the radiation field may vary only in the vertical direction and are independent of the horizontal coordinates. Taken together, these assumptions specify the so-called *standard problem* of atmospheric optics and provide a model relevant to a great variety of applications in diverse fields of science and technology. In this chapter we will not make any further assumptions and will derive several important equations describing the internal diffuse radiation field as well as the diffuse radiation exiting the medium.

10.1 The standard problem

Let us consider a plane-parallel layer extending in the vertical direction from $z = z_b$ to $z = z_t$, where the z-axis of the laboratory right-handed coordinate system is perpendicular to the boundaries of the medium and is directed upwards, and "b" and "t"

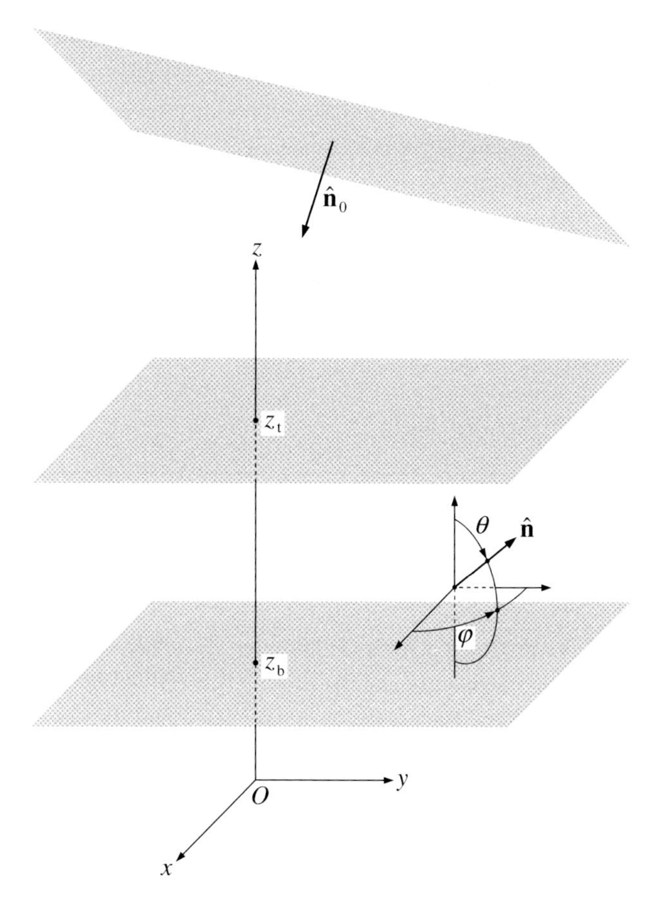

Figure 10.1.1. Plane-parallel scattering medium illuminated from above by a parallel quasi-monochromatic beam of light.

stand for "bottom" and "top", respectively (Fig. 10.1.1). A propagation direction $\hat{\mathbf{n}}$ at a point in space will be specified by a couplet $\{u, \varphi\}$, where $u = -\cos\theta \in [-1, +1]$ is the direction cosine, and θ and φ are the corresponding polar and azimuth angles with respect to the local coordinate system having the same spatial orientation as the laboratory coordinate system. It is also convenient to introduce a nonnegative quantity $\mu = |u| \in [0, 1]$. In order to make many formulas of this and the following chapters more compact, we will denote by $\hat{\mu}$ the pair of arguments $\{\mu, \varphi\}$ and by $-\hat{\mu}$ the pair of arguments $\{-\mu, \varphi\}$ (note that $\hat{\mu}$ and $-\hat{\mu}$ are not unit vectors). A $\hat{\mu}$ always corresponds to a downward direction and a $-\hat{\mu}$ always corresponds to an upward direction. We also denote

$$\int \mathrm{d}\hat{\mu} = \int_0^1 \mathrm{d}\mu \int_0^{2\pi} \mathrm{d}\varphi. \tag{10.1.1}$$

Let us assume that the scattering layer is illuminated from above by a plane electromagnetic wave or a parallel quasi-monochromatic beam of light propagating in the direction $\hat{\mathbf{n}}_0 = \{\mu_0, \varphi_0\}$. The uniformity and the infinite transverse extent of the wave or the beam ensure that all parameters of the internal radiation field and those of the radiation leaving the scattering layer are independent of the coordinates x and y. Therefore, Eq. (8.11.5) can be rewritten in the form

$$-u\frac{\mathrm{d}\tilde{\mathbf{I}}(z,\hat{\mathbf{n}})}{\mathrm{d}z} = -n_0(z)\mathbf{K}(z,\hat{\mathbf{n}})\tilde{\mathbf{I}}(z,\hat{\mathbf{n}}) + n_0(z)\int_{4\pi}\mathrm{d}\hat{\mathbf{n}}'\,\mathbf{Z}(z,\hat{\mathbf{n}},\hat{\mathbf{n}}')\tilde{\mathbf{I}}(z,\hat{\mathbf{n}}') \tag{10.1.2}$$

and must be supplemented by the boundary conditions

$$\tilde{\mathbf{I}}(z_t,\hat{\mu}) = \delta(\mu-\mu_0)\delta(\varphi-\varphi_0)\mathbf{I}_0, \tag{10.1.3}$$

$$\tilde{\mathbf{I}}(z_b,-\hat{\mu}) = \mathbf{0}, \tag{10.1.4}$$

where

$$\tilde{\mathbf{I}}(z,\hat{\mathbf{n}}) = \delta(\hat{\mathbf{n}}-\hat{\mathbf{n}}_0)\mathbf{I}_c(z) + \tilde{\mathbf{I}}_d(z,\hat{\mathbf{n}})$$

is the full specific intensity column vector including both the coherent and the diffuse components, $\mathbf{K}$ and $\mathbf{Z}$ are the extinction and the phase matrix, respectively, averaged over particle states (note that we have omitted the angular brackets for the sake of brevity), $\mathbf{I}_0$ is the Stokes column vector of the incident radiation, and $\mathbf{0}$ is a zero four-element column. As in Chapter 8, the tilde distinguishes specific intensity column vectors from Stokes column vectors. The boundary conditions follow directly from the integral form of the VRTE, Eq. (8.10.9), and mean that the downwelling radiation at the upper boundary of the layer consists only of the incident radiation and that there is no upwelling radiation at the lower boundary. Equations (10.1.2)–(10.1.4) collectively represent what we have called the *standard problem*.

Since $n_0(z)$ is a common factor in both terms on the right-hand side of Eq. (10.1.2), it is convenient to eliminate it by introducing a new vertical "coordinate" $\psi(z)$ according to $\mathrm{d}\psi = -n_0(z)\mathrm{d}z$ or

$$\psi(z) = \int_z^\infty \mathrm{d}z'\,n_0(z'). \tag{10.1.5}$$

Clearly, $\psi(z)$ has the dimension m^{-2} and is the number of particles in a vertical column having a unit cross section and extending from $z' = z$ to infinity. It is, therefore, natural to call it the "particle depth". Unlike the z-coordinate, which increases in the upward direction, the ψ-coordinate increases in the downward direction. We then have

$$u\frac{\mathrm{d}\tilde{\mathbf{I}}(\psi,\hat{\mathbf{n}})}{\mathrm{d}\psi} = -\mathbf{K}(\psi,\hat{\mathbf{n}})\tilde{\mathbf{I}}(\psi,\hat{\mathbf{n}}) + \int_{4\pi}\mathrm{d}\hat{\mathbf{n}}'\,\mathbf{Z}(\psi,\hat{\mathbf{n}},\hat{\mathbf{n}}')\tilde{\mathbf{I}}(\psi,\hat{\mathbf{n}}'), \tag{10.1.6}$$

$$\tilde{\mathbf{I}}(0,\hat{\mu}) = \delta(\mu-\mu_0)\delta(\varphi-\varphi_0)\mathbf{I}_0, \tag{10.1.7}$$

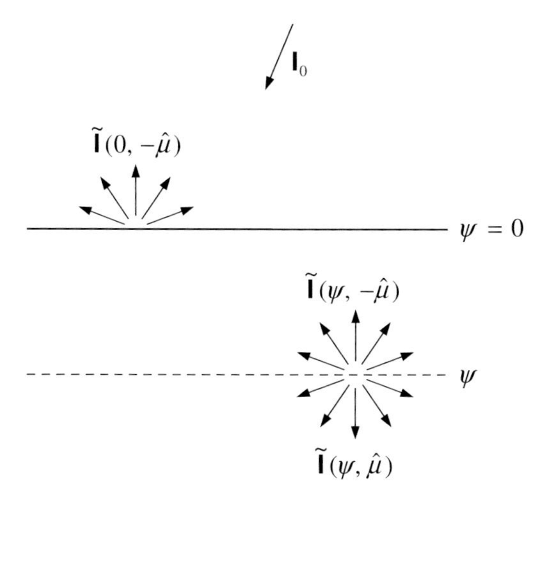

Figure 10.1.2. The standard problem.

$$\tilde{\mathbf{I}}(\Psi, -\hat{\mu}) = \mathbf{0}, \tag{10.1.8}$$

where $\Psi = \psi(z_{\mathrm{b}})$ is the “particle thickness” of the layer (Fig. 10.1.2).

10.2 The propagator

Before attempting to solve the full VRTE, let us first consider the solution of the homogenous differential transfer equation

$$\mu \frac{\mathrm{d}\tilde{\mathbf{I}}(\psi, \hat{\mu})}{\mathrm{d}\psi} = -\mathbf{K}(\psi, \hat{\mu})\,\tilde{\mathbf{I}}(\psi, \hat{\mu}), \qquad \psi \geq \psi_0 \tag{10.2.1}$$

supplemented by the initial condition

$$\tilde{\mathbf{I}}(\psi_0, \hat{\mu}) = \tilde{\mathbf{I}}_0. \tag{10.2.2}$$

It is convenient to express $\mathbf{I}(\psi, \hat{\mu})$ in terms of the solution of the following auxiliary initial-value problem:

$$\mu \frac{\mathrm{d}\mathbf{X}(\psi, \psi_0, \hat{\mu})}{\mathrm{d}\psi} = -\mathbf{K}(\psi, \hat{\mu})\mathbf{X}(\psi, \psi_0, \hat{\mu}), \qquad \psi \geq \psi_0, \tag{10.2.3}$$

$$\mathbf{X}(\psi_0, \psi_0, \hat{\mu}) = \mathbf{\Delta}, \tag{10.2.4}$$

where, as before, $\mathbf{\Delta}$ is the 4×4 unit matrix and $\mathbf{X}(\psi,\psi_0,\hat{\mu})$ is a 4×4 real matrix called the matrizant (Frazer *et al.*, 1957; Birkhoff and Rota, 1969), the evolution operator (Landi Degl'Innocenti and Landolfi, 2004), or the propagator (Flatau and Stephens, 1988). Specifically, if the propagator is known then the solution of Eqs. (10.2.1)–(10.2.2) is simply

$$\tilde{\mathbf{I}}(\psi,\hat{\mu}) = \mathbf{X}(\psi,\psi_0,\hat{\mu})\,\tilde{\mathbf{I}}_0. \tag{10.2.5}$$

The propagator has the obvious semi-group property

$$\mathbf{X}(\psi,\psi_0,\hat{\mu}) = \mathbf{X}(\psi,\psi_1,\hat{\mu})\mathbf{X}(\psi_1,\psi_0,\hat{\mu}), \tag{10.2.6}$$

where $\psi_0 \le \psi_1 \le \psi$. Indeed, since the matrix $\mathbf{Y}(\psi,\psi_0,\hat{\mu}) = \mathbf{X}(\psi,\psi_1,\hat{\mu}) \times \mathbf{X}(\psi_1,\psi_0,\hat{\mu})$ is the solution of the same differential matrix equation (10.2.3) with the same initial condition $\mathbf{Y}(\psi_0,\psi_0,\hat{\mu}) = \mathbf{\Delta}$, the property (10.2.6) follows from the well-known mathematical fact that the differential equation (10.2.3) has only one solution satisfying the initial condition (10.2.4).

If the scattering layer is homogeneous then $\mathbf{K}(\psi,\hat{\mu}) \equiv \mathbf{K}(\hat{\mu})$, and the propagator can be written in the form of a matrix exponential:

$$\mathbf{X}(\psi,\psi_0,\hat{\mu}) = \exp[-(\psi-\psi_0)\mathbf{K}(\hat{\mu})/\mu]. \tag{10.2.7}$$

If the layer is inhomogeneous, one should exploit the semi-group property (10.2.6) by subdividing the interval $[\psi_0,\psi]$ into a number N of equal subintervals $[\psi_0,\psi_1]$, ..., $[\psi_{n-1},\psi_n]$, ..., $[\psi_{N-1},\psi]$ and calculating the propagator in the limit $N\to\infty$:

$$\begin{aligned}
\mathbf{X}(\psi,\psi_0,\hat{\mu}) &= \lim_{N\to\infty} \{\exp[-(\Delta\psi/\mu)\mathbf{K}(\psi_{N-1}+\Delta\psi/2,\hat{\mu})]\cdots \\
&\qquad \times\exp[-(\Delta\psi/\mu)\mathbf{K}(\psi_{n-1}+\Delta\psi/2,\hat{\mu})]\cdots \\
&\qquad \times\exp[-(\Delta\psi/\mu)\mathbf{K}(\psi_0+\Delta\psi/2,\hat{\mu})]\} \\
&= \lim_{N\to\infty} \{[\mathbf{\Delta}-(\Delta\psi/\mu)\mathbf{K}(\psi_{N-1}+\Delta\psi/2,\hat{\mu})]\cdots \\
&\qquad \times[\mathbf{\Delta}-(\Delta\psi/\mu)\mathbf{K}(\psi_{n-1}+\Delta\psi/2,\hat{\mu})]\cdots \\
&\qquad \times[\mathbf{\Delta}-(\Delta\psi/\mu)\mathbf{K}(\psi_0+\Delta\psi/2,\hat{\mu})]\},
\end{aligned} \tag{10.2.8}$$

where $\Delta\psi = (\psi-\psi_0)/N$ and $\psi_n = \psi_0 + n\Delta\psi$.

Similarly, the solution of the equation

$$-\mu\frac{\mathrm{d}\tilde{\mathbf{I}}(\psi,-\hat{\mu})}{\mathrm{d}\psi} = -\mathbf{K}(\psi,-\hat{\mu})\,\tilde{\mathbf{I}}(\psi,-\hat{\mu}), \qquad \psi \le \psi_0 \tag{10.2.9}$$

supplemented by the initial condition

$$\tilde{\mathbf{I}}(\psi_0,-\hat{\mu}) = \tilde{\mathbf{I}}_0 \tag{10.2.10}$$

can be expressed in terms of the solution of the auxiliary initial-value problem

$$-\mu\frac{\mathrm{d}\mathbf{X}(\psi,\psi_0,-\hat{\mu})}{\mathrm{d}\psi} = -\mathbf{K}(\psi,-\hat{\mu})\mathbf{X}(\psi,\psi_0,-\hat{\mu}), \qquad \psi\le\psi_0, \tag{10.2.11}$$

$$\mathbf{X}(\psi_0, \psi_0, -\hat{\mu}) = \mathbf{\Delta} \tag{10.2.12}$$

as

$$\tilde{\mathbf{I}}(\psi, -\hat{\mu}) = \mathbf{X}(\psi, \psi_0, -\hat{\mu})\,\tilde{\mathbf{I}}_0. \tag{10.2.13}$$

The propagator $\mathbf{X}(\psi, \psi_0, -\hat{\mu})$ has the semi-group property

$$\mathbf{X}(\psi, \psi_0, -\hat{\mu}) = \mathbf{X}(\psi, \psi_1, -\hat{\mu})\mathbf{X}(\psi_1, \psi_0, -\hat{\mu}), \qquad \psi \le \psi_1 \le \psi_0 \tag{10.2.14}$$

and is given by

$$\mathbf{X}(\psi, \psi_0, -\hat{\mu}) = \exp[-(\psi_0 - \psi)\mathbf{K}(-\hat{\mu})/\mu] \tag{10.2.15}$$

if the layer is homogeneous and by

$$\begin{aligned}
\mathbf{X}(\psi, \psi_0, -\hat{\mu}) &= \lim_{N \to \infty} \{\exp[-(\Delta\psi/\mu)\mathbf{K}(\psi_{N-1} - \Delta\psi/2, -\hat{\mu})] \cdots \\
&\qquad \times \exp[-(\Delta\psi/\mu)\mathbf{K}(\psi_{n-1} - \Delta\psi/2, -\hat{\mu})] \cdots \\
&\qquad \times \exp[-(\Delta\psi/\mu)\mathbf{K}(\psi_0 - \Delta\psi/2, -\hat{\mu})]\} \\
&= \lim_{N \to \infty} \{[\mathbf{\Delta} - (\Delta\psi/\mu)\mathbf{K}(\psi_{N-1} - \Delta\psi/2, -\hat{\mu})] \cdots \\
&\qquad \times [\mathbf{\Delta} - (\Delta\psi/\mu)\mathbf{K}(\psi_{n-1} - \Delta\psi/2, -\hat{\mu})] \cdots \\
&\qquad \times [\mathbf{\Delta} - (\Delta\psi/\mu)\mathbf{K}(\psi_0 - \Delta\psi/2, -\hat{\mu})]\}
\end{aligned} \tag{10.2.16}$$

if the layer is inhomogeneous, where $\Delta\psi = (\psi_0 - \psi)/N$ and $\psi_n = \psi_0 - n\Delta\psi$.

10.3 The general problem

The standard problem (10.1.6)–(10.1.8) implies that the scattering layer is illuminated only from above by a monodirectional source of light. It is useful, however, to consider the following more general boundary values, which include the boundary conditions (10.1.7) and (10.1.8) as a particular case:

$$\tilde{\mathbf{I}}(0, \hat{\mu}) = \tilde{\mathbf{I}}_{\downarrow}(\hat{\mu}), \tag{10.3.1}$$

$$\tilde{\mathbf{I}}(\Psi, -\hat{\mu}) = \tilde{\mathbf{I}}_{\uparrow}(-\hat{\mu}), \tag{10.3.2}$$

where $\tilde{\mathbf{I}}_{\downarrow}(\hat{\mu})$ and $\tilde{\mathbf{I}}_{\uparrow}(-\hat{\mu})$ are arbitrary (Fig. 10.3.1). We will call Eqs. (10.1.6), (10.3.1), and (10.3.2) the *general problem*.

Let us now assume that the incident light is quasi-monochromatic, meaning that the specific intensity column vectors $\tilde{\mathbf{I}}_{\downarrow}(\hat{\mu})$ and $\tilde{\mathbf{I}}_{\uparrow}(-\hat{\mu})$ represent "bundles" of uncorrelated quasi-monochromatic beams with intensity and polarization state potentially varying with direction of incidence. The results of Section 8.15 allow us to express the radiation field $\tilde{\mathbf{I}}(\psi, \hat{\mathbf{n}})$ for $\psi \in [0, \Psi]$ in terms of the $\tilde{\mathbf{I}}_{\downarrow}(\hat{\mu})$ and $\tilde{\mathbf{I}}_{\uparrow}(-\hat{\mu})$ as follows:

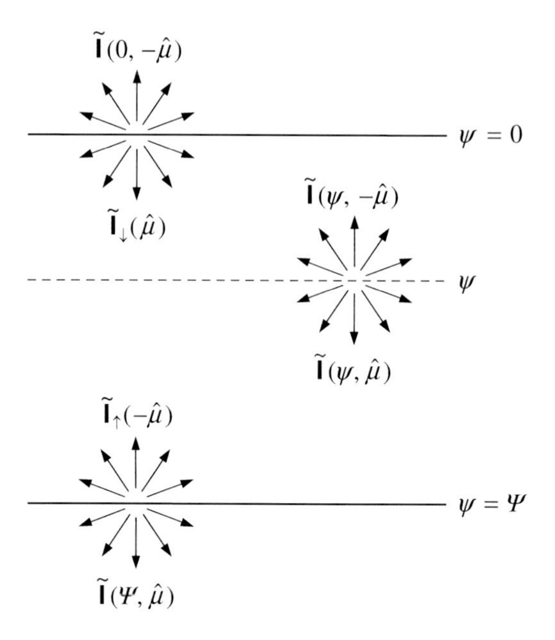

Figure 10.3.1. The general problem.

$$\tilde{\mathbf{I}}(\psi,\hat{\mu}) = \mathbf{X}(\psi,0,\hat{\mu})\tilde{\mathbf{I}}_{\downarrow}(\hat{\mu}) + \frac{1}{\pi}\int \mathrm{d}\hat{\mu}'\mu'\mathbf{D}(\psi,\hat{\mu},\hat{\mu}')\tilde{\mathbf{I}}_{\downarrow}(\hat{\mu}')$$
$$+ \frac{1}{\pi}\int \mathrm{d}\hat{\mu}'\mu'\mathbf{U}^{\dagger}(\psi,\hat{\mu},\hat{\mu}')\tilde{\mathbf{I}}_{\uparrow}(-\hat{\mu}'), \tag{10.3.3}$$

$$\tilde{\mathbf{I}}(\psi,-\hat{\mu}) = \mathbf{X}(\psi,\Psi,-\hat{\mu})\tilde{\mathbf{I}}_{\uparrow}(-\hat{\mu}) + \frac{1}{\pi}\int \mathrm{d}\hat{\mu}'\mu'\mathbf{U}(\psi,\hat{\mu},\hat{\mu}')\tilde{\mathbf{I}}_{\downarrow}(\hat{\mu}')$$
$$+ \frac{1}{\pi}\int \mathrm{d}\hat{\mu}'\mu'\mathbf{D}^{\dagger}(\psi,\hat{\mu},\hat{\mu}')\tilde{\mathbf{I}}_{\uparrow}(-\hat{\mu}'), \tag{10.3.4}$$

where the 4×4 matrices $\mathbf{U}$ and $\mathbf{D}$ describe the response of the scattering layer to the radiation incident on the upper boundary from above, while the 4×4 matrices $\mathbf{U}^{\dagger}$ and $\mathbf{D}^{\dagger}$ describe the response to the radiation illuminating the bottom boundary of the layer from below. The first terms on the right-hand side of Eqs. (10.3.3) and (10.3.4) describe the direct (coherent) propagation of the incident light, whereas the remaining terms describe the result of multiple scattering. The corresponding reflection and transmission matrices determine the Stokes parameters of the radiation exiting the layer and are defined as

$$\mathbf{R}(\hat{\mu},\hat{\mu}') = \mathbf{U}(0,\hat{\mu},\hat{\mu}'), \tag{10.3.5}$$

$$\mathbf{T}(\hat{\mu},\hat{\mu}') = \mathbf{D}(\Psi,\hat{\mu},\hat{\mu}'), \tag{10.3.6}$$

$$\mathbf{R}^{\dagger}(\hat{\mu},\hat{\mu}') = \mathbf{U}^{\dagger}(\Psi,\hat{\mu},\hat{\mu}'), \tag{10.3.7}$$

$$\mathbf{T}^{\dagger}(\hat{\mu},\hat{\mu}') = \mathbf{D}^{\dagger}(0,\hat{\mu},\hat{\mu}'). \tag{10.3.8}$$

The matrices $\mathbf{R}$ and $\mathbf{T}$ describe the response of the layer to the external radiation falling from above, whereas the matrices $\mathbf{R}^\dagger$ and $\mathbf{T}^\dagger$ describe the response to the external radiation falling from below.

It is easy to verify that the solution of the standard problem can now be expressed as

$$\tilde{\mathbf{I}}(\psi, \hat{\mu}) = \delta(\mu - \mu_0)\delta(\varphi - \varphi_0)\mathbf{X}(\psi, 0, \hat{\mu}_0)\mathbf{I}_0 + \frac{1}{\pi}\mu_0 \mathbf{D}(\psi, \hat{\mu}, \hat{\mu}_0)\mathbf{I}_0, \quad (10.3.9)$$

$$\tilde{\mathbf{I}}(\psi, -\hat{\mu}) = \frac{1}{\pi}\mu_0 \mathbf{U}(\psi, \hat{\mu}, \hat{\mu}_0)\mathbf{I}_0, \quad (10.3.10)$$

$$\tilde{\mathbf{I}}(\Psi, \hat{\mu}) = \delta(\mu - \mu_0)\delta(\varphi - \varphi_0)\mathbf{X}(\Psi, 0, \hat{\mu}_0)\mathbf{I}_0 + \frac{1}{\pi}\mu_0 \mathbf{T}(\hat{\mu}, \hat{\mu}_0)\mathbf{I}_0, \quad (10.3.11)$$

$$\tilde{\mathbf{I}}(0, -\hat{\mu}) = \frac{1}{\pi}\mu_0 \mathbf{R}(\hat{\mu}, \hat{\mu}_0)\mathbf{I}_0. \quad (10.3.12)$$

10.4 Adding equations

In this section we will describe an elegant mathematical scheme for computing the matrices $\mathbf{U}$, $\mathbf{D}$, $\mathbf{U}^\dagger$, $\mathbf{D}^\dagger$, $\mathbf{R}$, $\mathbf{T}$, $\mathbf{R}^\dagger$, and $\mathbf{T}^\dagger$ for an arbitrary scattering slab based on so-called adding equations. Let us divide the slab $[0, \Psi]$ into layers $[0, \psi]$ and $[\psi, \Psi]$ (Fig. 10.4.1). Applying Eqs. (10.3.3)–(10.3.8) to the two component layers and to the combined slab yields

$$\mathbf{U}(\psi, \hat{\mu}, \hat{\mu}') = \mathbf{R}_2(\hat{\mu}, \hat{\mu}')\mathbf{X}(\psi, 0, \hat{\mu}') + \frac{1}{\pi}\int \mathrm{d}\hat{\mu}''\mu''\mathbf{R}_2(\hat{\mu}, \hat{\mu}'')\mathbf{D}(\psi, \hat{\mu}'', \hat{\mu}'), \quad (10.4.1)$$

$$\mathbf{D}(\psi, \hat{\mu}, \hat{\mu}') = \mathbf{T}_1(\hat{\mu}, \hat{\mu}') + \frac{1}{\pi}\int \mathrm{d}\hat{\mu}''\mu''\mathbf{R}_1^\dagger(\hat{\mu}, \hat{\mu}'')\mathbf{U}(\psi, \hat{\mu}'', \hat{\mu}'), \quad (10.4.2)$$

$$\mathbf{U}^\dagger(\psi, \hat{\mu}, \hat{\mu}') = \mathbf{R}_1^\dagger(\hat{\mu}, \hat{\mu}')\mathbf{X}(\psi, \Psi, -\hat{\mu}') + \frac{1}{\pi}\int \mathrm{d}\hat{\mu}''\mu''\mathbf{R}_1^\dagger(\hat{\mu}, \hat{\mu}'')\mathbf{D}^\dagger(\psi, \hat{\mu}'', \hat{\mu}'), \quad (10.4.3)$$

$$\mathbf{D}^\dagger(\psi, \hat{\mu}, \hat{\mu}') = \mathbf{T}_2^\dagger(\hat{\mu}, \hat{\mu}') + \frac{1}{\pi}\int \mathrm{d}\hat{\mu}''\mu''\mathbf{R}_2(\hat{\mu}, \hat{\mu}'')\mathbf{U}^\dagger(\psi, \hat{\mu}'', \hat{\mu}'), \quad (10.4.4)$$

where the subscripts 1 and 2 denote the reflection and transmission matrices of isolated layers 1 and 2, respectively. Indeed, we can apply Eqs. (10.3.3), (10.3.6), and (10.3.7) to layer 1 and write

$$\tilde{\mathbf{I}}(\psi, \hat{\mu}) = \mathbf{X}(\psi, 0, \hat{\mu})\tilde{\mathbf{I}}_\downarrow(\hat{\mu}) + \frac{1}{\pi}\int \mathrm{d}\hat{\mu}'\mu'\mathbf{T}_1(\hat{\mu}, \hat{\mu}')\tilde{\mathbf{I}}_\downarrow(\hat{\mu}') + \frac{1}{\pi}\int \mathrm{d}\hat{\mu}'\mu'\mathbf{R}_1^\dagger(\hat{\mu}, \hat{\mu}')\tilde{\mathbf{I}}(\psi, -\hat{\mu}')$$

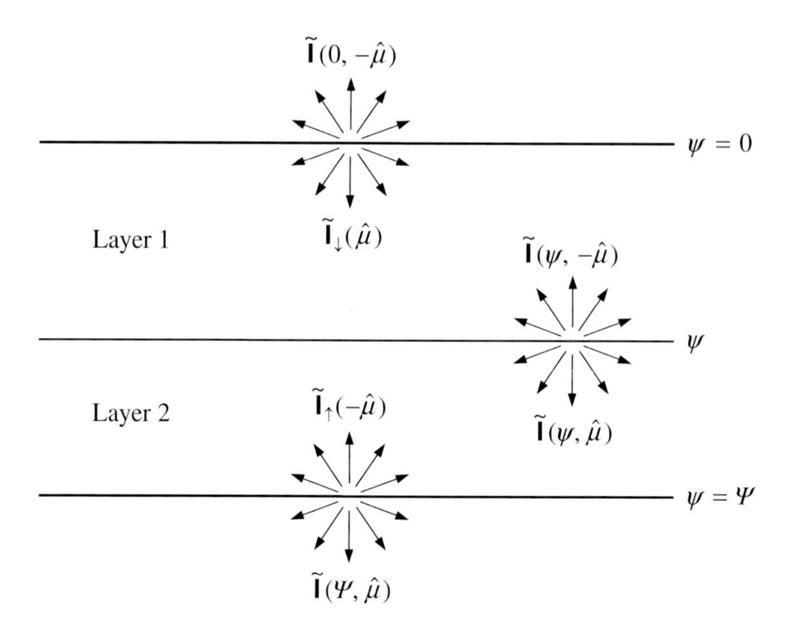

Figure 10.4.1. Illustration of the adding principle.

$$\begin{aligned}
&= \mathbf{X}(\psi, 0, \hat{\mu})\tilde{\mathbf{I}}_{\downarrow}(\hat{\mu}) + \frac{1}{\pi}\int d\hat{\mu}'\mu'\mathbf{T}_1(\hat{\mu}, \hat{\mu}')\tilde{\mathbf{I}}_{\downarrow}(\hat{\mu}') \\
&\quad + \frac{1}{\pi}\int d\hat{\mu}'\mu'\mathbf{R}_1^{\dagger}(\hat{\mu}, \hat{\mu}')\Bigg[\mathbf{X}(\psi, \Psi, -\hat{\mu}')\tilde{\mathbf{I}}_{\uparrow}(-\hat{\mu}') \\
&\qquad + \frac{1}{\pi}\int d\hat{\mu}''\mu''\mathbf{U}(\psi, \hat{\mu}', \hat{\mu}'')\tilde{\mathbf{I}}_{\downarrow}(\hat{\mu}'') \\
&\qquad + \frac{1}{\pi}\int d\hat{\mu}''\mu''\mathbf{D}^{\dagger}(\psi, \hat{\mu}', \hat{\mu}'')\tilde{\mathbf{I}}_{\uparrow}(-\hat{\mu}'')\Bigg],
\end{aligned} \tag{10.4.5}$$

which, after comparison with Eq. (10.3.3), gives Eqs. (10.4.2) and (10.4.3). Similarly, Eqs. (10.4.1) and (10.4.4) follow from

$$\begin{aligned}
\tilde{\mathbf{I}}(\psi, -\hat{\mu}) &= \mathbf{X}(\psi, \Psi, -\hat{\mu})\tilde{\mathbf{I}}_{\uparrow}(-\hat{\mu}) + \frac{1}{\pi}\int d\hat{\mu}'\mu'\mathbf{T}_2^{\dagger}(\hat{\mu}, \hat{\mu}')\tilde{\mathbf{I}}_{\uparrow}(-\hat{\mu}') \\
&\quad + \frac{1}{\pi}\int d\hat{\mu}'\mu'\mathbf{R}_2(\hat{\mu}, \hat{\mu}')\tilde{\mathbf{I}}(\psi, \hat{\mu}') \\
&= \mathbf{X}(\psi, \Psi, -\hat{\mu})\tilde{\mathbf{I}}_{\uparrow}(-\hat{\mu}) + \frac{1}{\pi}\int d\hat{\mu}'\mu'\mathbf{T}_2^{\dagger}(\hat{\mu}, \hat{\mu}')\tilde{\mathbf{I}}_{\uparrow}(-\hat{\mu}') \\
&\quad + \frac{1}{\pi}\int d\hat{\mu}'\mu'\mathbf{R}_2(\hat{\mu}, \hat{\mu}')\Bigg[\mathbf{X}(\psi, 0, \hat{\mu}')\tilde{\mathbf{I}}_{\downarrow}(\hat{\mu}')
\end{aligned}$$

$$+ \frac{1}{\pi}\int \mathrm{d}\hat{\mu}''\mu''\mathbf{D}(\psi,\hat{\mu}',\hat{\mu}'')\widetilde{\mathbf{I}}_{\downarrow}(\hat{\mu}'')$$
$$\left. + \frac{1}{\pi}\int \mathrm{d}\hat{\mu}''\mu''\mathbf{U}^{\dagger}(\psi,\hat{\mu}',\hat{\mu}'')\widetilde{\mathbf{I}}_{\uparrow}(-\hat{\mu}'')\right] \quad (10.4.6)$$

and Eq. (10.3.4). By analogy, one can derive

$$\mathbf{R}(\hat{\mu},\hat{\mu}') = \mathbf{R}_1(\hat{\mu},\hat{\mu}') + \mathbf{X}(0,\psi,-\hat{\mu})\mathbf{U}(\psi,\hat{\mu},\hat{\mu}') + \frac{1}{\pi}\int \mathrm{d}\hat{\mu}''\mu''\mathbf{T}_1^{\dagger}(\hat{\mu},\hat{\mu}'')\mathbf{U}(\psi,\hat{\mu}'',\hat{\mu}'), \quad (10.4.7)$$

$$\mathbf{T}(\hat{\mu},\hat{\mu}') = \mathbf{T}_2(\hat{\mu},\hat{\mu}')\mathbf{X}(\psi,0,\hat{\mu}') + \mathbf{X}(\Psi,\psi,\hat{\mu})\mathbf{D}(\psi,\hat{\mu},\hat{\mu}') + \frac{1}{\pi}\int \mathrm{d}\hat{\mu}''\mu''\mathbf{T}_2(\hat{\mu},\hat{\mu}'')\mathbf{D}(\psi,\hat{\mu}'',\hat{\mu}'), \quad (10.4.8)$$

$$\mathbf{R}^{\dagger}(\hat{\mu},\hat{\mu}') = \mathbf{R}_2^{\dagger}(\hat{\mu},\hat{\mu}') + \mathbf{X}(\Psi,\psi,\hat{\mu})\mathbf{U}^{\dagger}(\psi,\hat{\mu},\hat{\mu}') + \frac{1}{\pi}\int \mathrm{d}\hat{\mu}''\mu''\mathbf{T}_2(\hat{\mu},\hat{\mu}'')\mathbf{U}^{\dagger}(\psi,\hat{\mu}'',\hat{\mu}'), \quad (10.4.9)$$

$$\mathbf{T}^{\dagger}(\hat{\mu},\hat{\mu}') = \mathbf{T}_1^{\dagger}(\hat{\mu},\hat{\mu}')\mathbf{X}(\psi,\Psi,-\hat{\mu}') + \mathbf{X}(0,\psi,-\hat{\mu})\mathbf{D}^{\dagger}(\psi,\hat{\mu},\hat{\mu}') + \frac{1}{\pi}\int \mathrm{d}\hat{\mu}''\mu''\mathbf{T}_1^{\dagger}(\hat{\mu},\hat{\mu}'')\mathbf{D}^{\dagger}(\psi,\hat{\mu}'',\hat{\mu}'). \quad (10.4.10)$$

The interpretation of Eqs. (10.4.1)–(10.4.4) and (10.4.7)–(10.4.10) is rather transparent. For example, Eq. (10.4.1) indicates that the upwelling radiation at the interface between layers 1 and 2 in response to the beam incident on the combined slab from above is simply the result of reflection of the corresponding downwelling radiation by layer 2. This downwelling radiation consists of:

- The attenuated direct component represented by the propagator $\mathbf{X}(\psi,0,\hat{\mu}')$ (scattering path 1 in Fig. 10.4.2).
- The diffuse component represented by the matrix $\mathbf{D}(\psi,\hat{\mu}'',\hat{\mu}')$ (scattering path 2 in Fig. 10.4.2).

Similarly, Eq. (10.4.7) shows that the reflected radiation in response to the beam illuminating the combined slab from above consists of three components:

- The scattering paths that never reach the interface between layers 1 and 2 (the first term on the right-hand side of Eq. (10.4.7) and scattering path 1 in Fig. 10.4.3).
- The scattering paths "reflected" by layer 2 and "transmitted" by layer 1 without scattering (the second term on the right-hand side of Eq. (10.4.7) and scattering path 2 in Fig. 10.4.3).
- The scattering paths "reflected" by layer 2 and "diffusely transmitted" by layer 1 (the third term on the right-hand side of Eq. (10.4.7) and scattering path 3 in Fig. 10.4.3).

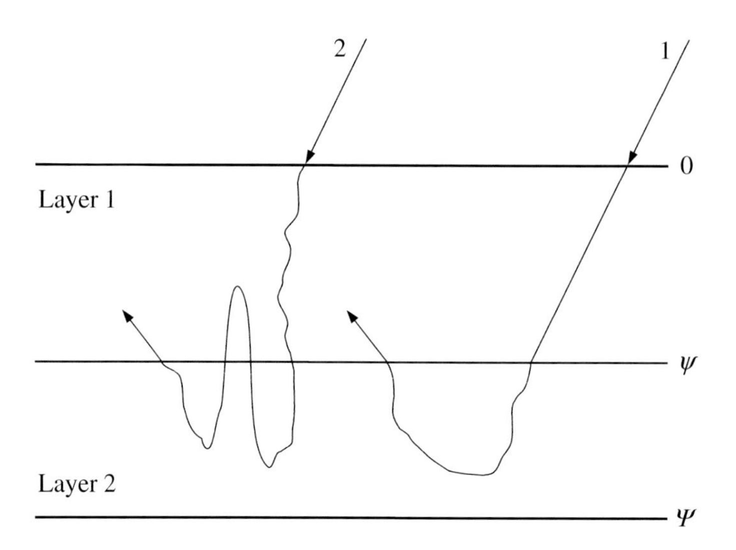

Figure 10.4.2. Physical interpretation of Eq. (10.4.1).

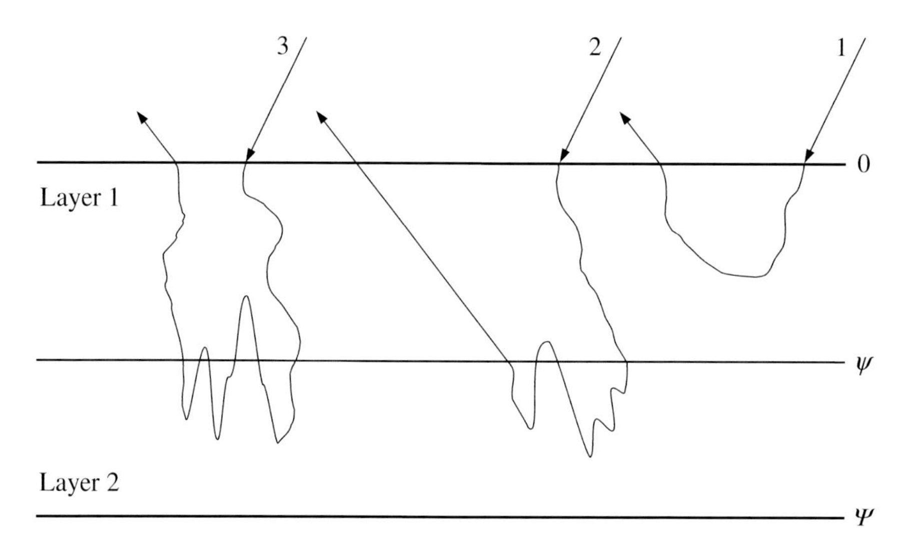

Figure 10.4.3. Physical interpretation of Eq. (10.4.7).

The reader may find it a useful exercise to give similar graphical interpretations of Eqs. (10.4.2)–(10.4.4) and (10.4.8)–(10.4.10).

Equations (10.4.1)–(10.4.4) and (10.4.7)–(10.4.10) are called adding equations because they allow one to compute the scattering properties of the combined slab provided that the scattering properties of each component layer are known. Indeed, if the matrices $\mathbf{R}_1$, $\mathbf{T}_1$, $\mathbf{R}_1^{\dagger}$, and $\mathbf{T}_1^{\dagger}$ for layer 1 in isolation from layer 2 and the matrices $\mathbf{R}_2$, $\mathbf{T}_2$, $\mathbf{R}_2^{\dagger}$, and $\mathbf{T}_2^{\dagger}$ for layer 2 in isolation from layer 1 are known then one can solve Eqs. (10.4.1)–(10.4.4) and find the matrices $\mathbf{U}$, $\mathbf{D}$, $\mathbf{U}^\dagger$, and $\mathbf{D}^\dagger$ describing the radiation field at the interface between the layers in the combined slab. A numerical implementation of this procedure can involve replacing the angular integrals by appropriate quadrature sums (Appendix D). For example, Eq. (10.4.1) becomes

$$\begin{aligned}\mathbf{U}(\psi;\mu_i,\varphi_j;\mu_k,\varphi_l) &= \mathbf{R}_2(\mu_i,\varphi_j;\mu_k,\varphi_l)\mathbf{X}(\psi,0;\mu_k,\varphi_l)\\ &\quad + \frac{1}{\pi}\sum_{m=1}^{N_\mu}\sum_{n=1}^{N_\varphi} w_m u_n \mu_m \mathbf{R}_2(\mu_i,\varphi_j;\mu_m,\varphi_n)\\ &\qquad\qquad \times \mathbf{D}(\psi;\mu_m,\varphi_n;\mu_k,\varphi_l),\end{aligned}$$

where μ_i and w_i $(i = 1,\ldots,N_\mu)$ are quadrature division points and weights on the interval $[0,1]$, and φ_i and u_i $(i = 1,\ldots,N_\varphi)$ are quadrature division points and weights on the interval $[0,2\pi]$. The resulting system of linear algebraic equations for the unknown values of the matrices $\mathbf{U}$, $\mathbf{D}$, $\mathbf{U}^\dagger$, and $\mathbf{D}^\dagger$ at the quadrature division points can be solved using one of the many available numerical techniques. After the matrices $\mathbf{U}$, $\mathbf{D}$, $\mathbf{U}^\dagger$, and $\mathbf{D}^\dagger$ at the quadrature division points have been found, the reflection and transmission matrices of the combined slab can be calculated using the discretized version of Eqs. (10.4.7)–(10.4.10). Adding two identical layers is traditionally called the doubling procedure.

Furthermore, let us assume that the matrices $\mathbf{U}_1$, $\mathbf{D}_1$, $\mathbf{U}_1^\dagger$, and $\mathbf{D}_1^\dagger$ for a level inside layer 1 are known, where the subscript 1 indicates that these matrices pertain to layer 1 taken in isolation from layer 2. Then the matrices $\mathbf{U}$, $\mathbf{D}$, $\mathbf{U}^\dagger$, and $\mathbf{D}^\dagger$ for the same level in the combined slab can also be easily calculated. Indeed, applying Eqs. (10.3.3) and (10.3.4) to each component layer and to the combined slab, we derive

$$\begin{aligned}\mathbf{U}(\psi',\hat{\mu},\hat{\mu}') &= \mathbf{U}_1(\psi',\hat{\mu},\hat{\mu}') + \mathbf{X}(\psi',\psi,-\hat{\mu})\mathbf{U}(\psi,\hat{\mu},\hat{\mu}')\\ &\quad + \frac{1}{\pi}\int \mathrm{d}\hat{\mu}''\mu''\mathbf{D}_1^\dagger(\psi',\hat{\mu},\hat{\mu}'')\mathbf{U}(\psi,\hat{\mu}'',\hat{\mu}'),\end{aligned} \tag{10.4.11}$$

$$\mathbf{D}(\psi',\hat{\mu},\hat{\mu}') = \mathbf{D}_1(\psi',\hat{\mu},\hat{\mu}') + \frac{1}{\pi}\int \mathrm{d}\hat{\mu}''\mu''\mathbf{U}_1^\dagger(\psi',\hat{\mu},\hat{\mu}'')\mathbf{U}(\psi,\hat{\mu}'',\hat{\mu}'), \tag{10.4.12}$$

$$\begin{aligned}\mathbf{U}^\dagger(\psi',\hat{\mu},\hat{\mu}') &= \mathbf{U}_1^\dagger(\psi',\hat{\mu},\hat{\mu}')\mathbf{X}(\psi,\Psi,-\hat{\mu}')\\ &\quad + \frac{1}{\pi}\int \mathrm{d}\hat{\mu}''\mu''\mathbf{U}_1^\dagger(\psi',\hat{\mu},\hat{\mu}'')\mathbf{D}^\dagger(\psi,\hat{\mu}'',\hat{\mu}'),\end{aligned} \tag{10.4.13}$$

$$\begin{aligned}\mathbf{D}^\dagger(\psi',\hat{\mu},\hat{\mu}') &= \mathbf{D}_1^\dagger(\psi',\hat{\mu},\hat{\mu}')\mathbf{X}(\psi,\Psi,-\hat{\mu}') + \mathbf{X}(\psi',\psi,-\hat{\mu})\mathbf{D}^\dagger(\psi,\hat{\mu},\hat{\mu}')\\ &\quad + \frac{1}{\pi}\int \mathrm{d}\hat{\mu}''\mu''\mathbf{D}_1^\dagger(\psi',\hat{\mu},\hat{\mu}'')\mathbf{D}^\dagger(\psi,\hat{\mu}'',\hat{\mu}')\end{aligned} \tag{10.4.14}$$

for $\psi' \in [0,\psi]$ (Fig. 10.4.4(a)). Similarly, if we know the matrices $\mathbf{U}_2$, $\mathbf{D}_2$, $\mathbf{U}_2^\dagger$, and $\mathbf{D}_2^\dagger$ for a level inside layer 2 taken in isolation from layer 1 then

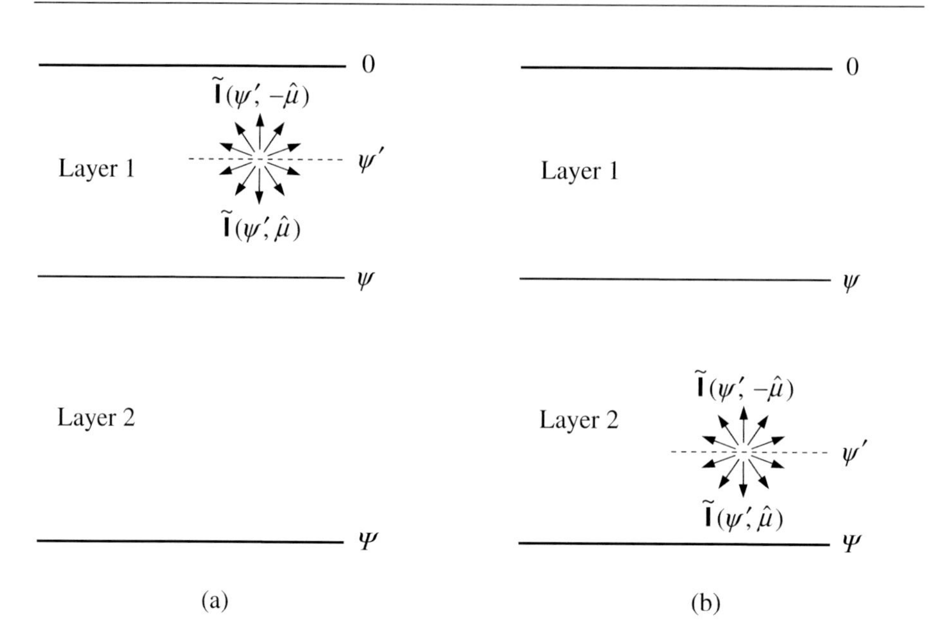

Figure 10.4.4. Internal radiation field.

$$\begin{aligned}\mathbf{U}(\psi', \hat{\mu}, \hat{\mu}') &= \mathbf{U}_2(\psi' - \psi, \hat{\mu}, \hat{\mu}')\mathbf{X}(\psi, 0, \hat{\mu}') \\ &\quad + \frac{1}{\pi}\int \mathrm{d}\hat{\mu}''\mu''\mathbf{U}_2(\psi' - \psi, \hat{\mu}, \hat{\mu}'')\mathbf{D}(\psi, \hat{\mu}'', \hat{\mu}'), \qquad (10.4.15)\end{aligned}$$

$$\begin{aligned}\mathbf{D}(\psi', \hat{\mu}, \hat{\mu}') &= \mathbf{D}_2(\psi' - \psi, \hat{\mu}, \hat{\mu}')\mathbf{X}(\psi, 0, \hat{\mu}') + \mathbf{X}(\psi', \psi, \hat{\mu})\mathbf{D}(\psi, \hat{\mu}, \hat{\mu}') \\ &\quad + \frac{1}{\pi}\int \mathrm{d}\hat{\mu}''\mu''\mathbf{D}_2(\psi' - \psi, \hat{\mu}, \hat{\mu}'')\mathbf{D}(\psi, \hat{\mu}'', \hat{\mu}'), \qquad (10.4.16)\end{aligned}$$

$$\begin{aligned}\mathbf{U}^{\dagger}(\psi', \hat{\mu}, \hat{\mu}') &= \mathbf{U}_2^{\dagger}(\psi' - \psi, \hat{\mu}, \hat{\mu}') + \mathbf{X}(\psi', \psi, \hat{\mu})\mathbf{U}^{\dagger}(\psi, \hat{\mu}, \hat{\mu}') \\ &\quad + \frac{1}{\pi}\int \mathrm{d}\hat{\mu}''\mu''\mathbf{D}_2(\psi' - \psi, \hat{\mu}, \hat{\mu}'')\mathbf{U}^{\dagger}(\psi, \hat{\mu}'', \hat{\mu}'), \qquad (10.4.17)\end{aligned}$$

$$\begin{aligned}\mathbf{D}^{\dagger}(\psi', \hat{\mu}, \hat{\mu}') &= \mathbf{D}_2^{\dagger}(\psi' - \psi, \hat{\mu}, \hat{\mu}') \\ &\quad + \frac{1}{\pi}\int \mathrm{d}\hat{\mu}''\mu''\mathbf{U}_2(\psi' - \psi, \hat{\mu}, \hat{\mu}'')\mathbf{U}^{\dagger}(\psi, \hat{\mu}'', \hat{\mu}') \qquad (10.4.18)\end{aligned}$$

for $\psi' \in [\psi, \Psi]$ (Fig. 10.4.4(b)).

The physical meaning of these formulas is rather transparent. For example, the first term on the right-hand side of Eq. (10.4.11) represents the contribution of scattering paths that never reach the interface between layers 1 and 2, as shown schematically by scattering path 1 in Fig. 10.4.5. The second term describes the contribution of the scattering paths that cross the interface at least once, exit layer 2 in the direction $\hat{\mu}$, and reach the level ψ' without scattering, as illustrated by scattering path 2 in Fig. 10.4.5. The last term gives the contribution of the scattering paths that cross the

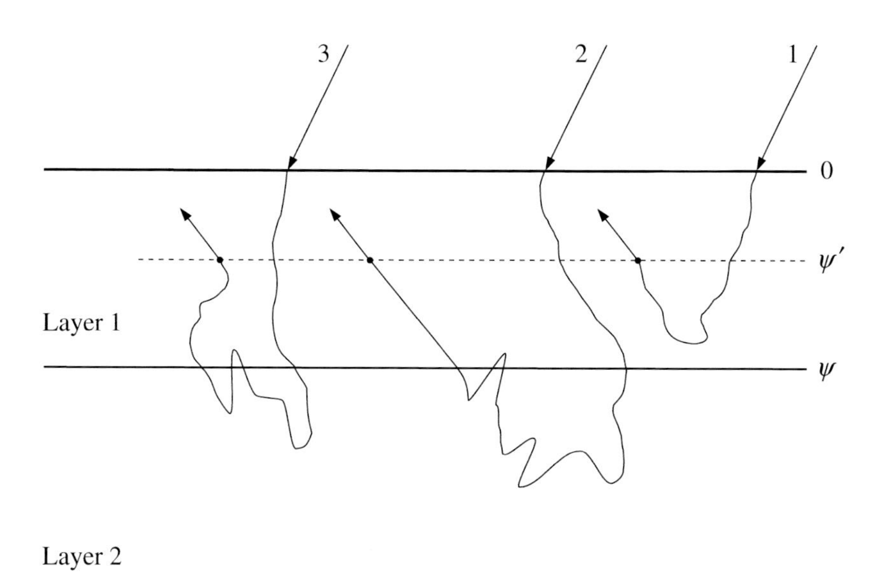

Figure 10.4.5. Physical interpretation of Eq. (10.4.11).

interface at least once and are "scattered" at least once inside layer 1 before they reach the level ψ' (scattering path 3 in Fig. 10.4.5).

A practical implementation of the adding method can involve the following three basic steps:

- A vertically inhomogeneous slab of particle thickness Ψ is approximated by a stack of N partial homogeneous layers having particle thicknesses $\Psi_1, ...,$ Ψ_N such that

$$\Psi = \sum_{n=1}^{N} \Psi_n$$

(Fig. 10.4.6). The number of partial layers and their particle thicknesses can depend on the degree of vertical inhomogeneity of the original slab as well as on the desired numerical accuracy of computations.

- The reflection and transmission matrices $\mathbf{R}_n$, $\mathbf{T}_n$, $\mathbf{R}_n^\dagger$, and $\mathbf{T}_n^\dagger$ of partial layer n in isolation from all other layers are computed by using the doubling method (Fig. 10.4.7). The doubling process can be started with a layer having a particle thickness $\Delta\Psi_n = \Psi_n / 2^{k_n}$ small enough that the reflection and transmission matrices for this layer can be computed by considering only the first order of scattering. Specifically, choosing the number of doubling events k_n sufficiently large that all elements of the matrices $\Delta\Psi_n \mathbf{Z}_n$ and $\Delta\Psi_n \mathbf{K}_n$ are much smaller than unity, using Eqs. (10.1.6), (10.2.7), (10.2.15), and

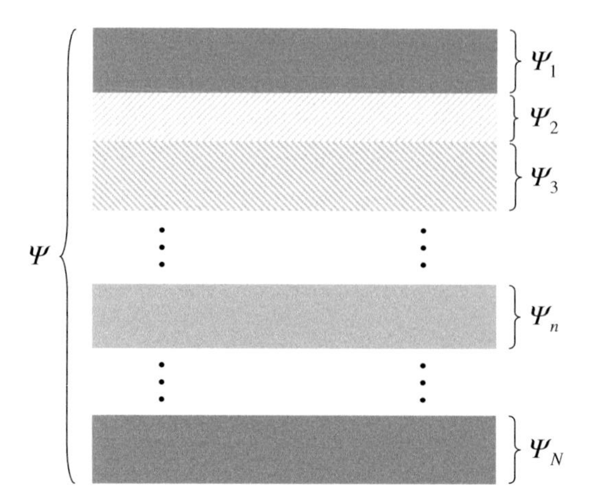

Figure 10.4.6. Representation of a vertically inhomogeneous scattering slab by a stack of N homogeneous layers.

(10.3.1)–(10.3.8), and neglecting all terms proportional to $(\Delta\Psi_n)^m$ with $m > 1$, we derive

$$\mathbf{X}_n(\Delta\Psi_n, 0, \hat{\mu}) = \mathbf{\Delta} - \frac{\Delta\Psi_n}{\mu}\mathbf{K}_n(\hat{\mu}), \tag{10.4.19}$$

$$\mathbf{X}_n(0, \Delta\Psi_n, -\hat{\mu}) = \mathbf{\Delta} - \frac{\Delta\Psi_n}{\mu}\mathbf{K}_n(-\hat{\mu}), \tag{10.4.20}$$

$$\mathbf{R}_{\Delta\Psi_n}(\hat{\mu}, \hat{\mu}') = \frac{\pi\Delta\Psi_n}{\mu\mu'}\mathbf{Z}_n(-\hat{\mu}, \hat{\mu}'), \tag{10.4.21}$$

$$\mathbf{T}_{\Delta\Psi_n}(\hat{\mu}, \hat{\mu}') = \frac{\pi\Delta\Psi_n}{\mu\mu'}\mathbf{Z}_n(\hat{\mu}, \hat{\mu}'), \tag{10.4.22}$$

$$\mathbf{R}^{\dagger}_{\Delta\Psi_n}(\hat{\mu}, \hat{\mu}') = \frac{\pi\Delta\Psi_n}{\mu\mu'}\mathbf{Z}_n(\hat{\mu}, -\hat{\mu}'), \tag{10.4.23}$$

$$\mathbf{T}^{\dagger}_{\Delta\Psi_n}(\hat{\mu}, \hat{\mu}') = \frac{\pi\Delta\Psi_n}{\mu\mu'}\mathbf{Z}_n(-\hat{\mu}, -\hat{\mu}'). \tag{10.4.24}$$

Obviously, the doubling procedure will also yield the matrices $\mathbf{U}_n$, $\mathbf{D}_n$, $\mathbf{U}^{\dagger}_n$, and $\mathbf{D}^{\dagger}_n$ at $2^{k_n} - 1$ equidistant levels inside the nth partial layer (Fig. 10.4.7).

- The N partial homogeneous layers are recursively added starting from layer 1 and moving down or starting from layer N and moving up. This process gives the reflection and transmission matrices of the combined slab and the matrices $\mathbf{U}$, $\mathbf{D}$, $\mathbf{U}^{\dagger}$, and $\mathbf{D}^{\dagger}$ at the $N-1$ interfaces between the partial layers as well as at the $\Sigma_{n=1}^{N}(2^{k_n} - 1)$ levels inside the partial layers rendered by the doubling procedure.

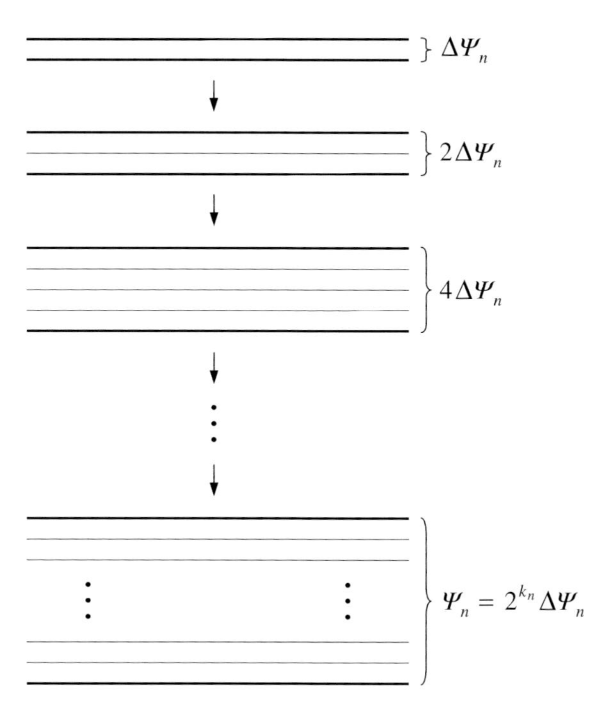

Figure 10.4.7. The doubling procedure.

10.5 Invariant imbedding equations

Adding computations can become inefficient if the specific vertical structure of the scattering slab necessitates partitioning the slab into a very large number N of homogeneous layers in order to ensure the requisite numerical accuracy. In such cases one may prefer to resort to solving numerically differential so-called invariant imbedding equations for the reflection and transmission matrices as functions of the particle thickness of the slab Ψ.

To derive the invariant imbedding equations, let us assume that ψ in Fig. 10.4.1 is so small that all terms proportional to ψ^m with $m > 1$ can be neglected. We then have

$$\mathbf{X}(\psi, 0, \hat{\mu}) = \mathbf{\Delta} - \frac{\psi}{\mu}\mathbf{K}(0, \hat{\mu}), \tag{10.5.1}$$

$$\mathbf{X}(0, \psi, -\hat{\mu}) = \mathbf{\Delta} - \frac{\psi}{\mu}\mathbf{K}(0, -\hat{\mu}), \tag{10.5.2}$$

$$\mathbf{R}_1(\hat{\mu}, \hat{\mu}') = \frac{\pi\psi}{\mu\mu'}\mathbf{Z}(0, -\hat{\mu}, \hat{\mu}'), \tag{10.5.3}$$

$$\mathbf{T}_1(\hat{\mu}, \hat{\mu}') = \frac{\pi\psi}{\mu\mu'}\mathbf{Z}(0, \hat{\mu}, \hat{\mu}'), \tag{10.5.4}$$

$$\mathbf{R}_1^{\dagger}(\hat{\mu},\hat{\mu}') = \frac{\pi\psi}{\mu\mu'}\mathbf{Z}(0,\hat{\mu},-\hat{\mu}'), \tag{10.5.5}$$

$$\mathbf{T}_1^{\dagger}(\hat{\mu},\hat{\mu}') = \frac{\pi\psi}{\mu\mu'}\mathbf{Z}(0,-\hat{\mu},-\hat{\mu}') \tag{10.5.6}$$

(cf. Eqs. (10.4.19)–(10.4.24)). Substituting these formulas in Eqs. (10.4.1)–(10.4.4), we obtain in the limit $\psi \to 0$

$$\begin{aligned}\mathbf{U}(\psi,\hat{\mu},\hat{\mu}') = {} & \mathbf{R}_2(\hat{\mu},\hat{\mu}') - \frac{\psi}{\mu'}\mathbf{R}(\hat{\mu},\hat{\mu}')\mathbf{K}(0,\hat{\mu}') \\ & + \frac{\psi}{\mu'}\int \mathrm{d}\hat{\mu}''\mathbf{R}(\hat{\mu},\hat{\mu}'')\mathbf{Z}(0,\hat{\mu}'',\hat{\mu}') \\ & + \frac{\psi}{\pi}\int \mathrm{d}\hat{\mu}''\int \mathrm{d}\hat{\mu}'''\mathbf{R}(\hat{\mu},\hat{\mu}'')\mathbf{Z}(0,\hat{\mu}'',-\hat{\mu}''')\mathbf{R}(\hat{\mu}''',\hat{\mu}'),\end{aligned} \tag{10.5.7}$$

$$\mathbf{D}(\psi,\hat{\mu},\hat{\mu}') = \frac{\pi\psi}{\mu\mu'}\mathbf{Z}(0,\hat{\mu},\hat{\mu}') + \frac{\psi}{\mu}\int \mathrm{d}\hat{\mu}''\mathbf{Z}(0,\hat{\mu},-\hat{\mu}'')\mathbf{R}(\hat{\mu}'',\hat{\mu}'), \tag{10.5.8}$$

$$\begin{aligned}\mathbf{U}^{\dagger}(\psi,\hat{\mu},\hat{\mu}') = {} & \frac{\pi\psi}{\mu\mu'}\mathbf{Z}(0,\hat{\mu},-\hat{\mu}')\mathbf{X}(0,\Psi,-\hat{\mu}') \\ & + \frac{\psi}{\mu}\int \mathrm{d}\hat{\mu}''\mathbf{Z}(0,\hat{\mu},-\hat{\mu}'')\mathbf{T}^{\dagger}(\hat{\mu}'',\hat{\mu}'),\end{aligned} \tag{10.5.9}$$

$$\begin{aligned}\mathbf{D}^{\dagger}(\psi,\hat{\mu},\hat{\mu}') = {} & \mathbf{T}_2^{\dagger}(\hat{\mu},\hat{\mu}') \\ & + \frac{\psi}{\mu'}\int \mathrm{d}\hat{\mu}''\mathbf{R}(\hat{\mu},\hat{\mu}'')\mathbf{Z}(0,\hat{\mu}'',-\hat{\mu}')\mathbf{X}(0,\Psi,-\hat{\mu}') \\ & + \frac{\psi}{\pi}\int \mathrm{d}\hat{\mu}''\int \mathrm{d}\hat{\mu}'''\mathbf{R}(\hat{\mu},\hat{\mu}'')\mathbf{Z}(0,\hat{\mu}'',-\hat{\mu}''')\mathbf{T}^{\dagger}(\hat{\mu}''',\hat{\mu}').\end{aligned} \tag{10.5.10}$$

Finally, substituting Eqs. (10.5.1)–(10.5.10) into Eqs. (10.4.7)–(10.4.10) yields

$$\begin{aligned}\frac{\partial\mathbf{R}(\hat{\mu},\hat{\mu}')}{\partial\Psi_{\uparrow}} = {} & -\frac{1}{\mu'}\mathbf{R}(\hat{\mu},\hat{\mu}')\mathbf{K}(0,\hat{\mu}') - \frac{1}{\mu}\mathbf{K}(0,-\hat{\mu})\mathbf{R}(\hat{\mu},\hat{\mu}') \\ & + \frac{\pi}{\mu\mu'}\mathbf{Z}(0,-\hat{\mu},\hat{\mu}') + \frac{1}{\mu'}\int \mathrm{d}\hat{\mu}''\mathbf{R}(\hat{\mu},\hat{\mu}'')\mathbf{Z}(0,\hat{\mu}'',\hat{\mu}') \\ & + \frac{1}{\mu}\int \mathrm{d}\hat{\mu}''\mathbf{Z}(0,-\hat{\mu},-\hat{\mu}'')\mathbf{R}(\hat{\mu}'',\hat{\mu}') \\ & + \frac{1}{\pi}\int \mathrm{d}\hat{\mu}''\int \mathrm{d}\hat{\mu}'''\mathbf{R}(\hat{\mu},\hat{\mu}'')\mathbf{Z}(0,\hat{\mu}'',-\hat{\mu}''')\mathbf{R}(\hat{\mu}''',\hat{\mu}'),\end{aligned} \tag{10.5.11a}$$

$$\frac{\partial \mathbf{T}(\hat{\mu},\hat{\mu}')}{\partial \Psi_{\uparrow}} = -\frac{1}{\mu'}\mathbf{T}(\hat{\mu},\hat{\mu}')\mathbf{K}(0,\hat{\mu}') + \frac{\pi}{\mu\mu'}\mathbf{X}(\Psi,0,\hat{\mu})\mathbf{Z}(0,\hat{\mu},\hat{\mu}')$$
$$+ \frac{1}{\mu'}\int \mathrm{d}\hat{\mu}''\,\mathbf{T}(\hat{\mu},\hat{\mu}'')\mathbf{Z}(0,\hat{\mu}'',\hat{\mu}')$$
$$+ \frac{1}{\mu}\mathbf{X}(\Psi,0,\hat{\mu})\int \mathrm{d}\hat{\mu}''\,\mathbf{Z}(0,\hat{\mu},-\hat{\mu}'')\mathbf{R}(\hat{\mu}'',\hat{\mu}')$$
$$+ \frac{1}{\pi}\int \mathrm{d}\hat{\mu}''\int \mathrm{d}\hat{\mu}'''\,\mathbf{T}(\hat{\mu},\hat{\mu}'')\mathbf{Z}(0,\hat{\mu}'',-\hat{\mu}''')\mathbf{R}(\hat{\mu}''',\hat{\mu}'), \tag{10.5.12a}$$

$$\frac{\partial \mathbf{R}^{\dagger}(\hat{\mu},\hat{\mu}')}{\partial \Psi_{\uparrow}} = \frac{\pi}{\mu\mu'}\mathbf{X}(\Psi,0,\hat{\mu})\mathbf{Z}(0,\hat{\mu},-\hat{\mu}')\mathbf{X}(0,\Psi,-\hat{\mu}')$$
$$+ \frac{1}{\mu}\mathbf{X}(\Psi,0,\hat{\mu})\int \mathrm{d}\hat{\mu}''\,\mathbf{Z}(0,\hat{\mu},-\hat{\mu}'')\mathbf{T}^{\dagger}(\hat{\mu}'',\hat{\mu}')$$
$$+ \frac{1}{\mu'}\int \mathrm{d}\hat{\mu}''\,\mathbf{T}(\hat{\mu},\hat{\mu}'')\mathbf{Z}(0,\hat{\mu}'',-\hat{\mu}')\mathbf{X}(0,\Psi,-\hat{\mu}')$$
$$+ \frac{1}{\pi}\int \mathrm{d}\hat{\mu}''\int \mathrm{d}\hat{\mu}'''\,\mathbf{T}(\hat{\mu},\hat{\mu}'')\mathbf{Z}(0,\hat{\mu}'',-\hat{\mu}''')\mathbf{T}^{\dagger}(\hat{\mu}''',\hat{\mu}'), \tag{10.5.13a}$$

$$\frac{\partial \mathbf{T}^{\dagger}(\hat{\mu},\hat{\mu}')}{\partial \Psi_{\uparrow}} = -\frac{1}{\mu}\mathbf{K}(0,-\hat{\mu})\mathbf{T}^{\dagger}(\hat{\mu},\hat{\mu}') + \frac{\pi}{\mu\mu'}\mathbf{Z}(0,-\hat{\mu},-\hat{\mu}')\mathbf{X}(0,\Psi,-\hat{\mu}')$$
$$+ \frac{1}{\mu}\int \mathrm{d}\hat{\mu}''\,\mathbf{Z}(0,-\hat{\mu},-\hat{\mu}'')\mathbf{T}^{\dagger}(\hat{\mu}'',\hat{\mu}')$$
$$+ \frac{1}{\mu'}\int \mathrm{d}\hat{\mu}''\,\mathbf{R}(\hat{\mu},\hat{\mu}'')\mathbf{Z}(0,\hat{\mu}'',-\hat{\mu}')\mathbf{X}(0,\Psi,-\hat{\mu}')$$
$$+ \frac{1}{\pi}\int \mathrm{d}\hat{\mu}''\int \mathrm{d}\hat{\mu}'''\,\mathbf{R}(\hat{\mu},\hat{\mu}'')\mathbf{Z}(0,\hat{\mu}'',-\hat{\mu}''')\mathbf{T}^{\dagger}(\hat{\mu}''',\hat{\mu}'), \tag{10.5.14a}$$

where the subscript $\uparrow$ indicates that the infinitesimally thin layer is added on top of the slab. Equations (10.5.11a)–(10.5.14a) are called invariant imbedding equations and must be supplemented by the initial conditions

$$\mathbf{R}(\hat{\mu},\hat{\mu}')\big|_{\Psi=0} = \mathbf{0}, \tag{10.5.15}$$

$$\mathbf{T}(\hat{\mu},\hat{\mu}')\big|_{\Psi=0} = \mathbf{0}, \tag{10.5.16}$$

$$\mathbf{R}^{\dagger}(\hat{\mu},\hat{\mu}')\big|_{\Psi=0} = \mathbf{0}, \tag{10.5.17}$$

$$\mathbf{T}^{\dagger}(\hat{\mu},\hat{\mu}')\big|_{\Psi=0} = \mathbf{0}, \tag{10.5.18}$$

where $\mathbf{0}$ is the 4×4 zero matrix.

In practice, the angular integrals in Eqs. (10.5.11a)–(10.5.14a) are replaced by appropriate quadrature sums, thereby yielding a system of ordinary differential equations. This system along with Eqs. (10.5.15)–(10.5.18) forms an initial-value problem which can be solved with one of the available numerical techniques.

Notice that Eqs. (10.5.11a) and (10.5.12a) are independent of Eqs. (10.5.13a) and (10.5.14a) and can be solved separately if only the matrices $\mathbf{R}$ and $\mathbf{T}$ are required. Furthermore, solving Eq. (10.5.11a) alone is sufficient if only the matrix $\mathbf{R}$ is required.

Although the invariant imbedding equations do not yield the internal radiation field directly, the latter can be found by combining the invariant imbedding equations and the adding method. Specifically, let us assume that one needs to find the radiation field at a level ψ inside a scattering slab having a particle thickness $\Psi > \psi$ (Fig. 10.4.1). One can first compute the reflection and transmission matrices of layers 1 and 2 by solving the invariant imbedding equations and then find the internal radiation field at the level ψ from Eqs. (10.4.1)–(10.4.4). This procedure is easily generalized if the internal radiation field is required at more than one level.

The reader may find it a useful exercise to derive a system of four companion invariant imbedding equations which have on the left-hand side the derivatives $\partial\mathbf{R}(\hat{\mu},\hat{\mu}')/\partial\Psi_{\downarrow}$, $\partial\mathbf{T}(\hat{\mu},\hat{\mu}')/\partial\Psi_{\downarrow}$, $\partial\mathbf{R}^{\dagger}(\hat{\mu},\hat{\mu}')/\partial\Psi_{\downarrow}$, and $\partial\mathbf{T}^{\dagger}(\hat{\mu},\hat{\mu}')/\partial\Psi_{\downarrow}$, where the subscript $\downarrow$ indicates that the infinitesimally thin layer is added to the bottom of the slab. Obviously, this is done by evaluating the limit $(\Psi-\psi)\to 0$. We will refer to these equations symbolically as Eqs. (10.5.11b)–(10.5.14b). If the scattering layer is homogeneous, it does not matter whether the infinitesimally thin layer is added to the top or to the bottom of the main layer. Therefore, by equating the right-hand sides of Eqs. (10.5.11a) and (10.5.11b), Eqs. (10.5.12a) and (10.5.12b), Eqs. (10.5.13a) and (10.5.13b), and Eqs. (10.5.14a) and (10.5.14b), one can obtain a system of four nonlinear integral equations for the matrices $\mathbf{R}$, $\mathbf{T}$, $\mathbf{R}^{\dagger}$, and $\mathbf{T}^{\dagger}$. Unfortunately, this system of equations allows an infinite continuous set of solutions, only one of which is physically relevant, and is very difficult to solve numerically (de Rooij and Domke, 1984).

10.6 Ambarzumian equation

If the scattering slab is homogeneous and semi-infinite, its reflection matrix for illumination from above must be independent of Ψ. Therefore, by equating the derivative $\partial\mathbf{R}(\hat{\mu},\hat{\mu}')/\partial\Psi_{\uparrow}$ in Eq. (10.5.11a) to zero, we obtain the following Ambarzumian nonlinear integral equation for the reflection matrix:

$$
\begin{aligned}
&\frac{1}{\mu'}\mathbf{R}(\hat{\mu},\hat{\mu}')\mathbf{K}(\hat{\mu}') + \frac{1}{\mu}\mathbf{K}(-\hat{\mu})\mathbf{R}(\hat{\mu},\hat{\mu}') \\
&\quad = \frac{\pi}{\mu\mu'}\mathbf{Z}(-\hat{\mu},\hat{\mu}') + \frac{1}{\mu'}\int \mathrm{d}\hat{\mu}''\mathbf{R}(\hat{\mu},\hat{\mu}'')\mathbf{Z}(\hat{\mu}'',\hat{\mu}')
\end{aligned}
$$

$$+ \frac{1}{\mu} \int \mathrm{d}\hat{\mu}'' \mathbf{Z}(-\hat{\mu}, -\hat{\mu}'') \mathbf{R}(\hat{\mu}'', \hat{\mu}')$$

$$+ \frac{1}{\pi} \int \mathrm{d}\hat{\mu}'' \int \mathrm{d}\hat{\mu}''' \mathbf{R}(\hat{\mu}, \hat{\mu}'') \mathbf{Z}(\hat{\mu}'', -\hat{\mu}''') \mathbf{R}(\hat{\mu}''', \hat{\mu}'). \tag{10.6.1}$$

This equation permits only a discrete set of solutions, and the physically relevant solution can be selected using a simple linear constraint (de Rooij and Domke, 1984). For example, when the semi-infinite slab is composed of nonabsorbing particles, the linear constraint can be derived from the obvious fact that the net flow of power through the boundary of the layer must be equal to zero: all electromagnetic energy entering the layer must eventually leave it. The actual numerical procedure involves replacing the integrals in Eq. (10.6.1) with appropriate quadrature sums and solving the resulting system of nonlinear algebraic equations using the method of iterations.

10.7 Reciprocity relations for the reflection and transmission matrices

Assuming that the solution of the initial value problem (10.5.11a)–(10.5.18) is unique, one can easily derive that the reflection and transmission matrices obey the following reciprocity relations:

$$\mathbf{R}(\mu', \varphi' + \pi; \mu, \varphi + \pi) = \mathbf{\Delta}_3 [\mathbf{R}(\mu, \varphi; \mu', \varphi')]^{\mathrm{T}} \mathbf{\Delta}_3, \tag{10.7.1}$$

$$\mathbf{R}^{\dagger}(\mu', \varphi' + \pi; \mu, \varphi + \pi) = \mathbf{\Delta}_3 [\mathbf{R}^{\dagger}(\mu, \varphi; \mu', \varphi')]^{\mathrm{T}} \mathbf{\Delta}_3, \tag{10.7.2}$$

$$\mathbf{T}^{\dagger}(\mu', \varphi' + \pi; \mu, \varphi + \pi) = \mathbf{\Delta}_3 [\mathbf{T}(\mu, \varphi; \mu', \varphi')]^{\mathrm{T}} \mathbf{\Delta}_3, \tag{10.7.3}$$

where, as before, $\mathbf{\Delta}_3 = \mathrm{diag}[1, 1, -1, 1]$. Equations (10.7.1)–(10.7.3) ultimately follow from the reciprocity relations for the phase and extinction matrices, Eqs. (3.7.31) and (3.8.16), which can be written in the form

$$\mathbf{Z}(-\mu', \varphi' + \pi; -\mu, \varphi + \pi) = \mathbf{\Delta}_3 [\mathbf{Z}(\mu, \varphi; \mu', \varphi')]^{\mathrm{T}} \mathbf{\Delta}_3, \tag{10.7.4}$$

$$\mathbf{K}(-\mu, \varphi + \pi) = \mathbf{\Delta}_3 [\mathbf{K}(\mu, \varphi)]^{\mathrm{T}} \mathbf{\Delta}_3. \tag{10.7.5}$$

Indeed, Eqs. (10.2.8) and (10.2.16) along with the matrix identity $(\mathbf{AB})^{\mathrm{T}} = \mathbf{B}^{\mathrm{T}} \mathbf{A}^{\mathrm{T}}$ yield

$$\mathbf{X}(0, \Psi; -\mu, \varphi + \pi) = \mathbf{\Delta}_3 [\mathbf{X}(\Psi, 0; \mu, \varphi)]^{\mathrm{T}} \mathbf{\Delta}_3. \tag{10.7.6}$$

The initial conditions (10.5.15)–(10.5.18) obviously satisfy the reciprocity relations (10.7.1)–(10.7.3). We can then add an infinitesimally thin layer to the "initial slab" of particle thickness zero and find out from Eqs. (10.5.11a)–(10.5.14a) that the reflection and transmission matrices of the resulting slab also satisfy the reciprocity relations (10.7.1)–(10.7.3). We can continue this recursive process of adding infinitesimally thin layers and find out that at each recursive step the resulting reflection and transmission matrices satisfy the reciprocity relations (10.7.1)–(10.7.3). Therefore, the reflection and transmission matrices of the final slab also satisfy these relations.

The reciprocity relations are fundamental properties of the reflection and transmission matrices and can be used in practice to check the accuracy of adding/doubling or invariant imbedding computer codes. Alternatively, they can be used to considerably shorten the requisite computer time by reducing the number of independent scattering geometries (i.e., the number of couplets $\{\hat{\mu}, \hat{\mu}'\}$ for which the reflection and transmission matrices are computed explicitly) by a factor of almost two.

10.8 Notes and further reading

The adding concept goes back to Stokes (1862), who analyzed the reflection and transmission of light by a stack of glass plates, and it was introduced to radiative transfer by van de Hulst (1963). Equations (10.3.3) and (10.3.4) generalize the interaction principle introduced by Redheffer (1962) and later used by Grant and Hunt (1969) to derive formulas of the so-called matrix operator method closely related to the adding method (see Hunt, 1971; Plass *et al.*, 1973). The invariant imbedding equations can be derived using heuristic so-called principles of invariance pioneered by Ambarzumian (1943) and Chandresekhar (1947b) (see also Chandrasekhar, 1950). Our derivation of the vector adding and invariant imbedding equations for vertically inhomogeneous scattering slabs containing arbitrarily oriented nonspherical particles follows that in Mishchenko (1990a).

Ishimaru *et al.* (1984) solved the boundary-value problem (10.1.6)–(10.1.8) for a homogeneous slab filled with spheroids having vertically aligned axes using the so-called eigenvalue-eigenvector technique (Ishimaru, 1978). Liou and Takano (2002) used the adding method to compute the reflectance of a slab comprising ice crystals randomly oriented in the horizontal plane. Picard *et al.* (2004) modeled the radar backscatter from forested areas by solving the plane-parallel VRTE with the so-called discrete ordinate method.

Chapter 11

Macroscopically isotropic and mirror-symmetric scattering media

An important particular type of discrete scattering medium is a macroscopically isotropic and mirror-symmetric medium (hereinafter isotropic and symmetric medium, or ISM). By definition, an ISM comprises spherically symmetric and/or randomly oriented nonspherical particles. Furthermore, each nonspherical particle must have a plane of symmetry and/or must be accompanied by a mirror counterpart.

Although this type of scattering medium might be thought to be a rather special case, it nonetheless provides a very good numerical description of the scattering properties of many particle collections encountered in practice and is by far the most often used theoretical model. Moreover, we shall see below that the assumption of microscopic isotropy and mirror symmetry leads to significant mathematical simplifications and allows one to develop efficient computer algorithms.

It turns out that a convenient concept in analyses of single and multiple scattering of light by ISMs is that of a scattering matrix. As we have seen before, the phase matrix is defined such that it relates the Stokes parameters of the incident and scattered waves defined relative to the meridional planes containing the incidence and scattering directions. In contrast, the scattering matrix $\mathbf{F}$ relates the Stokes parameters of the incident and scattered waves defined with respect to the scattering plane, that is, the plane through the unit vectors $\hat{\mathbf{n}}^{\text{inc}}$ and $\hat{\mathbf{n}}^{\text{sca}}$ (Perrin, 1942; van de Hulst, 1957).

A simple way to introduce the scattering matrix is to direct the z-axis of the reference frame along the incident beam and superpose the meridional plane with $\varphi = 0$ and the scattering plane (Fig. 11.0.1). Then the scattering matrix $\mathbf{F}$ can be defined as

$$\mathbf{F}(\theta^{\text{sca}}) = \mathbf{Z}(\theta^{\text{sca}}, \varphi^{\text{sca}} = 0; \theta^{\text{inc}} = 0, \varphi^{\text{inc}} = 0). \tag{11.0.1}$$

In general, all 16 elements of the scattering matrix are nonzero and depend on the

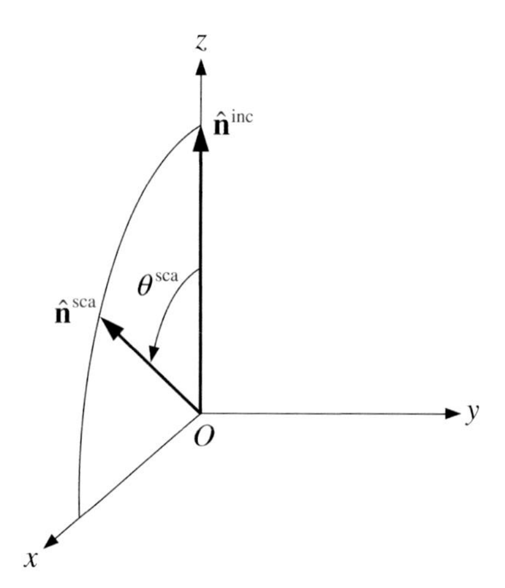

Figure 11.0.1. On the definition of the scattering matrix.

particle orientation with respect to the incident and scattered beams.

This choice of laboratory reference frame, with the z-axis along the incidence direction and the xz-half-plane with $x \geq 0$ coinciding with the scattering plane, can often be inconvenient because any change in the incidence direction and/or orientation of the scattering plane also changes the orientation of the scattering particle with respect to the coordinate system. However, we will show in this chapter that the notion of the scattering matrix can be very useful in application to ISMs because then the scattering matrix becomes independent of incidence direction and orientation of the scattering plane, depends only on the angle $\Theta = \arccos(\hat{\mathbf{n}}^{\text{inc}} \cdot \hat{\mathbf{n}}^{\text{sca}})$ between the incidence and scattering directions, and has a simple block-diagonal structure.

11.1 Symmetries of the Stokes scattering matrix

We begin by considering special symmetry properties of the amplitude scattering matrix that exist when both the incidence and the scattering directions lie in the xz-plane (van de Hulst, 1957). For the particle shown schematically in Fig. 11.1.1(a), let

$$\begin{bmatrix} S_{11} & S_{12} \\ S_{21} & S_{22} \end{bmatrix} \tag{11.1.1a}$$

be the amplitude scattering matrix that corresponds to the directions of incidence and scattering given by $\hat{\mathbf{n}}^{\text{inc}}$ and $\hat{\mathbf{n}}^{\text{sca}}$, respectively (Fig. 11.1.2). Rotating this particle by $180°$ about the bisectrix (i.e., the line in the scattering plane that bisects the angle $\pi - \Theta$ between the unit vectors $-\hat{\mathbf{n}}^{\text{inc}}$ and $\hat{\mathbf{n}}^{\text{sca}}$ in Fig. 11.1.2) puts it in the orientation schematically shown in Fig. 11.1.1(b). It is clear that the amplitude scattering matrix (11.1.1a) is also the amplitude scattering matrix for this rotated particle when

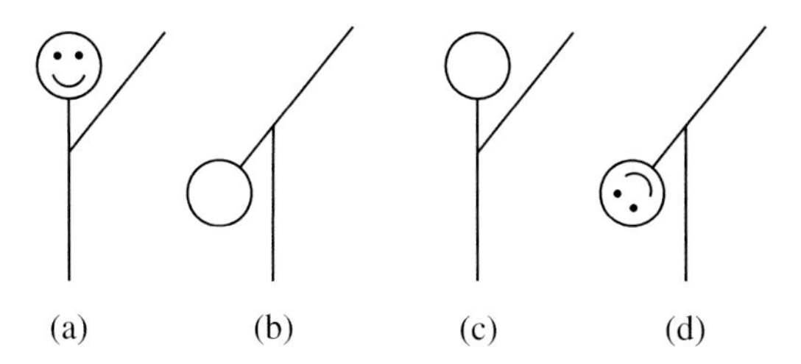

Figure 11.1.1. Two orientations of an arbitrary particle and two orientations of its mirror counterpart that give rise to certain symmetries in scattering patterns. (After van de Hulst 1957.)

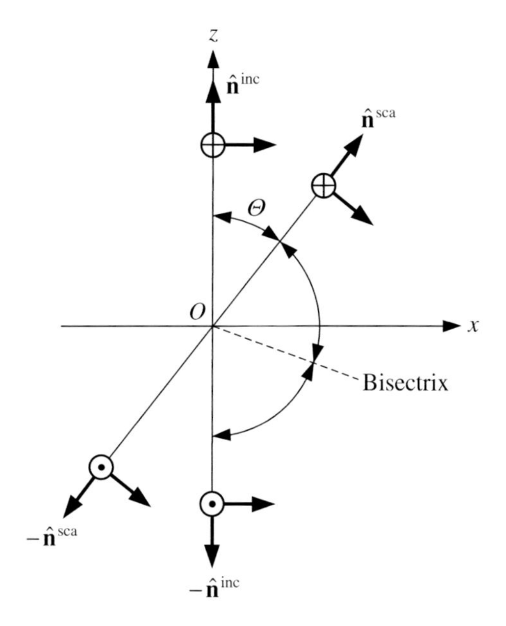

Figure 11.1.2. The *xz*-plane of the reference frame acts as the scattering plane. The arrows perpendicular to the unit $\hat{\mathbf{n}}$ vectors show the corresponding unit $\hat{\boldsymbol{\theta}}$ vectors. The symbols ⊕ and ⊙ indicate the corresponding unit $\hat{\boldsymbol{\varphi}}$ vectors, which are directed into and out of the paper, respectively.

the directions of incidence and scattering are given by $-\hat{\mathbf{n}}^{\text{sca}}$ and $-\hat{\mathbf{n}}^{\text{inc}}$, respectively. Therefore, the reciprocity relation (3.4.21) implies that the amplitude scattering matrix of the particle shown in Fig. 11.1.1(b) that corresponds to the original directions of incidence and scattering, $\hat{\mathbf{n}}^{\text{inc}}$ and $\hat{\mathbf{n}}^{\text{sca}}$, is simply

$$\begin{bmatrix} S_{11} & -S_{21} \\ -S_{12} & S_{22} \end{bmatrix}. \tag{11.1.1b}$$

Mirroring the original particle, Fig. 11.1.1(a), with respect to the scattering plane gives the particle shown in Fig. 11.1.1(c). If we also reversed the direction of the unit vectors $\hat{\boldsymbol{\varphi}}^{\text{inc}}$ and $\hat{\boldsymbol{\varphi}}^{\text{sca}}$ in Fig. (11.1.2), then we would have the same scattering problem as for the particle shown in Fig. 11.1.1(a). We may thus conclude that the amplitude scattering matrix for the particle shown in Fig. 11.1.1(c) that corresponds to the directions of incidence and scattering $\hat{\mathbf{n}}^{\text{inc}}$ and $\hat{\mathbf{n}}^{\text{sca}}$ is

$$\begin{bmatrix} S_{11} & -S_{12} \\ -S_{21} & S_{22} \end{bmatrix}. \tag{11.1.1c}$$

Finally, mirroring the original particle with respect to the bisectrix plane (i.e., the plane through the bisectrix and the y-axis) gives the particle shown in Fig. 11.1.1(d). Since this particle is simply the mirror-symmetric counterpart of the particle shown in Fig. 11.1.1(b), its amplitude scattering matrix corresponding to the directions of incidence and scattering $\hat{\mathbf{n}}^{\text{inc}}$ and $\hat{\mathbf{n}}^{\text{sca}}$ is

$$\begin{bmatrix} S_{11} & S_{21} \\ S_{12} & S_{22} \end{bmatrix}. \tag{11.1.1d}$$

It can be seen that any two of the three transformations shown in Figs. 11.1.1(b)–11.1.1(d) give the third.

We will now discuss the implications of Eqs. (11.1.1a)–(11.1.1d) for Stokes scattering matrices of collections of independently scattering particles, by considering the following four examples (van de Hulst, 1957):

1. Let us first assume that there is only one kind of particle and that each particle in a specific orientation, say Fig. 11.1.1(a), is accompanied by a particle in the reciprocal orientation, Fig. 11.1.1(b). It then follows from Eqs. (3.7.11)–(3.7.26), (11.0.1), (11.1.1a), and (11.1.1b) that the single-particle scattering matrix averaged over particle states has the following symmetry:

$$\begin{bmatrix} \langle F_{11}\rangle_\xi & \langle F_{12}\rangle_\xi & \langle F_{13}\rangle_\xi & \langle F_{14}\rangle_\xi \\ \langle F_{12}\rangle_\xi & \langle F_{22}\rangle_\xi & \langle F_{23}\rangle_\xi & \langle F_{24}\rangle_\xi \\ -\langle F_{13}\rangle_\xi & -\langle F_{23}\rangle_\xi & \langle F_{33}\rangle_\xi & \langle F_{34}\rangle_\xi \\ \langle F_{14}\rangle_\xi & \langle F_{24}\rangle_\xi & -\langle F_{34}\rangle_\xi & \langle F_{44}\rangle_\xi \end{bmatrix}. \tag{11.1.2}$$

   The number of independent matrix elements is 10.

2. As a second example, let us assume that for each particle in orientation (a) a mirror particle in orientation (c) is present (Fig. 11.1.1). This excludes, for example, scattering media composed of only right-handed or only left-handed helices. It is easy to verify that the resulting average scattering matrix involves eight independent elements and has the following structure:

$$\begin{bmatrix} \langle F_{11}\rangle_\xi & \langle F_{12}\rangle_\xi & 0 & 0 \\ \langle F_{21}\rangle_\xi & \langle F_{22}\rangle_\xi & 0 & 0 \\ 0 & 0 & \langle F_{33}\rangle_\xi & \langle F_{34}\rangle_\xi \\ 0 & 0 & \langle F_{43}\rangle_\xi & \langle F_{44}\rangle_\xi \end{bmatrix}. \tag{11.1.3}$$

3. As a third example, assume that any particle in orientation (a) is accompanied by a mirror counterpart in orientation (d), Fig. 11.1.1. The average scattering matrix becomes

$$\begin{bmatrix} \langle F_{11}\rangle_\xi & \langle F_{12}\rangle_\xi & \langle F_{13}\rangle_\xi & \langle F_{14}\rangle_\xi \\ \langle F_{12}\rangle_\xi & \langle F_{22}\rangle_\xi & \langle F_{23}\rangle_\xi & \langle F_{24}\rangle_\xi \\ \langle F_{13}\rangle_\xi & \langle F_{23}\rangle_\xi & \langle F_{33}\rangle_\xi & \langle F_{34}\rangle_\xi \\ -\langle F_{14}\rangle_\xi & -\langle F_{24}\rangle_\xi & -\langle F_{34}\rangle_\xi & \langle F_{44}\rangle_\xi \end{bmatrix} \tag{11.1.4}$$

and has 10 independent elements.

4. Finally, let us make any two of the preceding assumptions. The third assumption follows automatically, so that there are equal numbers of particles in orientations (a), (b), (c), and (d). The resulting average scattering matrix is

$$\begin{bmatrix} \langle F_{11}\rangle_\xi & \langle F_{12}\rangle_\xi & 0 & 0 \\ \langle F_{12}\rangle_\xi & \langle F_{22}\rangle_\xi & 0 & 0 \\ 0 & 0 & \langle F_{33}\rangle_\xi & \langle F_{34}\rangle_\xi \\ 0 & 0 & -\langle F_{34}\rangle_\xi & \langle F_{44}\rangle_\xi \end{bmatrix} \tag{11.1.5}$$

and has eight nonzero elements, of which only six are independent.

11.2 Macroscopically isotropic and mirror-symmetric scattering medium

Now we are ready to consider scattering by a medium containing *randomly oriented* particles. This means that there are many particles of each type and their orientation distribution is uniform (see Eq. (5.3.9)). In this case the assumptions of example 1 from the previous section are satisfied, and the average scattering matrix is given by Eq. (11.1.2). Furthermore, if particles and their mirror counterparts are present in equal numbers or each particle has a plane of symmetry, then the assumptions of example 4 are satisfied, and the resulting average scattering matrix is given by Eq. (11.1.5).

As a consequence of random particle orientation, the scattering medium is *macroscopically isotropic* (i.e., there is no preferred propagation direction and no preferred plane through the incidence direction). Therefore, the scattering matrix becomes independent of the incidence direction and the orientation of the scattering plane and depends only on the angle between the incidence and scattering directions, that is, the scattering angle

$$\Theta = \arccos(\hat{\mathbf{n}}^{\text{inc}} \cdot \hat{\mathbf{n}}^{\text{sca}}), \qquad \Theta \in [0, \pi].$$

Furthermore, the assumptions of example 4 ensure that the scattering medium is *mac-*

roscopically mirror-symmetric with respect to any plane and make the structure of the scattering matrix especially simple. Therefore, scattering media composed of equal numbers of randomly oriented particles and their mirror counterparts and/or of randomly oriented particles having a plane of symmetry can be called *macroscopically isotropic and mirror-symmetric*. To emphasize that the scattering matrix of an ISM depends only on the scattering angle, we rewrite Eq. (11.1.5) as

$$\langle \mathbf{F}(\Theta)\rangle_\xi = \begin{bmatrix} \langle F_{11}(\Theta)\rangle_\xi & \langle F_{12}(\Theta)\rangle_\xi & 0 & 0 \\ \langle F_{12}(\Theta)\rangle_\xi & \langle F_{22}(\Theta)\rangle_\xi & 0 & 0 \\ 0 & 0 & \langle F_{33}(\Theta)\rangle_\xi & \langle F_{34}(\Theta)\rangle_\xi \\ 0 & 0 & -\langle F_{34}(\Theta)\rangle_\xi & \langle F_{44}(\Theta)\rangle_\xi \end{bmatrix}. \tag{11.2.1}$$

As a direct consequence of Eqs. (3.7.29) and (3.7.30) we have the inequalities

$$\langle F_{11}\rangle_\xi \ge 0, \tag{11.2.2}$$

$$|\langle F_{ij}\rangle_\xi| \le \langle F_{11}\rangle_\xi, \qquad i, j = 1,\ldots,4. \tag{11.2.3}$$

Additional general inequalities for the elements of the scattering matrix (11.2.1) are as follows:

$$[\langle F_{33}\rangle_\xi + \langle F_{44}\rangle_\xi]^2 + 4[\langle F_{34}\rangle_\xi]^2 \le [\langle F_{11}\rangle_\xi + \langle F_{22}\rangle_\xi]^2 - 4[\langle F_{12}\rangle_\xi]^2, \tag{11.2.4}$$

$$|\langle F_{33}\rangle_\xi - \langle F_{44}\rangle_\xi| \le \langle F_{11}\rangle_\xi - \langle F_{22}\rangle_\xi, \tag{11.2.5}$$

$$|\langle F_{22}\rangle_\xi - \langle F_{12}\rangle_\xi| \le \langle F_{11}\rangle_\xi - \langle F_{12}\rangle_\xi, \tag{11.2.6}$$

$$|\langle F_{22}\rangle_\xi + \langle F_{12}\rangle_\xi| \le \langle F_{11}\rangle_\xi + \langle F_{12}\rangle_\xi. \tag{11.2.7}$$

The proof of these and other useful inequalities is given in Hovenier *et al.* (1986).

11.3 Phase matrix

Knowledge of the matrix $\langle \mathbf{F}(\Theta)\rangle_\xi$ can be used to calculate the average Stokes phase matrix for an ISM. Assume that $0 < \varphi^{\text{sca}} - \varphi^{\text{inc}} < \pi$ and consider the phase matrices $\langle \mathbf{Z}(\theta^{\text{sca}}, \varphi^{\text{sca}}; \theta^{\text{inc}}, \varphi^{\text{inc}})\rangle_\xi$ and $\langle \mathbf{Z}(\theta^{\text{sca}}, \varphi^{\text{inc}}; \theta^{\text{inc}}, \varphi^{\text{sca}})\rangle_\xi$. The second matrix involves the same polar angles of the incident and scattered beams as the first, but the azimuth angles are switched, as indicated in their respective scattering geometries; these are shown in Figs. 11.3.1(a) and (b). The phase matrix links the Stokes vectors of the incident and scattered beams, specified relative to their respective meridional planes. Therefore, to compute the Stokes vector of the scattered beam with respect to its meridional plane, we must:

- Calculate the Stokes vector of the incident beam with respect to the scattering plane.

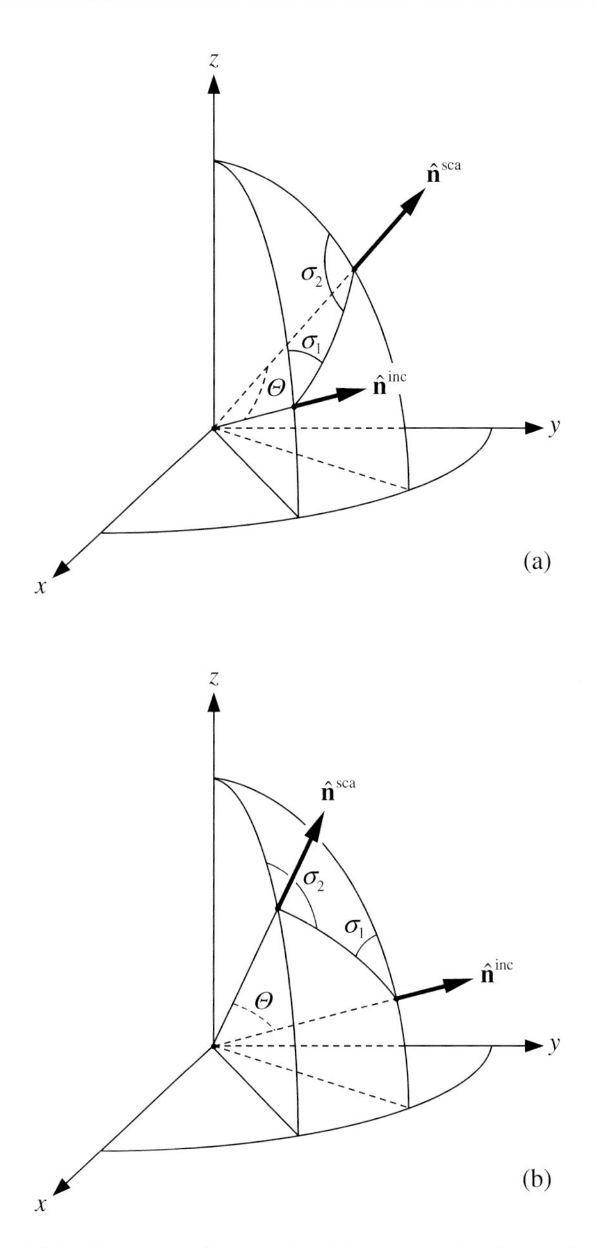

Figure 11.3.1. Illustration of the relationship between the phase and scattering matrices.

- Multiply it by the scattering matrix, thereby obtaining the Stokes vector of the scattered beam with respect to the scattering plane.
- Compute the Stokes vector of the scattered beam with respect to its meridional plane (Chandrasekhar, 1950).

This procedure involves two rotations of the reference plane, as shown in Figs.

11.3.1(a) and (b), and yields

$$\langle \mathbf{Z}(\theta^{\text{sca}},\varphi^{\text{sca}};\theta^{\text{inc}},\varphi^{\text{inc}})\rangle_\xi = \mathbf{L}(-\sigma_2)\langle \mathbf{F}(\Theta)\rangle_\xi \mathbf{L}(\pi-\sigma_1)$$

$$= \begin{bmatrix} \langle F_{11}(\Theta)\rangle_\xi & C_1\langle F_{12}(\Theta)\rangle_\xi & S_1\langle F_{12}(\Theta)\rangle_\xi & 0 \\ C_2\langle F_{12}(\Theta)\rangle_\xi & C_1C_2\langle F_{22}(\Theta)\rangle_\xi - S_1S_2\langle F_{33}(\Theta)\rangle_\xi & S_1C_2\langle F_{22}(\Theta)\rangle_\xi + C_1S_2\langle F_{33}(\Theta)\rangle_\xi & S_2\langle F_{34}(\Theta)\rangle_\xi \\ -S_2\langle F_{12}(\Theta)\rangle_\xi & -C_1S_2\langle F_{22}(\Theta)\rangle_\xi - S_1C_2\langle F_{33}(\Theta)\rangle_\xi & -S_1S_2\langle F_{22}(\Theta)\rangle_\xi + C_1C_2\langle F_{33}(\Theta)\rangle_\xi & C_2\langle F_{34}(\Theta)\rangle_\xi \\ 0 & S_1\langle F_{34}(\Theta)\rangle_\xi & -C_1\langle F_{34}(\Theta)\rangle_\xi & \langle F_{44}(\Theta)\rangle_\xi \end{bmatrix}, \tag{11.3.1}$$

$$\langle \mathbf{Z}(\theta^{\text{sca}},\varphi^{\text{inc}};\theta^{\text{inc}},\varphi^{\text{sca}})\rangle_\xi = \mathbf{L}(\sigma_2-\pi)\langle \mathbf{F}(\Theta)\rangle_\xi \mathbf{L}(\sigma_1)$$

$$= \begin{bmatrix} \langle F_{11}(\Theta)\rangle_\xi & C_1\langle F_{12}(\Theta)\rangle_\xi & -S_1\langle F_{12}(\Theta)\rangle_\xi & 0 \\ C_2\langle F_{12}(\Theta)\rangle_\xi & C_1C_2\langle F_{22}(\Theta)\rangle_\xi - S_1S_2\langle F_{33}(\Theta)\rangle_\xi & -S_1C_2\langle F_{22}(\Theta)\rangle_\xi - C_1S_2\langle F_{33}(\Theta)\rangle_\xi & -S_2\langle F_{34}(\Theta)\rangle_\xi \\ S_2\langle F_{12}(\Theta)\rangle_\xi & C_1S_2\langle F_{22}(\Theta)\rangle_\xi + S_1C_2\langle F_{33}(\Theta)\rangle_\xi & -S_1S_2\langle F_{22}(\Theta)\rangle_\xi + C_1C_2\langle F_{33}(\Theta)\rangle_\xi & C_2\langle F_{34}(\Theta)\rangle_\xi \\ 0 & -S_1\langle F_{34}(\Theta)\rangle_\xi & -C_1\langle F_{34}(\Theta)\rangle_\xi & \langle F_{44}(\Theta)\rangle_\xi \end{bmatrix}, \tag{11.3.2}$$

where

$$C_i = \cos 2\sigma_i, \qquad S_i = \sin 2\sigma_i, \qquad i = 1, 2, \tag{11.3.3}$$

and the rotation matrix $\mathbf{L}$ is defined by Eq. (2.8.4). (Recall that a rotation angle is positive if the rotation is performed in the clockwise direction when one is looking in the direction of propagation; see Section 2.8.) The scattering angle Θ and the angles σ_1 and σ_2 can be calculated from θ^{sca}, θ^{inc}, φ^{sca}, and φ^{inc} using spherical trigonometry:

$$\cos\Theta = \cos\theta^{\text{sca}}\cos\theta^{\text{inc}} + \sin\theta^{\text{sca}}\sin\theta^{\text{inc}}\cos(\varphi^{\text{sca}}-\varphi^{\text{inc}}), \tag{11.3.4}$$

$$\cos\sigma_1 = \frac{\cos\theta^{\text{sca}} - \cos\theta^{\text{inc}}\cos\Theta}{\sin\theta^{\text{inc}}\sin\Theta}, \tag{11.3.5}$$

$$\cos\sigma_2 = \frac{\cos\theta^{\text{inc}} - \cos\theta^{\text{sca}}\cos\Theta}{\sin\theta^{\text{sca}}\sin\Theta}. \tag{11.3.6}$$

Equations (11.3.1) and (11.3.3)–(11.3.6) demonstrate the obvious fact that the phase matrix of an ISM depends only on the difference between the azimuthal angles of the scattering and incidence directions rather than on their specific values. In particular,

$$\langle \mathbf{Z}(\theta^{\rm sca}, 2\pi - \varphi^{\rm inc}; \theta^{\rm inc}, 2\pi - \varphi^{\rm sca})\rangle_\xi = \langle \mathbf{Z}(\theta^{\rm sca}, \varphi^{\rm sca}; \theta^{\rm inc}, \varphi^{\rm inc})\rangle_\xi \tag{11.3.7}$$

or, formally allowing negative azimuth-angle values,

$$\langle \mathbf{Z}(\theta^{\rm sca}, -\varphi^{\rm inc}; \theta^{\rm inc}, -\varphi^{\rm sca})\rangle_\xi = \langle \mathbf{Z}(\theta^{\rm sca}, \varphi^{\rm sca}; \theta^{\rm inc}, \varphi^{\rm inc})\rangle_\xi. \tag{11.3.8}$$

Comparison of Eqs. (11.3.1) and (11.3.2) yields the symmetry relation (Hovenier, 1969):

$$\begin{aligned}\langle \mathbf{Z}(\theta^{\rm sca}, \varphi^{\rm inc}; \theta^{\rm inc}, \varphi^{\rm sca})\rangle_\xi &= \langle \mathbf{Z}(\theta^{\rm sca}, -\varphi^{\rm sca}; \theta^{\rm inc}, -\varphi^{\rm inc})\rangle_\xi \\ &= \mathbf{\Delta}_{34}\langle \mathbf{Z}(\theta^{\rm sca}, \varphi^{\rm sca}; \theta^{\rm inc}, \varphi^{\rm inc})\rangle_\xi \mathbf{\Delta}_{34},\end{aligned} \tag{11.3.9}$$

where

$$\mathbf{\Delta}_{34} = \mathbf{\Delta}_{34}^{\rm T} = \mathbf{\Delta}_{34}^{-1} = \begin{bmatrix} 1 & 0 & 0 & 0 \\ 0 & 1 & 0 & 0 \\ 0 & 0 & -1 & 0 \\ 0 & 0 & 0 & -1 \end{bmatrix}. \tag{11.3.10}$$

Obviously, Eq. (11.3.9) is a manifestation of mirror symmetry with respect to the meridional plane of the incidence direction (cf. Fig. 11.3.1) or, equivalently, with respect to the xz-half-plane with $x \geq 0$. It is also easy to see from either Eq. (11.3.1) or Eq. (11.3.2) that (Hovenier, 1969)

$$\langle \mathbf{Z}(\pi - \theta^{\rm sca}, \varphi^{\rm sca}; \pi - \theta^{\rm inc}, \varphi^{\rm inc})\rangle_\xi = \mathbf{\Delta}_{34}\langle \mathbf{Z}(\theta^{\rm sca}, \varphi^{\rm sca}; \theta^{\rm inc}, \varphi^{\rm inc})\rangle_\xi \mathbf{\Delta}_{34}, \tag{11.3.11}$$

which is a manifestation of mirror symmetry with respect to the xy-plane. Finally, we can verify that

$$\begin{aligned}&\langle \mathbf{Z}(\pi - \theta^{\rm inc}, \varphi^{\rm inc} + \pi; \pi - \theta^{\rm sca}, \varphi^{\rm sca} + \pi)\rangle_\xi \\ &\quad = \langle \mathbf{Z}(\pi - \theta^{\rm inc}, \varphi^{\rm inc}; \pi - \theta^{\rm sca}, \varphi^{\rm sca})\rangle_\xi \\ &\quad = \mathbf{\Delta}_3[\langle \mathbf{Z}(\theta^{\rm sca}, \varphi^{\rm sca}; \theta^{\rm inc}, \varphi^{\rm inc})\rangle_\xi]^{\rm T} \mathbf{\Delta}_3,\end{aligned} \tag{11.3.12}$$

where the matrix $\mathbf{\Delta}_3$ is given by Eq. (3.7.32). Obviously, this is the reciprocity relation (3.7.31). Other symmetry relations can be derived by forming combinations of Eqs. (11.3.9), (11.3.11), and (11.3.12). For example, combining Eqs. (11.3.9) and (11.3.11) yields

$$\langle \mathbf{Z}(\pi - \theta^{\rm sca}, \varphi^{\rm inc}; \pi - \theta^{\rm inc}, \varphi^{\rm sca})\rangle_\xi = \langle \mathbf{Z}(\theta^{\rm sca}, \varphi^{\rm sca}; \theta^{\rm inc}, \varphi^{\rm inc})\rangle_\xi. \tag{11.3.13}$$

Although Eq. (11.3.1) is valid only for $0 < \varphi^{\rm sca} - \varphi^{\rm inc} < \pi$, combining it with Eq.

(11.3.9) yields the phase matrix for all possible incidence and scattering directions. The symmetry relations (11.3.11) and (11.3.12) further reduce the range of independent scattering geometries and can be very helpful in theoretical calculations or consistency checks on measurements.

11.4 Forward-scattering direction and extinction matrix

By virtue of spatial isotropy, the extinction matrix of an ISM is independent of the direction of light propagation and orientation of the reference plane used to define the Stokes parameters. It also follows from Eqs. (3.8.10)–(3.8.13) and (11.1.1a)–(11.1.1d) that

$$\begin{aligned}\langle K_{13}\rangle_\xi &= \langle K_{14}\rangle_\xi = \langle K_{23}\rangle_\xi = \langle K_{24}\rangle_\xi \\ &= \langle K_{31}\rangle_\xi = \langle K_{32}\rangle_\xi = \langle K_{41}\rangle_\xi = \langle K_{42}\rangle_\xi = 0.\end{aligned}$$

Furthermore, we are about to show that the remaining off-diagonal elements of the average extinction matrix also vanish.

We will assume for simplicity that light is incident along the positive direction of the z-axis of the laboratory reference frame and will use the xz-half-plane with $x \geq 0$ as the meridional plane of the incident beam. Let us affix a reference frame to the particle and call it the particle reference frame. We will also assume that the initial orientation of a particle is such that the particle reference frame coincides with the laboratory reference frame. The forward-scattering amplitude matrix of the particle in the initial orientation computed in the laboratory reference frame is thus equal to the forward-scattering amplitude matrix computed in the particle reference frame. We will denote the latter as $\mathbf{S}_P$.

Let us now rotate the particle along with its reference frame through an Euler angle α about the z-axis in the clockwise direction as viewed in the positive z-direction (Figs. C.1 and 11.4.1) and denote the forward-scattering amplitude matrix of this rotated particle with respect to the laboratory reference frame as $\mathbf{S}_L^\alpha$. This matrix relates the column of the electric field vector components of the incident field to that of the field scattered in the exact forward direction:

$$\begin{bmatrix} E_{\theta L}^{\text{sca}} \\ E_{\varphi L}^{\text{sca}} \end{bmatrix} \propto \mathbf{S}_L^\alpha \begin{bmatrix} E_{\theta L}^{\text{inc}} \\ E_{\varphi L}^{\text{inc}} \end{bmatrix}, \tag{11.4.1}$$

where the subscript L indicates that all field components are computed in the laboratory reference frame. Figure 11.4.1 shows the directions of the respective unit θ- and φ-vectors for the incident and the forward-scattered beams. Simple trigonometry allows us to express the column of the electric vector components in the particle reference frame in terms of that in the laboratory reference frame by means of a trivial matrix multiplication (see Fig. 11.4.1):

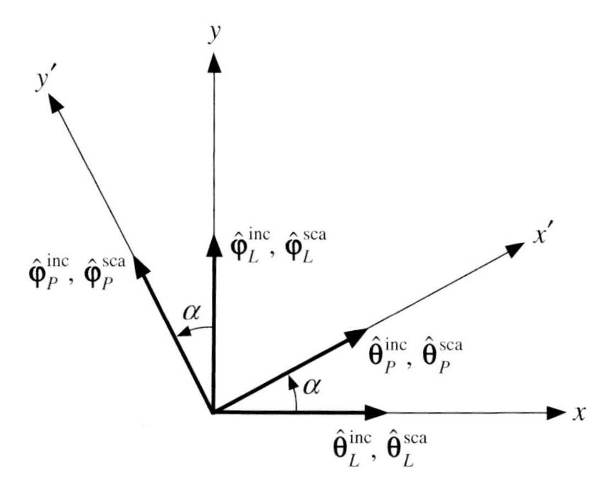

Figure 11.4.1. Rotation of the particle through an Euler angle α about the z-axis transforms the laboratory reference frame $L\{x, y, z\}$ into the particle reference frame $P\{x', y', z\}$. Since both the incident and the scattered beams propagate in the positive z-direction, their respective unit $\hat{\boldsymbol{\theta}}$ and $\hat{\boldsymbol{\varphi}}$ vectors are the same.

$$\begin{bmatrix} E_{\theta P}^{\rm inc} \\ E_{\varphi P}^{\rm inc} \end{bmatrix} = \begin{bmatrix} C & S \\ -S & C \end{bmatrix} \begin{bmatrix} E_{\theta L}^{\rm inc} \\ E_{\varphi L}^{\rm inc} \end{bmatrix}, \tag{11.4.2}$$

where $C = \cos\alpha$ and $S = \sin\alpha$. Conversely,

$$\begin{bmatrix} E_{\theta L}^{\rm sca} \\ E_{\varphi L}^{\rm sca} \end{bmatrix} = \begin{bmatrix} C & -S \\ S & C \end{bmatrix} \begin{bmatrix} E_{\theta P}^{\rm sca} \\ E_{\varphi P}^{\rm sca} \end{bmatrix}. \tag{11.4.3}$$

Rewriting Eq. (11.4.1) in the particle reference frame,

$$\begin{bmatrix} E_{\theta P}^{\rm sca} \\ E_{\varphi P}^{\rm sca} \end{bmatrix} \propto \mathbf{S}_P \begin{bmatrix} E_{\theta P}^{\rm inc} \\ E_{\varphi P}^{\rm inc} \end{bmatrix}, \tag{11.4.4}$$

and using Eqs. (11.4.2) and (11.4.3), we finally derive

$$\begin{aligned} \mathbf{S}_L^{\alpha} &= \begin{bmatrix} C & -S \\ S & C \end{bmatrix} \mathbf{S}_P \begin{bmatrix} C & S \\ -S & C \end{bmatrix} \\ &= \begin{bmatrix} C^2 S_{11P} - SCS_{12P} - SCS_{21P} + S^2 S_{22P} & SCS_{11P} + C^2 S_{12P} - S^2 S_{21P} - SCS_{22P} \\ SCS_{11P} - S^2 S_{12P} + C^2 S_{21P} - SCS_{22P} & S^2 S_{11P} + SCS_{12P} + SCS_{21P} + C^2 S_{22P} \end{bmatrix}. \end{aligned} \tag{11.4.5}$$

For $\alpha = 0$ and $\alpha = \pi/2$,

$$\mathbf{S}_L^0 = \begin{bmatrix} S_{11P} & S_{12P} \\ S_{21P} & S_{22P} \end{bmatrix}, \tag{11.4.6}$$

$$\mathbf{S}_L^{\pi/2} = \begin{bmatrix} S_{22P} & -S_{21P} \\ -S_{12P} & S_{11P} \end{bmatrix}. \tag{11.4.7}$$

Because we are assuming random orientation of the particles in the small volume element, for each particle in the initial orientation, $\alpha = 0$, there is always a particle of the same type but in the orientation corresponding to $\alpha = \pi/2$. It, therefore, follows from Eqs. (3.8.9), (3.8.14), (11.4.6), and (11.4.7) that

$$\langle K_{12}\rangle_\xi = \langle K_{21}\rangle_\xi = \langle K_{34}\rangle_\xi = \langle K_{43}\rangle_\xi = 0.$$

Finally, recalling Eq. (3.9.9), we conclude that the extinction matrix of a small volume element containing equal numbers of randomly oriented particles and their mirror-symmetric counterparts and/or randomly oriented particles having a plane of symmetry is diagonal:

$$\langle \mathbf{K}(\hat{\mathbf{n}})\rangle_\xi \equiv \langle \mathbf{K}\rangle_\xi = \langle C_{\text{ext}}\rangle_\xi \mathbf{\Delta}, \tag{11.4.8}$$

where $\langle C_{\text{ext}}\rangle_\xi$ is the average extinction cross section per particle which is now independent of the direction of propagation and polarization state of the incident light. This significant simplification is useful in many practical circumstances.

The scattering matrix also becomes simpler when $\Theta = 0$. From Eqs. (3.7.12), (3.7.15), (3.7.22), (3.7.25), (1.4.6), and (1.4.7), we find that

$$\langle F_{12}(0)\rangle_\xi = \langle F_{21}(0)\rangle_\xi = \langle F_{34}(0)\rangle_\xi = \langle F_{43}(0)\rangle_\xi = 0.$$

Equation (11.4.5) gives for $\alpha = \pi/4$:

$$\mathbf{S}_L^{\pi/4} = \frac{1}{2}\begin{bmatrix} S_{11P} - S_{12P} - S_{21P} + S_{22P} & S_{11P} + S_{12P} - S_{21P} - S_{22P} \\ S_{11P} - S_{12P} + S_{21P} - S_{22P} & S_{11P} + S_{12P} + S_{21P} + S_{22P} \end{bmatrix}. \tag{11.4.9}$$

Equations (3.7.16), (3.7.21), (11.4.6), and (11.4.9) and a considerable amount of algebra yield

$$\langle F_{22}(0)\rangle_\xi = \langle F_{33}(0)\rangle_\xi.$$

Thus, recalling Eq. (11.2.1), we find that the forward-scattering matrix for an ISM is diagonal and has only three independent elements:

$$\langle \mathbf{F}(0)\rangle_\xi = \begin{bmatrix} \langle F_{11}(0)\rangle_\xi & 0 & 0 & 0 \\ 0 & \langle F_{22}(0)\rangle_\xi & 0 & 0 \\ 0 & 0 & \langle F_{22}(0)\rangle_\xi & 0 \\ 0 & 0 & 0 & \langle F_{44}(0)\rangle_\xi \end{bmatrix} \tag{11.4.10}$$

(van de Hulst, 1957).

Rotationally-symmetric particles are obviously mirror-symmetric with respect to the plane through the direction of propagation and the axis of symmetry. Choosing this plane as the $x'z'$-plane of the particle reference frame, we see from Eq. (11.1.1c) that $S_{12P} = S_{21P} = 0$. This simplifies the amplitude scattering matrices (11.4.6) and (11.4.9) and ultimately yields

$$\langle F_{44}(0)\rangle_\xi = 2\langle F_{22}(0)\rangle_\xi - \langle F_{11}(0)\rangle_\xi, \tag{11.4.11}$$

$$0 \le \langle F_{22}(0)\rangle_\xi \le \langle F_{11}(0)\rangle_\xi \tag{11.4.12}$$

(Mishchenko and Travis, 1994a; Hovenier and Mackowski, 1998).

11.5 Backward scattering

Equation (11.0.1) provides an unambiguous definition of the scattering matrix in terms of the phase matrix, except for the exact backscattering direction. Indeed, the backscattering direction for an incidence direction $(\theta^{\mathrm{inc}}, \varphi^{\mathrm{inc}})$ is given by $(\pi - \theta^{\mathrm{inc}}, \varphi^{\mathrm{inc}} + \pi)$. Therefore, the complete definition of the scattering matrix should be as follows:

$$\mathbf{F}(\theta^{\mathrm{sca}}) = \begin{cases} \mathbf{Z}(\theta^{\mathrm{sca}}, 0; 0, 0) & \text{for } \theta^{\mathrm{sca}} \in [0, \pi), \\ \mathbf{Z}(\pi, \pi; 0, 0) & \text{for } \theta^{\mathrm{sca}} = \pi, \end{cases}$$

which seems to be different from Eq. (11.0.1). It is easy to see, however, that

$$\mathbf{Z}(\pi, 0; 0, 0) = \mathbf{L}(\pi)\mathbf{Z}(\pi, \pi; 0, 0) \equiv \mathbf{Z}(\pi, \pi; 0, 0),$$

see Eq. (2.8.3), which demonstrates the equivalence of the two definitions.

We are ready now to consider the case of scattering in the exact backward direction, using the complete definition of the scattering matrix and the backscattering theorem derived in Section 3.4. Let us assume that light is incident along the positive z-axis of the laboratory coordinate system and is scattered in the opposite direction; we use the xz half-plane with $x \ge 0$ as the meridional plane of the incident beam. As in the previous section, we consider two particle orientations relative to the laboratory reference frame:

- The initial orientation, when the particle reference frame coincides with the laboratory reference frame.
- The orientation obtained by rotating the particle about the z-axis through a positive Euler angle α.

Figure 11.5.1 shows the respective unit θ- and φ-vectors for the incident beam and the backscattered beam. Denote the backscattering amplitude matrix in the particle reference frame as $\mathbf{S}_P$ and the backscattering amplitude matrix in the laboratory reference frame for the rotated particle as $\mathbf{S}_L^\alpha$. A derivation similar to that in the previous section gives

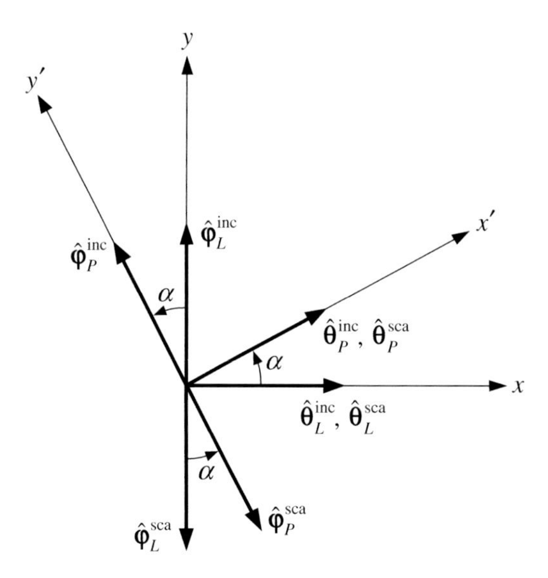

Figure 11.5.1. As in Fig. 11.4.1, but for the case of scattering in the exact backward direction.

$$\mathbf{S}_L^\alpha = \begin{bmatrix} C & S \\ -S & C \end{bmatrix} \mathbf{S}_P \begin{bmatrix} C & S \\ -S & C \end{bmatrix}$$

$$= \begin{bmatrix} C^2 S_{11P} - SCS_{12P} + SCS_{21P} - S^2 S_{22P} & SCS_{11P} + C^2 S_{12P} + S^2 S_{21P} + SCS_{22P} \\ -SCS_{11P} + S^2 S_{12P} + C^2 S_{21P} - SCS_{22P} & -S^2 S_{11P} - SCS_{12P} + SCS_{21P} + C^2 S_{22P} \end{bmatrix}. \quad (11.5.1)$$

This formula can be simplified, because the backscattering theorem (3.4.22) yields $S_{21P} = -S_{12P}$. Assuming that particles are randomly oriented and considering the cases $\alpha = 0$ and $\alpha = \pi/2$, we find that

$$\langle F_{12}(\pi)\rangle_\xi = \langle F_{21}(\pi)\rangle_\xi = \langle F_{34}(\pi)\rangle_\xi = \langle F_{43}(\pi)\rangle_\xi = 0.$$

Similarly, considering the cases $\alpha = 0$ and $\alpha = \pi/4$ yields

$$\langle F_{33}(\pi)\rangle_\xi = -\langle F_{22}(\pi)\rangle_\xi.$$

Finally, recalling Eqs. (3.7.38) and (11.2.1), we conclude that the backscattering matrix for an ISM is diagonal and has only two independent elements:

$$\langle \mathbf{F}(\pi)\rangle_\xi = \begin{bmatrix} \langle F_{11}(\pi)\rangle_\xi & 0 & 0 & 0 \\ 0 & \langle F_{22}(\pi)\rangle_\xi & 0 & 0 \\ 0 & 0 & -\langle F_{22}(\pi)\rangle_\xi & 0 \\ 0 & 0 & 0 & \langle F_{11}(\pi)\rangle_\xi - 2\langle F_{22}(\pi)\rangle_\xi \end{bmatrix} \quad (11.5.2)$$

(Mishchenko and Hovenier, 1995). According to Eq. (11.2.3) $\langle F_{44}\rangle_\xi \le \langle F_{11}\rangle_\xi$, so we always have

$$\langle F_{22}(\pi)\rangle_\xi \ge 0. \tag{11.5.3}$$

11.6 Scattering cross section and asymmetry parameter

Like all other macroscopic scattering characteristics, the average scattering cross section per particle for an ISM is independent of the direction of illumination. Therefore, we will evaluate the integral on the right-hand side of Eq. (3.9.10) assuming that the incident light propagates along the positive z-axis of the laboratory reference frame and that the xz-half-plane with $x \ge 0$ is the meridional plane of the incident beam. Figure 11.6.1 shows that in order to compute the Stokes column vector of the scattered beam with respect to its own meridional plane, we must rotate the reference frame of the incident light by the angle φ, thereby modifying the Stokes column vector of the incident light according to Eq. (2.8.3) with $\eta = \varphi$, and then multiply the new Stokes column vector of the incident light by the scattering matrix. Therefore, the average phase matrix is simply

$$\langle \mathbf{Z}(\hat{\mathbf{n}}^{\mathrm{sca}}, \hat{\mathbf{n}}^{\mathrm{inc}})\rangle_\xi = \langle \mathbf{F}(\theta)\rangle_\xi \mathbf{L}(\varphi)$$
$$= \begin{bmatrix} \langle F_{11}(\theta)\rangle_\xi & \langle F_{12}(\theta)\rangle_\xi \cos 2\varphi & -\langle F_{12}(\theta)\rangle_\xi \sin 2\varphi & 0 \\ \langle F_{12}(\theta)\rangle_\xi & \langle F_{22}(\theta)\rangle_\xi \cos 2\varphi & -\langle F_{22}(\theta)\rangle_\xi \sin 2\varphi & 0 \\ 0 & \langle F_{33}(\theta)\rangle_\xi \sin 2\varphi & \langle F_{33}(\theta)\rangle_\xi \cos 2\varphi & \langle F_{34}(\theta)\rangle_\xi \\ 0 & -\langle F_{34}(\theta)\rangle_\xi \sin 2\varphi & -\langle F_{34}(\theta)\rangle_\xi \cos 2\varphi & \langle F_{44}(\theta)\rangle_\xi \end{bmatrix}. \tag{11.6.1}$$

Substituting this formula in Eq. (3.9.10), we find that the average scattering cross section per particle is independent of the polarization state of the incident light and is given by

$$\langle C_{\mathrm{sca}}\rangle_\xi = 2\pi \int_0^\pi \mathrm{d}\theta \sin\theta \, \langle F_{11}(\theta)\rangle_\xi. \tag{11.6.2}$$

The corresponding asymmetry parameter must also be independent of $\hat{\mathbf{n}}^{\mathrm{inc}}$, and Eqs. (3.9.15), (3.9.19), and (11.6.1) yield

$$\langle \cos\Theta\rangle = \frac{2\pi}{\langle C_{\mathrm{sca}}\rangle_\xi} \int_0^\pi \mathrm{d}\theta \sin\theta \cos\theta \, \langle F_{11}(\theta)\rangle_\xi. \tag{11.6.3}$$

Obviously, $\langle \cos\Theta\rangle$ is polarization-independent. The average absorption cross section,

$$\langle C_{\mathrm{abs}}\rangle_\xi = \langle C_{\mathrm{ext}}\rangle_\xi - \langle C_{\mathrm{sca}}\rangle_\xi, \tag{11.6.4}$$

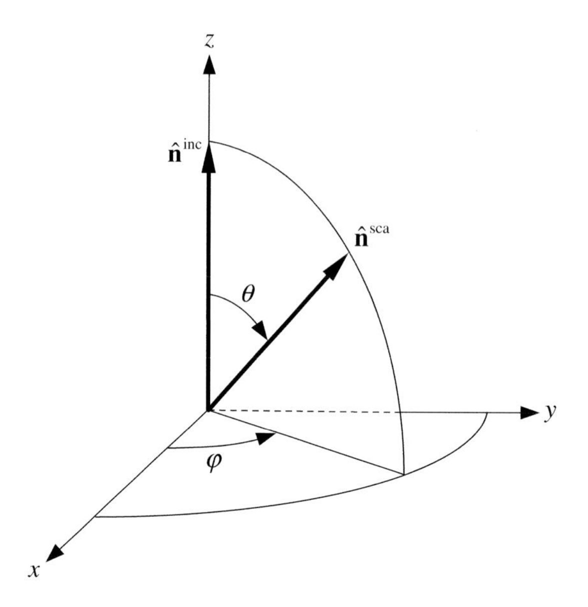

Figure 11.6.1. Illustration of the relationship between the phase and scattering matrices when the incident light propagates along the positive z-axis.

and the single-scattering albedo,

$$\varpi = \frac{\langle C_{\text{sca}} \rangle_\xi}{\langle C_{\text{ext}} \rangle_\xi}, \tag{11.6.5}$$

are also independent of the direction and polarization state of the incident beam. The same, of course, is true of the extinction, scattering, and absorption efficiency factors, defined as

$$Q_{\text{ext}} = \frac{\langle C_{\text{ext}} \rangle_\xi}{\langle G \rangle_\xi}, \qquad Q_{\text{sca}} = \frac{\langle C_{\text{sca}} \rangle_\xi}{\langle G \rangle_\xi}, \qquad Q_{\text{abs}} = \frac{\langle C_{\text{abs}} \rangle_\xi}{\langle G \rangle_\xi}, \tag{11.6.6}$$

respectively, where $\langle G \rangle_\xi$ is the average area of the particle projection.

11.7 Thermal emission

Because the ensemble-averaged emission Stokes column vector for an ISM must be independent of the emission direction, we will calculate the integral on the right-hand side of Eq. (3.13.6) for light emitted in the positive direction of the z-axis and will use the meridional plane $\varphi = 0$ as the reference plane for defining the emission Stokes column vector. It is then obvious from Fig. 11.7.1 that the corresponding average phase matrix can be calculated as

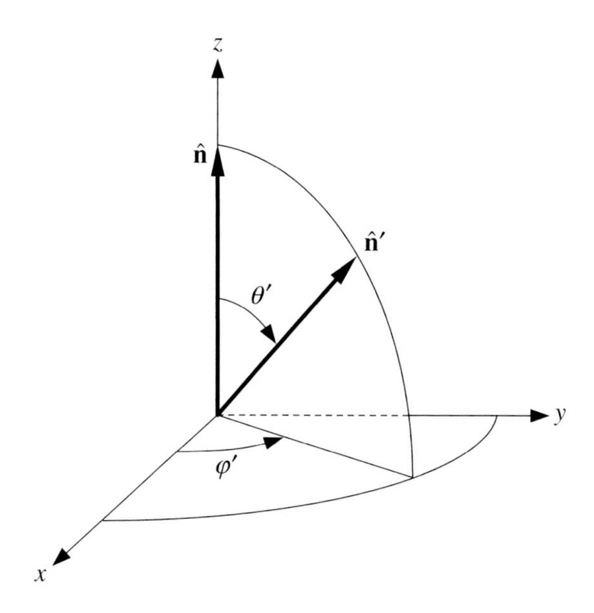

Figure 11.7.1. Illustration of the relationship between the phase and scattering matrices when the scattered light propagates along the positive z-axis.

$$\langle \mathbf{Z}(\hat{\mathbf{n}}, \hat{\mathbf{n}}') \rangle_\xi = \mathbf{L}(-\varphi') \langle \mathbf{F}(\theta') \rangle_\xi$$

$$= \begin{bmatrix} \langle F_{11}(\theta') \rangle_\xi & \langle F_{12}(\theta') \rangle_\xi & 0 & 0 \\ \langle F_{12}(\theta') \rangle_\xi \cos 2\varphi' & \langle F_{22}(\theta') \rangle_\xi \cos 2\varphi' & \langle F_{33}(\theta') \rangle_\xi \sin 2\varphi' & \langle F_{34}(\theta') \rangle_\xi \sin 2\varphi' \\ -\langle F_{12}(\theta') \rangle_\xi \sin 2\varphi' & -\langle F_{22}(\theta') \rangle_\xi \sin 2\varphi' & \langle F_{33}(\theta') \rangle_\xi \cos 2\varphi' & \langle F_{34}(\theta') \rangle_\xi \cos 2\varphi' \\ 0 & 0 & -\langle F_{34}(\theta') \rangle_\xi & \langle F_{44}(\theta') \rangle_\xi \end{bmatrix}. \tag{11.7.1}$$

Inserting this formula and Eqs. (11.4.8) and (11.6.2) in Eq. (3.13.6) yields

$$\langle \mathbf{K}_\mathrm{e}(\hat{\mathbf{n}}, T, \omega) \rangle_\xi \equiv \langle \mathbf{K}_\mathrm{e}(T, \omega) \rangle_\xi = \langle C_\mathrm{abs} \rangle_\xi \, \mathbf{I}_\mathrm{b}(T, \omega), \tag{11.7.2}$$

where $\langle C_\mathrm{abs} \rangle_\xi$ may depend on frequency and $\mathbf{I}_\mathrm{b}(T, \omega)$ is the blackbody Stokes column vector defined by Eq. (3.13.4). Thus, the emitted radiation is not only isotropic but also unpolarized. The first (and the only nonzero) element of the average emission Stokes column vector is simply equal to the product of the average absorption cross section and the Planck function.

11.8 Spherically symmetric particles

The structure of the scattering matrix simplifies further for spherically symmetric particles, that is, for homogeneous or radially inhomogeneous spherical bodies composed of optically isotropic materials. The refractive index inside such particles is a function of the distance from the particle center only. Irrespective of their "orientation" relative to the laboratory reference frame, spherically symmetric particles are obviously mirror-symmetric with respect to the xz-plane. Directing the incident light along the positive z-axis, restricting the scattering direction to the xz-half-plane with $x \geq 0$, and using this plane for reference, we find from Eqs. (11.1.1a) and (11.1.1c) that the amplitude scattering matrix is always diagonal: $S_{12} \equiv S_{21} \equiv 0$. Therefore, Eqs. (3.7.11), (3.7.16), (3.7.21), (3.7.26), and (11.2.1) yield

$$\langle \mathbf{F}(\Theta) \rangle_\xi = \begin{bmatrix} \langle F_{11}(\Theta) \rangle_\xi & \langle F_{12}(\Theta) \rangle_\xi & 0 & 0 \\ \langle F_{12}(\Theta) \rangle_\xi & \langle F_{11}(\Theta) \rangle_\xi & 0 & 0 \\ 0 & 0 & \langle F_{33}(\Theta) \rangle_\xi & \langle F_{34}(\Theta) \rangle_\xi \\ 0 & 0 & -\langle F_{34}(\Theta) \rangle_\xi & \langle F_{33}(\Theta) \rangle_\xi \end{bmatrix}. \tag{11.8.1}$$

A scattering matrix of this type appears in the standard Lorenz–Mie theory of light scattering by homogeneous isotropic spheres. The results of the previous sections on forward and backward scattering imply that

$$\langle F_{33}(0) \rangle_\xi = \langle F_{11}(0) \rangle_\xi, \tag{11.8.2}$$

$$\langle F_{33}(\pi) \rangle_\xi = -\langle F_{11}(\pi) \rangle_\xi. \tag{11.8.3}$$

11.9 Effects of nonsphericity and orientation

The previous discussion of symmetries enables us to summarize the most fundamental effects of particle nonsphericity and orientation on the average single-particle characteristics. If particles are not spherically symmetric and do not form an ISM, then, in general:

- The 4×4 extinction matrix does not degenerate to a direction- and polarization-independent scalar extinction cross section.
- The extinction, scattering, and absorption cross sections, the single-scattering albedo, and the asymmetry parameter depend on the direction and polarization state of the incident beam.
- All four elements of the emission vector are nonzero and orientation-dependent.
- The scattering matrix $\langle \mathbf{F} \rangle_\xi$ does not have the simple block-diagonal structure of Eq. (11.2.1): all 16 elements of the scattering matrix can be nonzero and depend on the incidence direction and the orientation of the scattering plane rather than only on the scattering angle.
- The phase matrix depends on the specific values of the azimuthal angles of the incidence and scattering directions rather than on their difference, it cannot be represented in the form of Eqs. (11.3.1) and (11.3.2), and it does not obey the symmetry relations (11.3.9) and (11.3.11).

Any of these effects can directly indicate the presence of oriented particles lacking spherical symmetry. For example, measurements of interstellar polarization are used in astrophysics to detect preferentially oriented dust grains causing different values of extinction for different polarization components of the transmitted starlight (Martin, 1978). Similarly, the depolarization of radiowave signals propagating through the Earth's atmosphere may indicate the presence of partially aligned nonspherical hydrometeors (Oguchi, 1983).

If nonspherical particles are randomly oriented and form an ISM, then:

- The extinction matrix reduces to the scalar extinction cross section, Eq. (11.4.8).
- All optical cross sections, the single-scattering albedo, and the asymmetry parameter become orientation- and polarization-independent.
- The emitted radiation becomes isotropic and unpolarized.
- The phase matrix depends only on the difference between the azimuthal angles of the incidence and scattering directions rather than on their specific values, has the structure specified by Eqs. (11.3.1) and (11.3.2), and obeys the symmetry relations (11.3.9) and (11.3.11).
- The scattering matrix becomes block-diagonal, Eq. (11.2.1), depends only on the scattering angle, and possesses almost the same structure as the Lorenz–Mie scattering matrix (11.8.1).

Despite the similarity of the matrices (11.2.1) and (11.8.1), the Lorenz–Mie identities $\langle F_{22}(\Theta)\rangle_\xi \equiv \langle F_{11}(\Theta)\rangle_\xi$ and $\langle F_{44}(\Theta)\rangle_\xi \equiv \langle F_{33}(\Theta)\rangle_\xi$ as well as Eqs. (11.8.2) and (11.8.3) do not hold, in general, for nonspherical particles. As a consequence, measurements of the linear backscattering depolarization ratio

$$\delta_{\mathrm{L}} = \frac{\langle F_{11}(\pi)\rangle_\xi - \langle F_{22}(\pi)\rangle_\xi}{\langle F_{11}(\pi)\rangle_\xi + \langle F_{22}(\pi)\rangle_\xi} \tag{11.9.1}$$

and the closely related circular backscattering depolarization ratio

$$\delta_{\mathrm{C}} = \frac{\langle F_{11}(\pi)\rangle_\xi + \langle F_{44}(\pi)\rangle_\xi}{\langle F_{11}(\pi)\rangle_\xi - \langle F_{44}(\pi)\rangle_\xi} = \frac{2\delta_{\mathrm{L}}}{1 - 2\delta_{\mathrm{L}}} \tag{11.9.2}$$

are, perhaps, the most reliable means of detecting particle nonsphericity (Mishchenko and Hovenier, 1995). Besides the above qualitative distinctions, which unequivocally distinguish randomly oriented nonspherical particles from spheres, there can be significant quantitative differences in specific scattering patterns. They will be discussed in some detail in Section 11.13.

11.10 Normalized scattering and phase matrices

It is convenient and customary in many types of application to use the so-called normalized scattering matrix

$$\tilde{\mathbf{F}}(\Theta) = \frac{4\pi}{\langle C_{\text{sca}} \rangle_\xi} \langle \mathbf{F}(\Theta) \rangle_\xi = \begin{bmatrix} a_1(\Theta) & b_1(\Theta) & 0 & 0 \\ b_1(\Theta) & a_2(\Theta) & 0 & 0 \\ 0 & 0 & a_3(\Theta) & b_2(\Theta) \\ 0 & 0 & -b_2(\Theta) & a_4(\Theta) \end{bmatrix}, \tag{11.10.1}$$

the elements of which are dimensionless. Similarly, the normalized phase matrix can be defined as

$$\tilde{\mathbf{Z}}(\theta^{\text{sca}}, \varphi^{\text{sca}}; \theta^{\text{inc}}, \varphi^{\text{inc}}) = \frac{4\pi}{\langle C_{\text{sca}} \rangle_\xi} \langle \mathbf{Z}(\theta^{\text{sca}}, \varphi^{\text{sca}}; \theta^{\text{inc}}, \varphi^{\text{inc}}) \rangle_\xi. \tag{11.10.2}$$

The (1,1) element of the normalized scattering matrix, $a_1(\Theta)$, is traditionally called the phase function and, as follows from Eqs. (11.6.2) and (11.10.1), satisfies the normalization condition:

$$\frac{1}{2} \int_0^\pi \mathrm{d}\Theta \sin\Theta \, a_1(\Theta) = 1. \tag{11.10.3}$$

Remember that we have already used the term "phase function" to name the quantity p defined by Eq. (3.9.17). It can be easily seen from Eqs. (3.9.15), (3.9.17), (11.0.1), and (11.10.1) that the differential scattering cross section $\mathrm{d}C_{\text{sca}}/\mathrm{d}\Omega$ reduces to $\langle F_{11} \rangle_\xi$, and so p reduces to a_1, when unpolarized incident light propagates along the positive z-axis and is scattered in the xz-half-plane with $x \geq 0$. Equations (11.6.3) and (11.10.1) yield

$$\langle \cos\Theta \rangle = \frac{1}{2} \int_0^\pi \mathrm{d}\Theta \sin\Theta \, a_1(\Theta) \cos\Theta. \tag{11.10.4}$$

The normalized scattering matrix possesses many properties of the regular scattering matrix, e.g.,

$$a_1 \geq 0, \tag{11.10.5}$$

$$|a_i| \leq a_1, \qquad i = 2, 3, 4, \tag{11.10.6}$$

$$|b_i| \leq a_1, \qquad i = 1, 2, \tag{11.10.7}$$

$$(a_3 + a_4)^2 + 4b_2^2 \leq (a_1 + a_2)^2 - 4b_1^2, \tag{11.10.8}$$

$$|a_3 - a_4| \leq a_1 - a_2, \tag{11.10.9}$$

$$|a_2 - b_1| \leq a_1 - b_1, \tag{11.10.10}$$

$$|a_2 + b_1| \leq a_1 + b_1, \tag{11.10.11}$$

$$\tilde{\mathbf{F}}(0) = \begin{bmatrix} a_1(0) & 0 & 0 & 0 \\ 0 & a_2(0) & 0 & 0 \\ 0 & 0 & a_2(0) & 0 \\ 0 & 0 & 0 & a_4(0) \end{bmatrix}, \tag{11.10.12}$$

$$\tilde{\mathbf{F}}(\pi) = \begin{bmatrix} a_1(\pi) & 0 & 0 & 0 \\ 0 & a_2(\pi) & 0 & 0 \\ 0 & 0 & -a_2(\pi) & 0 \\ 0 & 0 & 0 & a_4(\pi) \end{bmatrix}, \tag{11.10.13}$$

$$a_4(\pi) = a_1(\pi) - 2a_2(\pi), \qquad a_2(\pi) \ge 0. \tag{11.10.14}$$

Also,

$$a_4(0) = 2a_2(0) - a_1(0), \qquad 0 \le a_2(0) \le a_1(0) \tag{11.10.15}$$

for rotationally symmetric particles and

$$\tilde{\mathbf{F}}(\Theta) = \begin{bmatrix} a_1(\Theta) & b_1(\Theta) & 0 & 0 \\ b_1(\Theta) & a_1(\Theta) & 0 & 0 \\ 0 & 0 & a_3(\Theta) & b_2(\Theta) \\ 0 & 0 & -b_2(\Theta) & a_3(\Theta) \end{bmatrix}, \tag{11.10.16}$$

$$a_3(0) = a_1(0), \qquad a_3(\pi) = -a_1(\pi) \tag{11.10.17}$$

for spherically symmetric particles. Similarly, for $0 < \varphi^{\text{sca}} - \varphi^{\text{inc}} < \pi$ the normalized phase matrix is given by

$$\tilde{\mathbf{Z}}(\theta^{\text{sca}}, \varphi^{\text{sca}}; \theta^{\text{inc}}, \varphi^{\text{inc}})$$

$$= \begin{bmatrix} a_1(\Theta) & C_1 b_1(\Theta) & S_1 b_1(\Theta) & 0 \\ C_2 b_1(\Theta) & C_1 C_2 a_2(\Theta) - S_1 S_2 a_3(\Theta) & S_1 C_2 a_2(\Theta) + C_1 S_2 a_3(\Theta) & S_2 b_2(\Theta) \\ -S_2 b_1(\Theta) & -C_1 S_2 a_2(\Theta) - S_1 C_2 a_3(\Theta) & -S_1 S_2 a_2(\Theta) + C_1 C_2 a_3(\Theta) & C_2 b_2(\Theta) \\ 0 & S_1 b_2(\Theta) & -C_1 b_2(\Theta) & a_4(\Theta) \end{bmatrix} \tag{11.10.18}$$

(cf. Eq. (11.3.1)) and has the same symmetry properties as the regular phase matrix:

$$\tilde{\mathbf{Z}}(\theta^{\text{sca}}, -\varphi^{\text{inc}}; \theta^{\text{inc}}, -\varphi^{\text{sca}}) = \tilde{\mathbf{Z}}(\theta^{\text{sca}}, \varphi^{\text{sca}}; \theta^{\text{inc}}, \varphi^{\text{inc}}), \tag{11.10.19}$$

$$\tilde{\mathbf{Z}}(\theta^{\text{sca}}, -\varphi^{\text{sca}}; \theta^{\text{inc}}, -\varphi^{\text{inc}}) = \mathbf{\Delta}_{34} \tilde{\mathbf{Z}}(\theta^{\text{sca}}, \varphi^{\text{sca}}; \theta^{\text{inc}}, \varphi^{\text{inc}}) \mathbf{\Delta}_{34}, \tag{11.10.20}$$

$$\tilde{\mathbf{Z}}(\pi - \theta^{\text{sca}}, \varphi^{\text{sca}}; \pi - \theta^{\text{inc}}, \varphi^{\text{inc}}) = \mathbf{\Delta}_{34} \tilde{\mathbf{Z}}(\theta^{\text{sca}}, \varphi^{\text{sca}}; \theta^{\text{inc}}, \varphi^{\text{inc}}) \mathbf{\Delta}_{34}, \tag{11.10.21}$$

$$\tilde{\mathbf{Z}}(\pi - \theta^{\text{inc}}, \varphi^{\text{inc}}; \pi - \theta^{\text{sca}}, \varphi^{\text{sca}}) = \mathbf{\Delta}_3 [\tilde{\mathbf{Z}}(\theta^{\text{sca}}, \varphi^{\text{sca}}; \theta^{\text{inc}}, \varphi^{\text{inc}})]^{\text{T}} \mathbf{\Delta}_3. \tag{11.10.22}$$

11.11 Expansion in generalized spherical functions

A traditional way of specifying the elements of the normalized scattering matrix is to tabulate their numerical values in a representative grid of scattering angles (e.g., Deirmendjian, 1969). However, a more mathematically appealing and practically efficient way is to expand the scattering matrix elements in so-called generalized spherical functions $P^s_{mn}(\cos\Theta)$ or, equivalently, in Wigner d-functions $d^s_{mn}(\Theta) = \mathrm{i}^{n-m} \times P^s_{mn}(\cos\Theta)$ (see Appendix F):

$$a_1(\Theta) = \sum_{s=0}^{s_{\max}} \alpha_1^s P^s_{00}(\cos\Theta) = \sum_{s=0}^{s_{\max}} \alpha_1^s d^s_{00}(\Theta), \tag{11.11.1}$$

$$a_2(\Theta) + a_3(\Theta) = \sum_{s=0}^{s_{\max}} (\alpha_2^s + \alpha_3^s) P^s_{22}(\cos\Theta) = \sum_{s=0}^{s_{\max}} (\alpha_2^s + \alpha_3^s) d^s_{22}(\Theta), \tag{11.11.2}$$

$$a_2(\Theta) - a_3(\Theta) = \sum_{s=0}^{s_{\max}} (\alpha_2^s - \alpha_3^s) P^s_{2,-2}(\cos\Theta) = \sum_{s=0}^{s_{\max}} (\alpha_2^s - \alpha_3^s) d^s_{2,-2}(\Theta), \tag{11.11.3}$$

$$a_4(\Theta) = \sum_{s=0}^{s_{\max}} \alpha_4^s P^s_{00}(\cos\Theta) = \sum_{s=0}^{s_{\max}} \alpha_4^s d^s_{00}(\Theta), \tag{11.11.4}$$

$$b_1(\Theta) = \sum_{s=0}^{s_{\max}} \beta_1^s P^s_{02}(\cos\Theta) = -\sum_{s=0}^{s_{\max}} \beta_1^s d^s_{02}(\Theta), \tag{11.11.5}$$

$$b_2(\Theta) = \sum_{s=0}^{s_{\max}} \beta_2^s P^s_{02}(\cos\Theta) = -\sum_{s=0}^{s_{\max}} \beta_2^s d^s_{02}(\Theta) \tag{11.11.6}$$

(Siewert, 1981; de Haan *et al.*, 1987). According to Appendix F, these expansions always exist in the sense of either Eq. (F.3.5) or Eq. (F.6.6) provided that

$$\int_0^\pi \mathrm{d}\Theta \sin\Theta [a_i(\Theta)]^2 < \infty, \qquad i = 1, 2, 3, 4, \tag{11.11.7}$$

$$\int_0^\pi \mathrm{d}\Theta \sin\Theta [b_i(\Theta)]^2 < \infty, \qquad i = 1, 2. \tag{11.11.8}$$

In view of the inequalities (11.10.6) and (11.10.7), it is sufficient to require that

$$\int_0^\pi \mathrm{d}\Theta \sin\Theta [a_1(\Theta)]^2 < \infty. \tag{11.11.9}$$

There are no reasons to expect that the latter condition can be violated for real particles occurring in nature. According to Eqs. (F.1.8) and (F.1.10),

$$d^s_{2,-2}(0) = d^s_{02}(0) = 0 \tag{11.11.10}$$

and

$$d^s_{22}(\pi) = d^s_{02}(\pi) = 0. \tag{11.11.11}$$

Therefore, Eqs. (11.11.1), (11.11.2), (11.11.4), and (11.11.5) identically reproduce the specific structure of the normalized scattering matrix for the exact forward and backward directions as given by Eqs. (11.10.12) and (11.10.13).

The number of nonzero terms in the expansions (11.11.1)–(11.11.6) is in principle infinite. In practice, however, the expansions are truncated at $s = s_{\max}$, the $s_{\max}$ being chosen such that the corresponding finite sums differ from the respective scattering matrix elements on the entire interval $\Theta \in [0, \pi]$ of scattering angles within the requisite numerical accuracy. Since $d^s_{mn}(\Theta) \equiv 0$ for $s < \max(|m|, |n|)$, the coefficients α^0_2, α^1_2, α^0_3, α^1_3, β^0_1, β^1_1, β^0_2, and β^1_2 are not defined. However, it is often convenient to formally equate them to zero:

$$\alpha^0_2 = \alpha^1_2 = \alpha^0_3 = \alpha^1_3 = \beta^0_1 = \beta^1_1 = \beta^0_2 = \beta^1_2 = 0. \tag{11.11.12}$$

The angular behavior of several d-functions entering Eqs. (11.11.1)–(11.11.6) is illustrated in Fig. F.1.1. Equations (F.1.8), (F.1.10), (11.10.12), and (11.10.13) yield for the exact forward and exact backward directions:

$$a_1(0) = \sum_{s=0}^{s_{\max}} \alpha^s_1, \tag{11.11.13}$$

$$2a_2(0) = \sum_{s=0}^{s_{\max}} (\alpha^s_2 + \alpha^s_3), \tag{11.11.14}$$

$$a_4(0) = \sum_{s=0}^{s_{\max}} \alpha^s_4, \tag{11.11.15}$$

$$a_1(\pi) = \sum_{s=0}^{s_{\max}} (-1)^s \alpha^s_1, \tag{11.11.16}$$

$$2a_2(\pi) = \sum_{s=0}^{s_{\max}} (-1)^s (\alpha^s_2 - \alpha^s_3), \tag{11.11.17}$$

$$a_4(\pi) = \sum_{s=0}^{s_{\max}} (-1)^s \alpha^s_4. \tag{11.11.18}$$

The properties of the generalized spherical functions and the Wigner d-functions are summarized in Appendix F. For given m and n, either type of function with $s \geq \max(|m|, |n|)$, when multiplied by $(s + \tfrac{1}{2})^{1/2}$, forms a complete orthonormal set of functions of $\cos\Theta \in [-1, +1]$ (or $\Theta \in [0, \pi]$). Therefore, using the orthogonality relation (F.3.1), we obtain from Eqs. (11.11.1)–(11.11.6)

$$\alpha_1^s = (s+\tfrac{1}{2}) \int_0^\pi \mathrm{d}\Theta \sin\Theta \, a_1(\Theta) \, d_{00}^s(\Theta), \tag{11.11.19}$$

$$\alpha_2^s + \alpha_3^s = (s+\tfrac{1}{2}) \int_0^\pi \mathrm{d}\Theta \sin\Theta [a_2(\Theta) + a_3(\Theta)] \, d_{22}^s(\Theta), \tag{11.11.20}$$

$$\alpha_2^s - \alpha_3^s = (s+\tfrac{1}{2}) \int_0^\pi \mathrm{d}\Theta \sin\Theta [a_2(\Theta) - a_3(\Theta)] \, d_{2,-2}^s(\Theta), \tag{11.11.21}$$

$$\alpha_4^s = (s+\tfrac{1}{2}) \int_0^\pi \mathrm{d}\Theta \sin\Theta \, a_4(\Theta) \, d_{00}^s(\Theta), \tag{11.11.22}$$

$$\beta_1^s = -(s+\tfrac{1}{2}) \int_0^\pi \mathrm{d}\Theta \sin\Theta \, b_1(\Theta) \, d_{02}^s(\Theta), \tag{11.11.23}$$

$$\beta_2^s = -(s+\tfrac{1}{2}) \int_0^\pi \mathrm{d}\Theta \sin\Theta \, b_2(\Theta) \, d_{02}^s(\Theta) \tag{11.11.24}$$

(cf. Eq. (F.3.6)). These formulas suggest a simple, albeit not always the most elegant and efficient, way to compute the expansion coefficients by evaluating the integrals numerically using a suitable quadrature formula (de Rooij and van der Stap, 1984). Of course, this procedure assumes the knowledge of the scattering matrix elements at the quadrature division points.

Because the Wigner *d*-functions possess well-known and convenient mathematical properties and can be efficiently computed by using a simple and numerically stable recurrence relation, the expansions (11.11.1)–(11.11.6) offer substantial practical advantages. For example, if the expansion coefficients appearing in these expansions are known, then the elements of the normalized scattering matrix can be calculated easily for practically any number of scattering angles and with a minimal expenditure of computer time. Hence instead of tabulating the elements of the scattering matrix for a large number of scattering angles (cf. Deirmendjian, 1969) and resorting to interpolation in order to find the scattering matrix at intermediate points, one can provide a complete and accurate specification of the scattering matrix by tabulating a limited (and usually small) number of numerically significant expansion coefficients. This also explains why the expansion coefficients are especially convenient in averaging over particle states: instead of computing ensemble-averaged scattering matrix elements, one can average a (much) smaller number of expansion coefficients.

An additional advantage of expanding the scattering matrix elements in Wigner *d*-functions is that the latter obey an addition theorem, Eq. (F.7.8), and thereby provide an elegant analytical way of calculating the coefficients in a Fourier azimuthal decomposition of the normalized phase matrix (Kuščer and Ribarič, 1959; Domke, 1974; de Haan *et al.*, 1987). This Fourier decomposition is then used to handle the azimuthal dependence of the solution of the VRTE efficiently (Section 12.7).

Another important advantage offered by the expansions (11.11.1)–(11.11.6) is that using the (superposition) *T*-matrix method (Chapter 5 of MTL and Section 9.1), the expansion coefficients for certain types of nonspherical particles can be calculated analytically without computing the scattering matrix itself.

The expansion coefficients obey the general inequalities

$$|\alpha_i^s| \le 2s+1, \qquad i = 1,2,3,4, \tag{11.11.25}$$

$$|\beta_i^s| < \frac{2s+1}{\sqrt{2}}, \qquad i = 1,2. \tag{11.11.26}$$

These and other useful inequalities were derived by van der Mee and Hovenier (1990). Since, for each s, $d_{00}^s(\Theta)$ is also a Legendre polynomial $P_s(\cos\Theta)$, Eq. (11.11.1) is the well-known expansion of the phase function in Legendre polynomials (Chandrasekhar, 1950; Sobolev, 1975; van de Hulst, 1980). Equation (F.1.15) implies that $d_{00}^0(\Theta) \equiv 1$. Therefore, Eq. (11.11.19) and the normalization condition (11.10.3) yield the identity

$$\alpha_1^0 \equiv 1, \tag{11.11.27}$$

while the orthogonality property of the d-functions and Eq. (11.10.4) result in the relation

$$\langle\cos\Theta\rangle = \tfrac{1}{3}\alpha_1^1. \tag{11.11.28}$$

To illustrate the dependence of the expansion coefficients α_i^s and β_i^s on particle physical characteristics, Fig. 11.11.1 depicts them as a function of s for two polydisperse models of spherical particles, each described by a gamma distribution of particle radii given by Eq. (5.3.15). For both models, the relative refractive index is $m = 1.5$ and the effective variance is $v_{\rm eff} = 0.2$. The effective size parameter $x_{\rm eff} = k_1 r_{\rm eff}$ is equal to 5 for the first model and to 30 for the second model. Figure 11.11.2 visualizes the four independent elements of the normalized Lorenz–Mie scattering matrix for both models. The computations have been performed using the Lorenz–Mie code described in Section 5.10 of MTL.

Figure 11.11.1 reveals the typical behavior of the expansion coefficients α_i^s with increasing index s: they first grow in magnitude and then decay to absolute values below a reasonable numerical threshold. The greater the size of particles relative to the wavelength, the larger the maximum absolute value of the expansion coefficients and the slower their decay. This trend is largely explained by the rapid growth of the height of the forward-scattering peak in the elements $a_1(\Theta)$ and $a_3(\Theta)$ with increasing size parameter (see Fig. 11.11.2 and Eqs. (11.11.13)–(11.11.15)). The $|\beta_i^s|$ remain significantly smaller than the $|\alpha_i^s|$ and exhibit more pronounced oscillations. The former trait is obviously explained by the fact that the elements $b_1(\Theta)$ and $b_2(\Theta)$ vanish at $\Theta = 0$ rather than having a strong peak typical of the elements $a_1(\Theta)$ and $a_3(\Theta)$.

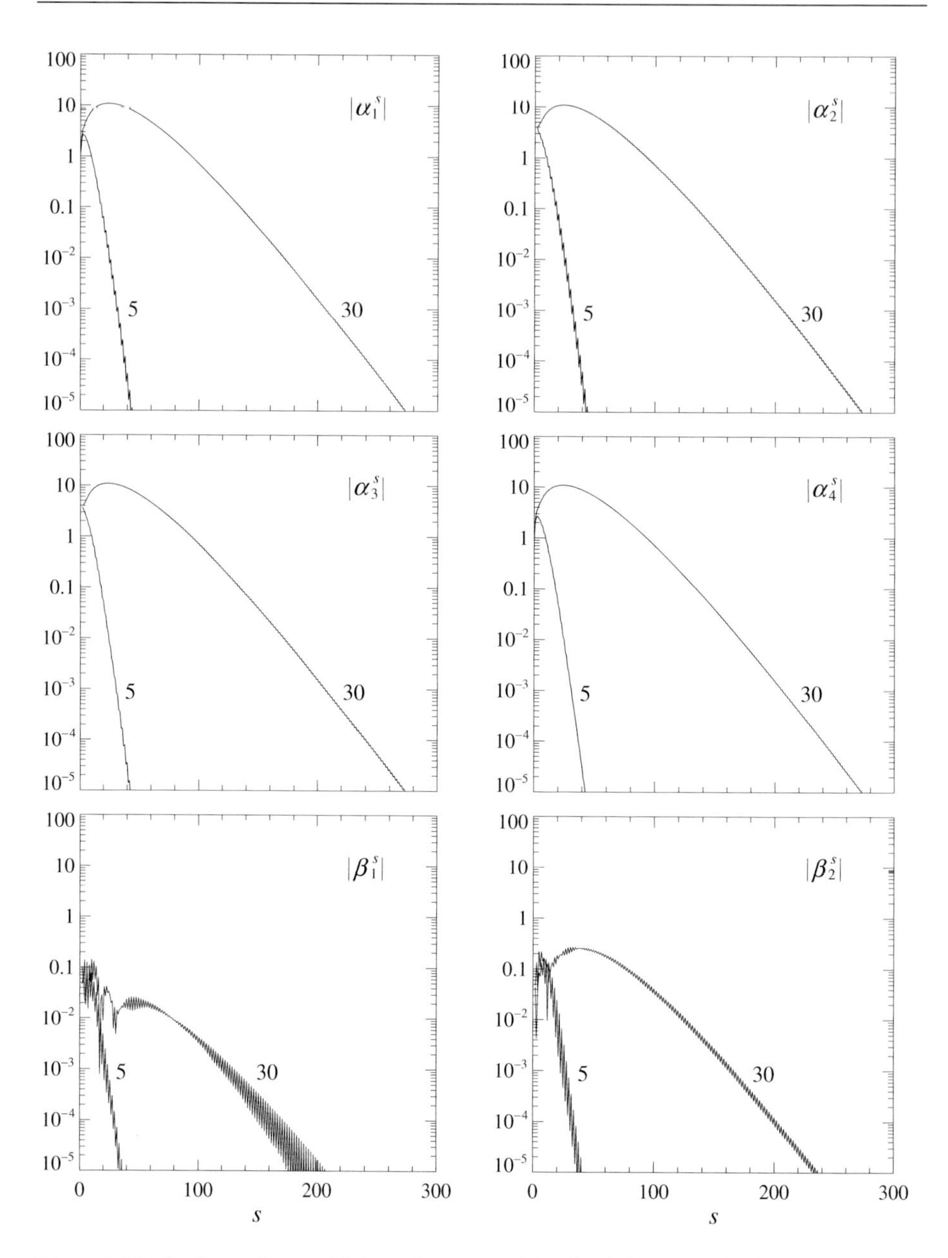

Figure 11.11.1. Expansion coefficients for two models of polydisperse spherical particles with effective size parameters $x_{\text{eff}} = 5$ and 30 (see text).

11.12 Circular-polarization representation

Equations (11.11.1)–(11.11.6) become more compact and their origin becomes more transparent if one uses the circular-polarization representation of the Stokes vector

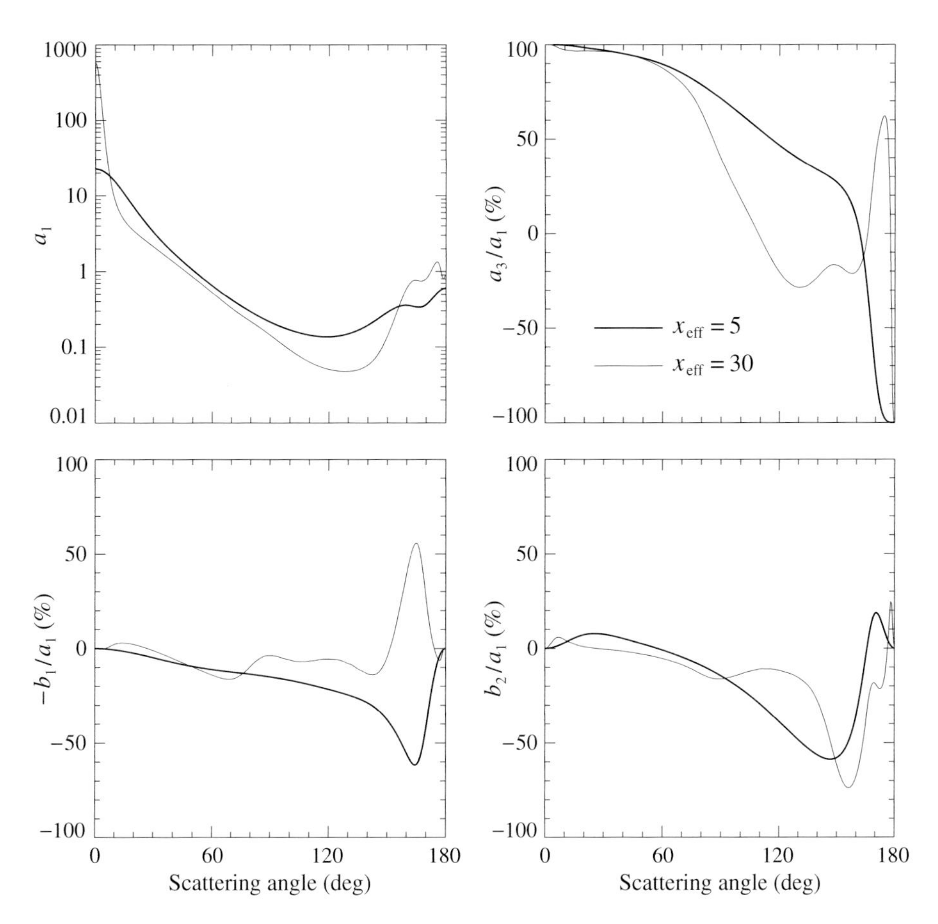

Figure 11.11.2. Elements of the normalized Stokes scattering matrix for two models of polydisperse spherical particles (see text).

(Kuščer and Ribarič, 1959; Domke, 1974; Hovenier and van der Mee, 1983). We begin by defining the circular components of a transverse electromagnetic wave as

$$\begin{bmatrix} E_+ \\ E_- \end{bmatrix} = \mathbf{q} \begin{bmatrix} E_\theta \\ E_\varphi \end{bmatrix}, \tag{11.12.1}$$

where

$$\mathbf{q} = \frac{1}{\sqrt{2}} \begin{bmatrix} 1 & \mathrm{i} \\ 1 & -\mathrm{i} \end{bmatrix}. \tag{11.12.2}$$

Using Eqs. (3.3.6) and (11.12.2), we find that the corresponding circular-polarization amplitude scattering matrix $\mathbf{C}$ is expressed in terms of the regular amplitude scattering matrix as

$$\mathbf{C} = \begin{bmatrix} C_{++} & C_{+-} \\ C_{-+} & C_{--} \end{bmatrix}$$

$$= \mathbf{q}\mathbf{S}\mathbf{q}^{-1}$$

$$= \frac{1}{2}\begin{bmatrix} S_{11} - \mathrm{i}S_{12} + \mathrm{i}S_{21} + S_{22} & S_{11} + \mathrm{i}S_{12} + \mathrm{i}S_{21} - S_{22} \\ S_{11} - \mathrm{i}S_{12} - \mathrm{i}S_{21} - S_{22} & S_{11} + \mathrm{i}S_{12} - \mathrm{i}S_{21} + S_{22} \end{bmatrix}, \quad (11.12.3)$$

where the arguments $(\hat{\mathbf{n}}^{\text{sca}}, \hat{\mathbf{n}}^{\text{inc}})$ are omitted for brevity and

$$\mathbf{q}^{-1} = \frac{1}{\sqrt{2}}\begin{bmatrix} 1 & 1 \\ -\mathrm{i} & \mathrm{i} \end{bmatrix}.$$

The usefulness of the circular electric vector components becomes clear from the simple formulas

$$I_2 = E_- E_+^*, \quad (11.12.4\text{a})$$

$$I_0 = E_+ E_+^*, \quad (11.12.4\text{b})$$

$$I_{-0} = E_- E_-^*, \quad (11.12.4\text{c})$$

$$I_{-2} = E_+ E_-^*, \quad (11.12.4\text{d})$$

which follow, after some algebra, from

$$\begin{bmatrix} E_\theta \\ E_\varphi \end{bmatrix} = \mathbf{q}^{-1}\begin{bmatrix} E_+ \\ E_- \end{bmatrix}$$

and Eqs. (2.6.4) and (2.6.10). It is easy to verify using the first equality of Eq. (11.12.3) and Eqs. (11.12.4) that the circular-polarization phase matrix is given by

$$\mathbf{Z}^{\text{CP}} = \left\| Z_{pq}^{\text{CP}} \right\| = \begin{bmatrix} C_{--}C_{++}^* & C_{-+}C_{++}^* & C_{--}C_{+-}^* & C_{-+}C_{+-}^* \\ C_{+-}C_{++}^* & C_{++}C_{++}^* & C_{+-}C_{+-}^* & C_{++}C_{+-}^* \\ C_{--}C_{-+}^* & C_{-+}C_{-+}^* & C_{--}C_{--}^* & C_{-+}C_{--}^* \\ C_{+-}C_{-+}^* & C_{++}C_{-+}^* & C_{+-}C_{--}^* & C_{++}C_{--}^* \end{bmatrix},$$

$$p, q = 2, 0, -0, -2. \quad (11.12.5)$$

Alternatively, it can be found from Eq. (3.7.28).

Consider now scattering by an ISM. The normalized scattering and phase matrices in the circular-polarization representation are defined by analogy with the corresponding Stokes matrices:

$$\widetilde{\mathbf{F}}^{\text{CP}}(\Theta) = \frac{4\pi}{\langle C_{\text{sca}} \rangle_\xi} \langle \mathbf{Z}^{\text{CP}}(\theta^{\text{sca}} = \Theta, \varphi^{\text{sca}} = 0; \theta^{\text{inc}} = 0, \varphi^{\text{inc}} = 0) \rangle_\xi, \quad (11.12.6)$$

$$\widetilde{\mathbf{Z}}^{\text{CP}}(\theta^{\text{sca}}, \varphi^{\text{sca}}; \theta^{\text{inc}}, \varphi^{\text{inc}}) = \frac{4\pi}{\langle C_{\text{sca}} \rangle_\xi} \langle \mathbf{Z}^{\text{CP}}(\theta^{\text{sca}}, \varphi^{\text{sca}}; \theta^{\text{inc}}, \varphi^{\text{inc}}) \rangle_\xi, \quad (11.12.7)$$

where $\langle \mathbf{Z}^{\mathrm{CP}}(\theta^{\mathrm{sca}}, \varphi^{\mathrm{sca}}; \theta^{\mathrm{inc}}, \varphi^{\mathrm{inc}})\rangle_\xi$ is the average circular-polarization phase matrix. From Eqs. (3.7.28), (11.10.1), (2.6.12), and (2.6.16) we have

$$\widetilde{\mathbf{F}}^{\mathrm{CP}} = \left\| \widetilde{F}^{\mathrm{CP}}_{pq} \right\| = \frac{1}{2}\begin{bmatrix} a_2 + a_3 & b_1 + \mathrm{i}b_2 & b_1 - \mathrm{i}b_2 & a_2 - a_3 \\ b_1 + \mathrm{i}b_2 & a_1 + a_4 & a_1 - a_4 & b_1 - \mathrm{i}b_2 \\ b_1 - \mathrm{i}b_2 & a_1 - a_4 & a_1 + a_4 & b_1 + \mathrm{i}b_2 \\ a_2 - a_3 & b_1 - \mathrm{i}b_2 & b_1 + \mathrm{i}b_2 & a_2 + a_3 \end{bmatrix},$$

$$p, q = 2, 0, -0, -2 \qquad (11.12.8)$$

(the argument Θ is omitted for the sake of brevity). Obviously, this matrix has several symmetry properties:

$$\widetilde{F}^{\mathrm{CP}}_{pq}(\Theta) = \widetilde{F}^{\mathrm{CP}}_{qp}(\Theta) = \widetilde{F}^{\mathrm{CP}}_{-p,-q}(\Theta), \qquad (11.12.9)$$

$$\widetilde{F}^{\mathrm{CP}}_{pp}(\Theta),\ \widetilde{F}^{\mathrm{CP}}_{p,-p}(\Theta) \text{ are real}, \qquad (11.12.10)$$

$$\widetilde{F}^{\mathrm{CP}}_{20}(\Theta) = [\widetilde{F}^{\mathrm{CP}}_{2,-0}(\Theta)]^*. \qquad (11.12.11)$$

An elegant and compact way to expand the elements $\widetilde{F}^{\mathrm{CP}}_{pq}$ is to use generalized spherical functions P^s_{pq}:

$$\widetilde{F}^{\mathrm{CP}}_{pq}(\Theta) = \sum_{s=0}^{s_{\max}} g^s_{pq} P^s_{pq}(\cos\Theta), \qquad p, q = 2, 0, -0, -2, \qquad (11.12.12)$$

which indicates the rationale for the specific choice of values for the p, q indices for the circular-polarization phase matrix and the corresponding Stokes vector component subscripts (cf. Eqs. (11.12.4)). Another justification for this choice of expansion functions comes from the consideration of certain properties of the rotation group (Domke, 1974). The expression for the expansion coefficients g^s_{pq} follows from Eqs. (11.12.12) and (F.6.8):

$$g^s_{pq} = (s + \tfrac{1}{2}) \int_{-1}^{+1} \mathrm{d}(\cos\Theta)\, \widetilde{F}^{\mathrm{CP}}_{pq}(\Theta)\, P^s_{pq}(\cos\Theta), \qquad p, q = 2, 0, -0, -2. \qquad (11.12.13)$$

Note that for $P^s_{pq}(\cos\Theta)$ no distinction is made between $p, q = 0$ and $p, q = -0$. For the values of p and q used here, all functions $P^s_{pq}(\cos\Theta)$ are real-valued (see Eq. (F.6.1)). Since $P^s_{pq}(\cos\Theta) \equiv 0$ for $s < \max(|p|, |q|)$, the corresponding expansion coefficients g^s_{pq} in Eq. (11.12.12) are not defined. However, it is convenient to complete the definition by equating them to zero, which is consistent with Eq. (11.11.12).

Using Eqs. (11.12.9)–(11.12.11), (11.12.13), and (F.6.2), we derive the following symmetry relations:

$$g^s_{pq} = g^s_{qp} = g^s_{-p,-q}, \qquad (11.12.14)$$

$$g^s_{pp},\ g^s_{p,-p} \text{ are real}, \qquad (11.12.15)$$

$$g_{20}^{s} = (g_{2,-0}^{s})^{*}. \tag{11.12.16}$$

Inserting Eq. (11.12.12) into Eq. (11.12.8) yields the expansions (11.11.1)–(11.11.6) with expansion coefficients

$$\alpha_1^s = g_{00}^s + g_{0,-0}^s, \tag{11.12.17}$$

$$\alpha_2^s = g_{22}^s + g_{2,-2}^s, \tag{11.12.18}$$

$$\alpha_3^s = g_{22}^s - g_{2,-2}^s, \tag{11.12.19}$$

$$\alpha_4^s = g_{00}^s - g_{0,-0}^s, \tag{11.12.20}$$

$$\beta_1^s = 2\,\mathrm{Re}\,g_{02}^s, \tag{11.12.21}$$

$$\beta_2^s = 2\,\mathrm{Im}\,g_{02}^s. \tag{11.12.22}$$

By analogy with Eq. (11.3.1) and using Eqs. (2.8.8) and (11.12.8), we find for $0 < \varphi^{\mathrm{sca}} - \varphi^{\mathrm{inc}} < \pi$:

$$\widetilde{\mathbf{Z}}^{\mathrm{CP}}(\theta^{\mathrm{sca}}, \varphi^{\mathrm{sca}}; \theta^{\mathrm{inc}}, \varphi^{\mathrm{inc}}) = \mathbf{L}^{\mathrm{CP}}(-\sigma_2)\widetilde{\mathbf{F}}^{\mathrm{CP}}(\Theta)\mathbf{L}^{\mathrm{CP}}(\pi - \sigma_1)$$

$$= \frac{1}{2}\begin{bmatrix} (a_2 + a_3)\mathrm{e}^{-\mathrm{i}2(\sigma_1+\sigma_2)} & (b_1 + \mathrm{i}b_2)\mathrm{e}^{-\mathrm{i}2\sigma_2} & (b_1 - \mathrm{i}b_2)\mathrm{e}^{-\mathrm{i}2\sigma_2} & (a_2 - a_3)\mathrm{e}^{\mathrm{i}2(\sigma_1-\sigma_2)} \\ (b_1 + \mathrm{i}b_2)\mathrm{e}^{-\mathrm{i}2\sigma_1} & a_1 + a_4 & a_1 - a_4 & (b_1 - \mathrm{i}b_2)\mathrm{e}^{\mathrm{i}2\sigma_1} \\ (b_1 - \mathrm{i}b_2)\mathrm{e}^{-\mathrm{i}2\sigma_1} & a_1 - a_4 & a_1 + a_4 & (b_1 + \mathrm{i}b_2)\mathrm{e}^{\mathrm{i}2\sigma_1} \\ (a_2 - a_3)\mathrm{e}^{\mathrm{i}2(\sigma_2-\sigma_1)} & (b_1 - \mathrm{i}b_2)\mathrm{e}^{\mathrm{i}2\sigma_2} & (b_1 + \mathrm{i}b_2)\mathrm{e}^{\mathrm{i}2\sigma_2} & (a_2 + a_3)\mathrm{e}^{\mathrm{i}2(\sigma_1+\sigma_2)} \end{bmatrix}, \tag{11.12.23}$$

where we have omitted the argument Θ in the as and bs. As is the case with the normalized Stokes phase matrix, the normalized circular-polarization phase matrix depends on the difference between the azimuth angles of the scattering and incidence directions rather than on their specific values. Applying the transformation rule (3.7.28) to Eqs. (11.10.19)–(11.10.22) yields, after some algebra, the following symmetry relations:

$$\widetilde{\mathbf{Z}}^{\mathrm{CP}}(\theta^{\mathrm{sca}}, -\varphi^{\mathrm{inc}}; \theta^{\mathrm{inc}}, -\varphi^{\mathrm{sca}}) = \widetilde{\mathbf{Z}}^{\mathrm{CP}}(\theta^{\mathrm{sca}}, \varphi^{\mathrm{sca}}; \theta^{\mathrm{inc}}, \varphi^{\mathrm{inc}}), \tag{11.12.24}$$

$$\begin{aligned} \widetilde{\mathbf{Z}}^{\mathrm{CP}}&(\theta^{\mathrm{sca}}, -\varphi^{\mathrm{sca}}; \theta^{\mathrm{inc}}, -\varphi^{\mathrm{inc}}) \\ &= \mathbf{A}\boldsymbol{\Delta}_{34}\mathbf{A}^{-1}\widetilde{\mathbf{Z}}^{\mathrm{CP}}(\theta^{\mathrm{sca}}, \varphi^{\mathrm{sca}}; \theta^{\mathrm{inc}}, \varphi^{\mathrm{inc}})\mathbf{A}\boldsymbol{\Delta}_{34}\mathbf{A}^{-1} \\ &= \boldsymbol{\Delta}^{\mathrm{CP}}\,\widetilde{\mathbf{Z}}^{\mathrm{CP}}(\theta^{\mathrm{sca}}, \varphi^{\mathrm{sca}}; \theta^{\mathrm{inc}}, \varphi^{\mathrm{inc}})\boldsymbol{\Delta}^{\mathrm{CP}}, \end{aligned} \tag{11.12.25}$$

$$\widetilde{\mathbf{Z}}^{\mathrm{CP}}(\pi - \theta^{\mathrm{sca}}, \varphi^{\mathrm{sca}}; \pi - \theta^{\mathrm{inc}}, \varphi^{\mathrm{inc}}) = \boldsymbol{\Delta}^{\mathrm{CP}}\,\widetilde{\mathbf{Z}}^{\mathrm{CP}}(\theta^{\mathrm{sca}}, \varphi^{\mathrm{sca}}; \theta^{\mathrm{inc}}, \varphi^{\mathrm{inc}})\boldsymbol{\Delta}^{\mathrm{CP}}, \tag{11.12.26}$$

$$\widetilde{\mathbf{Z}}^{\mathrm{CP}}(\pi - \theta^{\mathrm{inc}}, \varphi^{\mathrm{inc}}; \pi - \theta^{\mathrm{sca}}, \varphi^{\mathrm{sca}}) = [\widetilde{\mathbf{Z}}^{\mathrm{CP}}(\theta^{\mathrm{sca}}, \varphi^{\mathrm{sca}}; \theta^{\mathrm{inc}}, \varphi^{\mathrm{inc}})]^{\mathrm{T}}, \tag{11.12.27}$$

where

$$\mathbf{\Delta}^{\mathrm{CP}} = \begin{bmatrix} 0 & 0 & 0 & 1 \\ 0 & 0 & 1 & 0 \\ 0 & 1 & 0 & 0 \\ 1 & 0 & 0 & 0 \end{bmatrix}. \tag{11.12.28}$$

Equations (11.12.25) and (11.12.26) can also be written as follows:

$$\widetilde{Z}^{\mathrm{CP}}_{pq}(\theta^{\mathrm{sca}}, -\varphi^{\mathrm{sca}}; \theta^{\mathrm{inc}}, -\varphi^{\mathrm{inc}}) = \widetilde{Z}^{\mathrm{CP}}_{-p,-q}(\theta^{\mathrm{sca}}, \varphi^{\mathrm{sca}}; \theta^{\mathrm{inc}}, \varphi^{\mathrm{inc}}), \tag{11.12.29}$$

$$\widetilde{Z}^{\mathrm{CP}}_{pq}(\pi - \theta^{\mathrm{sca}}, \varphi^{\mathrm{sca}}; \pi - \theta^{\mathrm{inc}}, \varphi^{\mathrm{inc}}) = \widetilde{Z}^{\mathrm{CP}}_{-p,-q}(\theta^{\mathrm{sca}}, \varphi^{\mathrm{sca}}; \theta^{\mathrm{inc}}, \varphi^{\mathrm{inc}}). \tag{11.12.30}$$

11.13 Illustrative examples

Mishchenko *et al.* (2000a) and MTL provide a detailed discussion of extinction, scattering, and absorption properties of particles having diverse morphologies and compositions and encountered in various environments. Therefore, the limited purpose of the several illustrative examples given below is to highlight the most typical traits of the single-scattering patterns generated by small particles.

The bottom curve in Fig. 11.13.1 shows the extinction efficiency factor defined by Eq. (11.6.6) versus size parameter x for monodisperse spheres with a relative refractive index $m = 1.5$. The curve exhibits a succession of major low-frequency maxima and minima with superimposed high-frequency ripple consisting of sharp, irregularly spaced extrema some of which are super-narrow spike-like features. The major maxima and minima are called the "interference structure" since, as traditionally explained, they are the result of interference of light diffracted and transmitted by the particle. Unlike the interference structure, the ripple is caused by the resonance behavior of coefficients a_n and b_n appearing in the formulas of the Lorenz–Mie theory. The interference structure and ripple are typical attributes of all scattering characteristics of nonabsorbing monodisperse spheres.

The ripple structure rapidly weakens and then vanishes with increasing absorption, as the other curves in Fig. 11.13.1 demonstrate. Increasing m_{I} beyond 0.001 starts to affect and eventually eradicates the interference structure as well. However, the first interference maximum at $x \approx 4$ survives, albeit becomes significantly less pronounced, even at $m_{\mathrm{I}} = 0.1$.

A very similar smoothing effect on the interference and ripple structure is caused by particle polydispersity. Indeed, as Fig. 11.13.2 illustrates, increasing the width of the size distribution (see Fig. 11.13.3) first extinguishes the ripple and then eliminates the interference structure in Q_{ext}. It is interesting that as narrow a dispersion of sizes as that corresponding to $v_{\mathrm{eff}} = 0.01$ completely washes the ripple structure out. The first major maximum of the interference structure persists to much larger values of v_{eff}, but eventually fades away too.

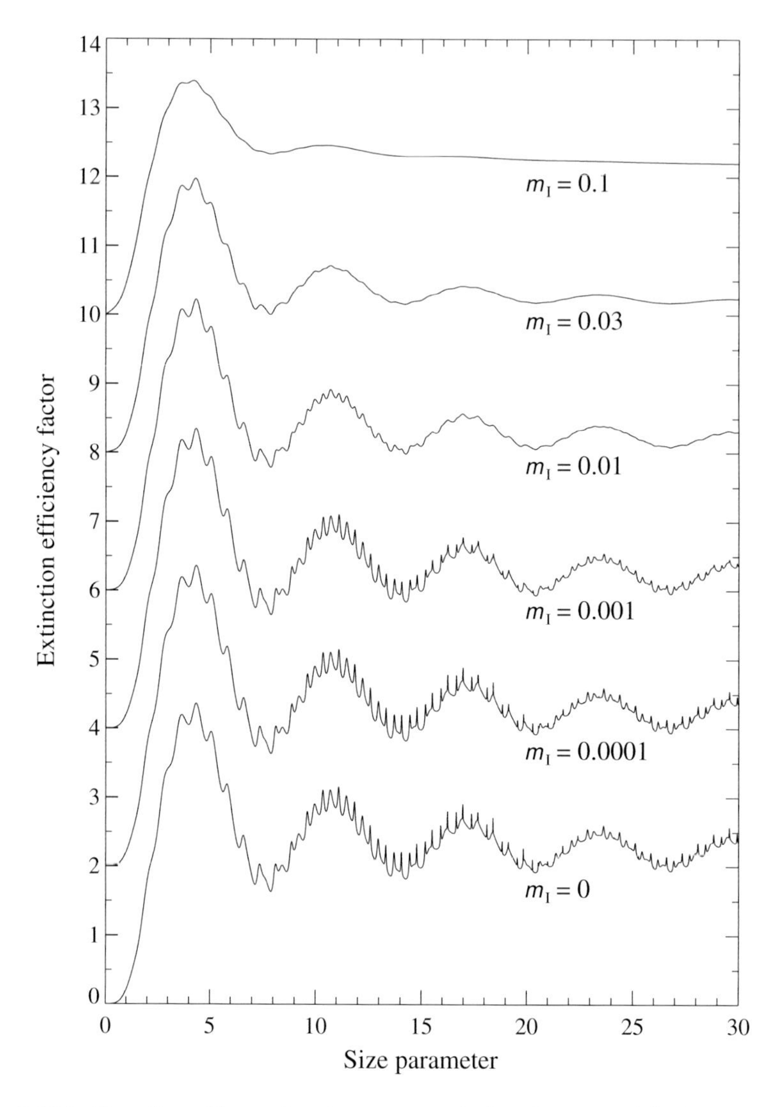

Figure 11.13.1. The effect of increasing absorption on the interference and ripple structure of the extinction efficiency factor for monodisperse spherical particles with the real part of the relative refractive index $m_R = 1.5$. The vertical axis scale applies to the curve with $m_I = 0$, the other curves being successively displaced upward by 2.

Plate 11.13.1 parallels Fig. 11.13.2 and shows the effect of increasing width of the size distribution on the degree of linear polarization of scattered light for unpolarized incident light,

$$P_Q(\Theta) = -\frac{Q^{\text{sca}}(\Theta)}{I^{\text{sca}}(\Theta)} = -\frac{b_1(\Theta)}{a_1(\Theta)}.$$

The case $v_{\text{eff}} = 0.01$ demonstrates that even a very narrow size distribution is suffi-

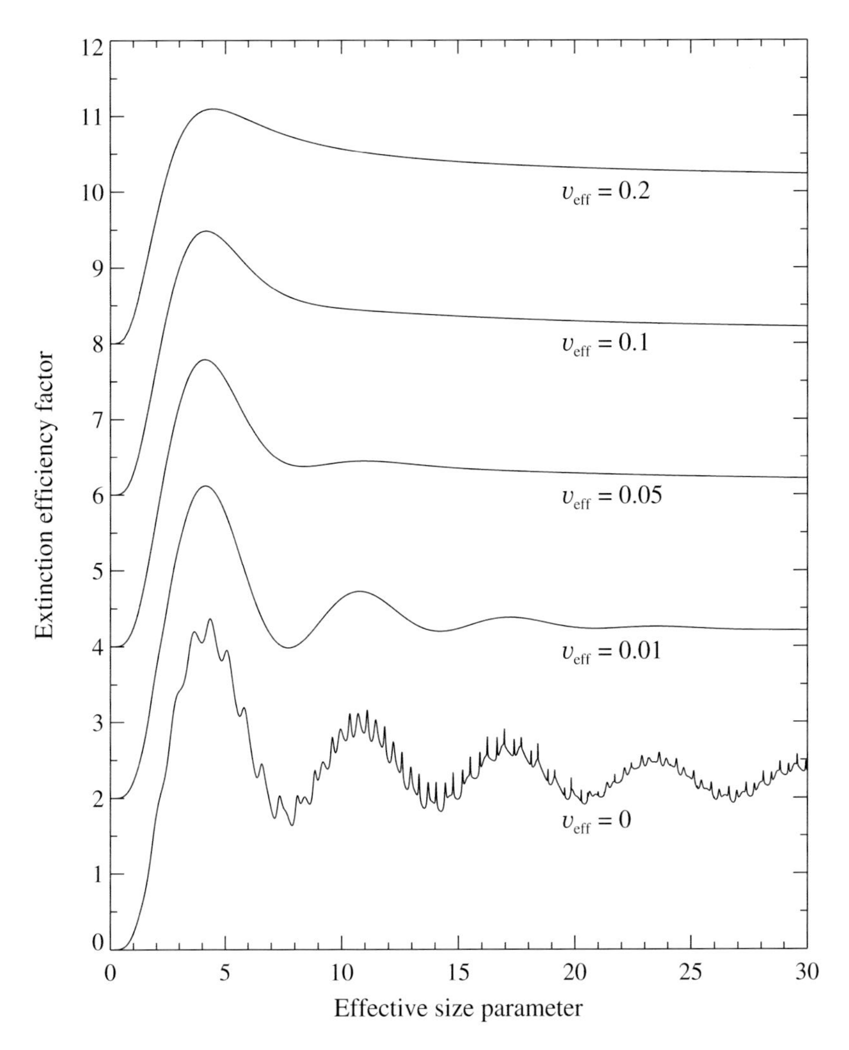

Figure 11.13.2. The effect of increasing width of the size distribution on the interference and ripple structure in Q_{ext} for nonabsorbing spherical particles with the relative refractive index 1.5 and effective size parameters $x_{eff} = k_1 r_{eff}$ ranging from 0 to 30. The particle size distributions used in these computations are depicted in Fig. 11.13.3. The vertical axis scale applies to the curve with $v_{eff} = 0$, the other curves being successively displaced upward by 2.

cient to extinguish most of the interference and resonance effects. With increasing effective variance, the maxima are smoothed out, the minima are filled in, and the polarization becomes more neutral. All these effects of broadening the size distribution are easy to understand qualitatively in terms of taking weighted averages along vertical lines of increasing length in the polarization diagram for monodisperse particles.

Figure 11.13.4 shows the extinction efficiency factor, the single-scattering albedo, and the asymmetry parameter versus effective size parameter $x_{eff} = k_1 r_{eff}$ for four

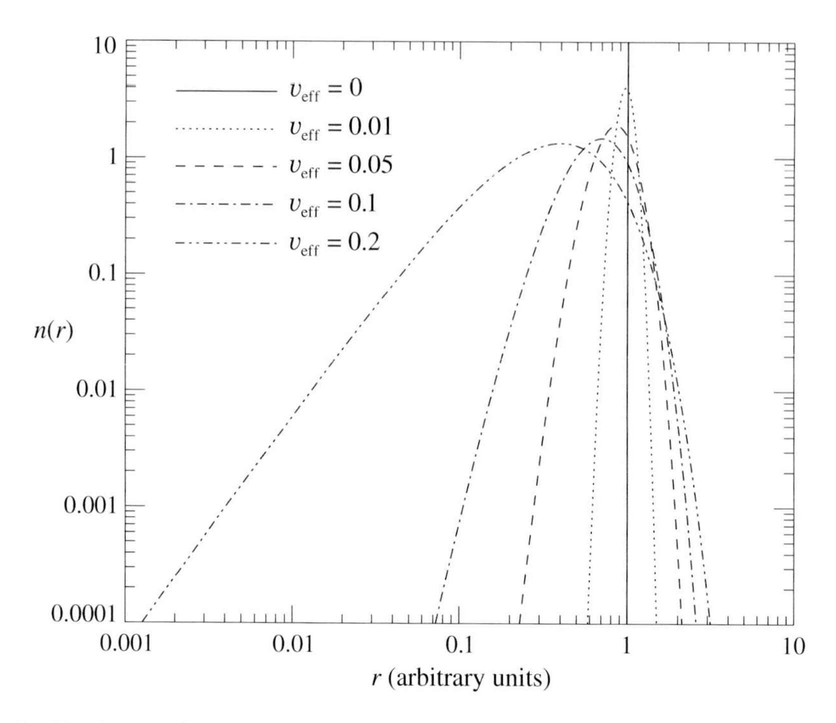

Figure 11.13.3. Gamma size distributions, Eq. (5.3.15), with $r_{\mathrm{min}} = 0$, $r_{\mathrm{max}} = \infty$, $r_{\mathrm{eff}} = 1$ (in arbitrary units of length) and $v_{\mathrm{eff}} = 0, 0.01, 0.05, 0.1$, and 0.2. The size distributions are normalized according to Eq. (5.3.8). The value $v_{\mathrm{eff}} = 0$ corresponds to monodisperse particles.

models of polydisperse spherical particles characterized by a moderately wide size distribution. It is seen that for nonabsorbing wavelength-sized particles ($x_{\mathrm{eff}} \sim 5$), the extinction cross section can exceed the particle geometrical cross section by more than a factor of 3. As the particle size becomes much larger, Q_{ext} tends to the asymptotic geometrical-optics value 2, with equal contributions from the rays striking the particle and the light diffracted by the particle projection (Section 9.2). For nonabsorbing particles much smaller than the wavelength,

$$Q_{\mathrm{ext}} = Q_{\mathrm{sca}} \underset{x \to 0}{\propto} \frac{1}{\lambda_1^4}, \tag{11.13.1}$$

as first demonstrated by Lord Rayleigh and hence called Rayleigh scattering. For absorbing particles, extinction in the Rayleigh limit is dominated by absorption and varies as

$$Q_{\mathrm{ext}} \approx Q_{\mathrm{abs}} \underset{x \to 0}{\propto} \frac{1}{\lambda_1}. \tag{11.13.2}$$

The single-scattering albedo is identically equal to unity for the nonabsorbing particles but is significantly smaller than unity for the absorbing spheres and vanishes in the Rayleigh limit in accordance with Eqs. (11.13.1) and (11.13.2). The $\langle \cos\Theta \rangle$ rap-

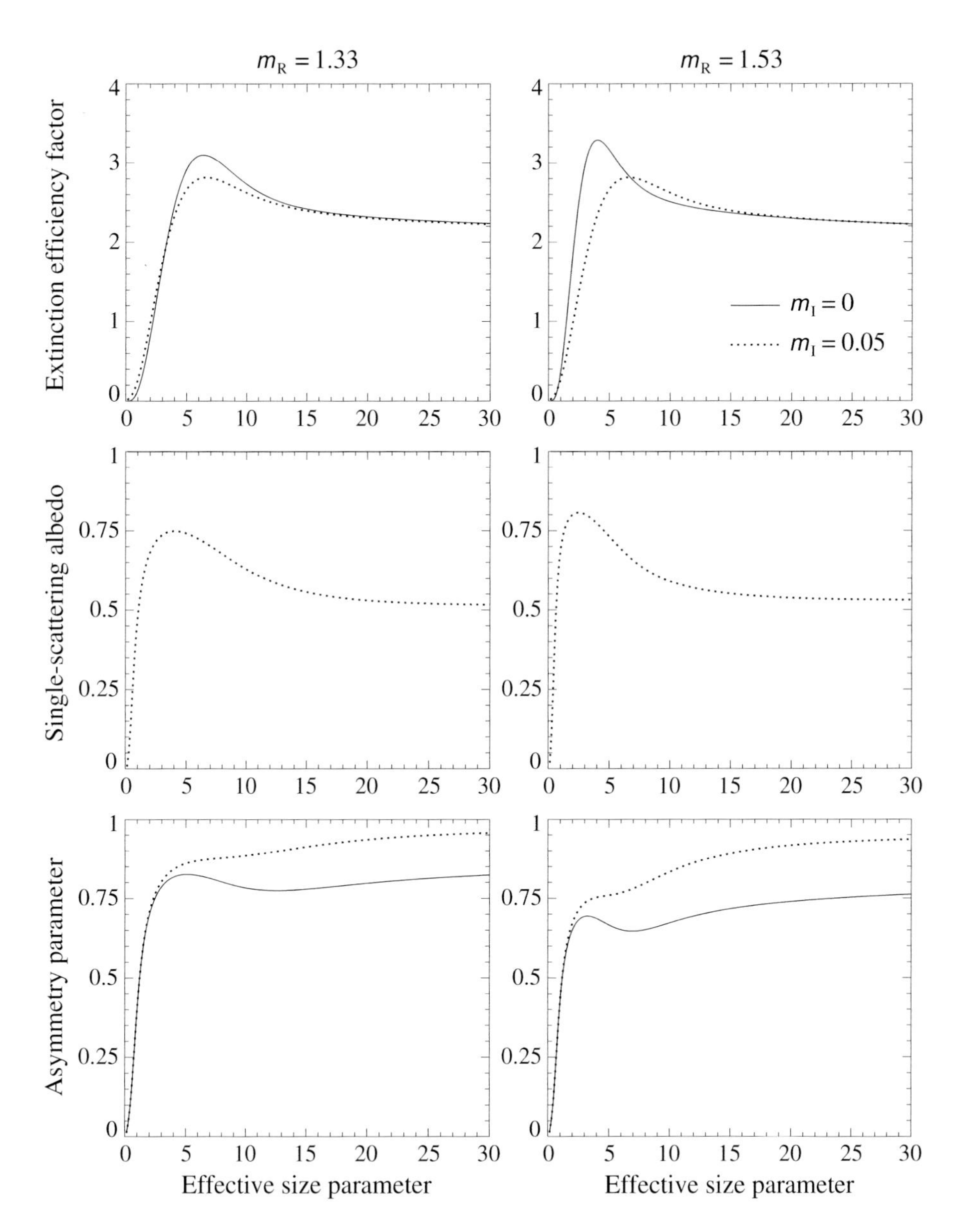

Figure 11.13.4. Q_{ext}, ϖ, and $\langle \cos\Theta \rangle$ versus x_{eff} for a gamma distribution of spherical particles with $v_{eff} = 0.15$, $m_R = 1.33$ and 1.53, and $m_I = 0$ and 0.05. Note that $\varpi \equiv 1$ for nonabsorbing particles with $m_I = 0$.

idly grows from zero to values exceeding 0.5 as x_{eff} increases from 0 to about 2. Then it remains positive, thereby indicating forward-scattering particles, and shows relatively little dependence on the particle size parameter.

The phase function in the Rayleigh limit (the upper left panel in Fig. 11.13.5) is nearly isotropic and is symmetric with respect to the scattering angle $\Theta = 90°$, thereby

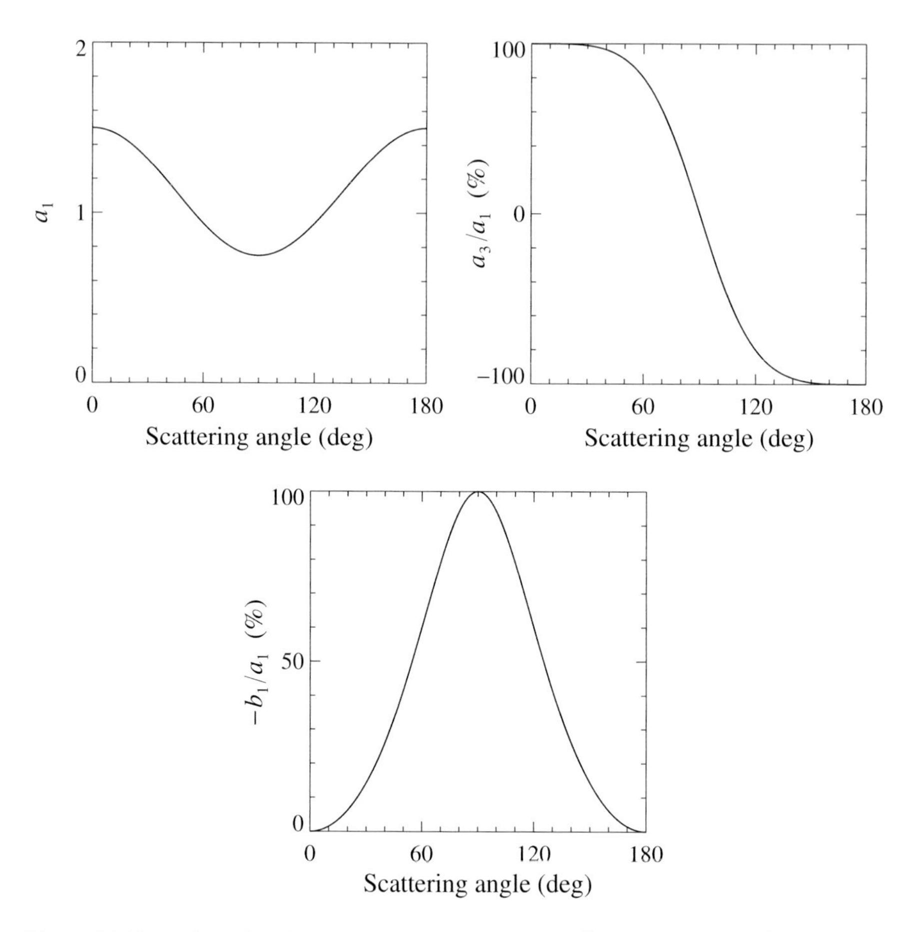

Figure 11.13.5. Phase function $a_1(\Theta)$ and the ratios $a_3(\Theta)/a_1(\Theta)$ and $-b_1(\Theta)/a_1(\Theta)$ versus scattering angle Θ for a spherically symmetric particle in the Rayleigh limit.

causing the asymmetry parameter to vanish:

$$\langle\cos\Theta\rangle \underset{x\to 0}{=} 0. \tag{11.13.3}$$

The ratio $-b_1(\Theta)/a_1(\Theta)$ (the lower panel in Fig. 11.13.5), is always positive, has the classical bell-like shape, and reaches 100% at the scattering angle $\Theta = 90°$.

The elements of the normalized Stokes scattering matrix exhibit significant variability in the intermediate (so-called resonance) region of size parameters ($1 \lesssim x_{\text{eff}} \lesssim 100$), but eventually start to develop features that can be explained through the concepts of geometrical optics applicable to particles much larger than the wavelength. Specifically, the concentration of light near $\Theta = 0°$ (Fig. 11.13.6) is caused by the diffraction of light on the particle projection. The diffraction peak rapidly grows in magnitude and becomes much narrower with increasing size parameter. The external reflection (see the ray-tracing diagram, Fig. 9.2.1) does not generate any distinctive feature, whereas the rays refracted twice cause a broad enhancement of the phase

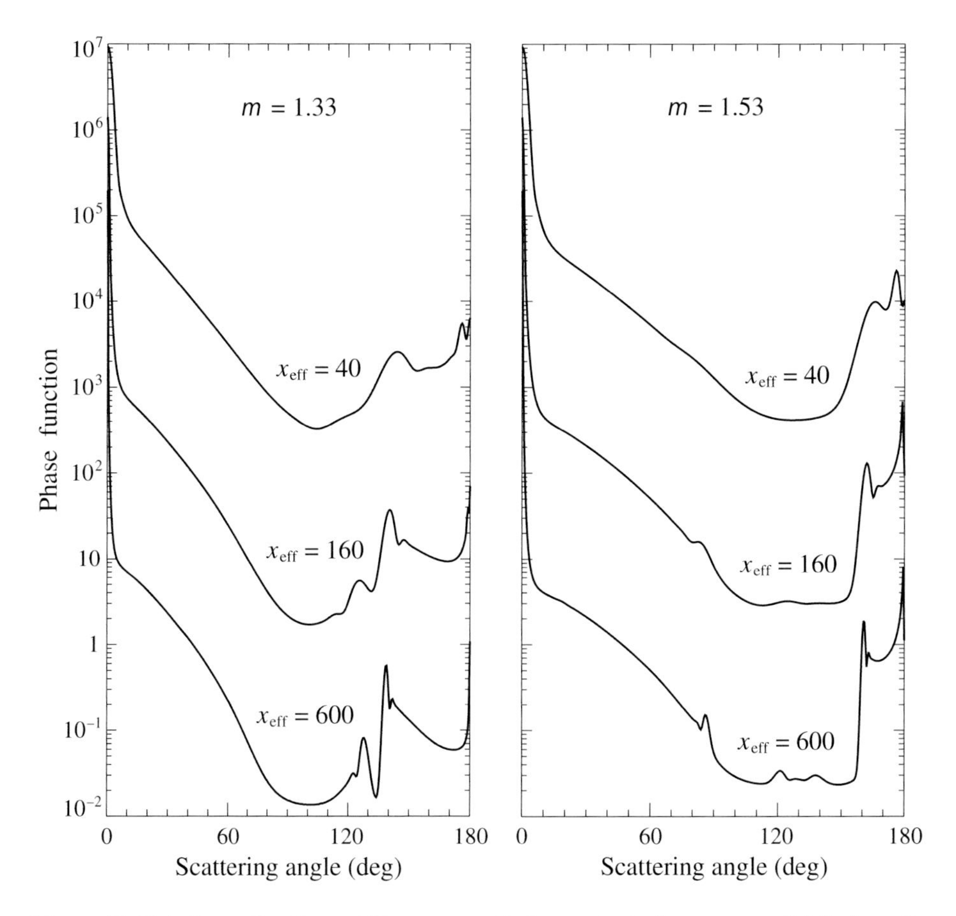

Figure 11.13.6. Phase function $a_1(\Theta)$ versus scattering angle Θ for a gamma distribution of homogeneous spheres with $v_{\text{eff}} = 0.07$, $x_{\text{eff}} = 40$, 160, and 600, and relative refractive indices $m_{\text{R}} = 1.33$ and 1.53. The vertical axis scale applies to the curves with $x_{\text{eff}} = 600$, the other curves being successively displaced upward by a factor of 100.

functions in the forward-scattering hemisphere.

The features at $\Theta \approx 137°$ and 130° for $m = 1.33$ (160° and 88° for $m = 1.53$) are the primary and secondary rainbows generated by the rays undergoing one and two internal reflections, respectively. The low-intensity zone (about 7° wide for m = 1.33 and 72° wide for m = 1.53) between the primary and secondary rainbows is called the Alexander dark band. Here the phase functions are mostly determined by the weak contribution from the externally reflected rays. The slight change of the rainbow angle with wavelength caused by dispersion (change of the relative refractive index with wavelength) gives rise to spectacular colorful rainbows often observed during showers illuminated by the sun at an altitude lower than about 40°.

The enhancement of intensity in the backscattering direction ($\Theta \approx 180°$) is called the glory and can be seen from an airplane as a series of colored rings around the shadow cast by the airplane on the cloud top. An obvious, but relatively insignificant contributor to the glory are central rays externally and internally reflected in the backscattering direction. Snell's law predicts that for real relative refractive indices in the range $2^{1/2} \le m \le 2$, a noncentral incident ray may emerge at $\Theta = 180°$ after just one internal reflection. However, this mechanism does not explain the pronounced glory generated by water droplets with $m = 1.33$ ($< 2^{1/2}$) and $x_{\rm eff} = 600$. The physical origin of the glory remains the subject of active research. In a recent paper called *Does the glory have a simple explanation?* Nussenzveig (2002) concludes that the glory is produced by near-peripheral incident light, optical resonances and van de Hulst's surface waves being the main contributors.

The ratio $-b_1(\Theta)/a_1(\Theta)$ (Fig. 11.13.7) is small at small scattering angles because of the predominance of unpolarized diffracted light (for unpolarized incident light). Most of the light scattered into the forward hemisphere is due to twice refracted rays and is negatively polarized, as follows from Fresnel's formulas. Externally reflected rays are strongly positively polarized at all scattering angles and cause the broad positive polarization at $\Theta \gtrsim 80°$, including the Alexander dark band. The primary and secondary rainbows cause pronounced peaks of positive polarization. It is interesting that the secondary rainbow remains visible in polarization to smaller size parameters than in intensity.

The dependence of all scattering and absorption characteristics on particle microphysical properties can become much more complex if particles are nonspherical and are partially or perfectly aligned. This is especially true of the interference structure and ripple, which now strongly depend on the particle orientation with respect to the incidence and scattering directions and on polarization of the incident light. However, averaging over orientations reinforces the effect of averaging over sizes and eradicates many resonance features, thereby making scattering patterns for randomly oriented, polydisperse nonspherical particles even smoother than those for surface- or volume-equivalent polydisperse spheres. In fact, it is not always easy to distinguish spherical and randomly oriented nonspherical particles based on qualitative differences in their scattering patterns (Section 11.9).

However, there can be significant quantitative differences in specific scattering patterns. As an example, Fig. 11.13.8 contrasts the elements of the normalized Stokes scattering matrix for polydisperse spheres and surface-equivalent, randomly oriented spheroids with a relative refractive index 1.53 + i0.008. The left-most top diagram of this figure shows the corresponding phase functions and reveals the following five distinct scattering-angle ranges:

$$\begin{array}{ll} \text{nonsphere} \approx \text{sphere} & \text{from } \Theta = 0° \text{ to } \Theta \sim 15°-20°; \\ \text{nonsphere} > \text{sphere} & \text{from } \Theta \sim 15°-20° \text{ to } \Theta \sim 35°; \\ \text{nonsphere} < \text{sphere} & \text{from } \Theta \sim 35° \text{ to } \Theta \sim 85°; \\ \text{nonsphere} \gg \text{sphere} & \text{from } \Theta \sim 85° \text{ to } \Theta \sim 150°; \\ \text{nonsphere} \ll \text{sphere} & \text{from } \Theta \sim 150° \text{ to } \Theta = 180°. \end{array} \quad (11.13.4)$$

Although the specific boundaries of these regions can be expected to shift with

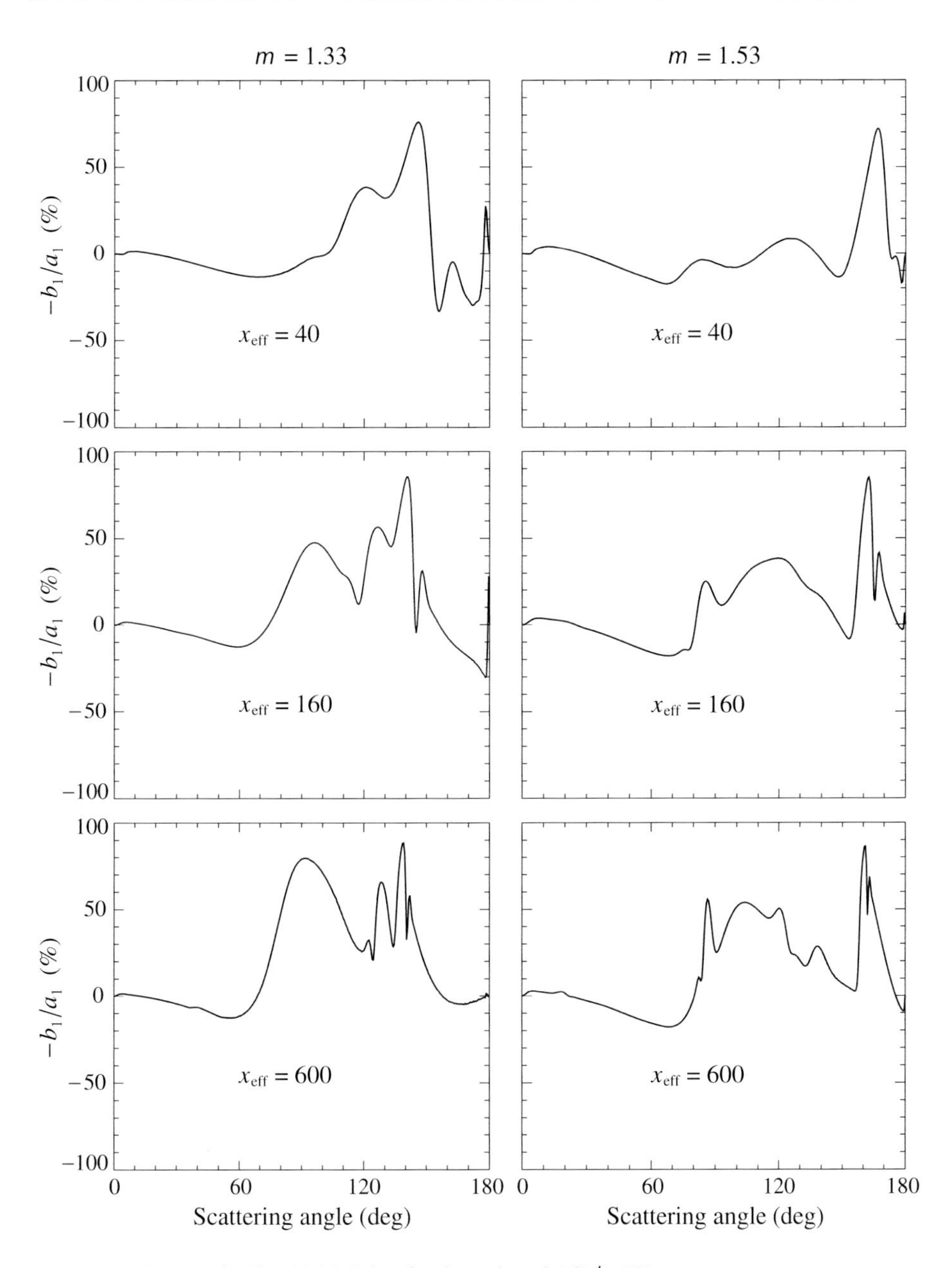

Figure 11.13.7. As in Fig. 11.13.6, but for the ratio $-b_1(\Theta)/a_1(\Theta)$.

changing particle shape and relative refractive index, the enhanced side-scattering and suppressed backscattering appear to be rather universal characteristics of nonspherical particles.

The degree of linear polarization for unpolarized incident light, $-b_1(\Theta)/a_1(\Theta)$, tends to be positive at scattering angles around 120° for the spheroids, but is negative at most scattering angles for the spheres. Whereas $a_2(\Theta)/a_1(\Theta) \equiv 1$ for spherically sym-

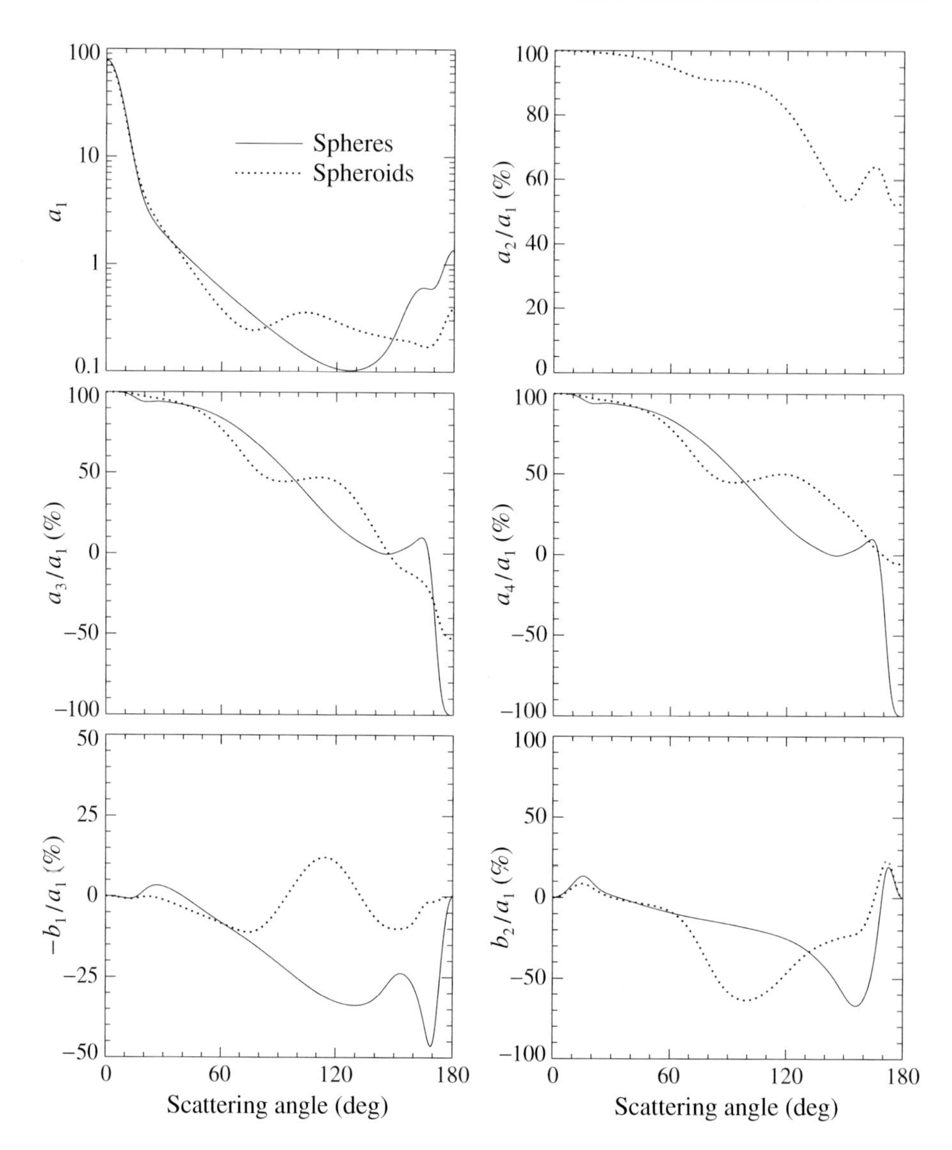

Figure 11.13.8. Elements of the normalized Stokes scattering matrix for a gamma distribution of spheres and surface-area-equivalent, randomly oriented oblate spheroids with $x_{\text{eff}} = 10$, $v_{\text{eff}} = 0.1$, and $m = 1.53 + \text{i}0.008$. The ratio of the larger to the smaller spheroid axes is 2.

metric scatterers, the $a_2(\Theta)/a_1(\Theta)$ curve for the spheroids significantly deviates from unity and leads to strong backscattering depolarization (see Eq. (11.9.1)). Similarly, $a_3(\Theta)/a_1(\Theta) \equiv a_4(\Theta)/a_1(\Theta)$ for spherically symmetric particles, whereas the $a_4(\Theta)/a_1(\Theta)$ for the spheroids tends to be greater than the $a_3(\Theta)/a_1(\Theta)$ at most scattering angles, especially in the backscattering direction. The ratios $b_2(\Theta)/a_1(\Theta)$ for the spheres and the spheroids also reveal significant quantitative differences at scattering angles exceeding 60°.

The corresponding optical cross sections, single-scattering albedos, and asymme-

Table 11.13.1. Efficiency factors, single-scattering albedo, and asymmetry parameter for a gamma distribution of spheres and surface-area-equivalent, randomly oriented oblate spheroids with $x_{\text{eff}} = 10$, $v_{\text{eff}} = 0.1$, and $m = 1.53 + \text{i}0.008$. The ratio of the larger to the smaller spheroid axes is 2.

Particles	Q_{ext}	Q_{sca}	Q_{abs}	ϖ	$\langle\cos\Theta\rangle$
Spheres	2.457	2.106	0.351	0.857	0.720
Spheroids	2.505	2.182	0.323	0.871	0.688

try parameters are listed in Table 11.13.1. Clearly, the nonspherical/spherical differences in the integral scattering and absorption characteristics are not nearly as significant as those in the scattering matrix elements, which appears to be another typical trait of scattering by nonspherical particles.

Chapter 12

Radiative transfer in plane-parallel, macroscopically isotropic and mirror-symmetric scattering media

Although the assumption of plane-parallel geometry made in Chapter 10 has greatly simplified the theoretical analysis of the VRTE, the latter still remains too cumbersome and laborious to find extensive practical applications. Therefore, in this chapter we will introduce an additional simplifying restriction that makes the problem manageable. Specifically, we will assume that the plane-parallel scattering medium is macroscopically isotropic and mirror-symmetric, thereby making applicable the results of the preceding chapter. We remind the reader that an ISM is composed of randomly oriented particles with a plane of symmetry and/or particles and their mirror counterparts in random orientation.

12.1 The standard problem

In view of Eq. (11.4.8), the integro-differential VRTE (10.1.2) can now be re-written in the form

$$u \frac{\mathrm{d}\tilde{\mathbf{I}}(\tau, \hat{\mathbf{n}})}{\mathrm{d}\tau} = -\tilde{\mathbf{I}}(\tau, \hat{\mathbf{n}}) + \frac{1}{C_{\mathrm{ext}}(\tau)} \int_{4\pi} \mathrm{d}\hat{\mathbf{n}}' \, \mathbf{Z}(\tau, \hat{\mathbf{n}}, \hat{\mathbf{n}}') \tilde{\mathbf{I}}(\tau, \hat{\mathbf{n}}'), \tag{12.1.1}$$

where

$$\tau(z) = \int_z^\infty \mathrm{d}z' n_0(z') C_{\mathrm{ext}}(z') \tag{12.1.2}$$

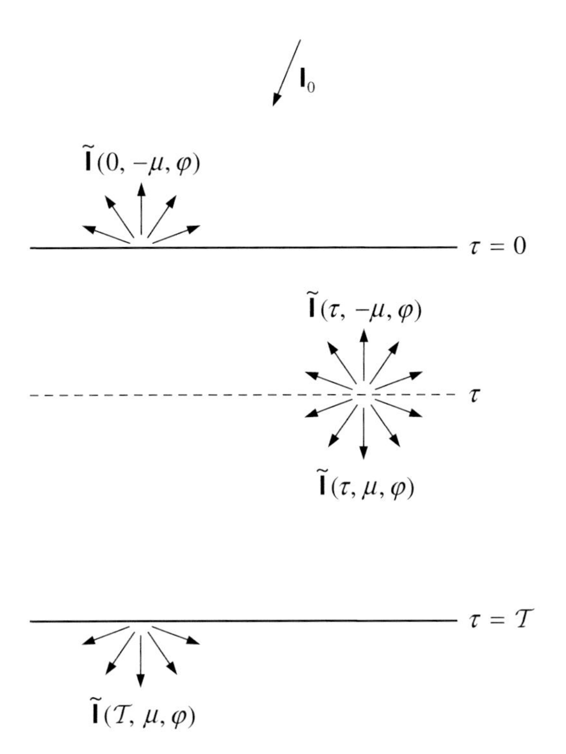

Figure 12.1.1. The standard problem.

is the so-called *optical depth*, which now replaces the geometrical depth z, and C_{ext} is the depth-dependent average extinction cross section.[1] Unlike the geometrical depth z and the particle depth ψ, the optical depth is dimensionless.

It is convenient to re-write Eq. (12.1.1) in the following form:

$$u \frac{\mathrm{d}\tilde{\mathbf{I}}(\tau, u, \varphi)}{\mathrm{d}\tau} = -\tilde{\mathbf{I}}(\tau, u, \varphi) + \frac{1}{C_{\text{ext}}(\tau)} \int_0^{2\pi} \mathrm{d}\varphi' \int_{-1}^{+1} \mathrm{d}u' \, \mathbf{Z}(\tau; u, \varphi; u', \varphi') \tilde{\mathbf{I}}(\tau, u', \varphi'), \tag{12.1.3}$$

where, as before, we use the notation

$$\tilde{\mathbf{I}}(\tau, u, \varphi) = \tilde{\mathbf{I}}(\tau, \theta, \varphi),$$
$$\mathbf{Z}(\tau; u, \varphi; u', \varphi') = \mathbf{Z}(\tau; \theta, \varphi; \theta', \varphi').$$

Finally, the phase matrix can be replaced by the normalized phase matrix according to Eq. (11.10.2), thereby yielding

[1] As in Chapter 10, we omit the angular brackets used previously to denote averages over particle states.

$$u\frac{\mathrm{d}\tilde{\mathbf{I}}(\tau,u,\varphi)}{\mathrm{d}\tau} = -\tilde{\mathbf{I}}(\tau,u,\varphi) + \frac{\varpi(\tau)}{4\pi}\int_0^{2\pi}\mathrm{d}\varphi'\int_{-1}^{+1}\mathrm{d}u'\,\tilde{\mathbf{Z}}(\tau;u,\varphi;u',\varphi')\,\tilde{\mathbf{I}}(\tau,u',\varphi'), \tag{12.1.4}$$

where

$$\varpi(\tau) = \frac{C_{\mathrm{sca}}(\tau)}{C_{\mathrm{ext}}(\tau)} \tag{12.1.5}$$

is the optical-depth-dependent single-scattering albedo. The standard problem is now reformulated by supplementing Eq. (12.1.4) with the boundary conditions

$$\tilde{\mathbf{I}}(0,\mu,\varphi) = \delta(\mu-\mu_0)\delta(\varphi-\varphi_0)\mathbf{I}_0, \tag{12.1.6}$$

$$\tilde{\mathbf{I}}(\mathcal{T},-\mu,\varphi) = \mathbf{0}, \tag{12.1.7}$$

where $\mathcal{T} = \tau(z_\mathrm{b})$ is the *optical thickness* of the layer (Fig. 12.1.1).

12.2 The general problem

By analogy with Section 10.3, the boundary values specifying the general problem read

$$\tilde{\mathbf{I}}(0,\mu,\varphi) = \tilde{\mathbf{I}}_\downarrow(\mu,\varphi), \tag{12.2.1}$$

$$\tilde{\mathbf{I}}(\mathcal{T},-\mu,\varphi) = \tilde{\mathbf{I}}_\uparrow(-\mu,\varphi) \tag{12.2.2}$$

(see Fig. 12.2.1). The fact that the extinction matrix has now reduced to a direction-independent scalar extinction cross section allows us to replace the matrix propagators (10.2.8) and (10.2.16) with simple scalar exponentials $\exp[-(\tau-\tau_0)/\mu]$ for $\tau \ge \tau_0$ and $\exp[-(\tau_0-\tau)/\mu]$ for $\tau < \tau_0$. As a consequence, the radiation field for $\tau \in [0,\mathcal{T}]$ can be expressed in terms of the specific intensity vectors of the external light as follows:

$$\tilde{\mathbf{I}}(\tau,\mu,\varphi) = \exp(-\tau/\mu)\tilde{\mathbf{I}}_\downarrow(\mu,\varphi) + \frac{1}{\pi}\int_0^{2\pi}\mathrm{d}\varphi'\int_0^{+1}\mathrm{d}\mu'\mu'\,\mathbf{D}(\tau;\mu,\varphi;\mu',\varphi')\,\tilde{\mathbf{I}}_\downarrow(\mu',\varphi') + \frac{1}{\pi}\int_0^{2\pi}\mathrm{d}\varphi'\int_0^{+1}\mathrm{d}\mu'\mu'\,\mathbf{U}^\dagger(\tau;\mu,\varphi;\mu',\varphi')\,\tilde{\mathbf{I}}_\uparrow(-\mu',\varphi'), \tag{12.2.3}$$

$$\tilde{\mathbf{I}}(\tau,-\mu,\varphi) = \exp[-(\mathcal{T}-\tau)/\mu]\,\tilde{\mathbf{I}}_\uparrow(-\mu,\varphi)$$

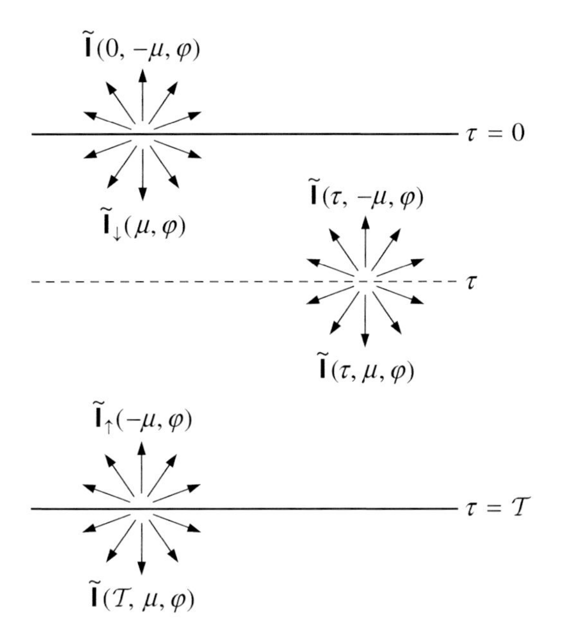

Figure 12.2.1. The general problem.

$$
\begin{aligned}
&+ \frac{1}{\pi}\int_0^{2\pi} \mathrm{d}\varphi' \int_0^{+1} \mathrm{d}\mu'\mu' \, \mathbf{U}(\tau; \mu, \varphi; \mu', \varphi')\, \tilde{\mathbf{I}}_{\downarrow}(\mu', \varphi') \\
&+ \frac{1}{\pi}\int_0^{2\pi} \mathrm{d}\varphi' \int_0^{+1} \mathrm{d}\mu'\mu' \, \mathbf{D}^{\dagger}(\tau; \mu, \varphi; \mu', \varphi')\, \tilde{\mathbf{I}}_{\uparrow}(-\mu', \varphi').
\end{aligned}
\tag{12.2.4}
$$

The corresponding reflection and transmission matrices are defined by

$$\mathbf{R}(\mu, \varphi; \mu', \varphi') = \mathbf{U}(0; \mu, \varphi; \mu', \varphi'), \tag{12.2.5}$$

$$\mathbf{T}(\mu, \varphi; \mu', \varphi') = \mathbf{D}(\mathcal{T}; \mu, \varphi; \mu', \varphi'), \tag{12.2.6}$$

$$\mathbf{R}^{\dagger}(\mu, \varphi; \mu', \varphi') = \mathbf{U}^{\dagger}(\mathcal{T}; \mu, \varphi; \mu', \varphi'), \tag{12.2.7}$$

$$\mathbf{T}^{\dagger}(\mu, \varphi; \mu', \varphi') = \mathbf{D}^{\dagger}(0; \mu, \varphi; \mu', \varphi'), \tag{12.2.8}$$

whereas the solution of the standard problem takes the form

$$
\begin{aligned}
\tilde{\mathbf{I}}(\tau, \mu, \varphi) = {} & \delta(\mu - \mu_0)\delta(\varphi - \varphi_0)\exp(-\tau/\mu_0)\mathbf{I}_0 \\
& + \frac{1}{\pi}\mu_0 \mathbf{D}(\tau; \mu, \varphi; \mu_0, \varphi_0)\mathbf{I}_0,
\end{aligned}
\tag{12.2.9}
$$

$$\tilde{\mathbf{I}}(\tau, -\mu, \varphi) = \frac{1}{\pi}\mu_0 \mathbf{U}(\tau; \mu, \varphi; \mu_0, \varphi_0)\mathbf{I}_0, \tag{12.2.10}$$

$$\tilde{\mathbf{I}}(\mathcal{T}, \mu, \varphi) = \delta(\mu - \mu_0)\delta(\varphi - \varphi_0)\exp(-\mathcal{T}/\mu_0)\mathbf{I}_0$$

$$+\frac{1}{\pi}\mu_0\mathbf{T}(\mu,\varphi;\mu_0,\varphi_0)\mathbf{I}_0, \tag{12.2.11}$$

$$\tilde{\mathbf{I}}(0,-\mu,\varphi) = \frac{1}{\pi}\mu_0\mathbf{R}(\mu,\varphi;\mu_0,\varphi_0)\mathbf{I}_0. \tag{12.2.12}$$

Analogously to the case for the phase matrix, the azimuthal symmetry of the medium causes the matrices $\mathbf{U}$, $\mathbf{D}$, $\mathbf{U}^\dagger$, $\mathbf{D}^\dagger$, $\mathbf{R}$, $\mathbf{T}$, $\mathbf{R}^\dagger$, and $\mathbf{T}^\dagger$ to depend on the difference $\varphi-\varphi'$ rather than on φ and φ' separately. In other words, if φ, φ', φ_1, and φ_1' are such that $\varphi-\varphi'=\varphi_1-\varphi_1'$ then $\mathbf{D}(\tau;\mu,\varphi_1;\mu',\varphi_1') = \mathbf{D}(\tau;\mu,\varphi;\mu',\varphi')$, and analogously for the other matrices. This is equivalent to the statement that each of these matrices has the property

$$\mathbf{Y}(\mu,\varphi+\Delta\varphi;\mu',\varphi'+\Delta\varphi) = \mathbf{Y}(\mu,\varphi;\mu',\varphi') \tag{12.2.13}$$

for any φ, φ', and $\Delta\varphi$. Although this property of azimuthal rotational invariance seems to be rather obvious, the more critical reader may appreciate the following formal proof. Recall first Eqs. (10.4.21)–(10.4.24) which suggest that the reflection and transmission matrices of optically thick layers do satisfy Eq. (12.2.13) because they are linearly expressed in terms of the phase matrix. The azimuthal rotational invariance of the matrices $\mathbf{U}$, $\mathbf{D}$, $\mathbf{U}^\dagger$, $\mathbf{D}^\dagger$, $\mathbf{R}$, $\mathbf{T}$, $\mathbf{R}^\dagger$, and $\mathbf{T}^\dagger$ for an arbitrary layer then follows by induction from the adding/doubling equations, the fact that all matrix propagators reduce to azimuth-independent scalars, and the fact that if 4×4 matrices $\mathbf{Y}_1(\mu,\varphi;\mu',\varphi')$ and $\mathbf{Y}_2(\mu,\varphi;\mu',\varphi')$ satisfy Eq. (12.2.13) then the matrix

$$\mathbf{Y}(\mu,\varphi;\mu',\varphi') = \int_0^{2\pi}\mathrm{d}\varphi''\int_0^1\mathrm{d}\mu''\mu''\,\mathbf{Y}_1(\mu,\varphi;\mu'',\varphi'')\mathbf{Y}_2(\mu'',\varphi'';\mu',\varphi') \tag{12.2.14}$$

also possesses the property of azimuthal rotational invariance. Indeed, Eqs. (10.4.1)–(10.4.4) can be solved by iteration using the first terms on the right-hand sides as an initial approximation. Therefore, if the reflection and transmission matrices of each layer satisfy Eq. (12.2.13) then the matrices $\mathbf{U}$, $\mathbf{D}$, $\mathbf{U}^\dagger$, and $\mathbf{D}^\dagger$ of the combined slab also satisfy this equation. Finally, Eqs. (10.4.7)–(10.4.10) show that the reflection and transmission matrices of the combined slab also satisfy Eq. (12.2.13).

12.3 Adding equations

The aim of this section is to show how the adding equations transform if one uses the simplifying assumption of macroscopic isotropy and mirror symmetry. Let us consider an arbitrary plane-parallel slab having an optical thickness $\mathcal{T}$ and divide it into layers $[0,\tau]$ and $[\tau,\mathcal{T}]$ (see Fig. 12.3.1). The reader can easily verify that the adding equations (10.4.1)–(10.4.4) and (10.4.7)–(10.4.10) now become

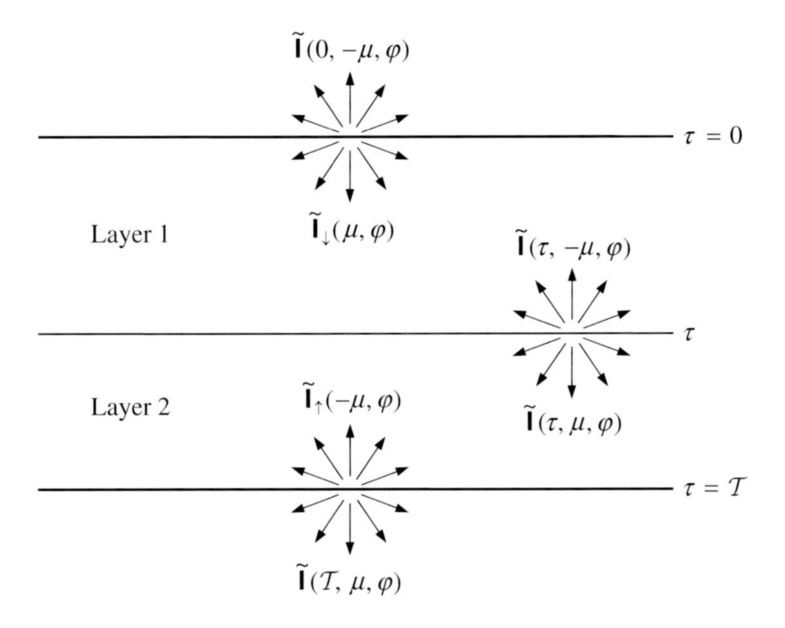

Figure 12.3.1. Illustration of the adding principle.

$$\mathbf{U}(\tau; \mu, \varphi; \mu', \varphi') = \exp(-\tau/\mu')\mathbf{R}_2(\mu, \varphi; \mu', \varphi') + \frac{1}{\pi}\int_0^{2\pi} \mathrm{d}\varphi'' \int_0^1 \mathrm{d}\mu''\mu''\, \mathbf{R}_2(\mu, \varphi; \mu'', \varphi'') \times \mathbf{D}(\tau; \mu'', \varphi''; \mu', \varphi'), \tag{12.3.1}$$

$$\mathbf{D}(\tau; \mu, \varphi; \mu', \varphi') = \mathbf{T}_1(\mu, \varphi; \mu', \varphi') + \frac{1}{\pi}\int_0^{2\pi} \mathrm{d}\varphi'' \int_0^1 \mathrm{d}\mu''\mu''\, \mathbf{R}_1^{\dagger}(\mu, \varphi; \mu'', \varphi'') \times \mathbf{U}(\tau; \mu'', \varphi''; \mu', \varphi'), \tag{12.3.2}$$

$$\mathbf{U}^{\dagger}(\tau; \mu, \varphi; \mu', \varphi') = \exp[-(\mathcal{T} - \tau)/\mu']\mathbf{R}_1^{\dagger}(\mu, \varphi; \mu', \varphi') + \frac{1}{\pi}\int_0^{2\pi} \mathrm{d}\varphi'' \int_0^1 \mathrm{d}\mu''\mu''\, \mathbf{R}_1^{\dagger}(\mu, \varphi; \mu'', \varphi'') \times \mathbf{D}^{\dagger}(\tau; \mu'', \varphi''; \mu', \varphi'), \tag{12.3.3}$$

$$\mathbf{D}^{\dagger}(\tau; \mu, \varphi; \mu', \varphi') = \mathbf{T}_2^{\dagger}(\mu, \varphi; \mu', \varphi') + \frac{1}{\pi}\int_0^{2\pi} \mathrm{d}\varphi'' \int_0^1 \mathrm{d}\mu''\mu''\, \mathbf{R}_2(\mu, \varphi; \mu'', \varphi'') \times \mathbf{U}^{\dagger}(\tau; \mu'', \varphi''; \mu', \varphi'), \tag{12.3.4}$$

$$\mathbf{R}(\mu, \varphi; \mu', \varphi') = \mathbf{R}_1(\mu, \varphi; \mu', \varphi') + \exp(-\tau/\mu)\mathbf{U}(\tau; \mu, \varphi; \mu', \varphi')$$

$$+\frac{1}{\pi}\int_0^{2\pi}\mathrm{d}\varphi''\int_0^1\mathrm{d}\mu''\mu''\,\mathbf{T}_1^\dagger(\mu,\varphi;\mu'',\varphi'')\times\mathbf{U}(\tau;\mu'',\varphi'';\mu',\varphi'), \tag{12.3.5}$$

$$\begin{aligned}\mathbf{T}(\mu,\varphi;\mu',\varphi') &= \exp(-\tau/\mu')\mathbf{T}_2(\mu,\varphi;\mu',\varphi') \\ &\quad+\exp[-(\mathcal{T}-\tau)/\mu]\mathbf{D}(\tau;\mu,\varphi;\mu',\varphi') \\ &\quad+\frac{1}{\pi}\int_0^{2\pi}\mathrm{d}\varphi''\int_0^1\mathrm{d}\mu''\mu''\,\mathbf{T}_2(\mu,\varphi;\mu'',\varphi'') \\ &\qquad\times\mathbf{D}(\tau;\mu'',\varphi'';\mu',\varphi'),\end{aligned} \tag{12.3.6}$$

$$\begin{aligned}\mathbf{R}^\dagger(\mu,\varphi;\mu',\varphi') &= \mathbf{R}_2(\mu,\varphi;\mu',\varphi') \\ &\quad+\exp[-(\mathcal{T}-\tau)/\mu]\mathbf{U}^\dagger(\tau;\mu,\varphi;\mu',\varphi') \\ &\quad+\frac{1}{\pi}\int_0^{2\pi}\mathrm{d}\varphi''\int_0^1\mathrm{d}\mu''\mu''\,\mathbf{T}_2(\mu,\varphi;\mu'',\varphi'') \\ &\qquad\times\mathbf{U}^\dagger(\tau;\mu'',\varphi'';\mu',\varphi'),\end{aligned} \tag{12.3.7}$$

$$\begin{aligned}\mathbf{T}^\dagger(\mu,\varphi;\mu',\varphi') &= \exp[-(\mathcal{T}-\tau)/\mu']\mathbf{T}_1^\dagger(\mu,\varphi;\mu',\varphi') \\ &\quad+\exp(-\tau/\mu)\mathbf{D}^\dagger(\tau;\mu,\varphi;\mu',\varphi') \\ &\quad+\frac{1}{\pi}\int_0^{2\pi}\mathrm{d}\varphi''\int_0^1\mathrm{d}\mu''\mu''\,\mathbf{T}_1^\dagger(\mu,\varphi;\mu'',\varphi'') \\ &\qquad\times\mathbf{D}^\dagger(\tau;\mu'',\varphi'';\mu',\varphi').\end{aligned} \tag{12.3.8}$$

Equations (10.4.11)–(10.4.18) describing the polarized radiation field at any optical depth τ' inside the combined slab (see Fig. 12.3.2) now take the form

$$\begin{aligned}\mathbf{U}(\tau';\mu,\varphi;\mu',\varphi') &= \mathbf{U}_1(\tau';\mu,\varphi;\mu',\varphi') \\ &\quad+\exp[-(\tau-\tau')/\mu]\mathbf{U}(\tau;\mu,\varphi;\mu',\varphi') \\ &\quad+\frac{1}{\pi}\int_0^{2\pi}\mathrm{d}\varphi''\int_0^1\mathrm{d}\mu''\mu''\,\mathbf{D}_1^\dagger(\tau';\mu,\varphi;\mu'',\varphi'') \\ &\qquad\times\mathbf{U}(\tau;\mu'',\varphi'';\mu',\varphi'),\end{aligned} \tag{12.3.9}$$

$$\begin{aligned}\mathbf{D}(\tau';\mu,\varphi;\mu',\varphi') &= \mathbf{D}_1(\tau';\mu,\varphi;\mu',\varphi') \\ &\quad+\frac{1}{\pi}\int_0^{2\pi}\mathrm{d}\varphi''\int_0^1\mathrm{d}\mu''\mu''\,\mathbf{U}_1^\dagger(\tau';\mu,\varphi;\mu'',\varphi'') \\ &\qquad\times\mathbf{U}(\tau;\mu'',\varphi'';\mu',\varphi'),\end{aligned} \tag{12.3.10}$$

$$\begin{aligned}\mathbf{U}^\dagger(\tau';\mu,\varphi;\mu',\varphi') &= \exp[-(\mathcal{T}-\tau)/\mu']\mathbf{U}_1^\dagger(\tau';\mu,\varphi;\mu',\varphi') \\ &\quad+\frac{1}{\pi}\int_0^{2\pi}\mathrm{d}\varphi''\int_0^1\mathrm{d}\mu''\mu''\,\mathbf{U}_1^\dagger(\tau';\mu,\varphi;\mu'',\varphi'') \\ &\qquad\times\mathbf{D}^\dagger(\tau;\mu'',\varphi'';\mu',\varphi'),\end{aligned} \tag{12.3.11}$$

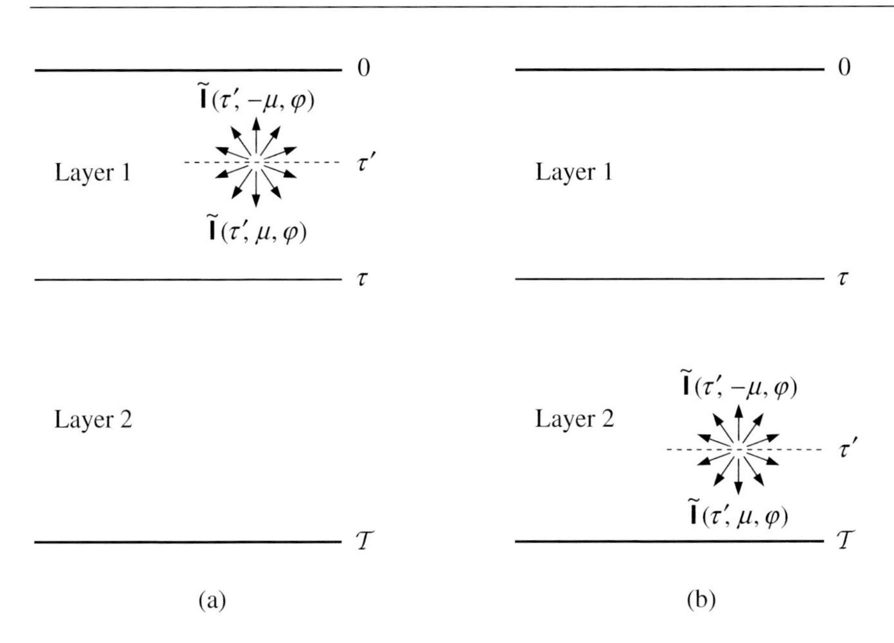

Figure 12.3.2. Internal radiation field.

$$\begin{aligned}\mathbf{D}^{\dagger}(\tau';\mu,\varphi;\mu',\varphi') &= \exp[-(\mathcal{T}-\tau)/\mu']\mathbf{D}_1^{\dagger}(\tau';\mu,\varphi;\mu',\varphi') \\ &\quad + \exp[-(\tau-\tau')/\mu]\mathbf{D}^{\dagger}(\tau;\mu,\varphi;\mu',\varphi') \\ &\quad + \frac{1}{\pi}\int_0^{2\pi}\mathrm{d}\varphi''\int_0^1 \mathrm{d}\mu''\mu''\,\mathbf{D}_1^{\dagger}(\tau';\mu,\varphi;\mu'',\varphi'') \\ &\quad\quad \times \mathbf{D}^{\dagger}(\tau;\mu'',\varphi'';\mu',\varphi') \end{aligned} \tag{12.3.12}$$

for $\tau' \in [0, \tau]$ and

$$\begin{aligned}\mathbf{U}(\tau';\mu,\varphi;\mu',\varphi') &= \exp(-\tau/\mu')\mathbf{U}_2(\tau'-\tau;\mu,\varphi;\mu',\varphi') \\ &\quad + \frac{1}{\pi}\int_0^{2\pi}\mathrm{d}\varphi''\int_0^1 \mathrm{d}\mu''\mu''\,\mathbf{U}_2(\tau'-\tau;\mu,\varphi;\mu'',\varphi'') \\ &\quad\quad \times \mathbf{D}(\tau;\mu'',\varphi'';\mu',\varphi'), \end{aligned} \tag{12.3.13}$$

$$\begin{aligned}\mathbf{D}(\tau';\mu,\varphi;\mu',\varphi') &= \exp(-\tau/\mu')\mathbf{D}_2(\tau'-\tau;\mu,\varphi;\mu',\varphi') \\ &\quad + \exp[-(\tau'-\tau)/\mu]\mathbf{D}(\tau;\mu,\varphi;\mu',\varphi') \\ &\quad + \frac{1}{\pi}\int_0^{2\pi}\mathrm{d}\varphi''\int_0^1 \mathrm{d}\mu''\mu''\,\mathbf{D}_2(\tau'-\tau;\mu,\varphi;\mu'',\varphi'') \\ &\quad\quad \times \mathbf{D}(\tau;\mu'',\varphi'';\mu',\varphi'), \end{aligned} \tag{12.3.14}$$

$$\begin{aligned}\mathbf{U}^{\dagger}(\tau';\mu,\varphi;\mu',\varphi') &= \mathbf{U}_2^{\dagger}(\tau'-\tau;\mu,\varphi;\mu',\varphi') \\ &\quad + \exp[-(\tau'-\tau)/\mu]\mathbf{U}^{\dagger}(\tau;\mu,\varphi;\mu',\varphi') \end{aligned}$$

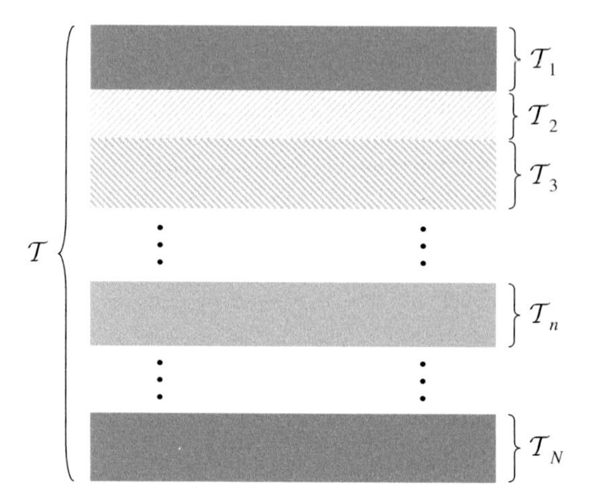

Figure 12.3.3. Representation of a vertically inhomogeneous scattering slab by a stack of N homogeneous layers.

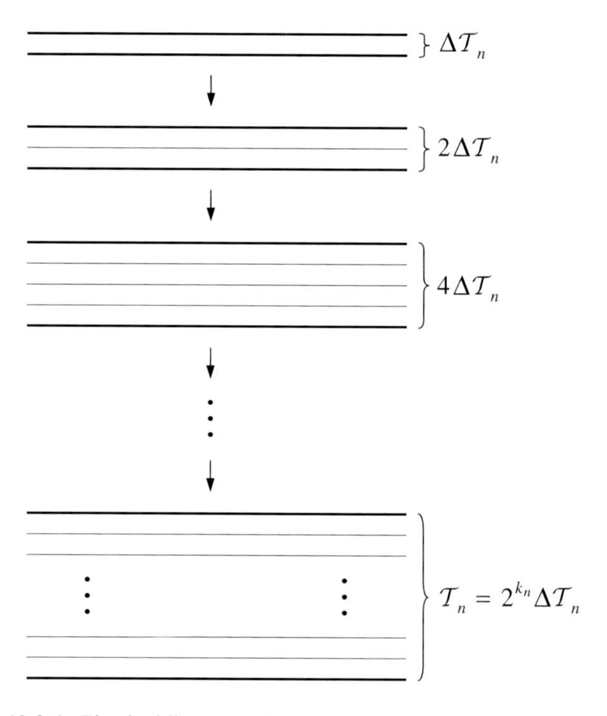

Figure 12.3.4. The doubling procedure.

$$+\frac{1}{\pi}\int_0^{2\pi}\mathrm{d}\varphi''\int_0^1\mathrm{d}\mu''\mu''\,\mathbf{D}_2(\tau'-\tau;\mu,\varphi;\mu'',\varphi'')$$
$$\times\mathbf{U}^{\dagger}(\tau;\mu'',\varphi'';\mu',\varphi'), \qquad (12.3.15)$$

$$\mathbf{D}^{\dagger}(\tau'; \mu, \varphi; \mu', \varphi') = \mathbf{D}_2^{\dagger}(\tau' - \tau; \mu, \varphi; \mu', \varphi') + \frac{1}{\pi}\int_0^{2\pi} \mathrm{d}\varphi'' \int_0^1 \mathrm{d}\mu''\mu''\, \mathbf{U}_2(\tau' - \tau; \mu, \varphi; \mu'', \varphi'') \times \mathbf{U}^{\dagger}(\tau; \mu'', \varphi''; \mu', \varphi') \tag{12.3.16}$$

for $\tau' \in [\tau, \mathcal{T}]$. The practical implementation of the adding scheme for a multi-layer scattering slab is now illustrated by Figs. 12.3.3 and 12.3.4. Finally, the reflection and transmission matrices of an infinitesimally thin layer, Fig. 12.3.4, are now given by

$$\mathbf{R}_{\Delta\mathcal{T}_n}(\mu, \varphi; \mu', \varphi') = \frac{\varpi\Delta\mathcal{T}_n}{4\mu\mu'}\widetilde{\mathbf{Z}}_n(-\mu, \varphi; \mu', \varphi'), \tag{12.3.17}$$

$$\mathbf{T}_{\Delta\mathcal{T}_n}(\mu, \varphi; \mu', \varphi') = \frac{\varpi\Delta\mathcal{T}_n}{4\mu\mu'}\widetilde{\mathbf{Z}}_n(\mu, \varphi; \mu', \varphi'), \tag{12.3.18}$$

$$\mathbf{R}^{\dagger}_{\Delta\mathcal{T}_n}(\mu, \varphi; \mu', \varphi') = \frac{\varpi\Delta\mathcal{T}_n}{4\mu\mu'}\widetilde{\mathbf{Z}}_n(\mu, \varphi; -\mu', \varphi'), \tag{12.3.19}$$

$$\mathbf{T}^{\dagger}_{\Delta\mathcal{T}_n}(\mu, \varphi; \mu', \varphi') = \frac{\varpi\Delta\mathcal{T}_n}{4\mu\mu'}\widetilde{\mathbf{Z}}_n(-\mu, \varphi; -\mu', \varphi'). \tag{12.3.20}$$

12.4 Invariant imbedding and Ambarzumian equations

It is also quite straightforward to rewrite the invariant imbedding equations and the Ambarzumian equation for the special case of a macroscopically isotropic and mirror-symmetric scattering slab. Specifically, Eqs. (10.5.11a)–(10.5.14a) now read

$$\begin{aligned}\frac{\partial \mathbf{R}(\mu, \varphi; \mu', \varphi')}{\partial \mathcal{T}_{\uparrow}} = {} & -\frac{\mu + \mu'}{\mu\mu'}\mathbf{R}(\mu, \varphi; \mu', \varphi') \\ & + \frac{\varpi}{4\mu\mu'}\widetilde{\mathbf{Z}}(0; -\mu, \varphi; \mu', \varphi') \\ & + \frac{\varpi}{4\pi\mu'}\int_0^{2\pi}\mathrm{d}\varphi''\int_0^1 \mathrm{d}\mu''\, \mathbf{R}(\mu, \varphi; \mu'', \varphi'') \\ & \qquad \times \widetilde{\mathbf{Z}}(0; \mu'', \varphi''; \mu', \varphi') \\ & + \frac{\varpi}{4\pi\mu}\int_0^{2\pi}\mathrm{d}\varphi''\int_0^1 \mathrm{d}\mu''\, \widetilde{\mathbf{Z}}(0; -\mu, \varphi; -\mu'', \varphi'') \\ & \qquad \times \mathbf{R}(\mu'', \varphi''; \mu', \varphi') \\ & + \frac{\varpi}{4\pi^2}\int_0^{2\pi}\mathrm{d}\varphi''\int_0^1 \mathrm{d}\mu''\int_0^{2\pi}\mathrm{d}\varphi'''\int_0^1 \mathrm{d}\mu''' \\ & \qquad \times \mathbf{R}(\mu, \varphi; \mu'', \varphi'')\widetilde{\mathbf{Z}}(0; \mu'', \varphi''; -\mu''', \varphi''') \\ & \qquad \times \mathbf{R}(\mu''', \varphi'''; \mu', \varphi'),\end{aligned} \tag{12.4.1}$$

$$\begin{aligned}
\frac{\partial \mathbf{T}(\mu,\varphi;\mu',\varphi')}{\partial \mathcal{T}_{\uparrow}} &= -\frac{1}{\mu'}\mathbf{T}(\mu,\varphi;\mu',\varphi') + \frac{\varpi \mathrm{e}^{-\mathcal{T}/\mu}}{4\mu\mu'}\widetilde{\mathbf{Z}}(0;\mu,\varphi;\mu',\varphi') \\
&+ \frac{\varpi}{4\pi\mu'}\int_0^{2\pi}\mathrm{d}\varphi''\int_0^1 \mathrm{d}\mu''\,\mathbf{T}(\mu,\varphi;\mu'',\varphi'') \\
&\qquad \times \mathbf{Z}(0;\mu'',\varphi'';\mu',\varphi') \\
&+ \frac{\varpi \mathrm{e}^{-\mathcal{T}/\mu}}{4\pi\mu}\int_0^{2\pi}\mathrm{d}\varphi''\int_0^1 \mathrm{d}\mu''\,\mathbf{Z}(0;\mu,\varphi;-\mu'',\varphi'') \\
&\qquad \times \mathbf{R}(\mu'',\varphi'';\mu',\varphi') \\
&+ \frac{\varpi}{4\pi^2}\int_0^{2\pi}\mathrm{d}\varphi''\int_0^1 \mathrm{d}\mu''\int_0^{2\pi}\mathrm{d}\varphi'''\int_0^1 \mathrm{d}\mu''' \\
&\quad \times \mathbf{T}(\mu,\varphi;\mu'',\varphi'')\mathbf{Z}(0;\mu'',\varphi'';-\mu''',\varphi''') \\
&\quad \times \mathbf{R}(\mu''',\varphi''';\mu',\varphi'),
\end{aligned} \tag{12.4.2}$$

$$\begin{aligned}
\frac{\partial \mathbf{R}^{\dagger}(\mu,\varphi;\mu',\varphi')}{\partial \mathcal{T}_{\uparrow}} &= -\frac{\varpi}{4\mu\mu'}\exp\left[-\mathcal{T}\left(\frac{1}{\mu}+\frac{1}{\mu'}\right)\right]\widetilde{\mathbf{Z}}(0;\mu,\varphi;-\mu',\varphi') \\
&+ \frac{\varpi \mathrm{e}^{-\mathcal{T}/\mu}}{4\pi\mu}\int_0^{2\pi}\mathrm{d}\varphi''\int_0^1 \mathrm{d}\mu''\,\widetilde{\mathbf{Z}}(0;\mu,\varphi;-\mu'',\varphi'') \\
&\qquad \times \mathbf{T}^{\dagger}(\mu'',\varphi'';\mu',\varphi') \\
&+ \frac{\varpi \mathrm{e}^{-\mathcal{T}/\mu'}}{4\pi\mu'}\int_0^{2\pi}\mathrm{d}\varphi''\int_0^1 \mathrm{d}\mu''\,\mathbf{T}(\mu,\varphi;\mu'',\varphi'') \\
&\qquad \times \widetilde{\mathbf{Z}}(0;\mu'',\varphi'';-\mu',\varphi') \\
&+ \frac{\varpi}{4\pi^2}\int_0^{2\pi}\mathrm{d}\varphi''\int_0^1 \mathrm{d}\mu''\int_0^{2\pi}\mathrm{d}\varphi'''\int_0^1 \mathrm{d}\mu''' \\
&\quad \times \mathbf{T}(\mu,\varphi;\mu'',\varphi'')\widetilde{\mathbf{Z}}(0;\mu'',\varphi'';-\mu''',\varphi''') \\
&\quad \times \mathbf{T}^{\dagger}(\mu''',\varphi''';\mu',\varphi'),
\end{aligned} \tag{12.4.3}$$

$$\begin{aligned}
\frac{\partial \mathbf{T}^{\dagger}(\mu,\varphi;\mu',\varphi')}{\partial \mathcal{T}_{\uparrow}} &= -\frac{1}{\mu}\mathbf{T}^{\dagger}(\mu,\varphi;\mu',\varphi') \\
&+ \frac{\varpi \mathrm{e}^{-\mathcal{T}/\mu'}}{4\mu\mu'}\widetilde{\mathbf{Z}}(0;-\mu,\varphi;-\mu',\varphi') \\
&+ \frac{\varpi}{4\pi\mu}\int_0^{2\pi}\mathrm{d}\varphi''\int_0^1 \mathrm{d}\mu''\,\widetilde{\mathbf{Z}}(0;-\mu,\varphi;-\mu'',\varphi'') \\
&\qquad \times \mathbf{T}^{\dagger}(\mu'',\varphi'';\mu',\varphi') \\
&+ \frac{\varpi \mathrm{e}^{-\mathcal{T}/\mu'}}{4\pi\mu'}\int_0^{2\pi}\mathrm{d}\varphi''\int_0^1 \mathrm{d}\mu''\,\mathbf{R}(\mu,\varphi;\mu'',\varphi'')
\end{aligned}$$

$$\times \widetilde{\mathbf{Z}}(0; \mu'', \varphi''; -\mu', \varphi')$$

$$+ \frac{\varpi}{4\pi^2} \int_0^{2\pi} \mathrm{d}\varphi'' \int_0^1 \mathrm{d}\mu'' \int_0^{2\pi} \mathrm{d}\varphi''' \int_0^1 \mathrm{d}\mu'''$$

$$\times \mathbf{R}(\mu, \varphi; \mu'', \varphi'') \mathbf{Z}(0; \mu'', \varphi''; -\mu''', \varphi''')$$

$$\times \mathbf{T}^\dagger(\mu''', \varphi'''; \mu', \varphi') \tag{12.4.4}$$

and are supplemented by the initial conditions

$$\mathbf{R}(\mu, \varphi; \mu', \varphi')\big|_{\mathcal{T}=0} = \mathbf{0}, \tag{12.4.5}$$

$$\mathbf{T}(\mu, \varphi; \mu', \varphi')\big|_{\mathcal{T}=0} = \mathbf{0}, \tag{12.4.6}$$

$$\mathbf{R}^\dagger(\mu, \varphi; \mu', \varphi')\big|_{\mathcal{T}=0} = \mathbf{0}, \tag{12.4.7}$$

$$\mathbf{T}^\dagger(\mu, \varphi; \mu', \varphi')\big|_{\mathcal{T}=0} = \mathbf{0}. \tag{12.4.8}$$

Similarly, the Ambarzumian nonlinear integral equation for the reflection matrix of a semi-infinite homogeneous slab now becomes

$$(\mu + \mu')\mathbf{R}(\mu, \varphi; \mu', \varphi') = \frac{\varpi}{4} \widetilde{\mathbf{Z}}(-\mu, \varphi; \mu', \varphi')$$

$$+ \frac{\varpi\mu}{4\pi} \int_0^{2\pi} \mathrm{d}\varphi'' \int_0^1 \mathrm{d}\mu'' \, \mathbf{R}(\mu, \varphi; \mu'', \varphi'')$$

$$\times \widetilde{\mathbf{Z}}(\mu'', \varphi''; \mu', \varphi')$$

$$+ \frac{\varpi\mu'}{4\pi} \int_0^{2\pi} \mathrm{d}\varphi'' \int_0^1 \mathrm{d}\mu'' \, \widetilde{\mathbf{Z}}(-\mu, \varphi; -\mu'', \varphi'')$$

$$\times \mathbf{R}(\mu'', \varphi''; \mu', \varphi')$$

$$+ \frac{\varpi\mu\mu'}{4\pi^2} \int_0^{2\pi} \mathrm{d}\varphi'' \int_0^1 \mathrm{d}\mu'' \int_0^{2\pi} \mathrm{d}\varphi''' \int_0^1 \mathrm{d}\mu'''$$

$$\times \mathbf{R}(\mu, \varphi; \mu'', \varphi'') \widetilde{\mathbf{Z}}(\mu'', \varphi''; -\mu''', \varphi''')$$

$$\times \mathbf{R}(\mu''', \varphi'''; \mu', \varphi'). \tag{12.4.9}$$

12.5 Successive orders of scattering

It is often convenient to expand either the specific intensity column vector or the 4×4 matrices $\mathbf{U}$, $\mathbf{D}$, $\mathbf{U}^\dagger$, $\mathbf{D}^\dagger$, $\mathbf{R}$, $\mathbf{T}$, $\mathbf{R}^\dagger$, and $\mathbf{T}^\dagger$ for a homogeneous slab in powers of the single-scattering albedo. Indeed, the physical interpretation of such Newmann series is very transparent: an nth term in the Newmann expansion of one of the above-mentioned quantities represents the contribution of light that was scattered n times.

Let us consider, for example, the reflection of light by a homogeneous slab and express the corresponding reflection matrix as follows:

$$\mathbf{R}(\mu, \varphi; \mu', \varphi') = \sum_{n=1}^{N} \varpi^n \mathbf{R}_n(\mu, \varphi; \mu', \varphi'). \tag{12.5.1}$$

In the case of an optically semi-infinite homogeneous slab, the Ambarzumian's equation (12.4.9) yields

$$\mathbf{R}_1(\mu, \varphi; \mu', \varphi') = \frac{1}{4(\mu + \mu')} \tilde{\mathbf{Z}}(-\mu, \varphi; \mu', \varphi'), \tag{12.5.2}$$

$$\begin{aligned}\mathbf{R}_2(\mu, \varphi; \mu', \varphi') = {} & \frac{\mu}{4\pi(\mu + \mu')} \int_0^{2\pi} \mathrm{d}\varphi'' \int_0^1 \mathrm{d}\mu'' \, \mathbf{R}_1(\mu, \varphi; \mu'', \varphi'') \\ & \times \tilde{\mathbf{Z}}(\mu'', \varphi''; \mu', \varphi') \\ & + \frac{\mu'}{4\pi(\mu + \mu')} \int_0^{2\pi} \mathrm{d}\varphi'' \int_0^1 \mathrm{d}\mu'' \, \tilde{\mathbf{Z}}(-\mu, \varphi; -\mu'', \varphi'') \\ & \times \mathbf{R}_1(\mu'', \varphi''; \mu', \varphi'),\end{aligned} \tag{12.5.3}$$

$$\begin{aligned}\mathbf{R}_n(\mu, \varphi; \mu', \varphi') = {} & \frac{\mu}{4\pi(\mu + \mu')} \int_0^{2\pi} \mathrm{d}\varphi'' \int_0^1 \mathrm{d}\mu'' \, \mathbf{R}_{n-1}(\mu, \varphi; \mu'', \varphi'') \\ & \times \tilde{\mathbf{Z}}(\mu'', \varphi''; \mu', \varphi') \\ & + \frac{\mu'}{4\pi(\mu + \mu')} \int_0^{2\pi} \mathrm{d}\varphi'' \int_0^1 \mathrm{d}\mu'' \, \tilde{\mathbf{Z}}(-\mu, \varphi; -\mu'', \varphi'') \\ & \times \mathbf{R}_{n-1}(\mu'', \varphi''; \mu', \varphi') \\ & + \sum_{n'=1}^{n-2} \frac{\mu\mu'}{4\pi^2(\mu + \mu')} \\ & \times \int_0^{2\pi} \mathrm{d}\varphi'' \int_0^1 \mathrm{d}\mu'' \int_0^{2\pi} \mathrm{d}\varphi''' \int_0^1 \mathrm{d}\mu''' \\ & \times \mathbf{R}_{n-n'-1}(\mu, \varphi; \mu'', \varphi'') \tilde{\mathbf{Z}}(\mu'', \varphi''; -\mu''', \varphi''') \\ & \times \mathbf{R}_{n'}(\mu''', \varphi'''; \mu', \varphi'), \quad n \geq 3\end{aligned} \tag{12.5.4}$$

(cf. Gross, 1962). Note that the upper summation limit N in Eq. (12.5.1) is infinite in general, but has to be a finite number in actual computer calculations. The practical value of N is usually determined from the requirement that the contribution of the $(N + 1)$th term be below a certain numerical threshold.

It is clear that the use of the recursion formula (12.5.4) can become rather onerous for $n \gg 1$. However, the quantities $\mathbf{R}_n(\mu, \varphi; \mu', \varphi')$ are independent of ϖ and are functions of only the phase matrix. Therefore, once they have been computed, Eq. (12.5.1) can be used to find the reflection matrix for any value of the single-scattering albedo provided that the phase matrix remains the same.

Various ways to implement the order-of-scattering approach are discussed by Hansen and Travis (1974), Hovenier *et al.* (2004), and Min and Duan (2004). For optically thin slabs ($\mathcal{T} < 1$), the order-of-scattering expansion converges rather rapidly with increasing N and can be truncated after a few (or several) terms, thereby yielding a viable alternative to the adding/doubling and invariant imbedding methods. For cases in which the contribution of higher-order terms is significant ($\mathcal{T} \gtrsim 1$, $\varpi \sim 1$), the order-of-scattering approach becomes (much) less efficient than the other techniques. However, there are several situations in which it is worthwhile to compute separately the contributions of at least a few orders of scattering (Hansen and Travis, 1974):

- In a Fourier series expansion in azimuth of the specific intensity column vector or the reflection and transmission matrices (see Section 12.7 below), the high frequency terms arise from light scattered a small number of times. Therefore, many Fourier terms can be computed accurately by including only one or a few orders of scattering.
- In the adding method, a significant saving of computer time can be achieved by computing two or three orders of scattering for the initial layer and taking a significantly larger $\Delta\mathcal{T}$ than that afforded by the first-order-scattering formulas (12.3.17)–(12.3.20).
- Equation (12.5.1) and its analogues for other characteristics of the radiation field show that for a homogeneous layer, the contribution of light scattered n times with conservative scattering ($\varpi = 1$) yields the solution for all other values of the single-scattering albedo upon multiplication by the factor ϖ^n. This property of the order-of-scattering solution can be useful for computing absorption line profiles provided that the phase matrix and the optical thickness remain nearly constant within the wavelength interval of the spectral line.
- The results for successive orders provide physical insight useful for understanding the process of multiple scattering.

12.6 Symmetry relations

12.6.1 Phase matrix

Using the notation introduced in Section 12.1, the four basic symmetry relations for the phase matrix, Eqs. (11.10.19)–(11.10.22), can be re-written as follows:

$$\widetilde{\mathbf{Z}}(\tau; u, -\varphi'; u', -\varphi) = \widetilde{\mathbf{Z}}(\tau; u, \varphi; u', \varphi'), \qquad (12.6.1)$$

$$\widetilde{\mathbf{Z}}(\tau; u, -\varphi; u', -\varphi') = \mathbf{\Delta}_{34} \widetilde{\mathbf{Z}}(\tau; u, \varphi; u', \varphi') \mathbf{\Delta}_{34}, \qquad (12.6.2)$$

$$\widetilde{\mathbf{Z}}(\tau; -u, \varphi; -u', \varphi') = \mathbf{\Delta}_{34}\widetilde{\mathbf{Z}}(\tau; u, \varphi; u', \varphi')\mathbf{\Delta}_{34}, \tag{12.6.3}$$

$$\widetilde{\mathbf{Z}}(\tau; -u', \varphi'; -u, \varphi) = \mathbf{\Delta}_{3}[\widetilde{\mathbf{Z}}(\tau; u, \varphi; u', \varphi')]^{\mathrm{T}}\mathbf{\Delta}_{3}. \tag{12.6.4}$$

Several more symmetry relations can be obtained by forming various combinations of these formulas. Specifically, combining Eqs. (12.6.2), and (12.6.4) yields

$$\widetilde{\mathbf{Z}}(\tau; -u', -\varphi'; -u, -\varphi) = \mathbf{\Delta}_{4}[\widetilde{\mathbf{Z}}(\tau; u, \varphi; u', \varphi')]^{\mathrm{T}}\mathbf{\Delta}_{4}, \tag{12.6.5}$$

where $\mathbf{\Delta}_4 = \mathrm{diag}[1, 1, 1, -1]$. Analogously,

$$\widetilde{\mathbf{Z}}(\tau; -u, \varphi'; -u', \varphi) = \widetilde{\mathbf{Z}}(\tau; u, \varphi; u', \varphi') \tag{12.6.6}$$

(see Eqs. (12.6.1)–(12.6.3)),

$$\widetilde{\mathbf{Z}}(\tau; u', \varphi'; u, \varphi) = \mathbf{\Delta}_{4}[\widetilde{\mathbf{Z}}(\tau; u, \varphi; u', \varphi')]^{\mathrm{T}}\mathbf{\Delta}_{4} \tag{12.6.7}$$

(see Eqs. (12.6.3) and (12.6.4)), and

$$\widetilde{\mathbf{Z}}(\tau; u', -\varphi'; u, -\varphi) = \mathbf{\Delta}_{3}[\widetilde{\mathbf{Z}}(\tau; u, \varphi; u', \varphi')]^{\mathrm{T}}\mathbf{\Delta}_{3} \tag{12.6.8}$$

(see Eqs. (12.6.3) and (12.6.5)).

12.6.2 Reflection and transmission matrices

The reciprocity relations for the reflection and transmission matrices, Eqs. (10.7.1)–(10.7.3), now take the form

$$\mathbf{R}(\mu', \varphi'; \mu, \varphi) = \mathbf{\Delta}_{3}[\mathbf{R}(\mu, \varphi; \mu', \varphi')]^{\mathrm{T}}\mathbf{\Delta}_{3}, \tag{12.6.9}$$

$$\mathbf{R}^{\dagger}(\mu', \varphi'; \mu, \varphi) = \mathbf{\Delta}_{3}[\mathbf{R}^{\dagger}(\mu, \varphi; \mu', \varphi')]^{\mathrm{T}}\mathbf{\Delta}_{3}, \tag{12.6.10}$$

$$\mathbf{T}^{\dagger}(\mu', \varphi'; \mu, \varphi) = \mathbf{\Delta}_{3}[\mathbf{T}(\mu, \varphi; \mu', \varphi')]^{\mathrm{T}}\mathbf{\Delta}_{3}. \tag{12.6.11}$$

Applying the derivation technique presented in Section 10.7 to the invariant imbedding equations (12.4.1)–(12.4.4) and using the symmetry relation (12.6.2), we further obtain

$$\mathbf{R}(\mu, -\varphi; \mu', -\varphi') = \mathbf{\Delta}_{34}\mathbf{R}(\mu, \varphi; \mu', \varphi')\mathbf{\Delta}_{34}, \tag{12.6.12}$$

$$\mathbf{T}(\mu, -\varphi; \mu', -\varphi') = \mathbf{\Delta}_{34}\mathbf{T}(\mu, \varphi; \mu', \varphi')\mathbf{\Delta}_{34}, \tag{12.6.13}$$

$$\mathbf{R}^{\dagger}(\mu, -\varphi; \mu', -\varphi') = \mathbf{\Delta}_{34}\mathbf{R}^{\dagger}(\mu, \varphi; \mu', \varphi')\mathbf{\Delta}_{34}, \tag{12.6.14}$$

$$\mathbf{T}^{\dagger}(\mu, -\varphi; \mu', -\varphi') = \mathbf{\Delta}_{34}\mathbf{T}^{\dagger}(\mu, \varphi; \mu', \varphi')\mathbf{\Delta}_{34}, \tag{12.6.15}$$

where we have taken into account that

$$\int_0^{2\pi} \mathrm{d}\varphi''\, \mathbf{R}(\mu, \varphi; \mu'', -\varphi'')\widetilde{\mathbf{Z}}(0, \mu'', -\varphi''; \mu', \varphi')$$

$$= \int_0^{2\pi} \mathrm{d}\varphi''\, \mathbf{R}(\mu, \varphi; \mu'', \varphi'')\widetilde{\mathbf{Z}}(0, \mu'', \varphi''; \mu', \varphi') \tag{12.6.16}$$

and analogously for other integrals of the same type. Combining Eqs. (12.6.9)–(12.6.15) yields three more symmetry relations:

$$\mathbf{R}(\mu', -\varphi'; \mu, -\varphi) = \mathbf{\Delta}_4[\mathbf{R}(\mu, \varphi; \mu', \varphi')]^{\mathrm{T}}\mathbf{\Delta}_4, \tag{12.6.17}$$

$$\mathbf{R}^{\dagger}(\mu', -\varphi'; \mu, -\varphi) = \mathbf{\Delta}_4[\mathbf{R}^{\dagger}(\mu, \varphi; \mu', \varphi')]^{\mathrm{T}}\mathbf{\Delta}_4, \tag{12.6.18}$$

$$\mathbf{T}^{\dagger}(\mu', -\varphi'; \mu, -\varphi) = \mathbf{\Delta}_4[\mathbf{T}(\mu, \varphi; \mu', \varphi')]^{\mathrm{T}}\mathbf{\Delta}_4. \tag{12.6.19}$$

If the scattering slab is homogeneous, then the only difference between the cases of illumination from above and illumination from below is the sense in which the azimuth angle is reckoned. Hence

$$\mathbf{R}^{\dagger}(\mu, \varphi; \mu', \varphi') = \mathbf{R}(\mu, -\varphi; \mu', -\varphi'), \tag{12.6.20}$$

$$\mathbf{T}^{\dagger}(\mu, \varphi; \mu', \varphi') = \mathbf{T}(\mu, -\varphi; \mu', -\varphi'). \tag{12.6.21}$$

Thus, for a homogeneous layer, $\mathbf{R}^{\dagger}$ and $\mathbf{T}^{\dagger}$ never need to be computed if $\mathbf{R}$ and $\mathbf{T}$ are known. By combining Eqs. (12.6.9)–(12.6.15) and (12.6.17)–(12.6.21), one can derive numerous additional symmetry relations; they are listed in Section 4.5 of Hovenier *et al.* (2004).

12.6.3 Matrices describing the internal field

By iterating the adding equations (12.3.1)–(12.3.4) (with the first term on the right-hand side of each equation serving as the initial approximation) and then applying the symmetry relations (12.6.12)–(12.6.15), it is straightforward to derive the symmetry relations for the matrices $\mathbf{U}$, $\mathbf{D}$, $\mathbf{U}^{\dagger}$, and $\mathbf{D}^{\dagger}$:

$$\mathbf{U}(\tau; \mu, -\varphi; \mu', -\varphi') = \mathbf{\Delta}_{34}\mathbf{U}(\tau; \mu, \varphi; \mu', \varphi')\mathbf{\Delta}_{34}, \tag{12.6.22}$$

$$\mathbf{D}(\tau; \mu, -\varphi; \mu', -\varphi') = \mathbf{\Delta}_{34}\mathbf{D}(\tau; \mu, \varphi; \mu', \varphi')\mathbf{\Delta}_{34}, \tag{12.6.23}$$

$$\mathbf{U}^{\dagger}(\tau; \mu, -\varphi; \mu', -\varphi') = \mathbf{\Delta}_{34}\mathbf{U}^{\dagger}(\tau; \mu, \varphi; \mu', \varphi')\mathbf{\Delta}_{34}, \tag{12.6.24}$$

$$\mathbf{D}^{\dagger}(\tau; \mu, -\varphi; \mu', -\varphi') = \mathbf{\Delta}_{34}\mathbf{D}^{\dagger}(\tau; \mu, \varphi; \mu', \varphi')\mathbf{\Delta}_{34} \tag{12.6.25}$$

(Hovenier *et al.*, 2004).

If the scattering slab is homogeneous, then the radiation field at an optical depth τ in response to illumination from above must be the essentially same as that at an optical depth $\mathcal{T} - \tau$ in response to illumination from below, the only difference being the sense in which the azimuth angle is reckoned. Hence

$$\mathbf{U}^{\dagger}(\mathcal{T} - \tau; \mu, \varphi; \mu', \varphi') = \mathbf{\Delta}_{34}\mathbf{U}(\tau; \mu, -\varphi; \mu', -\varphi')\mathbf{\Delta}_{34}, \tag{12.6.26}$$

$$\mathbf{D}^{\dagger}(\mathcal{T} - \tau; \mu, \varphi; \mu', \varphi') = \mathbf{\Delta}_{34}\mathbf{D}(\tau; \mu, -\varphi; \mu', -\varphi')\mathbf{\Delta}_{34}. \tag{12.6.27}$$

12.6.4 Perpendicular directions

Propagation directions perpendicular to the upper and lower boundaries of a plane-parallel scattering slab, both for the incident light ($\mu_0 = 1$) and for the scattered light ($u = \pm 1$), are rather special because they do not provide an implicit way to specify the meridional plane and azimuth and, thus, the plane of reference for defining the Stokes parameters. The singularity of the perpendicular directions gives rise to many additional symmetry relations for the phase matrix, the reflection and transmission matrices, and the matrices describing the internal radiation. These relations are studied in detail in Hovenier and de Haan (1985) and Hovenier *et al.* (2004) and are important for practical purposes because perpendicular propagation directions are often encountered in actual observations and measurements.

12.7 Fourier decomposition

The fact that the normalized phase matrix depends on the difference of azimuth angles of the scattering and incidence directions allows an efficient treatment of the azimuthal dependence of the specific intensity vector by means of a Fourier-series decomposition of the VRTE. Since the corresponding formulas are much simpler in the circular-polarization representation, we will discuss the Fourier expansion of only the circular-polarization specific intensity column vector. The extension of this formalism to the standard Stokes representation is described in detail in Hovenier *et al.* (2004).

12.7.1 Fourier decomposition of the VRTE

Recalling Eqs. (2.6.10), (2.6.12), (2.6.14), and (2.6.16), we easily derive from Eq. (12.1.4) the VRTE for the circular-polarization specific intensity column vector:

$$u\frac{\mathrm{d}\tilde{\mathbf{I}}^{\mathrm{CP}}(\tau,u,\varphi)}{\mathrm{d}\tau} = -\tilde{\mathbf{I}}^{\mathrm{CP}}(\tau,u,\varphi) + \frac{\varpi(\tau)}{4\pi}\int_0^{2\pi}\mathrm{d}\varphi'\int_{-1}^{+1}\mathrm{d}u'\,\tilde{\mathbf{Z}}^{\mathrm{CP}}(\tau;u,\varphi;u',\varphi')\,\tilde{\mathbf{I}}^{\mathrm{CP}}(\tau,u',\varphi'), \tag{12.7.1}$$

where

$$\tilde{\mathbf{I}}^{\mathrm{CP}}(\tau,u,\varphi) = \mathbf{A}\tilde{\mathbf{I}}(\tau,u,\varphi), \tag{12.7.2}$$

$$\tilde{\mathbf{Z}}^{\mathrm{CP}}(\tau;u,\varphi;u',\varphi') = \mathbf{A}\tilde{\mathbf{Z}}(\tau;u,\varphi;u',\varphi')\mathbf{A}^{-1}. \tag{12.7.3}$$

Let us first assume that $0 < \varphi - \varphi' < \pi$ and expand $\tilde{\mathbf{I}}^{\mathrm{CP}}$ and $\tilde{\mathbf{Z}}^{\mathrm{CP}}$ in the following Fourier series:

$$\tilde{\mathbf{I}}^{\mathrm{CP}}(\tau,u,\varphi) = \sum_{m=-\infty}^{\infty}\tilde{\mathbf{I}}^{\mathrm{CP},m}(\tau,u)\exp(\mathrm{i}m\varphi), \tag{12.7.4}$$

$$\widetilde{\mathbf{Z}}^{\mathrm{CP}}(\tau;u,\varphi;u',\varphi') = \sum_{m=-\infty}^{\infty} \widetilde{\mathbf{Z}}^{\mathrm{CP},m}(\tau;u,u')\exp[\mathrm{i}m(\varphi-\varphi')]. \tag{12.7.5}$$

Substituting Eqs. (12.7.4) and (12.7.5) in Eq. (12.7.1) and taking into account that

$$\int_0^{2\pi} \mathrm{d}\varphi \exp[\mathrm{i}(m-m')\varphi] = 2\pi\delta_{mm'}, \tag{12.7.6}$$

where $\delta_{mm'}$ is the Kronecker delta defined by Eq. (F.1.9), yields

$$u\frac{\mathrm{d}\widetilde{\mathbf{I}}^{\mathrm{CP},m}(\tau,u)}{\mathrm{d}\tau} = -\widetilde{\mathbf{I}}^{\mathrm{CP},m}(\tau,u) + \frac{\varpi(\tau)}{2}\int_{-1}^{+1}\mathrm{d}u'\,\widetilde{\mathbf{Z}}^{\mathrm{CP},m}(\tau;u,u')\,\widetilde{\mathbf{I}}^{\mathrm{CP},m}(\tau,u'),$$
$$m = 0, \pm 1, \pm 2, \ldots. \tag{12.7.7}$$

Thus each Fourier component of the circular-polarization specific intensity column vector is a function of only one angular variable and satisfies a separate simplified VRTE. Each term in the Fourier series (12.7.4) can be treated independently, thereby allowing a significant reduction in computer storage requirements. The matrices $\mathbf{U}$, $\mathbf{D}$, $\mathbf{U}^{\dagger}$, $\mathbf{D}^{\dagger}$, $\mathbf{R}$, $\mathbf{T}$, $\mathbf{R}^{\dagger}$, and $\mathbf{T}^{\dagger}$ can also be expanded in Fourier series, thereby resulting in separate adding, invariant-imbedding, and order-of-scattering equations for each Fourier index m. An additional advantage of the Fourier decomposition is that the numerical solution of these equations becomes much faster with increasing m.

12.7.2 Fourier components of the phase matrix

Perhaps the greatest advantage of the Fourier decomposition of the VRTE is that the Fourier components of the phase matrix can be computed very efficiently using simple analytical expressions (Kuščer and Ribarič, 1959). Let us first re-write Eq. (11.12.23) as follows:

$$\widetilde{Z}_{pq}^{\mathrm{CP}}(\tau;u,\varphi;u',\varphi') = \widetilde{F}_{pq}^{\mathrm{CP}}(\Theta)\exp[-\mathrm{i}(p\sigma_2+q\sigma_1)], \qquad p,q = 2,0,-0,-2, \tag{12.7.8}$$

where the angles Θ, σ_1, and σ_2 are shown in Fig. 12.7.1 (cf. Fig. 11.3.1(a)). Direct comparison with Fig. F.7.1 and the use of Eqs. (F.6.1)–(F.6.3) allows us to reformulate the addition theorem, Eq. (F.7.8), in terms of the generalized spherical functions:

$$P_{pq}^{s}(\cos\Theta)\exp[-\mathrm{i}(p\sigma_2+q\sigma_1)] = \sum_{m=-s}^{s}(-1)^m P_{pm}^{s}(u)P_{mq}^{s}(u')\exp[\mathrm{i}m(\varphi'-\varphi)],$$
$$p,q = 2,0,-0,-2. \tag{12.7.9}$$

Recalling now Eq. (11.12.12), we derive from Eq. (12.7.8)

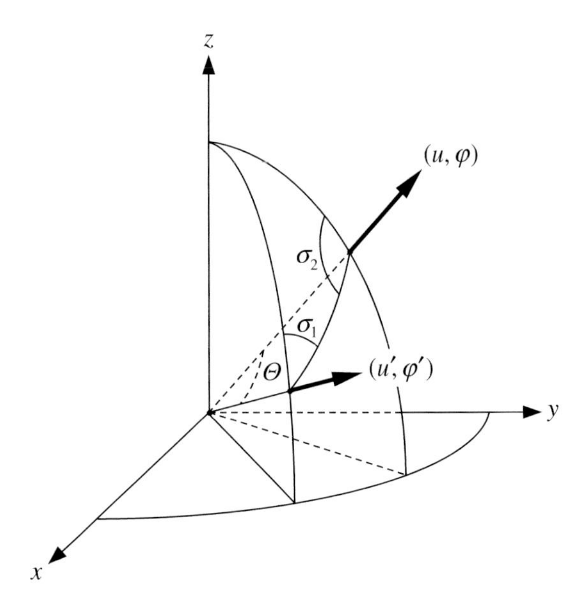

Figure 12.7.1. Illustration of Eq. (12.7.8).

$$\widetilde{Z}_{pq}^{\mathrm{CP}}(\tau; u, \varphi; u', \varphi') = \sum_{s=0}^{s_{\max}} g_{pq}^{s}(\tau)\, P_{pq}^{s}(\cos\Theta) \exp[-\mathrm{i}(p\sigma_2 + q\sigma_1)]$$
$$= \sum_{s=0}^{s_{\max}} g_{pq}^{s}(\tau) \sum_{m=-s}^{s} (-1)^m P_{pm}^{s}(u)\, P_{mq}^{s}(u') \exp[\mathrm{i}m(\varphi' - \varphi)],$$
$$p, q = 2, 0, -0, -2. \quad (12.7.10)$$

The final result follows from the comparison of Eqs. (12.7.5) and (12.7.10);

$$\widetilde{Z}_{pq}^{\mathrm{CP},-m}(\tau; u, u') = (-1)^m \sum_{s=|m|}^{s_{\max}} g_{pq}^{s}(\tau)\, P_{pm}^{s}(u)\, P_{mq}^{s}(u'),$$
$$p, q = 2, 0, -0, -2. \quad (12.7.11)$$

This remarkably simple and compact formula affords a very efficient numerical procedure for computing the Fourier components of the circular-polarization phase matrix. Although Eqs. (12.7.5) and (12.7.11) were derived assuming that $0 < \varphi - \varphi' < \pi$, the reader can easily verify that they exactly reproduce the symmetry relation (11.12.29),

$$\widetilde{Z}_{pq}^{\mathrm{CP}}(\tau; u, -\varphi; u', -\varphi') = \widetilde{Z}_{-p,-q}^{\mathrm{CP}}(\tau; u, \varphi; u', \varphi'), \quad (12.7.12)$$

and are, therefore, valid for any φ and φ'.

From Eqs. (12.7.11), (F.6.2), (F.6.3), and (11.12.14), one can easily derive the following symmetry relations:

$$\widetilde{\mathbf{Z}}^{\mathrm{CP},m}(\tau;-u,-u') = \widetilde{\mathbf{Z}}^{\mathrm{CP},-m}(\tau;u,u'), \tag{12.7.13}$$

$$\widetilde{\mathbf{Z}}^{\mathrm{CP},m}(\tau;-u',-u) = [\widetilde{\mathbf{Z}}^{\mathrm{CP},-m}(\tau;u,u')]^{\mathrm{T}}, \tag{12.7.14}$$

$$\widetilde{Z}_{pq}^{\mathrm{CP},-m}(\tau;u,u') = \widetilde{Z}_{-p,-q}^{\mathrm{CP},m}(\tau;u,u'). \tag{12.7.15}$$

These, in turn, give rise to symmetry relations for the Fourier components of the reflection and transmission matrices as well as for the matrices describing the internal radiation field; they are listed in Domke and Yanovitskij (1986).

12.8 Scalar approximation

Equation (12.1.4) can be further simplified by neglecting polarization and so replacing the specific intensity vector by its first element (i.e., the specific intensity) and the normalized phase matrix by its (1, 1) element (i.e., the phase function):

$$\begin{aligned} u\frac{\mathrm{d}\widetilde{I}(\tau,u,\varphi)}{\mathrm{d}\tau} &= -\widetilde{I}(\tau,u,\varphi) \\ &\quad + \frac{\varpi(\tau)}{4\pi}\int_0^{2\pi}\mathrm{d}\varphi'\int_{-1}^{+1}\mathrm{d}u'\,\widetilde{Z}_{11}(\tau;u,\varphi;u',\varphi')\widetilde{I}(\tau,u',\varphi') \\ &= -\widetilde{I}(\tau,u,\varphi) + \frac{\varpi(\tau)}{4\pi}\int_0^{2\pi}\mathrm{d}\varphi'\int_{-1}^{+1}\mathrm{d}u'\,a_1(\tau;\Theta)\widetilde{I}(\tau,u',\varphi'), \end{aligned} \tag{12.8.1}$$

where

$$\Theta = \arccos[uu' + (1-u^2)^{1/2}(1-u'^2)^{1/2}\cos(\varphi-\varphi')] \tag{12.8.2}$$

is the scattering angle (see Eqs. (11.3.4) and (11.10.18)). Although ignoring the vector nature of light and replacing the exact VRTE by its approximate scalar counterpart has no rigorous physical justification, this simplification is widely used when the medium is illuminated by unpolarized light and only the specific intensity of multiply scattered light needs to be computed. The main reason for doing that is a great saving of computer resources. We will discuss the accuracy of the scalar approximation in Section 13.1.

The reader may find it a useful exercise to derive the following formulas:

$$\widetilde{I}(\tau,u,\varphi) = \sum_{m=0}^{\infty}(2-\delta_{m0})\,\widetilde{I}^{m}(\tau,u)\cos m\varphi, \tag{12.8.3}$$

$$\widetilde{Z}_{11}(\tau;u,\varphi;u',\varphi') = \sum_{m=0}^{\infty}(2-\delta_{m0})\,\widetilde{Z}_{11}^{m}(\tau;u,u')\cos m(\varphi-\varphi'), \tag{12.8.4}$$

$$\widetilde{Z}_{11}^{m}(\tau;u,u') = (-1)^m\sum_{s=m}^{s_{\max}}\alpha_1^s(\tau)\,P_{m0}^s(u)\,P_{m0}^s(u')$$

$$= \sum_{s=m}^{s_{\max}} \alpha_1^s(\tau) \frac{(s-m)!}{(s+m)!} P_s^m(u) P_s^m(u'), \tag{12.8.5}$$

where the expansion coefficients $\alpha_1^s(\tau)$ are given by Eq. (11.12.17) and $P_s^m(u)$ are associated Legendre functions given by Eq. (F.5.2).

12.9 Notes and further reading

Classical texts on the scalar and vector theories of radiative transfer in plane-parallel ISMs are those by Chandrasekhar (1950), Kourganoff (1952), Davison (1958), Sobolev (1975), and van de Hulst (1980). Among more recent monographs are those by Minin (1988), Sushkevich *et al.* (1990), Yanovitskij (1997), and Hovenier *et al.* (2004).

Abstract mathematical aspects of the vector and scalar theories of radiative transfer in ISMs have been extensively studied by Vladimirov (1961), Case and Zweifel (1967), Maslennikov (1969, 1989), Williams (1971), van der Mee (1981), Kaper *et al.* (1982), Ershov and Shikhov (1985), Germogenova (1986), and Greenberg *et al.* (1987). Among many other issues, these publications address the problem of existence and uniqueness of solution of the RTE.

The adding/doubling method remains the most elegant, easy-to-implement, and efficient exact technique to solve numerically the scalar or the vector RTE for plane-parallel ISMs. Comprehensive descriptions of this technique and specific implementation issues can be found in Hansen and Travis (1974), de Haan *et al.* (1987), Stammes *et al.* (1989), and Hovenier *et al.* (2004). Useful discussions are also contained in Wiscombe (1976) and Takashima (1985).

Adams and Kattawar (1970) solved the scalar counterparts of the invariant imbedding equations (12.4.1) and (12.4.2) using the standard Runge-Kutta method. Sato *et al.* (1977) solved the scalar version of Eq. (12.4.1) by applying a more efficient numerical technique based on a special predictor–corrector scheme. They showed that for a slab with continuously varying optical properties, using this technique is more efficient than modeling the slab by a stack of a large number of homogeneous layers and applying the adding/doubling method. Mishchenko (1990b) extended their technique to the full vector case.

Efficient iterative solutions of the scalar version of the Ambarzumian equation (12.4.9) have been developed by Dlugach and Yanovitskij (1974) and Mishchenko *et al.* (1999b). The extension to the full vector case has been described by de Rooij (1985) and Mishchenko (1996). The corresponding computer codes are by far the fastest and most accurate means of calculating the (polarized) reflectivity of a semi-infinite homogeneous slab. The computer code described in Mishchenko *et al.* (1999b) is publicly available at http://www.giss.nasa.gov/~crmim/brf.

The monograph edited by Lenoble (1985) provides a comprehensive review of exact numerical solution techniques for the scalar and the full vector RTE in the case of plane-parallel ISMs. Among more recent publications we note those on the spherical harmonics solution (Garcia and Siewert, 1986), the discrete ordinate method (Stamnes *et al.*, 1988; Nakajima and King, 1992; Schulz *et al.*, 1999; Schulz and Stamnes, 2000; Siewert, 2000; Rozanov and Kokhanovsky, 2005), and the so-called F_N method (Garcia and Siewert, 1989).

Benchmark numbers with a guaranteed number of correct decimals have always been instrumental in checking the accuracy of existing computer codes or in the development and testing of new software. Extensive tabular material can be found in the books by van de Hulst (1980) and Lenoble (1985) as well as in the numerous publications referenced therein. Accurate numbers pertaining to multiple scattering of polarized light by polydisperse models of spherical and nonspherical particles have been tabulated in Garcia and Siewert (1986, 1989), Mishchenko (1990b, 1991b), Wauben and Hovenier (1992), and Siewert (2000).

Special effects arise when the scattering medium has a complex macroscopic morphology (e.g., Diner and Martonchik, 1984; Martonchik and Diner, 1985; Evans, 1993; Cairns *et al.*, 2000; Liou, 2002; Marshak and Davis, 2005) or when the incident light is neither a plane electromagnetic wave nor a uniform quasi-monochromatic parallel beam of light of infinite lateral extent (e.g., Mueller and Crosbie, 1997 and references therein). Although the validity of the RTE in many such situations has never been established, the RTT has been widely used to address specific practical problems. Not surprisingly, most applications have been based on the inefficient yet flexible and virtually universal Monte Carlo method. This technique was introduced a long time ago (e.g., Plass and Kattawar, 1968; Marchuk *et al.*, 1980) and remains very popular and often indispensable (e.g., Bruscaglioni *et al.*, 1993; Ambirajan and Look, 1997; Bartel and Hielscher, 2000; Hopcraft *et al.*, 2000; Ishimoto and Masuda, 2002; Barker *et al.*, 2003; Jaillon and Saint-Jalmes, 2003; Vaillon *et al.*, 2004; Postylyakov, 2004; and references therein).

Chapter 13

Illustrative applications of radiative transfer theory

There have been so many applications of radiative transfer in various areas of science and technology that listing even a small fraction of them would take too much space and would hardly be instructive. Therefore, in this chapter we will describe several selected illustrative examples which demonstrate the main qualitative and quantitative effects of multiple scattering and are expected to be of interest to a broad range of customers of the RTE.

13.1 Accuracy of the scalar approximation

We already mentioned in Section 12.8 that although the scalar approximation has no specific physical justification, it has been widely used in situations when the incident light is unpolarized (i.e., $Q_0 = U_0 = V_0 = 0$) and only the specific intensity of multiply scattered light needs to be computed. Comparison of Eqs. (12.3.17)–(12.3.20) and Eq. (11.3.1) shows that the scalar approximation becomes exact in the limit $\mathcal{T} \to 0$, i.e., when the first-order-scattering approximation is applicable. However, the process of multiple scattering engages all the other elements of the phase matrix and may result in significant differences between exact vector and approximate scalar computations of radiative transfer.

Numerical errors in the specific intensity of the reflected light resulting from the neglect of polarization were examined by Hansen (1971a) on the basis of accurate adding/doubling calculations of multiple scattering. He concluded that in most cases the errors of the scalar approximation should be less than or of the order of 1% for light reflected by a cloud of spherical particles with sizes of the order of or larger than the wavelength of light. On the other hand, it has been known since the pioneering work by Chandrasekhar (1950) that the errors can be much greater in the case of a semi-infinite atmosphere with pure Rayleigh scattering.

The aim of this section is to present a systematic survey of the errors induced by the neglect of polarization in radiance calculations for various types of scattering. We begin by considering the classical case of a homogeneous Rayleigh-scattering atmosphere (without and with depolarization) overlying a Lambertian surface with albedo ranging from zero to one (Mishchenko *et al.*, 1994). Then we consider multiple scattering by homogeneous slabs composed of spherical and randomly oriented nonspherical particles with sizes ranging from essentially zero (Rayleigh scatterers) to several times the incident wavelength.

13.1.1 Rayleigh-scattering slabs

The Rayleigh single-scattering law appears both in the RTE for a medium composed of discrete, macroscopic, widely separated particles with sizes much smaller than the wavelength (Section 11.13) and in the RTE derived for a continuous medium with random fluctuations of the refractive index (Papanicolaou and Burridge, 1975). Therefore, the results of this subsection will be applicable to both situations.

In order to take into account the potential anisotropy of molecules forming a continuous random medium, the normalized Stokes scattering matrix for a purely gaseous medium is usually parameterized in the following form:

$$\tilde{\mathbf{F}}(\Theta) = \Delta \begin{bmatrix} \frac{3}{4}(1+\cos^2\Theta) & -\frac{3}{4}\sin^2\Theta & 0 & 0 \\ -\frac{3}{4}\sin^2\Theta & \frac{3}{4}(1+\cos^2\Theta) & 0 & 0 \\ 0 & 0 & \frac{3}{2}\cos\Theta & 0 \\ 0 & 0 & 0 & \Delta'\frac{3}{2}\cos\Theta \end{bmatrix} + (1-\Delta)\begin{bmatrix} 1 & 0 & 0 & 0 \\ 0 & 0 & 0 & 0 \\ 0 & 0 & 0 & 0 \\ 0 & 0 & 0 & 0 \end{bmatrix}, \tag{13.1.1}$$

where

$$\Delta = \frac{1-\delta}{1+\delta/2}, \tag{13.1.2}$$

$$\Delta' = \frac{1-2\delta}{1-\delta}, \tag{13.1.3}$$

and δ is the so-called depolarization factor (Hansen and Travis, 1974). For pure Rayleigh scattering, the depolarization factor vanishes, whereas for most real gases it substantially deviates from zero. For example, δ is close to 0.03 for air and 0.09 for CO_2. The scattering matrix elements for the case of pure Rayleigh scattering are visualized in Fig. 11.13.5.

Let us consider a plane-parallel Rayleigh-scattering slab illuminated from above by an unpolarized parallel beam of light. The symmetries of multiple scattering in plane-parallel ISMs (Subsection 12.6.2) allow us, without loss of generality, to reduce the number of free parameters by fixing the azimuth angle of the incident beam at zero and to reduce by half the range of azimuth angles of the scattering direction. Thus the direction of incidence is specified by a couplet $\{\mu_0, \varphi_0 = 0\}$ and that of the reflected light by a couplet $\{-\mu, \varphi\}$, with $\mu_0 \in [0,1]$, $\mu \in [0,1]$, and $\varphi \in [0°, 180°]$.

Throughout this section, we will supply rigorously calculated "vector" quantities by the superscript "v" and approximate "scalar" quantities by the superscript "s". To quantify the errors in the reflected specific intensity resulting from the use of the scalar approximation, we will use several characteristics. The most general of them is the percent error defined as

$$\varepsilon(\mathcal{T}, \mu, \mu_0, \varphi) = \frac{\tilde{I}^{\mathrm{v}}(0,-\mu,\varphi) - \tilde{I}^{\mathrm{s}}(0,-\mu,\varphi)}{\tilde{I}^{\mathrm{v}}(0,-\mu,\varphi)} \times 100\%,$$

$$\mu_0 \in [0,1], \quad \mu \in [0,1], \quad \varphi \in [0°, 180°]. \tag{13.1.4}$$

Here, $\mathcal{T}$ is the optical thickness of the scattering slab,

$$\tilde{I}^{\mathrm{v}}(0,-\mu,\varphi) = \frac{1}{\pi}\mu_0 R_{11}(\mathcal{T}; \mu, \varphi; \mu_0, 0) I_0 \tag{13.1.5}$$

and

$$\tilde{I}^{\mathrm{s}}(0,-\mu,\varphi) = \frac{1}{\pi}\mu_0 R(\mathcal{T}; \mu, \varphi; \mu_0, 0) I_0 \tag{13.1.6}$$

are the vector and scalar specific intensities of the reflected light, respectively (see Eq. (12.2.12)), $R(\mathcal{T}; \mu, \varphi; \mu_0, \varphi_0)$, called the reflection coefficient, is the quantity replacing the reflection matrix $\mathbf{R}(\mathcal{T}; \mu, \varphi; \mu_0, \varphi_0)$ in the framework of the scalar approximation, and I_0 is the intensity of the incident parallel beam.

The degree to which the scalar approximation underestimates the reflected specific intensity for given $\mathcal{T}$, μ, and μ_0 is characterized by the local underestimation defined by

$$\varepsilon_{\mathrm{u}}(\mathcal{T}, \mu, \mu_0) = \max_{\varphi \in [0°, 180°]} \varepsilon(\mathcal{T}, \mu, \mu_0, \varphi). \tag{13.1.7}$$

Note that for some particular values of μ and μ_0, the local underestimation may be negative, thereby implying that the scalar approximation overestimates the specific intensity for all $\varphi \in [0°, 180°]$. Analogously, the local overestimation is defined as

$$\varepsilon_{\mathrm{o}}(\mathcal{T}, \mu, \mu_0) = \min_{\varphi \in [0°, 180°]} \varepsilon(\mathcal{T}, \mu, \mu_0, \varphi) \tag{13.1.8}$$

so that a positive local overestimation for some μ and μ_0 implies that the scalar approximation underestimates the specific intensity for all $\varphi \in [0°, 180°]$. The maximum underestimation and the maximum overestimation are given by

$$\varepsilon_{\mathrm{u}}^{\max}(\mathcal{T}) = \max_{\mu\in[0,1],\ \mu_0\in[0,1]} \varepsilon_{\mathrm{u}}(\mathcal{T},\mu,\mu_0), \tag{13.1.9}$$

$$\varepsilon_{\mathrm{o}}^{\max}(\mathcal{T}) = -\max_{\mu\in[0,1],\ \mu_0\in[0,1]} \varepsilon_{\mathrm{o}}(\mathcal{T},\mu,\mu_0). \tag{13.1.10}$$

The azimuth angles at which the local underestimation and the local overestimation are reached are denoted by $\varphi_{\mathrm{u}}(\mathcal{T},\mu,\mu_0)$ and $\varphi_{\mathrm{o}}(\mathcal{T},\mu,\mu_0)$, respectively, while the corresponding phase angles are denoted by $\alpha_{\mathrm{u}}(\mathcal{T},\mu,\mu_0)$ and $\alpha_{\mathrm{o}}(\mathcal{T},\mu,\mu_0)$. The phase angle is a quantity widely used in planetary astrophysics and defined as the angle between the reflection direction, $\{-\mu,\varphi\}$, and the direction opposite to the incidence direction, $\{-\mu_0,\pi\}$.

In both scalar and vector calculations of light reflection by finite atmospheres, we solved numerically the invariant imbedding equation (12.4.1) using the predictor–corrector scheme described in Sato *et al.* (1977) and Mishchenko (1990b). In particular, the reflection coefficient and the reflection matrix were Fourier-decomposed in azimuth, and the invariant imbedding equations for each Fourier mode were converted into a system of ordinary differential equations by replacing the remaining μ-integrals on the interval [0, 1] with 30-point Gaussian quadrature sums (see Section 10.4). The right-hand side of Eq. (13.1.4) was then evaluated at the respective Gaussian values of μ and μ_0 and at 61 equidistant azimuth angles φ = 0°, 3°, …, 177°, and 180°.

To model approximately more complex situations when a gaseous slab is bounded from below by a particulate surface such as a snow or a soil surface, the initial condition (12.4.5) is replaced by the initial condition

$$\mathbf{R}(\mathcal{T};\mu,\varphi;\mu_0,\varphi_0)\big|_{\mathcal{T}=0} \equiv \begin{bmatrix} A_{\mathrm{L}} & 0 & 0 & 0 \\ 0 & 0 & 0 & 0 \\ 0 & 0 & 0 & 0 \\ 0 & 0 & 0 & 0 \end{bmatrix}, \qquad A_{\mathrm{L}} \in [0,1]. \tag{13.1.11}$$

The above corresponds to an imaginary semi-infinite particulate medium reflecting light according to the isotropic so-called Lambert law.

In order to keep the discussion to a reasonable size, we will display only the most representative numerical results that illustrate our basic conclusions. Figures 13.1.1–13.1.6 depict the maximum underestimation $\varepsilon_{\mathrm{u}}^{\max}(\mathcal{T})$ and the maximum overestimation $\varepsilon_{\mathrm{o}}^{\max}(\mathcal{T})$ for different values of the single-scattering albedo ϖ, the depolarization factor δ, and the Lambertian albedo A_{L}. These plots can be used in practice to decide whether one may use the scalar approximation or must invoke the rigorous VRTE.

It is seen that the vector–scalar differences decrease with increasing depolarization factor and/or increasing surface albedo. For optically thin layers ($\mathcal{T} \lesssim 1$), the differences always decrease with decreasing single-scattering albedo. To explain this trait,

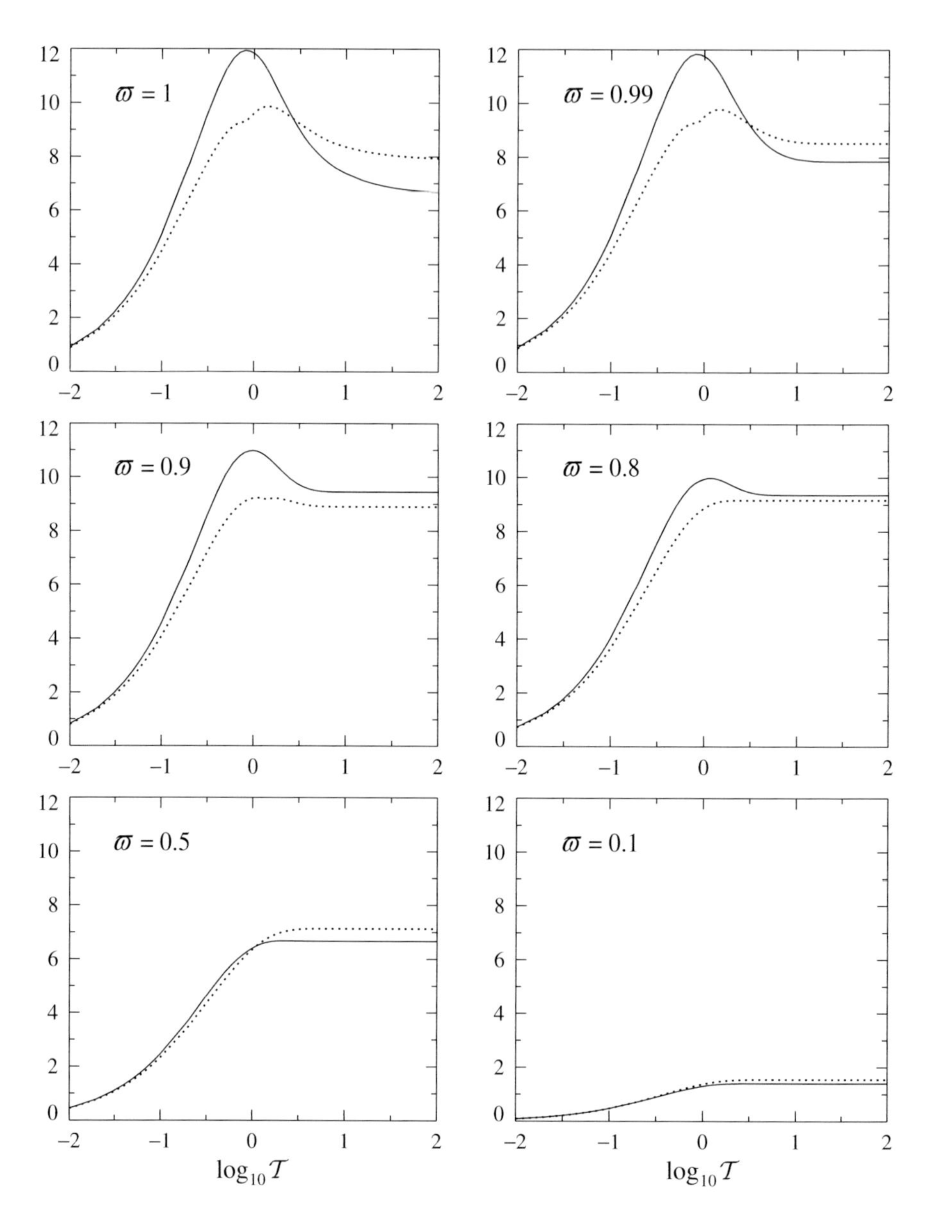

Figure 13.1.1. Maximum overestimation $\varepsilon_{\text{o}}^{\max}(\mathcal{T})$ (in %, solid curves) and maximum underestimation $\varepsilon_{\text{u}}^{\max}(\mathcal{T})$ (in %, dotted curves) versus optical thickness $\mathcal{T}$ for $\delta = 0$, $A_{\text{L}} = 0$, and $\varpi = 1$, 0.99, 0.9, 0.8, 0.5, and 0.1.

let us expand the "vector" and "scalar" reflected intensities in the Newmann order-of-scattering series following the approach outlined in Section 12.5:

$$\widetilde{I}^{\text{v}}(0, -\mu, \varphi) = \sum_{n=1}^{\infty} \varpi^n \widetilde{I}_n^{\text{v}}(0, -\mu, \varphi), \tag{13.1.12}$$

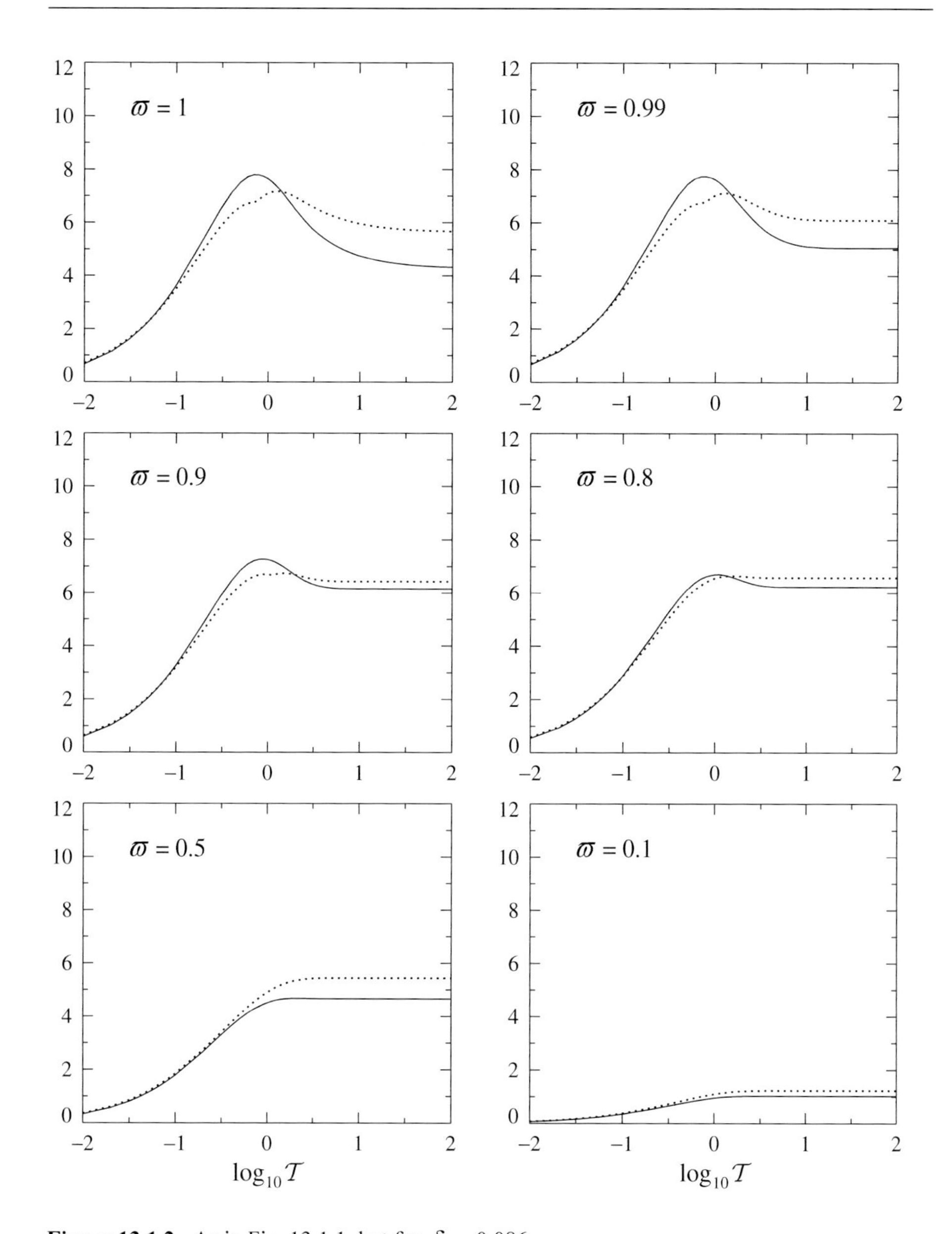

Figure 13.1.2. As in Fig. 13.1.1, but for $\delta = 0.086$.

$$\widetilde{I}^{\mathrm{s}}(0, -\mu, \varphi) = \sum_{n=1}^{\infty} \varpi^{n} \widetilde{I}_{n}^{\mathrm{s}}(0, -\mu, \varphi), \tag{13.1.13}$$

where $\widetilde{I}_n^{\mathrm{v}}(0, -\mu, \varphi)$ and $\widetilde{I}_n^{\mathrm{s}}(0, -\mu, \varphi)$ are independent of ϖ. In the limit $\mathcal{T} \to 0$, we may keep only the contributions of the first and second orders of scattering and rewrite Eq. (13.1.4) as follows:

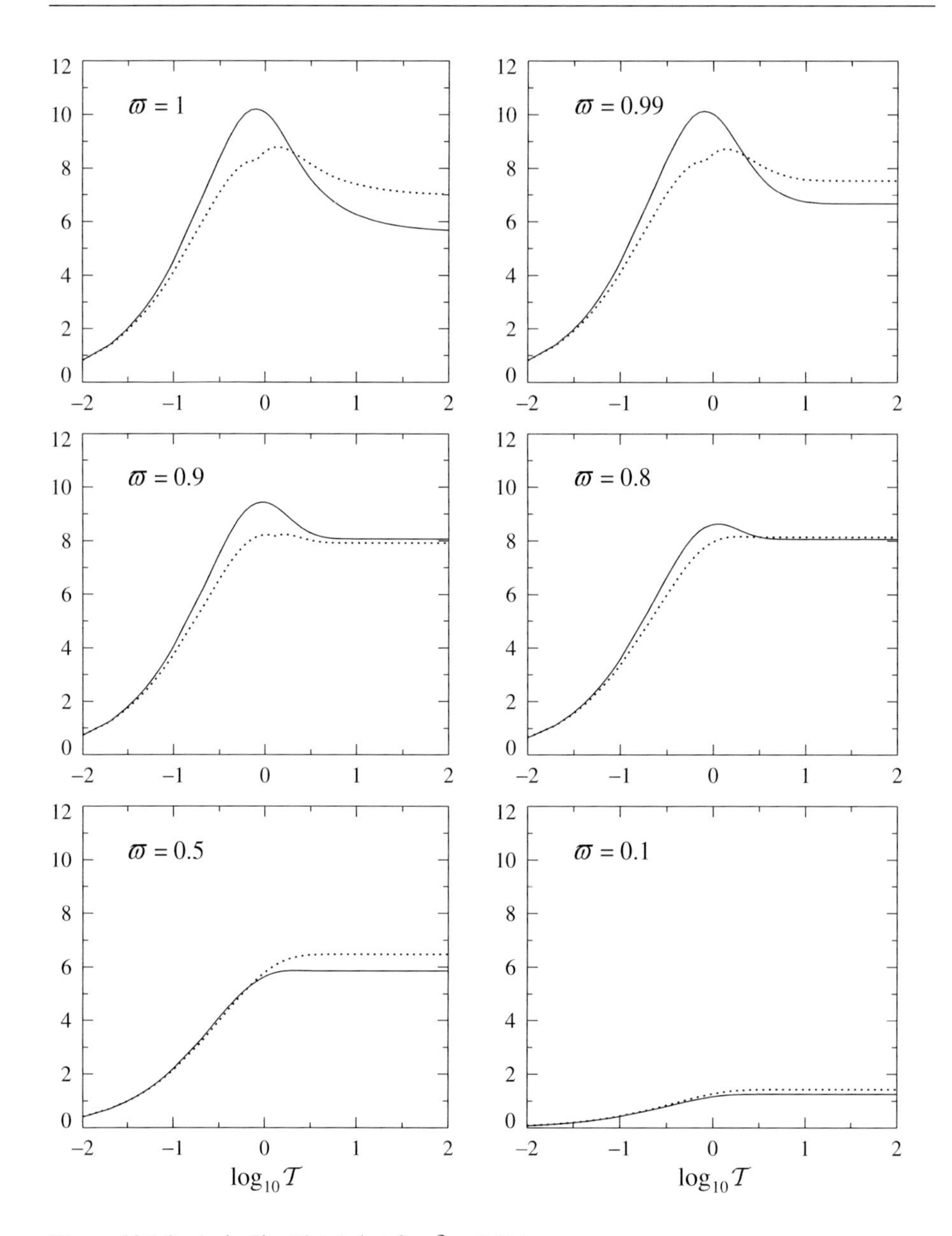

Figure 13.1.3. As in Fig. 13.1.1, but for $\delta = 0.031$.

$$\varepsilon = \frac{\varpi(\widetilde{I}_2^{\mathrm{v}} - \widetilde{I}_2^{\mathrm{s}})}{\widetilde{I}_1^{\mathrm{v}} + \varpi\widetilde{I}_2^{\mathrm{v}}} \times 100\%, \tag{13.1.14}$$

where we have taken into account that $\widetilde{I}_1^{\mathrm{s}} \equiv \widetilde{I}_1^{\mathrm{v}}$. It is then easy to see that Eq. (13.1.14) explains indeed the ϖ-dependence of the percent error for $\mathcal{T} \lesssim 1$.

For optically thick atmospheres ($\mathcal{T} \gg 1$), the vector–scalar differences first increase as the single-scattering albedo decreases from 1 to about 0.8, but then decrease

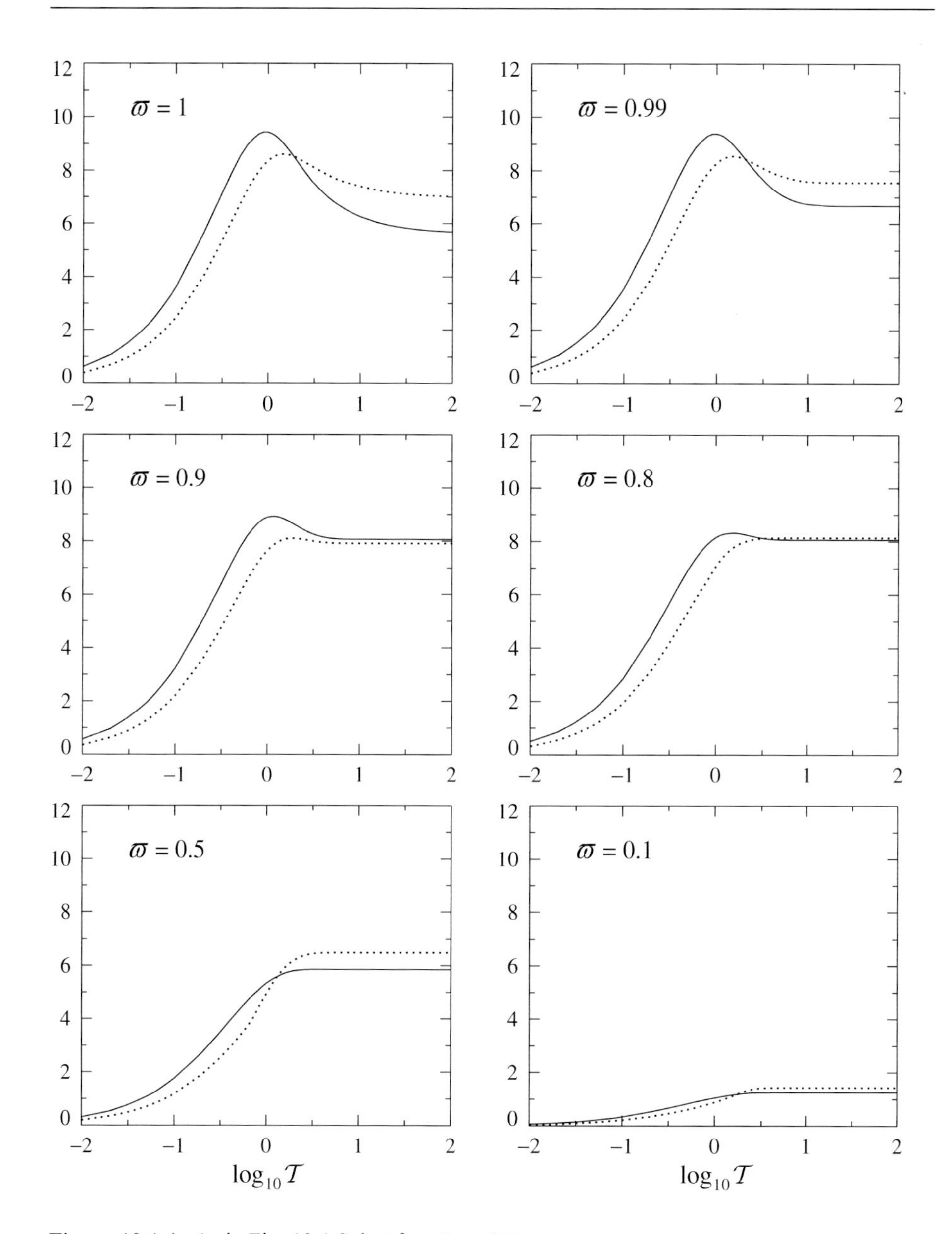

Figure 13.1.4. As in Fig. 13.1.3, but for $A_L = 0.1$.

with further decrease of ϖ. With $\mathcal{T} \to \infty$, all the curves tend to the corresponding asymptotic limits which are independent of the surface albedo and depend only on the single-scattering albedo and depolarization factor. For single-scattering albedos equal or close to one (conservative or nearly conservative scattering) and small surface albedos, the curves have a characteristic maximum at $\mathcal{T} \approx 1$ which disappears with increasing absorption and/or increasing surface albedo. Both errors displayed in Figs. 13.1.1–13.1.6 have roughly the same order of magnitude.

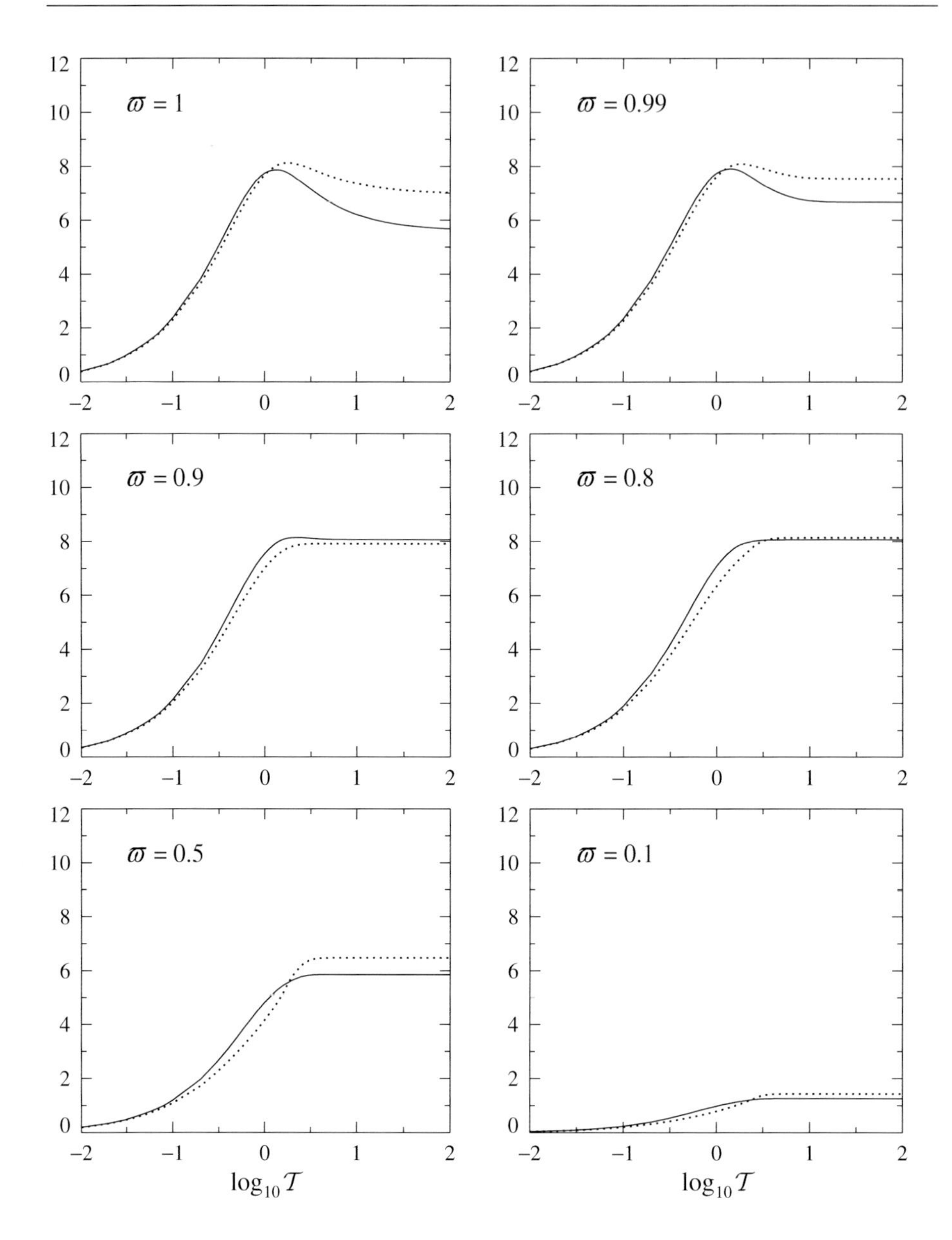

Figure 13.1.5. As in Fig. 13.1.3, but for $A_L = 0.4$.

Figures 13.1.7 and 13.1.8 illustrate the angular distribution of the vector–scalar differences for $\mathcal{T} = 1$, $\delta = 0.031$, $\varpi = 1$, and $A_L = 0$. Note, however, that the contour plots displayed are rather typical and, in conjunction with Figs. 13.1.1–13.1.6, give a general idea of what can be expected for other values of the optical thickness, single-scattering albedo, depolarization factor, and surface albedo. Moreover, we have found that, independently of the parameter values, the local underestimation is always reached in the azimuth plane 180°, i.e.,

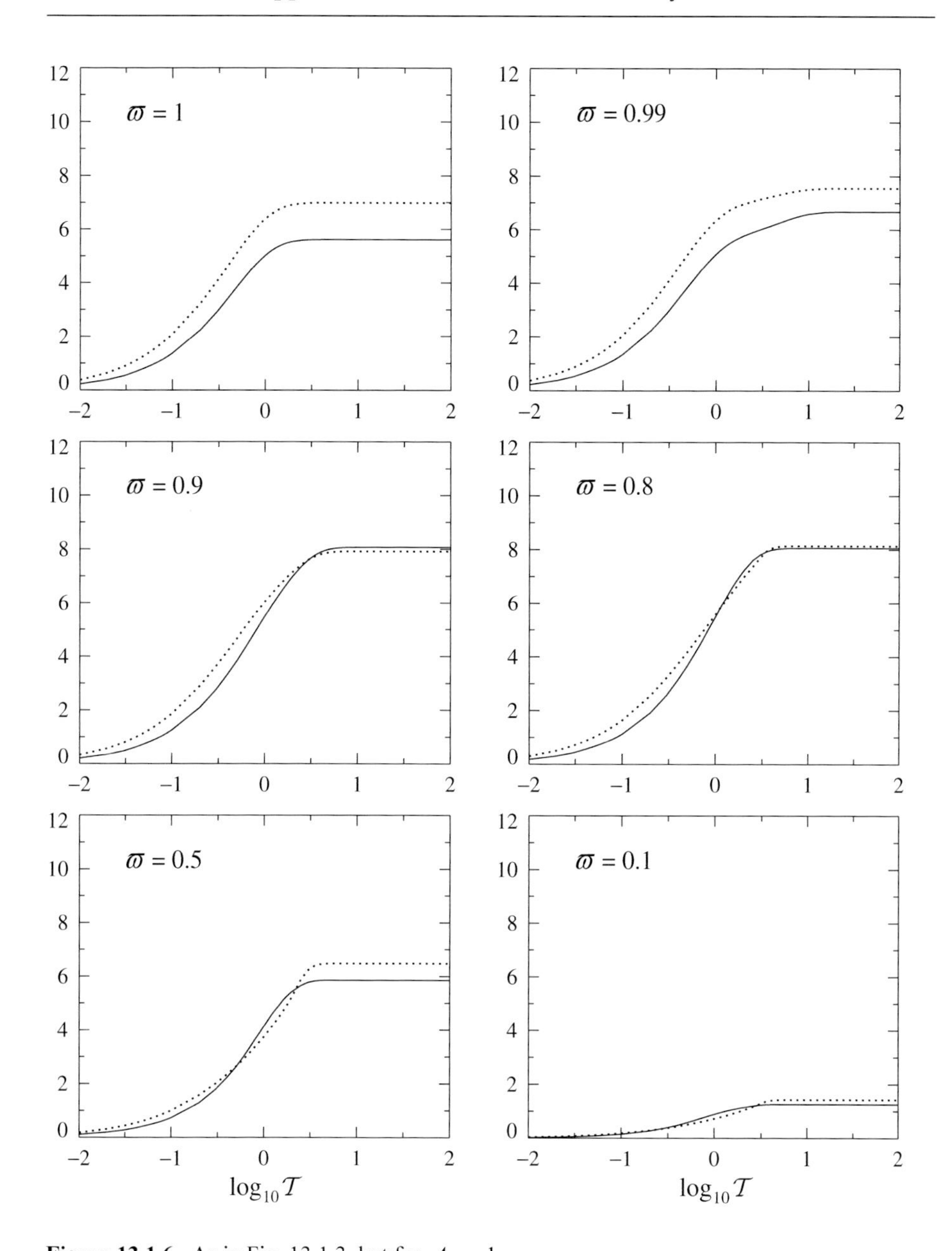

Figure 13.1.6. As in Fig. 13.1.3, but for $A_L = 1$.

$$\varphi_u(\mathcal{T}, \mu, \mu_0) \equiv 180°. \tag{13.1.15}$$

Note that the small step size in azimuth angle in our calculations of $\varphi_u(\mathcal{T}, \mu, \mu_0)$ made the identity (13.1.15) numerically very accurate. Therefore, the contour plot of $\alpha_u(\mathcal{T}, \mu, \mu_0)$ shown in panel (a) of Plate 13.1.1 is exactly the same for any $\mathcal{T}$, δ, ϖ, and A_L. Also, owing to the identity (13.1.15), the contour plot of the percent error $\varepsilon(\mathcal{T}, \mu, \mu_0, \varphi)$ shown in Fig. 13.1.8(a) is at the same time the contour plot of the

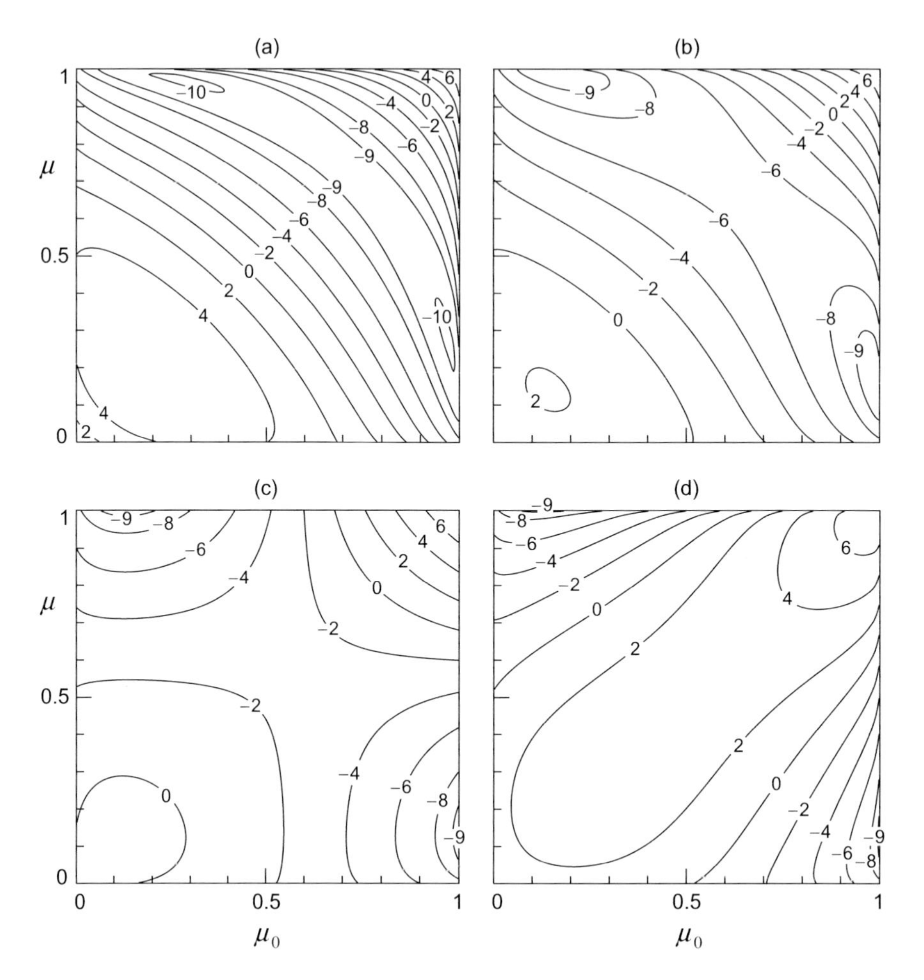

Figure 13.1.7. (a) Contour plot of percent error $\varepsilon(\mathcal{T}, \mu, \mu_0, \varphi)$ for $\mathcal{T} = 1$, $\delta = 0.031$, $\varpi = 1$, $A_{\rm L} = 0$, and $\varphi = 0°$. (b) As in (a), but for $\varphi = 60°$. (c) As in (a), but for $\varphi = 90°$. (d) As in (a), but for $\varphi = 120°$.

local underestimation $\varepsilon_{\rm u}(\mathcal{T}, \mu, \mu_0)$.

The symmetry relations (12.6.9) and (12.6.12) imply that the $(1, 1)$ element of the reflection matrix has the following symmetry property:

$$R_{11}(\mathcal{T}; \mu_0, \varphi; \mu, 0) = R_{11}(\mathcal{T}; \mu, \varphi; \mu_0, 0). \tag{13.1.16}$$

The scalar reflection coefficient possesses the same symmetry property:

$$R(\mathcal{T}; \mu, \varphi; \mu_0, 0) = R(\mathcal{T}; \mu_0, \varphi; \mu, 0). \tag{13.1.17}$$

Equations (13.1.16) and (13.1.17) explain why all panels of Figs. 13.1.7 and 13.1.8 are symmetric with respect to the diagonal $\mu = \mu_0$.

One sees from Fig. 13.1.7 and panel (a) of Fig. 13.1.8 that the vector–scalar differences are somewhat larger in the azimuth planes $\varphi = 0°$ and $\varphi = 180°$, although

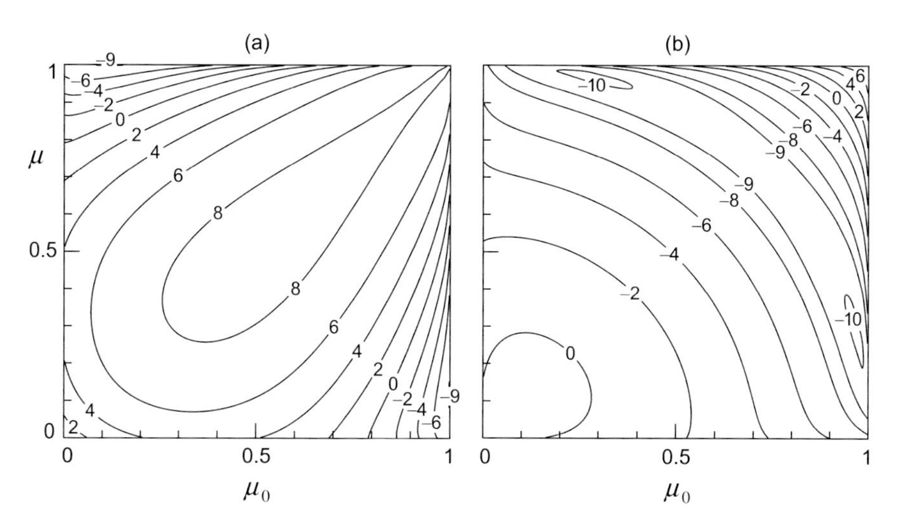

Figure 13.1.8. (a) As in Fig. 13.1.7(a), but for $\varphi = 180°$. This diagram is also a contour plot of local underestimation $\varepsilon_u(\mathcal{T}, \mu, \mu_0)$ (in percent). (b) Contour plot of local overestimation $\varepsilon_o(\mathcal{T}, \mu, \mu_0)$ (in percent) for $\mathcal{T} = 1$, $\delta = 0.031$, $\varpi = 1$, and $A_L = 0$.

they are rather significant in other azimuth planes as well. As was noted above, the local underestimation is always reached at $\varphi = 180°$. On the other hand, for a given μ_0 the azimuth plane of the local overestimation rotates from $0°$ for μ equal or close to one towards $90°$ (or, equivalently, $270°$) with $\mu \to 0$ (see panel (c) of Plate 13.1.1). This trend is illustrated schematically in Fig. 13.1.9.

The most intriguing result of our calculations is that the local overestimation reaches its maximum negative values at phase angles equal or close to $90°$ for all μ_0 (see panel (b) of Plate 13.1.1), whereas the local underestimation is maximal at phase angles equal or close to $0°$ for almost all μ_0 (see panel (a) of Plate 13.1.1). This remarkable trait, which is invariant for all $\mathcal{T}$, δ, ϖ, and A_L, may help to explain why the vector–scalar differences are so large in the case of Rayleigh scattering. We have already stated that for unpolarized incident light and any directions of incidence and reflection, the contribution of light scattered only once to the reflected specific intensity is exactly the same in both vector and scalar formulations and is proportional to $a_1(\pi - \alpha)$, where $a_1(\Theta)$ is the $(1,1)$ element of the normalized Stokes scattering matrix, Θ is the scattering angle given by Eq. (12.8.2), and $\alpha = \pi - \Theta$ is the phase angle. On the other hand, the light scattered in the atmosphere many times can be expected to become largely unpolarized (see Section 13.3). Therefore, it is reasonable to assume that it is light scattered a small number of times, but more than once, which carries the greatest differences between the approximate "scalar" and the rigorous "vector" intensity. Thus, our task is to explain why these differences are maximal at phase angles equal or close to $0°$ and $90°$.

In this explanation, we must take into account the following three factors which become significant when one considers an elementary Rayleigh scattering event with

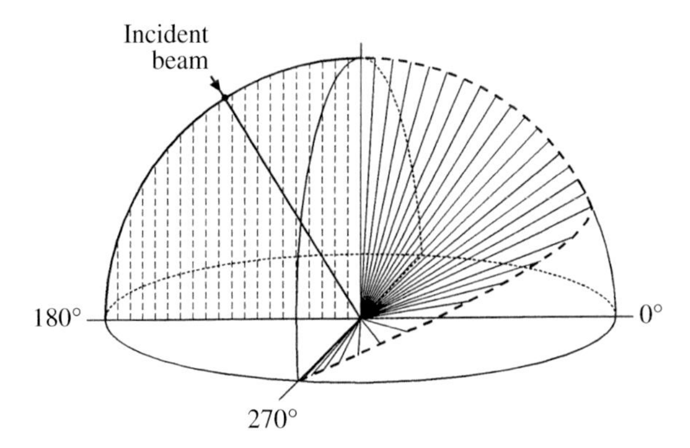

Figure 13.1.9. Rotation of the azimuth plane at which the local overestimation is reached. The incidence direction lies in the azimuth plane $\varphi_0 = 0°$. The dashed curve visualizes the rotation of the azimuth plane of local overestimation as μ decreases from 1 to 0. $\varphi_o(\mathcal{T}, \mu, \mu_0)$ is equal to zero for μ greater than some critical value and then rotates toward $270°$ (or, equivalently, toward $90°$) as μ approaches zero.

respect to the scattering plane rather than with respect to the meridional planes of the incidence and scattering directions:

- Two pronounced features of the Rayleigh scattering matrix are strong polarization at scattering angles near $90°$ (i.e., $-b_1(90°)/a_1(90°) \simeq 100\%$) and the nearly isotropic phase function with strong side scattering (see Fig. 11.13.5).
- Polarization of light directly affects the scattered intensity only through the second Stokes parameter, Q, and the element b_1 of the normalized scattering matrix. The effect is maximal when Q has the largest possible absolute value (i.e., when $|Q|$ is close or equal to I) and, in the case of Rayleigh scattering, when the scattering angle is close to $90°$ (see Eq. (13.1.1)).
- Unlike the intensity, the Stokes parameter Q changes not only due to light scattering, but also due to rotations of the reference plane. The maximum possible absolute change of Q due to a rotation of the reference plane occurs when the angle of rotation is $\pm 90°$ and Q changes its sign while not changing its absolute value (see Eq. (2.8.4)).

Thus, we can expect that low-order (and primarily second-order) light-scattering paths involving right scattering angles and right angles of rotation of the scattering plane will carry the greatest vector–scalar differences. Because the Rayleigh phase function is nearly isotropic, the contribution of such scattering paths to the reflected specific intensity will be rather significant, thus potentially explaining large vector–scalar differences in the case of Rayleigh scattering.

Two such second-order light-scattering paths are shown in Fig. 13.1.10. The scattering path in panel (a) involves two scattering events with $\Theta = 90°$, but does not involve rotations of the scattering plane so that the resulting phase angle is equal to zero. The scattering path shown in panel (b) involves not only two scattering events

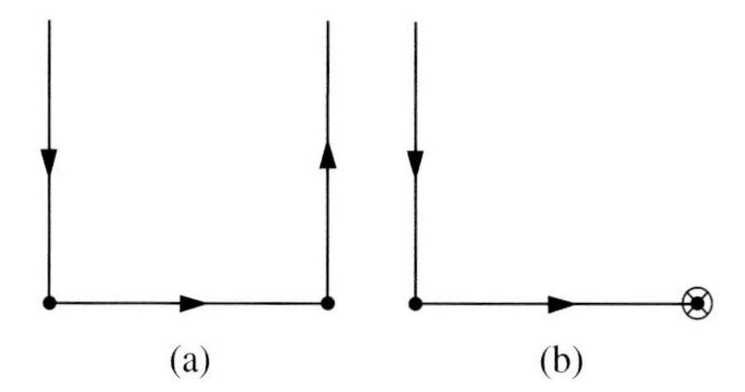

Figure 13.1.10. Two second-order light-scattering paths involving 90° scattering angles. The direction of propagation denoted by a crossed circle is into the paper. The path shown in panel (b) does whereas the path shown in panel (a) does not involve the right-angle rotation of the scattering plane.

with $\Theta = 90°$, but also the right-angle rotation of the scattering plane so that the resulting phase angle is equal to 90°. In the scalar approximation, the contribution of these two paths to the reflected specific intensity is the same and, in the absence of depolarization, is proportional to $[a_1(90°)]^2 = 9/16$. In the exact vector formulation, the contribution of the scattering path in panel (a) to the reflected specific intensity is proportional to the $(1, 1)$ element of the matrix $\widetilde{\mathbf{F}}(90°)\widetilde{\mathbf{F}}(90°)$, which is equal to $18/16$, whereas the contribution of the scattering path in panel (b) is proportional to the $(1, 1)$ element of the matrix $\widetilde{\mathbf{F}}(90°)\mathbf{L}(-90°)\widetilde{\mathbf{F}}(90°)$, which is equal to zero. Thus, in the first case (zero phase angle), the scalar approximation significantly underestimates the specific intensity, whereas in the second case (90° phase angle), it equally significantly overestimates the specific intensity, which is in full agreement with the results displayed in panels (a) and (b) of Plate 13.1.1.

The correctness of this theoretical explanation is corroborated by the results shown in Fig. 13.1.11. These data were calculated for a semi-infinite Rayleigh slab with $\varpi = 1$ and $\delta = 0$ by solving numerically the vector and scalar versions of the Ambarzumian nonlinear integral equation (12.4.9) using the iterative procedures described in Mishchenko (1996) and Mishchenko *et al.* (1999b). It is clearly seen that, in agreement with our theoretical analysis, the vector–scalar differences in the second-order-scattering contribution are much greater than those in the total specific intensity, and the errors are maximal at phase angles 0° and 90°.

Plate 13.1.2 demonstrates that the errors of the scalar approximation in diffuse transmission are quite comparable to those in reflection. The specific angular locations of maximal transmission errors can be explained easily by considering second-order scattering paths involving two scattering events with $\Theta = 90°$ and the resulting phase angles close to 90° and 180°.

13.1.2 Polydisperse spherical particles and spheroids

As we have already pointed out, the results of the previous subsection apply equally to continuous slabs with random fluctuations of the refractive index and to plane-

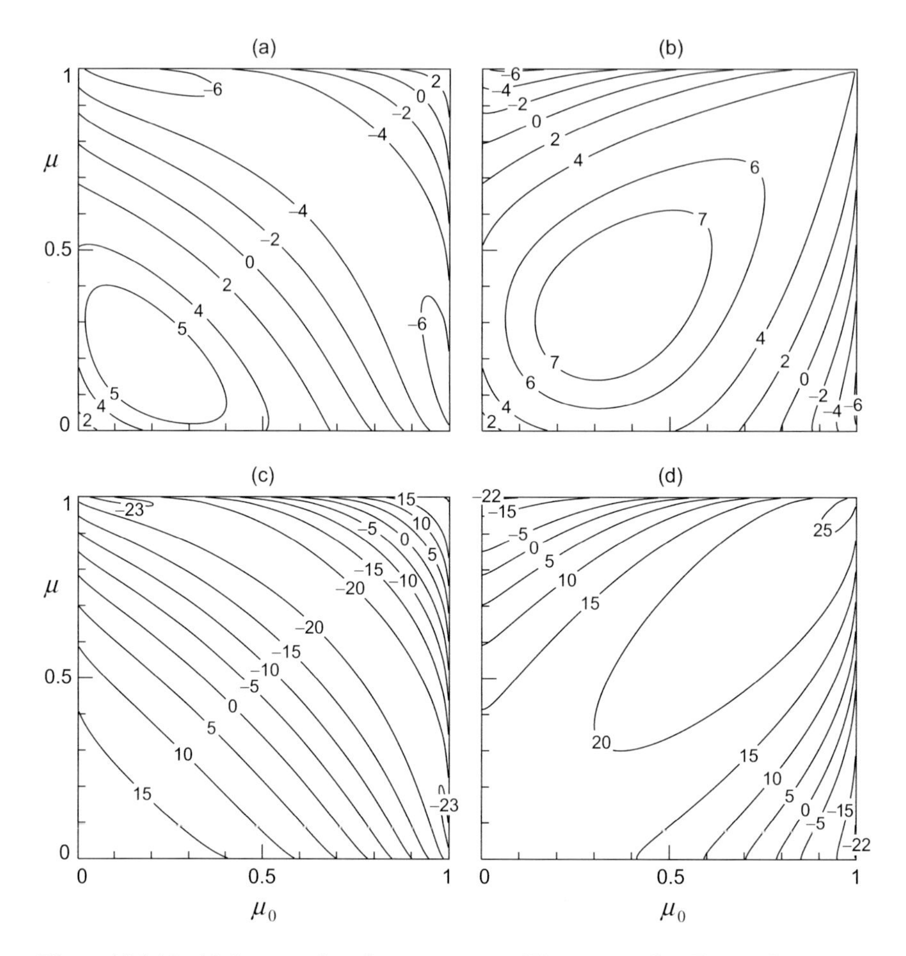

Figure 13.1.11. (a) Contour plot of percent error $\varepsilon(\mathcal{T}, \mu, \mu_0, \varphi)$ for $\mathcal{T} = \infty$, $\delta = 0$, $\varpi = 1$, and $\varphi = 0°$. (b) As in (a), but for $\varphi = 180°$. (c) As in (a), but for the second-order-scattering contribution to the reflected intensity. (d) As in (c), but for $\varphi = 180°$.

parallel discrete random media composed of very small, widely separated particles. Consider now what happens when the size of the discrete particles increases and exceeds the Rayleigh threshold. To determine the upper limit of the vector–scalar differences in the reflected specific intensity, we will consider only nonabsorbing particles with $m_{\text{I}} = 0$ and $\varpi = 1$ and assume that $A_{\text{L}} = 0$.

Figures 13.1.12 and 13.1.13 are analogous to Figs. 13.1.1–13.1.6 but show the results for polydisperse spheres distributed over sizes according to the gamma law (5.3.15) with $r_{\min} = 0$ and $r_{\max} = \infty$. The effective variance of the size distribution is fixed at 0.1, while the effective size parameter $x_{\text{eff}} = k_1 r_{\text{eff}}$ varies from 0.01 to 20. The relative refractive index is 1.33 and represents water droplets at visible wavelengths. Figure 13.1.14 depicts the corresponding phase functions and ratios $-b_1/a_1$. The single-scattering computations were performed using the public-domain Lorenz–Mie code posted at http://www.giss.nasa.gov/~crmim.

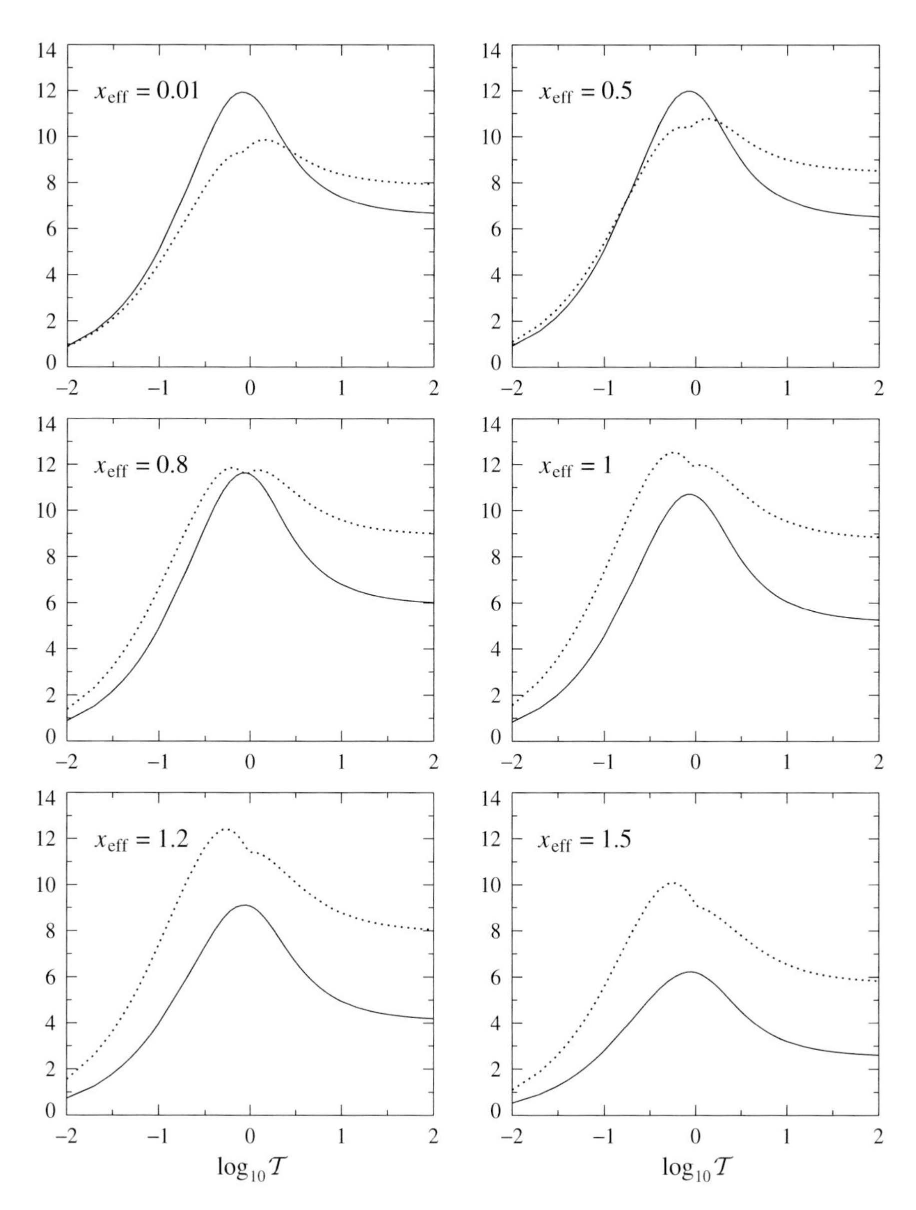

Figure 13.1.12. Maximum overestimation $\varepsilon_{\rm o}^{\rm max}(\mathcal{T})$ (in percent, solid curves) and maximum underestimation $\varepsilon_{\rm u}^{\rm max}(\mathcal{T})$ (in percent, dotted curves) versus optical thickness $\mathcal{T}$ for polydisperse spheres with effective size parameters ranging from 0.01 to 1.5.

Comparison of Fig. 13.1.14 with Fig. 11.13.5 shows that water droplets with $x_{\rm eff} = 0.01$ are still Rayleigh scatterers, which makes the upper left panel of Fig. 13.1.12 virtually indistinguishable from the upper left panel of Fig. 13.1.1. However, the particles with $x_{\rm eff} \gtrsim 2$ are already well outside the Rayleigh domain (see the dashed curves in the two middle panels of Fig. 13.1.14). In the interval of effective size parameters from 0 to about 1.5, the behavior of the maximum overestimation,

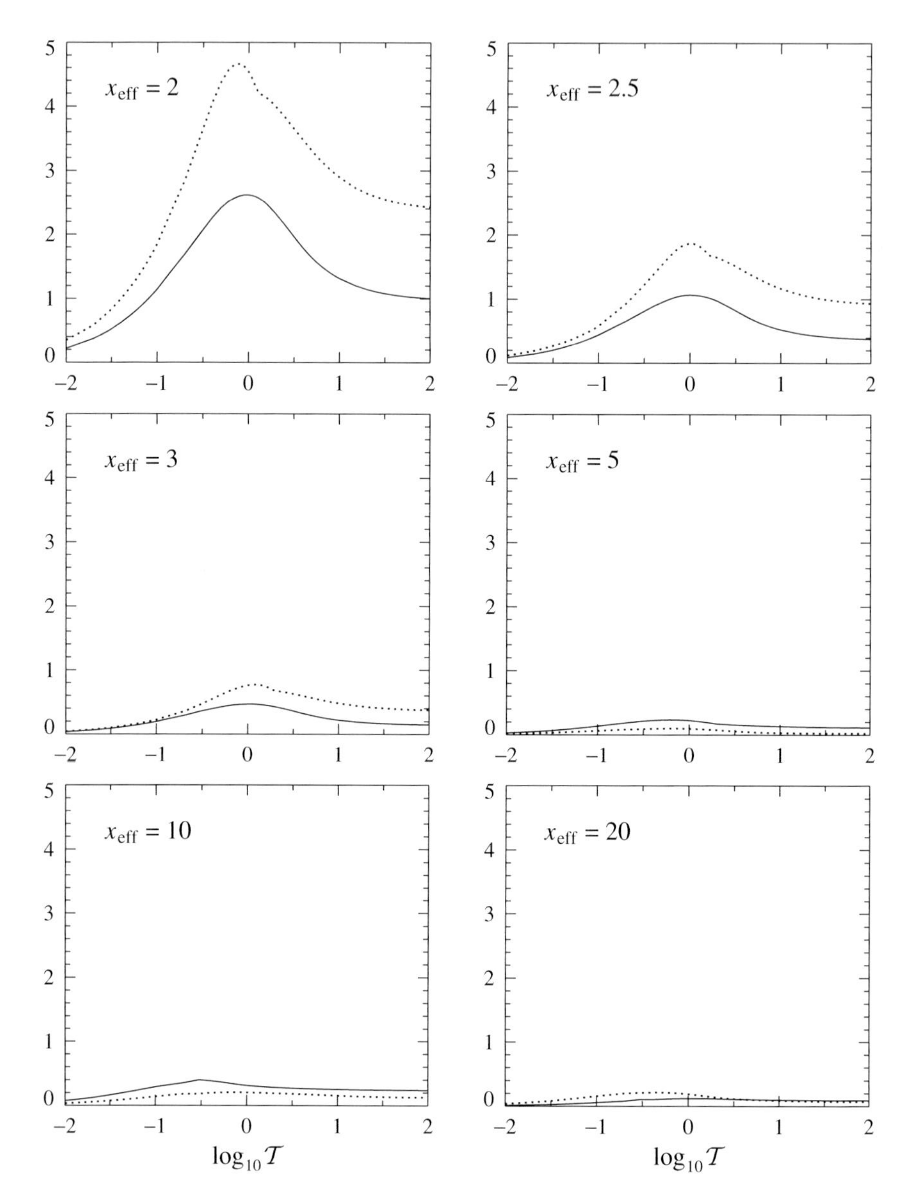

Figure 13.1.13. As in Fig. 13.1.12, but for effective size parameters ranging from 2 to 20.

$\varepsilon_o^{max}(\mathcal{T})$, and the maximum underestimation, $\varepsilon_u^{max}(\mathcal{T})$, is quite different (Fig. 13.1.12). While the $\varepsilon_o^{max}(\mathcal{T})$ decreases monotonically with increasing x_{eff}, the $\varepsilon_u^{max}(\mathcal{T})$ first increases, reaches its maximum values at $x_{eff} \simeq 1$, and only then starts to decline. In fact, the maximum $\varepsilon_u^{max}(\mathcal{T})$ value for $x_{eff} = 1$ is indicative of vector–scalar differences in the reflected specific intensity even exceeding those for pure Rayleigh scattering (cf. the solid and broken curves in the upper left panel of Fig. 13.1.1).

The interval $1 \lesssim x_{eff} \lesssim 3$ is a transition zone where the vector–scalar differences

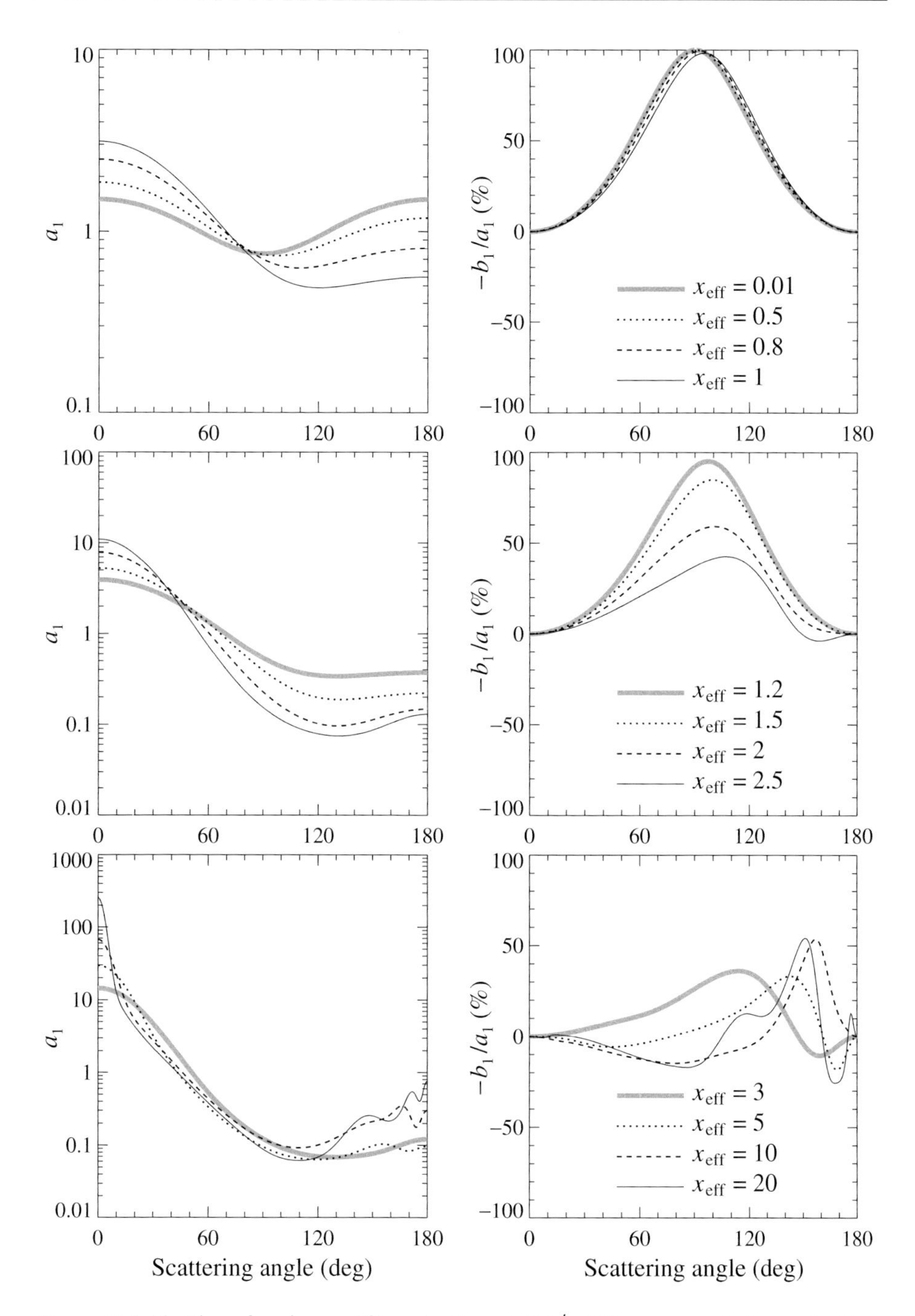

Figure 13.1.14. Phase functions $a_1(\Theta)$ and ratios $-b_1(\Theta)/a_1(\Theta)$ versus scattering angle Θ for polydisperse spheres with effective size parameters ranging from 0.01 to 20.

rapidly decrease to values below 1% (Figs. 13.1.12 and 13.1.13). At larger size parameters, the differences always remain smaller than 1%, thereby indicating that the scalar approximation provides results accurate enough for most practical applications. This result may also imply that large vector–scalar differences are only typical of scatterers exhibiting Rayleigh-like ratios $-b_1(\Theta)/a_1(\Theta)$ (see Fig. 13.1.14).

To explain the intriguing behavior of the maximum underestimation at small size parameters, let us recall Eq. (13.1.14) and Fig. 13.1.10 and analyze the consequences of the results shown in the top two panels of Fig. 13.1.14. It is seen from the latter that as the effective size parameter increases from 0.01 to 1, there is essentially no change in the ratio $-b_1/a_1$ and a very little change in the side-scattering phase function $a_1(90°)$. However, there is a significant decrease in the backscattering phase function $a_1(180°)$. This implies that:

- The contributions of the two second-order-scattering paths shown in Fig. 13.1.10 to the reflected specific intensity should be expected to remain approximately constant.
- The first-order-scattering contribution to the reflected specific intensity in situations corresponding to a 90° phase angle should also be expected to remain approximately constant.
- The first-order-scattering contribution to the reflected specific intensity in situations corresponding to a 0° phase angle should be expected to substantially decrease.

Since the local overestimation is maximal at phase angles close to 90°, it should be expected to remain almost the same, as Fig. 13.1.12 demonstrates indeed. However, the significant decrease of the first-order-scattering contribution in Eq. (13.1.14) should cause an increase in the percent error at phase angles close to 0° and, thus, an increase in the local underestimation. This explains the behavior of the dotted curves in the upper four panels of Fig. 13.1.12.

To analyze how the vector–scalar differences in the reflected specific intensity can be affected by particle nonsphericity, we present, in Figs. 13.1.15 and 13.1.16, the results computed for polydisperse, randomly oriented oblate spheroids with the refractive index 1.33 and with the ratio of the larger to the smaller axis equal to 2. The size of a spheroid is specified in terms of the radius of the sphere having the same surface area. To save computer time, we used the power law size distribution (5.3.14) with effective variance fixed at 0.1. The single-scattering computations for the spheroids were performed with the public-domain T-matrix code posted at http://www.giss.nasa.gov/~crmim.

Comparison of Figs. 13.1.14 and 13.1.17 shows that the Rayleigh domain extends to somewhat larger effective size parameters for the spheroids than for the surface-equivalent spheres, which is a well-known trait of single scattering by nonspherical particles (cf. Mishchenko and Travis, 1994b and references therein). Accordingly, the vector–scalar differences for the spheroids (Figs. 13.1.15 and 13.1.16) peak at $x_{\text{eff}} \approx 1.2$ rather than at $x_{\text{eff}} \approx 1$ and become insignificant at $x_{\text{eff}} \approx 5$ rather than at $x_{\text{eff}} \approx 3$. These effects of nonsphericity appear to be rather minor.

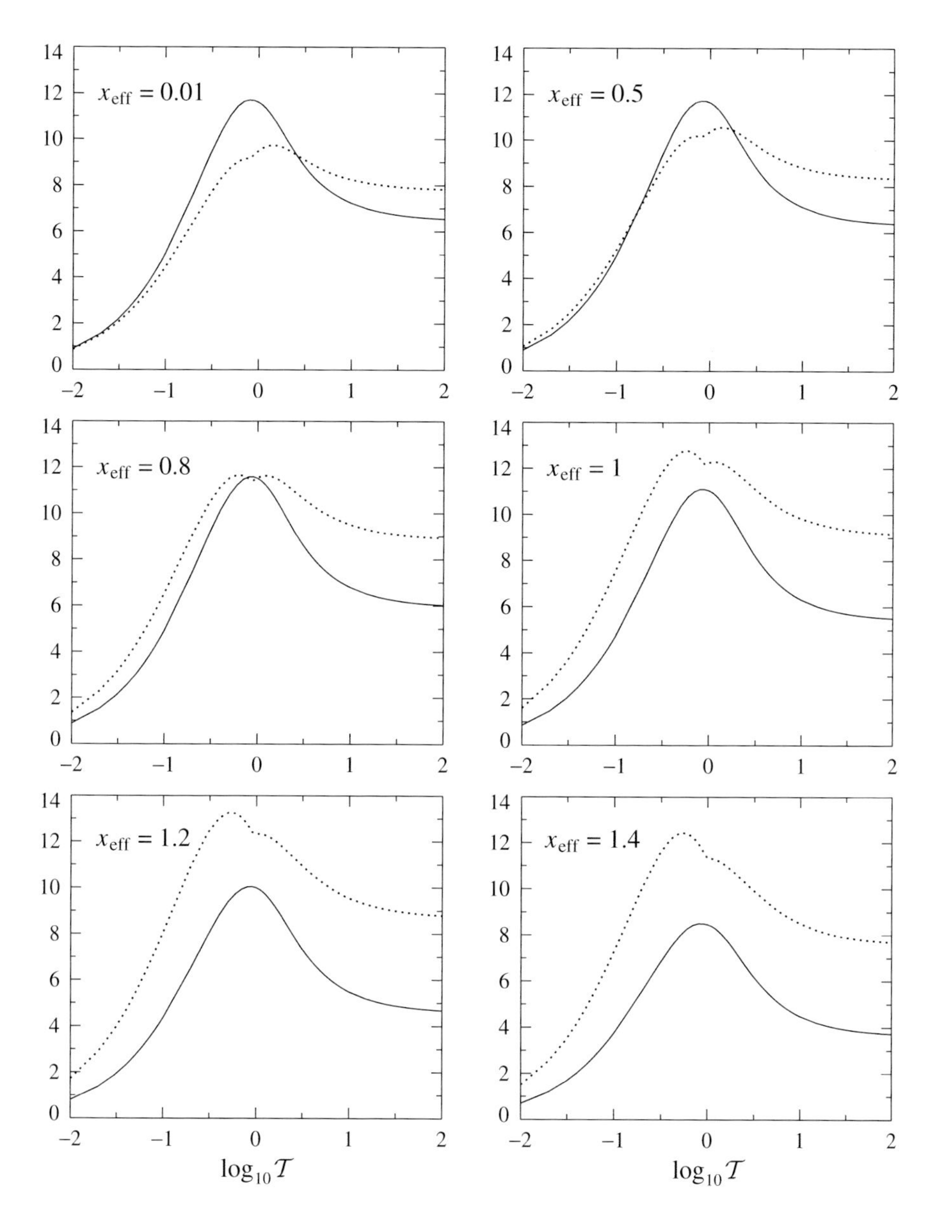

Figure 13.1.15. Maximum overestimation $\varepsilon_{\text{o}}^{\max}(\mathcal{T})$ (in percent, solid curves) and maximum underestimation $\varepsilon_{\text{u}}^{\max}(\mathcal{T})$ (in percent, dotted curves) versus optical thickness $\mathcal{T}$ for polydisperse, randomly oriented oblate spheroids with effective size parameters ranging from 0.01 to 1.4.

However, the situation can be completely different for nonspherical particles with extreme aspect ratios since in that case the Rayleigh domain can persist to significantly greater size parameters (Zakharova and Mishchenko, 2000, 2001). This is illustrated in Fig. 13.1.18 which shows the phase function and the ratio $-b_1/a_1$ for randomly oriented, monodisperse oblate spheroids with the ratio of the larger to the

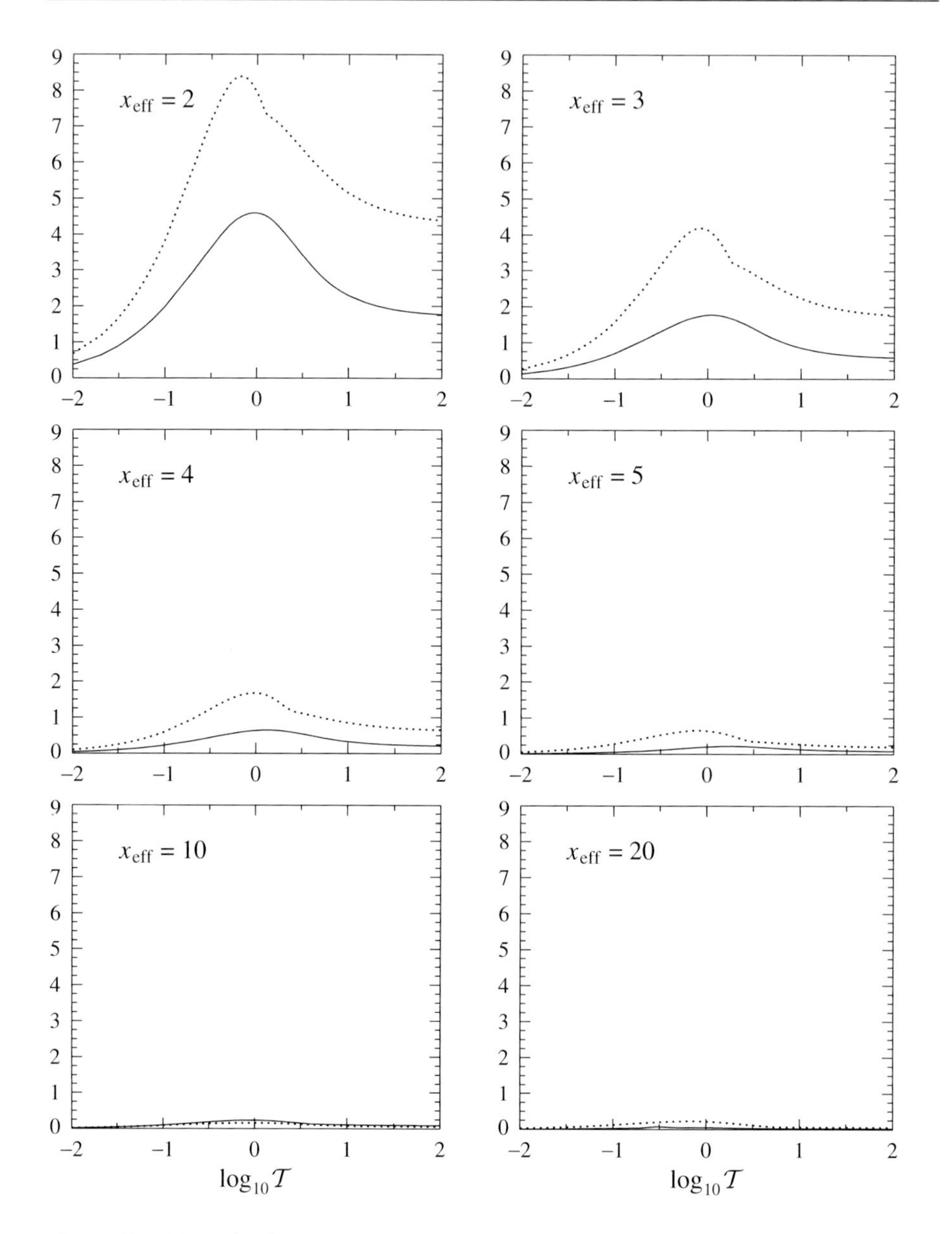

Figure 13.1.16. As in Fig. 13.1.15, but for effective size parameters ranging from 2 to 20.

smaller axis equal to 20. The refractive index is 1.31, which is a value typical of water ice at visible wavelengths. Figure 13.1.19 shows that the corresponding vector–scalar differences in the reflected specific intensity remain significant even for equal-surface-area-sphere size parameters as large as 12 and perhaps even larger. Unfortunately, the limitations of the T-matrix code used in the single-scattering computations did not allow us to determine the threshold size-parameter value above which the vector–scalar differences for these plate-like particles decrease to values below 1%.

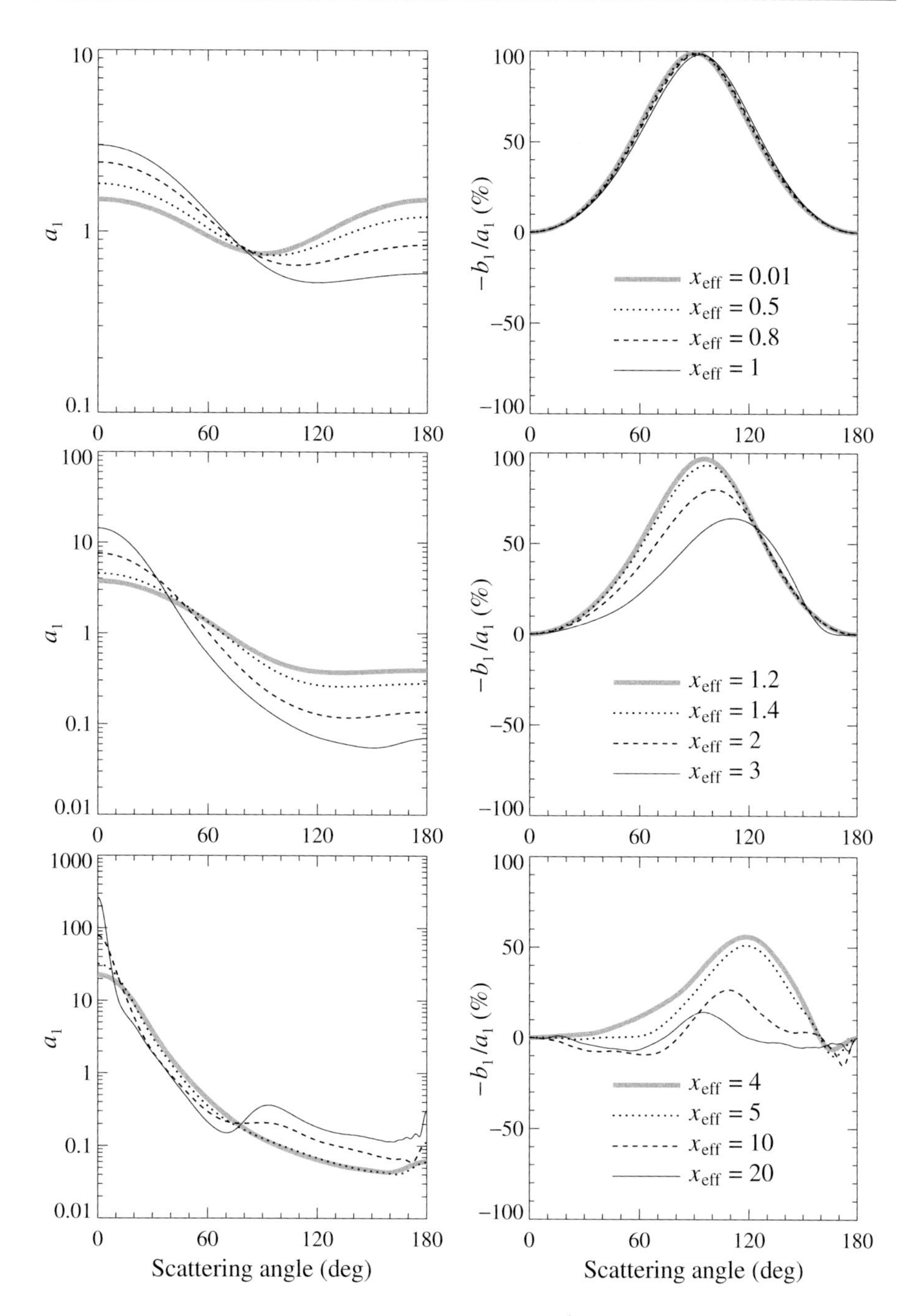

Figure 13.1.17. Phase functions $a_1(\Theta)$ and ratios $-b_1(\Theta)/a_1(\Theta)$ versus scattering angle Θ for polydisperse, randomly oriented oblate spheroids with effective size parameters ranging from 0.01 to 20.

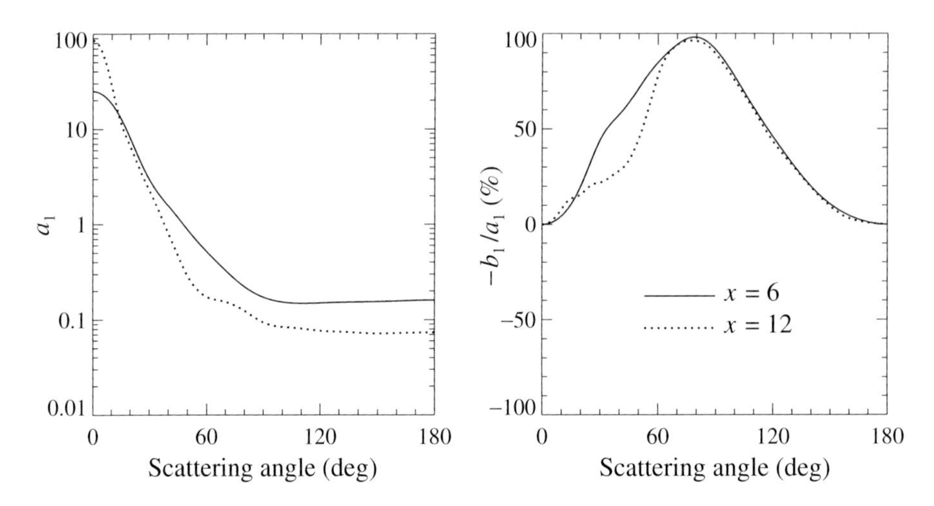

Figure 13.1.18. Phase functions $a_1(\Theta)$ and ratios $-b_1(\Theta)/a_1(\Theta)$ versus scattering angle Θ for randomly oriented oblate ice spheroids with an aspect ratio 20 and surface-equivalent-sphere size parameters 6 and 12.

Another class of particles possessing similar single-scattering properties are clusters composed of Rayleigh-sized monomers (West, 1991; Liu and Mishchenko, 2005). When such a cluster has an overall size comparable to or greater than the incident wavelength, its phase function develops a pronounced forward-scattering peak similar to those shown in the left-hand panel of Fig. 13.1.18, whereas the angular profile of the ratio $-b_1/a_1$ is similar to those shown on the right-hand panel of Fig. 13.1.18 and still closely resembles that of Rayleigh scattering. Therefore, by analogy with Fig. 13.1.19, one should expect large vector–scalar differences in the reflected

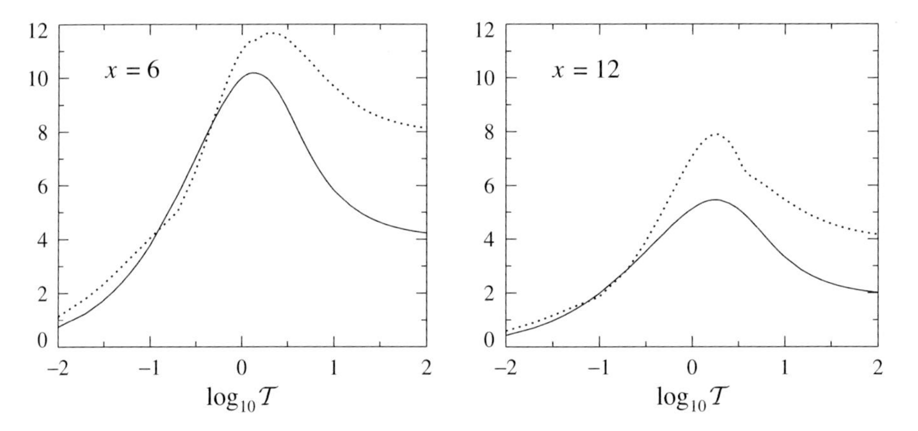

Figure 13.1.19. Maximum overestimation $\varepsilon_{\rm o}^{\rm max}(\mathcal{T})$ (in percent, solid curves) and maximum underestimation $\varepsilon_{\rm u}^{\rm max}(\mathcal{T})$ (in percent, dotted curves) versus optical thickness $\mathcal{T}$ for randomly oriented oblate ice spheroids with an aspect ratio 20 and surface-equivalent-sphere size parameters 6 and 12.

specific intensity for cluster size parameters significantly exceeding the corresponding threshold value for the surface- or volume-equivalent sphere.

13.2 Directional reflectance and spherical and plane albedos

In this section we will analyze how the various parameters of a scattering slab and the process of multiple scattering affect the angular distribution of the reflected specific intensity and its integral characteristics.

Figure 13.2.1 shows the phase function and the ratios a_3/a_1, $-b_1/a_1$, and b_2/a_1

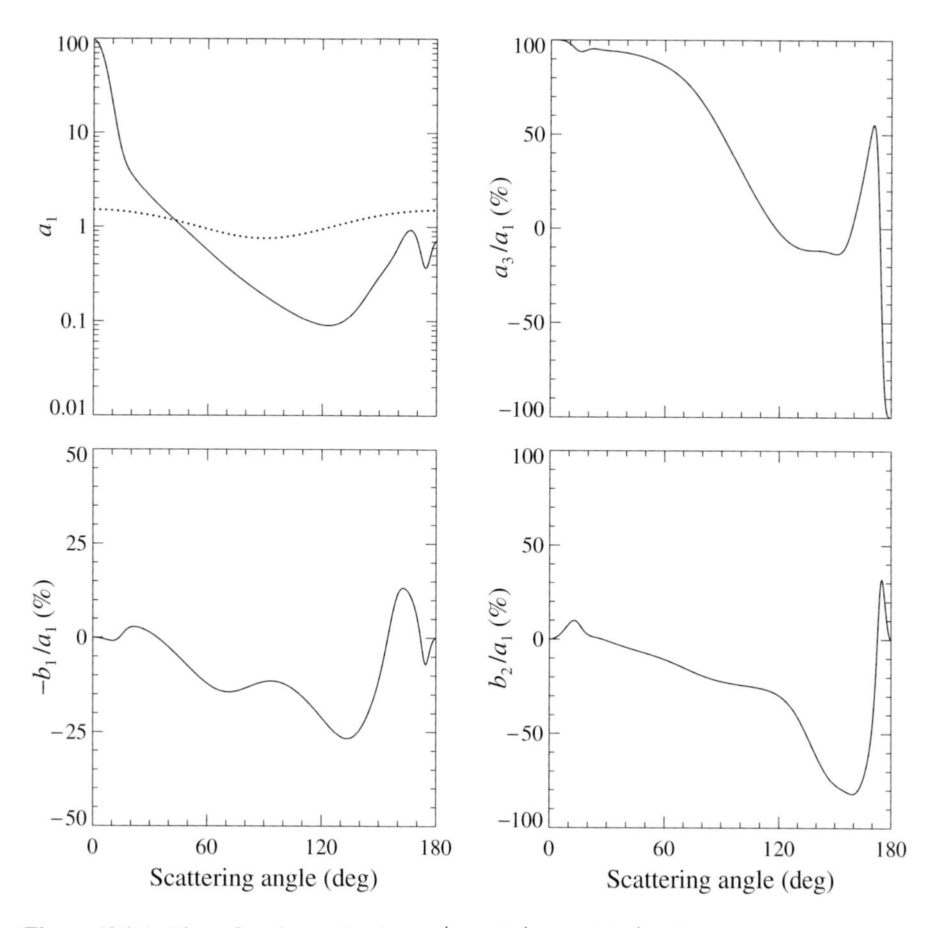

Figure 13.2.1. Phase function and ratios a_3/a_1, $-b_1/a_1$, and b_2/a_1 for a gamma size distribution of spherical particles with an effective radius $r_{\text{eff}} = 1.05$ μm, an effective variance $v_{\text{eff}} = 0.07$, and a relative refractive index $m = 1.44$. The incident wavelength is $\lambda_1 = 550$ nm. The dotted curve depicts the phase function for pure Rayleigh scattering (i.e., without depolarization).

for a gamma distribution of spherical particles given by Eq. (5.3.15) with $r_{\min} = 0$ and $r_{\max} = \infty$. The effective radius and the effective variance of the size distribution are $r_{\text{eff}} = 1.05\ \mu\text{m}$ and $v_{\text{eff}} = 0.07$, respectively, the particle relative refractive index is $m = 1.44$, and the wavelength of the incident light is $\lambda_1 = 550$ nm. These particle parameters characterize the sulfuric acid aerosols forming the main cloud in Venus' atmosphere (Hansen and Hovenier, 1974). Note that owing to $m_{\text{I}} = 0$ these particles are nonabsorbing (i.e., $\varpi = 1$).

Let us consider a homogeneous slab composed of such polydisperse spherical particles and assume that it is illuminated by an unpolarized parallel beam of light incident perpendicularly to the upper boundary of the slab. The intensity of the incident beam is $I_0 = \pi\ \text{W}\,\text{m}^{-2}$. Owing to the azimuthal symmetry of the illumination–reflection geometry, the reflected specific intensity given by

$$\tilde{I}(0, -\mu, \varphi) = R_{11}(\mathcal{T}; \mu, \varphi; 1, \varphi_0) \quad (\text{W}\,\text{m}^{-2}\,\text{sr}^{-1}) \tag{13.2.1}$$

(see Eq. (12.2.12)) is independent of the azimuth angles of the incidence and reflection directions, φ_0 and φ, and depends only on the polar angle of the reflection direction, θ. Figure 13.2.2 shows the reflected specific intensity versus $180° - \theta$ for five optical thickness values increasing from 0.01 to 100. Note that the angle $180° - \theta$ is equal to the scattering angle for the first-order scattering in the slab.

The comparison of the bottom curve in Fig. 13.2.2 and the solid curve in the upper left diagram of Fig. 13.2.1 reveals that the angular distribution of the specific intensity reflected by an optically thin layer closely follows that of single scattering. This is consistent with Eq. (12.3.17) which is convenient to rewrite here in the form

$$\mathbf{R}(\mathcal{T}; \mu, \varphi; \mu_0, \varphi_0) \underset{\mathcal{T} \to 0}{=} \frac{\varpi \mathcal{T}}{4\mu\mu_0} \tilde{\mathbf{Z}}(-\mu, \varphi; \mu_0, \varphi_0). \tag{13.2.2}$$

In particular, such phase-function features as the glory centered at $\Theta = 180°$ and the primary rainbow centered at $\Theta \sim 163°$ clearly show up in the reflected light. As the optical thickness grows from 0.01 to 100, the specific intensity increases by three orders of magnitude, the characteristic phase-function features become less pronounced, and the angular profile of the reflected specific intensity becomes very smooth.

All these effects are easy to understand qualitatively. Indeed, consider three scattering slabs with optical thicknesses $\mathcal{T}_1 < \mathcal{T}_2 < \mathcal{T}_3$, as shown in Fig. 13.2.3. Scattering path 1 contributes to the specific intensity reflected by all three slabs. Scattering path 2 contributes to the specific intensity reflected by layers 2 and 3, but not by layer 1. Finally, scattering path 3 contributes only to the specific intensity reflected by layer 3. Thus, increasing optical thickness affords more scattering paths contributing to the reflected light, which explains the increase of the reflected specific intensity in Fig. 13.2.2.

However, the scattering paths reaching large optical depths, such as scattering paths 2 and 3 in Fig. 13.2.3, are longer than those controlling the reflectivity of thin

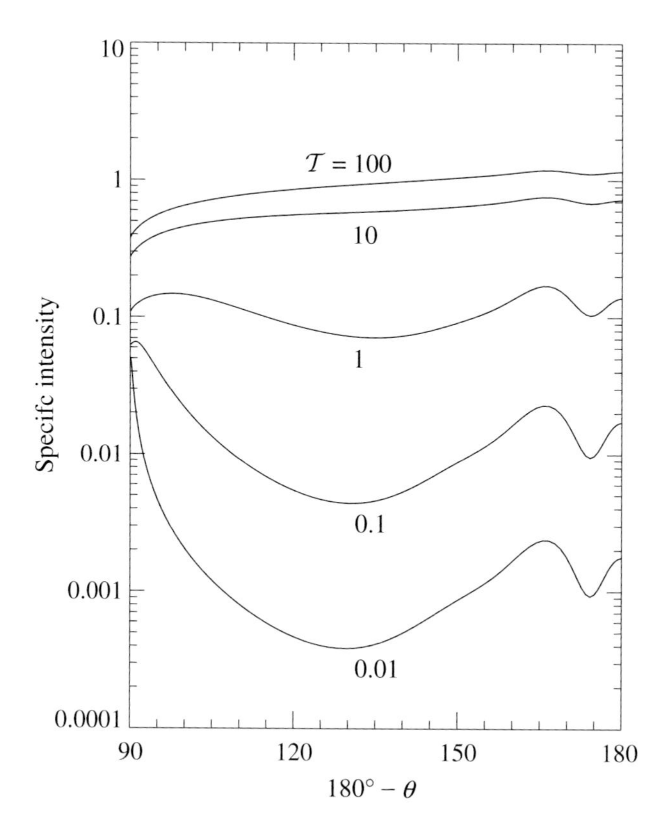

Figure 13.2.2. Angular distribution of the reflected specific intensity (in $\mathrm{W\,m^{-2}\,sr^{-2}}$) for a homogeneous slab composed of polydisperse spherical particles with an effective radius r_{eff} = 1.05 μm, an effective variance $v_{\mathrm{eff}} = 0.07$, and a relative refractive index $m = 1.44$. The wavelength of the incident light is λ_1 = 550 nm. The slab optical thickness increases from $\mathcal{T}$ = 0.01 to $\mathcal{T} = 100$. Unpolarized external beam is incident perpendicularly to the upper boundary of the slab, and its intensity is $I_0 = \pi \ \mathrm{W\,m^{-2}}$.

layers, such as path 1, and involve many more scattering events. The waves following long scattering paths "forget" the original incidence direction and are more likely to contribute equally to all reflection directions, thereby creating a more isotropic distribution of the reflected specific intensity than that typical of optically thin slabs. The latter is dominated by the first-order-scattering contribution and preserves pronounced single-scattering features such as the glory and the rainbow.

It is instructive to visualize the overall increase of the reflected specific intensity with increasing optical thickness by plotting the so-called spherical albedo defined by

$$A_{\mathrm{S}}(\mathcal{T}) = \frac{2}{\pi} \int_0^{2\pi} \mathrm{d}\varphi \int_0^1 \mathrm{d}\mu\,\mu \int_0^1 \mathrm{d}\mu_0\,\mu_0\, R_{11}(\mathcal{T};\mu,\varphi;\mu_0,\varphi_0) \tag{13.2.3}$$

as a function of $\mathcal{T}$. The term "spherical albedo" comes from planetary astrophysics where A_{S} represents the ratio of the electromagnetic energy reflected by the whole

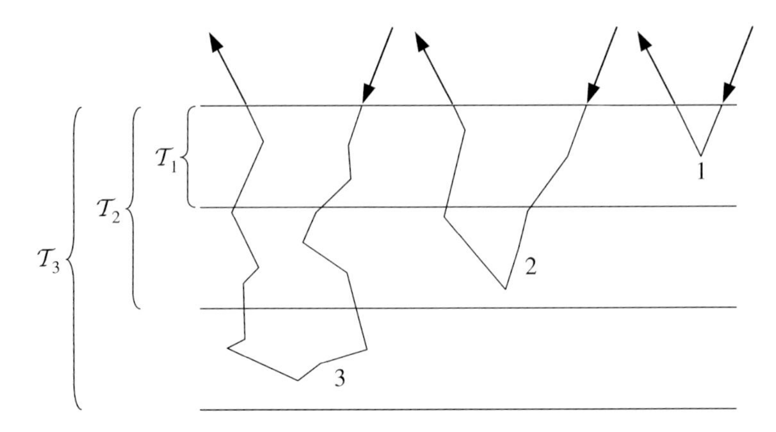

Figure 13.2.3. Various scattering paths contributing to the specific intensity of reflected light.

planet covered with a uniform cloud layer to the total (unpolarized) solar energy falling on the planet (Sobolev, 1975). The solid curve in the upper left panel of Fig. 13.2.4 depicts A_S versus $\mathcal{T}$ for a slab composed of the micron-sized sulfuric-acid aerosols. Taking into account that the $\mathcal{T}$ scale in Fig. 13.2.4 is logarithmic, one can clearly identify the following three regimes:

- A nearly linear growth of A_S on the interval $0 \lesssim \mathcal{T} \lesssim 1$, consistent with Eq. (13.2.2).
- A nearly logarithmic growth of A_S on the interval $1 \lesssim \mathcal{T} \lesssim 20$.
- The regime of slow saturation at $\mathcal{T} > 20$.

It is also seen that

$$\lim_{\mathcal{T} \to \infty} A_S(\mathcal{T})\bigg|_{\varpi = 1} = 1. \tag{13.2.4}$$

The above limit is an obvious manifestation of the energy conservation law: all light incident from above on a semi-infinite nonabsorbing scattering slab must eventually exit the slab through its only boundary.

The existence of the slow saturation regime for a slab composed of nonabsorbing particles is easy to understand. Indeed, the aerosol phase function, Fig. 13.2.1, has a pronounced forward-scattering peak, which implies that after the first scattering much of the incident light is still directed inside the slab. It then takes many scattering events before the light is redirected towards the upper boundary of the slab. Some of the multiple-scattering paths are so long and penetrate so deeply that they can still be terminated at the lower boundary of the slab even for very large $\mathcal{T}$ and cause the spherical albedo to be smaller than one. Therefore, it takes exceedingly large optical thicknesses to avoid the termination of the few extremely long scattering paths and thereby to prevent the loss of a few percent of the incident energy through the lower boundary of the slab.

The dotted curve in the same panel was computed for conservative Rayleigh

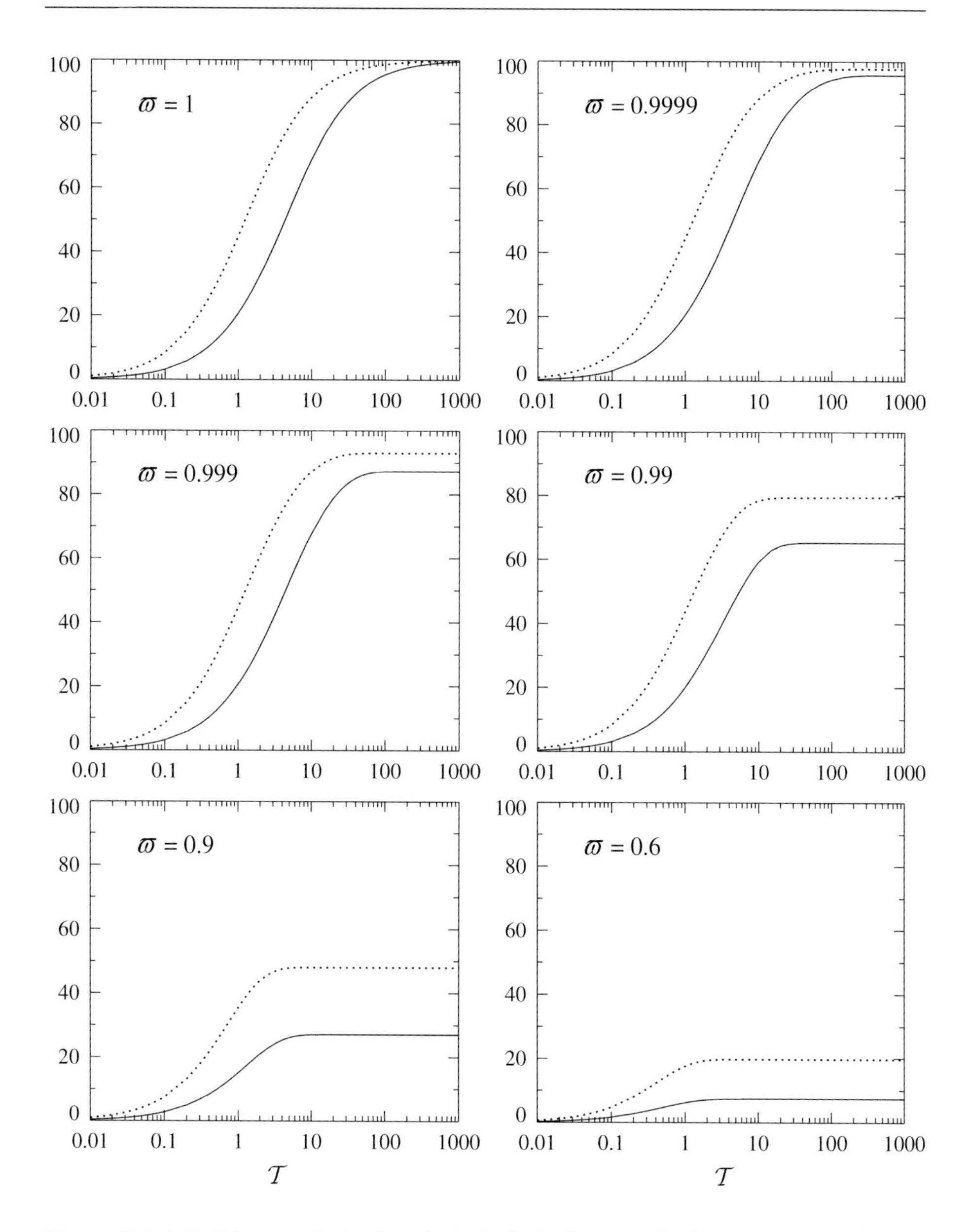

Figure 13.2.4. Solid curves depict the spherical albedo (in percent) of homogeneous slabs composed of polydisperse spherical particles. Dotted curves show the results for Rayleigh-scattering slabs.

scattering and shows spherical albedo values always exceeding those for optical-thickness-equivalent slabs composed of the micron-sized spheres. The overestimation is especially large at small optical thicknesses, where it exceeds a factor of two. The obvious reason for this overestimation is that the Rayleigh phase function, shown by the dotted curve in the upper left panel of Fig. 13.2.1, is nearly isotropic, lacks a forward-scattering peak, and has significantly larger side- and back-scattering values than the aerosol phase function. This means, in particular, that half of the single-scattered light is already directed towards the upper boundary of the slab, and the remaining half is distributed almost isotropically over the lower hemisphere of propagation directions. As a consequence, low-order scattering paths, which are less likely to get terminated at the lower boundary of the slab, provide a significantly stronger contribution to the reflected light than in the case of forward-scattering particles.

The presence of a lower boundary is not the only cause of termination of multiple-scattering paths. Another cause is absorption by gas molecules or particles. In the case of particles, absorption is caused by a nonzero imaginary part of the relative refractive index and results in single-scattering albedo values smaller than one. Although a nonzero m_{I} can also modify the elements of the normalized Stokes scattering matrix and, thus, the elements of the normalized phase matrix, it is instructive to examine the "pure" effect of absorption on the process of multiple scattering by simply varying the single-scattering albedo and keeping the phase matrix fixed. In particular, this approach will allow us to use the order-of-scattering terminology introduced in Section 12.5.

The corresponding numerical results for the micron-sized aerosols and for the pure Rayleigh scattering are also shown in Fig. 13.2.4. Not surprisingly, decreasing ϖ leads to progressively reduced spherical albedo values. This effect is much more pronounced at larger optical thicknesses where it causes an early saturation of A_{S} at values progressively smaller than one. To explain this result, let us recall that according to the order-of-scattering expansion (13.1.12) the contribution of an nth-order scattering path to the reflected specific intensity is proportional to the nth power of the single-scattering albedo. Therefore, although optically thick slabs can afford very long scattering paths, these paths get terminated simply because the factor ϖ^n becomes negligibly small. It is in fact remarkable that as small a deviation of the single-scattering albedo from the value one as 0.0001 (see the upper right panel of Fig. 13.2.4) already causes the loss of several percent of the incident energy in a slab with infinite optical thickness. The effect of absorption on spherical albedo is noticeably weaker for Rayleigh-scattering slabs, which can be explained by a larger contribution of low-order scattering paths to the reflected specific intensity and the fact that these paths are less affected by absorption.

As we have concluded above, one of the dominant effects of multiple scattering is to smear specific details of the single-scattering phase function and yield an increasingly uniform angular distribution of the reflected specific intensity. However, there are cases in which phase function features are so strong that they cannot be washed off completely even by multiple scattering in a semi-infinite nonabsorbing slab. As an example, Fig. 13.2.5 shows the phase functions computed for three ice particle models commonly used to represent cirrus cloud and snow crystals. Model 1 particles have highly irregular, random-fractal shapes introduced by Macke *et al.* (1996). Model 2 particles are homogeneous ice spheres. Model 3 particles are regular hex-

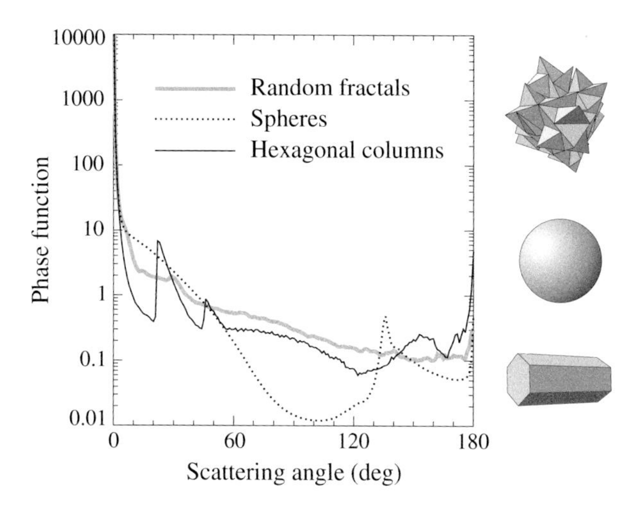

Figure 13.2.5. Phase functions for three ice-particle models.

agonal ice crystals with a length-to-diameter ratio of 2. The nonspherical model 1 and 3 particles are randomly oriented in three-dimensional space. For all three models we assumed the same power law distribution of projected-area-equivalent-sphere radii, Eq. (5.3.14), with an effective radius 50 μm and an effective variance 0.2. The wavelength of the incident light is $\lambda_1 = 650$ nm and the corresponding relative refractive index is $m = 1.311$. The phase functions were computed using the ray tracing technique coupled with the Kirchhoff approximation (Mishchenko and Macke, 1998) for models 1 and 3 and the Lorenz–Mie theory for model 2. As Fig. 13.2.5 shows, the three phase functions exhibit large differences exceeding an order of magnitude at some scattering angles. As discussed in Mishchenko *et al.* (1996), the model 1 and 3 particles may represent limiting cases of highly distorted and pristine ice crystals, respectively.

Plate 13.2.1 shows the reflected specific intensities for three homogeneous semi-infinite slabs composed of the model 1, 2, and 3 particles, while Plate 13.2.2 depicts the ratios 2/1, 3/1, and 3/2 of the reflected specific intensities for the respective models. Although the spherical albedo is equal to one for all three slabs owing to $m_I = 0$, it is clearly seen from Plates 13.2.1 and 13.2.2 that particle shape can indeed have a profound effect on directional reflectance even for semi-infinite particulate slabs. Specific intensity differences between the different models are moderate at nearly normal incidence ($\mu_0 = 0.9$), but increase significantly with decreasing μ_0 and can cause intensity ratios smaller than 0.2 or greater than 3 at grazing reflection directions corresponding to small values of μ. The latter trend is explained by increasing relative contribution of the first-order scattering to the reflection matrix, as follows from Eq. (12.4.9) (see Hovenier and Stam, 2006 for a comprehensive discussion), and the large phase-function differences seen in Fig. 13.2.5.

Hexagonal ice crystals (model 3) produce the most structured radiance field dominated by the backscattering peak and the primary ($\Theta \sim 22°$) and secondary

($\Theta \sim 46°$) halos in the corresponding phase function (cf. the thin solid curve in Fig. 13.2.5 and the two lower panels in the right-most column of Plate 13.2.1). These features clearly show up in the 3/1 and 3/2 intensity ratios as well (Plate 13.2.2). The spherical ice particles produce a noticeable enhancement of reflected specific intensity caused by the primary rainbow. This feature is particularly evident in the 2/1 ratio. The radiance field produced by the featureless phase function of irregular ice crystals (model 1) is by far the least structured (left-most column of Plate 13.2.1). These results illustrate the importance of accurate treatment of single-scattering phase functions for realistic cirrus cloud and snow particle models in various remote-sensing and atmospheric radiation applications.

A widespread practice in many applied science and engineering disciplines is to replace the actual phase function by an asymmetry-parameter-equivalent so-called Henyey–Greenstein (HG) phase function given by

$$P_{\mathrm{HG}}(\Theta) = \frac{1 - g^2}{(1 - 2g\cos\Theta + g^2)^{3/2}}, \tag{13.2.5}$$

where $g \equiv \langle \cos\Theta \rangle$ is the asymmetry parameter (e.g., Sobolev, 1975; Tuchin, 2002). Although not a solution of the Maxwell equations, this "unphysical" phase function has several attractive properties:

- It is given by a simple analytical expression and is specified fully by only one model parameter, g.
- It is always normalized according to Eq. (11.10.3).
- It is defined in the entire theoretical range of asymmetry-parameter values $-1 \le g \le 1$ and can, therefore, be used to model forward-scattering, isotropically scattering, and backward-scattering particles.
- It has a forward-scattering peak, for $g > 0$, reminiscent of the diffraction peak typical of wavelength-sized and larger particles.
- The Legendre expansion coefficients appearing in Eq. (11.11.1) are given by the following simple formula:

$$\alpha_1^s = (2s + 1)g^s. \tag{13.2.6}$$

Figure 13.2.6 compares Lorenz–Mie phase functions computed for two models of polydisperse spherical particles with their $\langle \cos\Theta \rangle$-equivalent HG counterparts. For both models we assumed the gamma size distribution (5.3.15) with an effective radius $r_{\mathrm{eff}} = a = 10$ μm and an effective variance $v_{\mathrm{eff}} = b = 0.1$. The model 1 and 2 values of the relative refractive index were 1.55 + i0.001 and 1.55 + i0.004, respectively, whereas the incident wavelength was fixed at 0.63 μm. The corresponding values of the asymmetry parameter were 0.838 and 0.901.

Plate 13.2.3 shows the angular distribution of the reflected specific intensity for a semi-infinite homogeneous slab composed of model 1 particles as well as for its HG counterpart. Two obvious features of the reflected specific intensity distributions shown in the left-hand column are the backscattering enhancement caused by the pri-

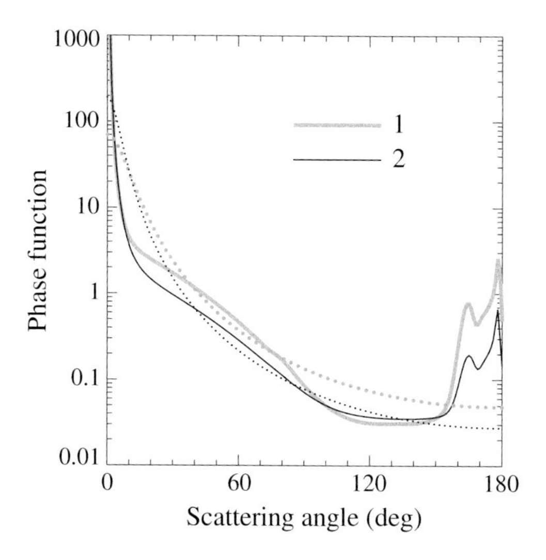

Figure 13.2.6. Phase functions for model 1 and 2 polydisperse spherical particles (solid curves) and their Henyey–Greenstein counterparts (dotted curves).

mary glory in the Lorenz–Mie phase function (gray solid curve in Fig. 13.2.6) and the strong near-forward scattering for the cases of grazing and near-grazing incidence (μ_0 equal or close to zero) caused by the diffraction peak. The reflectance patterns for the asymmetry-parameter-equivalent HG phase function lack the first feature, which is explained by the absence of the backscattering phase function peak similar to the glory. The right-hand column in Plate 13.2.3 shows that errors in the reflected specific intensity caused by the use of the approximate HG phase function can be very large and can, in fact, exceed a factor of 20 at backscattering geometries and a factor of 3 at near-forward-scattering geometries. These errors can be unequivocally attributed to the large phase-function differences. Thus, Plate 13.2.3 makes a strong case against using the HG phase function in directional reflectance computations even for semi-infinite slabs.

The upper panel of Fig. 13.2.7 depicts the so-called plane albedo given by

$$A_{\mathrm{P}}(\mathcal{T}, \mu_0) = \frac{1}{\pi} \int_0^{2\pi} \mathrm{d}\varphi \int_0^1 \mathrm{d}\mu\, \mu\, R_{11}(\mathcal{T}; \mu, \varphi; \mu_0, \varphi_0) \tag{13.2.7}$$

as a function of μ_0 for two semi-infinite homogeneous slabs composed of the model 1 and 2 particles, respectively. In general, the plane albedo characterizes situations when a slab is illuminated by an unpolarized parallel beam of light incident from above and is defined as the ratio of the radiant energy reflected by the slab per unit area of the upper boundary per unit time to the incident energy per unit area of the upper boundary per unit time (Sobolev, 1975). Comparison of Eqs. (13.2.3) and (13.2.7) reveals a close connection between the spherical and plane albedos. Specifically,

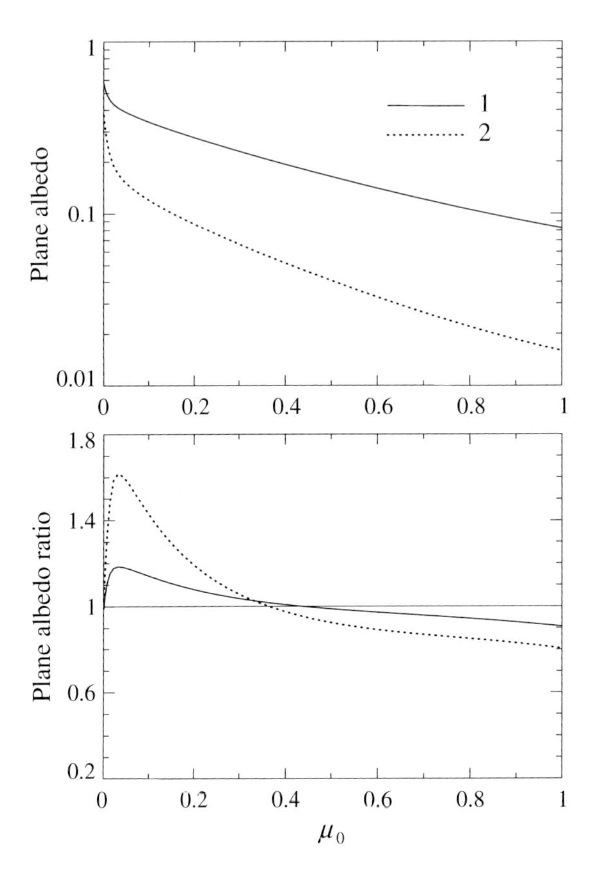

Figure 13.2.7. Top panel: plane albedo versus μ_0 for two homogeneous semi-infinite slabs composed of model 1 and 2 spherical particles, respectively. Bottom panel: plane albedos computed for HG phase functions relative to their Lorenz–Mie counterparts.

$$A_S(\mathcal{T}) = 2\int_0^1 d\mu_0\, \mu_0\, A_P(\mathcal{T}, \mu_0). \tag{13.2.8}$$

We also computed the plane albedo using the asymmetry-parameter-equivalent HG phase functions. The ratios of these approximate plane-albedo values to the respective exact ones are shown in the bottom panel of Fig. 13.2.7.

Not surprisingly, the plane albedos shown in the upper panel of Fig. 13.2.7 decrease with increasing m_I and, thus, with decreasing single-scattering albedo. The asymmetry-parameter-equivalent HG phase functions cause significant plane-albedo errors, especially for grazing illumination. The use of the HG phase functions overestimates $A_P(\mathcal{T}, \mu_0)$ for small μ_0 and underestimates it for μ_0 close to one, which is naturally explained by the angular pattern of the phase-function differences seen in Fig. 13.2.6. The plane-albedo errors increase significantly with increasing absorption. This trend is caused by the increasing relative contribution of the first-order scattering coupled with the large phase-function differences.

The respective spherical albedo ratios (0.99 for model 1 and 0.98 for model 2) are much closer to unity. This much better accuracy of the HG estimate of the spherical albedo can be explained in terms of cancellation of the plane-albedo errors upon integration over μ_0 in Eq. (13.2.8).

13.3 Polarization as an effect and as a particle characterization tool

The widespread use of the scalar approximation has caused an equally widespread ignorance of an important scattering effect called polarization. This term refers to the situation when an initially unpolarized incident light becomes polarized upon scattering. This means that at least one of the elements of the specific intensity column vector other than the specific intensity acquires a nonzero value.

This effect is illustrated in Fig. 13.3.1 which parallels Fig. 13.2.2 but shows the absolute value of the second element of the specific intensity vector of the reflected light. Owing to the particular illumination geometry and to the incident beam being unpolarized, the third and fourth elements of the specific intensity column vector are equal to zero, whereas the second one is independent of the azimuth angle of the reflection direction.

There are two striking differences between the results shown in Fig. 13.2.2 and in Fig. 13.3.1. First, the overall growth of $|\widetilde{Q}|$ as $\mathcal{T}$ increases from 0.01 to 100 is more than an order of magnitude smaller than that of $\widetilde{I}$. Second, the saturation of $|\widetilde{Q}|$ occurs at smaller values of the optical thickness than that of $\widetilde{I}$. In particular, the $|\widetilde{Q}|$ curves for $\mathcal{T} = 10$ and $\mathcal{T} = 100$ are hardly distinguishable, while the overall growth of $|\widetilde{Q}|$ as $\mathcal{T}$ increases from 1 to 100 is less than a factor of 2. These results suggest unequivocally that the main contribution to $\widetilde{Q}$ comes from the first few orders of scattering, the first-order scattering being the prime contributor, whereas light scattered many times becomes largely unpolarized.

These conclusions are corroborated by Fig. 13.3.2 which shows the corresponding signed degree of linear polarization of the reflected light (cf. Eq. (2.9.23)). One can see indeed that the ratio $-\widetilde{Q}/\widetilde{I}$ for small $\mathcal{T}$ essentially replicates the ratio $-b_1/a_1$ of the elements of the normalized Stokes scattering matrix (see the bottom left-hand panel in Fig. 13.2.1), whereas the growth of $\mathcal{T}$ only serves to make the polarization more neutral.

Note that the deep spikes in the curves shown in Fig. 13.3.1 correspond to so-called inversion (or neutral) points, i.e., scattering angles at which the signed degree of linear polarization switches sign. The remarkable constancy of both inversion angles with increasing $\mathcal{T}$ in Fig. 13.3.2 indicates too that the main contribution to polarization comes from the first-order scattering.

The above conclusions regarding $\widetilde{Q}$ apply also to the third element of the specific intensity column vector, $\widetilde{U}$, in cases when the scattering geometry does not cause the

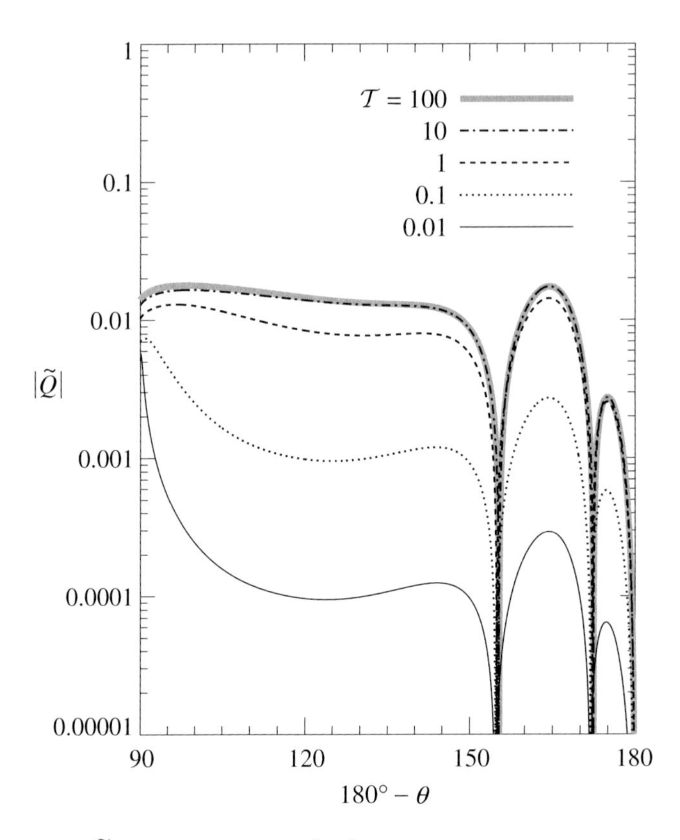

Figure 13.3.1. $|\tilde{Q}(0, -\mu, \varphi)|$ (in $\mathrm{W\,m^{-2}\,sr^{-2}}$) versus $180° - \arccos\mu$ for a homogeneous slab composed of polydisperse spherical particles with an effective radius $r_{\mathrm{eff}} = 1.05$ µm, an effective variance $v_{\mathrm{eff}} = 0.07$, and a relative refractive index $m = 1.44$. The wavelength of the incident light is $\lambda_1 = 550$ nm. The slab optical thickness varies from $\mathcal{T} = 0.01$ to $\mathcal{T} = 100$. Unpolarized external beam is incident perpendicularly to the upper boundary of the slab, and its intensity is $I_0 = \pi$ $\mathrm{Wm^{-2}}$.

latter to vanish. The first-order scattering of unpolarized incident light in an ISM does not contribute to the fourth element of the specific intensity column vector, $\tilde{V}$, as follows from the comparison of Eq. (13.2.2) with Eqs. (11.3.1) and (11.3.2) (Hansen, 1971b). Therefore, the main contribution to this element comes from several low-order-scattering events except first-order scattering. This explains the small magnitude of $\tilde{V}/\tilde{I}$ for all $\mathcal{T}$ in most cases and the fact that the corresponding inversion points can move considerably with increasing $\mathcal{T}$ (Hansen and Travis, 1974).

The majority of particle characterization techniques in disciplines such as terrestrial and planetary remote sensing, astrophysics, and biomedicine are based on intensity measurements (e.g., Stephens, 1994; Liou, 2002; Tuchin, 2002). However, there are two major factors that can make polarimetry a much more sensitive particle characterization tool.

First, the absolute accuracy of intensity measurements is typically of order $0.05I$

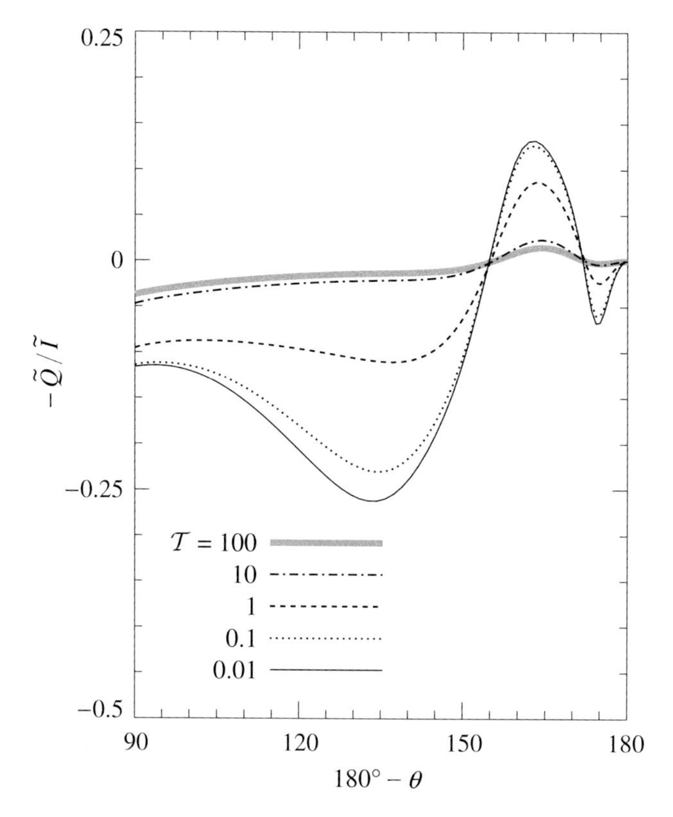

Figure 13.3.2. As in Fig. 13.3.1, but for $-\widetilde{Q}(0,-\mu,\varphi)/\widetilde{I}(0,-\mu,\varphi)$.

and can become significantly worse for weak signals, whereas the absolute accuracy of measurements of the ratios Q/I, U/I, and V/I can be as good as 0.001 and sometimes is much better (see, e.g., Table 1.3 of Tinbergen (1996) which lists the best accuracies obtained in various types of astrophysical polarimetric observations). Theoretically, these ratios can vary between −1 and +1 (see the inequality (2.9.16)). However, the 0.001 absolute accuracy even makes informative data spanning a significantly narrower range, as, for example, in Fig. 13.3.2.

Second, since a major contribution to $\widetilde{Q}$ and $\widetilde{U}$ comes from light scattered once, these quantities preserve more information content of the Stokes scattering matrix than the specific intensity. Furthermore, the single-scattering polarization $-b_1(\Theta)/a_1(\Theta)$ exhibits a much stronger variability with particle size, shape, and relative refractive index than the phase function $a_1(\Theta)$, which makes the former a more sensitive indicator of particle microphysical characteristics (see Chapters 9 and 10 of MTL and references therein).

A classical example of the use of polarimetry in remote sensing is the analysis of ground-based polarization observations of Venus by Hansen and Hovenier (1974). Figures 13.3.3 and 13.3.4 show the results of measurements of the signed degree of linear polarization of sunlight reflected by the entire planet as a function of scattering

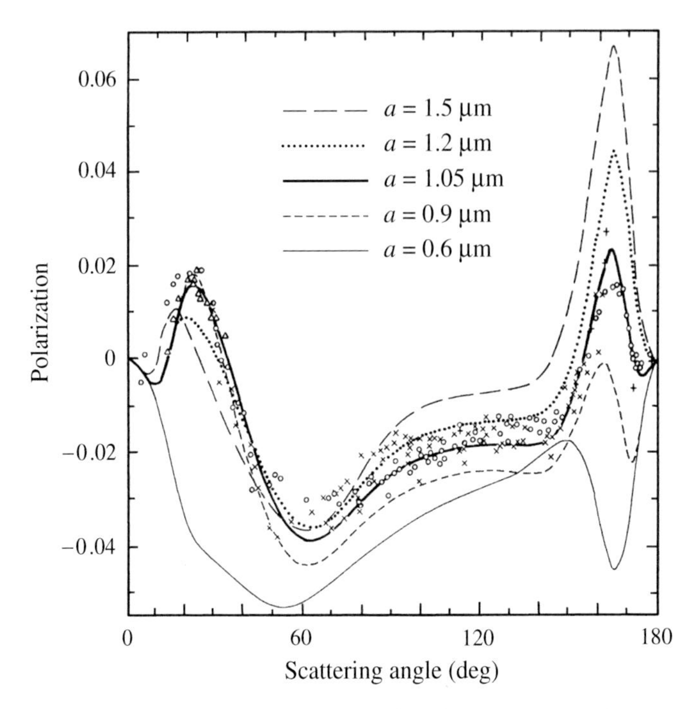

Figure 13.3.3. Observations of the polarization of sunlight reflected by Venus in the visual wavelength region (symbols) and theoretical computations at 0.55 μm wavelength (curves). The theoretical results are based on a model of nonabsorbing spherical particles with fixed relative refractive index ($m = 1.44$) and fixed effective variance of the size distribution ($v_{\rm eff} = 0.07$). The different curves show the influence of the effective radius $a \equiv r_{\rm eff}$ on the polarization. (After Hansen and Hovenier, 1974.)

angle at wavelengths 0.55 and 0.99 μm, respectively. The scattering angle refers to first-order scattering only, i.e., it is the angle between the unti-solar direction and the direction towards the Earth as viewed from Venus. The curves depict the results of theoretical calculations based on a simple model of the Venus atmosphere in the form of a homogeneous, optically semi-infinite, locally plane-parallel cloud layer uniformly covering the entire planet. The cloud particles were assumed to be spherical, and their single-scattering properties were modeled using the Lorenz–Mie theory. The computations of multiple scattering of light in the atmosphere were based on the adding/doubling method. Hansen and Hovenier used the simple gamma distribution (5.3.15) to represent analytically the distribution of cloud particles over sizes and found the parameters a $(= r_{\rm eff})$ and b $(= v_{\rm eff})$ of this distribution, as well as the relative refractive index, by minimizing the differences between the observational data and the results of model computations.

From the comparisons between the computed and observed quantities, Hansen and Hovenier deduced the following:

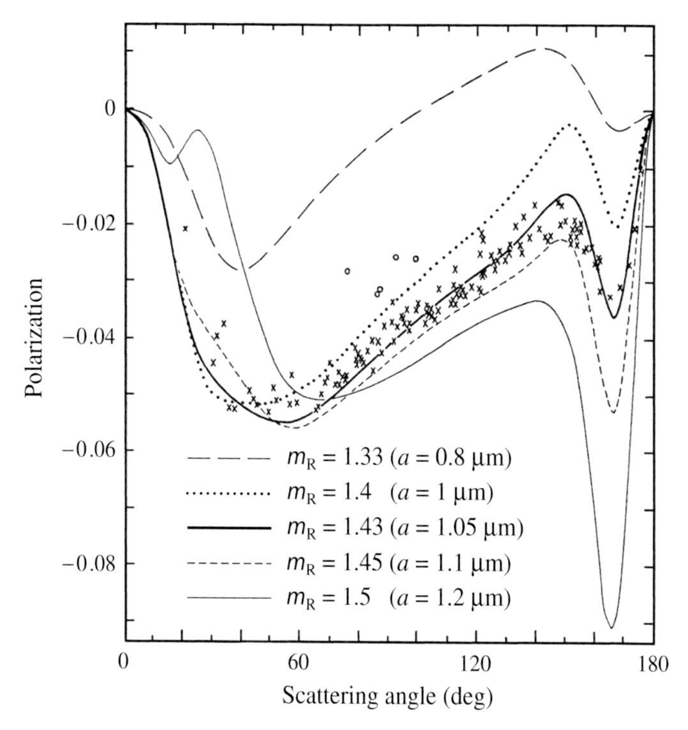

Figure 13.3.4. Observations (symbols) and theoretical computations (curves) of the polarization of the sunlight reflected by Venus at 0.99 μm wavelength. The different theoretical curves are for various relative refractive indices, the effective radius being selected in each case to yield the closest agreement with the observations. The effective variance of the cloud-droplet size distribution is fixed at 0.07. (After Hansen and Hovenier, 1974.)

- The observations can indeed be reproduced quantitatively using a model of nonabsorbing spherical particles. The measurements at visible wavelengths contain a clear signature of the spherical particle shape, such as the primary rainbow at $\Theta \sim 160°$ and an anomalous diffraction feature at $\Theta \sim 25°$ (cf. Fig. 13.3.3 and the lower left panel of Fig. 13.2.1). This interpretation is confirmed by the spectral variation of the observed polarization.
- The effective radius of the cloud droplets is 1.05 ± 0.10 μm. This value is consistent with the large observed variations of the rainbow and anomalous diffraction features with changing wavelength in precise agreement with the Lorenz–Mie theory, including a changeover toward Rayleigh scattering at longer wavelengths (and thus smaller size parameters; see the bottom left-hand panel of Plate 11.13.1). The large difference in the observed polarization between visible and near-infrared wavelengths is a direct indication that the size of the cloud droplets must be of the order of the wavelength in this spectral region. In particular, the complete disappearance of the rainbow and anomalous diffraction features at wavelengths $\gtrsim 0.9$ μm and their obvious

prominence at wavelengths $\lesssim 0.55$ μm allows one to determine the particle size with extreme precision.

- The particle size distribution is narrow, with an effective variance $v_{\text{eff}} = 0.07 \pm 0.02$. The upper limit on the effective-variance value follows from the fact that the anomalous diffraction feature can only be observed for $v_{\text{eff}} \lesssim 0.07$ (see Plate 11.13.1). The lower limit is consistent with the measurements at wavelengths ~ 1 μm which show negative polarization at all scattering angles (see Fig. 13.3.4 and Plate 11.13.1).
- The cloud-particle relative refractive index has a normal spectral dispersion, decreasing from 1.46 ± 0.015 at a wavelength 0.365 μm to 1.43 ± 0.015 at a wavelength 0.99 μm. The extreme sensitivity of polarization measurements to the relative refractive index is well illustrated by Fig. 13.3.4.

Based on the spectral dependence of the refractive index, Hansen and Hovenier concluded that the cloud particles consist of a concentrated (76% by weight) aqueous solution of sulfuric acid ($H_2SO_4 - H_2O$). This remarkable result has been confirmed by subsequent *in situ* measurements and observations from Venus-orbiting satellites (e.g., Sato *et al.*, 1996 and references on page 139 of Hovenier *et al.*, 2004).

13.4 Depolarization

Another important effect of multiple scattering is depolarization. This term refers to the situation when an initially completely polarized (either linearly or circularly) beam becomes partially polarized or even completely unpolarized upon scattering. Most natural sources of light are not completely polarized and some of them, such as the sun, are unpolarized. Therefore, depolarization is usually observed in measurements involving an artificial source of illumination such as a lidar or a transmitting antenna.

The effect of depolarization is especially relevant to analyses of monostatic lidar and radar observations involving the measurement of two or more Stokes parameters of light reflected by a particulate medium in the exact backscattering direction (i.e., towards the source of illumination). Such observations can also be influenced quite significantly by the effect of coherent backscattering. Therefore, we will postpone the discussion of various manifestations of depolarization and its practical usage as a particle characterization tool until Chapter 14.

13.5 Further reading

Diverse applications of radiative transfer in atmospheric radiation and terrestrial remote sensing are discussed in the monographs by Ulaby *et al.* (1986), Asrar (1989),

Goody and Yung (1989), Liou (1992, 2002), Janssen (1993), Jin (1993), Lenoble (1993), Fung (1994), Natsuyama *et al.* (1998), Thomas and Stamnes (1999), Guzzi (2003), Kokhanovsky (2003), and Sharkov (2003). The subject of ocean optics is well covered in the books by Shifrin (1988), Mobley (1994), and Spinrad *et al.* (1994). Planetary remote sensing and various applications of radiative transfer in astrophysics are described in Gehrels (1974), Dolginov *et al.* (1995), Hanel *et al.* (2003), Morozhenko (2004), and Videen *et al.* (2004a,b). Applications in engineering and biomedicine are discussed thoroughly in the monographs by Bayvel and Jones (1981), Tuchin (2002), and Modest (2003) as well as in the special journal issues edited by Mengüç *et al.* (2002, 2005).

Radiance errors resulting from the use of the scalar approximation in radiative transfer computations for realistic atmosphere–surface models and their practical implications are studied in Kattawar and Adams (1990), Stammes (1994), Petropavlovskikh *et al.* (2000), Oikarinen (2001), Hasekamp *et al.* (2002), Landgraf *et al.* (2004), Levy *et al.* (2004), Loughman *et al.* (2004), Sromovsky (2005), and Stam and Hovenier (2005). Detailed discussions of polarized radiative transfer in Rayleigh-scattering slabs, including specific phenomena like points of neutral polarization and neutral lines, can be found in the monographs by Coulson (1988) and Viik (1989) as well as in the numerous publications referenced therein.

Dubovik *et al.* (2002) discussed the retrieval of microphysical properties of aerosols using multi-wavelength measurements of extinction and sky radiances. The characterization of water-cloud droplets and spherical aerosol particles using radiance measurements from space is reviewed by King *et al.* (1992), Martonchik *et al.* (1998), King *et al.* (1999), and Rossow and Schiffer (1999).

The tutorial papers by Mishchenko and Travis (1997a,b) and Mishchenko *et al.* (2004c) provide a systematic sensitivity analysis of various passive aerosol retrieval algorithms based on intensity and/or polarization measurements from aircraft or space platforms and demonstrate the great superiority of the algorithms utilizing polarimetric data. Specific applications of polarimetry in remote sensing of terrestrial aerosols and clouds have been documented in Brogniez *et al.* (1992), Buriez *et al.* (1997), Bréon and Goloub (1998), Deuzé *et al.* (2000), Masuda *et al.* (2000), Sano and Mukai (2000), Knap *et al.* (2005), and Chowdhary *et al.* (2001, 2002, 2005). Numerous other applications of the vector RTT in terrestrial and planetary remote sensing are listed in Section 5.1 of Hovenier *et al.* (2004). They range from the interpretation of ground-based polarimetric observations of Venus to potential detection and characterization of extrasolar planets (Stam *et al.*, 2004). New approaches to retrieval algorithm development based on the application of perturbation procedures to the VRTE are discussed in Polonsky and Box (2002), Postylyakov (2004), and Hasekamp and Landgraf (2005) as well as in the earlier publications referenced therein.

There is a rapidly growing number of publications in which numerical solutions of the RTE are used to model directional reflectance and transmittance characteristics of various particulate surfaces (see, e.g., Aoki *et al.*, 1999; Leroux *et al.*, 1999; Mishchenko *et al.*, 1999b; Petrova *et al.*, 2001; Kokhanovsky, 2004; Liang, 2004; Okin and Painter, 2004; and references therein). The formal applicability of the RTT rests on the assumption that scattering particles are located in each-other's far-field zones. The violation of this assumption in the case of particulate surfaces can lead to specific high-density effects (e.g., Kumar and Tien, 1990; Mishchenko, 1994; Garg *et al.*, 1998; Nashashibi and Sarabandi, 1999; Shinde *et al.*, 1999; Tsang and Kong, 2001; Loiko and Miskevich, 2004; and references therein). Therefore, it is important to analyze both theoretically and experimentally to what extent the classical RTE can be applied to densely packed particulate media. Some progress in this direction has been reported by Sergent *et al.* (1998), Hespel *et al.* (2003), Li and Zhou (2004), Painter and Dozier (2004), and Zhang and Voss (2005).

Chapter 14

Coherent backscattering

The main advantage of the microphysical approach to radiative transfer is that it establishes a direct link between the macroscopic Maxwell equations and the RTE via a sequence of eight unambiguously defined and physically realizable assumptions and approximations summarized in Section 8.11. This link ensures that all parameters entering the RTE are well-defined and measurable physical quantities and thereby enables direct quantitative comparisons of RTT results with results of controlled laboratory experiments and full-closure field experiments.

The above statement can be rephrased as follows. Suppose one performs a detailed set of measurements of various characteristics of electromagnetic radiation multiply scattered in a sparse discrete random medium and supplements them by comprehensive accurate measurements of macro- and microphysical parameters of the scattering medium. Suppose also that the optical measurements are sufficiently accurate and comprehensive and can be reproduced quantitatively by the RTE when the latter is applied to the measured macro- and microphysical parameters. Then one may conclude that, in all likelihood, the observed scattering process falls in the realm of radiative transfer.

By keeping assumptions 1 through 7 from Section 8.11 but relaxing approximation 8, one can extend the microphysical approach and establish a similar direct link between the macroscopic Maxwell equations and the effect of coherent backscattering of light by sparse discrete random media. Specifically, one can supplement the computation of the ladder component of the coherency dyadic, $\ddot{C}_{\mathrm{L}}(\mathbf{r})$, with the computation of the so-called "cyclical" component, $\ddot{C}_{\mathrm{C}}(\mathbf{r})$. The latter is the sum of all so-called maximally crossed diagrams in the diagrammatic representation of the coherency dyadic. As pointed out in Sections 1.8 and 8.11, the sum of the corresponding specific coherency dyadics can be expected to provide a better representation of the properties of radiation scattered by the medium in directions exactly or approximately opposite to the illumination direction.

It should be recognized from the outset that CB is not an independent physical phenomenon. It is implicitly contained in the exact solution of the Maxwell equations but "falls through the cracks" when one resorts to the ladder approximation in order to simplify the computation. Therefore, one may characterize CB as the difference between the exact solution of the Maxwell equations for a sparse discrete random medium and the ladder approximation, although this characterization may not be quite accurate since it still neglects the existence of light-scattering paths that go through a particle more than once.

CB can manifest itself in several different ways. Apparently, the first laboratory observation of CB, in the form of the so-called polarization opposition effect (Mishchenko, 1993), was reported by Lyot (1929). Oetking (1966) observed CB in the form of a narrow intensity peak centered at the exact backscattering direction. However, neither Lyot nor Oetking offered a correct theoretical explanation of their experimental results.

The first theoretical prediction of the potential presence of CB in multiply scattered light was made by K. M. Watson with a reference to a private communication from R. Ruffine (Watson, 1969). Barabanenkov (1973) introduced the concept of maximally crossed (or cyclical) diagrams which proved to be a very useful interpretation tool widely accepted in the multiple-scattering theory.

The first true laboratory demonstrations of CB accompanied by a correct theoretical interpretation should be credited to Kuga and Ishimaru (1984), Tsang and Ishimaru (1984), Van Albada and Lagendijk (1985), and Wolf and Maret (1985). Since then, CB has been the subject of active theoretical and experimental research and has been the centerpiece of many applications of electromagnetic scattering in remote sensing and particle characterization.

14.1 Specific coherency dyadic

Consider again a scattering object in the form of a large group of discrete, randomly and sparsely distributed particles (Fig. 14.1.1). The object is illuminated by a plane electromagnetic wave propagating in the direction of a unit vector $\hat{\mathbf{n}}_0$,

$$\mathbf{E}^{\mathrm{inc}}(\mathbf{r},t) = \mathbf{E}_0^{\mathrm{inc}} \exp(\mathrm{i}k_1\hat{\mathbf{n}}_0 \cdot \mathbf{r} - \mathrm{i}\omega t), \qquad \mathbf{E}_0^{\mathrm{inc}} \cdot \hat{\mathbf{n}}_0 = 0. \tag{14.1.1}$$

The reader may recall that the RTE was derived in Chapter 8 by neglecting all diagrams with crossing connectors in the diagrammatic representation of the coherency dyadic. Following the line of reasoning outlined in Section 8.11, one may indeed conclude that upon statistical averaging the contribution of all the diagrams of the type illustrated in Fig. 14.1.2 must vanish at observation points located either inside the object (observation point 1 in Fig. 14.1.1) or outside the object (observation point 2).

Figure 14.1.1. Scattering of a plane electromagnetic wave by a volume of sparse, discrete random medium.

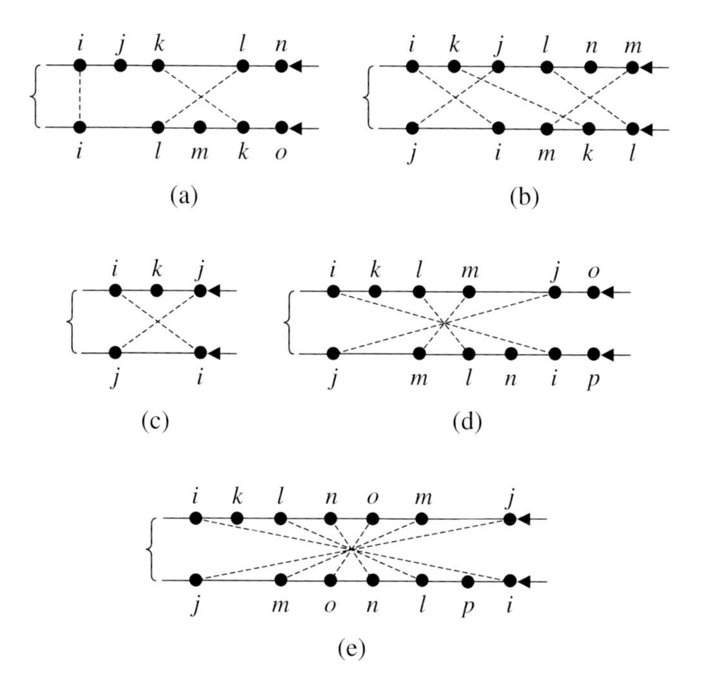

Figure 14.1.2. Diagrams with crossing connectors.

However, there is an exception corresponding to the situation when the observation point is very far (ideally, infinitely far) from the scattering object and is located within its "back-shadow" (observation point 3). Then the class of diagrams illustrated by panels (c)–(e) in Fig. 14.1.2 gives a nonzero contribution that causes CB. These diagrams are called maximally crossed since they can be drawn in such a way that all connectors cross at one point.

The expression for the cumulative contribution of all maximally crossed (or cycli-

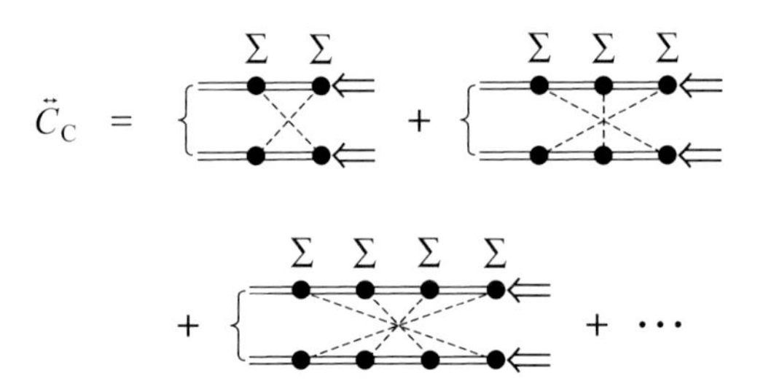

Figure 14.1.3. The cyclical part of the coherency dyadic.

cal) diagrams to the coherency dyadic at an observation point can be derived using the diagrammatic technique introduced in Chapter 8. The reader can verify that the final result can be summarized by the diagrammatic expression shown in Fig. 14.1.3. The symbol Σ has the usual meaning and denotes both the summation over all appropriate particles and the statistical averaging over the particle states and positions, whereas the double lines account for the effect of coherent attenuation and, possibly, dichroism. It is very instructive to compare Fig. 14.1.3 with Fig. 8.6.1 since this comparison reveals quite vividly the morphological difference between the participating diagrams.

In order to simplify further discussion, we will assume that the scattering medium is a plane-parallel slab of infinite horizontal extent, as shown in Fig. 14.1.4. It will be convenient for our purposes to express the coherency dyadic at a remote observation point as the sum of the coherent (subscript c), single-scattering (subscript 1), diffuse multiple-scattering (subscript M), and cyclical (subscript C) components:

$$\begin{aligned}\ddot{C} &= \ddot{C}_L + \ddot{C}_C \\ &= \ddot{C}_c + \ddot{C}_1 + \ddot{C}_M + \ddot{C}_C. \end{aligned} \tag{14.1.2}$$

The corresponding diagrammatic expressions for these components are shown in Figs. 14.1.3 and 14.1.5 (cf. Fig. 8.6.1).

According to subsections 8.14.1 and 8.14.2, the coherent, single-scattering, and diffuse multiple-scattering components of the coherency dyadic at the remote observation point can be expressed in terms of the respective specific coherency dyadics:

$$\ddot{C}_c = \int_{4\pi} d\hat{\mathbf{n}}\, \ddot{\Sigma}_c(\hat{\mathbf{n}}), \tag{14.1.3}$$

$$\ddot{C}_1 = \int_{4\pi} d\hat{\mathbf{n}}\, \ddot{\Sigma}_1(\hat{\mathbf{n}}), \tag{14.1.4}$$

$$\ddot{C}_M = \int_{4\pi} d\hat{\mathbf{n}}\, \ddot{\Sigma}_M(\hat{\mathbf{n}}), \tag{14.1.5}$$

where the unit vector $\hat{\mathbf{n}}$ specifies the direction of the incoming scattered light. The specific coherency dyadics characterize the angular distribution of electromagnetic radiation entering the observation point. Both $\ddot{\Sigma}_1(\hat{\mathbf{n}})$ and $\ddot{\Sigma}_M(\hat{\mathbf{n}})$ vanish if the unit

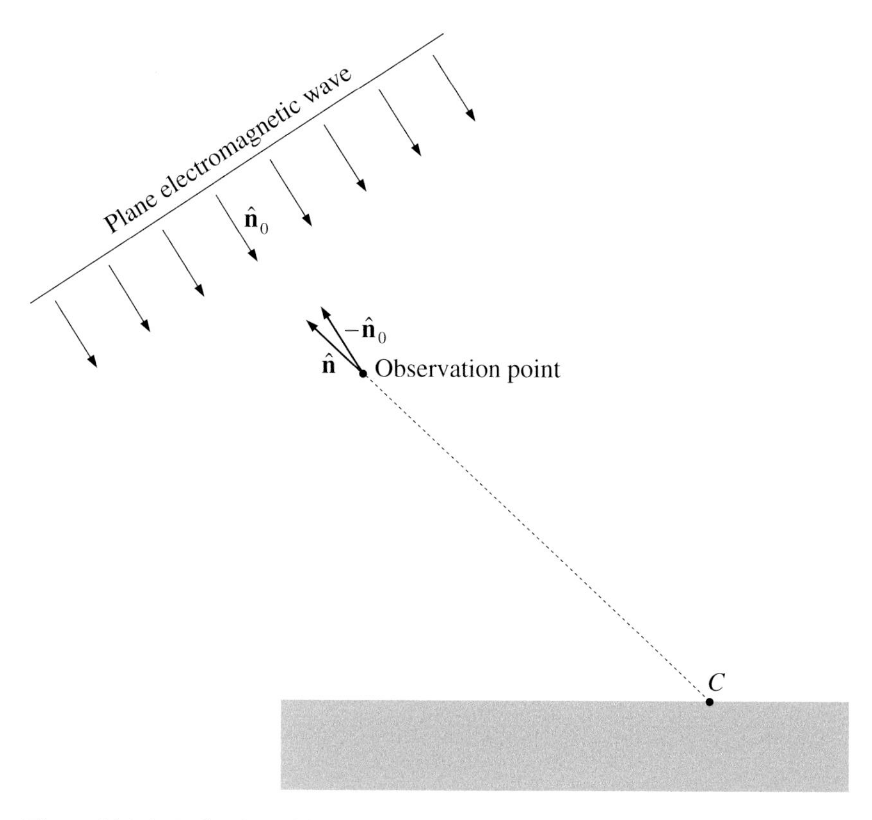

Figure 14.1.4. Reflection of light by a plane-parallel slab of sparse, discrete random medium.

vector $\hat{\mathbf{n}}$ does not specify an upward direction. The coherent specific coherency dyadic is given by

$$\ddot{\Sigma}_c(\hat{\mathbf{n}}) = \delta(\hat{\mathbf{n}} - \hat{\mathbf{n}}_0)\,\ddot{\rho}^{\text{inc}}, \tag{14.1.6}$$

where

$$\ddot{\rho}^{\text{inc}} = \mathbf{E}_0^{\text{inc}} \otimes (\mathbf{E}_0^{\text{inc}})^* \tag{14.1.7}$$

is the coherency dyad of the incident plane wave. Obviously, $\ddot{\Sigma}_c(\hat{\mathbf{n}})$ vanishes unless $\hat{\mathbf{n}} = \hat{\mathbf{n}}_0$. The sum of the single-scattering and diffuse multiple-scattering specific coherency dyadics is equal to the diffuse specific coherency dyadic $\ddot{\Sigma}_d(\hat{\mathbf{n}})$ given by Eq. (8.14.10):

$$\ddot{\Sigma}_d(\hat{\mathbf{n}}) = \ddot{\Sigma}_1(\hat{\mathbf{n}}) + \ddot{\Sigma}_M(\hat{\mathbf{n}}). \tag{14.1.8}$$

According to subsection 8.14.2, $\ddot{\Sigma}_d(\hat{\mathbf{n}})$ is equal to the internal diffuse specific coherency dyadic at the boundary point C where the line drawn through the observation point in the direction $-\hat{\mathbf{n}}$ enters the scattering slab (Fig. 14.1.4). As such it can be computed by solving the RTE.

It is rather straightforward to show that an expression similar to Eq. (14.1.5) can

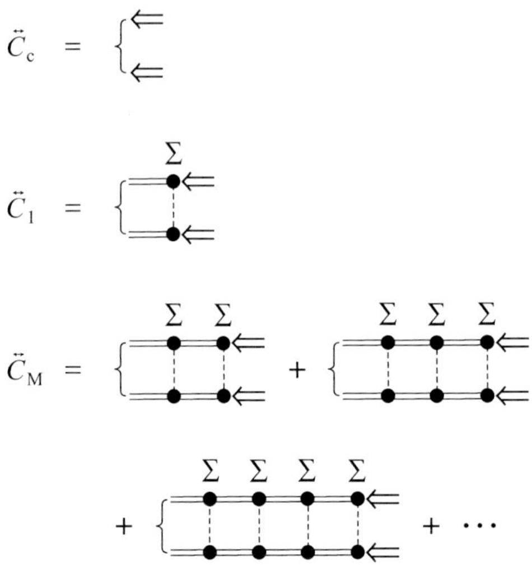

Figure 14.1.5. Various components of the ladder coherency dyadic.

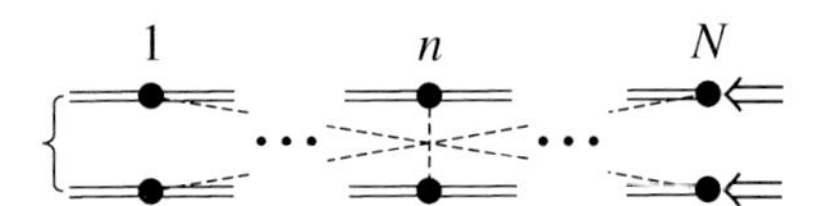

Figure 14.1.6. A cyclical diagram involving N connected particles.

be written for the cyclical component of the coherency dyadic:

$$\ddot{C}_{\mathrm{C}} = \int_{4\pi} \mathrm{d}\hat{\mathbf{n}}\, \ddot{\Sigma}_{\mathrm{C}}(\hat{\mathbf{n}}). \tag{14.1.9}$$

Indeed, the upper and lower scattering paths of each cyclical diagram involve the same group of particles but taken in opposite order. This is illustrated in Fig. 14.1.6 which shows a cyclical diagram involving N connected particles. The light propagating along the upper path arrives at the observation point in the form of a spherical wavelet centered at particle 1, whereas that propagating along the lower path arrives at the observation point in the form of a spherical wavelet centered at particle N. However, if the distance d_{1N} between particles 1 and N is much smaller than the distance d from the slab to the observation point,

$$d_{1N} \ll d, \tag{14.1.10}$$

then the direction of propagation of both wavelets at the observation point will be essentially the same. As a consequence, the expanded analytical expression for the cyclical component of the coherency dyadic analogous to Eq. (8.14.10) leads to Eq.

(14.1.9).

The above results show that the full coherency dyadic can be expressed in terms of the full specific coherency dyadic,

$$\ddot{C} = \int_{4\pi} \mathrm{d}\hat{\mathbf{n}}\, \ddot{\Sigma}(\hat{\mathbf{n}}). \tag{14.1.11}$$

The latter, in turn, can be expressed as the sum of the coherent, single-scattering, diffuse multiple-scattering, and cyclical components:

$$\begin{aligned}\ddot{\Sigma}(\hat{\mathbf{n}}) &= \ddot{\Sigma}_{\mathrm{L}}(\hat{\mathbf{n}}) + \ddot{\Sigma}_{\mathrm{C}}(\hat{\mathbf{n}}) \\ &= \ddot{\Sigma}_{\mathrm{c}}(\hat{\mathbf{n}}) + \ddot{\Sigma}_{1}(\hat{\mathbf{n}}) + \ddot{\Sigma}_{\mathrm{M}}(\hat{\mathbf{n}}) + \ddot{\Sigma}_{\mathrm{C}}(\hat{\mathbf{n}}).\end{aligned} \tag{14.1.12}$$

14.2 Reflected light

From this point on, we will often specify propagation directions using the notation introduced in Section 10.1. The expanded analytical versions of the diagrammatic formulas in Figs. 14.1.3 and 14.1.5 show that for light reflected in an upward direction $\hat{\mathbf{n}}$,

$$\hat{\mathbf{n}} \cdot \ddot{\Sigma}_{1}(-\hat{\mu}) = \ddot{\Sigma}_{1}(-\hat{\mu}) \cdot \hat{\mathbf{n}} = \mathbf{0}, \tag{14.2.1}$$

$$\hat{\mathbf{n}} \cdot \ddot{\Sigma}_{\mathrm{M}}(-\hat{\mu}) = \ddot{\Sigma}_{\mathrm{M}}(-\hat{\mu}) \cdot \hat{\mathbf{n}} = \mathbf{0}, \tag{14.2.2}$$

$$\hat{\mathbf{n}} \cdot \ddot{\Sigma}_{\mathrm{C}}(-\hat{\mu}) = \ddot{\Sigma}_{\mathrm{C}}(-\hat{\mu}) \cdot \hat{\mathbf{n}} = \mathbf{0}. \tag{14.2.3}$$

Furthermore, they show that the single-scattering, diffuse multiple-scattering, and cyclical components of the specific coherency dyadic are linearly expressed in the coherency dyad of the incident plane electromagnetic wave given by Eq. (14.1.7). Consequently, we may do two important things. First, we may define the corresponding specific coherency column vectors,

$$\tilde{\mathbf{J}}(-\hat{\mu}) = \tilde{\mathbf{J}}^{1}(-\hat{\mu}) + \tilde{\mathbf{J}}^{\mathrm{M}}(-\hat{\mu}) + \tilde{\mathbf{J}}^{\mathrm{C}}(-\hat{\mu}), \tag{14.2.4}$$

$$\tilde{\mathbf{J}}^{1}(-\hat{\mu}) = \frac{1}{2}\sqrt{\frac{\epsilon_1}{\mu_0}} \begin{bmatrix} \hat{\boldsymbol{\theta}}(\hat{\mathbf{n}}) \cdot \ddot{\Sigma}_{1}(-\hat{\mu}) \cdot \hat{\boldsymbol{\theta}}(\hat{\mathbf{n}}) \\ \hat{\boldsymbol{\theta}}(\hat{\mathbf{n}}) \cdot \ddot{\Sigma}_{1}(-\hat{\mu}) \cdot \hat{\boldsymbol{\varphi}}(\hat{\mathbf{n}}) \\ \hat{\boldsymbol{\varphi}}(\hat{\mathbf{n}}) \cdot \ddot{\Sigma}_{1}(-\hat{\mu}) \cdot \hat{\boldsymbol{\theta}}(\hat{\mathbf{n}}) \\ \hat{\boldsymbol{\varphi}}(\hat{\mathbf{n}}) \cdot \ddot{\Sigma}_{1}(-\hat{\mu}) \cdot \hat{\boldsymbol{\varphi}}(\hat{\mathbf{n}}) \end{bmatrix}, \tag{14.2.5}$$

$$\tilde{\mathbf{J}}^{\mathrm{M}}(-\hat{\mu}) = \frac{1}{2}\sqrt{\frac{\epsilon_1}{\mu_0}} \begin{bmatrix} \hat{\boldsymbol{\theta}}(\hat{\mathbf{n}}) \cdot \ddot{\Sigma}_{\mathrm{M}}(-\hat{\mu}) \cdot \hat{\boldsymbol{\theta}}(\hat{\mathbf{n}}) \\ \hat{\boldsymbol{\theta}}(\hat{\mathbf{n}}) \cdot \ddot{\Sigma}_{\mathrm{M}}(-\hat{\mu}) \cdot \hat{\boldsymbol{\varphi}}(\hat{\mathbf{n}}) \\ \hat{\boldsymbol{\varphi}}(\hat{\mathbf{n}}) \cdot \ddot{\Sigma}_{\mathrm{M}}(-\hat{\mu}) \cdot \hat{\boldsymbol{\theta}}(\hat{\mathbf{n}}) \\ \hat{\boldsymbol{\varphi}}(\hat{\mathbf{n}}) \cdot \ddot{\Sigma}_{\mathrm{M}}(-\hat{\mu}) \cdot \hat{\boldsymbol{\varphi}}(\hat{\mathbf{n}}) \end{bmatrix}, \tag{14.2.6}$$

$$\widetilde{\mathbf{J}}^{\mathrm{C}}(-\hat{\mu}) = \frac{1}{2}\sqrt{\frac{\epsilon_1}{\mu_0}} \begin{bmatrix} \hat{\boldsymbol{\theta}}(\hat{\mathbf{n}}) \cdot \ddot{\Sigma}_{\mathrm{C}}(-\hat{\mu}) \cdot \hat{\boldsymbol{\theta}}(\hat{\mathbf{n}}) \\ \hat{\boldsymbol{\theta}}(\hat{\mathbf{n}}) \cdot \ddot{\Sigma}_{\mathrm{C}}(-\hat{\mu}) \cdot \hat{\boldsymbol{\varphi}}(\hat{\mathbf{n}}) \\ \hat{\boldsymbol{\varphi}}(\hat{\mathbf{n}}) \cdot \ddot{\Sigma}_{\mathrm{C}}(-\hat{\mu}) \cdot \hat{\boldsymbol{\theta}}(\hat{\mathbf{n}}) \\ \hat{\boldsymbol{\varphi}}(\hat{\mathbf{n}}) \cdot \ddot{\Sigma}_{\mathrm{C}}(-\hat{\mu}) \cdot \hat{\boldsymbol{\varphi}}(\hat{\mathbf{n}}) \end{bmatrix}. \tag{14.2.7}$$

Second, we may define the corresponding 4×4 coherency reflection matrix $\mathcal{R}^{J}(\hat{\mu}, \hat{\mu}_0)$ according to

$$\widetilde{\mathbf{J}}(-\hat{\mu}) = \frac{1}{\pi}\mu_0 \mathcal{R}^{J}(\hat{\mu}, \hat{\mu}_0)\mathbf{J}_0 \tag{14.2.8}$$

and represent it as a sum of the corresponding single-scattering, diffuse multiple-scattering, and cyclical components:

$$\mathcal{R}^{J}(\hat{\mu}, \hat{\mu}_0) = \mathcal{R}^{J1}(\hat{\mu}, \hat{\mu}_0) + \mathcal{R}^{J\mathrm{M}}(\hat{\mu}, \hat{\mu}_0) + \mathcal{R}^{J\mathrm{C}}(\hat{\mu}, \hat{\mu}_0), \tag{14.2.9}$$

where $\mathbf{J}_0$ is the coherency column vector of the incident plane wave:

$$\mathbf{J}_0 = \frac{1}{2}\sqrt{\frac{\epsilon_1}{\mu_0}} \begin{bmatrix} E_{0\theta}^{\mathrm{inc}}(E_{0\theta}^{\mathrm{inc}})^* \\ E_{0\theta}^{\mathrm{inc}}(E_{0\varphi}^{\mathrm{inc}})^* \\ E_{0\varphi}^{\mathrm{inc}}(E_{0\theta}^{\mathrm{inc}})^* \\ E_{0\varphi}^{\mathrm{inc}}(E_{0\varphi}^{\mathrm{inc}})^* \end{bmatrix}. \tag{14.2.10}$$

In the Stokes-vector representation, we have

$$\widetilde{\mathbf{I}}(-\hat{\mu}) = \widetilde{\mathbf{I}}^{1}(-\hat{\mu}) + \widetilde{\mathbf{I}}^{\mathrm{M}}(-\hat{\mu}) + \widetilde{\mathbf{I}}^{\mathrm{C}}(-\hat{\mu}), \tag{14.2.11}$$

$$\widetilde{\mathbf{I}}^{1}(-\hat{\mu}) = \mathbf{D}\widetilde{\mathbf{J}}^{1}(-\hat{\mu}), \tag{14.2.12}$$

$$\widetilde{\mathbf{I}}^{\mathrm{M}}(-\hat{\mu}) = \mathbf{D}\widetilde{\mathbf{J}}^{\mathrm{M}}(-\hat{\mu}), \tag{14.2.13}$$

$$\widetilde{\mathbf{I}}^{\mathrm{C}}(-\hat{\mu}) = \mathbf{D}\widetilde{\mathbf{J}}^{\mathrm{C}}(-\hat{\mu}), \tag{14.2.14}$$

$$\widetilde{\mathbf{I}}(-\hat{\mu}) = \frac{1}{\pi}\mu_0 \mathcal{R}(\hat{\mu}, \hat{\mu}_0)\mathbf{I}_0, \tag{14.2.15}$$

$$\mathcal{R}(\hat{\mu}, \hat{\mu}_0) = \mathcal{R}^{1}(\hat{\mu}, \hat{\mu}_0) + \mathcal{R}^{\mathrm{M}}(\hat{\mu}, \hat{\mu}_0) + \mathcal{R}^{\mathrm{C}}(\hat{\mu}, \hat{\mu}_0), \tag{14.2.16}$$

$$\mathcal{R}^{1}(\hat{\mu}, \hat{\mu}_0) = \mathbf{D}\mathcal{R}^{J1}(\hat{\mu}, \hat{\mu}_0)\mathbf{D}^{-1}, \tag{14.2.17}$$

$$\mathcal{R}^{\mathrm{M}}(\hat{\mu}, \hat{\mu}_0) = \mathbf{D}\mathcal{R}^{J\mathrm{M}}(\hat{\mu}, \hat{\mu}_0)\mathbf{D}^{-1}, \tag{14.2.18}$$

$$\mathcal{R}^{\mathrm{C}}(\hat{\mu}, \hat{\mu}_0) = \mathbf{D}\mathcal{R}^{J\mathrm{C}}(\hat{\mu}, \hat{\mu}_0)\mathbf{D}^{-1}, \tag{14.2.19}$$

$$\mathbf{I}_0 = \mathbf{D}\mathbf{J}_0. \tag{14.2.20}$$

It is obvious that the sum of the single-scattering and diffuse multiple-scattering components of the reflection matrix,

$$\mathbf{R}(\hat{\mu}, \hat{\mu}_0) = \mathcal{R}^{1}(\hat{\mu}, \hat{\mu}_0) + \mathcal{R}^{\mathrm{M}}(\hat{\mu}, \hat{\mu}_0), \tag{14.2.21}$$

yields the Stokes reflection matrix obtained by solving the RTE (cf. Eq. (10.3.12)).

Consider now the response of a well-collimated polarization-sensitive detector of electromagnetic energy located at the distant observation point, Fig. 14.1.4. Let us imagine that the detector scans a range of upward propagation directions $\hat{\mathbf{n}}$ including the exact backscattering direction given by $\hat{\mathbf{n}} = -\hat{\mathbf{n}}_0$. According to the previous discussion of CB, it is useful to consider the following three particular situations:

- The incoming propagation direction $\hat{\mathbf{n}}$ is far from the exact backscattering direction $-\hat{\mathbf{n}}_0$. Then the cyclical specific coherency dyadic vanishes, and the detector response is fully determined by the diffuse specific coherency dyadic $\ddot{\Sigma}_{\rm d}(\hat{\mathbf{n}})$. This means that the response of the detector can be fully quantified in terms of the reflection matrix $\mathbf{R}(\hat{\mu}, \hat{\mu}_0)$ obtained by solving the RTE.

- The detector registers light propagating in the exact backscattering direction. Then the effects of CB can be expected to be maximal and must be taken into account. We shall demonstrate in the following section that the Saxon's reciprocity relation (3.4.19) can be used to derive an exact analytical expression of the cyclical component of the backscattering reflection matrix $\mathcal{R}^{\rm C}(\hat{\mu}_0, \hat{\mu}_0)$ in terms of the diffuse multiple-scattering component $\mathcal{R}^{\rm M}(\hat{\mu}_0, \hat{\mu}_0)$. This fundamental result also allows one to fully quantify the response of the detector centered at the exact backscattering direction in terms of the solution of the RTE.

- As the detector axis deviates more and more from the exact backscattering direction the effects of CB can be expected to weaken and gradually disappear. The computation of the angular profile of the detector response in this transition region of incoming directions is a difficult task and will be discussed in Section 14.7.

14.3 Exact backscattering direction

We will now discuss the computation of the cyclical specific coherency dyadic for the case of the exact backscattering direction, $\hat{\mathbf{n}} = -\hat{\mathbf{n}}_0$. Let us consider first the simplest two-particle diagrams depicted in Fig. 14.3.1. All four diagrams involve the same particles, 1 and 2, and an observation point D located at a very large distance from the scattering medium, as shown schematically in Fig. 14.3.2. Let $\mathbf{r}_A$ be the position vector of the boundary point A. Then the corresponding time-independent part of the electric field is

$$\mathbf{E}_A^{\rm inc} = \mathbf{E}_0^{\rm inc} \exp(\mathrm{i}k_1 \hat{\mathbf{n}}_0 \cdot \mathbf{r}_A). \tag{14.3.1}$$

It is easy to see that the time-independent part of the electric field at the boundary point B is

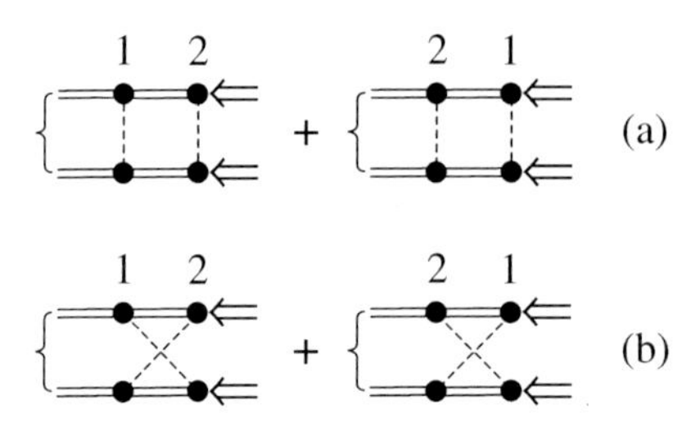

Figure 14.3.1. Conjugate pairs of two-particle ladder and cyclical diagrams.

$$\mathbf{E}_B^{\mathrm{inc}} = \mathbf{E}_A^{\mathrm{inc}} \exp(\mathrm{i}k_1 \overline{BC}), \tag{14.3.2}$$

where $\overline{BC}$ is the distance from point C to point B.

Let us introduce the following two dyadics:

$$\ddot{Y} = \frac{\exp(\mathrm{i}k_1 \overline{DA})\, \ddot{\eta}(-\hat{\mathbf{n}}_0, \overline{A1})}{\overline{D1}} \cdot \ddot{A}_1(-\hat{\mathbf{n}}_0, \hat{\mathbf{R}}_{12}) \cdot \frac{\ddot{\eta}(\hat{\mathbf{R}}_{12}, \overline{12})}{\overline{12}}$$
$$\cdot \ddot{A}_2(\hat{\mathbf{R}}_{12}, \hat{\mathbf{n}}_0) \cdot \ddot{\eta}(\hat{\mathbf{n}}_0, \overline{2B}) \exp(\mathrm{i}k_1 \overline{BC}), \tag{14.3.3}$$

$$\ddot{Y}' = \frac{\exp(\mathrm{i}k_1 \overline{DC}) \exp(\mathrm{i}k_1 \overline{CB})\, \ddot{\eta}(-\hat{\mathbf{n}}_0, \overline{B2})}{\overline{D2}} \cdot \ddot{A}_2(-\hat{\mathbf{n}}_0, \hat{\mathbf{R}}_{21})$$
$$\cdot \frac{\ddot{\eta}(\hat{\mathbf{R}}_{21}, \overline{21})}{\overline{21}} \cdot \ddot{A}_1(\hat{\mathbf{R}}_{21}, \hat{\mathbf{n}}_0) \cdot \ddot{\eta}(\hat{\mathbf{n}}_0, \overline{1A}), \tag{14.3.4}$$

where $\hat{\mathbf{R}}_{12}$ and $\hat{\mathbf{R}}_{21} = -\hat{\mathbf{R}}_{12}$ are unit vectors shown in Fig. 14.3.2 and $\ddot{\eta}$ is the coherent transmission dyadic given by Eq. (8.3.13). It is easy to see that the contribution of the two ladder diagrams shown in Fig. 14.3.1(a) to $\ddot{\Sigma}_{\mathrm{M}}(-\hat{\mathbf{n}}_0)$ is described by the dyadic

$$\ddot{M} = (\ddot{Y} \cdot \mathbf{E}_A^{\mathrm{inc}}) \otimes (\ddot{Y} \cdot \mathbf{E}_A^{\mathrm{inc}})^* + (\ddot{Y}' \cdot \mathbf{E}_A^{\mathrm{inc}}) \otimes (\ddot{Y}' \cdot \mathbf{E}_A^{\mathrm{inc}})^*$$
$$= \ddot{Y} \cdot \ddot{\rho}^{\mathrm{inc}} \cdot \ddot{Y}^{\mathrm{T}*} + \ddot{Y}' \cdot \ddot{\rho}^{\mathrm{inc}} \cdot (\ddot{Y}')^{\mathrm{T}*}. \tag{14.3.5}$$

Analogously, the contribution of the two cyclical diagrams shown in Fig. 14.3.1(b) to $\ddot{\Sigma}_{\mathrm{C}}(-\hat{\mathbf{n}}_0)$ is described by the dyadic

$$\ddot{C} = (\ddot{Y} \cdot \mathbf{E}_A^{\mathrm{inc}}) \otimes (\ddot{Y}' \cdot \mathbf{E}_A^{\mathrm{inc}})^* + (\ddot{Y}' \cdot \mathbf{E}_A^{\mathrm{inc}}) \otimes (\ddot{Y} \cdot \mathbf{E}_A^{\mathrm{inc}})^*$$
$$= \ddot{Y} \cdot \ddot{\rho}^{\mathrm{inc}} \cdot (\ddot{Y}')^{\mathrm{T}*} + \ddot{Y}' \cdot \ddot{\rho}^{\mathrm{inc}} \cdot \ddot{Y}^{\mathrm{T}*}. \tag{14.3.6}$$

It is now the right time to recall the reciprocity relations (3.4.19) and (8.3.28). The application of these relations along with the equality $DA = DC$ and the approximate equality

$$\frac{1}{\overline{D1}} \approx \frac{1}{\overline{D2}} \tag{14.3.7}$$

to Eqs. (14.3.3) and (14.3.4) yields

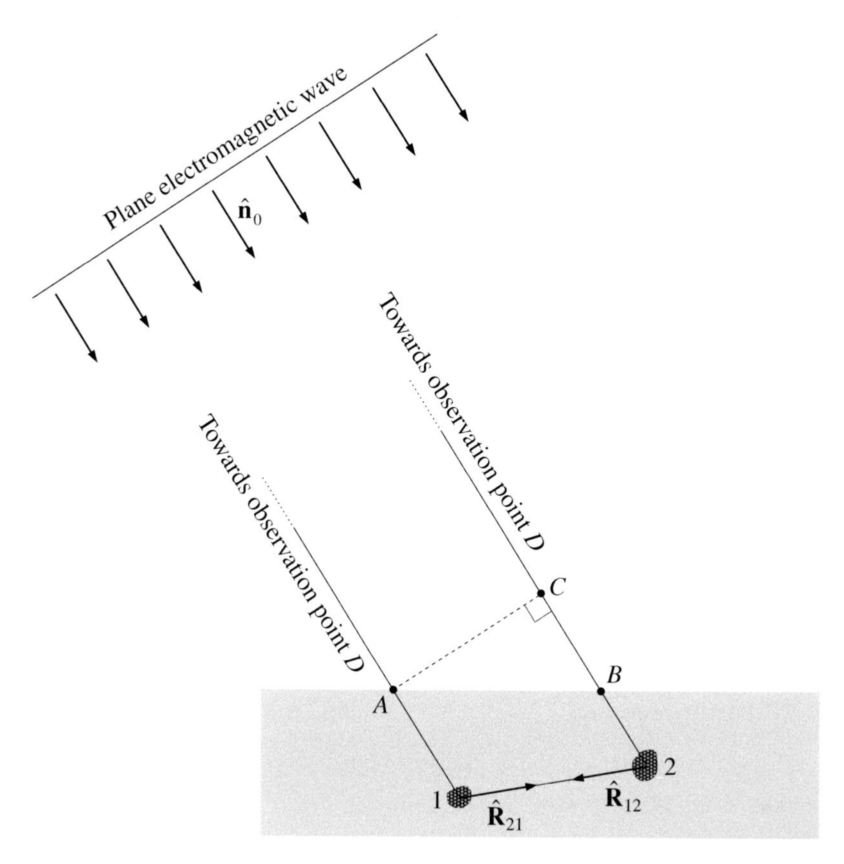

Figure 14.3.2. Derivation of Eqs. (14.3.3) and (14.3.4).

$$\vec{Y}' = \vec{Y}^{\mathrm{T}}, \tag{14.3.8}$$

whereas the substitution of Eq. (14.3.8) in Eqs. (14.3.5) and (14.3.6) gives

$$\vec{M} = \vec{Y} \cdot \vec{\rho}^{\,\mathrm{inc}} \cdot \vec{Y}^{\mathrm{T}*} + \vec{Y}^{\mathrm{T}} \cdot \vec{\rho}^{\,\mathrm{inc}} \cdot \vec{Y}^{*}, \tag{14.3.9}$$

$$\vec{C} = \vec{Y} \cdot \vec{\rho}^{\,\mathrm{inc}} \cdot \vec{Y}^{*} + \vec{Y}^{\mathrm{T}} \cdot \vec{\rho}^{\,\mathrm{inc}} \cdot \vec{Y}^{\mathrm{T}*}. \tag{14.3.10}$$

The next step is to introduce the following four scalars:

$$a = \hat{\boldsymbol{\theta}}(\hat{\mathbf{n}}_0) \cdot \vec{Y} \cdot \hat{\boldsymbol{\theta}}(\hat{\mathbf{n}}_0), \tag{14.3.11}$$

$$b = \hat{\boldsymbol{\theta}}(\hat{\mathbf{n}}_0) \cdot \vec{Y} \cdot \hat{\boldsymbol{\varphi}}(\hat{\mathbf{n}}_0), \tag{14.3.12}$$

$$c = \hat{\boldsymbol{\varphi}}(\hat{\mathbf{n}}_0) \cdot \vec{Y} \cdot \hat{\boldsymbol{\theta}}(\hat{\mathbf{n}}_0), \tag{14.3.13}$$

$$d = \hat{\boldsymbol{\varphi}}(\hat{\mathbf{n}}_0) \cdot \vec{Y} \cdot \hat{\boldsymbol{\varphi}}(\hat{\mathbf{n}}_0). \tag{14.3.14}$$

Using Eq. (14.3.9) in Eq. (14.2.6) evaluated for $\hat{\mathbf{n}} = -\hat{\mathbf{n}}_0$ and taking into account Eqs. (3.4.20), (A.6), (A.12), and (14.2.10) shows that the contribution of the two lad-

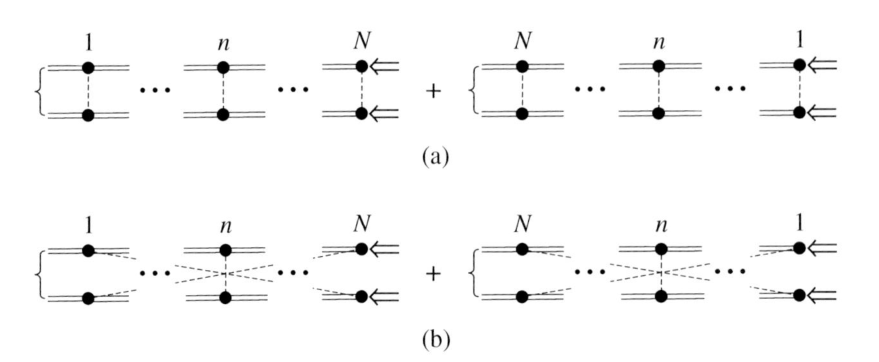

Figure 14.3.3. Conjugate pairs of N-particle ladder and cyclical diagrams.

der diagrams shown in Fig. 14.3.1(a) to the diffuse multiple-scattering specific coherency column vector is given by

$$\begin{bmatrix} 2aa^* & ab^* + ac^* & ba^* + ca^* & bb^* + cc^* \\ -ab^* - ac^* & -2ad^* & -bc^* - cb^* & -bd^* - cd^* \\ -ba^* - ca^* & -bc^* - cb^* & -2da^* & -db^* - dc^* \\ bb^* + cc^* & bd^* + cd^* & db^* + dc^* & 2dd^* \end{bmatrix} \mathbf{J}_0. \tag{14.3.15}$$

Analogously, using Eq. (14.3.10) in Eq. (14.2.7) evaluated for $\hat{\mathbf{n}} = -\hat{\mathbf{n}}_0$ shows that the contribution of the two cyclical diagrams shown in Fig. 14.3.1(b) to the coherent specific coherency column vector is given by

$$\begin{bmatrix} 2aa^* & ab^* + ac^* & ba^* + ca^* & bc^* + cb^* \\ -ab^* - ac^* & -2ad^* & -bb^* - cc^* & -bd^* - cd^* \\ -ba^* - ca^* & -bb^* - cc^* & -2da^* & -db^* - dc^* \\ bc^* + cb^* & bd^* + cd^* & db^* + dc^* & 2dd^* \end{bmatrix} \mathbf{J}_0. \tag{14.3.16}$$

Although Eqs. (14.3.15) and (14.3.16) were derived for sums of two-particle ladder and cyclical diagrams, respectively, it is easy to see that they are valid for sums of ladder and cyclical diagrams involving any N particles, Fig. 14.3.3. Therefore, Eqs. (14.2.8), (14.2.9), (14.3.15) and (14.3.16) imply that the backscattering coherent matrix $\boldsymbol{\mathcal{R}}^{\mathrm{JC}}(\hat{\mu}_0^{\pi}, \hat{\mu}_0)$ can be expressed in terms of the backscattering diffuse multiple-scattering matrix $\boldsymbol{\mathcal{R}}^{\mathrm{JM}}(\hat{\mu}_0^{\pi}, \hat{\mu}_0)$, where $\hat{\mu}_0^{\pi}$ denotes the couplet $\{\mu_0, \varphi_0 + \pi\}$. Specifically,

$$\boldsymbol{\mathcal{R}}^{\mathrm{JC}} = \begin{bmatrix} \mathcal{R}_{11}^{\mathrm{JM}} & \mathcal{R}_{12}^{\mathrm{JM}} & \mathcal{R}_{13}^{\mathrm{JM}} & -\mathcal{R}_{23}^{\mathrm{JM}} \\ \mathcal{R}_{21}^{\mathrm{JM}} & \mathcal{R}_{22}^{\mathrm{JM}} & -\mathcal{R}_{14}^{\mathrm{JM}} & \mathcal{R}_{24}^{\mathrm{JM}} \\ \mathcal{R}_{31}^{\mathrm{JM}} & -\mathcal{R}_{14}^{\mathrm{JM}} & \mathcal{R}_{33}^{\mathrm{JM}} & \mathcal{R}_{34}^{\mathrm{JM}} \\ -\mathcal{R}_{23}^{\mathrm{JM}} & \mathcal{R}_{42}^{\mathrm{JM}} & \mathcal{R}_{43}^{\mathrm{JM}} & \mathcal{R}_{44}^{\mathrm{JM}} \end{bmatrix} \tag{14.3.17}$$

(Mishchenko, 1992a). The importance of this rigorous relationship is difficult to overstate. Indeed, it demonstrates that although the RTT is based on the neglect of all cyclical diagrams, all observable characteristics of CB at the *exact backscattering direction* can still be calculated by solving the RTE. Specific applications of this fundamental result will be described in Sections 14.5 and 14.6.

In what follows, we will simplify the discussion by assuming that the scattering medium is macroscopically isotropic and mirror-symmetric. Equations (12.6.9) and (12.6.12) imply that the backscattering Stokes matrix $\boldsymbol{\mathcal{R}}^{\mathrm{M}}(\hat{\mu}_0^{\pi}, \hat{\mu}_0)$ has the following block-diagonal structure:

$$\boldsymbol{\mathcal{R}}^{\mathrm{M}} = \begin{bmatrix} \mathcal{R}_{11}^{\mathrm{M}} & \mathcal{R}_{12}^{\mathrm{M}} & 0 & 0 \\ \mathcal{R}_{12}^{\mathrm{M}} & \mathcal{R}_{22}^{\mathrm{M}} & 0 & 0 \\ 0 & 0 & \mathcal{R}_{33}^{\mathrm{M}} & \mathcal{R}_{34}^{\mathrm{M}} \\ 0 & 0 & -\mathcal{R}_{34}^{\mathrm{M}} & \mathcal{R}_{44}^{\mathrm{M}} \end{bmatrix}. \tag{14.3.18}$$

Equation (14.2.18) then yields for the backscattering coherency matrix $\boldsymbol{\mathcal{R}}^{J\mathrm{M}}(\hat{\mu}_0^{\pi}, \hat{\mu}_0)$:

$$\boldsymbol{\mathcal{R}}^{J\mathrm{M}} = \begin{bmatrix} \mathcal{R}_{11}^{J\mathrm{M}} & 0 & 0 & \mathcal{R}_{14}^{J\mathrm{M}} \\ 0 & \mathcal{R}_{22}^{J\mathrm{M}} & \mathcal{R}_{23}^{J\mathrm{M}} & 0 \\ 0 & \mathcal{R}_{23}^{J\mathrm{M}} & \mathcal{R}_{33}^{J\mathrm{M}} & 0 \\ \mathcal{R}_{14}^{J\mathrm{M}} & 0 & 0 & \mathcal{R}_{44}^{J\mathrm{M}} \end{bmatrix}. \tag{14.3.19}$$

As a consequence, Eq. (14.3.17) becomes considerably simpler:

$$\boldsymbol{\mathcal{R}}^{J\mathrm{C}} = \begin{bmatrix} \mathcal{R}_{11}^{J\mathrm{M}} & 0 & 0 & -\mathcal{R}_{23}^{J\mathrm{M}} \\ 0 & \mathcal{R}_{22}^{J\mathrm{M}} & -\mathcal{R}_{14}^{J\mathrm{M}} & 0 \\ 0 & -\mathcal{R}_{14}^{J\mathrm{M}} & \mathcal{R}_{33}^{J\mathrm{M}} & 0 \\ -\mathcal{R}_{23}^{J\mathrm{M}} & 0 & 0 & \mathcal{R}_{44}^{J\mathrm{M}} \end{bmatrix}. \tag{14.3.20}$$

Equations (4.2.18) and (4.2.19) finally yield for the backscattering Stokes matrix $\boldsymbol{\mathcal{R}}^{\mathrm{C}}(\hat{\mu}_0^{\pi}, \hat{\mu}_0)$:

$$\boldsymbol{\mathcal{R}}^{\mathrm{C}} = \begin{bmatrix} \mathcal{R}_{11}^{\mathrm{C}} & \mathcal{R}_{12}^{\mathrm{M}} & 0 & 0 \\ \mathcal{R}_{12}^{\mathrm{M}} & \mathcal{R}_{22}^{\mathrm{C}} & 0 & 0 \\ 0 & 0 & \mathcal{R}_{33}^{\mathrm{C}} & \mathcal{R}_{34}^{\mathrm{M}} \\ 0 & 0 & -\mathcal{R}_{34}^{\mathrm{M}} & \mathcal{R}_{44}^{\mathrm{C}} \end{bmatrix}, \tag{14.3.21}$$

where

$$\mathcal{R}_{11}^{\mathrm{C}}(\hat{\mu}_0^{\pi}, \hat{\mu}_0) = \tfrac{1}{2}[\mathcal{R}_{11}^{\mathrm{M}}(\hat{\mu}_0^{\pi}, \hat{\mu}_0) + \mathcal{R}_{22}^{\mathrm{M}}(\hat{\mu}_0^{\pi}, \hat{\mu}_0) - \mathcal{R}_{33}^{\mathrm{M}}(\hat{\mu}_0^{\pi}, \hat{\mu}_0) + \mathcal{R}_{44}^{\mathrm{M}}(\hat{\mu}_0^{\pi}, \hat{\mu}_0)], \tag{14.3.22}$$

$$\mathcal{R}_{22}^{\mathrm{C}}(\hat{\mu}_0^{\pi}, \hat{\mu}_0) = \tfrac{1}{2}[\mathcal{R}_{11}^{\mathrm{M}}(\hat{\mu}_0^{\pi}, \hat{\mu}_0) + \mathcal{R}_{22}^{\mathrm{M}}(\hat{\mu}_0^{\pi}, \hat{\mu}_0) + \mathcal{R}_{33}^{\mathrm{M}}(\hat{\mu}_0^{\pi}, \hat{\mu}_0) - \mathcal{R}_{44}^{\mathrm{M}}(\hat{\mu}_0^{\pi}, \hat{\mu}_0)], \tag{14.3.23}$$

$$\mathcal{R}^{\mathrm{C}}_{33}(\hat{\mu}_0^{\pi}, \hat{\mu}_0) = \tfrac{1}{2}[-\mathcal{R}^{\mathrm{M}}_{11}(\hat{\mu}_0^{\pi}, \hat{\mu}_0) + \mathcal{R}^{\mathrm{M}}_{22}(\hat{\mu}_0^{\pi}, \hat{\mu}_0) + \mathcal{R}^{\mathrm{M}}_{33}(\hat{\mu}_0^{\pi}, \hat{\mu}_0) + \mathcal{R}^{\mathrm{M}}_{44}(\hat{\mu}_0^{\pi}, \hat{\mu}_0)], \tag{14.3.24}$$

$$\mathcal{R}^{\mathrm{C}}_{44}(\hat{\mu}_0^{\pi}, \hat{\mu}_0) = \tfrac{1}{2}[\mathcal{R}^{\mathrm{M}}_{11}(\hat{\mu}_0^{\pi}, \hat{\mu}_0) - \mathcal{R}^{\mathrm{M}}_{22}(\hat{\mu}_0^{\pi}, \hat{\mu}_0) + \mathcal{R}^{\mathrm{M}}_{33}(\hat{\mu}_0^{\pi}, \hat{\mu}_0) + \mathcal{R}^{\mathrm{M}}_{44}(\hat{\mu}_0^{\pi}, \hat{\mu}_0)]. \tag{14.3.25}$$

The actual application of the above formulas requires the knowledge of the matrices $\boldsymbol{\mathcal{R}}^{1}(\hat{\mu}_0^{\pi}, \hat{\mu}_0)$ and $\boldsymbol{\mathcal{R}}^{\mathrm{M}}(\hat{\mu}_0^{\pi}, \hat{\mu}_0)$. According to Eq. (14.2.21), the diffuse multiple-scattering component can be found by subtracting the first-order scattering component from the reflection matrix $\mathbf{R}(\hat{\mu}, \hat{\mu}_0)$ obtained by solving the VRTE:

$$\boldsymbol{\mathcal{R}}^{\mathrm{M}}(\hat{\mu}, \hat{\mu}_0) = \mathbf{R}(\hat{\mu}, \hat{\mu}_0) - \boldsymbol{\mathcal{R}}^{1}(\hat{\mu}, \hat{\mu}_0). \tag{14.3.26}$$

A simple formula for the first-order-scattering contribution $\boldsymbol{\mathcal{R}}^{1}(\hat{\mu}, \hat{\mu}_0)$ can be derived from the invariant imbedding equation (12.4.1) using the order-of-scattering expansion (12.5.1) and by taking into account that

$$\boldsymbol{\mathcal{R}}^{1}(\hat{\mu}, \hat{\mu}_0) = \varpi \mathbf{R}_1(\hat{\mu}, \hat{\mu}_0). \tag{14.3.27}$$

Indeed, assuming that the scattering slab is homogeneous and substituting Eq. (12.5.1) in Eq. (12.4.1) yields

$$\frac{\partial \boldsymbol{\mathcal{R}}^{1}(\mu, \varphi; \mu_0, \varphi_0)}{\partial \mathcal{T}_{\uparrow}} = -\frac{\mu + \mu_0}{\mu \mu_0} \boldsymbol{\mathcal{R}}^{1}(\mu, \varphi; \mu_0, \varphi_0) + \frac{\varpi}{4 \mu \mu_0} \widetilde{\mathbf{Z}}(-\mu, \varphi; \mu_0, \varphi_0). \tag{14.3.28}$$

The solution of this equation satisfying the initial condition (12.4.5) is

$$\boldsymbol{\mathcal{R}}^{1}(\mu, \varphi; \mu_0, \varphi_0) = \frac{\varpi}{4} \frac{1 - \exp\left[-\mathcal{T}\left(\dfrac{1}{\mu} + \dfrac{1}{\mu_0}\right)\right]}{\mu + \mu_0} \widetilde{\mathbf{Z}}(-\mu, \varphi; \mu_0, \varphi_0). \tag{14.3.29}$$

In the case of a homogeneous semi-infinite slab,

$$\boldsymbol{\mathcal{R}}^{1}(\mu, \varphi; \mu_0, \varphi_0)\Big|_{\mathcal{T} \to \infty} = \frac{\varpi}{4(\mu + \mu_0)} \widetilde{\mathbf{Z}}(-\mu, \varphi; \mu_0, \varphi_0). \tag{14.3.30}$$

Since

$$\widetilde{\mathbf{Z}}(-\mu_0, \varphi_0 + \pi; \mu_0, \varphi_0) \equiv \widetilde{\mathbf{F}}(\pi) = \begin{bmatrix} a_1(\pi) & 0 & 0 & 0 \\ 0 & \dfrac{a_1(\pi) - a_4(\pi)}{2} & 0 & 0 \\ 0 & 0 & -\dfrac{a_1(\pi) - a_4(\pi)}{2} & 0 \\ 0 & 0 & 0 & a_4(\pi) \end{bmatrix}, \tag{14.3.31}$$

Eqs. (14.3.29) and (14.3.30) yield

$$\mathcal{R}^1(\hat{\mu}_0^\pi, \hat{\mu}_0) = \frac{\varpi[1 - \exp(-2\mathcal{T}/\mu_0)]}{8\mu_0} \widetilde{\mathbf{F}}(\pi), \tag{14.3.32}$$

$$\mathcal{R}^1(\hat{\mu}_0^\pi, \hat{\mu}_0)\Big|_{\mathcal{T}\to\infty} = \frac{\varpi}{8\mu_0} \widetilde{\mathbf{F}}(\pi). \tag{14.3.33}$$

The matrix $\mathbf{R}(\hat{\mu}, \hat{\mu}_0)$ for a finite slab can be found as a solution of the invariant imbedding equation (12.4.1) supplemented by the initial condition (12.4.5). Alternatively, it can be calculated using the doubling method. In the case of a homogeneous semi-infinite slab it is more efficient to compute the matrix $\mathbf{R}(\hat{\mu}, \hat{\mu}_0)$ by solving iteratively the Ambarzumian equation (12.4.9). The actual numerical data discussed in Subsection 14.5.6 and Section 14.6 were obtained with computer codes described in Mishchenko (1990b, 1996).

Note that by virtue of representing the solution of the VRTE for a plane-parallel slab of infinite horizontal extent, the matrices $\mathcal{R}^1(\hat{\mu}, \hat{\mu}_0)$ and $\mathcal{R}^{\mathrm{M}}(\hat{\mu}, \hat{\mu}_0)$ are independent of the distance from the upper boundary of the slab to the observation point. Equations (14.3.21)–(14.3.25) imply that this is also true of the cyclical matrix $\mathcal{R}^{\mathrm{C}}(\hat{\mu}_0^\pi, \hat{\mu}_0)$.

14.4 Other types of illumination

The above discussion of CB was explicitly based on the assumption that the incident light is a plane electromagnetic wave. However, we could have made the discussion more general by using the terminology introduced in Sections 3.10 and 8.15. For example, we could have assumed that the discrete random medium is illuminated by a parallel quasi-monochromatic beam and that significant changes of the transformation dyadic occur much more slowly than the random oscillations of the electric field amplitude. It is then straightforward to show that the above equations remain valid provided that the coherency dyad of the incident plane wave, Eq. (14.1.7), is replaced by the time average of the coherency dyad of the quasi-monochromatic beam, $\langle \vec{\rho}^{\mathrm{inc}} \rangle_t = \langle \mathbf{E}_0^{\mathrm{inc}}(t) \otimes [\mathbf{E}_0^{\mathrm{inc}}(t)]^* \rangle_t$. This result allows one to study CB of partially polarized and even unpolarized incident light.

Similarly, if the medium is illuminated by N quasi-monochromatic beams then it can be shown that the total angular distribution of the reflected light is obtained by adding the individual angular distributions computed for each incident quasi-monochromatic beam separately. In particular, if the N beams are incident in different directions then there will be N separate CB patterns centered at the respective backscattering directions. This conclusion remains valid if the medium is illuminated by several plane electromagnetic waves provided that all of them have different angular frequencies.

14.5 Photometric and polarimetric characteristics of coherent backscattering

The range of reflection directions affected by CB is usually rather small (see, for example, Fig. 1.8.2). Therefore, Eqs. (14.2.16) and (14.3.21)–(14.3.25) can be used to define several useful observable quantities relating characteristics of light reflected in the exact backscattering direction and in nearby directions not affected by CB. We will introduce these quantities under the assumption that the incident light is either unpolarized, fully linearly polarized, or fully circularly polarized. These polarization states of the incident light are most often encountered in practice. Typical examples are unpolarized sunlight and linearly or circularly polarized electromagnetic radiation emitted by lasers or radio antennas.

14.5.1 Unpolarized incident light

In this case the specific intensity reflected by a plane-parallel slab in the exact backscattering direction is given by

$$\begin{aligned}\widetilde{I}(-\mu_0,\varphi_0+\pi) &= \frac{1}{\pi}\mu_0\mathcal{R}_{11}(\hat{\mu}_0^\pi,\hat{\mu}_0)I_0\\ &= \frac{1}{\pi}\mu_0[\mathcal{R}_{11}^1(\hat{\mu}_0^\pi,\hat{\mu}_0)+\mathcal{R}_{11}^{\mathrm{M}}(\hat{\mu}_0^\pi,\hat{\mu}_0)+\mathcal{R}_{11}^{\mathrm{C}}(\hat{\mu}_0^\pi,\hat{\mu}_0)]I_0\\ &= \frac{1}{\pi}\mu_0\{\mathcal{R}_{11}^1(\hat{\mu}_0^\pi,\hat{\mu}_0)+\mathcal{R}_{11}^{\mathrm{M}}(\hat{\mu}_0^\pi,\hat{\mu}_0)+\tfrac{1}{2}[\mathcal{R}_{11}^{\mathrm{M}}(\hat{\mu}_0^\pi,\hat{\mu}_0)\\ &\qquad +\mathcal{R}_{22}^{\mathrm{M}}(\hat{\mu}_0^\pi,\hat{\mu}_0)-\mathcal{R}_{33}^{\mathrm{M}}(\hat{\mu}_0^\pi,\hat{\mu}_0)+\mathcal{R}_{44}^{\mathrm{M}}(\hat{\mu}_0^\pi,\hat{\mu}_0)]\}I_0,\end{aligned} \quad (14.5.1)$$

where I_0 is the incident intensity. Assuming that both $\boldsymbol{\mathcal{R}}^1(\hat{\mu},\hat{\mu}_0)$ and $\boldsymbol{\mathcal{R}}^{\mathrm{M}}(\hat{\mu},\hat{\mu}_0)$ do not change significantly over the range of reflection directions affected by CB, we can also write for the intensity of the surrounding "incoherent" (or "diffuse") background (see Fig. 14.5.1):

$$\widetilde{I}^{\,\mathrm{diff}}(-\mu_0,\varphi_0+\pi) = \frac{1}{\pi}\mu_0[\mathcal{R}_{11}^1(\hat{\mu}_0^\pi,\hat{\mu}_0)+\mathcal{R}_{11}^{\mathrm{M}}(\hat{\mu}_0^\pi,\hat{\mu}_0)]I_0. \quad (14.5.2)$$

The above equations can be used to define the corresponding "unpolarized" enhancement factor as the ratio of the total intensity reflected in the exact backscattering direction to that of the incoherent background:

$$\begin{aligned}\zeta_I = \frac{\widetilde{I}}{\widetilde{I}^{\,\mathrm{diff}}} &= \frac{\mathcal{R}_{11}^1+\mathcal{R}_{11}^{\mathrm{M}}+\tfrac{1}{2}(\mathcal{R}_{11}^{\mathrm{M}}+\mathcal{R}_{22}^{\mathrm{M}}-\mathcal{R}_{33}^{\mathrm{M}}+\mathcal{R}_{44}^{\mathrm{M}})}{\mathcal{R}_{11}^1+\mathcal{R}_{11}^{\mathrm{M}}}\\ &= 1+\frac{\mathcal{R}_{11}^{\mathrm{M}}+\mathcal{R}_{22}^{\mathrm{M}}-\mathcal{R}_{33}^{\mathrm{M}}+\mathcal{R}_{44}^{\mathrm{M}}}{2(\mathcal{R}_{11}^1+\mathcal{R}_{11}^{\mathrm{M}})},\end{aligned} \quad (14.5.3)$$

where we have omitted the angular arguments for the sake of brevity.

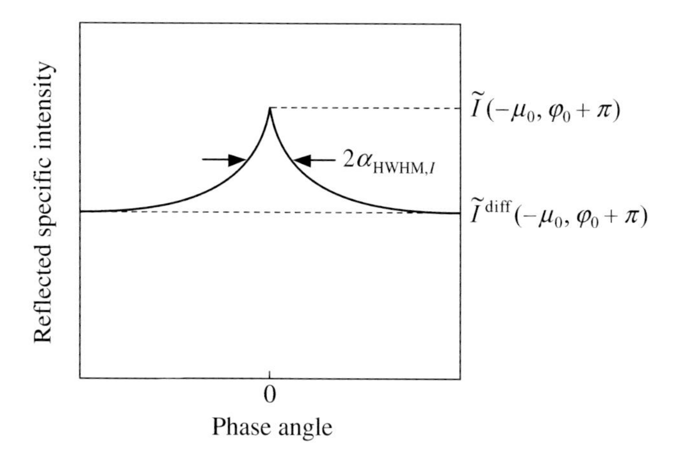

Figure 14.5.1. Coherent enhancement of backscattered intensity. The phase angle is defined as the angle between the unit vectors $\hat{\mathbf{n}}$ and $-\hat{\mathbf{n}}_0$.

It is interesting that the formula for the enhancement factor in the case of unpolarized incident light involves all four diagonal elements of the diffuse multiple-scattering component of the reflection matrix rather than only the (1, 1) element. This makes Eq. (14.5.3) quite different from the corresponding formula obtained in the scalar approximation (Tsang and Ishimaru, 1985):

$$\zeta = \frac{\mathcal{R}^1 + 2\mathcal{R}^{\mathrm{M}}}{\mathcal{R}^1 + \mathcal{R}^{\mathrm{M}}}, \tag{14.5.4}$$

where $\mathcal{R}^1$ and $\mathcal{R}^{\mathrm{M}}$ are the single-scattering and diffuse multiple-scattering components of the scalar reflection coefficient, respectively. In fact, Mishchenko and Dlugach (1992a) demonstrated on the basis of numerically exact radiative transfer calculations that the approximate "scalar" formula (14.5.4) can cause errors far exceeding the errors of the scalar approximation in radiance calculations discussed in Section 13.1.

14.5.2 Linearly polarized incident light

Let us now assume that the incident light is linearly polarized in the vertical direction, so that $\mathbf{I}_0 = [I_0 \; I_0 \; 0 \; 0]^{\mathrm{T}}$. The vertically and horizontally polarized components of the backscattered light are given by

$$\tilde{I}_{\mathrm{v}} = \tfrac{1}{2}(\tilde{I} + \tilde{Q}) \tag{14.5.5}$$

and

$$\tilde{I}_{\mathrm{h}} = \tfrac{1}{2}(\tilde{I} - \tilde{Q}), \tag{14.5.6}$$

respectively (see Eq. (2.6.9)). It is easy to see that the corresponding co-polarized and cross-polarized enhancement factors can be defined as follows:

$$\zeta_{\mathrm{vv}} = \frac{\tilde{I}_{\mathrm{v}}}{\tilde{I}_{\mathrm{v}}^{\mathrm{diff}}} = \frac{\mathcal{R}_{11}^{1} + \mathcal{R}_{22}^{1} + 2\mathcal{R}_{11}^{\mathrm{M}} + 4\mathcal{R}_{12}^{\mathrm{M}} + 2\mathcal{R}_{22}^{\mathrm{M}}}{\mathcal{R}_{11}^{1} + \mathcal{R}_{22}^{1} + \mathcal{R}_{11}^{\mathrm{M}} + 2\mathcal{R}_{12}^{\mathrm{M}} + \mathcal{R}_{22}^{\mathrm{M}}}$$

$$= 2 - \frac{\mathcal{R}_{11}^{1} + \mathcal{R}_{22}^{1}}{\mathcal{R}_{11}^{1} + \mathcal{R}_{22}^{1} + \mathcal{R}_{11}^{\mathrm{M}} + 2\mathcal{R}_{12}^{\mathrm{M}} + \mathcal{R}_{22}^{\mathrm{M}}}, \quad (14.5.7)$$

$$\zeta_{\mathrm{hv}} = \frac{\tilde{I}_{\mathrm{h}}}{\tilde{I}_{\mathrm{h}}^{\mathrm{diff}}} = \frac{\mathcal{R}_{11}^{1} - \mathcal{R}_{22}^{1} + \mathcal{R}_{11}^{\mathrm{M}} - \mathcal{R}_{22}^{\mathrm{M}} - \mathcal{R}_{33}^{\mathrm{M}} + \mathcal{R}_{44}^{\mathrm{M}}}{\mathcal{R}_{11}^{1} - \mathcal{R}_{22}^{1} + \mathcal{R}_{11}^{\mathrm{M}} - \mathcal{R}_{22}^{\mathrm{M}}}. \quad (14.5.8)$$

It is also useful to define the linear polarization ratio and its diffuse counterpart as the ratios of the corresponding cross-polarized and co-polarized backscattered specific intensities:

$$\mu_{\mathrm{L}} = \frac{\tilde{I}_{\mathrm{h}}}{\tilde{I}_{\mathrm{v}}} = \frac{\mathcal{R}_{11}^{1} - \mathcal{R}_{22}^{1} + \mathcal{R}_{11}^{\mathrm{M}} - \mathcal{R}_{22}^{\mathrm{M}} - \mathcal{R}_{33}^{\mathrm{M}} + \mathcal{R}_{44}^{\mathrm{M}}}{\mathcal{R}_{11}^{1} + \mathcal{R}_{22}^{1} + 2\mathcal{R}_{11}^{\mathrm{M}} + 4\mathcal{R}_{12}^{\mathrm{M}} + 2\mathcal{R}_{22}^{\mathrm{M}}}, \quad (14.5.9)$$

$$\mu_{\mathrm{L}}^{\mathrm{diff}} = \frac{\tilde{I}_{\mathrm{h}}^{\mathrm{diff}}}{\tilde{I}_{\mathrm{v}}^{\mathrm{diff}}} = \frac{\mathcal{R}_{11}^{1} - \mathcal{R}_{22}^{1} + \mathcal{R}_{11}^{\mathrm{M}} - \mathcal{R}_{22}^{\mathrm{M}}}{\mathcal{R}_{11}^{1} + \mathcal{R}_{22}^{1} + \mathcal{R}_{11}^{\mathrm{M}} + 2\mathcal{R}_{12}^{\mathrm{M}} + \mathcal{R}_{22}^{\mathrm{M}}}. \quad (14.5.10)$$

The physical meaning of the linear polarization ratios will be discussed in the following section.

14.5.3 Circularly polarized incident light

Let us now assume that the incident light is circularly polarized in the anti-clockwise sense as viewed by an observer looking in the direction of propagation, so that $\mathbf{I}_0 = [I_0 \ 0 \ 0 \ I_0]^{\mathrm{T}}$. The "same-helicity" and "opposite-helicity" components of the backscattered light are given by

$$\tilde{I}_{\mathrm{sh}} = \tfrac{1}{2}(\tilde{I} + \tilde{V}) \quad (14.5.11)$$

and

$$\tilde{I}_{\mathrm{oh}} = \tfrac{1}{2}(\tilde{I} - \tilde{V}), \quad (14.5.12)$$

respectively (see Eq. (2.6.10)). The corresponding "helicity-preserving" and "opposite-helicity" enhancement factors are given by

$$\zeta_{\mathrm{hp}} = \frac{\tilde{I}_{\mathrm{sh}}}{\tilde{I}_{\mathrm{sh}}^{\mathrm{diff}}} = \frac{\mathcal{R}_{11}^{1} + \mathcal{R}_{44}^{1} + 2\mathcal{R}_{11}^{\mathrm{M}} + 2\mathcal{R}_{44}^{\mathrm{M}}}{\mathcal{R}_{11}^{1} + \mathcal{R}_{44}^{1} + \mathcal{R}_{11}^{\mathrm{M}} + \mathcal{R}_{44}^{\mathrm{M}}}, \quad (14.5.13)$$

$$\zeta_{\mathrm{oh}} = \frac{\tilde{I}_{\mathrm{oh}}}{\tilde{I}_{\mathrm{oh}}^{\mathrm{diff}}} = \frac{\mathcal{R}_{11}^{1} - \mathcal{R}_{44}^{1} + \mathcal{R}_{11}^{\mathrm{M}} + \mathcal{R}_{22}^{\mathrm{M}} - \mathcal{R}_{33}^{\mathrm{M}} - \mathcal{R}_{44}^{\mathrm{M}}}{\mathcal{R}_{11}^{1} - \mathcal{R}_{44}^{1} + \mathcal{R}_{11}^{\mathrm{M}} - \mathcal{R}_{44}^{\mathrm{M}}}. \quad (14.5.14)$$

The circular polarization ratio and its diffuse counterpart are defined as the ratios of the corresponding same-helicity and opposite-helicity backscattered specific intensities:

$$\mu_{\mathrm{C}} = \frac{\widetilde{I}_{\mathrm{sh}}}{\widetilde{I}_{\mathrm{oh}}} = \frac{\mathcal{R}^1_{11} + \mathcal{R}^1_{44} + 2\mathcal{R}^{\mathrm{M}}_{11} + 2\mathcal{R}^{\mathrm{M}}_{44}}{\mathcal{R}^1_{11} - \mathcal{R}^1_{44} + \mathcal{R}^{\mathrm{M}}_{11} + \mathcal{R}^{\mathrm{M}}_{22} - \mathcal{R}^{\mathrm{M}}_{33} - \mathcal{R}^{\mathrm{M}}_{44}}, \tag{14.5.15}$$

$$\mu_{\mathrm{C}}^{\mathrm{diff}} = \frac{\widetilde{I}_{\mathrm{sh}}^{\mathrm{diff}}}{\widetilde{I}_{\mathrm{oh}}^{\mathrm{diff}}} = \frac{\mathcal{R}^1_{11} + \mathcal{R}^1_{44} + \mathcal{R}^{\mathrm{M}}_{11} + \mathcal{R}^{\mathrm{M}}_{44}}{\mathcal{R}^1_{11} - \mathcal{R}^1_{44} + \mathcal{R}^{\mathrm{M}}_{11} - \mathcal{R}^{\mathrm{M}}_{44}}. \tag{14.5.16}$$

14.5.4 General properties of the enhancement factors and polarization ratios

According to Subsection 4.5.2 of Hovenier *et al.* (2004), the matrices $\boldsymbol{\mathcal{R}}^1(\hat{\mu}, \hat{\mu}_0)$ and $\boldsymbol{\mathcal{R}}^{\mathrm{M}}(\hat{\mu}, \hat{\mu}_0)$ belong to the class of matrices called "sums of pure Mueller matrices". As such, they satisfy linear inequalities listed in Subsection A.1.3 of Hovenier *et al.* which can be used to derive several fundamental inequalities for the enhancement factors and polarization ratios introduced above. Specifically,

$$0 \le \zeta_I \le 2, \tag{14.5.17}$$

$$1 \le \zeta_{\mathrm{vv}} \le 2, \tag{14.5.18}$$

$$0 \le \zeta_{\mathrm{hv}} \le 2, \tag{14.5.19}$$

$$1 \le \zeta_{\mathrm{hp}} \le 2, \tag{14.5.20}$$

$$0 \le \zeta_{\mathrm{oh}} \le 2, \tag{14.5.21}$$

$$\mu_{\mathrm{L}} \ge 0, \tag{14.5.22}$$

$$\mu_{\mathrm{L}}^{\mathrm{diff}} \ge 0, \tag{14.5.23}$$

$$\mu_{\mathrm{C}} \ge 0, \tag{14.5.24}$$

$$\mu_{\mathrm{C}}^{\mathrm{diff}} \ge 0. \tag{14.5.25}$$

Note that these general inequalities do not require the unpolarized, cross-polarized, and opposite-helicity enhancement factors to be always greater than one. In other words, they do not exclude the possibility that CB can cause suppression rather than enhancement of the corresponding backscattered signal(s).

We will assume in this subsection that, in general, $a_4(\pi) \neq -a_1(\pi)$ in Eq. (14.3.31), which implies that

$$\mathcal{R}^1_{22}(\hat{\mu}_0^{\pi}, \hat{\mu}_0) \neq \mathcal{R}^1_{11}(\hat{\mu}_0^{\pi}, \hat{\mu}_0) \quad \text{and} \quad \mathcal{R}^1_{44}(\hat{\mu}_0^{\pi}, \hat{\mu}_0) \neq -\mathcal{R}^1_{11}(\hat{\mu}_0^{\pi}, \hat{\mu}_0). \tag{14.5.26}$$

Recalling the order-of-scattering expansion (12.5.1) and Eqs. (14.3.26) and (14.3.27), we can conclude that in the limit $\varpi \to 0$ the diffuse multiple-scattering component of the reflection matrix must vanish much faster than the single-scattering component. As a consequence, Eqs. (14.5.3), (14.5.7)–(14.5.10), and (14.5.13)–(14.5.16) yield

$$\lim_{\varpi \to 0} \zeta_I = \lim_{\varpi \to 0} \zeta_{\rm vv} = \lim_{\varpi \to 0} \zeta_{\rm hv} = \lim_{\varpi \to 0} \zeta_{\rm hp} = \lim_{\varpi \to 0} \zeta_{\rm oh} = 1, \tag{14.5.27}$$

$$\lim_{\varpi \to 0} \mu_{\rm L} = \lim_{\varpi \to 0} \mu_{\rm L}^{\rm diff} = \delta_{\rm L}, \tag{14.5.28}$$

$$\lim_{\varpi \to 0} \mu_{\rm C} = \lim_{\varpi \to 0} \mu_{\rm C}^{\rm diff} = \delta_{\rm C}, \tag{14.5.29}$$

where $\delta_{\rm L}$ and $\delta_{\rm C}$ are the linear and circular backscattering depolarization ratios defined by Eqs. (11.9.1) and (11.9.2), respectively.

In the limit $\mu_0 \to 0$, the diffuse reflection matrix $\mathbf{R}(\hat{\mu}_0^\pi, \hat{\mu}_0)$ also reduces to the first-order-scattering component:

$$\mu_0 \mathbf{R}(\hat{\mu}_0^\pi, \hat{\mu}_0)\Big|_{\mu_0 \to 0} = \mu_0 \mathcal{R}^1(\hat{\mu}_0^\pi, \hat{\mu}_0)\Big|_{\mu_0 \to 0}, \tag{14.5.30}$$

where

$$\mu_0 \mathcal{R}^1(\hat{\mu}_0^\pi, \hat{\mu}_0)\Big|_{\mu_0 \to 0} = \frac{\varpi \widetilde{\mathbf{F}}(\pi)}{8}. \tag{14.5.31}$$

In the case $\mathcal{T} = \infty$, these formulas follow directly from the Ambarzumian equation (12.4.9). However, they remain valid for any $\mathcal{T}$ since the amount of multiple scattering in a slab with a finite optical thickness cannot exceed that in a semi-infinite slab with the same single-scattering albedo and phase matrix (see Hovenier and Stam, 2006 for a rigorous proof). This result means that in the limit $\mu = \mu_0 \to 0$, the diffuse multiple-scattering component of the reflection matrix becomes negligible in comparison with the single-scattering component. As a consequence,

$$\lim_{\mu_0 \to 0} \zeta_I = \lim_{\mu_0 \to 0} \zeta_{\rm vv} = \lim_{\mu_0 \to 0} \zeta_{\rm hv} = \lim_{\mu_0 \to 0} \zeta_{\rm hp} = \lim_{\mu_0 \to 0} \zeta_{\rm oh} = 1, \tag{14.5.32}$$

$$\lim_{\mu_0 \to 0} \mu_{\rm L} = \lim_{\mu_0 \to 0} \mu_{\rm L}^{\rm diff} = \delta_{\rm L}, \tag{14.5.33}$$

$$\lim_{\mu_0 \to 0} \mu_{\rm C} = \lim_{\mu_0 \to 0} \mu_{\rm C}^{\rm diff} = \delta_{\rm C}. \tag{14.5.34}$$

Finally, Eqs. (13.2.2) and (14.3.32) yield

$$\mathbf{R}(\hat{\mu}_0^\pi, \hat{\mu}_0)\Big|_{\mathcal{T} \to 0} = \mathcal{R}^1(\hat{\mu}_0^\pi, \hat{\mu}_0)\Big|_{\mathcal{T} \to 0}, \tag{14.5.35}$$

where

$$\mathcal{R}^1(\hat{\mu}_0^\pi, \hat{\mu}_0)\Big|_{\mathcal{T} \to 0} = \frac{\varpi \mathcal{T}}{4\mu_0^2} \widetilde{\mathbf{F}}(\pi). \tag{14.5.36}$$

Equation (14.3.26) then implies that $\mathcal{R}^{\rm M}(\hat{\mu}_0^\pi, \hat{\mu}_0)$ vanishes and, consequently,

$$\lim_{\mathcal{T} \to 0} \zeta_I = \lim_{\mathcal{T} \to 0} \zeta_{\rm vv} = \lim_{\mathcal{T} \to 0} \zeta_{\rm hv} = \lim_{\mathcal{T} \to 0} \zeta_{\rm hp} = \lim_{\mathcal{T} \to 0} \zeta_{\rm oh} = 1, \tag{14.5.37}$$

$$\lim_{\mathcal{T} \to 0} \mu_{\rm L} = \lim_{\mathcal{T} \to 0} \mu_{\rm L}^{\rm diff} = \delta_{\rm L}, \tag{14.5.38}$$

$$\lim_{\mathcal{T} \to 0} \mu_{\rm C} = \lim_{\mathcal{T} \to 0} \mu_{\rm C}^{\rm diff} = \delta_{\rm C}. \tag{14.5.39}$$

The physical interpretation of Eqs. (4.5.27)–(14.5.29), (4.5.32)–(14.5.34), and (4.5.37)–(14.5.39) is very transparent: either limit $\varpi \to 0$, $\mu = \mu_0 \to 0$, or $\mathcal{T} \to 0$ eliminates multiple scattering and, consequently, any manifestation of CB.

14.5.5 Spherically symmetric particles

Equations (11.10.17), (14.3.31), and (14.3.32) imply that for spherically symmetric particles,

$$\mathcal{R}^1_{22}(\hat{\mu}_0^\pi, \hat{\mu}_0) \equiv \mathcal{R}^1_{11}(\hat{\mu}_0^\pi, \hat{\mu}_0) \quad \text{and} \quad \mathcal{R}^1_{44}(\hat{\mu}_0^\pi, \hat{\mu}_0) \equiv -\mathcal{R}^1_{11}(\hat{\mu}_0^\pi, \hat{\mu}_0). \tag{14.5.40}$$

As a consequence, Eqs. (14.5.7)–(14.5.10) and (14.5.13)–(14.5.16) can be simplified as follows:

$$\zeta_{\text{vv}} = 2\left[1 - \frac{\mathcal{R}^1_{11}}{2\mathcal{R}^1_{11} + \mathcal{R}^{\text{M}}_{11} + 2\mathcal{R}^{\text{M}}_{12} + \mathcal{R}^{\text{M}}_{22}}\right], \tag{14.5.41}$$

$$\zeta_{\text{hv}} = \frac{\mathcal{R}^{\text{M}}_{11} - \mathcal{R}^{\text{M}}_{22} - \mathcal{R}^{\text{M}}_{33} + \mathcal{R}^{\text{M}}_{44}}{\mathcal{R}^{\text{M}}_{11} - \mathcal{R}^{\text{M}}_{22}}, \tag{14.5.42}$$

$$\mu_{\text{L}} = \frac{\mathcal{R}^{\text{M}}_{11} - \mathcal{R}^{\text{M}}_{22} - \mathcal{R}^{\text{M}}_{33} + \mathcal{R}^{\text{M}}_{44}}{2(\mathcal{R}^1_{11} + \mathcal{R}^{\text{M}}_{11} + 2\mathcal{R}^{\text{M}}_{12} + \mathcal{R}^{\text{M}}_{22})}, \tag{14.5.43}$$

$$\mu_{\text{L}}^{\text{diff}} = \frac{\mathcal{R}^{\text{M}}_{11} - \mathcal{R}^{\text{M}}_{22}}{2\mathcal{R}^1_{11} + \mathcal{R}^{\text{M}}_{11} + 2\mathcal{R}^{\text{M}}_{12} + \mathcal{R}^{\text{M}}_{22}}, \tag{14.5.44}$$

$$\zeta_{\text{hp}} \equiv 2, \tag{14.5.45}$$

$$\zeta_{\text{oh}} = \frac{2\mathcal{R}^1_{11} + \mathcal{R}^{\text{M}}_{11} + \mathcal{R}^{\text{M}}_{22} - \mathcal{R}^{\text{M}}_{33} - \mathcal{R}^{\text{M}}_{44}}{2\mathcal{R}^1_{11} + \mathcal{R}^{\text{M}}_{11} - \mathcal{R}^{\text{M}}_{44}}, \tag{14.5.46}$$

$$\mu_{\text{C}} = \frac{2(\mathcal{R}^{\text{M}}_{11} + \mathcal{R}^{\text{M}}_{44})}{2\mathcal{R}^1_{11} + \mathcal{R}^{\text{M}}_{11} + \mathcal{R}^{\text{M}}_{22} - \mathcal{R}^{\text{M}}_{33} - \mathcal{R}^{\text{M}}_{44}}, \tag{14.5.47}$$

$$\mu_{\text{C}}^{\text{diff}} = \frac{\mathcal{R}^{\text{M}}_{11} + \mathcal{R}^{\text{M}}_{44}}{2\mathcal{R}^1_{11} + \mathcal{R}^{\text{M}}_{11} - \mathcal{R}^{\text{M}}_{44}}. \tag{14.5.48}$$

Now the enhancement factors ζ_{hv} and ζ_{hp} do not satisfy the limits (14.5.27), (14.5.32), and (14.5.37). Since the elements of the single-scattering reflection matrix do not contribute to the numerators of the linear and circular polarization ratios, we also have

$$\lim_{\varpi \to 0} \mu_{\text{L}} = \lim_{\varpi \to 0} \mu_{\text{L}}^{\text{diff}} = \lim_{\varpi \to 0} \mu_{\text{C}} = \lim_{\varpi \to 0} \mu_{\text{C}}^{\text{diff}} = 0, \tag{14.5.49}$$

$$\lim_{\mu_0 \to 0} \mu_{\text{L}} = \lim_{\mu_0 \to 0} \mu_{\text{L}}^{\text{diff}} = \lim_{\mu_0 \to 0} \mu_{\text{C}} = \lim_{\mu_0 \to 0} \mu_{\text{C}}^{\text{diff}} = 0, \tag{14.5.50}$$

$$\lim_{\mathcal{T} \to 0} \mu_{\text{L}} = \lim_{\mathcal{T} \to 0} \mu_{\text{L}}^{\text{diff}} = \lim_{\mathcal{T} \to 0} \mu_{\text{C}} = \lim_{\mathcal{T} \to 0} \mu_{\text{C}}^{\text{diff}} = 0. \tag{14.5.51}$$

Table 14.5.1. Enhancement factors and polarization ratios for homogeneous slabs consisting of nonabsorbing Rayleigh scatterers.

	$\mathcal{T} = 1$	$\mathcal{T} = \infty$
ζ_I	1.465559	1.5368115
ζ_{vv}	1.453065	1.7520882
ζ_{hv}	1.553473	1.1201587
ζ_{hp}	2	2
ζ_{oh}	1.337506	1.2509893
$\mu_{\mathrm{L}}^{\mathrm{diff}}$	0.142105	0.5166812
μ_{L}	0.151925	0.3303286
$\mu_{\mathrm{C}}^{\mathrm{diff}}$	0.239602	0.6170752
μ_{C}	0.358281	0.9865395

14.5.6 Benchmark results for Rayleigh scattering

Since Rayleigh scattering is the simplest physically realizable type of scattering, it has become a classical model in vector radiative transfer and has been used in many publications intended to provide benchmark numerical results (see, for example, Chandrasekhar, 1950; van de Hulst, 1980; Mishchenko, 1990b; and references therein). Given the exact nature of Eqs. (14.3.21)–(14.3.25) and the widespread acceptance of the Rayleigh scattering law as a canonical template, it is appropriate to end this section with Table 14.5.1 listing benchmark values of the enhancement factors and polarization ratios for homogeneous slabs composed of nonabsorbing ($\varpi = 1$) Rayleigh scatterers. The high numerical accuracy of the techniques described in Mishchenko (1990b, 1996) allows us to expect that these numbers are accurate to one or two units in the last decimals given. The numbers for $\mathcal{T} = \infty$ are consistent (to $\pm 1\times 10^{-6}$) with those calculated by Amic *et al.* (1997) on the basis of a completely independent solution technique.

14.6 Numerical results for polydisperse spheres and polydisperse, randomly oriented spheroids

Plates 14.6.1–14.6.8 are composed of color contour diagrams of the photometric and polarimetric characteristics of CB computed for homogeneous semi-infinite slabs consisting of polydisperse spherical particles. The helicity-preserving enhancement factor is not shown since it is identically equal to two, Eq. (14.5.45). The enhancement factors and polarization ratios are plotted as functions of the particle effective size parameter $x_{\rm eff} = 2\pi r_{\rm eff}/\lambda_1 \in [0, 30]$ and minus the cosine of the incidence polar angle $\mu_0 \in [0, 1]$ for a set of values of the real ($m_{\rm R}$ = 1.2, 1.4, 1.6, 1.8, and 2) and imaginary ($m_{\rm I}$ = 0, 0.002, 0.01, and 0.3) parts of the relative refractive index. The Lorenz–Mie computations were performed for the gamma size distribution (5.3.15) with a fixed effective variance $v_{\rm eff} = 0.1$. Plates 14.6.1–14.6.8 are accompanied by Figs. 14.6.1 and 14.6.2 which depict the corresponding single-scattering albedo ϖ and backscattering phase function $a_1(\pi)$ versus effective size parameter.

The results shown in Plates 14.6.1, 14.6.2, and 14.6.4 illustrate well the limit (14.5.32) and, for $x_{\rm eff} \lesssim 1$, the limit (14.5.27). For larger particles, the limit (14.5.27) cannot be verified because with increasing $m_{\rm I}$ the single-scattering albedo does not tend to zero (see Fig. 14.6.1). Similarly, the data displayed in Plates 14.6.5–14.6.8 illustrate well the limits (4.5.49) and (4.5.50) for all four polarization ratios. Although the general inequalities (14.5.17), (14.5.19), and (14.5.21) do not rule out values of ζ_I, $\zeta_{\rm hv}$, and $\zeta_{\rm oh}$ smaller than one, our numerical results suggest that the value one may be a more appropriate lower boundary for the unpolarized, cross-polarized, and opposite-helicity enhancement factors. Unfortunately, we have not been able to prove this analytically.

As follows from Eq. (14.5.41), the co-polarized enhancement factor $\zeta_{\rm vv}$ cannot reach the value two since, for real scattering particles, $\mathcal{R}^1_{11}(\hat{\mu}_0^\pi, \hat{\mu}_0)$ is never exactly zero but rather is a positive number. The degree of deviation of the co-polarized enhancement factor from the value two strongly depends on the value of the backscattering phase function $a_1(\pi)$ (cf. Eqs. (14.5.41), (14.3.33), and (14.3.31)). For nonabsorbing particles with $m_{\rm R} = 1.2$ and effective size parameters in the range $x_{\rm eff} \in [2, 20]$ and for nearly normal incidence, $\mathcal{R}^1_{11}(\hat{\mu}_0^\pi, \hat{\mu}_0)$ is small because $a_1(\pi)$ is small (Fig. 14.6.2) and μ_0 is close to one, whereas the multiple-scattering contribution is large. As a consequence, the deviation of $\zeta_{\rm vv}$ from the value two is small (Plate 14.6.2, top left-most diagram). In contrast, the (much) larger values of the backscattering phase function for nonabsorbing particles with larger relative refractive indices (Fig. 14.6.2) cause co-polarized enhancement factors significantly smaller than two (Plate 14.6.2, left-most column). According to Eq. (14.3.33), the single-scattering contribution to the backscattering reflection matrix decreases with increasing incidence polar angle, thereby causing $\zeta_{\rm vv}$ to be a monotonically increasing function of μ_0 in most cases.

Not surprisingly, the effective size parameters at which ζ_I and $\zeta_{\rm vv}$ reach maximal values are close to those at which $a_1(\pi)$ and, thus, $\mathcal{R}^1_{11}(\hat{\mu}_0^\pi, \hat{\mu}_0)$ have a minimum (cf. Eqs. (14.5.3) and (14.5.41)). Increasing $m_{\rm I}$ causes a decrease in the single-scattering albedo and, thus, a reduced diffuse multiple-scattering contribution to the backscattering reflection matrix $\boldsymbol{\mathcal{R}}^{\rm M}(\hat{\mu}_0^\pi, \hat{\mu}_0)$. This explains the significant reduction of the co-polarized enhancement factor as $m_{\rm I}$ increases from 0 to 0.01 (Plate 14.6.2). However, increasing $m_{\rm I}$ also makes the minima in the backscattering phase function for $m_{\rm R}$ = 1.2, 1.4, and 1.6 deeper (Fig. 14.6.2), which appears to explain why ζ_I

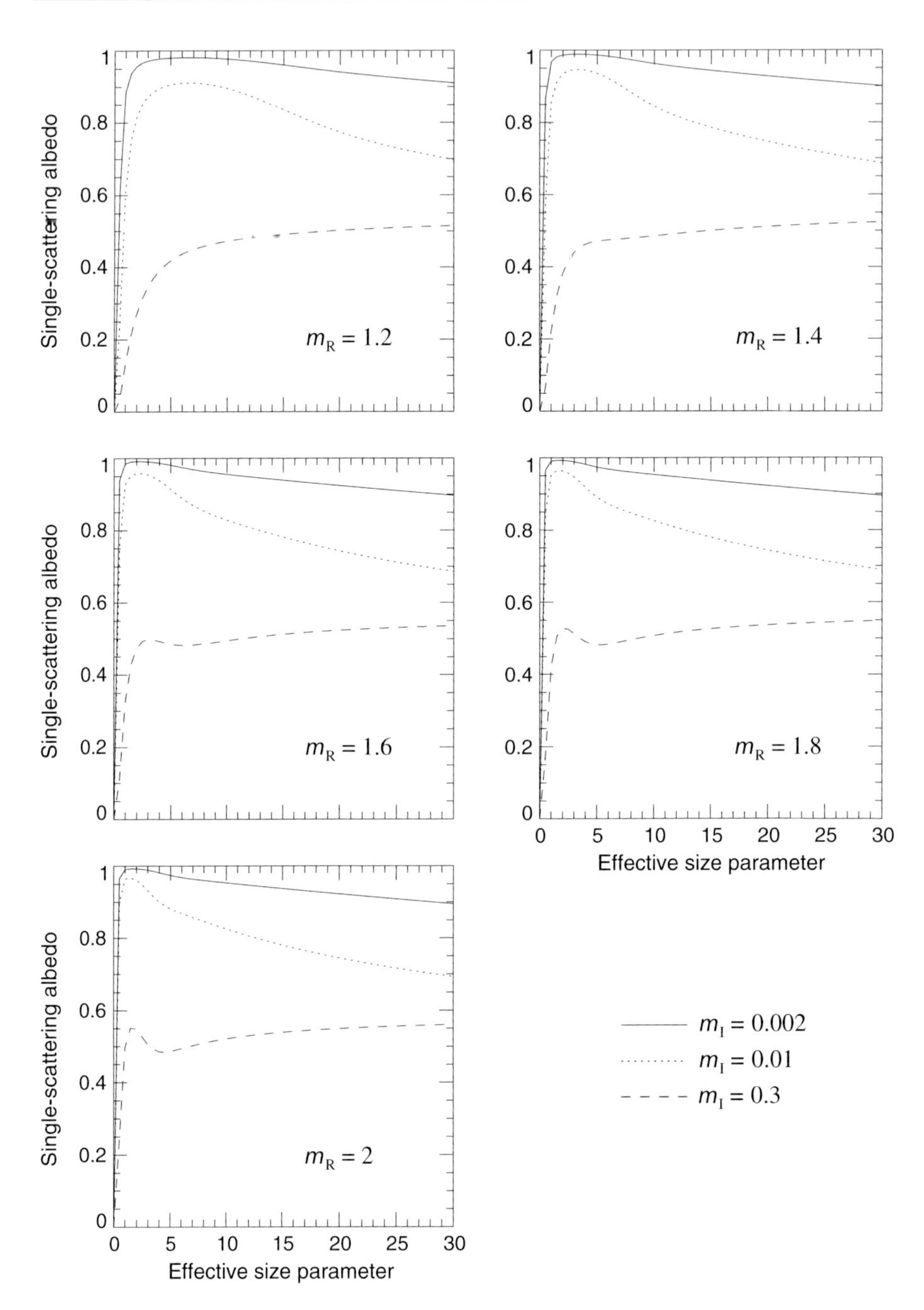

Figure 14.6.1. Single-scattering albedo ϖ versus effective size parameter x_{eff} for a gamma size distribution of spherical particles with $m_R = 1.2, 1.4, 1.6, 1.8$, and 2 and $m_I = 0.002$, 0.01, and 0.3. The effective variance of the size distribution is fixed at 0.1. Note that $\varpi \equiv 1$ for nonabsorbing particles with $m_I = 0$.

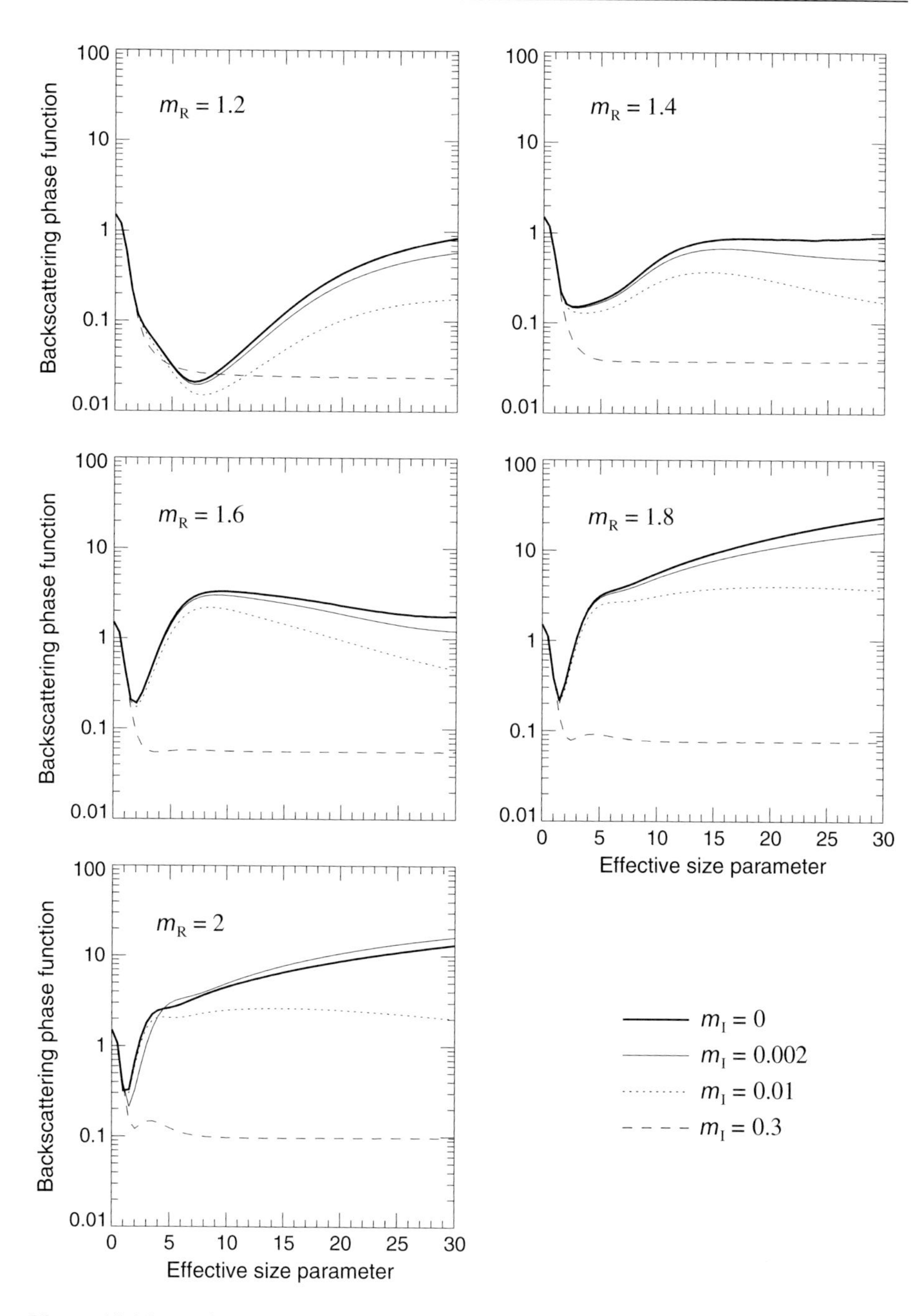

Figure 14.6.2. Backscattering phase function $a_1(\pi)$ versus effective size parameter x_{eff} for a gamma size distribution of spherical particles with $m_R = 1.2, 1.4, 1.6, 1.8$, and 2 and $m_I = 0$, 0.002, 0.01, and 0.3. The effective variance of the size distribution is fixed at 0.1.

often increases as $m_{\rm I}$ increases from 0 to 0.01. Plates 14.6.1 and 14.6.2 also show that an interesting cumulative result of the effects of increasing absorption on the single-scattering albedo and backscattering phase function can be an increase in ζ_I and $\zeta_{\rm vv}$ as $m_{\rm I}$ increases from 0.01 to 0.3, especially at larger effective size parameters.

Plate 14.6.3 shows that the cross-polarized enhancement factor $\zeta_{\rm hv}$ always increases with decreasing multiple-scattering contribution, i.e., with increasing absorption and/or decreasing μ_0, and can reach values very close to two. In contrast, the opposite-helicity enhancement factor $\zeta_{\rm oh}$ can either decrease or increase with increasing $m_{\rm I}$, Plate 14.6.4. This is explained by the presence of elements of both the matrix $\mathcal{R}^1(\hat{\mu}_0^\pi, \hat{\mu}_0)$ and the matrix $\mathcal{R}^{\rm M}(\hat{\mu}_0^\pi, \hat{\mu}_0)$ on the right-hand side of Eq. (14.5.46) and the presence of elements of only the matrix $\mathcal{R}^{\rm M}(\hat{\mu}_0^\pi, \hat{\mu}_0)$ on the right-hand side of Eq. (14.5.42).

It should be noted that accurate numerical computations of the cross-polarized enhancement factor become difficult for strongly absorbing particles and/or for grazing illumination directions since both the numerator and denominator of the right-hand side of Eq. (14.5.42) do not have a first-order-scattering component and become very small with vanishing multiple-scattering contribution. Therefore, such computations require the use of at least double-precision floating point variables with the general inequality (14.5.19) serving as an important basic check on numerical accuracy.

Equations (14.5.43) and (14.5.44) suggest that in the absence of multiple scattering the backscattered light must be fully linearly polarized in the vertical direction. In other words, single scattering by spherical particles in the exact backscattering direction does not depolarize linearly polarized incident light. However, upon multiple scattering both the linear polarization ratio $\mu_{\rm L}$ and the linear polarization ratio of the diffuse background $\mu_{\rm L}^{\rm diff}$ can become nonzero. This means that the backscattered light acquires a nonzero intensity component polarized in the horizontal direction. Thus, the backscattered light becomes partially polarized. Referring to Section 13.4, this effect can be called linear depolarization.

Similarly, let us consider the illumination of a slab by light circularly polarized in the anti-clockwise sense when looking in the direction of incidence $\hat{\mathbf{n}}_0$. Then Eqs. (14.5.47) and (14.5.48) indicate that single scattering by spherical particles in the exact backscattering direction produces light which is fully circularly polarized in the clockwise sense with respect to the unit vector of the backscattering direction $\hat{\mathbf{n}} = -\hat{\mathbf{n}}_0$. This means that the backscattered signal is fully circularly polarized in the anti-clockwise sense as viewed by an observer looking in the original incidence direction $\hat{\mathbf{n}}_0$. In the radar literature, this situation is usually characterized by saying that there is no backscattering depolarization. In terms of the polarization definitions used in this book, there is a complete switch in the sense of circular polarization for the backscattered light. Upon multiple scattering, both the circular polarization ratio $\mu_{\rm C}$ and the circular polarization ratio of the diffuse background $\mu_{\rm C}^{\rm diff}$ can become nonzero, resulting in partially polarized backscattered light. This phenomenon can be called circular depolarization.

Plate 14.6.6 shows that $\mu_L^{\rm diff}$ always decreases with increasing absorption. According to Eq. (14.5.44), the first-order-scattering contribution to $\mu_L^{\rm diff}$ in the case of spherically symmetric particles is identically equal to zero. Therefore, the decrease of $\mu_L^{\rm diff}$ with increasing m_I might be explained by a reduced diffuse multiple-scattering matrix $\mathcal{R}^{\rm M}(\hat{\mu}_0^\pi, \hat{\mu}_0)$. However, this simple qualitative explanation does not quite work in the case of $\mu_C^{\rm diff}$ (Plate 14.6.8), even though the first-order-scattering contribution to $\mu_C^{\rm diff}$ is also equal to zero (Eq. (14.5.48)). Indeed, the diagrams for m_R = 1.2 and 1.4 in Plate 14.6.8 show that $\mu_C^{\rm diff}$ can first increase with increasing m_I and then rapidly decreases. This different behavior of $\mu_L^{\rm diff}$ and $\mu_C^{\rm diff}$ with increasing absorption can only be explained by the fact that different elements of the matrix $\mathcal{R}^{\rm M}(\hat{\mu}_0^\pi, \hat{\mu}_0)$ are involved and that the process of multiple scattering of polarized light is extremely complex.

The numerical data displayed in Plates 14.6.6 and 14.6.8 cover a representative, albeit restricted, range of effective size parameters and real and imaginary parts of the relative refractive index. As such they may suggest that $\mu_L^{\rm diff}$ is always less than or equal to one,

$$\mu_L^{\rm diff} \le 1, \tag{14.6.1}$$

whereas $\mu_C^{\rm diff}$ is always greater than or equal to $\mu_L^{\rm diff}$:

$$\mu_L^{\rm diff} \le \mu_C^{\rm diff}. \tag{14.6.2}$$

Unfortunately, we were unable to give a general analytical proof of these inequalities.

The full linear and circular polarization ratios μ_L and μ_C are shown in Plates 14.6.5 and 14.6.7, respectively, and are more similar to each other than the corresponding diffuse ratios. In particular, the diagrams for m_R = 1.2 and 1.4 show that both μ_L and μ_C can first increase with increasing absorption and then rapidly decrease.

Using formulas of Section 14.5, we can express the full polarization ratios in terms of the diffuse polarization ratios and the enhancement factors as follows:

$$\mu_L = \frac{\zeta_{\rm hv}}{\zeta_{\rm vv}} \mu_L^{\rm diff}, \tag{14.6.3}$$

$$\mu_C = \frac{\zeta_{\rm hp}}{\zeta_{\rm oh}} \mu_C^{\rm diff}. \tag{14.6.4}$$

The enhancement factor $\zeta_{\rm hp}$ for spherical particles is identically equal to two, whereas the enhancement factors $\zeta_{\rm vv}$, $\zeta_{\rm hv}$, and $\zeta_{\rm oh}$ exhibit a rather complex dependence on the effective size parameter and real and imaginary parts of the relative refractive index. In view of Eqs. (14.5.21) and (14.5.45), we may conclude that the opposite-helicity enhancement factor is always smaller than or equal to the helicity-preserving enhancement factor,

$$\zeta_{\rm oh} \le \zeta_{\rm hp}. \tag{14.6.5}$$

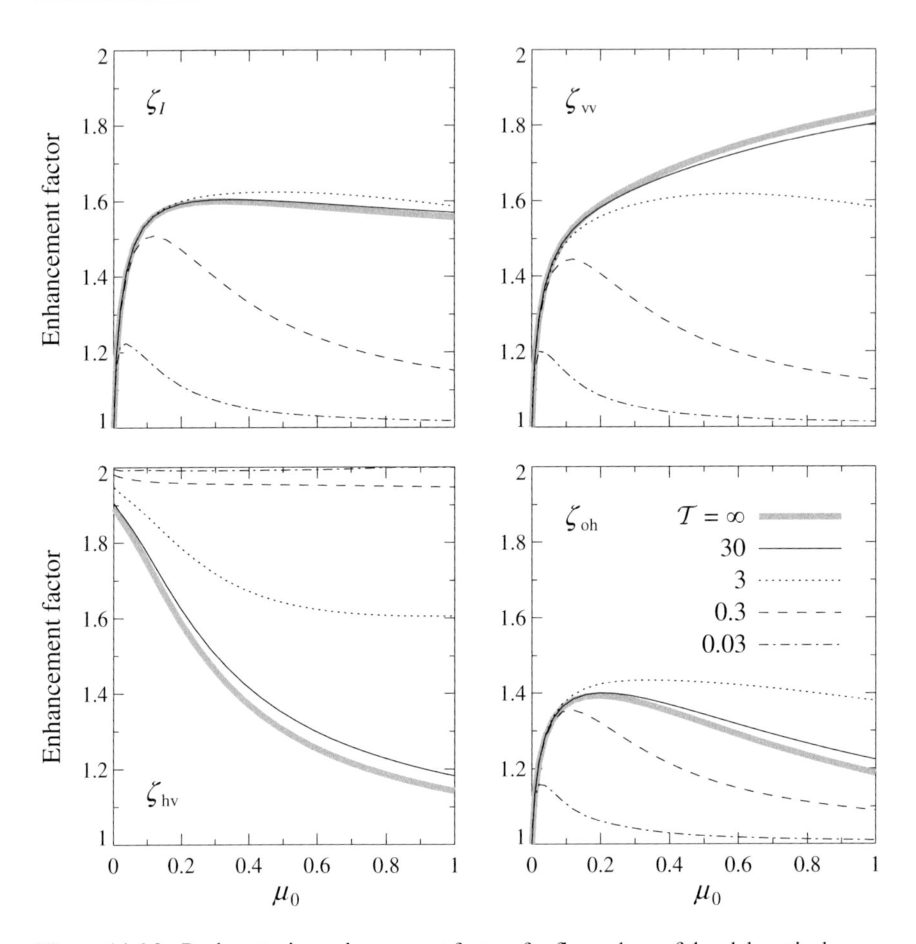

Figure 14.6.3. Backscattering enhancement factors for five values of the slab optical thickness.

Equation (14.6.4) then implies that the effect of CB is always to increase (or at least not to change) the circular polarization ratio:

$$\mu_C^{\text{diff}} \le \mu_C. \tag{14.6.6}$$

In contrast, CB can either increase or decrease the linear polarization ratio depending on the ratio of the cross-polarized to co-polarized enhancement factors. Plates 14.6.2 and 14.6.3 show that this ratio can be quite variable.

The computations displayed in Plates 14.6.5 and 14.6.7 may suggest that the inequality (14.6.1) applies also to the full linear polarization ratio,

$$\mu_L \le 1, \tag{14.6.7}$$

and that μ_L is always smaller than or equal to μ_C:

$$\mu_L \le \mu_C. \tag{14.6.8}$$

Again, it remains unclear whether these inequalities are general and can be proven

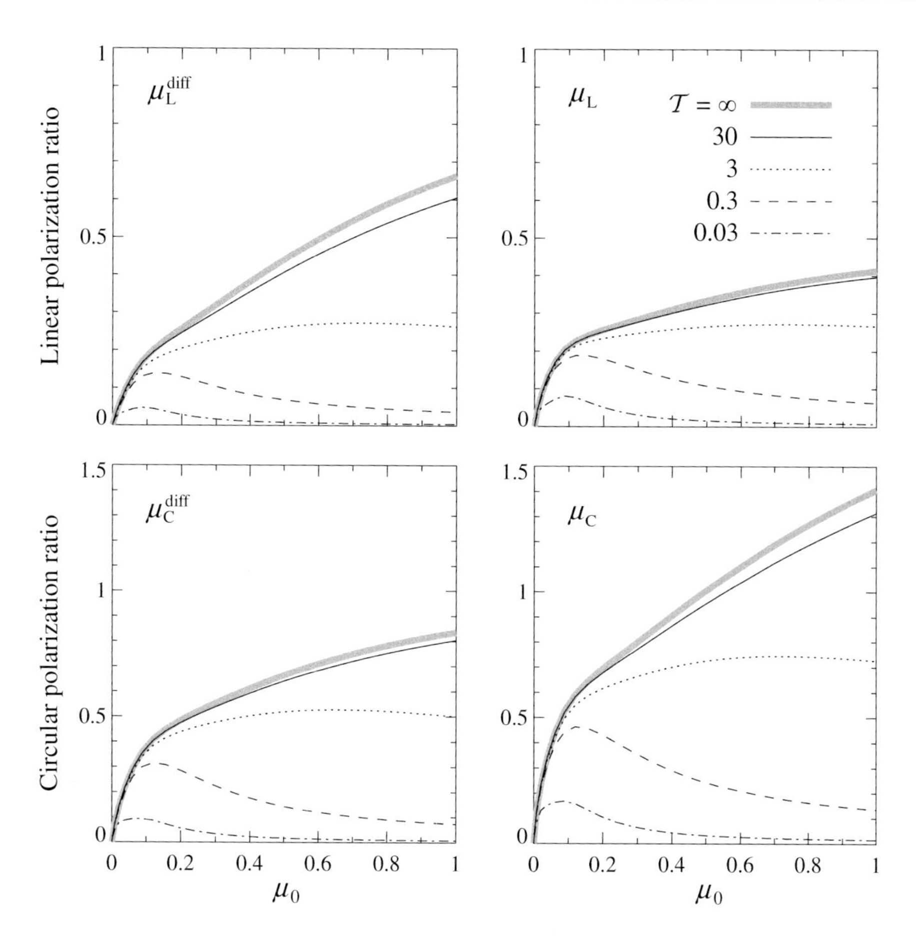

Figure 14.6.4. Polarization ratios for five values of the slab optical thickness.

analytically.

Figures 14.6.3 and 14.6.4 depict the backscattering enhancement factors and polarization ratios for finite and infinite homogeneous slabs consisting of polydisperse spherical particles. The results were computed for a relative refractive index $m = 1.5$ and the power law size distribution (5.3.14) with an effective size parameter $x_{\text{eff}} = 15$ and an effective variance $v_{\text{eff}} = 0.1$. One can see that all four polarization ratios decrease monotonically with decreasing $\mathcal{T}$ and obey the limit (14.5.51). Furthermore, the polarization ratios satisfy the inequalities (14.6.1), (14.6.2), and (14.6.6)–(14.6.8). The enhancement factors ζ_I, ζ_{vv}, and ζ_{oh} obey the limit (14.5.37), whereas the cross-polarized enhancement factor ζ_{hv} appears to tend to the value two as $\mathcal{T}$ approaches zero. As $\mathcal{T}$ decreases, the co-polarized enhancement factor monotonically decreases and the cross-polarized enhancement factor monotonically increases. The behavior of the unpolarized and opposite-helicity enhancement factors with decreasing optical thickness is different from that of either ζ_{vv} or ζ_{hv}: they first increase, reach a maximum, then start to decrease, and eventually vanish.

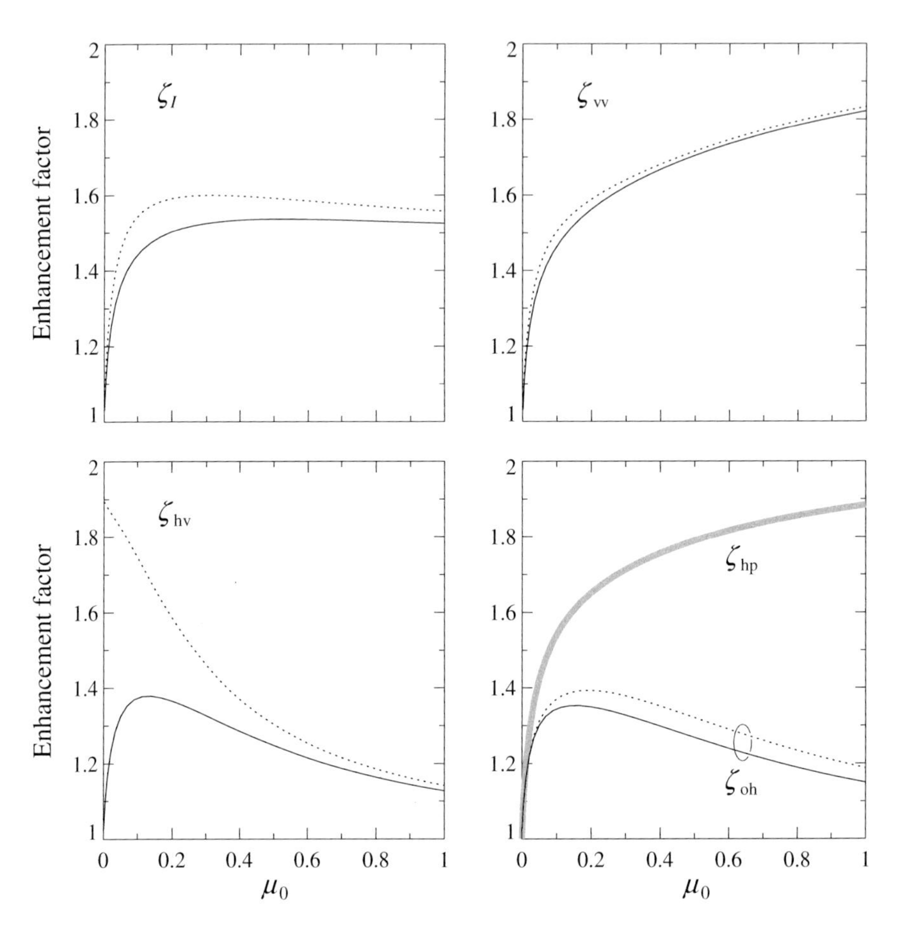

Figure 14.6.5. Backscattering enhancement factors for homogeneous semi-infinite slabs composed of randomly oriented polydisperse spheroids (solid curves) and surface-equivalent spheres (dotted curves). The thick gray curve depicts the helicity-preserving enhancement factor for the spheroids, whereas that for the spheres is identically equal to two.

Figures 14.6.5 and 14.6.6 compare the results for a semi-infinite slab composed of the above spherical particles with those for a semi-infinite slab consisting of surface-equivalent randomly oriented oblate spheroids. The ratio of the larger to the smaller spheroid semi-axes is equal to 1.4. The most notable difference is that the helicity-preserving enhancement factor for the spheroids deviates from the value two and, in fact, tends to the value one in the limit $\mu_0 \to 0$. All five enhancement factors for the spheroids satisfy the limit (14.5.32), whereas the polarization ratios satisfy the limits (14.5.33) and (14.5.34) with $\delta_L \approx 0.289$ and $\delta_C \approx 0.813$. The inequalities (14.6.1), (14.6.2), and (14.6.6)–(14.6.8) appear to be valid for nonspherical particles as well as for spheres.

Thus, the above results show that the helicity-preserving enhancement factor can be strongly affected by particle nonsphericity. One can expect, of course, that the actual deviation of ζ_{hp} from the value two will depend on the degree of particle as-

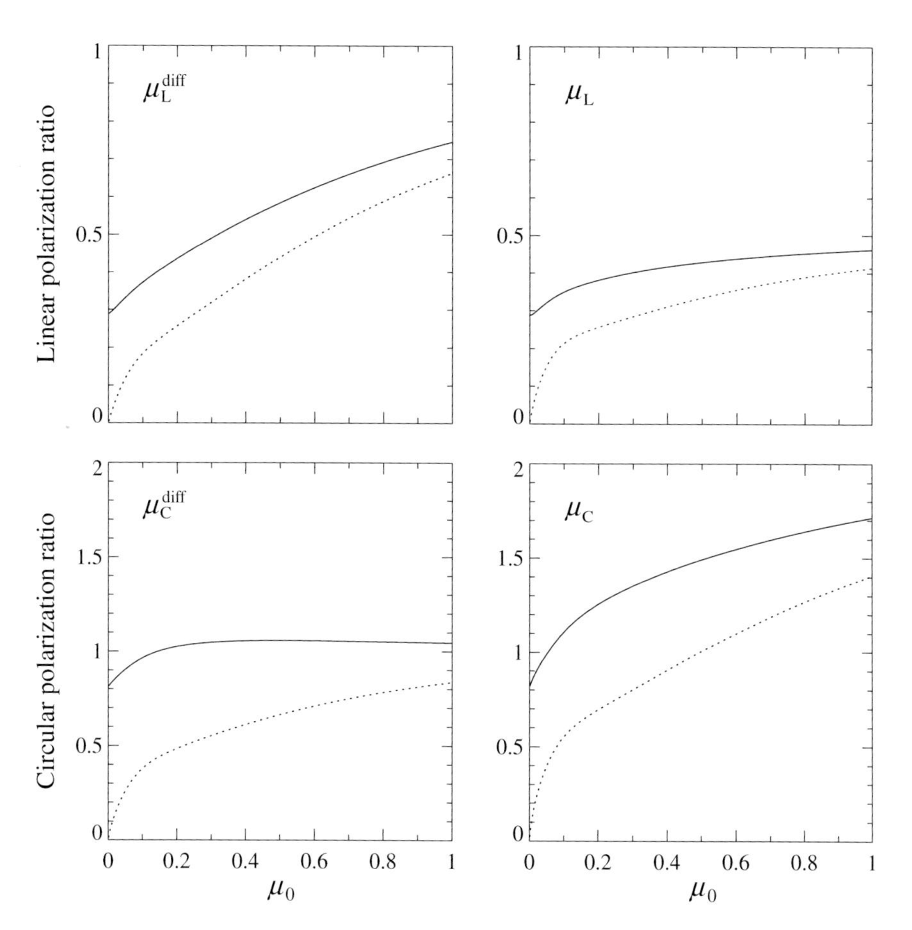

Figure 14.6.6. Polarization ratios for homogeneous semi-infinite slabs composed of randomly oriented polydisperse spheroids (solid curves) and surface-equivalent spheres (dotted curves).

phericity and particle size parameter and relative refractive index as well as on the optical thickness of the scattering slab. However, even the limited amount of numerical data shown in Fig. 14.6.5 suggests that the deviation of the helicity-preserving enhancement factor from the value two observed experimentally by Wiersma *et al.* (1995a) may be accounted for, at least partly, by the fact that the particles that formed their scattering samples were nonspherical. The results of more extensive computations of ζ_{hp} for polydisperse spheroids have recently been reported by Dlugach and Mishchenko (2006).

14.7 Angular profile of coherent backscattering

The approach described in Section 14.5 is quite general in that it allows accurate and efficient computer calculations for a wide range of physical models specified in terms of slab optical thickness and particle size distribution, shape, and relative refractive index. However, its major limitation is that it works only in the exact backscattering direction as well as outside the range of reflection directions affected by CB. Consequently, it does not allow one to determine the angular profile of specific manifestations of CB which may carry diagnostic information supplementary to that contained in the enhancement factors and polarization ratios.

The general vector theory of CB is much more involved than the vector RTT (cf. Tishkovets, 2002; Tishkovets *et al.*, 2002; Tishkovets and Mishchenko, 2004) and is still at an early stage of development and computer implementation. An approximate vector theory of CB for the canonical case of Rayleigh scattering was developed by Stephen and Cwilich (1986). Exact solutions of this problem have been obtained by Ozrin (1992b) and Amic *et al.* (1997), but they are applicable only to a semi-infinite homogeneous slab composed of nonabsorbing Rayleigh scatterers and illuminated by light incident perpendicularly to the slab boundary. The exact scalar theory of CB is much less complicated (Tsang and Ishimaru, 1985; Akkermans *et al.*, 1988; van der Mark *et al.*, 1988; Gorodnichev *et al.*, 1990; Ozrin, 1992a; Amic *et al.*, 1996), whereas the scalar diffusion approximation offers further simplifications and provides easy-to-use closed-form analytical expressions (Akkermans *et al.*, 1988; Barabanenkov and Ozrin, 1991). However, neither one can be applied directly to multiple scattering of electromagnetic waves by discrete random media composed of realistic particles and cannot be used to compute polarization characteristics of CB. As usual, the Monte Carlo technique (Iwai *et al.*, 1995; Lenke and Maret, 2000a; Lenke *et al.*, 2002; Muinonen, 2004) provides a rather general recourse. However, it can be very computer-intensive, especially for particles distributed over sizes, shapes, and/or orientations.

We will not attempt here a comprehensive account of these solution approaches. The interested reader can find detailed information in the publications cited above as well as in the reviews by Barabanenkov *et al.* (1991), Kuz'min and Romanov (1996), Lagendijk and van Tiggelen (1996), van Rossum and Nieuwenhuizen (1999), and Lenke and Maret (2000a). The objective of this section is much more limited. Specifically, we will use the exact vector solution for a semi-infinite Rayleigh slab derived by Amic *et al.* (1997) in order to illustrate the angular distribution of reflected intensity and polarization in the vicinity of the exact backscattering direction and will give a simple physical interpretation of these results.

Let us assume that a homogeneous semi-infinite slab composed of nonabsorbing Rayleigh scatterers is illuminated by an unpolarized quasi-monochromatic beam incident perpendicularly to the slab boundary. Figure 14.7.1 depicts the corresponding unpolarized backscattering enhancement factor $\zeta_I(q)$ as a function of the dimensionless so-called angular parameter q defined as

$$q = k_1 l \alpha, \tag{14.7.1}$$

where α is the phase angle and l is the mean free path of light in the scattering me-

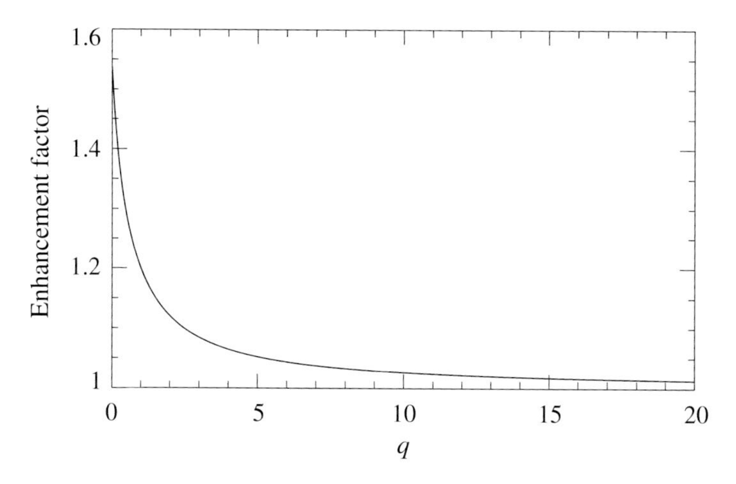

Figure 14.7.1. Theoretical angular profile of the unpolarized backscattering enhancement factor for a half-space of nonabsorbing Rayleigh particles illuminated by an unpolarized beam of light incident normally to the boundary of the scattering medium. (After Mishchenko *et al.*, 2000c.)

dium. As before, the phase angle is defined as the angle between the unit vectors $\hat{\mathbf{n}}$ and $-\hat{\mathbf{n}}_0$; in this case it is equal to the polar angle of the reflection direction. The mean free path is defined by

$$l = \frac{1}{n_0 \langle C_{\text{ext}} \rangle_\xi}, \tag{14.7.2}$$

where, as before, n_0 is the particle number density and $\langle C_{\text{ext}} \rangle_\xi$ is the extinction cross section per particle averaged over particle states. In other words, l is the geometrical distance corresponding to a unit optical path-length.

Figure 14.7.1 demonstrates the renowned coherent intensity peak centered at exactly the opposition. The amplitude of the peak is $\zeta_I(0) \approx 1.537$ (cf. Table 14.5.1) and its half-width at half-maximum is $q_{\text{HWHM},I} = k_1 l \alpha_{\text{HWHM},I} \approx 0.597$ (cf. Fig. 14.5.1). Thus the relationship between the half-width at half-maximum of the backscattering intensity peak and the mean free path for the case of conservative Rayleigh scattering and unpolarized normal illumination is given by

$$\alpha_{\text{HWHM},I} \approx \frac{0.597}{k_1 l}. \tag{14.7.3}$$

The appearance of the dimensionless angular parameter q as the primary angular variable in the theory of CB is not surprising. Indeed, let us consider two conjugate light-scattering paths involving a group of N particles (Fig. 14.7.2). The phase difference between the path shown by the broken lines and that shown by the solid lines is

$$\varDelta = k_1 \mathbf{R}_{N1} \cdot (\hat{\mathbf{n}}_0 + \hat{\mathbf{n}}), \tag{14.7.4}$$

where $\mathbf{R}_{N1} = \mathbf{r}_N - \mathbf{r}_1$ is the vector connecting the origins of particles 1 and N. Since

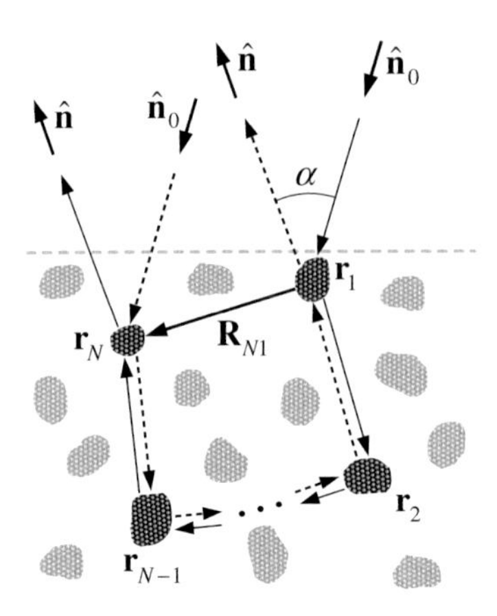

Figure 14.7.2. Derivation of Eq. (14.7.5).

$|\hat{\mathbf{n}}_0 + \hat{\mathbf{n}}| = 2\sin(\alpha/2)$, we can estimate the size of the angular region where the two light-scattering paths are fully or approximately coherent by requiring that Δ be smaller than about one:

$$2k_1 |\mathbf{R}_{N1}| \sin(\alpha/2) \cos\varsigma \lesssim 1, \tag{14.7.5}$$

where ς is the angle between the vectors $\mathbf{R}_{N1}$ and $\hat{\mathbf{n}}_0 + \hat{\mathbf{n}}$. In a discrete random medium, both N, $|\mathbf{R}_{N1}|$, and ς are random variables. Therefore, taking into account that $\alpha \ll 1$, we can conclude that the angular size of the coherence region for the entire medium can be found from $\gamma k_1 L \alpha_{\mathrm{HWHM}} \sim 1$, where L is an average distance between the end particles of a light-scattering path and γ is a quantity of order one which depends on such factors as type of particles, optical thickness of the medium, incidence direction, and polarization state of incident and reflected light. Approximating L by the mean free path l finally yields

$$\alpha_{\mathrm{HWHM}} \sim \frac{1}{\gamma k_1 l}. \tag{14.7.6}$$

For nonabsorbing media composed of anisotropically scattering particles with $\langle \cos\Theta \rangle > 0$, a better approximation for L is the so-called transport mean free path given by

$$l_{\mathrm{tr}} = \frac{l}{1 - \langle \cos\Theta \rangle} \tag{14.7.7}$$

(Amic *et al.*, 1996).

As we have already seen, in the case of normal incidence of unpolarized incident light, $1/\gamma \approx 0.60$. In the case of circularly polarized light,

$$\alpha_{\mathrm{HWHM,hp}} \approx \frac{0.34}{k_1 l}. \tag{14.7.8}$$

For linearly polarized incident light, the width of the co-polarized coherent peak

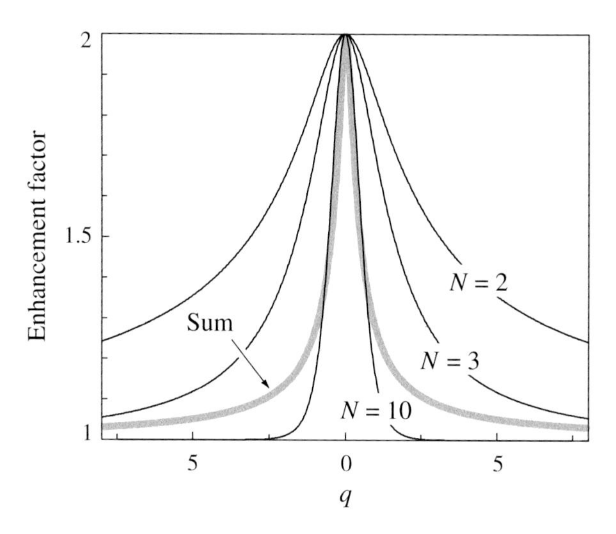

Figure 14.7.3. Scalar enhancement factors in different scattering orders N of light reflected by a homogeneous plane-parallel slab of optical thickness $\mathcal{T} = 12$. The slab is composed of nonabsorbing, isotropically scattering particles and is illuminated by light incident normally to the upper boundary of the slab. The thick gray curve shows the enhancement factor for the sum of the scattering orders 2 through 80. (After Labeyrie *et al.*, 2000.)

depends on the azimuth angles of the incidence and reflection directions. For example, if $\varphi_0 = 0°$ then

$$\alpha_{\text{HWHM,vv}}\Big|_{\varphi=0°} \approx \frac{0.65}{k_1 l}, \tag{14.7.9}$$

$$\alpha_{\text{HWHM,vv}}\Big|_{\varphi=90°} \approx \frac{0.48}{k_1 l}. \tag{14.7.10}$$

This azimuthal asymmetry has been observed experimentally (e.g., van Albada *et al.*, 1988) and will be discussed later in this section. The peak in the cross-polarized channel is significantly wider:

$$\alpha_{\text{HWHM,hv}}\Big|_{\varphi=0°} \approx \frac{1.33}{k_1 l}. \tag{14.7.11}$$

The backscattering intensity peak in Fig. 14.7.1 has a characteristic triangular vertex. The explanation of this shape is that light-scattering paths of increasing order yield narrower peaks (cf. Eq. (14.7.5)). This is illustrated by the different curves in Fig. 14.7.3 computed on the basis of the exact scalar theory of coherent backscattering by isotropically scattering particles (van der Mark *et al.*, 1988). In a semi-infinite medium composed of nonabsorbing scatterers, the cumulative contribution of all Nth-order scattering paths to the total intensity decreases rather slowly, as $N^{-3/2}$. As a result, even the longest paths with infinitesimal angular widths keep raising the tip of the backscattering peak so that it does not becomes rounded even at $\alpha = 0$.

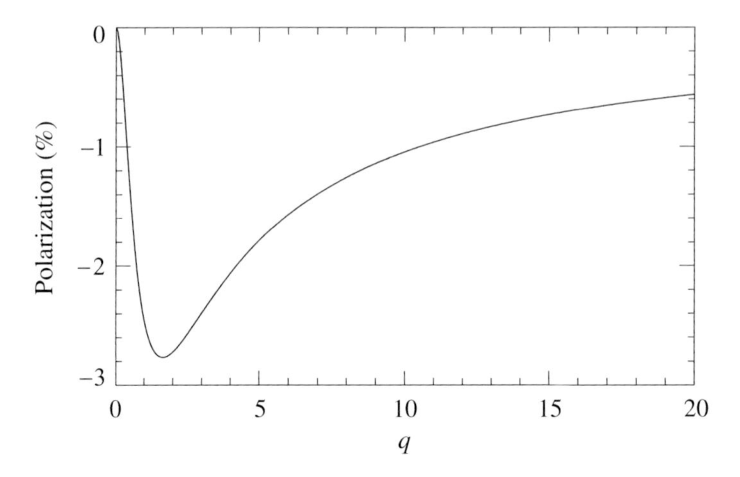

Figure 14.7.4. Theoretical angular profile of the signed degree of linear polarization of the reflected light for a half-space of nonabsorbing Rayleigh particles illuminated by an unpolarized beam of light incident normally to the boundary of the scattering medium. (After Mishchenko *et al.*, 2000c.)

It is evident from Fig. 14.7.3 that any factor leading to suppression of long multiple-scattering paths must cause a rounded backscattering peak. This effect has indeed been demonstrated experimentally for media with absorption ($\varpi \neq 1$) and finite optical thickness ($\mathcal{T} < \infty$) (Etemad *et al.*, 1987).

The signed degree of linear polarization of the reflected light is equal to minus the ratio of the second element of the total reflected specific intensity column vector to the total reflected specific intensity. In the case of unpolarized incident light,

$$P_Q(q) = -\frac{\widetilde{Q}(q)}{\widetilde{I}(q)} = -\frac{\mathcal{R}^1_{21}(q) + \mathcal{R}^\mathrm{M}_{21}(q) + \mathcal{R}^\mathrm{C}_{21}(q)}{\mathcal{R}^1_{11}(q) + \mathcal{R}^\mathrm{M}_{11}(q) + \mathcal{R}^\mathrm{C}_{11}(q)}. \tag{14.7.12}$$

Both $\mathcal{R}^1_{21}(0)$, $\mathcal{R}^\mathrm{M}_{21}(0)$, and $\mathcal{R}^\mathrm{C}_{21}(0)$ vanish, the latter two quantities as a consequence of azimuthal symmetry in the case of normal illumination and unpolarized incident light. Furthermore, both $\mathcal{R}^1_{11}(q)$, $\mathcal{R}^\mathrm{M}_{11}(q)$, $\mathcal{R}^1_{21}(q)$, and $\mathcal{R}^\mathrm{M}_{21}(q)$ change with reflection direction much more slowly than $\mathcal{R}^\mathrm{C}_{11}(q)$ and $\mathcal{R}^\mathrm{C}_{21}(q)$ and, thus, can be neglected within the range of reflection directions affected by CB. Consequently,

$$P_Q(q) \approx -\frac{\mathcal{R}^\mathrm{C}_{21}(q)}{\mathcal{R}^1_{11}(0) + \mathcal{R}^\mathrm{M}_{11}(0) + \mathcal{R}^\mathrm{C}_{11}(q)}. \tag{14.7.13}$$

This quantity is shown in Fig. 14.7.4. It is seen indeed that the reflected polarization is zero at the exact backscattering direction. However, with increasing q, polarization becomes negative, rapidly grows in absolute value, and reaches its minimal value $P_{Q,\min} \approx -2.765\%$ at a reflection direction very close to opposition ($q_P \approx 1.68$). The peak of negative polarization is highly asymmetric so that the half-minimal polarization value -1.383% is first reached at $q_{P,1} \approx 0.498$, which is even smaller than the

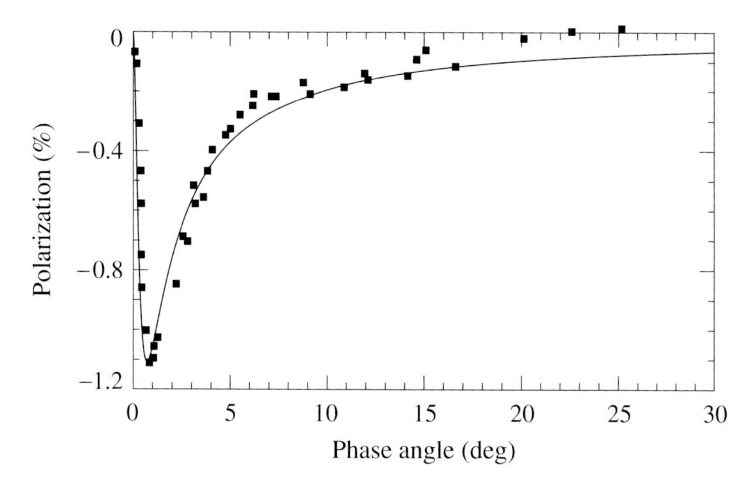

Figure 14.7.5. Polarization measurements for a particulate surface composed of microscopic magnesia particles (squares) and best-fit theoretical results (solid curve). (After Mishchenko *et al.*, 2000c.)

value $q_{\mathrm{HWHM},I} \approx 0.597$ corresponding to the half-width at half-maximum of the backscattering intensity peak, and then at a much larger $q_{P,2} \approx 7.10$. This unusual behavior of polarization at near-backscattering angles was called in Mishchenko (1993) the polarization opposition effect (POE).

Because lasers are the most frequently used sources of illumination and usually generate linearly or circularly polarized light, explicit laboratory demonstrations of the POE have been rare. Apparently the first laboratory observation of the POE was made by Lyot as long ago as in the 1920s (Lyot, 1929), although the physical origin of this effect was, of course, unknown at that time. Figure 14.7.5 shows Lyot's polarization measurements for a particulate surface obtained by burning a tape of magnesium under a glass plate until the deposit on the plate was completely opaque. Lyot described the observed phase curve of polarization as "puzzling" and attributed it to the very small size of magnesia grains. Unfortunately, he did not measure the actual size of the grains and their packing density and thus did not provide the information necessary to compute the mean free path l. Furthermore, the minimal measured polarization value is only −1.11%, compared with the theoretical value $P_{Q,\min} \approx -2.765\%$ computed for nonabsorbing Rayleigh particles. However, by assuming that the latter difference is explained by the finite particle size in Lyot's experiment, multiplying the theoretical polarization by a factor of 0.4, and assuming that the actual $k_1 l$ was close to a realistic value of 132, Mishchenko *et al.* (2000c) were able to almost perfectly reproduce the angular profile of the measured polarization up to phase angles of about 15° (solid curve in Fig. 14.7.5). At larger phase angles, the assumption of negligible single-scattering and diffuse multiple-scattering contributions, $\mathcal{R}^{1}_{21}(q)$ and $\mathcal{R}^{\mathrm{M}}_{21}(q)$, is no longer valid and causes a significant deviation of the theoretical curve from the measurements.

An unpolarized beam can be represented as a superposition of two linearly polarized beams with orthogonal vibration directions of the electric field vector (Section 2.9). Therefore, it is easy to see that the POE has the same physical origin as the azimuthal asymmetry of the backscattering intensity peak in the case of linearly polarized incident beam. Both effects appear to be a specific consequence of the Rayleigh single-scattering law (see Tishkovets *et al.*, 2002) and seem to disappear with increasing particle size parameter (e.g., van Albada *et al.*, 1988).

14.8 Further discussion of theoretical and practical aspects of coherent backscattering

It is instructive at this point to recapitulate the assumptions and approximations made in the unified microphysical theory of radiative transfer and coherent backscattering:

1. It is assumed that the scattering medium is illuminated by either: (i) one plane electromagnetic wave; (ii) several plane electromagnetic waves with different angular frequencies and arbitrary propagation directions; and/or (iii) one or several quasi-monochromatic beams with arbitrary propagation directions.
2. It is assumed that each particle is located in the far-field zones of all the other particles and that the observation point is also located in the far-field zones of all the particles forming the scattering medium.
3. All scattering paths going through a particle two or more times are neglected (the Twersky approximation).
4. It is assumed that the scattering system is ergodic and that averaging over time can be replaced by averaging over particle positions and states.
5. It is assumed that: (i) the position and state of each particle are statistically independent of each other and of those of all the other particles; and (ii) the spatial distribution of the particles throughout the medium is random and statistically uniform.
6. It is assumed that: (i) the scattering medium is a plane-parallel layer of infinite horizontal extent; and (ii) the observation point is located infinitely far from the layer.
7. It is assumed that the number of particles N forming the scattering medium is very large.
8. Only the ladder and maximally crossed diagrams are kept in the diagrammatic expansion of the coherency dyadic.
9. It is assumed that the Saxon's reciprocity relation (3.4.19) is valid.

Assumption 1 implies that the incident light is fully coherent across any plane perpendicular to the incidence direction. The use of low-coherence or converging incident beams may cause special effects studied in Tomita and Ikari (1991), Dogariu and Boreman (1996), Okamoto and Asakura (1996), and Kim *et al.* (2005).

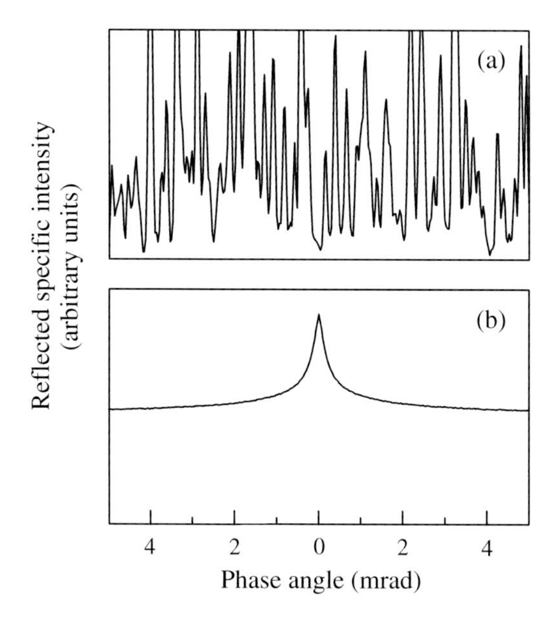

Figure 14.8.1. (a) Typical plot of the angular distribution of backscattered specific intensity for one configuration of a particulate sample. (b) The angular distribution of backscattered specific intensity after configurational averaging. (After Labeyrie *et al.*, 2000.)

Assumption 2 is, perhaps, the most critical in that it makes possible the characterization of the scattering properties of each particle in terms of its far-field scattering dyadic and, thereby, allows the introduction of the very concept of ladder and maximally crossed diagrams. Thus this assumption as well as the Twersky approximation are at the core of separating the total solution of the Maxwell's equations for a sparse discrete random medium into the radiative-transfer and coherent-backscattering components.

Assumptions 4 and 5 are naturally realized for particle suspensions provided that the measurement is taken over a sufficiently long period of time. However, experiments involving particulate surfaces and coherent sources of illumination such as lasers require special care (Etemad *et al.*, 1986; Kaveh *et al.*, 1986). Indeed, as we have seen in Section 1.4 and illustrate again in Fig. 14.8.1(a), a fixed particulate sample generates a speckle pattern with frequent irregular oscillations of the reflected intensity. To smooth out the speckle structure and separate the backscattering peak, one needs to rotate or vibrate the sample and accumulate the signal over a significant period of time (Fig. 14.8.1(b)).

Assumption 6(i) is by no means mandatory. However, it makes the theoretical analysis much simpler, especially vis-à-vis the concept of angular profile of CB. To estimate how far the observation point should actually be from the scattering slab, let us assume that it lies on the straight line going through particle 1 and point A in Fig. 14.3.2. Then the requirement that the phase difference between the two reciprocal paths going through particles 1 and 2 and arriving at the observation point D be much smaller than one yields

$$\frac{k_1(\overline{AC})^2}{2\,\overline{D2}} \ll 1.$$

Thus, in general, the distance d from the scattering medium to the observation point must satisfy the following inequality:

$$d \gg \tfrac{1}{2}k_1 l_{\rm tr}^2, \tag{14.8.1}$$

where $l_{\rm tr}$ is the transport mean free path defined by Eq. (14.7.7). For an optically semi-infinite medium composed of nonabsorbing, wavelength-sized or larger particles with $\langle\cos\Theta\rangle$ close to one, the requirement (14.8.1) can be rather demanding.

The Saxon's reciprocity relation is valid only in the far-field zone of a scatterer. Therefore, assumption 9 reinforces the main implication of assumption 2 according to which the very concept of CB can be directly applied to discrete random media only if the scattering particles are distributed sparsely.

The reciprocity relation becomes invalid if the scatterers and/or the host medium are made of naturally optically active materials or consist of magneto-optic materials and are subject to an external magnetic field (see, for example, the review by Potton, 2004). The natural optical activity and magnetic-field effects on CB have been studied both experimentally and theoretically in Golubentsev (1984), MacKintosh and John (1988), Martinez and Maynard (1994), van Tiggelen *et al.* (1996), Lacoste and van Tiggelen (2000), Lenke and Maret (2000b), and Lenke *et al.* (2000).

One theoretical problem that still awaits its solution is energy conservation. As we have seen in Section 8.13, the RTE ensures energy conservation by itself. Including the maximally crossed diagrams appears to destroy energy conservation by adding the energy contained in the coherent backscattering peak. It remains unclear whether this additional energy is "taken" from the far wings of the backscattering peak, which would imply that the contribution of the maximally crossed diagrams to the specific intensity at certain reflection directions may be negative.

14.9 Applications and further reading

The theory of CB and its applications are extensively discussed in the books edited by Sheng (1990), Fouque (1999), Sebbah (2001), and van Tiggelen and Skipetrov (2003). The effect of CB has been the basis of various optical characterization techniques and has been observed for various types of scatterers and scattering media including biological tissues (Yoo *et al.*, 1990; Yoon *et al.*, 1993; Kim *et al.*, 2005), randomized laser materials (Wiersma *et al.*, 1995b), nematic liquid crystals (Kuzmin *et al.*, 1996; Sapienza *et al.*, 2004), industrial materials (Schirrer *et al.*, 1997), atoms (Labeyrie *et al.*, 1999; Jonckheere *et al.*, 2000; Labeyrie *et al.*, 2000; Müller *et al.*, 2001), and photonic crystals (Koenderink *et al.*, 2000; Huang *et al.*, 2001).

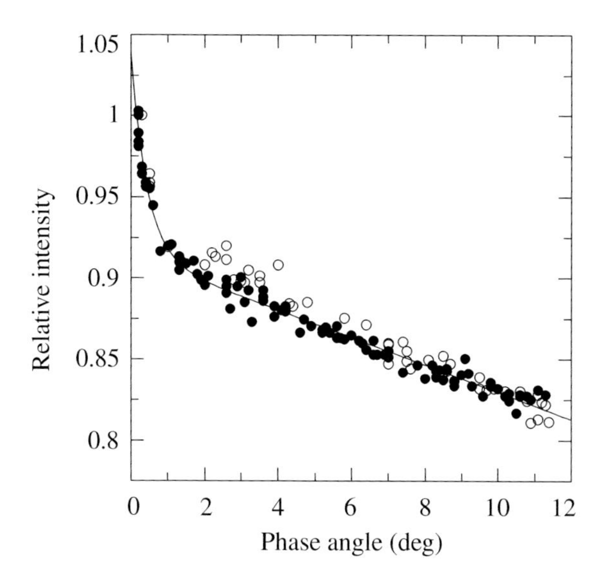

Figure 14.9.1. The brightness opposition effect for Europa (after Thompson and Lockwood, 1992).

Owing to its small angular width, the effect of CB is rather difficult to observe and measure. Yet it is often considered to be a precursor to an even more elusive effect called strong localization of light (Genack and Garcia, 1991; Wiersma *et al.*, 1997; Scheffold *et al.*, 1999; Wiersma *et al.*, 1999; Chabanov and Genack, 2003).

One of the more exotic manifestations of CB can be found in planetary astrophysics. Shkuratov (1988, 1989) and Muinonen (1990) were the first to suggest that CB might be responsible for some of the optical effects exhibited by atmosphereless solar system bodies at small and moderate phase angles. As an example, Fig. 14.9.1 shows the observed brightness of one of the large satellites of Jupiter, Europa, as a function of phase angle. These ground-based telescopic data as well as those collected from the Galileo spacecraft (Helfenstein *et al.*, 1998) clearly indicate the presence of an anomalously narrow backscattering peak centered at exactly the opposition. Similarly narrow intensity peaks have been observed for Saturn's A and B rings (Franklin and Cook, 1965), large satellites of Uranus and Neptune (Goguen *et al.*, 1989), and asteroids 44 Nysa and 64 Angelina (Harris *et al.*, 1989). All these objects have very high albedos and are believed to be covered with a layer of very small, weakly absorbing ice and/or silicate particles. Amazingly, the polarimetric phase curve for Europa exhibits a spike of negative polarization at a phase angle smaller than 1° (Fig. 14.9.2) which closely resembles the POE caused by CB. A similar feature has been observed for Saturn's rings (see Mishchenko, 1993 and references therein), for other Galilean satellites of Jupiter (Rosenbush and Kiselev, 2005), and for the high-albedo asteroid 64 Angelina (Rosenbush *et al.*, 2005). The angular characteristics of these intensity and polarization features are such that it may be impossible to reproduce them with an optical mechanism other than CB (Mishchenko and Dlugach, 1992b,

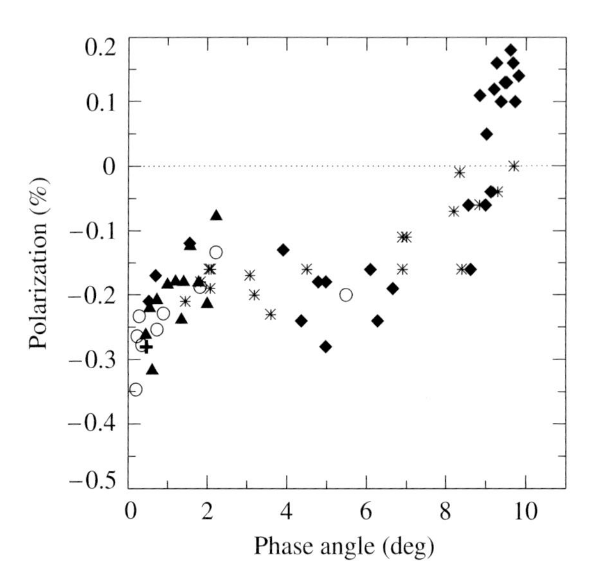

Figure 14.9.2. The polarization opposition effect for Europa (after Rosenbush *et al.*, 2002).

1993; Mishchenko, 1993; Rosenbush *et al.*, 1997), even though the formal applicability of the latter to densely distributed particles may be questionable.

Monostatic radars use the same antenna to transmit and receive electromagnetic waves. Therefore, radar measurements of particulate media are inevitably affected by CB. Several solar system objects have been found to generate radar returns quite uncharacteristic of bare solid surfaces (Ostro, 1993). For example, the icy Galilean satellites of Jupiter exhibit both high radar reflectivities and circular polarization ratios exceeding one. These measurements have been interpreted in terms of multiple scattering, including CB, of electromagnetic waves by voids or rocks imbedded in a transparent layer of ice (Hapke, 1990; Mishchenko, 1992b, 1996).

What is perhaps the most unexpected, unusually strong and highly depolarized echoes have been detected in radar observations of Mercury's poles (Harmon *et al.*, 1994). The fact that these echoes come from areas associated with deep craters provides a strong support for the hypothesis that water ice may have survived in permanently shadowed polar areas of Mercury despite the extreme proximity of this planet to the sun. Nozette *et al.* (2001) suggested that bistatic radar observations of the moon using the Clementine spacecraft may indicate the presence of similar deposits of ice at lunar poles.

Appendix A

Dyads and dyadics

The result of a dyadic operating on a vector is another vector. This operation may be thought of as a 3×3 matrix representing the dyadic multiplying a column matrix consisting of the initial vector components, thereby producing another column matrix consisting of the resulting vector components. The components of both vectors must be specified in the same coordinate system.

From another, coordinate-free, standpoint a dyadic can be introduced as a sum of so-called dyads, each dyad being the result of a dyadic product of two vectors $\mathbf{a} \otimes \mathbf{b}$ such that the operation $(\mathbf{a} \otimes \mathbf{b}) \cdot \mathbf{c}$ yields the vector $\mathbf{a}(\mathbf{b} \cdot \mathbf{c})$ and the operation $\mathbf{c} \cdot (\mathbf{a} \otimes \mathbf{b})$ yields the vector $(\mathbf{c} \cdot \mathbf{a})\mathbf{b}$. Note that the sum of two dyads is not necessarily a dyad. Any dyadic can be represented as a sum of at most nine dyads. For example, in Cartesian coordinates any dyadic $\overleftrightarrow{A}$ can be expressed as

$$\begin{aligned} \overleftrightarrow{A} = \; & A_{xx}\,\hat{\mathbf{x}} \otimes \hat{\mathbf{x}} + A_{xy}\,\hat{\mathbf{x}} \otimes \hat{\mathbf{y}} + A_{xz}\,\hat{\mathbf{x}} \otimes \hat{\mathbf{z}} \\ & + A_{yx}\,\hat{\mathbf{y}} \otimes \hat{\mathbf{x}} + A_{yy}\,\hat{\mathbf{y}} \otimes \hat{\mathbf{y}} + A_{yz}\,\hat{\mathbf{y}} \otimes \hat{\mathbf{z}} \\ & + A_{zx}\,\hat{\mathbf{z}} \otimes \hat{\mathbf{x}} + A_{zy}\,\hat{\mathbf{z}} \otimes \hat{\mathbf{y}} + A_{zz}\,\hat{\mathbf{z}} \otimes \hat{\mathbf{z}}, \end{aligned} \tag{A.1}$$

where $\hat{\mathbf{x}}$, $\hat{\mathbf{y}}$, and $\hat{\mathbf{z}}$ are the unit vectors along the x-, y-, and z-axis, respectively, and the coefficients A_{ij} can be thought of as the elements of the matrix representing the dyadic.

The vector product $(\mathbf{a} \otimes \mathbf{b}) \times \mathbf{c}$ is defined as a dyad $\mathbf{a} \otimes (\mathbf{b} \times \mathbf{c})$, and $\mathbf{c} \times (\mathbf{a} \otimes \mathbf{b})$ yields $(\mathbf{c} \times \mathbf{a}) \otimes \mathbf{b}$. The dot product of two dyads $\mathbf{a} \otimes \mathbf{b}$ and $\mathbf{c} \otimes \mathbf{d}$ yields the dyad $(\mathbf{b} \cdot \mathbf{c})(\mathbf{a} \otimes \mathbf{d})$.

The transpose of a dyadic $\overleftrightarrow{A}$ is a dyadic $\overleftrightarrow{A}^{\mathrm{T}}$ such that

$$\overleftrightarrow{A} \cdot \mathbf{a} = \mathbf{a} \cdot \overleftrightarrow{A}^{\mathrm{T}}$$

for any $\mathbf{a}$. One may easily verify that transposing a dyadic is equivalent to transposing the matrix representing the dyadic in a coordinate system. Obviously,

$$(\overleftrightarrow{A}^{\mathrm{T}})^{\mathrm{T}} = \overleftrightarrow{A}$$

and

$$\mathbf{a} \cdot \overleftrightarrow{A} = \overleftrightarrow{A}^{\mathrm{T}} \cdot \mathbf{a}.$$

A dyadic $\overleftrightarrow{A}$ is symmetric if

$$\overleftrightarrow{A}^{\mathrm{T}} = \overleftrightarrow{A}$$

and is Hermitian if

$$\overleftrightarrow{A}^{\mathrm{T}} = \overleftrightarrow{A}^{*}.$$

It is straightforward to show that

$$(\mathbf{a} \cdot \overleftrightarrow{A}) \cdot \mathbf{b} = \mathbf{a} \cdot (\overleftrightarrow{A} \cdot \mathbf{b}) = \mathbf{a} \cdot \overleftrightarrow{A} \cdot \mathbf{b}, \tag{A.2}$$

$$(\mathbf{a} \cdot \overleftrightarrow{A}) \cdot \overleftrightarrow{B} = \mathbf{a} \cdot (\overleftrightarrow{A} \cdot \overleftrightarrow{B}) = \mathbf{a} \cdot \overleftrightarrow{A} \cdot \overleftrightarrow{B}, \tag{A.3}$$

$$\overleftrightarrow{A} \cdot (\overleftrightarrow{B} \cdot \mathbf{a}) = (\overleftrightarrow{A} \cdot \overleftrightarrow{B}) \cdot \mathbf{a} = \overleftrightarrow{A} \cdot \overleftrightarrow{B} \cdot \mathbf{a}, \tag{A.4}$$

$$(\overleftrightarrow{A} \cdot \overleftrightarrow{B}) \cdot \overleftrightarrow{C} = \overleftrightarrow{A} \cdot (\overleftrightarrow{B} \cdot \overleftrightarrow{C}) = \overleftrightarrow{A} \cdot \overleftrightarrow{B} \cdot \overleftrightarrow{C}, \tag{A.5}$$

$$\mathbf{a} \cdot \overleftrightarrow{A} \cdot \mathbf{b} = \mathbf{b} \cdot \overleftrightarrow{A}^{\mathrm{T}} \cdot \mathbf{a}, \tag{A.6}$$

$$(\overleftrightarrow{A} \cdot \overleftrightarrow{B})^{\mathrm{T}} = \overleftrightarrow{B}^{\mathrm{T}} \cdot \overleftrightarrow{A}^{\mathrm{T}}, \tag{A.7}$$

$$(\overleftrightarrow{A} \cdot \mathbf{a}) \otimes (\overleftrightarrow{B} \cdot \mathbf{b}) = \overleftrightarrow{A} \cdot (\mathbf{a} \otimes \mathbf{b}) \cdot \overleftrightarrow{B}^{\mathrm{T}}. \tag{A.8}$$

The identity dyadic $\overleftrightarrow{I}$ is defined by the relations

$$\overleftrightarrow{I} \cdot \mathbf{a} = \mathbf{a} \cdot \overleftrightarrow{I} = \mathbf{a} \tag{A.9}$$

for any **a**. As a consequence,

$$\overleftrightarrow{I} \cdot \overleftrightarrow{A} = \overleftrightarrow{A} \cdot \overleftrightarrow{I} = \overleftrightarrow{A} \tag{A.10}$$

for any $\overleftrightarrow{A}$. Obviously,

$$\overleftrightarrow{I} = \hat{\mathbf{x}} \otimes \hat{\mathbf{x}} + \hat{\mathbf{y}} \otimes \hat{\mathbf{y}} + \hat{\mathbf{z}} \otimes \hat{\mathbf{z}} \tag{A.11}$$

in Cartesian coordinates and

$$\overleftrightarrow{I} = \hat{\mathbf{r}} \otimes \hat{\mathbf{r}} + \hat{\boldsymbol{\theta}} \otimes \hat{\boldsymbol{\theta}} + \hat{\boldsymbol{\varphi}} \otimes \hat{\boldsymbol{\varphi}} \tag{A.12}$$

in spherical polar coordinates, where $\hat{\mathbf{r}}$, $\hat{\boldsymbol{\theta}}$, and $\hat{\boldsymbol{\varphi}}$ are the corresponding unit vectors.

A useful compendium of formulas from dyadic algebra and dyadic analysis can be found in Appendix 3 of Van Bladel (1964).

Appendix B

Spherical wave expansion of a plane wave in the far-field zone

In this appendix we derive Eq. (3.4.15) following the approach described by Saxon (1955b). We begin with the well-known expansion of a plane wave in scalar spherical harmonics (Jackson, 1998, p. 471):

$$\exp(\mathrm{i}\mathbf{r}\cdot\mathbf{r}') = 4\pi \sum_{l=0}^{\infty} \mathrm{i}^l j_l(rr') \sum_{m=-l}^{l} Y_{lm}^*(\hat{\mathbf{r}}) Y_{lm}(\hat{\mathbf{r}}'), \tag{B.1}$$

where $\hat{\mathbf{r}} = \mathbf{r}/r$, $\hat{\mathbf{r}}' = \mathbf{r}'/r'$,

$$j_l(y) = y^l \left(-\frac{1}{y}\frac{\mathrm{d}}{\mathrm{d}y}\right)^l \left(\frac{\sin y}{y}\right) \tag{B.2}$$

are spherical Bessel functions of the first kind, and $Y_{lm}(\hat{\mathbf{r}})$ are scalar spherical harmonics. The latter are defined as

$$Y_{lm}(\hat{\mathbf{r}}) = \sqrt{\frac{(2l+1)(l-m)!}{4\pi(l+m)!}}\, P_l^m(\cos\theta)\exp(\mathrm{i}m\varphi), \tag{B.3}$$

where θ and φ are spherical angular coordinates of the unit vector $\hat{\mathbf{r}}$ and the P_l^m are associated Legendre functions defined in terms of Legendre polynomials P_l as follows:

$$P_l^m(x) = (-1)^m (1-x^2)^{m/2} \frac{\mathrm{d}^m}{\mathrm{d}x^m} P_l(x), \tag{B.4}$$

$$P_l(x) = \frac{1}{2^l l!}\frac{\mathrm{d}^l}{\mathrm{d}x^l}(x^2-1)^l \tag{B.5}$$

with $x \in [-1, 1]$. Using the asymptotic form (Arfken and Weber, 2001, p. 726),

$$j_l(y) \underset{y \to \infty}{=} \frac{1}{y} \sin\left(y - \frac{l\pi}{2}\right), \tag{B.6}$$

we have

$$j_l(rr') \underset{rr' \to \infty}{=} \frac{\mathrm{i}}{2rr'}\left[\exp\left(-\mathrm{i}rr' + \frac{\mathrm{i}l\pi}{2}\right) - \exp\left(\mathrm{i}rr' - \frac{\mathrm{i}l\pi}{2}\right)\right]. \tag{B.7}$$

Substituting this expression in Eq. (B.1) and making use of the completeness relation for spherical harmonics (Jackson, 1998, p. 108)

$$\sum_{l=0}^{\infty}\sum_{m=-l}^{l} Y_{lm}^{*}(\hat{\mathbf{r}})Y_{lm}(\hat{\mathbf{r}}') = \delta(\hat{\mathbf{r}} - \hat{\mathbf{r}}') \tag{B.8}$$

and the symmetry relation

$$Y_{lm}(-\hat{\mathbf{r}}') = (-1)^l Y_{lm}(\hat{\mathbf{r}}'), \tag{B.9}$$

we finally derive, after simple algebra,

$$\exp(\mathrm{i}rr'\hat{\mathbf{r}} \cdot \hat{\mathbf{r}}') \underset{rr' \to \infty}{=} \frac{\mathrm{i}2\pi}{r'}\left[\delta(\hat{\mathbf{r}} + \hat{\mathbf{r}}')\frac{\exp(-\mathrm{i}rr')}{r} - \delta(\hat{\mathbf{r}} - \hat{\mathbf{r}}')\frac{\exp(\mathrm{i}rr')}{r}\right], \tag{B.10}$$

where

$$\delta(\hat{\mathbf{r}} - \hat{\mathbf{r}}') = \delta(\cos\theta - \cos\theta')\delta(\varphi - \varphi') \tag{B.11}$$

is the solid-angle delta function.

A direct consequence of Eq. (B.10) is the so-called Jones' lemma (see, for example, Appendix XII of Born and Wolf, 1999), which states the following:

$$\frac{1}{R}\oint_S \mathrm{d}S\, f(\mathbf{r})\exp(\mathrm{i}k\hat{\mathbf{n}} \cdot \mathbf{r}) \underset{kR \to \infty}{=} \frac{\mathrm{i}2\pi}{k}[f(-R\hat{\mathbf{n}})\exp(-\mathrm{i}kR) - f(R\hat{\mathbf{n}})\exp(\mathrm{i}kR)], \tag{B.12}$$

where S is the surface of a sphere centered at the origin, R is the sphere radius, the position vector $\mathbf{r}$ connects the origin and a point on the surface, $f(\mathbf{r})$ is a "well-behaved" function of the position vector, and $\hat{\mathbf{n}}$ is a constant unit vector.

Appendix C

Euler rotation angles

Consider right-handed Cartesian coordinate systems $\{x, y, z\}$ and $\{x', y', z'\}$ having a common origin. It is often convenient to specify the orientation of the coordinate system $\{x', y', z'\}$ relative to the coordinate system $\{x, y, z\}$ in terms of three Euler rotation angles α, β, and γ which transform the coordinate system $\{x, y, z\}$ into the coordinate system $\{x', y', z'\}$, as shown in Fig. C.1. Specifically, the three consecutive Euler rotations are performed as follows:

- Rotation of the coordinate system $\{x, y, z\}$ about the z-axis through an angle $\alpha \in [0, 2\pi)$, reorienting the y-axis in such a way that it coincides with the line of nodes (i.e., the line formed by the intersection of the xy- and the $x'y'$-plane).

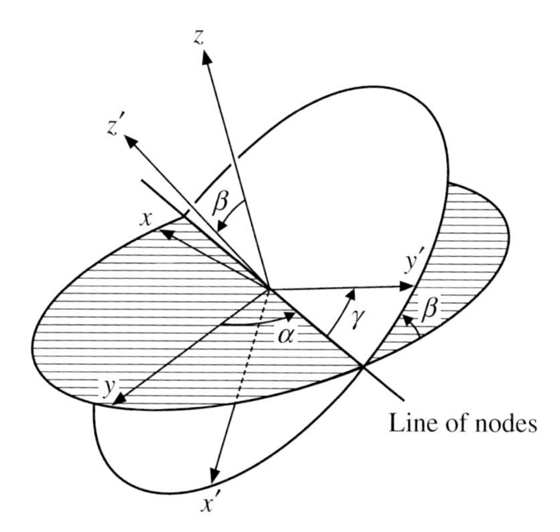

Figure C.1. Euler angles of rotation α, β, and γ transforming the coordinate system $\{x, y, z\}$ into the coordinate system $\{x', y', z'\}$.

- Rotation about the new y-axis through an angle $\beta \in [0, \pi]$.
- Rotation about the z'-axis through an angle $\gamma \in [0, 2\pi)$.

An angle of rotation is positive if the rotation is performed in the *clockwise* direction when one is looking in the positive direction of the rotation axis.

Appendix D

Integration quadrature formulas

Practical numerical evaluation of definite integrals is often based on using one of the so-called quadrature formulas. Assume, for example, that one needs to compute numerically the integral

$$I = \int_a^b \mathrm{d}x\, f(x), \tag{D.1}$$

where $f(x)$ is a real-valued function. The simplest approach is to use the (extended) trapezoidal rule

$$I \approx \frac{b-a}{N-1}\left[\frac{f(x_1)+f(x_N)}{2} + \sum_{n=2}^{N-1} f(x_n)\right], \tag{D.2}$$

where

$$x_n = a + \frac{(n-1)(b-a)}{N-1} \tag{D.3}$$

are the equidistant so-called division points such that $x_1 = a$ and $x_N = b$. By increasing the number of division points N, the result can be made arbitrarily accurate. However, this can lead to a rapid increase in computer time.

Usually a much more accurate and fast result can be obtained by using so-called quadrature formulas of the highest algebraic degree of precision (Krylov, 1962; Press *et al.*, 1992). Perhaps the most important example is the Gauss quadrature formula, which reads

$$\int_{-1}^{1} \mathrm{d}x\, f(x) \approx \sum_{n=1}^{N} w_n f(x_n), \tag{D.4}$$

where the nth quadrature division point x_n is the nth zero of the Nth degree Legendre

Table D.1. Gaussian division points and weights for $N = 9$.

n	x_n	w_n
1	−0.968160239507626	0.081274388361574
2	−0.836031107326636	0.180648160694857
3	−0.613371432700590	0.260610696402935
4	−0.324253423403809	0.312347077040003
5	0.000000000000000	0.330239355001260
6	0.324253423403809	0.312347077040003
7	0.613371432700590	0.260610696402935
8	0.836031107326636	0.180648160694857
9	0.968160239507626	0.081274388361574

polynomial $P_N(x)$ defined by Eq. (B.5), and the quadrature weights w_n are given by

$$w_n = \frac{2}{(1 - x_n^2)[P_N'(x_n)]^2}. \tag{D.5}$$

The Gauss quadrature formula is exact for all functions that can be represented by a polynomial of degree smaller than or equal to $2N - 1$. Substituting $f(x) \equiv 1$ yields a useful numerical check on the quadrature weights:

$$\sum_{n=1}^{N} w_n = 2. \tag{D.6}$$

As an illustration, Table D.1 lists the Gaussian division points and weights for $N =$ 9. Notice that $x_n = -x_{N-n+1}$ and $w_n = w_{N-n+1}$ so that the middle division point of the Gauss quadrature of any odd order is always zero.

The Gauss formula for an arbitrary integration interval $[a, b]$ follows from Eq. (D.4):

$$\int_a^b \mathrm{d}y\, f(y) \approx \sum_{n=1}^{N} u_n f(y_n), \tag{D.7}$$

where the corresponding division points and weights are now given by

$$y_n = \frac{b-a}{2} x_n + \frac{b+a}{2}, \tag{D.8}$$

$$u_n = \frac{b-a}{2} w_n. \tag{D.9}$$

For given a, b, and N, the division points of the Gauss quadrature are chosen

Table D.2. Division points and weights of the Markov quadrature formula on the interval [–1, 1] with $N = 9$ and one prescribed division point $x_9 = 1$.

n	x_n	w_n
1	–0.964440169705273	0.090714504923282
2	–0.817352784200412	0.200553298024552
3	–0.571383041208738	0.286386696357232
4	–0.256135670833455	0.337693966975930
5	0.090373369606853	0.348273002772967
6	0.426350485711139	0.316843775670438
7	0.711267485915709	0.247189378204593
8	0.910732089420060	0.147654019046315
9	1.000000000000000	0.024691358024692

automatically so that the formula is exact for polynomials of the highest possible degree. As a consequence, one has no direct control over the exact location of the division points, the middle division point $(b+a)/2$ for an odd N being the only exception. However, it is often convenient to have an integration formula that has one or more prescribed division points and still provides the highest possible degree of precision. If the number of prescribed division points is M and the total number of points is N then this so-called Markov quadrature formula (Krylov, 1962) is exact for all polynomials of degree smaller than or equal to $2N - M - 1$. The Markov quadrature with $M = 1$ and $x_1 = a$ is often called the Radau formula, whereas that with $M = 2$, $x_1 = a$, and $x_N = b$ is called the Lobatto formula. As an example, Table D.2 lists the division points and weights for the Markov quadrature formula on the interval [–1, 1] with $N = 9$ and one prescribed division point $x_9 = 1$.

Tables D.1 and D.2 were computed using FORTRAN subroutines GAUSS and MARK included in the code refl.f available at http://www.giss.nasa.gov/~crmim/brf/.

Appendix E

Stationary phase evaluation of a double integral

Consider the double integral

$$I = \int_{-\infty}^{+\infty} \mathrm{d}x \int_{-\infty}^{+\infty} \mathrm{d}y\, A(x, y) \frac{\exp[\mathrm{i}f(x, y)]}{r_1 r_2}, \tag{E.1}$$

where

$$f(x, y) = k(r_1 + r_2), \tag{E.2}$$

$$r_1 = \sqrt{x^2 + y^2 + z^2}, \tag{E.3}$$

$$r_2 = \sqrt{x^2 + y^2 + (Z - z)^2}. \tag{E.4}$$

The exponential $\exp[\mathrm{i}f(x, y)]$ is a rapidly oscillating function everywhere except in the region in which $f(x, y) \approx$ constant. Therefore, if $A(x, y)$ is a slowly varying function of x and y then the only significant contribution to I arises from the nearest vicinity of the stationary phase point determined from

$$\frac{\partial f(x, y)}{\partial x} = 0 \quad \text{and} \quad \frac{\partial f(x, y)}{\partial y} = 0 \tag{E.5}$$

and given by $x = 0$ and $y = 0$. Expanding $f(x, y)$ in a Taylor series about this point, we have

$$\begin{aligned} f(x, y) \approx\ & f(0, 0) + x \left.\frac{\partial f(x, y)}{\partial x}\right|_{x=0,\, y=0} + y \left.\frac{\partial f(x, y)}{\partial y}\right|_{x=0,\, y=0} \\ & + \frac{1}{2}x^2 \left.\frac{\partial^2 f(x, y)}{\partial x^2}\right|_{x=0,\, y=0} + \frac{1}{2}y^2 \left.\frac{\partial^2 f(x, y)}{\partial y^2}\right|_{x=0,\, y=0} \\ & + xy \left.\frac{\partial^2 f(x, y)}{\partial x \partial y}\right|_{x=0,\, y=0} \end{aligned}$$

$$= k(|z| + |Z - z|) + \frac{k}{2}\left(\frac{1}{|z|} + \frac{1}{|Z - z|}\right)(x^2 + y^2). \tag{E.6}$$

Finally, approximating

$$A(x, y) \approx A(0, 0), \tag{E.7}$$

$$\frac{1}{r_1 r_2} \approx \frac{1}{|z||Z - z|}, \tag{E.8}$$

substituting Eq. (E.6) in Eq. (E.1), and taking into account that

$$\int_{-\infty}^{+\infty} \mathrm{d}x \int_{-\infty}^{+\infty} \mathrm{d}y \exp[\mathrm{i}a(x^2 + y^2)] = \frac{\mathrm{i}\pi}{a}, \qquad a > 0 \tag{E.9}$$

yields

$$I = \frac{\mathrm{i}2\pi}{k} \frac{\exp[\mathrm{i}k(|z| + |Z - z|)]}{|z| + |Z - z|} A(0, 0). \tag{E.10}$$

Appendix F

Wigner functions, Jacobi polynomials, and generalized spherical functions

Jacobi polynomials, Wigner functions, and generalized spherical functions are closely related special functions which were introduced in classical analysis (Szegő, 1959), quantum theory of angular momentum (Wigner, 1959), and the theory of representations of the rotation group (Gelfand *et al.*, 1963), respectively. Because differences in notational conventions in various publications may often lead to confusion, we give in this appendix a short consistent summary of the main properties of these functions and their relationships.

F.1 Wigner *d*-functions

Wigner *d*-functions are defined as

$$\begin{aligned} d^s_{mn}(\theta) &= \sqrt{(s+m)!(s-m)!(s+n)!(s-n)!} \\ &\quad \times \sum_k (-1)^k \frac{(\cos\frac{1}{2}\theta)^{2s-2k+m-n}(\sin\frac{1}{2}\theta)^{2k-m+n}}{k!(s+m-k)!(s-n-k)!(n-m+k)!}, \end{aligned} \tag{F.1.1}$$

where s, m, and n are integers, $0 \le \theta \le \pi$, and the sum is taken over all integer values of k that lead to nonnegative factorials. Thus the summation index runs from $k_{\min} = \max(0, m-n)$ to $k_{\max} = \min(s+m, s-n)$. Therefore, $d^s_{mn}(\theta) = 0$ unless $k_{\max} \ge k_{\min}$, which is equivalent to requiring that $s \ge 0$ and $-s \le m, n \le s$. By making the substitutions $k \to s-n-k$, $k \to s+m-k$, and $k \to m-n+k$, respectively, we derive the following alternative expressions:

$$d^s_{mn}(\theta) = (-1)^{s-n}\sqrt{(s+m)!(s-m)!(s+n)!(s-n)!}$$
$$\times\sum_k(-1)^k\frac{(\cos\frac{1}{2}\theta)^{m+n+2k}(\sin\frac{1}{2}\theta)^{2s-m-n-2k}}{k!(s-m-k)!(s-n-k)!(m+n+k)!}, \tag{F.1.2}$$

$$d^s_{mn}(\theta) = (-1)^{s+m}\sqrt{(s+m)!(s-m)!(s+n)!(s-n)!}$$
$$\times\sum_k(-1)^k\frac{(\cos\frac{1}{2}\theta)^{2k-m-n}(\sin\frac{1}{2}\theta)^{2s+m+n-2k}}{k!(s+m-k)!(s+n-k)!(k-m-n)!}, \tag{F.1.3}$$

$$d^s_{mn}(\theta) = (-1)^{m-n}\sqrt{(s+m)!(s-m)!(s+n)!(s-n)!}$$
$$\times\sum_k(-1)^k\frac{(\cos\frac{1}{2}\theta)^{2s-2k-m+n}(\sin\frac{1}{2}\theta)^{2k+m-n}}{k!(s-m-k)!(s+n-k)!(m-n+k)!}. \tag{F.1.4}$$

As follows from the definition (F.1.1), the d-functions are real. It is also straightforward to verify, using Eqs. (F.1.1)–(F.1.4), that they have the following symmetry properties:

$$d^s_{mn}(\theta) = (-1)^{m-n}d^s_{-m,-n}(\theta) = (-1)^{m-n}d^s_{nm}(\theta) = d^s_{-n,-m}(\theta), \tag{F.1.5}$$

$$d^s_{mn}(-\theta) = (-1)^{m-n}d^s_{mn}(\theta) = d^s_{nm}(\theta), \tag{F.1.6}$$

$$d^s_{mn}(\pi-\theta) = (-1)^{s-n}d^s_{-mn}(\theta) = (-1)^{s+m}d^s_{m,-n}(\theta). \tag{F.1.7}$$

Furthermore,

$$d^s_{mn}(0) = \delta_{mn}, \tag{F.1.8}$$

where δ_{mn} is the Kronecker delta:

$$\delta_{mn} = \begin{cases}1 & \text{if } m=n,\\ 0 & \text{if } m\neq n.\end{cases} \tag{F.1.9}$$

Indeed, a nonzero term in the summation on the right-hand side of Eq. (F.1.1) must correspond to $2k = m-n$, while the requisite nonnegativity of $n-m+k$ implies that $n \geq m$. However, having $n > m$ would yield a negative k, which is disallowed. We thus have $n = m$ and $k = 0$, which leads to Eq. (F.1.8). Equations (F.1.7) and (F.1.8) imply that

$$d^s_{mn}(\pi) = (-1)^{s-n}\delta_{-mn}. \tag{F.1.10}$$

Denoting $x = \cos\theta$ and requiring that $\theta \in [0,\pi]$ yields

$$\cos\tfrac{1}{2}\theta = \sqrt{\tfrac{1}{2}(1+x)}, \qquad \sin\tfrac{1}{2}\theta = \sqrt{\tfrac{1}{2}(1-x)}. \tag{F.1.11}$$

Substituting Eq. (F.1.11) in Eq. (F.1.1) and modifying the resulting formula, we obtain

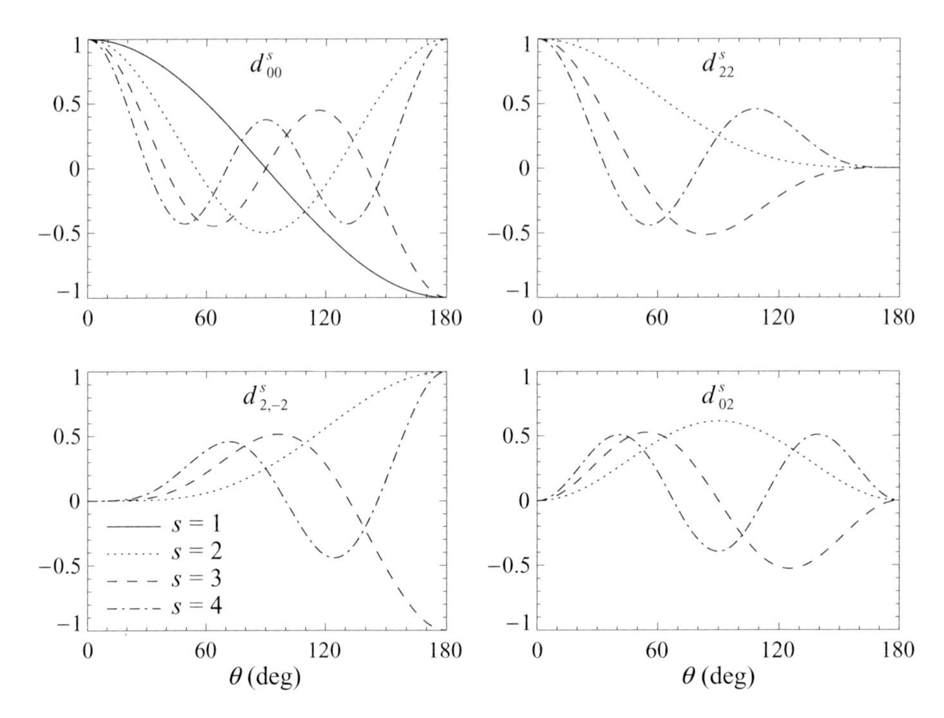

Figure F.1.1. Wigner *d*-functions.

$$\begin{aligned} d^s_{mn}(\theta) &= A^s_{mn}(1-x)^{(m-n)/2}(1+x)^{-(m+n)/2} \\ &\quad \times \sum_k (-1)^{s-n-k} \frac{(s+m)!(s-m)!(s-n)!}{k!(s+m-k)!(s-n-k)!(k-m+n)!} \\ &\quad \times (1+x)^{s+m-k}(1-x)^{k-m+n}, \end{aligned} \tag{F.1.12}$$

where

$$A^s_{mn} = \frac{(-1)^{s-n}}{2^s}\left[\frac{(s+n)!}{(s+m)!(s-m)!(s-n)!}\right]^{1/2}. \tag{F.1.13}$$

Finally, recalling the Leibniz rule,

$$\left(\frac{\mathrm{d}}{\mathrm{d}x}\right)^N [f(x)g(x)] = \sum_{k=0}^N \frac{N!}{k!(N-k)!}\left[\left(\frac{\mathrm{d}}{\mathrm{d}x}\right)^k f(x)\right]\left[\left(\frac{\mathrm{d}}{\mathrm{d}x}\right)^{N-k} g(x)\right], \tag{F.1.14}$$

and applying it to $f(x) = (1+x)^{s+m}$, $g(x) = (1-x)^{s-m}$, and $N = s-n$ we can rewrite Eq. (F.1.12) in the form

$$d^s_{mn}(\theta) = A^s_{mn}(1-x)^{(m-n)/2}(1+x)^{-(m+n)/2}\left(\frac{\mathrm{d}}{\mathrm{d}x}\right)^{s-n}[(1+x)^{s+m}(1-x)^{s-m}]. \tag{F.1.15}$$

Table F.1.1. Wigner d-functions appearing in the expansions (11.11.1)–(11.11.6).

s	$d^s_{00}(\theta)$	$d^s_{22}(\theta)$	$d^s_{2,-2}(\theta)$	$d^s_{02}(\theta)$
0	1	0	0	0
1	$\cos\theta$	0	0	0
2	$\frac{1}{2}(3\cos^2\theta - 1)$	$\frac{1}{4}(1+\cos\theta)^2$	$\frac{1}{4}(1-\cos\theta)^2$	$\frac{1}{2}\sqrt{\frac{3}{2}}\sin^2\theta$
3	$-\frac{1}{2}\cos\theta(3-5\cos^2\theta)$	$-\frac{1}{4}(1+\cos\theta)^2(2-3\cos\theta)$	$\frac{1}{4}(1-\cos\theta)^2(2+3\cos\theta)$	$\sqrt{\frac{15}{8}}\sin^2\theta\cos\theta$
4	$\frac{1}{8}(3-30\cos^2\theta+35\cos^4\theta)$	$\frac{1}{4}(1+\cos\theta)^2(1-7\cos\theta+7\cos^2\theta)$	$\frac{1}{4}(1-\cos\theta)^2(1+7\cos\theta+7\cos^2\theta)$	$-\sqrt{\frac{5}{32}}\sin^2\theta(1-7\cos^2\theta)$

Figure F.1.1 and Table F.1.1 illustrate the d-functions used in the expansions (11.11.1)–(11.11.6).

F.2 Jacobi polynomials

The Jacobi polynomial of degree q is given by Eq. (4.3.1) of Szegő (1959):

$$P_q^{(a,b)}(x) = (1-x)^{-a}(1+x)^{-b}\frac{(-1)^q}{2^q q!}\left(\frac{\mathrm{d}}{\mathrm{d}x}\right)^q [(1+x)^{q+b}(1-x)^{q+a}], \qquad \text{(F.2.1)}$$

where q is a nonnegative integer and $a > -1$ and $b > -1$ are real. Comparing Eq. (F.1.15) with Eq. (F.2.1), we obtain the following expression of the Wigner d-functions in terms of the Jacobi polynomials for $n \geq |m|$:

$$d_{mn}^s(\theta) = \frac{\xi_{mn}}{2^{(a+b)/2}}\left[\frac{q!(q+a+b)!}{(q+a)!(q+b)!}\right]^{1/2}(1-x)^{a/2}(1+x)^{b/2}P_q^{(a,b)}(x), \quad \text{(F.2.2)}$$

where $a = n-m$, $b = n+m$, $q = s-n$, and $\xi_{mn} = 1$. The condition $n \geq |m|$ ensures that $a \geq 0$ and $b \geq 0$, thereby preventing singularities for $x = \pm 1$. Using the symmetry relations (F.1.5), it is straightforward to show that Eq. (F.2.2) can be used for arbitrary m and n, provided that

$$a = |m-n|, \qquad b = |m+n|, \qquad q = s - \tfrac{1}{2}(a+b), \qquad \text{(F.2.3)}$$

$$\xi_{mn} = \begin{cases} 1 & \text{for } n \geq m, \\ (-1)^{m-n} & \text{for } n < m. \end{cases} \qquad \text{(F.2.4)}$$

F.3 Orthogonality and completeness

The orthogonality property of the Jacobi polynomials (Eq. (4.3.3) of Szegő, 1959) and Eqs. (F.2.2)–(F.2.4) lead to the following orthogonality property of the d-functions:

$$\int_0^\pi \mathrm{d}\theta \sin\theta \, d_{mn}^s(\theta) d_{mn}^{s'}(\theta) = \frac{2}{2s+1}\delta_{ss'}. \qquad \text{(F.3.1)}$$

The completeness property of the Jacobi polynomials (Szegő, 1959) and Eqs. (F.2.2) and (F.3.1) imply that functions $\sqrt{s+\frac{1}{2}}\, d_{mn}^s(\theta)$ with $s = s_{\min}, s_{\min}+1, \ldots$ form a complete orthonormal system of functions on $[0, \pi]$, where

$$s_{\min} = \max(|m|, |n|). \qquad \text{(F.3.2)}$$

This means that if a real-valued function $f(\theta)$ defined on the closed interval $[0, \pi]$ is square integrable on this interval, i.e., if

$$\int_0^\pi d\theta \sin\theta \, [f(\theta)]^2 < \infty \tag{F.3.3}$$

then there exists a unique set of coefficients η_s $(s \geq s_{\min})$ such that the series expansion

$$f(\theta) = \sum_{s=s_{\min}}^{\infty} \eta_s \, d^s_{mn}(\theta), \qquad \theta \in [0, \pi] \tag{F.3.4}$$

holds in the following sense:

$$\lim_{S\to\infty} \int_0^\pi d\theta \sin\theta \left| f(\theta) - \sum_{s=s_{\min}}^{S} \eta_s \, d^s_{mn}(\theta) \right|^2 = 0. \tag{F.3.5}$$

Conversely, if a real-valued function $f(\theta)$ on $[0, \pi]$ admits the expansion (F.3.4) in the sense of Eq. (F.3.5), then it is square integrable on $[0, \pi]$ and the expansion coefficients are given by

$$\eta_s = (s + \tfrac{1}{2}) \int_0^\pi d\theta \sin\theta \, f(\theta) \, d^s_{mn}(\theta). \tag{F.3.6}$$

The latter formula follows directly from Eqs. (F.3.4) and (F.3.1).

F.4 Recurrence relations

Using Eq. (4.5.1) of Szegő (1959) and Eq. (F.2.2), we obtain the following recurrence relation for the Wigner d-functions:

$$d^{s+1}_{mn}(\theta) = \frac{1}{s\sqrt{(s+1)^2 - m^2}\sqrt{(s+1)^2 - n^2}} \{(2s+1)[s(s+1)x - mn] d^s_{mn}(\theta)$$
$$- (s+1)\sqrt{s^2 - m^2}\sqrt{s^2 - n^2}\, d^{s-1}_{mn}(\theta)\}, \qquad s \geq s_{\min}. \tag{F.4.1}$$

The simplest way to derive this formula is to consider first the case $n \geq |m|$, which corresponds to $a = n - m$, $b = n + m$, $q = s - n$, and then to use the symmetry relations (F.1.5) in order to verify that Eq. (F.4.1) is correct for arbitrary m and n. The initial values are given by

$$d^{s_{\min}-1}_{mn}(\theta) = 0, \tag{F.4.2}$$

$$d^{s_{\min}}_{mn}(\theta) = \xi_{mn} 2^{-s_{\min}} \left[\frac{(2s_{\min})!}{(|m-n|)!(|m+n|)!} \right]^{1/2} (1-x)^{|m-n|/2} (1+x)^{|m+n|/2}, \tag{F.4.3}$$

where ξ_{mn} is given by Eq. (F.2.4). Equation (F.4.3) follows directly from Eq. (F.1.15) if $n \geq |m|$, and it is extended to arbitrary m and n using Eq. (F.1.5). From Eq. (F.1.15), we easily derive

$$\frac{\mathrm{d}}{\mathrm{d}\theta} d^s_{mn}(\theta) = \frac{m - n\cos\theta}{\sin\theta} d^s_{mn}(\theta) + \sqrt{(s+n)(s-n+1)}\, d^s_{m,n-1}(\theta) \qquad \text{(F.4.4)}$$

$$= -\frac{m - n\cos\theta}{\sin\theta} d^s_{mn}(\theta) - \sqrt{(s-n)(s+n+1)}\, d^s_{m,n+1}(\theta). \qquad \text{(F.4.5)}$$

Alternatively, we have from Eq. (4.5.5) of Szegő (1959) and Eq. (F.4.1)

$$\frac{\mathrm{d}}{\mathrm{d}\theta} d^s_{mn}(\theta) = \frac{1}{\sin\theta}\left[-\frac{(s+1)\sqrt{(s^2-m^2)(s^2-n^2)}}{s(2s+1)} d^{s-1}_{mn}(\theta) - \frac{mn}{s(s+1)} d^s_{mn}(\theta) + \frac{s\sqrt{(s+1)^2-m^2}\sqrt{(s+1)^2-n^2}}{(s+1)(2s+1)} d^{s+1}_{mn}(\theta) \right]. \qquad \text{(F.4.6)}$$

F.5 Legendre polynomials and associated Legendre functions

The Wigner d-functions with $m = 0$ and $n = 0$ are equivalent to the usual Legendre polynomials (cf. Eqs. (F.1.15) and (B.5)):

$$d^s_{00}(\theta) = P_s(x). \qquad \text{(F.5.1)}$$

For $n = 0$, we obtain

$$d^s_{m0}(\theta) = \sqrt{\frac{(s-m)!}{(s+m)!}}\, P^m_s(x), \qquad \text{(F.5.2)}$$

where $P^m_s(x)$ are associated Legendre functions defined by Eq. (B.4). Equations (F.4.1) and (F.5.2) give a simple recurrence relation for the associated Legendre functions:

$$(s-m+1)\, P^m_{s+1}(x) = (2s+1)\, x\, P^m_s(x) - (s+m)\, P^m_{s-1}(x). \qquad \text{(F.5.3)}$$

Despite its simplicity, the use of this relation in computer calculations for large s and $|m|$ results in overflows, whereas the original recurrence relation for the functions $d^s_{m0}(\theta)$ remains stable and accurate. Furthermore, the functions $d^s_{m0}(\theta)$ have simpler symmetry properties than the $P^m_s(x)$. It is, therefore, advisable to use the d-functions instead of the associated Legendre functions from both the analytical and the numerical standpoint.

F.6 Generalized spherical functions

The generalized spherical functions $P^s_{mn}(x)$ are complex-valued functions related to the Wigner d-functions by

$$P^s_{mn}(x) = \mathrm{i}^{m-n}\, d^s_{mn}(\theta). \tag{F.6.1}$$

Using Eqs. (F.1.5)–(F.1.7), we easily derive the following symmetry relations:

$$P^s_{mn}(x) = P^s_{nm}(x) = P^s_{-m,-n}(x) = (-1)^{m+n}\,[P^s_{mn}(x)]^*, \tag{F.6.2}$$

$$P^s_{mn}(-x) = (-1)^{s+m-n}\, P^s_{-mn}(x). \tag{F.6.3}$$

The corresponding orthogonality and normalization condition follows directly from Eq. (F.3.1):

$$\begin{aligned}\int_{-1}^{+1} \mathrm{d}x\, P^s_{mn}(x)\, P^{s'}_{mn}(x) &= (-1)^{m+n} \int_{-1}^{+1} \mathrm{d}x\, P^s_{mn}(x)[P^{s'}_{mn}(x)]^* \\ &= (-1)^{m+n}\,\frac{2}{2s+1}\,\delta_{ss'}.\end{aligned} \tag{F.6.4}$$

It is straightforward to show that the generalized spherical functions form a complete set of complex functions on the interval $x \in [-1,+1]$. This means that any complex-valued function $f(x)$, defined and square-integrable on the interval $x \in [-1,+1]$, can be uniquely expanded in the functions $P^s_{mn}(x)$ with $s = s_{\min}, s_{\min}+1, \ldots$ In other words, if

$$\int_{-1}^{+1} \mathrm{d}x\, |f(x)|^2 < \infty \tag{F.6.5}$$

then there exists a unique set of coefficients η_s $(s \ge s_{\min})$ such that

$$\lim_{S\to\infty} \int_{-1}^{+1} \mathrm{d}x \left| f(x) - \sum_{s=s_{\min}}^{S} \eta_s\, P^s_{mn}(x) \right|^2 = 0. \tag{F.6.6}$$

Conversely, if a complex-valued function $f(x)$ on $[-1,+1]$ admits the expansion

$$f(x) = \sum_{s=s_{\min}}^{\infty} \eta_s\, P^s_{mn}(x) \tag{F.6.7}$$

in the sense of Eq. (F.6.6), then it is square integrable on $[-1,+1]$ and the expansion coefficients are given by

$$\eta_s = (-1)^{m+n}(s+\tfrac{1}{2}) \int_{-1}^{+1} \mathrm{d}x\, f(x)\, P^s_{mn}(x) \tag{F.6.8}$$

(cf. Eqs. (F.6.4) and (F.6.7)).

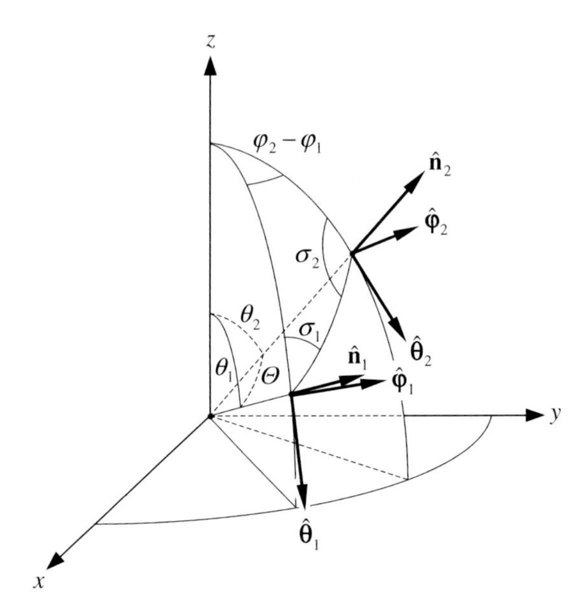

Figure F.7.1. Illustration of the addition theorem for Wigner *d*-functions.

F.7 Wigner *D*-functions, addition theorem, and unitarity

The Wigner *D*-functions are defined as

$$D^n_{mm'}(\alpha, \beta, \gamma) = e^{-im\alpha} d^n_{mm'}(\beta) e^{-im'\gamma}, \tag{F.7.1}$$

where

$$0 \le \alpha < 2\pi, \qquad 0 \le \beta \le \pi, \qquad 0 \le \gamma < 2\pi. \tag{F.7.2}$$

If the sets of Euler angles $(\alpha_1, \beta_1, \gamma_1)$ and $(\alpha_2, \beta_2, \gamma_2)$ (Appendix C) describe two consecutive rotations of a coordinate system and the set (α, β, γ) describes the resulting rotation, then the addition theorem for the *D*-functions reads

$$D^n_{mm'}(\alpha, \beta, \gamma) = \sum_{m''=-n}^{n} D^n_{mm''}(\alpha_1, \beta_1, \gamma_1) D^n_{m''m'}(\alpha_2, \beta_2, \gamma_2) \tag{F.7.3}$$

(see Eq. (2) in Section 4.7 of Varshalovich *et al.*, 1988). A direct consequence of the addition theorem is the unitarity condition

$$\sum_{m''=-n}^{n} D^n_{mm''}(\alpha, \beta, \gamma) D^n_{m''m'}(-\gamma, -\beta, -\alpha) = D^n_{mm'}(0,0,0) = \delta_{mm'} \tag{F.7.4}$$

(cf. Eq. (F.1.8)).

Using Eq. (F.7.3), we can derive the addition theorem for the Wigner *d*-functions. Consider the geometry shown in Fig. F.7.1, where the angles θ_1, θ_2, $\varphi_2 - \varphi_1$, σ_1, σ_2, and Θ are nonnegative and are related by

$$\cos\Theta = \cos\theta_2\cos\theta_1 + \sin\theta_2\sin\theta_1\cos(\varphi_2-\varphi_1), \tag{F.7.5}$$

$$\cos\sigma_1 = \frac{\cos\theta_2 - \cos\theta_1\cos\Theta}{\sin\theta_1\sin\Theta}, \tag{F.7.6}$$

$$\cos\sigma_2 = \frac{\cos\theta_1 - \cos\theta_2\cos\Theta}{\sin\theta_2\sin\Theta} \tag{F.7.7}$$

(cf. Eqs. (11.3.4)–(11.3.6)). The reference frame formed by the unit vectors $(\hat{\mathbf{n}}_1, \hat{\boldsymbol{\theta}}_1, \hat{\boldsymbol{\varphi}}_1)$ can be transformed into the reference frame formed by the unit vectors $(\hat{\mathbf{n}}_2, \hat{\boldsymbol{\theta}}_2, \hat{\boldsymbol{\varphi}}_2)$ in two ways:

- Via a single rotation through Euler angles $(\pi-\sigma_1, \Theta, -\sigma_2)$.
- Via two consecutive rotations through Euler angles $(0, -\theta_1, \varphi_2-\varphi_1)$ and $(0, \theta_2, 0)$.

We, therefore, derive from Eqs. (F.7.1) and (F.7.3)

$$e^{im\sigma_1} d^n_{mm'}(\Theta) e^{im'\sigma_2} = \sum_{m''=-n}^{n} (-1)^{m''} d^n_{mm''}(\theta_1) d^n_{m''m'}(\theta_2) e^{-im''(\varphi_2-\varphi_1)}. \tag{F.7.8}$$

Consider two special cases of Eq. (F.7.8). If $\varphi_2 - \varphi_1 = 0$ and $\theta_1 \ge \theta_2$ then $\sigma_1 = 0$, $\sigma_2 = \pi$, and $\Theta = \theta_1 - \theta_2$. We thus obtain

$$d^n_{mm'}(\theta_1-\theta_2) = \sum_{m''=-n}^{n} (-1)^{m''-m'} d^n_{mm''}(\theta_1) d^n_{m''m'}(\theta_2). \tag{F.7.9}$$

In particular, when $\theta_1 = \theta_2$, Eqs. (F.1.8) and (F.7.9) yield the unitarity condition for the d-functions:

$$\begin{aligned}\sum_{m''=-n}^{n} (-1)^{m''-m'} d^n_{mm''}(\theta) d^n_{m''m'}(\theta) &= \sum_{m''=-n}^{n} d^n_{mm''}(\theta) d^n_{m''m'}(-\theta)\\ &= \sum_{m''=-n}^{n} d^n_{mm''}(\theta) d^n_{m'm''}(\theta)\\ &= \delta_{mm'}.\end{aligned} \tag{F.7.10}$$

This formula can also be derived directly from the unitarity condition for the D-functions, Eq. (F.7.4), by substituting $\alpha = \gamma = 0$. If $\varphi_2 - \varphi_1 = \pi$, then $\sigma_1 = \sigma_2 = 0$ and $\Theta = \theta_1 + \theta_2$, and we have

$$d^n_{mm'}(\theta_1+\theta_2) = \sum_{m''=-n}^{n} d^n_{mm''}(\theta_1) d^n_{m''m'}(\theta_2). \tag{F.7.11}$$

Equation (F.7.10) yields

$$\sum_{m'=-n}^{n} [d^n_{mm'}(\theta)]^2 = 1, \tag{F.7.12}$$

which implies that

$$|d^n_{mm'}(\theta)| \le 1, \qquad \theta \in [0, \pi], \tag{F.7.13}$$

$$|P^n_{mm'}(x)| \le 1, \qquad x \in [-1, 1]. \tag{F.7.14}$$

F.8 Further reading

Detailed accounts of Jacobi polynomials, Wigner d-functions, and generalized spherical functions are given in Rose (1957), Szegő (1959), Gelfand *et al.* (1963), Vilenkin (1968), Biedenharn and Louck (1981), Varshalovich *et al.* (1988), Brink and Satchler (1993), and Edmonds (1996). Our definition of the d-functions is consistent with that of Rose (1957), Biedenharn and Louck (1981), Hovenier and van der Mee (1983), Varshalovich *et al.* (1988), and Brink and Satchler (1993). Vilenkin (1968) uses functions $P^s_{mn}(x)$ related to the Wigner d-functions by $P^s_{mn}(x) = \mathrm{i}^{n-m} d^s_{mn}(\theta)$. Edmonds (1996) uses a function $d^{(s)}_{mn}(\theta)$ which is related to $d^s_{mn}(\theta)$ by $d^{(s)}_{mn}(\theta) = (-1)^{m+n} d^s_{mn}(\theta)$.

Appendix G

Système International units

The system of physical units adopted in this book is the internationally accepted form of the metric system known as the Système International (SI). The SI is formed by base units, supplementary units, and units derived from the base units. The table below lists only those derived SI units that are used in this book.

Quantity	Unit	Symbol	Definition
Base units			
length	meter	m	
time	second	s	
mass	kilogram	kg	
electric current	ampere	A	
temperature	kelvin	K	
amount of substance	mole	mol	
Supplementary units			
plane angle	radian	rad	
solid angle	steradian	sr	
Derived units			
energy	joule	J	$\mathrm{N\,m = kg\,m^2\,s^{-2}}$
electric charge	coulomb	C	$\mathrm{A\,s}$
electric potential	volt	V	$\mathrm{W\,A^{-1} = kg\,m^2\,s^{-3}\,A^{-1}}$
electric capacitance	farad	F	$\mathrm{C\,V^{-1} = kg^{-1}\,m^{-2}\,s^4\,A^2}$
electric resistance	ohm	Ω	$\mathrm{V\,A^{-1} = kg\,m^2\,s^{-3}\,A^{-2}}$
magnetic flux	weber	Wb	$\mathrm{V\,s = kg\,m^2\,s^{-2}\,A^{-1}}$
inductance	henry	H	$\mathrm{Wb\,A^{-1} = kg\,m^2\,s^{-2}\,A^{-2}}$

Quantity	Unit	Symbol	Definition
frequency	hertz	Hz	s^{-1}
power	watt	W	$J\,s^{-1} = kg\,m^2\,s^{-3}$
force	newton	N	$kg\,m\,s^{-2}$

Appendix H

Abbreviations

Abbreviation	Definition	Introduced in Section
ADA	anomalous diffraction approximation	9.2
CB	coherent backscattering	1.8
DDA	discrete dipole approximation	9.1
DWS	diffusing wave spectroscopy	1.4
EBCM	extended boundary condition method	9.1
FDTDM	finite-difference time-domain method	9.1
FEM	finite-element method	9.1
FIEM	Fredholm integral equation method	9.1
FOSA	first-order-scattering approximation	7.7
GOA	geometrical optics approximation	9.2
GPMM	generalized PMM	9.1
HG	Henyey–Greenstein (phase function)	13.2
ISM	macroscopically isotropic and mirror-symmetric scattering medium	Introduction to Chapter 11
KA	Kirchhoff approximation	9.2
LDM	laser Doppler method	1.10
ME-GPMM	multiple-expansion GPMM	9.1
MOM	method of moments	9.1
MTL	Mishchenko *et al.* (2002)	Preface
MUSSA	modified uncorrelated single-scattering approximation	7.3
OCT	optical coherence tomography	1.10

Abbreviation	Definition	Introduced in Section
PCS	photon correlation spectroscopy	1.4
PMM	point-matching method	9.1
POE	polarization opposition effect	14.7
RGA	Rayleigh–Gans approximation	3.1
RTE	radiative transfer equation	Preface
RTT	radiative transfer theory	Preface
SM	superposition method	9.1
SSA	single-scattering approximation	7.1
SVM	separation of variables method	9.1
TMM	T-matrix method	9.1
USSA	uncorrelated single-scattering approximation	7.3
VIEM	volume integral equation method	9.1
VRTE	vector RTE	8.10
VSWF	vector spherical wave function	9.1

Appendix I

Glossary of symbols

This glossary lists the most important symbols whose specific definitions may not appear in their immediate context. Each entry gives the physical dimension of the corresponding quantity (if applicable) as well as the section in which the symbol is defined.

Symbol	Definition and dimension in SI units	Introduced in Section
English symbols		
$a_j(\Theta),\ j = 1, \ldots, 4$	diagonal elements of the normalized Stokes scattering matrix [–]	11.10
$\mathbf{A}$	4×4 transformation matrix [–]	2.6
$\overleftrightarrow{A}$	scattering dyadic [m]	3.3
$\langle \overleftrightarrow{A} \rangle_\xi$	scattering dyadic averaged over particle states [m]	8.2
$b_j(\Theta),\ j = 1, 2$	off-diagonal elements of the normalized Stokes scattering matrix [–]	11.10
$\mathbf{B}$	complex magnetic induction [Wb m^{-2}]	2.3
$\mathbf{B}$	4×4 transformation matrix [–]	2.6
$\boldsymbol{\mathcal{B}}$	magnetic induction [Wb m^{-2}]	2.1
$\overleftrightarrow{B}$	dimensionless scattering dyadic [–]	8.1
c	speed of light in a vacuum [m s^{-1}]	2.5
C_{abs}	absorption cross section [m^2]	3.9
C_{ext}	extinction cross section [m^2]	3.6
C_{sca}	scattering cross section [m^2]	3.9
$\overleftrightarrow{C}$	coherency dyadic [V^2 m^{-2}]	1.7

Symbol	Definition and dimension in SI units	Introduced in Section
$\ddot{C}_1$	single-scattering component of the coherency dyadic $[\mathrm{V}^2\ \mathrm{m}^{-2}]$	14.1
$\ddot{C}_\mathrm{c}$	coherent part of the coherency dyadic $[\mathrm{V}^2\ \mathrm{m}^{-2}]$	8.6
$\ddot{C}_\mathrm{C}$	cyclical component of the coherency dyadic $[\mathrm{V}^2\ \mathrm{m}^{-2}]$	Introduction to Chapter 14
$\ddot{C}_\mathrm{L}$	ladder component of the coherency dyadic $[\mathrm{V}^2\ \mathrm{m}^{-2}]$	8.6
$\ddot{C}_\mathrm{M}$	diffuse multiple-scattering component of the coherency dyadic $[\mathrm{V}^2\ \mathrm{m}^{-2}]$	14.1
$\langle\cos\Theta\rangle$	asymmetry parameter [–]	3.9
$\langle C_\mathrm{abs}\rangle_\xi$	absorption cross section per particle averaged over particle states $[\mathrm{m}^2]$	7.3
$\langle C_\mathrm{ext}\rangle_\xi$	extinction cross section per particle averaged over particle states $[\mathrm{m}^2]$	7.3
$\langle C_\mathrm{sca}\rangle_\xi$	scattering cross section per particle averaged over particle states $[\mathrm{m}^2]$	7.3
d	differential sign	
$d^s_{mn}(\theta)$	Wigner d-functions [–]	Appendix F
D	diameter of the sensitive surface of a detector [m]	3.6
$\mathbf{D}$	complex electric displacement $[\mathrm{C\ m}^{-2}]$	2.3
D	4×4 transformation matrix [–]	2.6
D	4×4 matrix describing internal radiation field [–]	10.3
$\mathcal{D}$	electric displacement $[\mathrm{C\ m}^{-2}]$	2.1
$\mathsf{D}^\dagger$	4×4 matrix describing internal radiation field [–]	10.3
e	base of natural logarithms [–]	
$\mathbf{E}$	complex electric field $[\mathrm{V\ m}^{-1}]$	1.1
E	two-component column formed by the θ- and φ-components of the electric field vector $[\mathrm{V\ m}^{-1}]$	3.3
$\boldsymbol{\mathcal{E}}$	electric field $[\mathrm{V\ m}^{-1}]$	1.1
$\mathbf{E}_\mathrm{c}$	coherent electric field $[\mathrm{V\ m}^{-1}]$	8.2
E_c	two-component column of the coherent electric field $[\mathrm{V\ m}^{-1}]$	8.3
E_θ, E_φ	spherical coordinate components of the electric field vector $[\mathrm{V\ m}^{-1}]$	2.6

Symbol	Definition and dimension in SI units	Introduced in Section
$\mathbf{F}$	4×4 Stokes scattering matrix [m^2]	Introduction to Chapter 11
$\widetilde{\mathbf{F}}$	4×4 normalized Stokes scattering matrix [–]	11.10
$\widetilde{\mathbf{F}}^{CP}$	4×4 normalized circular-polarization scattering matrix [–]	11.12
$\widetilde{F}^{CP}_{pq}$	elements of the normalized circular-polarization scattering matrix [–]	11.12
$\langle\mathbf{F}\rangle_\xi$	4×4 Stokes scattering matrix per particle averaged over particle states [m^2]	11.1
$\langle F_{ij}\rangle_\xi$	elements of the Stokes scattering matrix per particle averaged over particle states [m^2]	11.1
g^s_{pq}	expansion coefficients [–]	11.12
$G(r)$	scalar Green function [m^{-1}]	4.2
$\ddot{G}$	free space dyadic Green function [m^{-1}]	3.1
$\mathbf{h}$	2×2 coherent transmission amplitude matrix [–]	8.3
$\mathbf{H}$	complex magnetic field [$A\ m^{-1}$]	1.1
$\mathbf{H}$	4×4 coherent transmission Stokes matrix [–]	8.4
$\mathcal{H}$	magnetic field [$A\ m^{-1}$]	1.1
i	$\sqrt{-1}$ [–]	1.1
I	intensity [$W\ m^{-2}$]	2.5
I	first Stokes parameter [$W\ m^{-2}$]	2.6
$\mathbf{I}$	4×1 Stokes column vector [$W\ m^{-2}$]	2.6
Im	imaginary part	
I_2, I_0, I_{-0}, I_2	elements of the circular-polarization column vector [$W\ m^{-2}$]	2.6
I_b	Planck blackbody energy distribution [$W\ m^{-2}\ sr^{-1}\ rad^{-1}\ s$]	3.13
I_v, I_h	first and second elements of the modified Stokes column vector [$W\ m^{-2}$]	2.6
$\ddot{I}$	identity dyadic [–]	Appendix A
$\widetilde{I}$	first element of the specific intensity column vector (specific intensity) [$W\ m^{-2}\ sr^{-1}$]	8.10
$\mathbf{I}_b$	4×1 blackbody Stokes column vector [$W\ m^{-2}\ sr^{-1}\ rad^{-1}\ s$]	3.13
$\mathbf{I}_c$	4×1 Stokes column vector of the coherent electric field [$W\ m^{-2}$]	8.4

Symbol	Definition and dimension in SI units	Introduced in Section
$\mathbf{I}^{\mathrm{CP}}$	4×1 circular-polarization column vector [W m^{-2}]	2.6
$\mathbf{I}^{\mathrm{MS}}$	4×1 modified Stokes column vector [W m^{-2}]	2.6
$\tilde{\mathbf{I}}$	4×1 specific intensity column vector [W m^{-2} sr^{-1}]	8.10
$\tilde{\mathbf{I}}_{\mathrm{d}}$	4×1 diffuse specific intensity column vector [W m^{-2} sr^{-1}]	8.10
$\mathbf{J}$	complex current density [A m^{-2}]	2.3
$\mathbf{J}$	4×1 coherency column vector [W m^{-2}]	2.6
$\mathcal{J}$	current density [A m^{-2}]	2.1
$\mathbf{J}_{\mathrm{c}}$	4×1 coherency column vector of the coherent electric field [W m^{-2}]	8.4
$\tilde{\mathbf{J}}_{\mathrm{d}}$	4×1 diffuse coherency column vector [W m^{-2} sr^{-1}]	8.9
$k = k_{\mathrm{R}} + \mathrm{i}k_{\mathrm{I}}$	(complex) wave number [m^{-1}]	2.5
$\mathbf{k} = \mathbf{k}_{\mathrm{R}} + \mathrm{i}\mathbf{k}_{\mathrm{I}}$	(complex) wave vector [m^{-1}]	1.1, 2.5
$\mathbf{k}$	2×2 matrix propagation constant [m^{-1}]	8.3
k_1	wave number in the exterior region [m^{-1}]	1.8
$\mathbf{L}, \mathbf{L}^{\mathrm{MS}}, \mathbf{L}^{\mathrm{CP}}$	4×4 rotation matrices [–]	2.8
$m = m_{\mathrm{R}} + \mathrm{i}m_{\mathrm{I}}$	(complex) refractive index relative to vacuum or surrounding medium [–]	2.5, 3.1
$\mathcal{M}$	magnetization [A m^{-1}]	2.1
$n(r)$	size distribution function [m^{-1}]	5.3
n_0	particle number density [m^{-3}]	5.1
$O(x)$	quantity of the order of x	
$\boldsymbol{O}(x)$	vector with elements of the order of x	3.4
$\mathbf{O}(x)$	matrix with elements of the order of x	3.8
$p(x)$	probability density function [dimension is that of x^{-1}]	1.5
P	degree of (elliptical) polarization [–]	2.9
$\mathcal{P}$	electric polarization [C m^{-2}]	2.1
P_Q	signed degree of linear polarization [–]	2.9
$P_l(x)$	Legendre polynomials [–]	Appendix B
$P_l^m(x)$	associated Legendre functions [–]	Appendix B
$P_{mn}^s(x)$	generalized spherical functions [–]	Appendix F
Q	second Stokes parameter [W m^{-2}]	2.6
Q_{abs}	efficiency factor for absorption [–]	3.9
Q_{ext}	efficiency factor for extinction [–]	3.9

Symbol	Definition and dimension in SI units	Introduced in Section
Q_{sca}	efficiency factor for scattering [–]	3.9
$\tilde{Q}$	second element of the specific intensity column vector [W m^{-2} sr^{-1}]	8.10
r_{eff}	effective radius of a size distribution [m]	5.3
r_{max}	maximal radius of a size distribution [m]	5.3
r_{min}	minimal radius of a size distribution [m]	5.3
$\mathbf{r}$	radius (position) vector [m]	1.1
$\hat{\mathbf{r}}$	unit vector in the direction of the radius vector [–]	2.11
R	4×4 reflection matrix [–]	10.3
$\boldsymbol{\mathcal{R}}$	4×4 reflection matrix [–]	14.2
Re	real part	
$\Re^3$	entire three-dimensional space	1.1
$\boldsymbol{\mathcal{R}}^1$	4×4 single-scattering reflection matrix [–]	14.2
$\boldsymbol{\mathcal{R}}^{\text{C}}$	4×4 cyclical reflection matrix [–]	14.2
$\boldsymbol{\mathcal{R}}^{\text{M}}$	4×4 diffuse multiple-scattering reflection matrix [–]	14.2
$\mathsf{R}^{\dagger}$	4×4 reflection matrix [–]	10.3
S_{ij}	elements of the amplitude scattering matrix [m]	3.3
S	complex Poynting vector [W m^{-2}]	2.4
S	2×2 amplitude scattering matrix [m]	3.3
$\boldsymbol{\mathcal{S}}$	Poynting vector [W m^{-2}]	2.4
$\langle\boldsymbol{\mathcal{S}}\rangle_t$	time-averaged Poynting vector [W m^{-2}]	2.4
t	time [s]	1.1
$\mathcal{T}$	optical thickness of a scattering particulate layer [–]	12.1
T	4×4 transmission matrix [–]	10.3
$\ddot{T}$	dyadic transition operator [m^{-5}]	3.1
$\ddot{\mathcal{T}}$	transformation dyadic [–]	3.10
$\mathsf{T}^{\dagger}$	4×4 transmission matrix [–]	10.3
u	$-\cos\theta$ [–]	10.1
U	third Stokes parameter [W m^{-2}]	2.6
$\mathcal{U}$	electromagnetic energy density [J m^{-3}]	2.4
U	4×4 matrix describing internal radiation field [–]	10.3
$\tilde{U}$	third element of the specific intensity column vector [W m^{-2} sr^{-1}]	8.10

Symbol	Definition and dimension in SI units	Introduced in Section
$\langle \mathcal{U} \rangle_t$	time-averaged electromagnetic energy density [J m^{-3}]	2.5
$\mathbf{U}^{\dagger}$	4×4 matrix describing internal radiation field [–]	10.3
v_{eff}	effective variance of a size distribution [–]	5.3
V	fourth Stokes parameter [W m^{-2}]	2.6
V_{EXT}	exterior region [m^3]	3.1
V_{INT}	interior region [m^3]	3.1
$\tilde{V}$	fourth element of the specific intensity column vector [W m^{-2} sr^{-1}]	8.10
W	power [W]	2.4
x	size parameter [–]	3.2
x_{eff}	effective size parameter of a size distribution [–]	11.13
$\mathbf{X}$	4×4 matrix propagator [–]	10.2
$\mathbf{Z}$	4×4 Stokes phase matrix [m^2]	3.7
Z_{ij}	elements of the Stokes phase matrix [m^2]	3.7
$\tilde{Z}^{\text{CP}}_{pq}$	elements of the normalized circular-polarization phase matrix [–]	11.12
$\mathbf{Z}^{\text{CP}}$	4×4 circular-polarization phase matrix [m^2]	3.7
$\mathbf{Z}^{J}$	4×4 coherency phase matrix [m^2]	3.7
$\mathbf{Z}^{\text{MS}}$	4×4 modified Stokes phase matrix [m^2]	3.7
$\tilde{\mathbf{Z}}$	4×4 normalized Stokes phase matrix [–]	11.10
$\tilde{\mathbf{Z}}^{\text{CP}}$	4×4 normalized circular-polarization phase matrix [–]	11.12
$\langle \mathbf{Z} \rangle_{\xi}$	4×4 Stokes phase matrix per particle averaged over particle states [m^2]	6.1

Greek symbols

Symbol	Definition and dimension in SI units	Introduced in Section
α	phase angle [–]	13.1.1
α^s_j	expansion coefficients [–]	11.11
β	ellipticity angle of the polarization ellipse [rad]	2.7
β^s_j	expansion coefficients [–]	11.11
$\delta(x)$	Dirac delta function [dimension is that of x^{-1}]	3.1
$\delta(\mathbf{r})$	three-dimensional Dirac delta function [m^{-3}]	3.1
$\delta(\hat{\mathbf{n}})$	solid-angle Dirac delta function [–]	Appendix B
δ_{mn}	Kronecker delta [–]	Appendix F
δ_{C}	circular backscattering depolarization ratio [–]	11.9

Symbol	Definition and dimension in SI units	Introduced in Section
δ_L	linear backscattering depolarization ratio [–]	11.9
ΔS	area of the sensitive surface of a detector [m^2]	3.6
$\Delta\Omega$	angular aperture of a detector [sr]	3.6
$\mathbf{\Delta}$	4×4 unit matrix [–]	8.4
$\mathbf{\Delta}_{23}$	4×4 transformation matrix [–]	3.7
$\mathbf{\Delta}_3$	4×4 transformation matrix [–]	3.7
$\mathbf{\Delta}_{34}$	4×4 transformation matrix [–]	11.3
$\mathbf{\Delta}^{CP}$	4×4 transformation matrix [–]	11.12
$\mathbf{\Delta}^{MS}$	4×4 transformation matrix [–]	3.7
ε	complex permittivity [F m^{-1}]	2.3
$\in$	complex electric permittivity [F m^{-1}]	2.3
ϵ	electric permittivity [F m^{-1}]	2.1
ϵ_0	electric permittivity of free space [F m^{-1}]	2.1
ϵ_1	electric permittivity of the surrounding medium [F m^{-1}]	3.1
ζ	orientation angle of the polarization ellipse [rad]	2.7
ζ_{hp}	helicity-preserving enhancement factor [–]	14.5
ζ_{hv}	cross-polarized enhancement factor [–]	14.5
ζ_{oh}	opposite-helicity enhancement factor [–]	14.5
ζ_{vv}	co-polarized enhancement factor [–]	14.5
ζ_I	unpolarized enhancement factor [–]	14.5
$\vec{\eta}$	coherent transmission dyadic [–]	8.3
θ	polar (zenith) angle [rad]	2.6
$\hat{\boldsymbol{\theta}}$	unit vector in the θ direction [–]	2.6
Θ	scattering angle [rad]	3.9
$\vec{\kappa}$	dyadic propagation constant [m^{-1}]	8.3
$\mathbf{K}$	4×4 Stokes extinction matrix [m^2]	3.8
K_{ij}	elements of the Stokes extinction matrix [m^2]	3.8
$\mathbf{K}_e$	4×1 Stokes emission column vector [W sr^{-1} rad^{-1} s]	3.13
$\mathbf{K}^{CP}$	4×4 circular-polarization extinction matrix [m^2]	3.8
$\mathbf{K}^J$	4×4 coherency extinction matrix [m^2]	3.8
$\mathbf{K}^{MS}$	4×4 modified Stokes extinction matrix [m^2]	3.8
$\langle\mathbf{K}\rangle_\xi$	4×4 Stokes extinction matrix per particle averaged over particle states [m^2]	6.1
λ_1	wavelength in the surrounding medium [m]	3.5
μ	complex magnetic permeability [H m^{-1}]	2.3
μ	magnetic permeability [H m^{-1}]	2.1

Symbol	Definition and dimension in SI units	Introduced in Section
μ	$\lvert u \rvert$	10.1
μ_0	magnetic permeability of free space [H m^{-1}]	2.1
μ_C	circular polarization ratio [–]	14.5
μ_L	linear polarization ratio [–]	14.5
μ_C^{diff}	diffuse circular polarization ratio [–]	14.5
μ_L^{diff}	diffuse linear polarization ratio [–]	14.5
$\hat{\mu}$	couplet $\{\mu, \varphi\}$	10.1
$-\hat{\mu}$	couplet $\{-\mu, \varphi\}$	10.1
$\hat{\mu}_0^{\pi}$	couplet $\{\mu_0, \varphi + \pi\}$	14.3
π	pi [–]	
ϖ	single-scattering albedo [–]	3.9
$\mathrm{\rho}$	complex charge density [C m^{-3}]	2.3
ρ	charge density [C m^{-3}]	2.1
$\ddot{\rho}$	coherency dyad [V^2 m^{-2}]	2.12
$\boldsymbol{\rho}$	2×2 coherency (density) matrix [W m^{-2}]	2.6
$\boldsymbol{\rho}_c$	2×2 coherent coherency matrix [W m^{-2}]	8.8
$\tilde{\boldsymbol{\rho}}_d$	2×2 diffuse specific coherency matrix [W m^{-2} sr^{-1}]	8.8
$\mathrm{\sigma}$	complex conductivity [Ω^{-1} m^{-1}]	2.3
σ	conductivity [Ω^{-1} m^{-1}]	2.1
$\ddot{\Sigma}$	specific coherency dyadic [V^2 m^{-2} sr^{-1}]	1.7
$\ddot{\Sigma}_1$	single-scattering specific coherency dyadic [V^2 m^{-2} sr^{-1}]	14.1
$\ddot{\Sigma}_c$	coherent specific coherency dyadic [V^2 m^{-2} sr^{-1}]	14.1
$\ddot{\Sigma}_d$	diffuse specific coherency dyadic [V^2 m^{-2} sr^{-1}]	8.7
$\ddot{\Sigma}_C$	cyclical specific coherency dyadic [V^2 m^{-2} sr^{-1}]	14.1
$\ddot{\Sigma}_L$	ladder specific coherency dyadic [V^2 m^{-2} sr^{-1}]	8.6
$\ddot{\Sigma}_M$	diffuse multiple-scattering specific coherency dyadic [V^2 m^{-2} sr^{-1}]	14.1
τ	optical depth [–]	12.1
φ	azimuth angle [rad]	2.6
$\hat{\boldsymbol{\varphi}}$	unit vector in the φ direction [–]	2.6
χ	electric susceptibility [–]	2.1
ψ	particle depth [m^{-2}]	10.1
Ψ	particle thickness of a scattering particulate layer [m^{-2}]	10.1
ω	angular frequency [rad s^{-1}]	1.1

Symbol	Definition and dimension in SI units	Introduced in Section
Miscellaneous symbols		
$\mathbf{a}\cdot\mathbf{b}$	dot (scalar) product of vectors **a** and **b**	
$\mathbf{a}\times\mathbf{b}$	vector product of vectors **a** and **b**	
$\mathbf{a}\otimes\mathbf{b}$	dyadic product of vectors **a** and **b**	1.7, Appendix A
(a, b)	open interval $a < x < b$	
$[a, b]$	closed interval $a \le x \le b$	
$[a, b)$	semi-open interval $a \le x < b$	
A^{-1}	inverse of matrix A	2.6
A^{T}	transpose of matrix A	2.7
$\ddot{B}^{\mathrm{T}}$	transpose of dyad(ic) $\ddot{B}$	Appendix A
$\mathrm{diag}[a, b]$	diagonal matrix with diagonal elements a and b	
$\exp\mathsf{B}$	matrix exponential	8.3
$\exp\ddot{B}$	dyadic exponential	8.3
$\hat{\mathbf{n}}$	unit vector [–]	1.7
x^*	complex-conjugate value of x	2.3
$\lvert x\rvert$	absolute value of x	
$\langle x\rangle_t$	average of x over time	1.5
$\langle x\rangle_\xi$	average of x over particle states	6.1
$\langle x\rangle_{\mathbf{R}}$	average of x over particle coordinates	6.1
$\ddot{0}$	zero dyad	3.1
$\mathbf{0}$	zero vector	2.2
$\mathsf{0}$	zero matrix	8.14
∇	gradient [m^{-1}]	
$\in$	element of	
$\cup$	union of sets	

References

Abramowitz, M., and Stegun, I. A., eds. (1964). *Handbook of Mathematical Functions with Formulas, Graphs and Mathematical Tables* (National Bureau of Standards, Washington, DC).

Adams, C. N., and Kattawar, G. W. (1970). Solutions of the equations of radiative transfer by an invariant imbedding approach. *J. Quant. Spectrosc. Radiat. Transfer* **10**, 341–356.

Akhiezer, A. I., and Berestetskii, V. B. (1965). *Quantum Electrodynamics* (Wiley Interscience, New York).

Akkermans, E., Wolf, P. E., Maynard, R., and Maret, G. (1988). Theoretical study of the coherent backscattering of light by disordered media. *J. Phys. (France)* **49**, 77–98.

Albrecht, H.-E., Damaschke, N., Borys, M., and Tropea, C. (2003). *Laser Doppler and Phase Doppler Measurement Techniques* (Springer-Verlag, Berlin).

Ambarzumian, V. A. (1943). Diffuse reflection of light by a foggy medium. *Comptes Rendus (Doklady) l'Acad. Sci. l'URSS* **38**, 229–232.

Ambirajan, A., and Look, D. C. (1997). A backward Monte Carlo study of the multiple scattering of a polarized laser beam. *J. Quant. Spectrosc. Radiat. Transfer* **58**, 171–192.

Amic, E., Luck, J. M., and Nieuwenhuizen, Th. M. (1996). Anisotropic multiple scattering in diffusive media. *J. Phys. A: Math. Gen.* **29**, 4915–4955.

Amic, E., Luck, J. M., and Nieuwenhuizen, Th. M. (1997). Multiple Rayleigh scattering of electromagnetic waves. *J. Phys. I (France)* **7**, 445–483.

Aoki, T., Aoki, T., Fukabori, M., and Uchiyama, A. (1999). Numerical simulation of the atmospheric effects on snow albedo with a multiple scattering radiative transfer model for the atmosphere-snow system. *J. Meteorol. Soc. Japan* **77**, 595–614.

Apresyan, L. A., and Kravtsov, Yu. A. (1996). *Radiation Transfer. Statistical and Wave Aspects* (Gordon and Breach, Basel).

Arfken, G. B., and Weber, H. J. (2001). *Mathematical Methods for Physicists* (Academic

Press, San Diego).

Arnott, W. P., Dong, Y. Y., Hallett, J., and Poellot, M. R. (1994). Role of small ice crystals in radiative properties of cirrus: a case study, FIRE II, November 22, 1991. *J. Geophys. Res.* **99**, 1371–1381.

Arons, A. B., and Peppard, M. B. (1965). Einstein's proposal of the photon concept – a translation of the *Annalen der Physik* paper of 1905. *Am. J. Phys.* **33**, 367–374.

Asano, S., and Yamamoto, G. (1975). Light scattering by a spheroidal particle. *Appl. Opt.* **14**, 29–49.

Asrar, G., ed. (1989). *Theory and Applications of Optical Remote Sensing* (John Wiley & Sons, New York).

Azzam, R. M. A., and Bashara, N. M. (1977). *Ellipsometry and Polarized Light* (North Holland, Amsterdam).

Babenko, V. A., Astafyeva, L. G., and Kuzmin, V. N. (2003). *Electromagnetic Scattering in Disperse Media: Inhomogeneous and Anisotropic Particles* (Praxis Publishing, Chichester, UK).

Bangs, L. B., and Meza, M. (1995). Microspheres. 1: Selection, cleaning, and characterization. *IVD Technol.* **17**(3), 18–26.

Barabanenkov, Yu. N. (1973). Wave corrections to the transfer equation for "back" scattering. *Radiophys. Quantum Electron.* **16**, 65–71.

Barabanenkov, Yu. N. (1975). Multiple scattering of waves by ensembles of particles and the theory of radiation transport. *Sov. Phys.–Usp.* **18**, 673–689.

Barabanenkov, Yu. N., and Ozrin, V. D. (1991). Diffusion approximation in the theory of weak localization of radiation in a discrete random medium. *Radio Sci.* **26**, 747–750.

Barabanenkov, Yu. N., Vinogradov, A. G., Kravtsov, Yu. A., and Tatarskii, V. I. (1972). Application of the theory of multiple scattering of waves to the derivation of the radiation transfer equation for a statistically inhomogeneous medium. *Radiophys. Quantum Electron.* **15**, 1420–1425.

Barabanenkov, Yu. N., Kravtsov, Yu. A., Ozrin, V. D., and Saichev, A. I. (1991). Enhanced backscattering in optics. *Progr. Opt.* **29**, 65–197.

Barber, P. W., and Hill, S. C. (1990). *Light Scattering by Particles: Computational Methods* (World Scientific, Singapore).

Barker, H. W., Goldstein, R. K., and Stevens, D. E. (2003). Monte Carlo simulation of solar reflectances for cloudy atmospheres. *J. Atmos. Sci.* **60**, 1881–1894.

Bartel, S., and Hielscher, A. H. (2000). Monte Carlo simulations of the diffuse backscattering Mueller matrix for highly scattering media. *Appl. Opt.* **39**, 1580–1588.

Battaglia, A., and Mantovani, S. (2005). Forward Monte Carlo computations of fully polarized microwave radiation in non-isotropic media. *J. Quant. Spectrosc. Radiat. Transfer* **95**, 285–308.

Bayvel, L. P., and Jones, A. R. (1981). *Electromagnetic Scattering and its Applications* (Applied Science Publishers, London).

Berne, B. J., and Pecora, R. (1976). *Dynamic Light Scattering with Applications to Chemistry, Biology, and Physics* (John Wiley & Sons, New York). (Also: Dover Publications, Mineola, NY, 2000.)

Biedenharn, L. C., and Louck, J. D. (1981). *Angular Momentum in Quantum Physics: Theory and Application* (Addison-Wesley, Reading, MA).

Birkhoff, G., and Rota, G.-C. (1969). *Ordinary Differential Equations* (John Wiley & Sons, New York).

Bohm, D. (1951). *Quantum Theory* (Prentice-Hall, Englewood Cliffs, NJ).

Bohren, C. F., and Huffman, D. R. (1983). *Absorption and Scattering of Light by Small Particles* (John Wiley & Sons, New York).

Borghese, F., Denti, P., and Saija, R. (2003). *Scattering from Model Nonspherical Particles. Theory and Applications to Environmental Physics* (Springer-Verlag, Berlin).

Born, M., and Wolf, E. (1999). *Principles of Optics* (Cambridge University Press, Cambridge).

Borovoy, A. G. (1966). Method of iterations in multiple scattering: the transfer equation. *Izv. Vuzov Fizika*, No. 6, 50–54.

Bouma, B. E., and Tearney, G. J., eds. (2002). *Handbook of Optical Coherence Tomography* (Marcel Dekker, New York).

Bréon, F.-M., and Goloub, P. (1998). Cloud droplet effective radius from spaceborne polarization measurements. *Geophys. Res. Lett.* **25**, 1879–1882.

Brink, D. M., and Satchler, G. R. (1993). *Angular Momentum* (Oxford University Press, Oxford).

Brogniez, C., Santer, R., Diallo, B. S., *et al.* (1992). Comparative observations of stratospheric aerosols by ground-based lidar, balloon-borne polarimeter, and satellite solar occultation. *J. Geophys. Res.* **97**, 20 805–20 823.

Brosseau, C. (1998). *Polarized Light: A Statistical Optics Approach* (John Wiley & Sons, New York).

Brown, W., ed. (1993). *Dynamic Light Scattering: The Method and Some Applications* (Oxford University Press, Oxford).

Brown, W., ed. (1996). *Light Scattering: Principles and Development* (Oxford University Press, Oxford).

Bruning, J. H., and Lo, Y. T. (1971a). Multiple scattering of EM waves by spheres. I. Multipole expansion and ray-optical solutions. *IEEE Trans. Antennas Propag.* **19**, 378–390.

Bruning, J. H., and Lo, Y. T. (1971b). Multiple scattering of EM waves by spheres. II. Numerical and experimental results. *IEEE Trans. Antennas Propag.* **19**, 391–400.

Bruscaglioni, P., Zaccanti, G., and Wei, Q. (1993). Transmission of a pulsed polarized light beam through thick turbid media: numerical results. *Appl. Opt.* **32**, 6142–6150.

Buriez, J. C., Vanbauce, C., Parol, F., *et al.* (1997). Cloud detection and derivation of cloud properties from POLDER. *Int. J. Remote Sens.* **18**, 2785–2813.

Cairns, B., Lacis, A. A., and Carlson, B. E. (2000). Absorption within inhomogeneous clouds and its parameterization in general circulation models. *J. Atmos. Sci.* **57**, 700–714.

Case, K. M., and Zweifel, P. F. (1967). *Linear Transport Theory* (Addison-Wesley, Reading, MA).

Chabanov, A. A., and Genack, A. Z. (2003). Photon localization in resonant media. In *Wave Scattering in Complex Media: From Theory to Applications*, eds. B. A. van Tiggelen and S. E. Skipetrov, pp. 203–212 (Kluwer Academic Publishers, Dordrecht, The Netherlands).

Chamaillard, K., Jennings, S. G., Kleefeld, C., *et al.* (2003). Light backscattering and scattering by nonspherical sea-salt aerosols. *J. Quant. Spectrosc. Radiat. Transfer* **79–80**, 577–597.

Chandrasekhar, S. (1947a). On the radiative equilibrium of a stellar atmosphere. XV. *Astrophys. J.* **105**, 424–434.

Chandrasekhar, S. (1947b). On the radiative equilibrium of a stellar atmosphere. XVII. *Astrophys. J.* **105**, 441–460.

Chandrasekhar, S. (1950). *Radiative Transfer* (Oxford University Press, Oxford). (Also: Dover Publications, New York, 1960.)

Chang, P. C. Y., Hopcraft, K. I., Jakeman, E., and Walker, J. G. (2002). Optimum configuration for polarization photon correlation spectroscopy. *Meas. Sci. Technol.* **13**, 341–348.

Chowdhary, J., Cairns, B., Mishchenko, M., and Travis, L. (2001). Retrieval of aerosol properties over the ocean using multispectral and multiangle photopolarimetric measurements from the Research Scanning Polarimeter. *Geophys. Res. Lett.* **28**, 243–246.

Chowdhary, J., Cairns, B., and Travis, L. D. (2002). Case studies of aerosol retrievals over the ocean from multiangle, multispectral photopolarimetric remote sensing data. *J. Atmos. Sci.* **59**, 383–397.

Chowdhary, J., Cairns, B., Mishchenko, M. I., *et al.* (2005). Retrieval of aerosol scattering and absorption properties from photopolarimetric observations over the ocean during the CLAMS experiment. *J. Atmos. Sci.* **62**, 1093–1117.

Chu, B. (1991). *Laser Light Scattering. Basic Principles and Practice* (Academic Press, Boston).

Chwolson, O. (1889). Grundzüge einer mathematischen Theorie der inneren Diffusion des Lichtes. *Bull. l'Acad. Impériale Sci. St. Pétersbourg* **33**, 221–256.

Ciric, I. R., and Cooray, F. R. (2000). Separation of variables for electromagnetic scattering by spheroidal particles. In *Light Scattering by Nonspherical Particles: Theory, Measurements, and Applications*, eds. M. I. Mishchenko, J. W. Hovenier, and L. D. Travis, pp. 89–130 (Academic Press, San Diego).

Clarke, D., and Grainger, J. F. (1971). *Polarized Light and Optical Measurement* (Pergamon Press, Oxford).

Cloude, S. R., and Pottier, E. (1996). A review of target decomposition theorems in radar polarimetry. *IEEE Trans. Geosci. Remote Sens.* **34**, 498–518.

Collett, E. (1992). *Polarized Light: Fundamentals and Applications* (Marcel Dekker, New York).

Colton, D., and Kress, R. (1998). *Inverse Acoustic and Electromagnetic Scattering Theory* (Springer-Verlag, Berlin).

Cooray, M. F. R., and Ciric, I. R. (1992). Scattering of electromagnetic waves by a coated dielectric spheroid. *J. Electromagn. Waves Appl.* **6**, 1491–1507.

Coulson, K. L. (1988). *Polarization and Intensity of Light in the Atmosphere* (A. Deepak Publishing, Hampton, VA).

Crease, R. P. (2004). The greatest equations ever. *Phys. World* **17**(10), 14–15.

Crosignani, B., Di Porto, P., and Bertolotti, M. (1975). *Statistical Properties of Scattered Light* (Academic Press, New York).

Cummins, H. Z., and Pike, E. R., eds. (1974). *Photon Correlation and Light Beating Spec-*

troscopy (Plenum Press, New York).
Cummins, H. Z., and Pike, E. R., eds. (1977). *Photon Correlation Spectroscopy and Velocimetry* (Plenum Press, New York).
Dainty, J. C., ed. (1984). *Laser Speckle and Related Phenomena* (Springer-Verlag, Berlin).
Davis, E. J., and Schweiger, G. (2002). *The Airborne Microparticle: Its Physics, Chemistry, Optics, and Transport Phenomena* (Springer-Verlag, Berlin).
Davison, B. (1958). *Neutron Transport Theory* (Oxford University Press, London).
de Boer, J. F., and Milner, T. E. (2002). Review of polarization sensitive optical coherence tomography and Stokes vector determination. *J. Biomed. Opt.* **7**, 359–371.
de Haan, J. F., Bosma, P. B., and Hovenier, J. W. (1987). The adding method for multiple scattering calculations of polarized light. *Astron. Astrophys.* **183**, 371–391.
Deirmendjian, D. (1969). *Electromagnetic Scattering on Spherical Polydispersions* (Elsevier, New York).
de Rooij, W. A. (1985). Reflection and transmission of polarized light by planetary atmospheres. Ph. D. dissertation, Free University, Amsterdam.
de Rooij, W. A., and Domke, H. (1984). On the nonuniqueness of solutions for nonlinear integral equations in radiative transfer theory. *J. Quant. Spectrosc. Radiat. Transfer* **31**, 285–299.
de Rooij, W. A., and van der Stap, C. C. A. H. (1984). Expansion of Mie scattering matrices in generalized spherical functions. *Astron. Astrophys.* **131**, 237–248.
Deuzé, J. L., Goloub, P., Herman, M., *et al.* (2000). Estimate of the aerosol properties over the ocean with POLDER. *J. Geophys. Res.* **105**, 15 329–15 346.
Diner, D. J., and Martonchik, J. V. (1984). Atmospheric transfer of radiation above an inhomogeneous non-Lambertian reflective ground. I. Theory. *J. Quant. Spectrosc. Radiat. Transfer* **31**, 97–125.
Dlugach, J. M., and Mishchenko, M. I. (2006). Enhanced backscattering of polarized light: effect of particle nonsphericity on the helicity-preserving enhancement factor. *J. Quant. Spectrosc. Radiat. Transfer* (in press).
Dlugach, J. M., and Yanovitskij, E. G. (1974). The optical properties of Venus and the Jovian planets. II. Methods and results of calculations of the intensity of radiation diffusely reflected from semi-infinite homogeneous atmospheres. *Icarus* **22**, 66–81.
Dogariu, A., and Boreman, G. D. (1996). Enhanced backscattering in a converging-beam configuration. *Opt. Lett.* **21**, 1718–1720.
Doicu, A., Eremin, Yu., and Wriedt, T. (2000). *Acoustic and Electromagnetic Scattering Analysis Using Discrete Sources* (Academic Press, San Diego).
Dolginov, A. Z., Gnedin, Yu. N., and Silant'ev, N. A. (1970). Photon polarization and frequency change in multiple scattering. *J. Quant. Spectrosc. Radiat. Transfer* **10**, 707–754 (1970).
Dolginov, A. Z., Gnedin, Yu. N., and Silant'ev, N. A. (1995). *Propagation and Polarization of Radiation in Cosmic Media* (Gordon and Breach, Basel).
Domke, H. (1974). The expansion of scattering matrices for an isotropic medium in generalized spherical functions. *Astrophys. Space Sci.* **29**, 379–386.
Domke, H., and Yanovitskij, E. G. (1986). Principles of invariance applied to the computation

of internal polarized radiation in multilayered atmospheres. *J. Quant. Spectrosc. Radiat. Transfer* **36**, 175–186 (1986).

Draine, B. T. (2000). The discrete dipole approximation for light scattering by irregular targets. In *Light Scattering by Nonspherical Particles: Theory, Measurements, and Applications*, eds. M. I. Mishchenko, J. W. Hovenier, and L. D. Travis, pp. 131–145 (Academic Press, San Diego).

Dubovik, O., Holben, B. N., Lapyonok, T., *et al.* (2002). Non-spherical aerosol retrieval method employing light scattering by spheroids. *Geophys. Res. Lett.* **29**, 10.1029/2001GL014506.

Ebert, M., Weinbruch, S., Rausch, A., *et al.* (2002). Complex refractive index of aerosols during LACE 98 as derived from the analysis of individual particles. *J. Geophys. Res.* **107**, 8121.

Edmonds, A. R. (1996). *Angular Momentum in Quantum Mechanics* (Princeton University Press, Princeton, NJ).

Elfouhaily, T. M., and Guérin, C.-A. (2004). A critical survey of approximate scattering wave theories from random rough surfaces. *Waves Random Media* **14**, R1–R40.

Emde, C., Buehler, S. A., Davis, C., *et al.* (2004). A polarized discrete ordinate scattering model for simulations of limb and nadir long-wave measurements in 1-D/3-D spherical atmospheres. *J. Geophys. Res.* **109**, D24207.

Ershov, Yu. I., and Shikhov, S. B. (1985). *Mathematical Foundations of the Transfer Theory*, Vols. 1 and 2 (Energoatomizdat, Moscow) (in Russian).

Etemad, S., Thompson, R., and Andrejco, M. J. (1986). Weak localization of photons: universal fluctuations and ensemble averaging. *Phys. Rev. Lett.* **57**, 575–578.

Etemad, S., Thompson, R., Andrejco, M. J., *et al.* (1987). Weak localization of photons: termination of coherent random walks by absorption and confined geometry. *Phys. Rev. Lett.* **59**, 1420–1423.

Evans, K. F. (1993). Two-dimensional radiative transfer in cloudy atmospheres: the spherical harmonic spatial grid method. *J. Atmos. Sci.* **50**, 3111–3124.

Fabelinskii, I. L. (1968). *Molecular Scattering of Light* (Plenum Press, New York).

Fante, R. L. (1981). Relationship between radiative-transport theory and Maxwell's equations in dielectric media. *J. Opt. Soc. Am.* **71**, 460–468.

Farafonov, V. G., Voshchinnikov, N. V., and Somsikov, V. V. (1996). Light scattering by a core-mantle spheroidal particle. *Appl. Opt.* **35**, 5412–5426.

Farquhar, I. E. (1964). *Ergodic Theory in Statistical Mechanics* (John Wiley & Sons, London).

Fearn, H., and Lamb, W. E., Jr. (1991). Corrections to the golden rule. *Phys. Rev. A* **43**, 2124–2128.

Fercher, A. F., Drexler, W., Hitzenberger, C. K., and Lasser, T. (2003). Optical coherence tomography – principles and applications. *Rep. Progr. Phys.* **66**, 239–303.

Fernández, J. E., Hubbell, J. H., Hanson, A. L., and Spencer, L. V. (1993). Polarization effects of multiple scattering gamma transport. *Radiat. Phys. Chem.* **41**, 579–630.

Flatau, P. J., and Stephens, G. L. (1988). On the fundamental solution of the radiative transfer equation. *J. Geophys. Res.* **93**, 11 037–11 050.

Foldy, L. L. (1945). The multiple scattering of waves. *Phys. Rev.* **67**, 107–119.

Fouque, J.-P., ed. (1999). *Diffuse Waves in Complex Media* (Kluwer Academic Publishers, Dordrecht, The Netherlands).

Franklin, F. A., and Cook, A. F. (1965). Optical properties of Saturn's rings. II. Two-color phase curves of the two bright rings. *Astron. J.* **70**, 704–720.

Frazer, R. A., Duncan, W. J., and Collar, A. R. (1957). *Elementary Matrices and Some Applications to Dynamics and Differential Equations* (Cambridge University Press, Cambridge).

Fuller, K. A. (1995). Scattering and absorption cross sections of compounded spheres. III. Spheres containing arbitrarily located spherical inhomogeneities. *J. Opt. Soc. Am. A* **12**, 893–904.

Fuller, K. A., and Mackowski, D. W. (2000). Electromagnetic scattering by compounded spherical particles. In *Light Scattering by Nonspherical Particles: Theory, Measurements, and Applications*, eds. M. I. Mishchenko, J. W. Hovenier, and L. D. Travis, pp. 225–272 (Academic Press, San Diego).

Fung, A. K. (1994). *Microwave Scattering and Emission Models and Their Applications* (Artech House, Boston).

Gans, R. (1924). Die Farbe des Meeres. *Ann. Phys.* **75**, 1–22.

Garcia, R. D. M., and Siewert, C. E. (1986). A generalized spherical harmonics solution for radiative transfer models that include polarization effects. *J. Quant. Spectrosc. Radiat. Transfer* **36**, 401–423. (Errata: **40**, 83–84, 1988.)

Garcia, R. D. M., and Siewert, C. E. (1989). The F_N method for radiative transfer models that include polarization effects. *J. Quant. Spectrosc. Radiat. Transfer* **41**, 117–145.

Garg, R., Prud'homme, R. K., Aksay, I. A., *et al.* (1998). Optical transmission in highly concentrated dispersions. *J. Opt. Soc. Am. A* **15**, 932–935.

Gehrels, T., ed. (1974). *Planets, Stars and Nebulae Studied with Photopolarimetry* (University of Arizona Press, Tucson, AZ).

Gelfand, I. M., Minlos, R. A., and Shapiro, Z. Ya. (1963). *Representations of the Rotation and Lorentz Groups and their Applications* (Pergamon Press, New York). (Original Russian edition: Gosudarstvennoe Izdatel'stvo Fiziko-Matematicheskoy Literatury, Moscow, 1958.)

Genack, A. Z., and Garcia, N. (1991). Observation of photon localization in a three-dimensional disordered system. *Phys. Rev. Lett.* **66**, 2064–2067.

Germogenova, T. A. (1986). *Local Properties of Solutions of the Transfer Equation* (Nauka, Moscow) (in Russian).

Goguen, J. D., Hammel, H. B., and Brown, R. H. (1989). V photometry of Titania, Oberon, and Triton. *Icarus* **77**, 239–247.

Golubentsev, A. A. (1984). Interference correction to the albedo of a strongly gyrotropic medium with random inhomogeneities. *Radiophys. Quantum Electron.* **27**, 506–516.

Goody, R. M., and Yung, Y. L. (1989). *Atmospheric Radiation: Theoretical Basis* (Oxford University Press, New York).

Gorodnichev, E. E., Dudarev, S. L., and Rogozkin, D. B. (1990). Coherent wave backscattering by random medium. Exact solution of the albedo problem. *Phys. Lett. A* **144**, 48–54.

Grant, I. P., and Hunt, G. E. (1969). Discrete space theory of radiative transfer. I. Fundamentals. *Proc. Roy. Soc. London A* **313**, 183–197.

Greenberg, W., van der Mee, C., and Protopopescu, V. (1987). *Boundary Value Problems in Abstract Kinetic Theory* (Birkhäuser Verlag, Basel).

Gross, K. I. (1962). Discussion of an iterative solution of an equation of radiative transfer. *J. Math. & Phys.* **41**, 53–61.

Gustafson, B. Å. S. (2000). Microwave analog to light-scattering measurements. In *Light Scattering by Nonspherical Particles: Theory, Measurements, and Applications*, eds. M. I. Mishchenko, J. W. Hovenier, and L. D. Travis, pp. 367–390 (Academic Press, San Diego).

Guzzi, R., ed. (2003). *Exploring the Atmosphere by Remote Sensing Techniques* (Springer-Verlag, Berlin).

Haferman, J. L., Smith, T. F., and Krajewski, W. F. (1997). A multi-dimensional discrete-ordinates method for polarized radiative transfer. I. Validation for randomly oriented axisymmetric particles. *J. Quant. Spectrosc. Radiat. Transfer* **58**, 379–398. (Corrigendum: **60**, I (1998).)

Hanel, R. A., Conrath, B. J., Jennings, D. E., and Samuelson, R. E. (2003). *Exploration of the Solar System by Infrared Remote Sensing* (Cambridge University Press, Cambridge).

Hansen, J. E. (1971a). Multiple scattering of polarized light in planetary atmospheres. II. Sunlight reflected by terrestrial water clouds. *J. Atmos. Sci.* **28**, 1400–1426.

Hansen, J. E. (1971b). Circular polarization of sunlight reflected by clouds. *J. Atmos. Sci.* **28**, 1515–1516.

Hansen, J. E., and Hovenier, J. W. (1974). Interpretation of the polarization of Venus. *J. Atmos. Sci.* **31**, 1137–1160.

Hansen, J. E., and Travis, L. D. (1974). Light scattering in planetary atmospheres. *Space Sci. Rev.* **16**, 527–610.

Hapke, B. (1990). Coherent backscatter and the radar characteristics of outer planet satellites. *Icarus* **88**, 407–417.

Harmon, J. K., Slade, M. A., Vélez, R. A., *et al.* (1994). Radar mapping of Mercury's polar anomalies. *Nature* **369**, 213–215.

Harris, A. W., Young, J. W., Contreiras, L., *et al.* (1989). Phase relations of high albedo asteroids: the unusual opposition brightening of 44 Nysa and 64 Angelina. *Icarus* **81**, 365–374.

Hasekamp, O. P., and Landgraf, J. (2005). Linearization of vector radiative transfer with respect to aerosol properties and its use in satellite remote sensing. *J. Geophys. Res.* **110**, D04203.

Hasekamp, O. P., Landgraf, J., and van Oss, R. (2002). The need of polarization modeling for ozone profile retrieval from backscattered sunlight. *J. Geophys. Res.* **107**, 4692.

Helfenstein, P., Currier, N., Clark, B. E., *et al.* (1998). Galileo observations of Europa's opposition effect. *Icarus* **135**, 41–63.

Hespel, L., Mainguy, S., and Greffet, J.-J. (2003). Radiative properties of scattering and absorbing dense media: theory and experimental study. *J. Quant. Spectrosc. Radiat. Transfer* **77**, 193–210.

Holt, A. R. (1982). The scattering of electromagnetic waves by single hydrometeors. *Radio Sci.* **17**, 929–945.

Holt, A. R., Uzunoglu, N. K., and Evans, B. G. (1978). An integral equation solution to the scattering of electromagnetic radiation by dielectric spheroids and ellipsoids. *IEEE Trans. Antennas Propag.* **26**, 706–712.

Hopcraft, K. I., Chang, P. C. Y., Walker, J. G., and Jakeman, E. (2000). Properties of a polarized light-beam multiply scattered by a Rayleigh medium. In *Light Scattering from Microstructures*, eds. F. Moreno and F. González, pp. 135–158 (Springer-Verlag, Berlin).

Hopcraft, K. I., Chang, P. C. Y., Jakeman, E., and Walker, J. G. (2004). Polarization fluctuation spectroscopy. In *Photopolarimetry in Remote Sensing*, eds. G. Videen, Ya. Yatskiv, and M. Mishchenko, pp. 137–174 (Kluwer Academic Publishers, Dordrecht, The Netherlands).

Hovenier, J. W. (1969). Symmetry relationships for scattering of polarized light in a slab of randomly oriented particles. *J. Atmos. Sci.* **26**, 488–499.

Hovenier, J. W., ed. (1996). Special issue on light scattering by non-spherical particles. *J. Quant. Spectrosc. Radiat. Transfer* **55**, 535–694.

Hovenier, J. W. (2000). Measuring scattering matrices of small particles at optical wavelengths. In *Light Scattering by Nonspherical Particles: Theory, Measurements, and Applications*, eds. M. I. Mishchenko, J. W. Hovenier, and L. D. Travis, pp. 355–365 (Academic Press, San Diego).

Hovenier, J. W., and de Haan, J. F. (1985). Polarized light in planetary atmospheres for perpendicular directions. *Astron. Astrophys.* **146**, 185–191.

Hovenier, J. W., and Mackowski, D. W. (1998). Symmetry relations for forward and backward scattering by randomly oriented particles. *J. Quant. Spectrosc. Radiat. Transfer* **60**, 483–492.

Hovenier, J. W., and Stam, D. M. (2006). A peculiar discontinuity in the intensity of light reflected by a plane-parallel atmosphere. *J. Quant. Spectrosc. Radiat. Transfer* (in press).

Hovenier, J. W., and van der Mee, C. V. M. (1983). Fundamental relationships relevant to the transfer of polarized light in a scattering atmosphere. *Astron. Astrophys.* **128**, 1–16.

Hovenier, J. W., and van der Mee, C. V. M. (1996). Testing scattering matrices: A compendium of recipes. *J. Quant. Spectrosc. Radiat. Transfer* **55**, 649–661.

Hovenier, J. W., and van der Mee, C. V. M. (2000). Basic relationships for matrices describing scattering by small particles. In *Light Scattering by Nonspherical Particles: Theory, Measurements, and Applications*, eds. M. I. Mishchenko, J. W. Hovenier, and L. D. Travis, pp. 61–85 (Academic Press, San Diego).

Hovenier, J. W., van de Hulst, H. C., and van der Mee, C. V. M. (1986). Conditions for the elements of the scattering matrix. *Astron. Astrophys.* **157**, 301–310.

Hovenier, J. W., van der Mee, C., and Domke, H. (2004). *Transfer of Polarized Light in Planetary Atmospheres—Basic Concepts and Practical Methods* (Kluwer Academic Publishers, Dordrecht, The Netherlands).

Huang, J., Eradat, N., Raikh, M. E., *et al.* (2001). Anomalous coherent backscattering of light from opal photonic crystals. *Phys. Rev. Lett.* **86**, 4815–4818.

Huffman, D. R. (1988). The applicability of bulk optical constants to small particles. In *Optical Effects Associated with Small Particles*, eds. P. W. Barber and R. K. Chang, pp. 279–324 (World Scientific, Singapore).

Hunt, G. E. (1971). A review of computational techniques for analyzing the transfer of radiation through a model cloudy atmosphere. *J. Quant. Spectrosc. Radiat. Transfer* **11**, 655–690.

Hunt, A. J., and Huffman, D. R. (1973). A new polarization-modulated light scattering instrument. *Rev. Sci. Instrum.* **44**, 1753–1762.

Ishimaru, A. (1978). *Wave Propagation and Scattering in Random Media* (Academic Press, New York). (Also: IEEE Press, New York, 1997.)

Ishimaru, A., Lesselier, D., and Yeh, C. (1984). Multiple scattering calculations for nonspherical particles based on the vector radiative transfer theory. *Radio Sci.* **19**, 1356–1366.

Ishimoto, H., and Masuda, K. (2002). A Monte Carlo approach for the calculation of polarized light: application to an incident narrow beam. *J. Quant. Spectrosc. Radiat. Transfer* **72**, 467–483.

Ivanov, V. V. (1973). *Transfer of Radiation in Spectral Lines* (National Bureau of Standards, Washington, DC).

Ivanov, V. V. (1994). Making of radiative transfer theory. *Trudy (Proc.) Astron. Observ. St. Petersburg Univ.* **44**, 6–29 (in Russian).

Ivanov, A. P., Khairullina, A. Ya., and Kharkova, T. N. (1970). Experimental detection of cooperative effects in a scattering volume. *Opt. Spectrosc.* **28**, 204–207.

Iwai, T., Furukawa, H., and Asakura, T. (1995). Numerical analysis on enhanced backscattering of light based on Rayleigh–Debye scattering theory. *Opt. Rev.* **2**, 413–419.

Jackson, J. D. (1998). *Classical Electrodynamics* (John Wiley & Sons, New York).

Jaillon, F., and Saint-Jalmes, H. (2003). Description and time reduction of a Monte Carlo code to simulate propagation of polarized light through scattering media. *Appl. Opt.* **42**, 3290–3296.

Jakeman, E. (2000). Polarization fluctuations in light scattered by small particles. In *Light Scattering from Microstructures*, eds. F. Moreno and F. González, pp. 179–189 (Springer-Verlag, Berlin).

Janssen, M. A., ed. (1993). *Atmospheric Remote Sensing by Microwave Radiometry* (John Wiley & Sons, New York).

Jin, Y.-Q. (1993). *Electromagnetic Scattering Modelling for Quantitative Remote Sensing* (World Scientific, Singapore).

Jin, J. (2002). *The Finite Element Method in Electromagnetics* (John Wiley & Sons, New York).

Jonckheere, T., Müller, C. A., Kaiser, R., *et al.* (2000). Multiple scattering of light by atoms in the weak localization regime. *Phys. Rev. Lett.* **85**, 4269–4272.

Joosten, J. G. H., Geladé, E. T. F., and Pusey, P. N. (1990). Dynamic light scattering by nonergodic media: Brownian particles trapped in polyacrylamide gels. *Phys. Rev. A* **42**, 2161–2175.

Kahnert, F. M. (2003). Numerical methods in electromagnetic scattering theory. *J. Quant. Spectrosc. Radiat. Transfer* **79–80**, 775–824.

Kaper, H. G., Lekkerkerker, C. G., and Hejtmanek, J. (1982). *Spectral Methods in Linear Transport Theory* (Birkhäuser Verlag, Basel).

Kattawar, G. W., and Adams, C. N. (1990). Errors in radiance calculations induced by using scalar rather than Stokes vector theory in a realistic atmosphere–ocean system. *Proc. SPIE* **1302**, 2–12.

Kaveh, M., Rosenbluh, M., Edrei, I., and Freund, I. (1986). Weak localization and light scattering from disordered solids. *Phys. Rev. Lett.* **57**, 2049–2052.

Kerker, M. (1969). *The Scattering of Light and Other Electromagnetic Radiation* (Academic

Press, New York).

Khinchin, A. I. (1949). *Mathematical Foundations of Statistical Mechanics* (Dover Publications, New York).

Khlebtsov, N. G. (1992). Orientational averaging of light-scattering observables in the *T*-matrix approach. *Appl. Opt.* **31**, 5359–5365.

Kidd, R., Ardini, J., and Anton, A. (1989). Evolution of the modern photon. *Am. J. Phys.* **57**, 27–35.

Kim, Y. L., Liu, Y., Wali, R. K., *et al.* (2005). Low-coherent backscattering spectroscopy for tissue characterization. *Appl. Opt.* **44**, 366–377.

King, M. D., Kaufman, Y. J., Menzel, W. P., and Tanré, D. (1992). Remote sensing of cloud, aerosol, and water vapor properties from the Moderate Resolution Imaging Spectrometer (MODIS). *IEEE Trans. Geosci. Remote Sens.* **30**, 2–27.

King, M. D., Kaufman, Y. J., Tanré, D., and Nakajima, T. (1999). Remote sensing of tropospheric aerosols from space: past, present, and future. *Bull. Am. Meteorol. Soc.* **80**, 2229–2259.

Kittel, C. (1963). *Quantum Theory of Solids* (John Wiley & Sons, New York).

Kliger, D. S., Lewis, J. W., and Randall, C. E. (1990). *Polarized Light in Optics and Spectroscopy* (Academic Press, San Diego).

Klyatskin, V. I. (2004). Electromagnetic wave propagation in a randomly inhomogeneous medium as a problem in mathematical statistical physics. *Phys.–Uspekhi* **47**, 169–186.

Knap, W. H., C.-Labonnote, L., Brogniez, G., and Stammes, P. (2005). Modeling total and polarized reflectances of ice clouds: evaluation by means of POLDER and ATSR-2 measurements. *Appl. Opt.* **44**, 4060–4073.

Knyazikhin, Y., Marshak, A., Wiscombe, W. J., *et al.* (2002). A missing solution to the transport equation and its effect on estimation of cloud absorptive properties. *J. Atmos. Sci.* **59**, 3572–3585.

Koenderink, A. F., Megens, M., van Soest, G., *et al.* (2000). Enhanced backscattering from photonic crystals. *Phys. Lett. A* **268**, 104–111.

Kokhanovsky, A. A. (2003). *Polarization Optics of Random Media* (Praxis Publishing, Chichester, UK).

Kokhanovsky, A. A. (2004). *Light Scattering Media Optics: Problems and Solutions* (Praxis Publishing, Chichester, UK).

Kolokolova, L., Gustafson, B. Å. S., Mishchenko, M. I., and Videen, G., eds. (2003). Special issue on electromagnetic and light scattering by nonspherical particles 2002. *J. Quant. Spectrosc. Radiat. Transfer* **79–80**, 491–1198.

Kong, J. A. (2000). *Electromagnetic Wave Theory* (EMW Publishing, Cambridge, MA).

Kourganoff, V. (1952). *Basic Methods in Transfer Problems* (Clarendon Press, Oxford).

Kourganoff, V. (1969). *Introduction to the General Theory of Particle Transfer* (Gordon and Breach, New York).

Kramers, H. A. (1957). *Quantum Mechanics* (North-Holland, Amsterdam).

Krylov, V. I. (1962). *Approximate Calculation of Integrals* (MacMillan, New York).

Kuga, Y., and Ishimaru, A. (1984). Retroreflectance from a dense distribution of spherical particles. *J. Opt. Soc. Am. A* **1**, 831–835.

Kumar, S., and Tien, C. L. (1990). Dependent absorption and extinction of radiation by small particles. *ASMI J. Heat Transfer* **112**, 178–185.

Kunz, K. S., and Luebbers, R. J. (1993). *Finite Difference Time Domain Method for Electromagnetics* (CRC Press, Boca Raton, FL).

Kuščer, I., and Ribarič, M. (1959). Matrix formalism in the theory of diffusion of light. *Opt. Acta* **6**, 42–51 (1959).

Kusmartseva, O., and Smith, P. R. (2002). Robust method for non-sphere detection by photon correlation between scattered polarization states. *Meas. Sci. Technol.* **13**, 336–340.

Kuz'min, V. L., and Romanov, V. P. (1996). Coherent phenomena in light scattering from disordered systems. *Phys.–Uspekhi* **39**, 231–260.

Kuz'min, V. L., Romanov, V. P., and Zubkov, L. A. (1994). Propagation and scattering of light in fluctuating media. *Phys. Rep.* **248**, 71–368.

Kuzmin, V. L., Romanov, V. P., and Zubkov, L. A. (1996). Coherent backscattering of light from anisotropic scatterers. *Phys. Rev. E* **54**, 6798–6801.

Labeyrie, G., de Tomasi, F., Bernard, J.-C., *et al.* (1999). Coherent backscattering of light by cold atoms. *Phys. Rev. Lett.* **83**, 5266–5269.

Labeyrie, G., Müller, C. A., Wiersma, D. S., *et al.* (2000). Observation of coherent backscattering of light by cold atoms. *J. Opt. B: Quantum Semiclass. Opt.* **2**, 672–685.

Lacis, A. A., Chowdhary, J., Mishchenko, M. I., and Cairns, B. (1998). Modeling errors in diffuse-sky radiation: vector *vs.* scalar treatment. *Geophys. Res. Lett.* **25**, 135–138.

Lacoste, D., and van Tiggelen, B. A. (1999). Stokes parameters for light scattering from a Faraday-active sphere. *J. Quant. Spectrosc. Radiat. Transfer* **63**, 305–319.

Lacoste, D., and van Tiggelen, B. A. (2000). Coherent backscattering of light in a magnetic field. *Phys. Rev. E* **61**, 4556–4565.

Lagendijk, A., and van Tiggelen, B. A. (1996). Resonant multiple scattering of light. *Phys. Rep.* **270**, 143–215.

Lakhtakia, A., and Mulholland, G. W. (1993). On two numerical techniques for light scattering by dielectric agglomerated structures. *J. Res. Natl. Inst. Stand. Technol.* **98**, 699–716.

Lamb, W. E., Jr. (1995). Anti-photon. *Appl. Phys. B* **60**, 77–84.

Landgraf, J., Hasekamp, O. P., van Deelen, R., and Aben, I. (2004). Rotational Raman scattering of polarized light in the Earth atmosphere: a vector radiative transfer model using the radiative transfer perturbation theory approach. *J. Quant. Spectrosc. Radiat. Transfer* **87**, 399–433.

Landi Degl'Innocenti, E., and Landolfi, M. (2004). *Polarization in Spectral Lines* (Kluwer Academic Publishers, Dordrecht, The Netherlands).

Lax, M. (1951). Multiple scattering of waves. *Rev. Mod. Phys*. **23**, 287–310.

Lenke, R., and Maret, G. (2000a). Multiple scattering of light: coherent backscattering and transmission. In *Scattering in Polymeric and Colloidal Systems*, eds. W. Brown and K. Mortensen, pp. 1–73 (Gordon and Breach, Amsterdam).

Lenke, R., and Maret, G. (2000b). Magnetic field effects on coherent backscattering of light. *Eur. Phys. J. B* **17**, 171–185.

Lenke, R., Lehner, R., and Maret, G. (2000). Magnetic-field effects on coherent backscattering of light in case of Mie spheres. *Europhys. Lett*. **52**, 620–626.

Lenke, R., Tweer, R., and Maret, G. (2002). Coherent backscattering of turbid samples containing large Mie spheres. *J. Opt. A: Pure Appl. Opt.* **4**, 293–298.

Lenoble, J., ed. (1985). *Radiative Transfer in Scattering and Absorbing Atmospheres: Standard Computational Procedures* (A. Deepak Publishing, Hampton, VA).

Lenoble, J. (1993). *Atmospheric Radiative Transfer* (A. Deepak Publishing, Hampton, VA).

Leroux, C., Lenoble, J., Brogniez, G., *et al.* (1999). A model for the bidirectional polarized reflectance of snow. *J. Quant. Spectrosc. Radiat. Transfer* **61**, 273–285.

Levy, R. C., Remer, L. A., and Kaufman, Y. J. (2004). Effects of neglecting polarization on the MODIS aerosol retrieval over land. *IEEE Trans. Geosci. Remote Sens.* **42**, 2576–2583.

Li, S., and Zhou, X. (2004). Modeling and measuring the spectral bidirectional reflectance factor of snow-covered sea ice: an intercomparison study. *Hydrol. Process.* **18**, 3559–3581.

Li, L.-W., Kang, X.-K., and Leong, M.-S. (2002). *Spheroidal Wave Functions in Electromagnetic Theory* (John Wiley & Sons, New York).

Liang, S. (2004). *Quantitative Remote Sensing of Land Surfaces* (John Wiley & Sons, Hoboken, NJ).

Liou, K. N. (1992). *Radiation and Cloud Processes in the Atmosphere: Theory, Observation, and Modeling* (Oxford University Press, New York).

Liou, K. N. (2002). *An Introduction to Atmospheric Radiation* (Academic Press, San Diego).

Liou, K. N., and Takano, Y. (2002). Interpretation of cirrus cloud polarization measurements from radiative transfer theory. *Geophys. Res. Lett.* **29**, doi:10.1029/2001GL 014613.

Liou, K. N., Takano, Y., and Yang, P. (2000). Light scattering and radiative transfer in ice crystal clouds: applications to climate research. In *Light Scattering by Nonspherical Particles: Theory, Measurements, and Applications*, eds. M. I. Mishchenko, J. W. Hovenier, and L. D. Travis, pp. 3–27 (Academic Press, San Diego).

Lippmann, B. A., and Schwinger, J. (1950). Variational principles for scattering processes. *Phys. Rev.* **79**, 469–480.

Lipson, S. G., Lipson, H., and Tannhauser, D. S. (2001). *Optical Physics* (Cambridge University Press, Cambridge).

Liu, L., and Mishchenko, M. I. (2005). Effects of aggregation on scattering and radiative properties of soot aerosols. *J. Geophys. Res.* **110**, D11211.

Loiko, V. A., and Miskevich, A. (2004). The adding method for coherent transmittance and reflectance of a densely packed layer. *J. Quant. Spectrosc. Radiat. Transfer* **88**, 125–138.

Lommel, E. (1887). Die Photometrie der diffusen Zurückwerfung. *Sitzber. Acad. Wissensch. München* **17**, 95–124. (Reprinted in *Ann Phys. und Chem. (N.F.)* **36**, 473–502, 1889.)

Loughman, R. P., Griffioen, E., Oikarinen, L., *et al.* (2004). Comparison of radiative transfer models for limb-viewing scattered sunlight measurements. *J. Geophys. Res.* **109**, D06303.

Lumme, K., ed. (1998). Special issue on light scattering by non-spherical particles. *J. Quant.*

Spectrosc. Radiat. Transfer **60**, 301–500.
Lyot, B. (1929). Recherches sur la polarisation de la lumière des planetes et de quelques substances terrestres. *Ann. Observ. Paris, Sect. Meudon* **8**, No. 1. (English translation: Research on the polarization of light from planets and from some terrestrial substances, NASA Tech. Transl. NASA TT F-187, Washington, DC, 1964.)
Macke, A., Mueller, J., and Raschke, E. (1996). Scattering properties of atmospheric ice crystals. *J. Atmos. Sci.* **53**, 2813–2825.
MacKintosh, F. C., and John, S. (1988). Coherent backscattering of light in the presence of time-reversal-noninvariant and parity-nonconserving media. *Phys. Rev. B* **37**, 1884–1897.
Mackowski, D. W. (1994). Calculation of total cross sections of multiple-sphere clusters. *J. Opt. Soc. Am. A* **11**, 2851–2861.
Mackowski, D. W., and Mishchenko, M. I. (1996). Calculation of the *T* matrix and the scattering matrix for ensembles of spheres. *J. Opt. Soc. Am. A* **13**, 2266–2278.
Mandel, L., and Wolf, E. (1995). *Optical Coherence and Quantum Optics* (Cambridge University Press, Cambridge).
Marchuk, G. I., Mikhailov, G. A., Nazaraliev, M. A., *et al.* (1980). *The Monte Carlo Methods in Atmospheric Optics* (Springer-Verlag, Berlin). (Original Russian edition: Nauka, Novosibirsk, 1976.)
Maret, G., and Wolf, P. E. (1987). Multiple light scattering from disordered media. The effect of Brownian motion of scatterers. *Z. Phys. B* **65**, 409–413.
Marshak, A., and Davis, A. B., eds. (2005). *3D Radiative Transfer in Cloudy Atmospheres* (Springer-Verlag, Berlin).
Martin, P. G. (1978). *Cosmic Dust* (Oxford University Press, Oxford).
Martinez, A. S., and Maynard, R. (1994). Faraday effect and multiple scattering of light. *Phys. Rev. B* **50**, 3714–3732.
Martonchik, J. V., and Diner, D. J. (1985). Three-dimensional radiative transfer using a Fourier-transform matrix-operator method. *J. Quant. Spectrosc. Radiat. Transfer* **34**, 133–148.
Martonchik, J. V., Diner, D. J., Kahn, R. A., *et al.* (1998). Techniques for the retrieval of aerosol properties over land and ocean using multiangle imaging. *IEEE Trans. Geosci. Remote Sens.* **36**, 1212–1227.
Maslennikov, M. V. (1969). *The Milne Problem with Anisotropic Scattering* (American Mathematical Society, Providence, RI). (Original Russian edition: Nauka, Moscow, 1968.)
Maslennikov, M. V. (1989). *Axiomatic Model of Particle Transport Phenomena* (Nauka, Moscow) (in Russian).
Masuda, K., Takashima, T., Kawata, Y., *et al.* (2000). Retrieval of aerosol optical properties over the ocean using multispectral polarization measurements from space. *Appl. Math. Comput.* **116**, 103–114.
Maxwell, J. C. (1891). *A Treatise on Electricity and Magnetism* (Clarendon Press, Oxford). (Also: Dover Publications, New York, 1954).
Mengüç, M. P., Selçuk, N., Howell, J. R., and Sacadura, J.-F., eds. (2002). Special issue on Third International Symposium on Radiative Transfer. *J. Quant. Spectrosc. Radiat. Transfer* **73**, 129–528.

Mengüç, M. P., Selçuk, N., Webb, B. W., and Lemonnier, D., eds. (2005). Special issue on Fourth International Symposium on Radiative Transfer. *J. Quant. Spectrosc. Radiat. Transfer* **93**, 1–395.

Meyer, W. V., Smart, A. E., Brown, R. G. W., and Anisimov, M. A., eds. (1997). Feature issue on photon correlation and scattering. *Appl. Opt.* **36**, 7477–7677.

Meyer, W. V., Smart, A. E., and Brown, R. G. W., eds. (2001). Feature issue on photon correlation and scattering. *Appl. Opt.* **40**, 3965–4242.

Meystre, P., and Sargent, M., III (1999). *Elements of Quantum Optics* (Springer-Verlag, Berlin).

Miller, E. K., Medgyesi-Mitschang, L. N., and Newman, E. H. (1991). *Computational Electromagnetics: Frequency Domain Method of Moments* (IEEE Press, New York).

Min, Q., and Duan, M. (2004). A successive order of scattering model for solving vector radiative transfer in the atmosphere. *J. Quant. Spectrosc. Radiat. Transfer* **87**, 243–259.

Minin, I. N. (1988). *Theory of Radiative Transfer in Planetary Atmospheres* (Nauka, Moscow) (in Russian).

Mishchenko, M. I. (1990a). Multiple scattering of polarized light in anisotropic plane-parallel media. *Transp. Theory Statist. Phys.* **19**, 293–316.

Mishchenko, M. I. (1990b). The fast invariant imbedding method for polarized light: computational aspects and numerical results for Rayleigh scattering. *J. Quant. Spectrosc. Radiat. Transfer* **43**, 163–171.

Mishchenko, M. I. (1991a). Light scattering by randomly oriented axially symmetric particles. *J. Opt. Soc. Am. A* **8**, 871–882. (Errata: **9**, 497 (1992).)

Mishchenko, M. I. (1991b). Reflection of polarized light by plane-parallel slabs containing randomly-oriented, nonspherical particles. *J. Quant. Spectrosc. Radiat. Transfer* **46**, 171–181.

Mishchenko, M. I. (1992a). Enhanced backscattering of polarized light from discrete random media: calculations in exactly the backscattering direction. *J. Opt. Soc. Am. A* **9**, 978–982.

Mishchenko, M. I. (1992b). Polarization characteristics of the coherent backscatter opposition effect. *Earth Moon Planets* **58**, 127–144.

Mishchenko, M. I. (1993). On the nature of the polarization opposition effect exhibited by Saturn's rings. *Astrophys. J.* **411**, 351–361.

Mishchenko, M. I. (1994). Asymmetry parameters of the phase function for densely packed scattering grains. *J. Quant. Spectrosc. Radiat. Transfer* **52**, 95–110.

Mishchenko, M. I. (1996). Diffuse and coherent backscattering by discrete random media. I. Radar reflectivity, polarization ratios, and enhancement factors for a half-space of polydisperse, nonabsorbing and absorbing spherical particles. *J. Quant. Spectrosc. Radiat. Transfer* **56**, 673–702.

Mishchenko, M. I. (2002). Vector radiative transfer equation for arbitrarily shaped and arbitrarily oriented particles: a microphysical derivation from statistical electromagnetics. *Appl. Opt.* **41**, 7114–7134.

Mishchenko, M. I. (2003). Microphysical approach to polarized radiative transfer: extension to the case of an external observation point. *Appl. Opt.* **42**, 4963–4967.

Mishchenko, M. I. (2006). Scale invariance rule in electromagnetic scattering. *J. Quant. Spectrosc. Radiat. Transfer* (in press).

Mishchenko, M. I., and Dlugach, J. M. (1992a). The amplitude of the opposition effect due to weak localization of photons in discrete disordered media. *Astrophys. Space Sci.* **189**, 151–154.

Mishchenko, M. I., and Dlugach, J. M. (1992b). Can weak localization of photons explain the opposition effect of Saturn's rings? *Mon. Not. R. Astron. Soc.* **254**, 15P–18P.

Mishchenko, M. I., and Dlugach, J. M. (1993). Coherent backscatter and the opposition effect for E-type asteroids. *Planet. Space Sci.* **41**, 173–181.

Mishchenko, M. I., and Hovenier, J. W. (1995). Depolarization of light backscattered by randomly oriented nonspherical particles. *Opt. Lett.* **20**, 1356–1358.

Mishchenko, M. I., and Macke, A. (1998). Incorporation of physical optics effects and computation of the Legendre expansion for ray-tracing phase functions involving δ-function transmission. *J. Geophys. Res.* **103**, 1799–1805.

Mishchenko, M. I., and Mackowski, D. W. (1994). Light scattering by randomly oriented bispheres. *Opt. Lett.* **19**, 1604–1606.

Mishchenko, M. I., and Travis, L. D. (1994a). Light scattering by polydispersions of randomly oriented spheroids with sizes comparable to wavelengths of observation. *Appl. Opt.* **33**, 7206–7225.

Mishchenko, M. I., and Travis, L. D. (1994b). Light scattering by polydisperse, rotationally symmetric nonspherical particles: linear polarization. *J. Quant. Spectrosc. Radiat. Transfer* **51**, 759–778.

Mishchenko, M. I., and Travis, L. D. (1997a). Satellite retrieval of aerosol properties over the ocean using measurements of reflected sunlight: effect of instrumental errors and aerosol absorption. *J. Geophys. Res.* **102**, 13 543–13 553.

Mishchenko, M. I., and Travis, L. D. (1997b). Satellite retrieval of aerosol properties over the ocean using polarization as well as intensity of reflected sunlight. *J. Geophys. Res.* **102**, 16 989–17 013.

Mishchenko, M. I., Dlugach, J. M., and Yanovitskij, E. G. (1992). Multiple light scattering by polydispersions of randomly distributed, perfectly-aligned, infinite Mie cylinders illuminated perpendicularly to their axes. *J. Quant. Spectrosc. Radiat. Transfer* **47**, 401–410.

Mishchenko, M. I., Lacis, A. A., and Travis, L. D. (1994). Errors induced by the neglect of polarization in radiance calculations for Rayleigh-scattering atmospheres. *J. Quant. Spectrosc. Radiat. Transfer* **51**, 491–510.

Mishchenko, M. I., Rossow, W. B., Macke, A., and Lacis, A. A. (1996). Sensitivity of cirrus cloud albedo, bidirectional reflectance and optical thickness retrieval accuracy to ice particle shape. *J. Geophys. Res.* **101**, 16 973–16 985.

Mishchenko, M. I., Hovenier, J. W., and Travis, L. D., eds. (1999a). Special issue on light scattering by nonspherical particles '98. *J. Quant. Spectrosc. Radiat. Transfer* **63**, 127–738.

Mishchenko, M. I., Dlugach, J. M., Yanovitskij, E. G., and Zakharova, N. T. (1999b). Bidirectional reflectance of flat, optically thick particulate layers: an efficient radiative transfer solution and applications to snow and soil surfaces. *J. Quant. Spectrosc. Radiat. Transfer* **63**, 409–432.

Mishchenko, M. I., Hovenier, J. W., and Travis, L. D., eds. (2000a). *Light Scattering by Nonspherical Particles: Theory, Measurements, and Applications* (Academic Press,

San Diego).

Mishchenko, M. I., Hovenier, J. W., and Travis, L. D. (2000b). Concepts, terms, notation. In *Light Scattering by Nonspherical Particles: Theory, Measurements, and Applications*, eds. M. I. Mishchenko, J. W. Hovenier, and L. D. Travis, pp. 3–27 (Academic Press, San Diego).

Mishchenko, M. I., Luck, J.-M., and Nieuwenhuizen, T. M. (2000c). Full angular profile of the coherent polarization opposition effect. *J. Opt. Soc. Am. A* **17**, 888–891.

Mishchenko, M. I., Travis, L. D., and Lacis, A. A. (2002). *Scattering, Absorption, and Emission of Light by Small Particles* (Cambridge University Press, Cambridge). (Available in the .pdf format at http://www.giss.nasa.gov/~crmim/books.html.)

Mishchenko, M. I., Videen, G., Babenko, V. A., *et al.* (2004a). *T*-matrix theory of electromagnetic scattering by particles and its applications: a comprehensive reference database. *J. Quant. Spectrosc. Radiat. Transfer* **88**, 357–406.

Mishchenko, M. I., Hovenier, J. W., and Mackowski, D. W. (2004b). Singe scattering by a small volume element. *J. Opt. Soc. Am. A* **21**, 71–87.

Mishchenko, M. I., Cairns, B., Hansen, J. E., *et al.* (2004c). Monitoring of aerosol forcing of climate from space: analysis of measurement requirements. *J. Quant. Spectrosc. Radiat. Transfer* **88**, 149–161.

Mobley, C. D. (1994). *Light and Water: Radiative Transfer in Natural Waters* (Academic Press, San Diego).

Mobley, J., and Vo-Dinh, T. (2003). Optical properties of tissue. In *Biomedical Photonics Handbook*, ed. T. Vo-Dinh, pp. **2**-1–**2**-75 (CRC Press, Boca Raton, FL).

Modest, M. (2003). *Radiative Heat Transfer* (Academic Press, San-Diego).

Morozhenko, O. V. (2004). *Techniques and Results of Remote Sensing of Planetary Atmospheres* (Naukova Dumka, Kyiv) (in Ukrainian).

Morrison, J. A., and Cross, M.-J. (1974). Scattering of a plane electromagnetic wave by axisymmetric raindrops. *Bell Syst. Tech. J.* **53**, 955–1019.

Mueller, D. W., Jr., and Crosbie, A. L. (1997). Three-dimensional radiative transfer with polarization in a multiple scattering medium exposed to spatially varying radiation. *J. Quant. Spectrosc. Radiat. Transfer* **57**, 81–105.

Muinonen, K. (1989). Scattering of light by crystals: a modified Kirchhoff approximation. *Appl. Opt.* **28**, 3044–3050.

Muinonen, K. (1990). Light scattering by inhomogeneous media: backward enhancement and reversal of linear polarization. Ph.D. dissertation, University of Helsinki.

Muinonen, K. (2004). Coherent backscattering of light by complex random media of spherical scatterers: numerical solution. *Waves Random Media* **14**, 365–388.

Müller, C. (1969). *Foundations of the Mathematical Theory of Electromagnetic Waves* (Springer-Verlag, Berlin). (Original German edition: 1957.)

Müller, C. A., Jonckheere, T., Miniatura, C., and Delande, D. (2001). Weak localization of light by cold atoms: the impact of quantum internal structure. *Phys. Rev. A* **64**, 53804.

Nakajima, T., and King, M. D. (1992). Asymptotic theory for optically thick layers: application to the discrete ordinates method. *Appl. Opt.* **31**, 7669–7683.

Nashashibi, A., and Sarabandi, K. (1999). Experimental characterization of the effective propagation constant of dense random media. *IEEE Trans. Antennas Propag.* **47**, 1454–1462.

Natsuyama, H. H., Ueno, S., and Wang, A. P. (1998). *Terrestrial Radiative Transfer: Modeling, Computation, and Data Analysis* (Springer-Verlag, Tokyo).

Newton, R. G. (1982). *Scattering Theory of Waves and Particles* (Springer-Verlag, New York). (Also: Dover Publications, Mineola, NY, 2002.)

Nisato, G., Hébraud, P., Munch, J.-P., and Candau, S. J. (2000). Diffusing-wave-spectroscopy investigation of latex particle motion in polymer gels. *Phys. Rev. E* **61**, 2879–2887.

Nozette, S., Spudis, P. D., Robinson, M. S., *et al.* (2001). Integration of lunar polar remote-sensing data sets: evidence for ice at the lunar south pole. *J. Geophys. Res.* **106**, 23 253–23 266.

Nussenzveig, H. M. (2002). Does the glory have a simple explanation? *Opt. Lett.* **27**, 1379–1381.

Oetking, P. (1966). Photometric studies of diffusely reflecting surfaces with applications to the brightness of the Moon. *J. Geophys. Res.* **71**, 2505–2513.

Oguchi, T. (1973). Scattering properties of oblate raindrops and cross polarization of radio waves due to rain: calculations at 19.3 and 34.8 GHz. *J. Radio Res. Lab. Japan* **20**, 79–118.

Oguchi, T. (1983). Electromagnetic wave propagation and scattering in rain and other hydrometeors. *Proc. IEEE* **71**, 1029–1078.

Oikarinen, L. (2001). Polarization of light in UV-visible limb radiance measurements. *J. Geophys. Res.* **106**, 1533–1544.

Okamoto, T., and Asakura, T. (1996). Enhanced backscattering of partially coherent light. *Opt. Lett.* **21**, 369–371.

Okin, G. S., and Painter, T. H. (2004). Effect of grain size on remotely sensed spectral reflectance of sandy desert surfaces. *Remote Sens. Environ.* **89**, 272–280.

Onaka, T. (1980). Light scattering by spheroidal grains. *Ann. Tokyo Astron. Observ.* **18**, 1–54.

Ostro, S. J. (1993). Planetary radar astronomy. *Rev. Mod. Phys.* **65**, 1235–1279.

Ozrin, V. D. (1992a). Exact solution for coherent backscattering from a semi-infinite random medium of anisotropic scatterers. *Phys. Lett. A* **162**, 341–345.

Ozrin, V. D. (1992b). Exact solution for coherent backscattering of polarized light from a random medium of Rayleigh scatterers. *Waves Random Media* **2**, 141–164.

Painter, T. H., and Dozier, J. (2004). Measurements of the hemispherical-directional reflectance of snow at fine spectral and angular resolution. *J. Geophys. Res.* **109**, D18115.

Papanicolaou, G. C., and Burridge, R. (1975). Transport equations for the Stokes parameters from Maxwell's equations in a random medium. *J. Math. Phys.* **16**, 2074–2085.

Pecora, R., ed. (1985). *Dynamic Light Scattering: Applications of Photon Correlation Spectroscopy* (Plenum Press, New York).

Penttilä, A., and Lumme, K. (2004). The effect of particle shape on scattering – a study with a collection of axisymmetric particles and sphere clusters. *J. Quant. Spectrosc. Radiat. Transfer* **89**, 303–310.

Perrin, F. (1942). Polarization of light scattered by isotropic opalescent media. *J. Chem. Phys.* **10**, 415–427.

Peterson, B., and Ström, S. (1973). *T* matrix for electromagnetic scattering from an arbitrary number of scatterers and representations of E(3)*. *Phys. Rev. D* **8**, 3661–3678.

Peterson, B., and Ström, S. (1974). *T*-matrix formulation of electromagnetic scattering from multilayered scatterers. *Phys. Rev. D* **10**, 2670–2684.

Peterson, A. F., Ray, S. L., and Mittra, R. (1998). *Computational Methods for Electromagnetics* (IEEE Press, New York).

Petropavlovskikh, I., Loughman, R., DeLuisi, J., and Herman, B. (2000). A comparison of UV intensities calculated by spherical-atmosphere radiation transfer codes: application to the aerosol corrections. *J. Geophys. Res.* **105**, 14 737–14 746.

Petrova, E. V., Markiewicz, W. J., and Keller, H. U. (2001). Regolith surface reflectance: a new attempt to model. *Solar Syst. Res.* **35**, 278–290.

Picard, G., Le Toan, T., and Quegan, S. (2004). Radiative transfer modeling of cross-polarized backscatter from a pine forest using the discrete ordinate and eigenvalue method. *IEEE Trans. Geosci. Remote Sens.* **42**, 1720–1730.

Pike, E. R., and Abbiss, J. B., eds. (1997). *Light Scattering and Photon Correlation Spectroscopy* (Kluwer Academic Publishers, Dordrecht, The Netherlands).

Pike, R., and Sabatier, P., eds. (2001). *Scattering: Scattering and Inverse Scattering in Pure and Applied Science*, Vols. 1 and 2 (Academic Press, San Diego).

Pine, D. J., Weitz, D. A., Chaikin, P. M., and Herbolzheimer, E. (1988). Diffusing-wave spectroscopy. *Phys. Rev. Lett.* **60**, 1134–1137.

Pine, D. J., Weitz, D. A., Maret, G., *et al.* (1990). Dynamical correlations of multiply scattered light. In *Scattering and Localization of Classical Waves in Random Media*, ed. P. Sheng, pp. 312–372 (World Scientific, Singapore).

Pitter, M. C., Hopcraft, K. I., Jakeman, E., and Walker, J. G. (1999). Structure of polarization fluctuations and their relation to particle shape. *J. Quant. Spectrosc. Radiat. Transfer* **63**, 433–444.

Plass, G. N., and Kattawar, G. W. (1968). Monte Carlo calculations of light scattering from clouds. *Appl. Opt.* **7**, 415–419.

Plass, G. N., Kattawar, G. W., and Catchings, F. E. (1973). Matrix operator theory of radiative transfer. 1: Rayleigh scattering. *Appl. Opt.* **12**, 314–329.

Poincaré, H. (1890). Sur le problème des trois corps et les équations de la Dynamique. *Acta Math.* **13**, 1–270.

Poincaré, H. (1892). *Théorie Mathématique de la Lumière*, Vol. 2 (Georges Carré, Paris).

Polonsky, I. N., and Box, M. A. (2002). General perturbation technique for the calculation of radiative effects in scattering and absorbing media. *J. Opt. Soc. Am. A* **19**, 2281–2292.

Pomraning, G. C. (1991). *Linear Kinetic Theory and Particle Transport in Stochastic Mixtures* (World Scientific, Singapore).

Postylyakov, O. V. (2004). Linearized vector radiative transfer model MCC++ for a spherical atmosphere. *J. Quant. Spectrosc. Radiat. Transfer* **88**, 297–317.

Potton, R. J. (2004). Reciprocity in optics. *Rep. Progr. Phys.* **67**, 717–754.

Power, E. A. (1964). *Introductory Quantum Electrodynamics* (Longmans, London).

Preisendorfer, R. W. (1965). *Radiative Transfer on Discrete Spaces* (Pergamon Press, Oxford).

Press, W. H., Teukolsky, S. A., Vetterling, W. T., and Flannery, B. P. (1992). *Numerical Recipes in FORTRAN* (Cambridge University Press, Cambridge).

Prishivalko, A. P., Babenko, V. A., and Kuzmin, V. N. (1984). *Scattering and Absorption of Light by Inhomogeneous and Anisotropic Spherical Particles* (Nauka i Tekhnika,

Minsk, USSR) (in Russian).

Purcell, E. M., and Pennypacker, C. R. (1973). Scattering and absorption of light by non-spherical dielectric grains. *Astrophys. J.* **186**, 705–714.

Pusey, P. N., and van Megen, W. (1989). Dynamic light scattering by non-ergodic media. *Physica A* **157**, 705–741.

Pye, D. (2001). *Polarised Light in Science and Nature* (Institute of Physics Publishing, Bristol, UK).

Ravey, J.-C., and Mazeron, P. (1982). Light scattering in the physical optics approximation: application to large spheroids. *J. Opt. (Paris)* **13**, 273–282.

Rayleigh, Lord (1897). On the incidence of aerial and electric waves upon small obstacles in the form of ellipsoids or elliptic cylinders, and on the passage of electric waves through a circular aperture in a conducting screen. *Philos. Mag.* **44**, 28–52.

Redheffer, R. (1962). On the relation of transmission-line theory to scattering and transfer. *J. Math. & Phys.* **41**, 1–41.

Rose, M. E. (1957). *Elementary Theory of Angular Momentum* (John Wiley & Sons, New York. (Also: Dover Publications, New York, 1995.)

Rosenbush, V. K., and Kiselev, N. N. (2005). Polarization opposition effect for the Galilean satellites of Jupiter. *Icarus* **179**, 490–496.

Rosenbush, V. K., Avramchuk, V. V., Rosenbush, A. E., and Mishchenko, M. I. (1997). Polarization properties of the Galilean satellites of Jupiter: observations and preliminary analysis. *Astrophys. J.* **487**, 402–414.

Rosenbush, V., Kiselev, N., Avramchuk, V., and Mishchenko, M. (2002). Photometric and polarimetric opposition phenomena exhibited by solar system bodies. In *Optics of Cosmic Dust*, eds. G. Videen and M. Kocifaj, pp. 191–224 (Kluwer Academic Publishers, Dordrecht, The Netherlands).

Rosenbush, V. K., Kiselev, N. N., Shevchenko, V. G., *et al.* (2005). Polarization and brightness opposition effects for the E-type asteroid 64 Angelina. *Icarus* **178**, 222–234.

Rossow, W. B., and Schiffer, R. A. (1999). Advances in understanding clouds from ISCCP. *Bull. Am. Meteorol. Soc.* **80**, 2261–2287.

Roux, L., Mareschal, P., Vukadinovic, N., *et al.* (2001). Scattering by a slab containing randomly located cylinders: comparison between radiative transfer and electromagnetic simulation. *J. Opt. Soc. Am. A* **18**, 374–384.

Roychoudhuri, C., and Roy, R., eds. (2003). The nature of light: what is a photon? *OPN Trends* **3**, S1–S35.

Rozanov, V. V., and Kokhanovsky, A. A. (2005). The solution of the vector radiative transfer equation using the discrete ordinates technique: selected applications. *Atmos. Res.* (in press).

Rozenberg, G. V. (1955). Stokes vector-parameter. *Uspekhi Fiz. Nauk* **56**(1), 77–110 (in Russian).

Saillard, M., and Sentenac, A. (2001). Rigorous solutions for electromagnetic scattering from rough surfaces. *Waves Random Media* **11**, R103–R137.

Sano, I., and Mukai, S. (2000). Algorithm description of system flow for global aerosol distribution. *Appl. Math. Comput.* **116**, 79–91.

Sapienza, R., Mujumdar, S., Cheung, C., *et al.* (2004). Anisotropic weak localization of light. *Phys. Rev. Lett.* **92**, 033903.

Sappes, P. (2002). *Representation and Invariance of Scientific Structures* (CSLI Publications, Stanford, CA).

Sato, M., Kawabata, K., and Hansen, J. E. (1977). A fast invariant imbedding method for multiple scattering calculations and an application to equivalent widths of CO_2 lines on Venus. *Astrophys. J.* **216**, 947–962.

Sato, M., Travis, L. D., and Kawabata, K. (1996). Photopolarimetry analysis of the Venus atmosphere in polar regions. *Icarus* **124**, 569–585.

Saxon, D. S. (1955a). Tensor scattering matrix for the electromagnetic field. *Phys. Rev.* **100**, 1771–1775.

Saxon, D. S. (1955b). Lectures on the scattering of light (Scientific Report No. 9, Department of Meteorology, University of California at Los Angeles).

Scheffold, F., Lenke, R., Tweer, R., and Maret, G. (1999). Localization or classical diffusion of light? *Nature* **398**, 206–207.

Scheffold, F., Skipetrov, S. E., Romer, S., and Schurtenberger, P. (2001). Diffusing-wave spectroscopy of nonergodic media. *Phys. Rev. E* **63**, 061404.

Schiff, L. I. (1968). *Quantum Mechanics* (McGraw-Hill, New York).

Schirrer, R., Lenke, R., and Boudouaz, J. (1997). Study of mechanical damage in rubber-toughened poly(methyl methacrylate) by single and multiple scattering of light. *Polym. Eng. Sci.* **37**, 1748–1760.

Schmitt, J. M. (1999). Optical coherence tomography (OCT): a review. *IEEE J. Select. Topics Quantum Electron.* **5**, 1205–1215.

Schmitt, J. M., and Xiang, S. H. (1998). Cross-polarized backscatter in optical coherence tomography of biological tissue. *Opt. Lett.* **23**, 1060–1062.

Schmitz, K. S. (1990). *An Introduction to Dynamic Light Scattering by Macromolecules* (Academic Press, Boston).

Schulz, F. M., and Stamnes, K. (2000). Angular distribution of the Stokes vector in a plane-parallel, vertically inhomogeneous medium in the vector discrete ordinate radiative transfer (VDISORT) model. *J. Quant. Spectrosc. Radiat. Transfer* **65**, 609–620.

Schulz, F. M., Stamnes, K., and Weng, F. (1999). VDISORT: an improved and generalized discrete ordinate method for polarized (vector) radiative transfer. *J. Quant. Spectrosc. Radiat. Transfer* **61**, 105–122.

Schuster, A. (1905). Radiation through a foggy atmosphere. *Astrophys. J.* **21**, 1–22.

Sebbah, P., ed. (2001). *Waves and Imaging through Complex Media* (Kluwer Academic Publishers, Dordrecht, The Netherlands).

Sergent, C., Leroux, C., Pougatch, E., and Guirado, F. (1998). Hemispherical–directional reflectance measurements of natural snow in the 0.9–1.45 μm spectral range: comparison with adding–doubling modelling. *Ann. Glaciol.* **26**, 59–63.

Sharkov, E. A. (2003). *Passive Microwave Remote Sensing of the Earth: Physical Foundations* (Praxis Publishing, Chichester, UK).

Shchegrov, A. V., Maradudin, A. A., and Méndez, E. R. (2004). Multiple scattering of light from randomly rough surfaces. *Progr. Opt.* **46**, 117–241.

Sheng, P., ed. (1990). *Scattering and Localization of Classical Waves in Random Media* (World Scientific, Singapore).

Shifrin, K. S. (1968). *Scattering of Light in a Turbid Medium* (NASA Technical Translation NASA TT F-477). (Original Russian edition: Gostehteorizdat, Moscow, 1951.)

Shifrin, K. S. (1988). *Physical Optics of Ocean Water* (AIP Press, New York). (Original Russian edition: Gidrometeoizdat, Leningrad, 1983.)

Shinde, R., Balgi, G., Richter, S., *et al.* (1999). Investigation of static structure factor in dense suspensions by use of multiply scattered light. *Appl. Opt.* **38**, 197–204.

Shkuratov, Yu. G. (1988). A diffraction mechanism for the formation of the opposition effect of the brightness of surfaces having a complex structure. *Kinem. Fiz. Nebes. Tel* **4**(4), 33–39 (in Russian).

Shkuratov, Yu. G. (1989). A new mechanism for the negative polarization of light scattered by the solid surfaces of cosmic bodies. *Astron. Vestnik* **23**, 176–180 (in Russian).

Shurcliff, W. A. (1962). *Polarized Light: Production and Use* (Harvard University Press, Cambridge, MA).

Siewert, C. E. (1981). On the equation of transfer relevant to the scattering of polarized light. *Astrophys. J.* **245**, 1080–1086.

Siewert, C. E. (2000). Discrete-ordinates solution for radiative-transfer models that include polarization effects. *J. Quant. Spectrosc. Radiat. Transfer* **64**, 227–254.

Silver, S. (1949). Radiation from current distributions. In *Microwave Antenna Theory and Design*, ed. S. Silver, pp. 61–106 (McGraw-Hill, New York).

Silvester, P. P., and Ferrari, R. L. (1996). *Finite Elements for Electrical Engineers* (Cambridge University Press, New York).

Smith, P. R., Kusmartseva, O., and Naimimohasses, R. (2001). Evidence for particle-shape sensitivity in the correlation between polarization states of light scattering. *Opt. Lett.* **26**, 1289–1291.

Sobolev, V. V. (1975). *Light Scattering in Planetary Atmospheres* (Pergamon Press, Oxford). (Original Russian edition: Nauka, Moscow, 1972.)

Spinrad, R. W., Carder, K. L., and Perry, M. J., eds. (1994). *Ocean Optics* (Oxford University Press, New York).

Sromovsky, L. A. (2005). Effects of Rayleigh-scattering polarization on reflected intensity: a fast and accurate approximation method for atmospheres with aerosols. *Icarus* **173**, 284–294.

Stam, D. M., and Hovenier, J. W. (2005). Errors in calculated planetary phase functions and albedos due to neglecting polarization. *Astron. Astrophys.* **444**, 275–286.

Stam, D. M., Hovenier, J. W., and Waters, L. B. F. M. (2004). Using polarimetry to detect and characterize Jupiter-like extrasolar planets. *Astron. Astrophys.* **428**, 663–672.

Stammes, P. (1994). Errors in UV reflectivity and albedo calculations due to neglecting polarization. *SPIE Proc.* **2311**, 227–235.

Stammes, P., de Haan, J. F., and Hovenier, J. W. (1989). The polarized internal radiation field of a planetary atmosphere. *Astron. Astrophys.* **225**, 239–259.

Stamnes, K., Tsay, S.-C., Wiscombe, W., and Jayaweera, K. (1988). Numerically stable algorithm for discrete-ordinate-method radiative transfer in multiple scattering and emitting layered media. *Appl. Opt.* **27**, 2502–2509.

Stephen, M. J., and Cwilich, G. (1986). Rayleigh scattering and weak localization: effects of polarization. *Phys. Rev. B* **34**, 7564–7572.

Stephens, G. L. (1994). *Remote Sensing of the Lower Atmosphere* (Oxford University Press, New York).

Stokes, G. G. (1852). On the composition and resolution of streams of polarized light from different sources. *Trans. Cambridge Philos. Soc.* **9**, 399–416.

Stokes, G. G. (1862). On the intensity of the light reflected from or transmitted through a pile of plates. *Proc. R. Soc. London* **11**, 545–556.

Stratton, J. A. (1941). *Electromagnetic Theory* (McGraw Hill, New York).

Sushkevich, T. A., Strelkov, S. A., and Ioltukhovsky, A. A. (1990). *Method of Characteristics in Problems of Atmospheric Optics* (Nauka, Moscow) (in Russian).

Szegő, G. (1959). *Orthogonal Polynomials* (American Mathematical Society, New York).

Taflove, A., and Hagness, S. C. (2000). *Computational Electrodynamics: The Finite-Difference Time-Domain Method* (Artech House, Boston).

Takashima, T. (1985). Polarization effect on radiative transfer in Chandrasekhar's planetary problem. *Appl. Math. Comput.* **17**, 185–227.

Thomas, G. E., and Stamnes, K. (1999). *Radiative Transfer in the Atmosphere and Ocean* (Cambridge University Press, Cambridge).

Thompson, D. T., and Lockwood, G. W. (1992). Photoelectric photometry of Europa and Callisto 1976–1991. *J. Geophys. Res.* **97**, 14 761–14 772.

Tinbergen, J. (1996). *Astronomical Polarimetry* (Cambridge University Press, Cambridge).

Tishkovets, V. P. (2002). Multiple scattering of light by a layer of discrete random medium: backscattering. *J. Quant. Spectrosc. Radiat. Transfer* **72**, 123–137.

Tishkovets, V. P., and Mishchenko, M. I. (2004). Coherent backscattering of light by a layer of discrete random medium. *J. Quant. Spectrosc. Radiat. Transfer* **86**, 161–180.

Tishkovets, V. P., Litvinov, P. V., and Lyubchenko, M. V. (2002). Coherent opposition effect for semi-infinite discrete random medium in the double-scattering approximation. *J. Quant. Spectrosc. Radiat. Transfer* **72**, 803–811.

Tomita, M., and Ikari, H. (1991). Influence of finite coherence length of incoming light on enhanced backscattering. *Phys. Rev. B* **43**, 3716–3719.

Tsang, L., and Ishimaru, A. (1984). Backscattering enhancement of random discrete scatterers. *J. Opt. Soc. Am. A* **1**, 836–839.

Tsang, L., and Ishimaru, A. (1985). Theory of backscattering enhancement of random discrete isotropic scatterers based on the summation of all ladder and cyclical terms. *J. Opt. Soc. Am. A* **2**, 1331–1338.

Tsang, L., and Kong, J. A. (2001). *Scattering of Electromagnetic Waves: Advanced Topics* (John Wiley & Sons, New York).

Tsang, L., Kong, J. A., and Shin, R. T. (1985). *Theory of Microwave Remote Sensing* (John Wiley & Sons, New York).

Tsang, L., Kong, J. A., and Ding, K.-H. (2000). *Scattering of Electromagnetic Waves: Theories and Applications* (John Wiley & Sons, New York).

Tsang, L., Kong, J. A., Ding, K.-H., and Ao, C. O. (2001). *Scattering of Electromagnetic Waves: Numerical Simulations* (John Wiley & Sons, New York).

Tuchin, V. V., ed. (2002). *Handbook of Optical Biomedical Diagnostics* (SPIE Press, Bellingham, WA).

Tuchin, V. V., ed. (2004). *Coherent-Domain Optical Methods: Biomedical Diagnostics, Environmental and Material Science*, Vols. 1 and 2 (Kluwer Academic Publishers, Boston).

Twersky, V. (1964). On propagation in random media of discrete scatterers. *Proc. Symp. Appl. Math.* **16**, 84–116.

Uhlenbeck, G. E., and Ford, G. W. (1963). *Lectures in Statistical Mechanics* (American Mathematical Society, Providence, RI).

Ulaby, F. T., and Elachi, C., eds. (1990). *Radar Polarimetry for Geoscience Applications* (Artech House, Norwood, MA).

Ulaby, F. T., Moore, R. K., and Fung, A. K. (1986). *Microwave Remote Sensing: Active and Passive. Vol. III. From Theory to Applications* (Artech House, Norwood, MA).

Vaillon, R., Wong, B. T., and Mengüç, M. P. (2004). Polarized radiative transfer in a particle-laden semi-transparent medium via a vector Monte Carlo method. *J. Quant. Spectrosc. Radiat. Transfer* **84**, 383–394.

Van Albada, M. P., and Lagendijk, A. (1985). Observation of weak localization of light in a random medium. *Phys. Rev. Lett.* **55**, 2692–2695.

van Albada, M. P., van der Mark, M. B., and Lagendijk, A. (1987). Observation of weak localization of light in a finite slab: anisotropy effects and light-path classification. *Phys. Rev. Lett.* **58**, 361–364.

van Albada, M. P., van der Mark, M. B., and Lagendijk, A. (1988). Polarization effects in weak localization of light. *J. Phys. D: Appl. Phys.* **21**, S28–S31.

Van Bladel, J. (1964). *Electromagnetic Fields* (McGraw-Hill, New York).

van de Hulst, H. C. (1957). *Light Scattering by Small Particles* (John Wiley & Sons, New York). (Also: Dover Publications, New York, 1981.)

van de Hulst, H. C. (1963). A new look at multiple scattering (technical report, NASA Institute for Space Studies, New York).

van de Hulst, H. C. (1980). *Multiple Light Scattering. Tables, Formulas, and Applications*, Vols. 1 and 2 (Academic Press, San Diego).

van der Mark, M. B., van Albada, M. P., and Lagendijk, A. (1988). Light scattering in strongly scattering media: multiple scattering and weak localization. *Phys. Rev. B* **37**, 3575–3592.

van der Mee, C. V. M. (1981). *Semigroup and Factorization Methods in Transport Theory* (Mathematisch Centrum, Amsterdam).

van der Mee, C. V. M., and Hovenier, J. W. (1990). Expansion coefficients in polarized light transfer. *Astron. Astrophys.* **228**, 559–568.

van Rossum, M. C. W., and Nieuwenhuizen, Th. M. (1999). Multiple scattering of classical waves: microscopy, mesoscopy, and diffusion. *Rev. Mod. Phys.* **71**, 313–371.

van Tiggelen, B. A., and Skipetrov, S. E., eds. (2003). *Wave Scattering in Complex Media: From Theory to Applications* (Kluwer Academic Publishers, Dordrecht, The Netherlands).

van Tiggelen, B., and Stark, H. (2000). Nematic liquid crystals as a new challenge for radiative transfer. *Rev. Mod. Phys.* **72**, 1017–1039.

van Tiggelen, B. A., Maynard, R., and Nieuwenhuizen, Th. M. (1996). Theory for multiple light scattering from Rayleigh scatterers in magnetic fields. *Phys. Rev. E* **53**, 2881–2908.

Varadan, V. K., and Varadan, V. V., eds. (1980). *Acoustic, Electromagnetic and Elastic Wave Scattering – Focus on the T-Matrix Approach* (Pergamon Press, New York).

Varshalovich, D. A., Moskalev, A. N., and Khersonskii, V. K. (1988). *Quantum Theory of Angular Momentum* (World Scientific, Singapore). (Original Russian edition: Nauka, Leningrad, 1975.)

Videen, G., Ngo, D., Chýlek, P., and Pinnick, R. G. (1995). Light scattering from a sphere with an irregular inclusion. *J. Opt. Soc. Am. A* **12**, 922–928.

Videen, G., Fu, Q., and Chýlek, P., eds. (2001). Special issue on light scattering by nonspherical particles. *J. Quant. Spectrosc. Radiat. Transfer* **70**, 373–831.

Videen, G., Yatskiv, Ya., and Mishchenko, M., eds. (2004a). *Photopolarimetry in Remote Sensing* (Kluwer Academic Publishers, Dordrecht, The Netherlands).

Videen, G., Yatskiv, Ya. S., and Mishchenko, M. I., eds. (2004b). Special issue on photopolarimetry in remote sensing. *J. Quant. Spectrosc. Radiat. Transfer* **88**, 1–406.

Viik, T. (1989). *Rayleigh Scattering in a Homogeneous Plane-Parallel Atmosphere* (Valgus, Tallin) (in Russian).

Vilenkin, N. Ja. (1968). *Special Functions and the Theory of Group Representations* (American Mathematical Society, Providence, RI). (Original Russian edition: Nauka, Moscow, 1965.)

Vladimirov, V. S. (1961). *Mathematical Problems in the One-Velocity Theory of Particle Transport. Proceedings of the V. A. Steklov Mathematical Institute, Vol. 61* (Nauka, Moscow). (English translation: Report AECL-1661, Atomic Energy of Canada Limited, Chalk River, Ontario, Canada, 1963.)

Volakis, J. L., Chatterjee, A., and Kempel, L. C. (1998). *Finite Element Method for Electromagnetics* (IEEE Press, New York).

Volten, H., Muñoz, O., Rol, E., *et al.* (2001). Scattering matrices of mineral aerosol particles at 441.6 nm and 632.8 nm. *J. Geophys. Res.* **106**, 17 375–17 402.

Voshchinnikov, N. V., and Farafonov, V. G. (1993). Optical properties of spheroidal particles. *Astrophys. Space Sci.* **204**, 19–86.

Waterman, P. C. (1971). Symmetry, unitarity, and geometry in electromagnetic scattering. *Phys. Rev. D* **3**, 825–839.

Watson, K. M. (1969). Multiple scattering of electromagnetic waves in an underdense plasma. *J. Math. Phys.* **10**, 688–702.

Wauben, W. M. F., and Hovenier, J. W. (1992). Polarized radiation of an atmosphere containing randomly-oriented spheroids. *J. Quant. Spectrosc. Radiat. Transfer* **47**, 491–504.

West, R. A. (1991). Optical properties of aggregate particles whose outer diameter is comparable to the wavelength. *Appl. Opt.* **30**, 5316–5324.

West, R., Gibbs, D., Tsang, L., and Fung, A. K. (1994). Comparison of optical scattering experiments and the quasi-crystalline approximation for dense media. *J. Opt. Soc. Am. A* **11**, 1854 –1858.

Whittaker, E. (1987). *A History of the Theories of Aether and Electricity*, Vols. I and II (American Institute of Physics, New York).

Wiersma, D. S., van Albada, M. P., van Tiggelen, B. A., and Lagendijk, A. (1995a). Experimental evidence for recurrent multiple scattering events of light in disordered media. *Phys. Rev. Lett.* **74**, 4193–4196.

Wiersma, D. S., van Albada, M. P., and Lagendijk, A. (1995b). Coherent backscattering of light from amplifying random media. *Phys. Rev. Lett.* **75**, 1739–1742.

Wiersma, D. S., Bartolini, P., Lagendijk, A., and Righini, R. (1997). Localization of light in a disordered medium. *Nature* **390**, 671–673.

Wiersma, D. S., Rivas, J. G., Bartolini, P., *et al.* (1999). Localization or classical diffusion of light? Reply. *Nature* **398**, 207.

Wigner, E. P. (1959). *Group Theory and its Application to the Quantum Mechanics of Atomic Spectra* (Academic Press, New York).

Williams, M. M. R. (1971). *Mathematical Methods in Particle Transport Theory* (John Wiley & Sons, New York).

Wiscombe, W. J. (1976). On initialization, error and flux conservation in the doubling method. *J. Quant. Spectrosc. Radiat. Transfer* **16**, 637–658.

Wolf, E. (1978). Coherence and radiometry. *J. Opt. Soc. Am.* **68**, 6–17.

Wolf, P.-E., and Maret, G. (1985). Weak localization and coherent backscattering of photons in disordered media. *Phys. Rev. Lett.* **55**, 2696–2699.

Wriedt, T., ed. (1999). *Generalized Multipole Techniques for Electromagnetic and Light Scattering* (Elsevier, Amsterdam).

Wriedt, T. (2002). Using the T-matrix method for light scattering computations by non-axisymmetric particles: superellipsoids and realistically shaped particles. *Part. Part. Syst. Charact.* **19**, 256–268.

Wriedt, T., ed. (2004). Special issue on VII electromagnetic and light scattering by non-spherical particles: theory, measurement and applications. *J. Quant. Spectrosc. Radiat. Transfer* **89**, 1–460.

Xue, J.-Z., Pine, D. J., Milner, S. T., *et al.* (1992). Nonergodicity and light scattering from polymer gels. *Phys. Rev. A* **46**, 6550–6563.

Yang, P., and Liou, K. N. (2000). Finite difference time domain method for light scattering by nonspherical and inhomogeneous particles. In *Light Scattering by Nonspherical Particles: Theory, Measurements, and Applications*, eds. M. I. Mishchenko, J. W. Hovenier, and L. D. Travis, pp. 173–221 (Academic Press, San Diego).

Yanovitskij, E. G. (1997). *Light Scattering in Inhomogeneous Atmospheres* (Springer-Verlag, Berlin).

Yee, K. S. (1966). Numerical solution of initial boundary value problems involving Maxwell's equations in isotropic media. *IEEE Trans. Antennas Propag.* **14**, 302–307.

Yoo, K. M., Tang, G. C., and Alfano, R. R. (1990). Coherent backscattering of light from biological tissues. *Appl. Opt.* **29**, 3237–3239.

Yoon, G., Roy, D. N. G., and Straight, R. C. (1993). Coherent backscattering in biological media: measurement and estimation of optical properties. *Appl. Opt.* **32**, 580–585.

Zakharova, N. T., and Mishchenko, M. I. (2000). Scattering properties of needlelike and platelike ice spheroids with moderate size parameters. *Appl. Opt.* **39**, 5052–5057.

Zakharova, N. T., and Mishchenko, M. I. (2001) Scattering by randomly oriented thin ice disks with moderate equivalent-sphere size parameters. *J. Quant. Spectrosc. Radiat. Transfer* **70**, 465–471.

Zhang, H., and Voss, K. J. (2005). Comparisons of bidirectional reflectance distribution function measurements on prepared particulate surfaces and radiative-transfer models. *Appl.*

Opt. **44**, 597–610.

Zimnyakov, D. A., Oh, J.-T., Sinichkin, Yu. P., *et al.* (2004). Polarization-sensitive speckle spectroscopy of scattering media beyond the diffusion limit. *J. Opt. Soc. Am. A* **21**, 59–70.

Index

absorption, 1
absorption coefficient, 35
absorption cross section, *see* cross section, absorption
adding equations, 247–55, 306–11
adding method, 250–5
aerosol remote sensing, 363
Alexander dark band, 297–8
Ambarzumian equation, 258–9, 313, 379
amplitude scattering matrix, 79
  backscattering, 273–4
  backscattering theorem for, 82–3
  circular-polarization, 287–8
  forward-scattering, 179, 270–2
  reciprocity relation for, 82
  symmetry properties of, 262–4
angle
  azimuth, 37
  phase, 15, 327, 396
  polar, 37
  scattering, 237–9, 262, 265
  zenith, *see* angle, polar
angular frequency, 4, 26
anomalous diffraction approximation, *see* approximation, anomalous diffraction
approximation
  anomalous diffraction, 234
  Born, *see* approximation, Rayleigh–Gans
  far-field, 6, 74
    criteria of, 75–8
  first-order-scattering, *see* first-order-scattering approximation
  geometrical optics, 235–6
  Kirchhoff, 236, 353
  ray-tracing, *see* approximation, geometrical optics
  Rayleigh, 234
  Rayleigh–Debye, *see* approximation, Rayleigh–Gans
  Rayleigh–Gans, 71, 234
  scalar, *see* scalar approximation
  single-scattering, *see* single-scattering approximation
associated Legendre functions, 409, 424
asymmetry parameter, 105, 275, 354
  for spherical particles, 293–5, 301
attenuation coefficient, *see* extinction coefficient
averaging
  configurational, 125–6
  orientation, 126–8
    analytical, 126, 233
  over particle states, 125
  shape, 126–7
  size, 126–30
    effects of, 291–4

statistical, 123–30
azimuth angle, *see* angle, azimuth

backscattering depolarization ratio
circular, 279
linear, 279
backscattering theorem, 82
Bessel functions, spherical, 409
blackbody
energy distribution, 112
Stokes column vector, 113
Born approximation, *see* approximation, Rayleigh–Gans
Bouguer–Beer law, 206, 222
boundary conditions, 23–6
absorbing, 229–30
bulk matter, 67–8
optical constants of, 67–8

charge density, 21
surface, 24
circular-polarization column vector, 40
additivity of, 52
rotation transformation rule for, 48
circularly polarized light, *see* polarization, circular
cirrus cloud crystals, 352, 354
cloud remote sensing, 354, 363
clusters, 6–7, 233, 346
of spheres, 232
two-sphere, 147–9, 153–6
coherency column vector, 38–9, 61–2
of the coherent field, 180
transfer equation for, 180
coherency dyad of the electric field, 62–4, 105–9
average, 108–10
coherency dyadic, 13, 17, 191–2
at a remote observation point, 368, 371
at an external observation point, 211–2
coherent, 193, 212, 368
cyclical, 365, 368, 370
diffuse multiple-scattering, 368
ladder, 211, 365, 370
ladder approximation for, 192–3
single-scattering, 368
coherency matrix, 38, 61
additivity of, 52
coherency reflection matrix, 372
cyclical, 372
for the exact backscattering direction, 377
diffuse multiple scattering, 372
single-scattering, 372
coherent backscattering, 14–7, 203, 365–406
angular profile of, 395–402
benchmark results for, 386
by polydisperse spheres, 386–93
by polydisperse spheroids, 394–5
by solar system bodies, 405–6
half-width at half-maximum of, 397–9
scalar theory of, 396, 399
coherent field, 171–8
at an external point, 210–1
transfer equation for, 180
transversality condition for, 177–8
Twersky expansion of, 173
coherent intensity, 207
physical meaning of, 207
coherent Stokes column vector, *see* Stokes column vector, of the coherent field
coherent transmission amplitude matrix, 179
reciprocity relation for, 180
coherent transmission dyadic, 177
reciprocity relation for, 180
coherent transmission Stokes matrix, 181
completely polarized light, *see* polarization, full
complex permittivity, 27
computer codes, xiii, 229, 233, 234, 342
conductivity, 21
constitutive relations, 21, 22, 27
continuity equation, 21
integral form of, 23
coordinate system
Cartesian, 37, 126, 411
laboratory, 126, 270–1, 273
particle, 126, 270–1, 273
right-handed, 37
spherical, 37
cosine integral, 189
coupled dipole method, *see* discrete dipole

approximation
cross section
absorption, 102–3,
average, 147, 275
differential scattering, 104–5
extinction, 91, 102–4
average, 147, 272
scattering, 102–3, 105
average, 147, 275
for macroscopically isotropic and mirror-symmetric media, 275
current density, 21
surface, 25
cyclical diagrams, 15, 17, 366–8, 370, 402

delta function, *see* Dirac delta function
density matrix, *see* coherency matrix
depolarization, 99, 279, 362, 390
depolarization factor, 325
dichroism, 1, 6, 101, 206
differential equation methods, 227–30
differential volume element, 140, 221–2
diffraction, 296
diffusing wave spectroscopy, 8, 17–9
diffusion approximation, 396
Dirac delta function, 69
solid-angle, 410
three-dimensional, 69
discrete dipole approximation, 231, 233
discrete ordinate method, 260, 323
Doppler effect, 4, 18
doubling method, 251, 253–5
dyad, 407
zero, 70
dyadic, 407–8
Hermitian, 63, 408
identity, *see* identity dyadic
symmetric, 408
transpose of, 407
dyadic correlation function, 181
ladder approximation for, 191
dyadic exponential, 176
dyadic propagation constant, 177
dyadic transition operator, 71, 78, 117
integral equation for, 71
dynamic light scattering, *see* scattering, dynamic

effective radius, 129
effective size parameter, *see* size parameter, effective
effective variance, 129
efficiency factor
for absorption, 104, 276
for spherical particles, 301
for extinction, 104, 276
for spherical particles, 291–5, 301
for scattering, 104, 276
for spherical particles, 301
electric displacement, 21
electric energy density, 30
electric field, 2, 21
electric permittivity, 21
frequency-dependent, 27
of free space, 21
electric polarization, 21
electric susceptibility, 21
electromagnetic energy density, 29
time-averaged, 35, 61, 91
electromagnetic scattering, 1
electromagnetic scattering problems, classification of, 16–8
electromagnetic wave
plane, 1, 31–6
circularly polarized, 45
homogeneous, 33
inhomogeneous, 33
linearly polarized, 45
spherical, 58–62
incoming, 62
outgoing, 59
transverse, 5, 33, 59, 62
circular components of, 287
elementary volume element, *see* differential volume element
elliptically polarized light, *see* polarization, elliptical
emission Stokes column vector, 112, 114
average, 224, 276–7
emission, thermal, 4, 112, 224
energy conservation, 60, 69–70, 151, 208–9, 404
enhancement factor

co-polarized, 381–2
general properties of, 383–5
cross-polarized, 382
general properties of, 383–5
helicity-preserving, 382
general properties of, 383–5
opposite-helisity, 382
general properties of, 383–5
scalar, 381
unpolarized, 380
general properties of, 383–5
equilibrium, thermal, 112, 114
ergodic hypothesis, 10, 19, 123
ergodicity, 9–10, 19, 181, 201, 208, 402
Euler angles, 126, 270–1, 273, 411–2, 426–7
extended boundary condition method, 233
iterative, 233
extinction, 1
extinction coefficient, 206
extinction cross section, *see* cross section, extinction
extinction matrix, 6, 99–102
circular-polarization, 102
reciprocity relation for, 102
coherency, 100
average, 198
reciprocity relation for, 102
modified Stokes, 102
reciprocity relation for, 102
Stokes, 100
average, 135, 147, 162, 199
for macroscopically isotropic and mirror-symmetric media, 272
reciprocity relation for, 101
symmetry property of, 101

far-field approximation, *see* approximation, far-field
far-field scattering, *see* scattering, far-field
far-field zone, 5–6, 13, 73, 119, 122, 222, 229
finite-difference time-domain method, 230, 233
finite-element method, 229–30, 233
first-order-scattering approximation, 158–62, 216–7
conditions of applicability of, 164
fluorescence, 4
flux density vector, 208–9
F_N method, 323
Foldy approximation, 177
Foldy–Lax equations, 6–7, 11, 13, 18, 115–8
far-field version of, 118–22, 165
forward-scattering direction, 89–90, 99
Fredholm integral equation method, 232
Fresnel formulas, 235, 298
fully polarized light, *see* polarization, full

Gauss theorem, 23
general problem in radiative transfer, 245, 304
generalized spherical functions, 282, 425
completeness of, 425
normalization of, 425
symmetry relations for, 425
geometrical optics approximation, *see* approximation, geometrical optics
glory, 297–8, 348, 355
Green's function
dyadic, free space, 69–70
differential equation for, 69
scalar, 70
differential equation for, 70

halos, 353–4
Henyey–Greenstein phase function, 354–6
hydrometeors, nonspherical and partially aligned, 279

identity dyadic, 69, 408
incident field, 2, 66, 69, 71
independent scattering, 201
integral equation methods, 227
intensity, 35, 60, 62
interaction principle, 260
interference of light, 14–5
forward-scattering, 146–51
interference structure, 291
interstellar dust grains, 206
interstellar polarization, 206, 279
invariance principles, 260

invariant imbedding equations, 255–8, 311–3, 327, 378, 379
irradiance, *see* intensity

Jacobi polynomials, 422
Jones lemma, 410

Kirchhoff approximation, *see* approximation, Kirchhoff
Kronecker delta, 419

ladder approximation, 191–2, 201, 366
ladder diagrams, 13, 17, 191
Lambert law, 327
Lambertian albedo, 327
laser Doppler methods, 19
Legendre functions, associated, *see* associated Legendre functions
Legendre polynomials, 409, 413–4, 424
Leibniz rule, 195, 420
levitation, 131
  electrostatic, 7
  optical, 7
lidar observations, 362
linearly polarized light, *see* polarization, linear
Lippmann–Schwinger equation, 71, 84
Lorenz–Mie
  coefficients, 291
    resonance behavior of, 291
  computer code, 229
  identities, 279
  scattering matrix, 278
  theory, 227–9, 278, 291, 353

magnetic energy density, 30
magnetic field, 2, 21
magnetic induction, 21
magnetic permeability, 21
  of free space, 21
magnetization, 21
matrix exponential, 179, 181
matrix operator method, 260
matrix propagation constant, 178
maximally crossed diagrams, *see* cyclical diagrams
Maxwell equations, macroscopic, 3, 20
  curl, 68
  integral form of, 22–3, 27–8
  linearity of, 80
  plane-wave solution of, 31–6
  spherical-wave solution of, 58–62
mean free path, 397
measurement techniques for scattering, 237–9
  using microwaves, 238–9
  using visible and infrared light, 237–8
medium
  absorbing, 34
  lossless, 34
  nonabsorbing, 34
  nondispersive, 34
  time-dispersive, 22, 27
meridional plane, 37
method of moments, 231, 233
microwave analog technique, 7, 86, 238–9
modified Stokes column vector, 40
  additivity of, 52
  rotation transformation rule for, 48
modulator, electro-optic, 237–8
monochromatic light, 1, 26, 33
Monte Carlo method, 323, 396

natural light, *see* unpolarized light
Newman expansion, *see* order-of-scattering series
nonsphericity, effects of, 278–9, 298–301

observable, 5, 109
optical coherence tomography, 19
optical depth, 303
optical equivalence principle, 39–40, 52
optical theorem, 90, 101, 178
optical thickness, 304
order-of-scattering approach, 315
order-of-scattering series, 313, 328
orientation angle of the polarization ellipse, *see* polarization ellipse, orientation of
orientation averaging, *see* averaging, orientation
orientation distribution, 127
  axially symmetric, 127–8
  random, 127

orientation of the scattering object, *see* particle orientation
outgoing wave, 59, 70

parallel beam of light, 1, 33
partially polarized light, *see* polarization, partial
particle collection, random and tenuous, 17, 140
particle depth, 242
particle number density, 125
particle orientation, 126
particle thickness, 243
particles
  hexagonal, 352–3
  independently scattering, *see* independent scattering
  irregular, 352
  random-fractal, 352–3
  spherically symmetric, 277–8
phase angle, *see* angle, phase
phase function, 105, 280, 321
  backscattering, 387, 389
  expansion in Legendre polynomials, 285
  for polydisperse spherical particles, 296–8, 347–8, 354–5
  for randomly oriented spheroids, 298
  Henyey–Greenstein, *see* Henyey–Greenstein phase function
  normalization condition for, 280
  Rayleigh, 295–6
phase matrix, 6
  circular-polarization, 96, 288
    average, 289
    normalized, 288
      Fourier decomposition in azimuth of, 318–21
      symmetry properties of, 290
    reciprocity relation for, 98
  coherency, 95
    average, 198
    reciprocity relation for, 98
  modified Stokes, 96
    reciprocity relation for, 98
  normalized Stokes, 280–1
    symmetry properties of, 281, 315–6
  Stokes, 95–6
    average, 135, 147, 162, 199
    backscattering, 98
    expression in terms of the scattering matrix, 266–8
    for macroscopically isotropic and mirror-symmetric media, 266–70
    inequalities for, 97
    reciprocity relation for, 97
    symmetry relations for, 269
phase velocity, 31, 34, 59
phenomenological radiative transfer theory, *see* radiative transfer theory, phenomenological
photon, 223–6
photon correlation spectroscopy, 8, 17–9
photon distribution function, 223
photon gas, 223
physical optics approximation, *see* approximation, Kirchhoff
plane albedo, 355–6
plane wave, expansion in spherical waves, 409–10
Poincaré recurrence theorem, 19
Poincaré sphere, 64
point matching method, 230
  generalized, 230
    multiple-expansion, 230, 233
polar angle, *see* angle, polar
polarization, 99, 357
  circular, 43, 45, 99, 390
    degree of, 45, 53
  complete, *see* polarization, full
  degree of, 52
  elliptical, 43, 52, 99
    degree of, 52
  full, 51–3, 99
  left-handed, 42, 53
  linear, 43, 45, 99, 390
    degree of, 52–3
      for spherical particles, 292–3, 298–9
      for spheroidal particles, 299
      signed, 53–4, 400
  of multiply scattered light, 357–9
  partial, 51–3, 99, 390

right-handed, 42, 53
polarization analyzer, 237–8
polarization ellipse, 42–5, 51–3
ellipticity of, 42, 52–3
orientation of, 42, 52–3
polarization modulation technique, 238
polarization opposition effect, 366, 401–2, 405
polarization ratio
linear, 382
diffuse, 382
general properties of, 383–5
circular, 382–3
diffuse, 382–3
general properties of, 383–5
polarizer, 54–8, 237–9
position vector, 2
potential dyadic, 117
potential function, 116
Poynting vector, 28–30
complex, 30
time-averaged, 30, 34–5, 60, 88–9
probability density function, 124
normalization condition for, 124
propagator, 243–5

quadrature division points, *see* quadrature formula
quadrature formula, 251, 413–5
Gauss, 413–4
Lobatto, 415
Markov, 415
Radau, 415
quadrature sum, *see* quadrature formula
quadrature weights, *see* quadrature formula
quarter-wave plate, 237–8
quasi-monochromatic light, 4, 49–54, 97, 101, 108, 218, 245, 379, 402

radar observations, 362
of solar system bodies, 406
radiance, *see* specific intensity
radiation condition, 231
radiative transfer
in particulate surfaces, 363–4
in plane-parallel media, 240–60, 302–64
in plane-parallel, macroscopically isotropic and mirror-symmetric media, 302–64
in Rayleigh-scattering slabs, 324–37, 363
radiative transfer equation, 13, 165, 199, 365
for macroscopically isotropic and mirror-symmetric media, 302
scalar, 321
Fourier decomposition of, 321–2
vector, 199
Fourier decomposition of, 318–9
radiative transfer theory, 14, 16–7
phenomenological, 218–25
radius, 127
effective, *see* effective radius
equivalent-sphere, 127
radius vector, *see* position vector
rainbows, 297–8, 348
ray-tracing approximation, *see* approximation, geometrical optics
Rayleigh approximation, *see* approximation, Rayleigh
Rayleigh–Debye approximation, *see* approximation, Rayleigh–Gans
Rayleigh–Gans approximation, *see* approximation, Rayleigh–Gans
Rayleigh hypothesis, 230, 233
Rayleigh scattering, 294–6, 325, 386, 396
reciprocity, 15, 80–4
reference plane, 47
rotations of, 47
reflection coefficient, 326
reflection matrix, 246, 305, 372
cyclical, 372
for the exact backscattering direction, 377–8
diffuse multiple-scattering, 372
reciprocity relations for, 259
single-scattering, 372
symmetry properties of, 316–7
refractive index, 3, 34
relative, 68
relativity theory, 64
retarder, 56–8

ripple, high-frequency, 291
rotation matrix
  for circular-polarization representation, 48
  for modified Stokes column vector, 48
  Stokes, 48

scalar approximation, 321, 324, 381
  accuracy of, 324–5, 363, 381
    for Rayleigh-scattering slabs, 325–37
    for polydisperse spheres, 337–42
    for polydisperse spheroids, 342–7
scale invariance rule, 84–7, 97, 101, 104, 238
scattered field, 2, 66, 69–71
scattering
  Brillouin, 4
  by a fixed finite object, 1–5
  by random particles, 10–2, 131–9
  by variable objects, 110–1
  dynamic, 9, 17
  elastic, 1
  electromagnetic, *see* electromagnetic scattering
  far-field, 5, 11, 71–8, 131–6
  forward, 270–3
  independent, *see* independent scattering
  inelastic, 18
  multiple, 12–4
  Raman, 4
  single, 10–2
  static, 9, 17
scattering angle, *see* angle, scattering
scattering cross section, *see* cross section, scattering
scattering dyadic, 78
  average, 173
  forward-scattering, 178
  reciprocity relation for, 82
scattering matrix, circular-polarization, normalized, 288–9
  expansion in generalized spherical functions, 289–90
  symmetry properties of, 289
scattering matrix, Stokes, 261–2
  average, 264
  effects of nonsphericity on, 278–9
  for backward scattering, 273–5
  for forward scattering, 272–3
  for macroscopically isotropic and mirror-symmetric media, 266
  for rotationally symmetric particles, 272–3
  for spherically symmetric particles, 278
  inequalities for, 266, 275
  normalized, 280
    expansion in generalized spherical functions, 282
    expansion in Wigner *d*-functions, 282–6
    for polydisperse spheres, 285, 287, 298–300, 347–8
    for randomly oriented spheroids, 298–300
    for rotationally symmetric particles, 281
    for spherically symmetric particles, 281
    properties of, 280–1
  symmetries of, 264–5
scattering medium
  macroscopically isotropic, 265
  macroscopically isotropic and mirror-symmetric, 261, 266
  macroscopically mirror-symmetric, 265–6
scattering plane, 261–3
scattering tensor, 80
  reciprocity condition for, 81
secondary waves, 2
separation of variables method for spheroids, 229, 233
shape averaging, *see* averaging, shape
sine integral, 189
single-scattering albedo, 103–4
  average, 276
  for spherical particles, 293–5, 301, 387–8
single-scattering approximation, 11, 12, 141–57
  far-field, 142–5

modified uncorrelated, 145–7, 222
conditions of validity of, 151–7, 163
uncorrelated, 145–7
effects of forward-scattering interference on, 147–50
size averaging, *see* averaging, size
size distribution, 127–30
gamma, 128–30, 294, 338, 348, 354, 387–9
log normal, 128–30
modified bimodal log normal, 128–9
modified gamma, 128–9
modified power law, 128, 130
power law, 128
size parameter, 86
effective, 293, 338
Snell law, 235
snow crystals, 352, 354
solid angle element, differential, 61
specific coherency column vector, 371
cyclical, 372
diffuse, 198
integral equation for, 198
integro-differential equation for, 198
multiple-scattering, 371
single-scattering, 371
specific coherency dyadic, 13, 366–71
at an external observation point, 211–3, 368
coherent, 368–9, 371
cyclical, 370–1
for the exact backscattering direction, 373
diffuse, 195, 369
integral equation for, 195
integro-differential equation for, 196–7
multiple-scattering, 368–9, 371
ladder, 13, 194–5
integral equation for, 194
single-scattering, 368–9
specific coherency matrix
diffuse, 197
integral equation for, 197
integro-differential equation for, 197
specific intensity, diffuse, 207
physical meaning of, 207
specific intensity column vector, 13
at an external observation point, 213–4
diffuse, 199
integral equation for, 199
integro-differential equation for, 199
full, 200
integral equation for, 200
integro-differential equation for, 200
physical meaning of, 203–8
speckle pattern, 8, 403
speed of light
in a nonabsorbing medium, 35, 61
in a vacuum, 34
spherical albedo, 349–52, 357
spherical harmonics, scalar, 409
completeness relation for, 410
spherical harmonics solution, 323
spherical wave functions, vector, 228, 230, 232, 233
standard problem in radiative transfer, 240–3, 247, 302–6
static light scattering, *see* scattering, static
stationary phase method, 188, 416–7
Stokes column vector, 5, 39, 61–2
emission, *see* emission Stokes column vector
modified, *see* modified Stokes column vector
of the coherent field, 180, 199
at an external observation point, 211
physical meaning of, 203–8
transfer equation for, 180, 199
Stokes identity, 40
Stokes parameters, 5, 39
ellipsometric interpretation of, 41–5
for quasi-monochromatic light, 49
additivity of, 52
quadratic inequality for, 50–1
measurement of, 54–8
rotation transformation rule for, 48
Stokes theorem, 22

strong localization of light, 405
successive orders of scattering method, 313–5
superposition method, 232–3
superposition principle, 28, 71
Système International, 20, 429–30

T matrix, 232–3
 for a cluster, 233
T-matrix computer codes, 233
T-matrix method, 126, 232–3, 284
 superposition, 18, 147, 233, 284
temperature, absolute, 4, 112, 224
thermal emission, *see* emission, thermal
thermal equilibrium, *see* equilibrium, thermal
time, 2, 21
time-harmonic factor, 5, 36, 171
time-harmonic field, 3, 26
total field, 2, 7, 66, 71, 118
transformation dyadic, 108–11
 of a multi-particle group, 217
transition matrix, *see T* matrix
translation addition theorem, 232–3
transmission matrix, 246, 305
 reciprocity relation for, 259
 symmetry properties of, 316–7
transport mean free path, 398, 404
Twersky approximation, 13, 169–71, 201, 402–3

unpolarized light, 51–3, 99, 321, 326, 348, 357, 380

vector spherical wave functions, *see* spherical wave functions, vector
Venus clouds, 348, 359–62
volume integral equation, 70–1
volume integral equation method, 231

wave equation, vector, 68
wave number, 34, 59
wave vector, 4, 31
wavelength
 free-space, 35
 in a nonabsorbing medium, 36, 86
weak localization of electromagnetic waves, 14
Wigner *d*-functions, 282–4, 418–22
 addition theorem for, 426–7
 completeness of, 422
 orthogonality of, 422
 recurrence relations for, 423–4
 symmetry properties of, 419
 unitarity condition for, 427
Wigner *D*-functions, 426
 addition theorem for, 426
 unitarity condition for, 426

zenith angle, *see* angle, polar

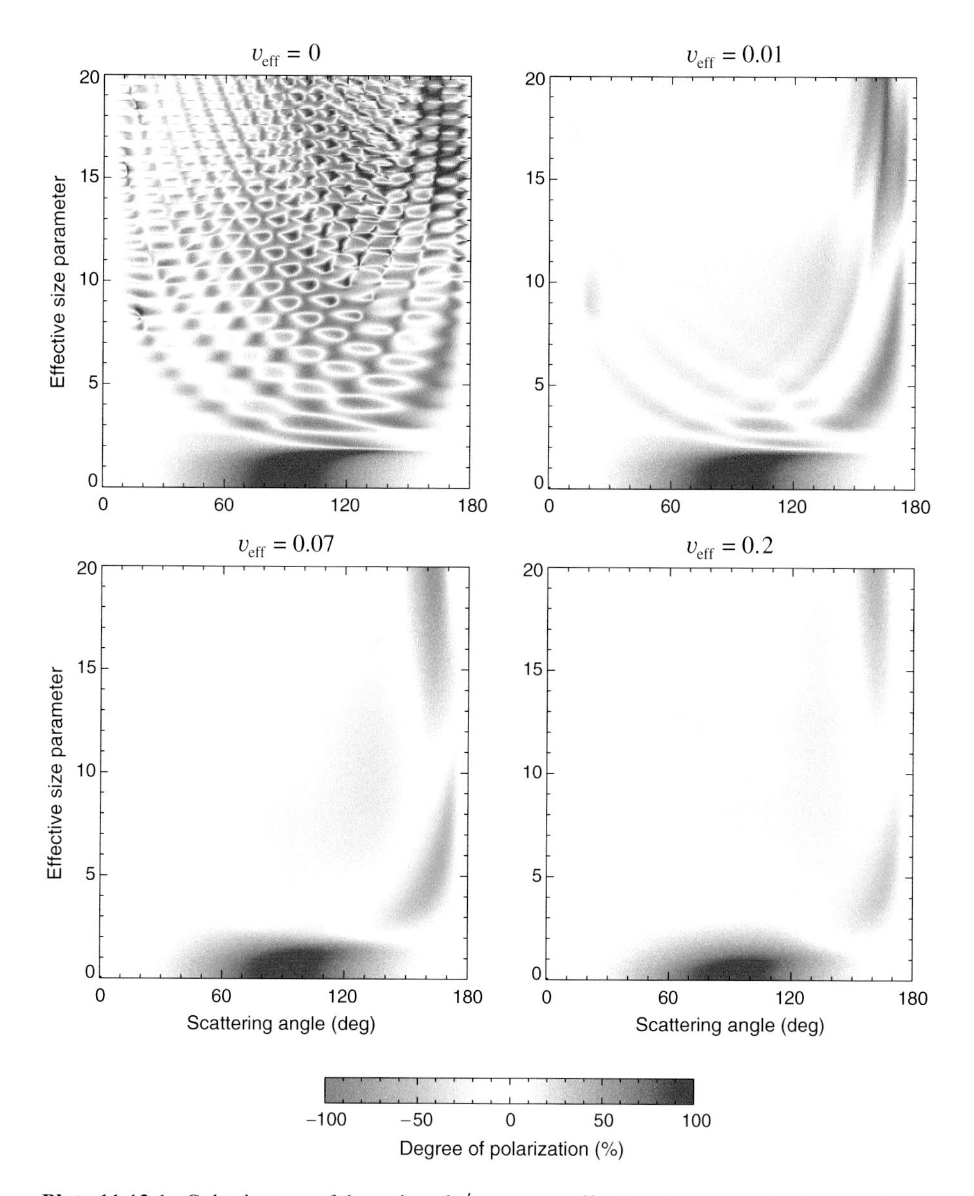

Plate 11.13.1. Color images of the ratio $-b_1/a_1$ versus effective size parameter $k_1 r_{eff}$ and scattering angle Θ, for the gamma distribution of spherical particles with $v_{eff} = 0$ (the value for monodisperse particles), 0.01, 0.07, and 0.2. The relative refractive index is fixed at 1.44.

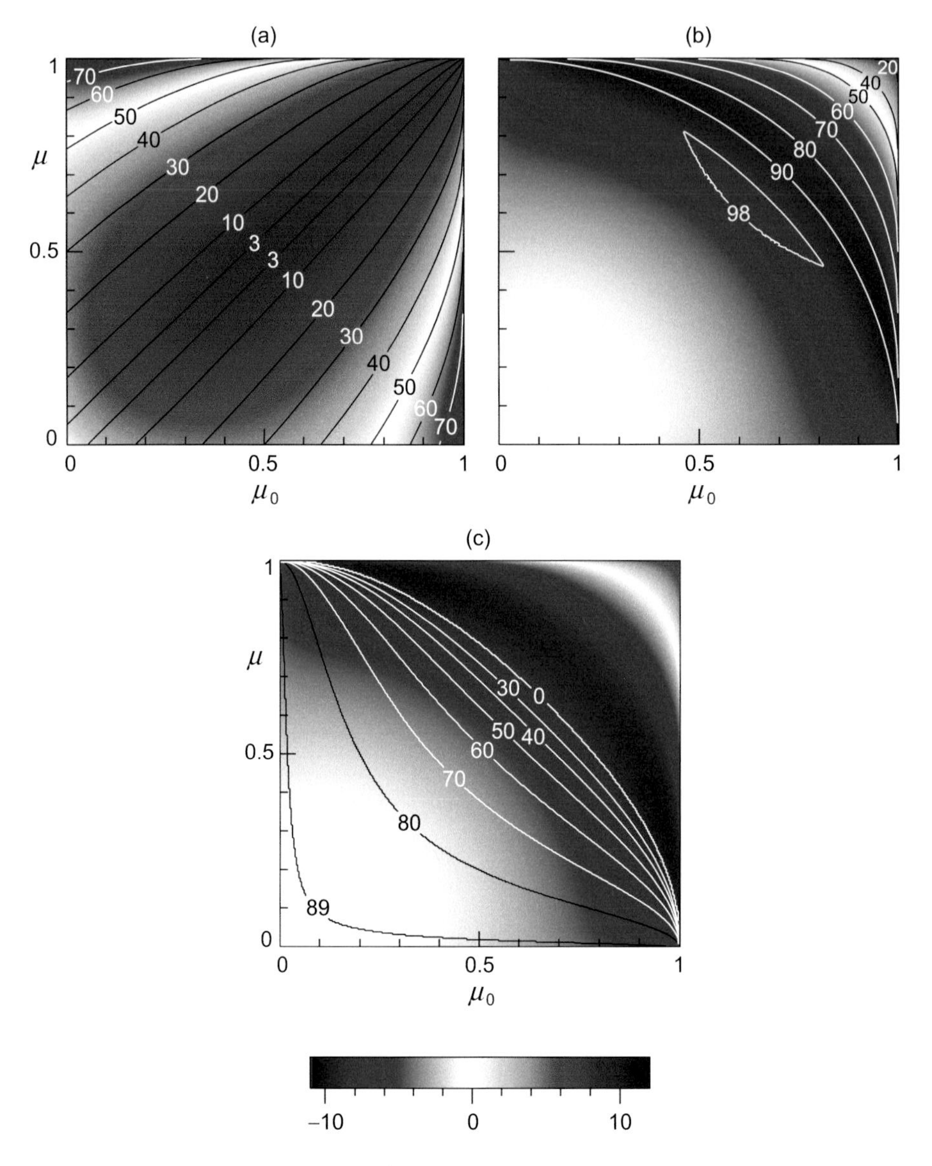

Plate 13.1.1. (a) Contour plot of $\alpha_u(\mathcal{T}, \mu, \mu_0)$ (in degrees) superimposed on the color diagram of local underestimation $\varepsilon_u(\mathcal{T}, \mu, \mu_0)$ (in percent) for $\mathcal{T} = 1$, $\delta = 0.031$, $\varpi = 1$, and $A_L = 0$. (b) Contour plot of $\alpha_o(\mathcal{T}, \mu, \mu_0)$ (in degrees) superimposed on the color diagram of local overestimation $\varepsilon_o(\mathcal{T}, \mu, \mu_0)$ (in percent) for $\mathcal{T} = 1$, $\delta = 0.031$, $\varpi = 1$, and $A_L = 0$. Note that below and to the left of the contour labeled by 90, $\alpha_o(\mathcal{T}, \mu, \mu_0)$ is a continuous function of μ and μ_0, is always greater than $90°$, and has a maximum at around $\mu = \mu_0 = 0.64$. (c) Contour plot of $\varphi_o(\mathcal{T}, \mu, \mu_0)$ (in degrees) superimposed on the color diagram of local overestimation $\varepsilon_o(\mathcal{T}, \mu, \mu_0)$ (in percent) for $\mathcal{T} = 1$, $\delta = 0.031$, $\varpi = 1$, and $A_L = 0$. Note that $\varphi_o(\mathcal{T}, \mu, \mu_0)$ is zero above and to the right of the contour labeled by 0.

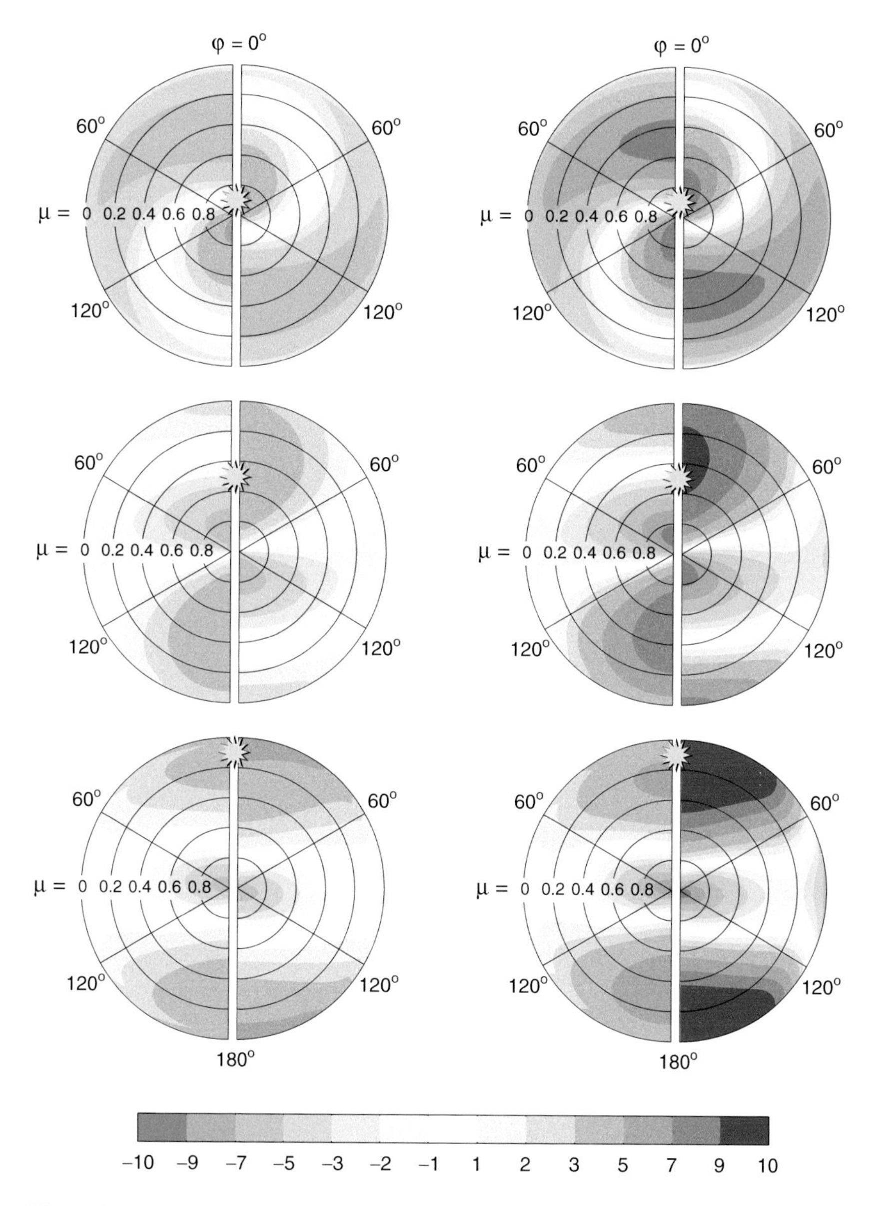

Plate 13.1.2. Scalar approximation specific intensity errors for Rayleigh-scattering slabs with $\delta = 0.031$ and $A_{\mathrm{L}} = 0$. The split-hemisphere polar diagrams display the errors for reflection in the left-hand hemispheres and for diffuse transmission in the right-hand hemispheres. The errors are defined by Eq. (13.1.4) for reflection and by a similar formula for diffuse transmission. The left-hand and right-hand columns shows results for $\mathcal{T} = 0.2$ and $\mathcal{T} = 0.8$, respectively. Illumination polar angles are for $\mu_0 = 0.9$, 0.5, and 0.1, respectively, as indicated by the yellow stars. The azimuth angle of the unpolarized incident beam is zero. (After Lacis *et al*., 1998.)

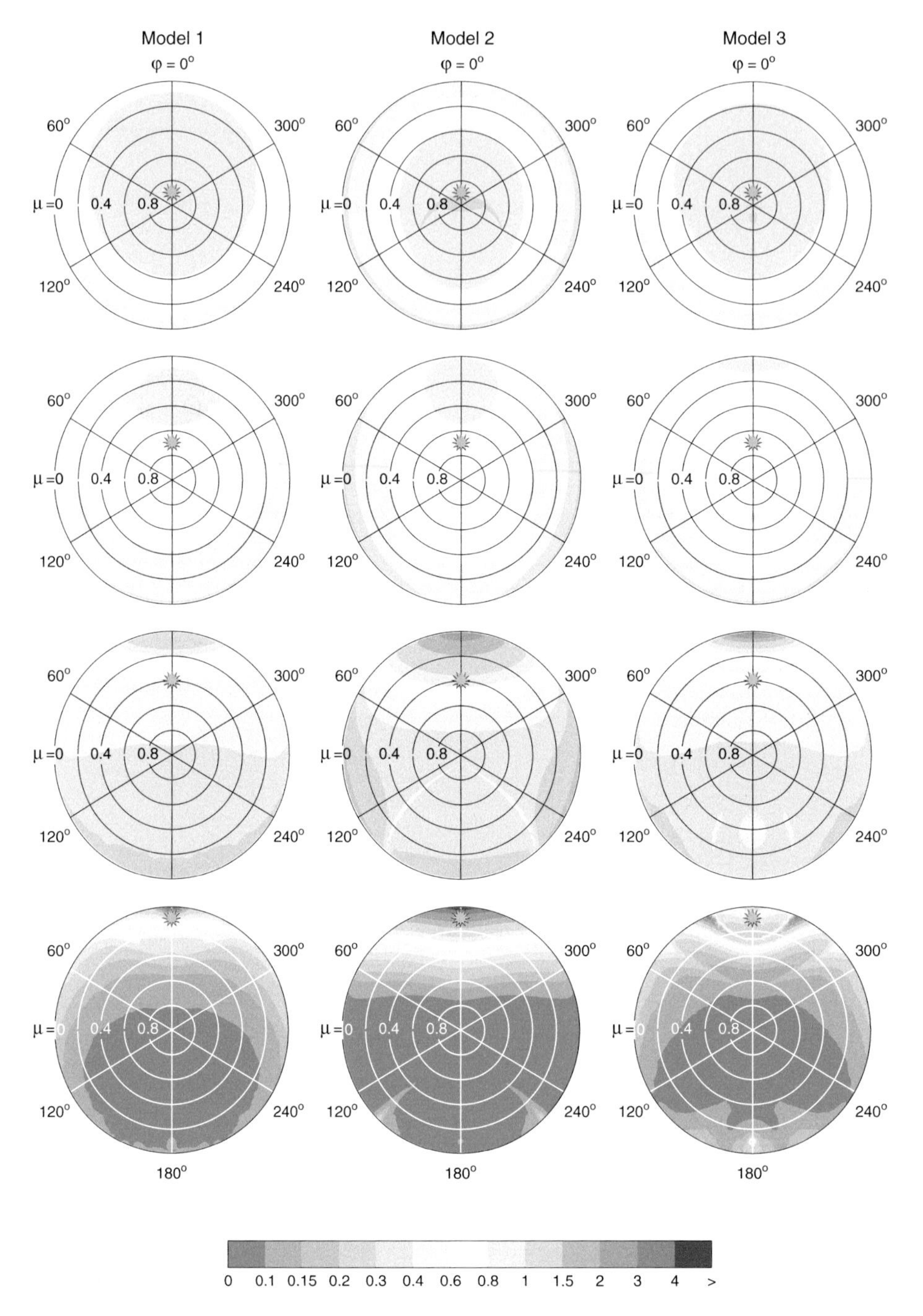

Plate 13.2.1. Reflected specific intensity (in $\mathrm{W\,m^{-2}\,sr^{-2}}$) versus μ and φ for semi-infinite slabs composed of model 1, 2, and 3 ice particles. The four values of minus the cosine of the illumination zenith angle $\mu_0 = 0.9$, 0.7, 0.4, and 0.1 are indicated by the yellow stars. The azimuth angle of the unpolarized incident beam is zero and its intensity is $I_0 = \pi\ \mathrm{Wm^{-2}}$.

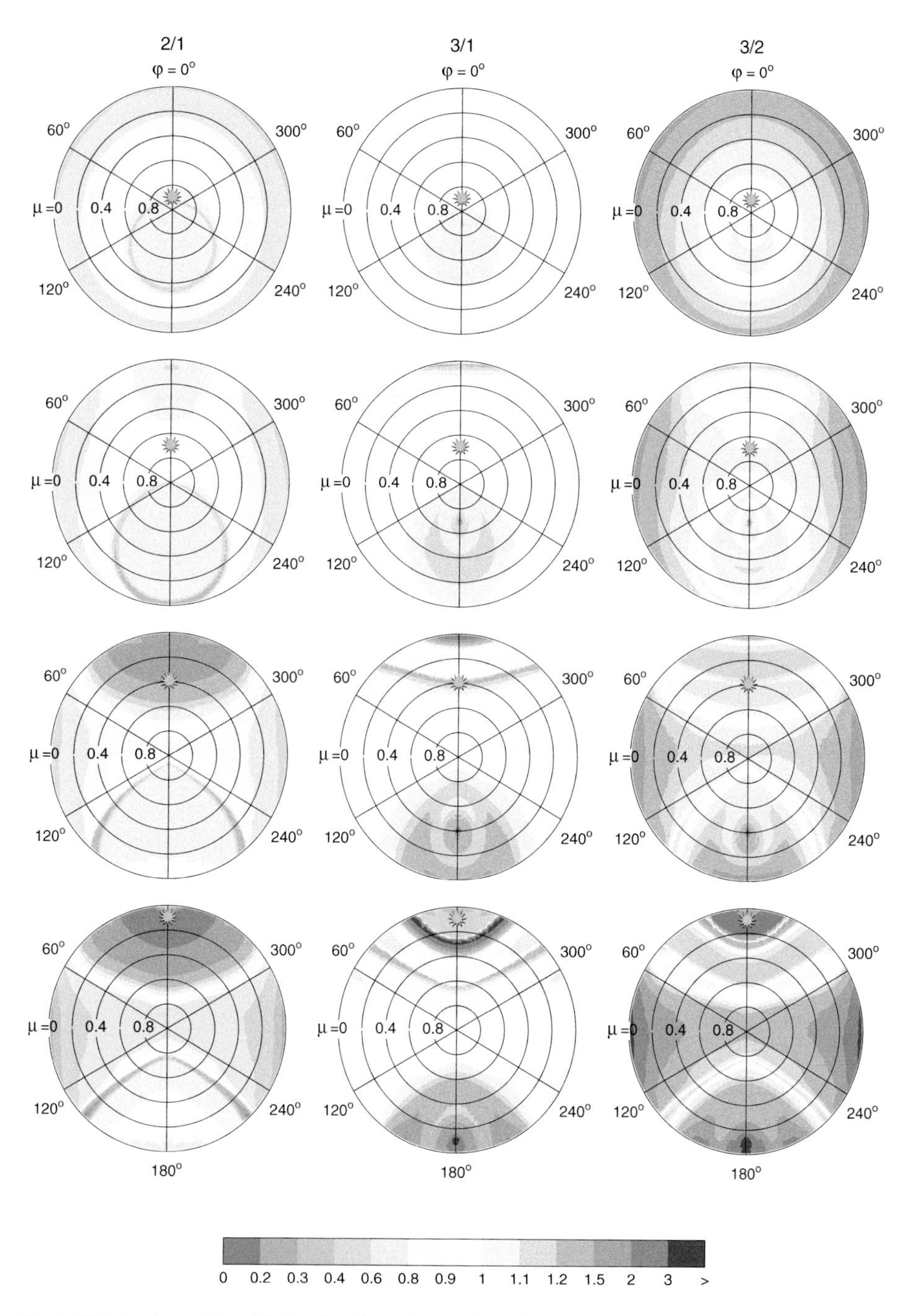

Plate 13.2.2. As in Plate 13.2.1, but for ratios of the reflected specific intensities.

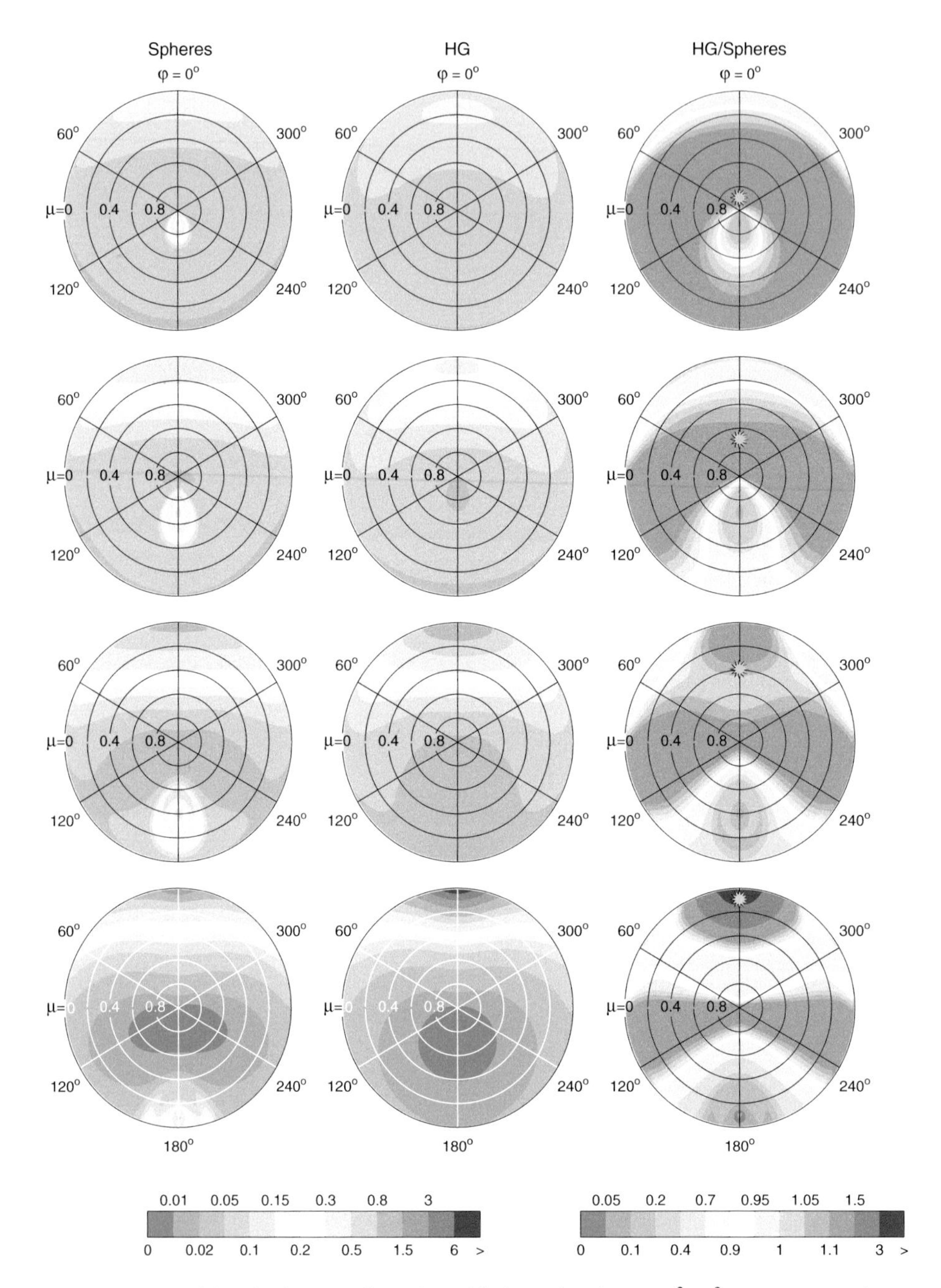

Plate 13.2.3. Left-hand column: reflected specific intensity (in $\mathrm{W\,m^{-2}\,sr^{-2}}$) versus μ and φ for a semi-infinite slab composed of model 1 spherical particles. The four values of minus the cosine of the illumination zenith angle μ_0 = 0.9, 0.7, 0.4, and 0.1 are indicated by the yellow stars in the right-hand column. The azimuth angle of the unpolarized incident beam is zero and its intensity is $I_0 = \pi$ $\mathrm{Wm^{-2}}$. Central column: as in the left-hand column, but for the asymmetry-parameter-equivalent HG phase function. Right-hand column: the ratio of the specific intensities shown in the central and left-hand columns.

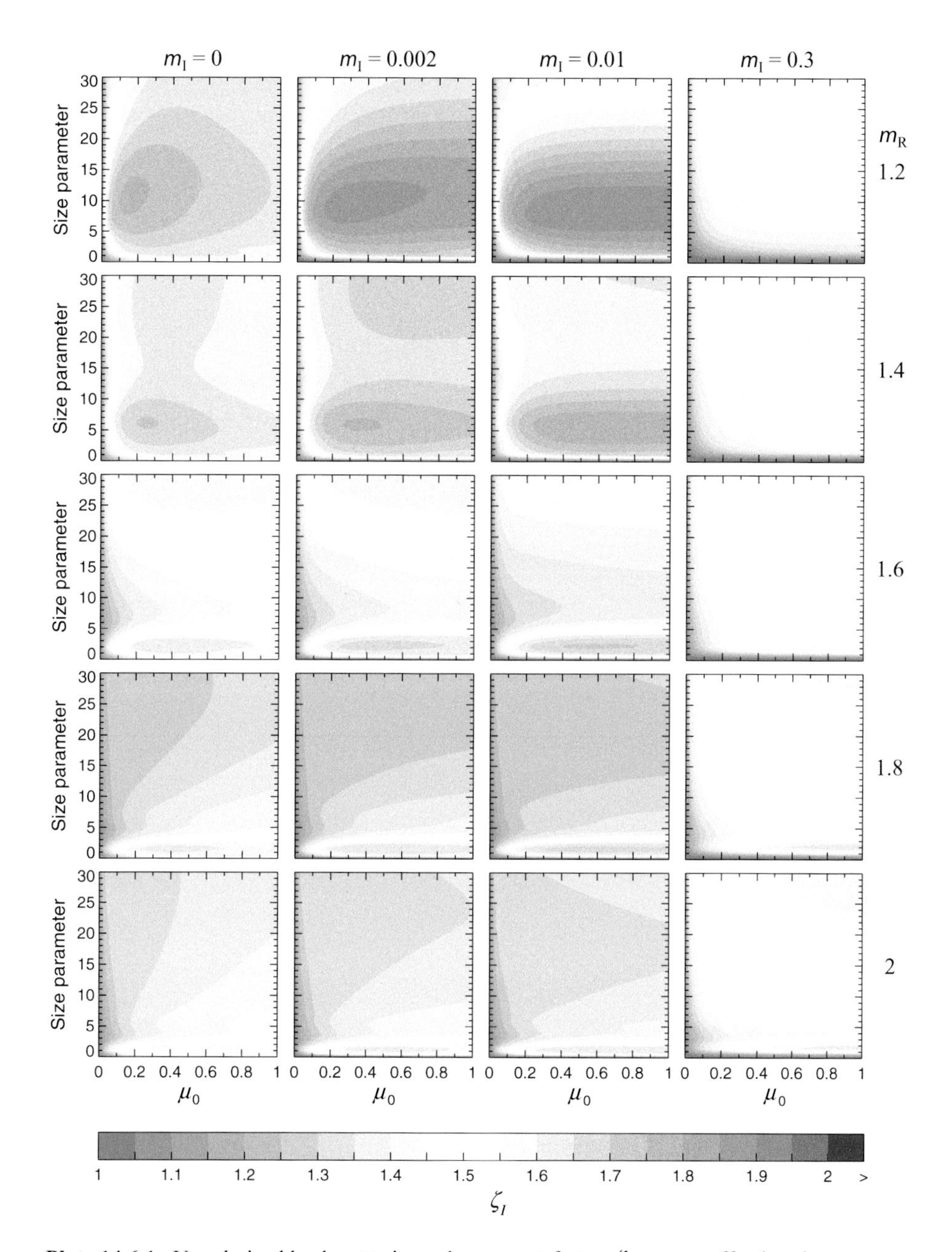

Plate 14.6.1. Unpolarized backscattering enhancement factor ζ_I versus effective size parameter x_{eff} and μ_0 for a homogeneous semi-infinite slab composed of polydisperse spherical particles with $m_R = 1.2$, 1.4, 1.6, 1.8, and 2 and $m_I = 0$, 0.002, 0.01, and 0.3.

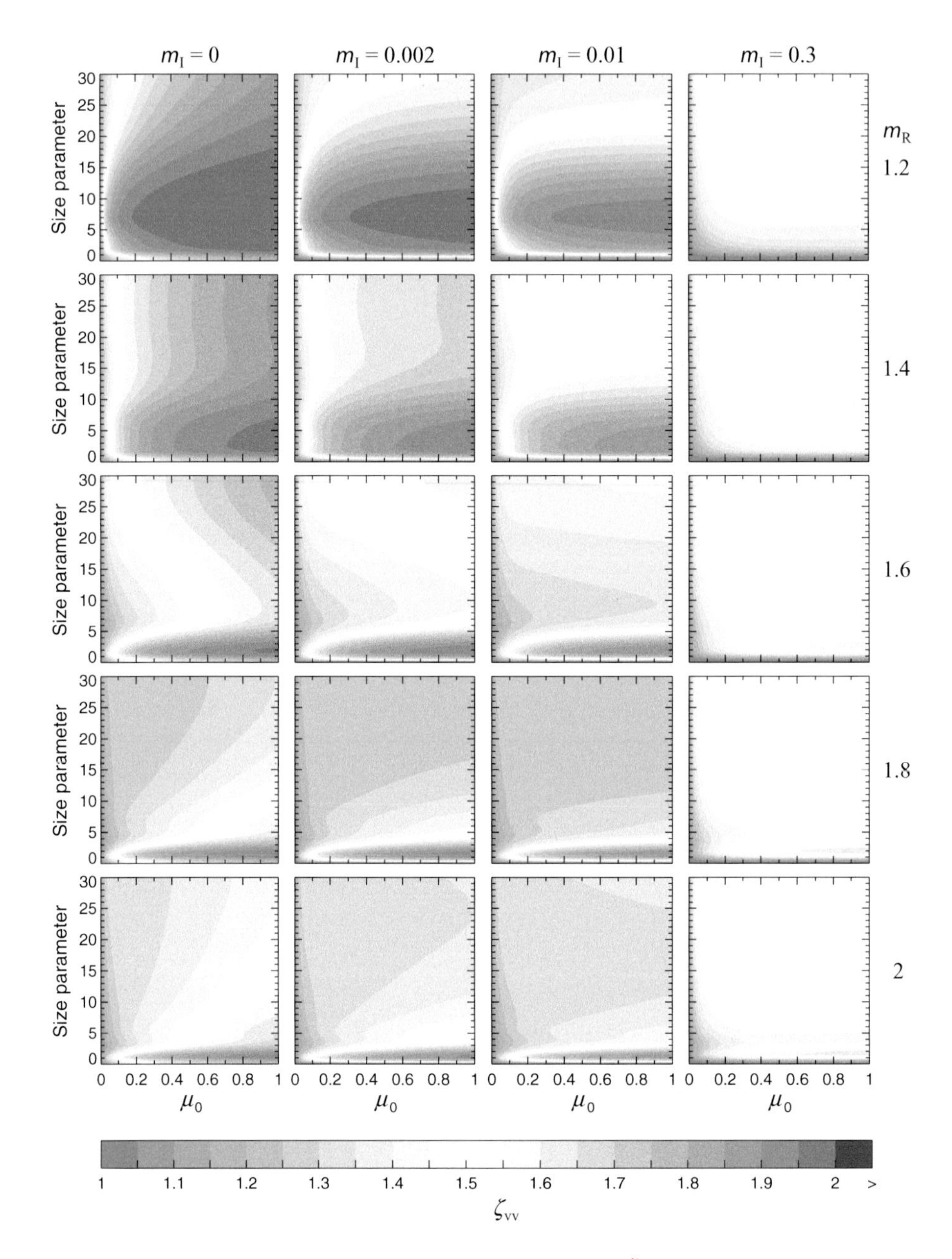

Plate 14.6.2. Co-polarized backscattering enhancement factor ζ_{vv} versus effective size parameter x_{eff} and μ_0 for a homogeneous semi-infinite slab composed of polydisperse spherical particles with m_R = 1.2, 1.4, 1.6, 1.8, and 2 and m_I = 0, 0.002, 0.01, and 0.3.

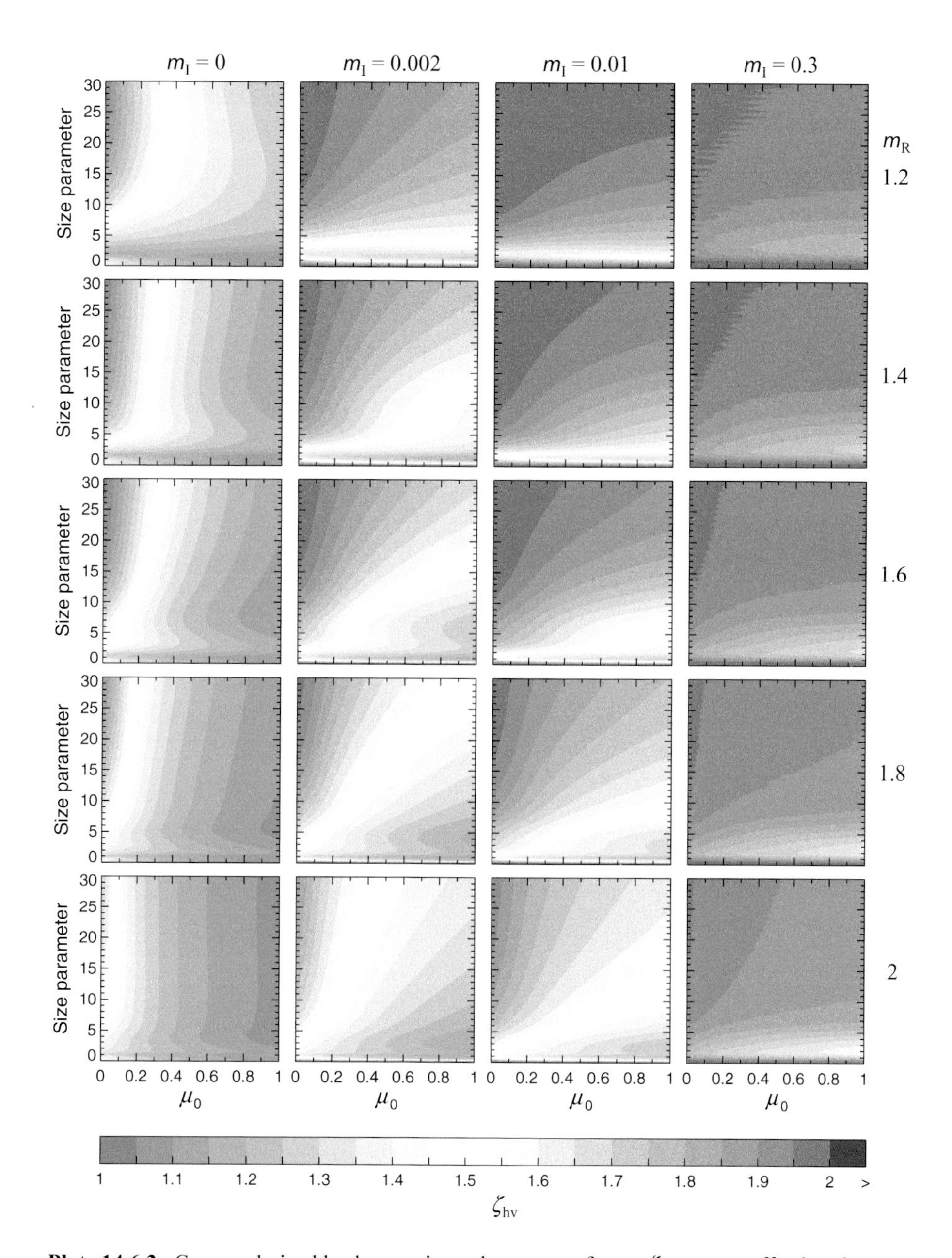

Plate 14.6.3. Cross-polarized backscattering enhancement factor ζ_{hv} versus effective size parameter x_{eff} and μ_0 for a homogeneous semi-infinite slab composed of polydisperse spherical particles with $m_R = 1.2$, 1.4, 1.6, 1.8, and 2 and $m_I = 0$, 0.002, 0.01, and 0.3.

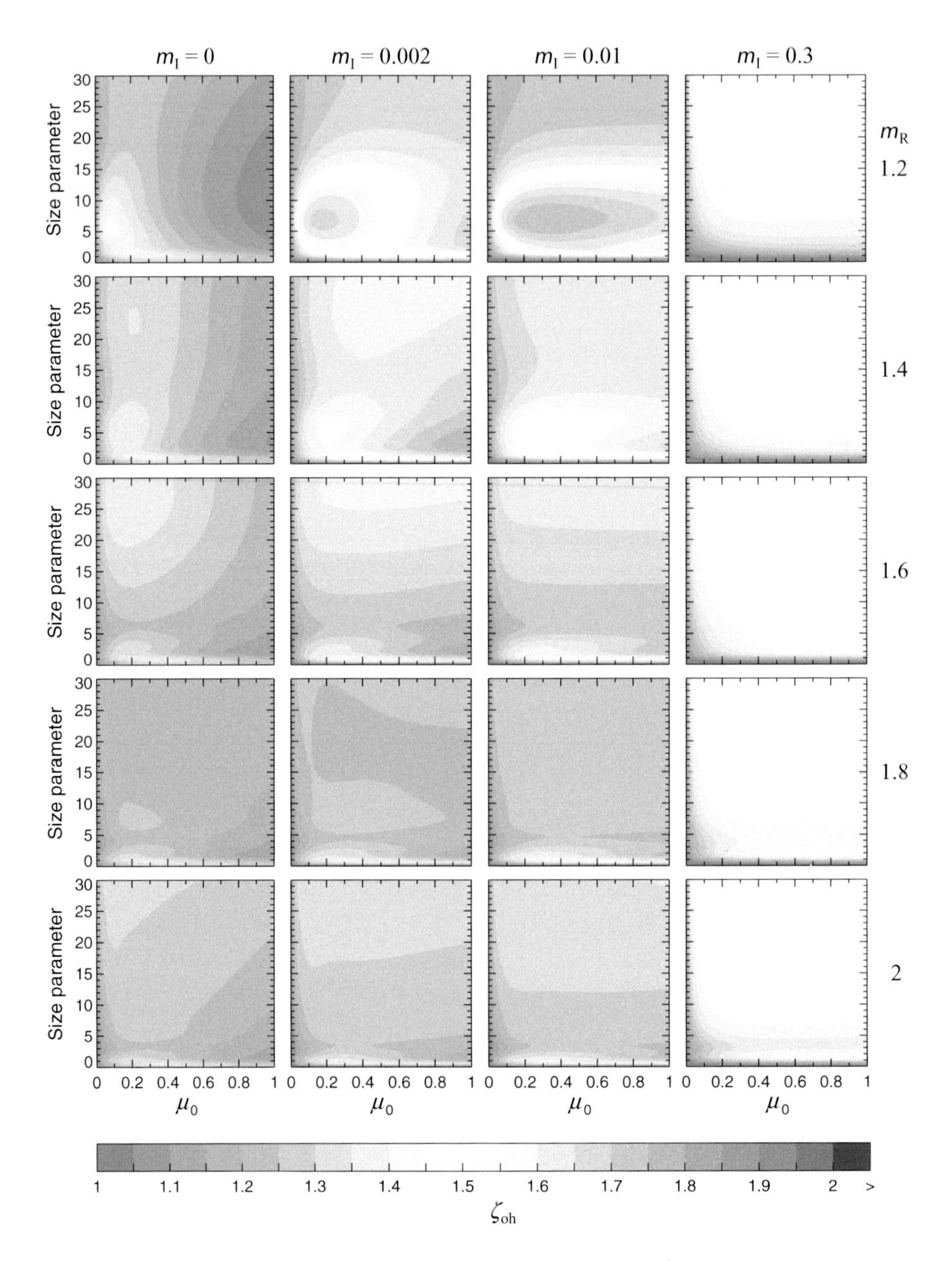

Plate 14.6.4. Opposite-helicity backscattering enhancement factor ζ_{oh} versus effective size parameter x_{eff} and μ_0 for a homogeneous semi-infinite slab composed of polydisperse spherical particles with m_R = 1.2, 1.4, 1.6, 1.8, and 2 and m_I = 0, 0.002, 0.01, and 0.3.

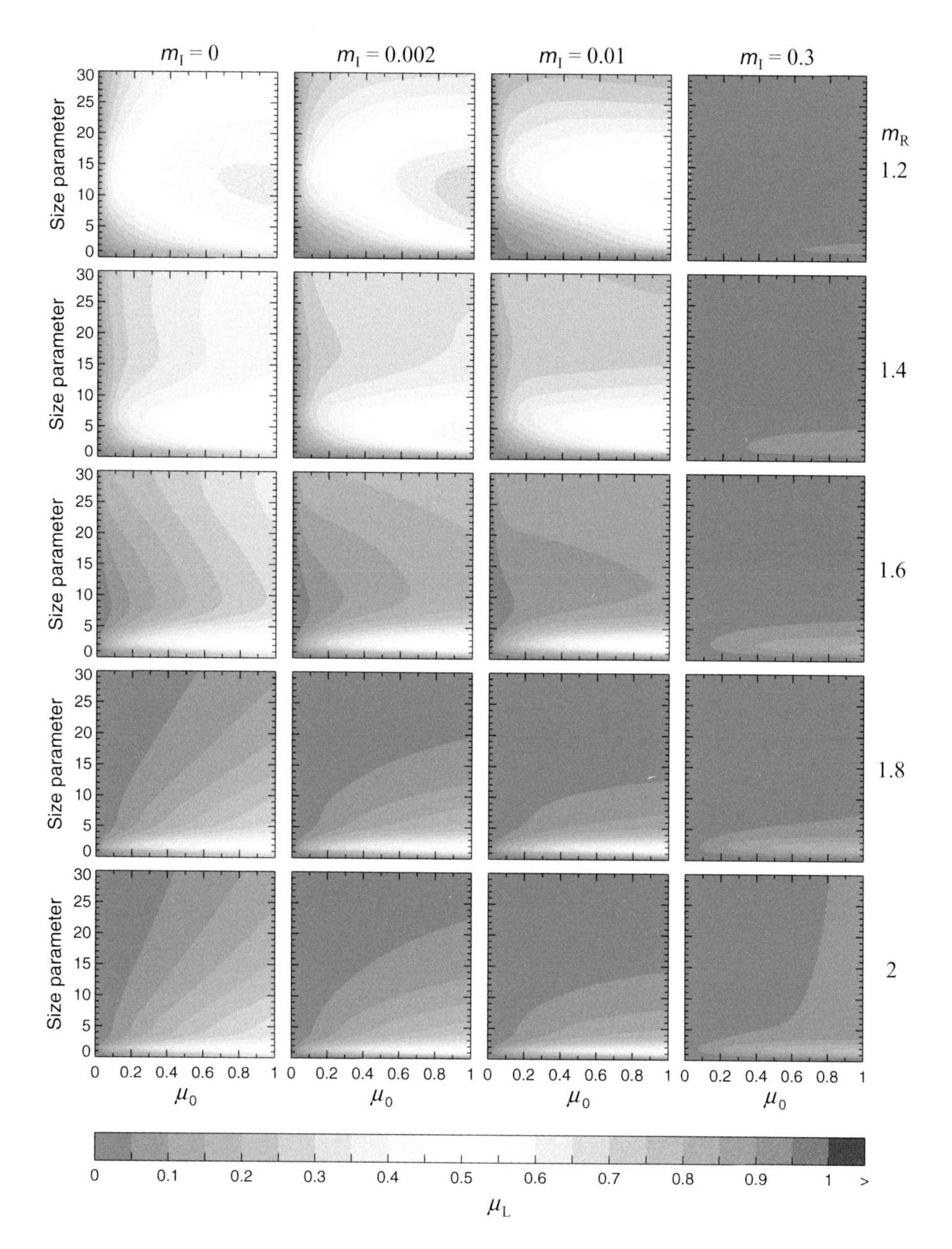

Plate 14.6.5. Linear polarization ratio μ_L versus effective size parameter x_{eff} and μ_0 for a homogeneous semi-infinite slab composed of polydisperse spherical particles with $m_R = 1.2$, 1.4, 1.6, 1.8, and 2 and $m_I = 0$, 0.002, 0.01, and 0.3.

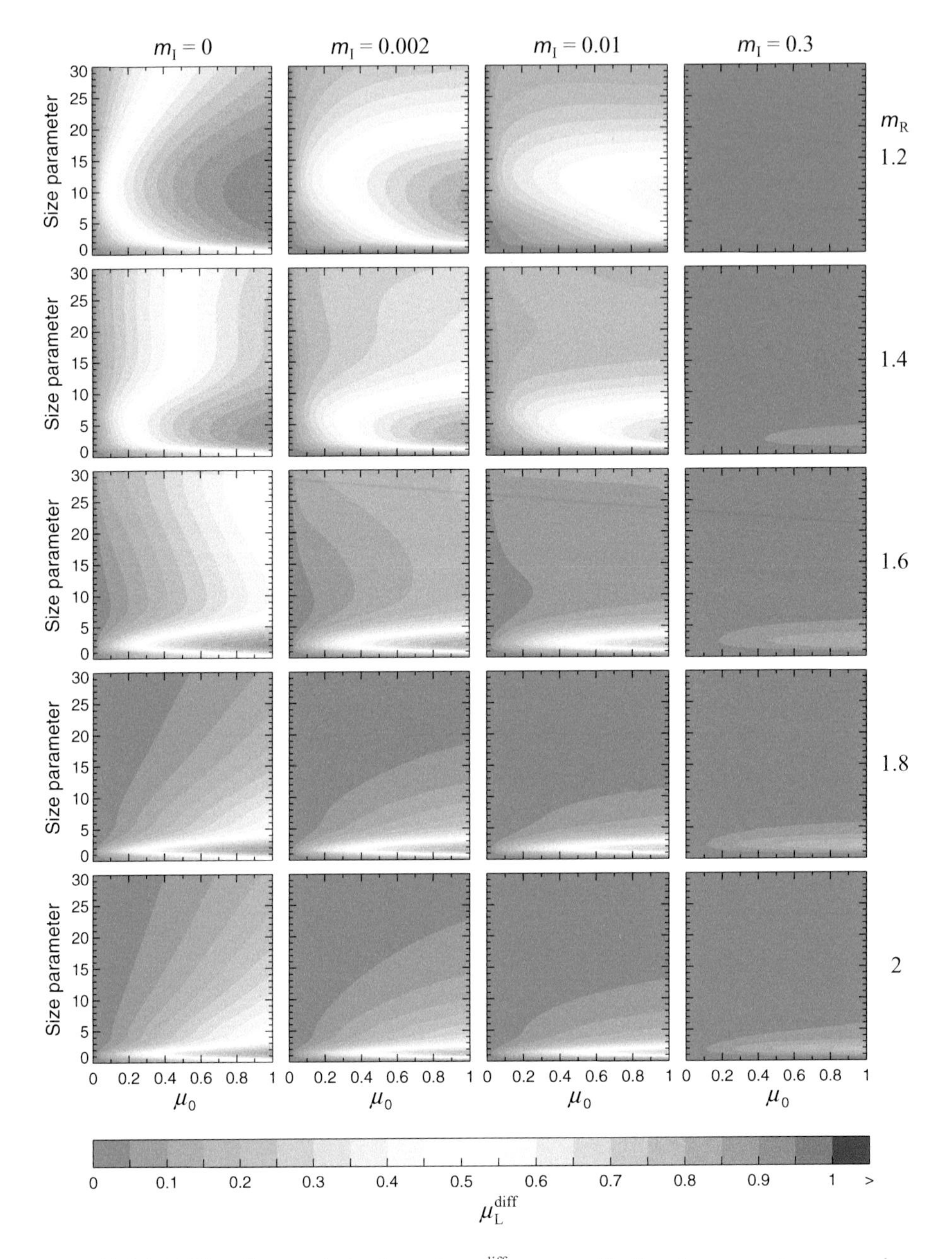

Plate 14.6.6. Diffuse linear polarization ratio μ_L^{diff} versus effective size parameter x_{eff} and μ_0 for a homogeneous semi-infinite slab composed of polydisperse spherical particles with m_R = 1.2, 1.4, 1.6, 1.8, and 2 and m_I = 0, 0.002, 0.01, and 0.3.

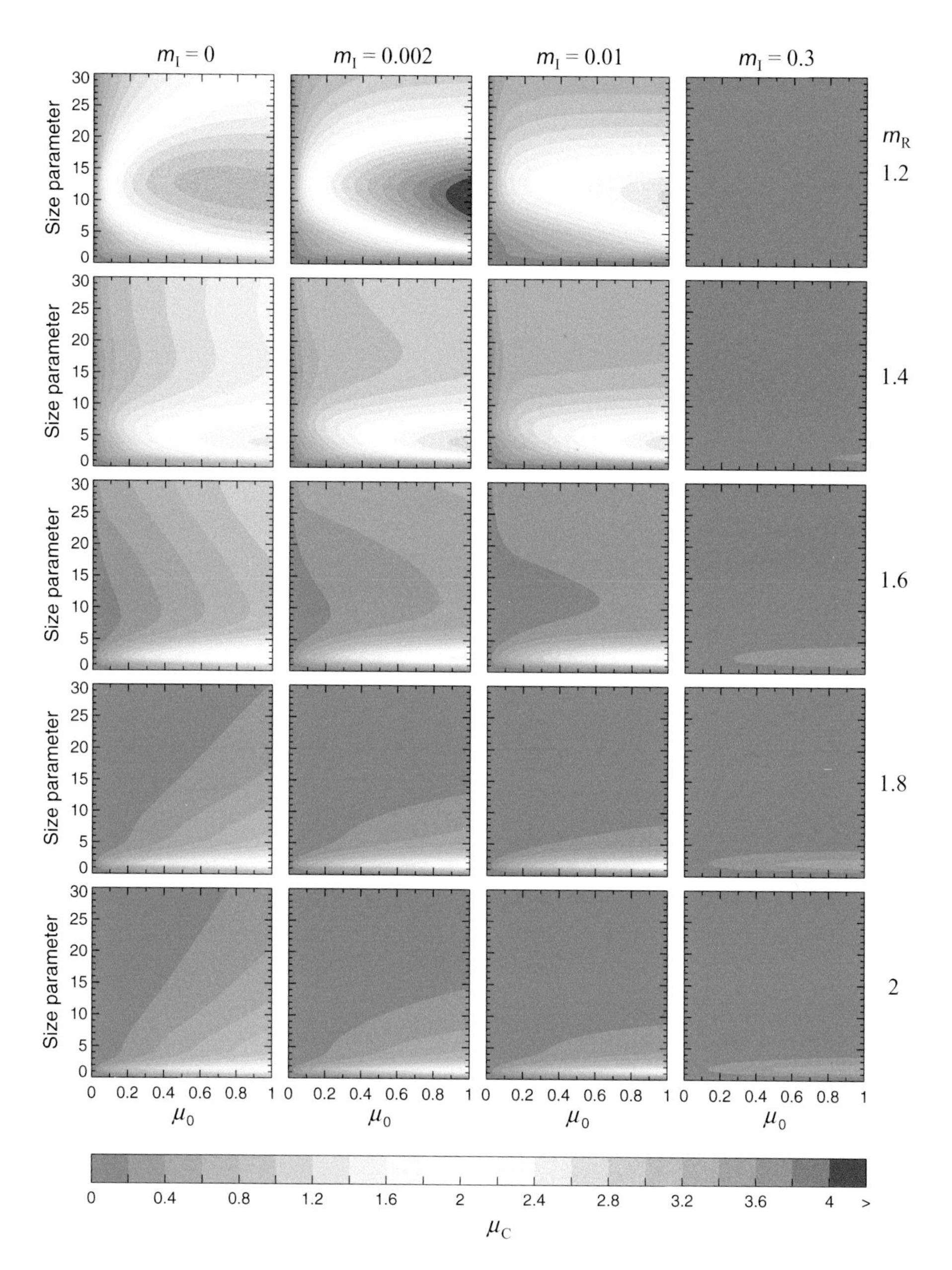

Plate 14.6.7. Circular polarization ratio μ_C versus effective size parameter x_{eff} and μ_0 for a homogeneous semi-infinite slab composed of polydisperse spherical particles with m_R = 1.2, 1.4, 1.6, 1.8, and 2 and m_I = 0, 0.002, 0.01, and 0.3.

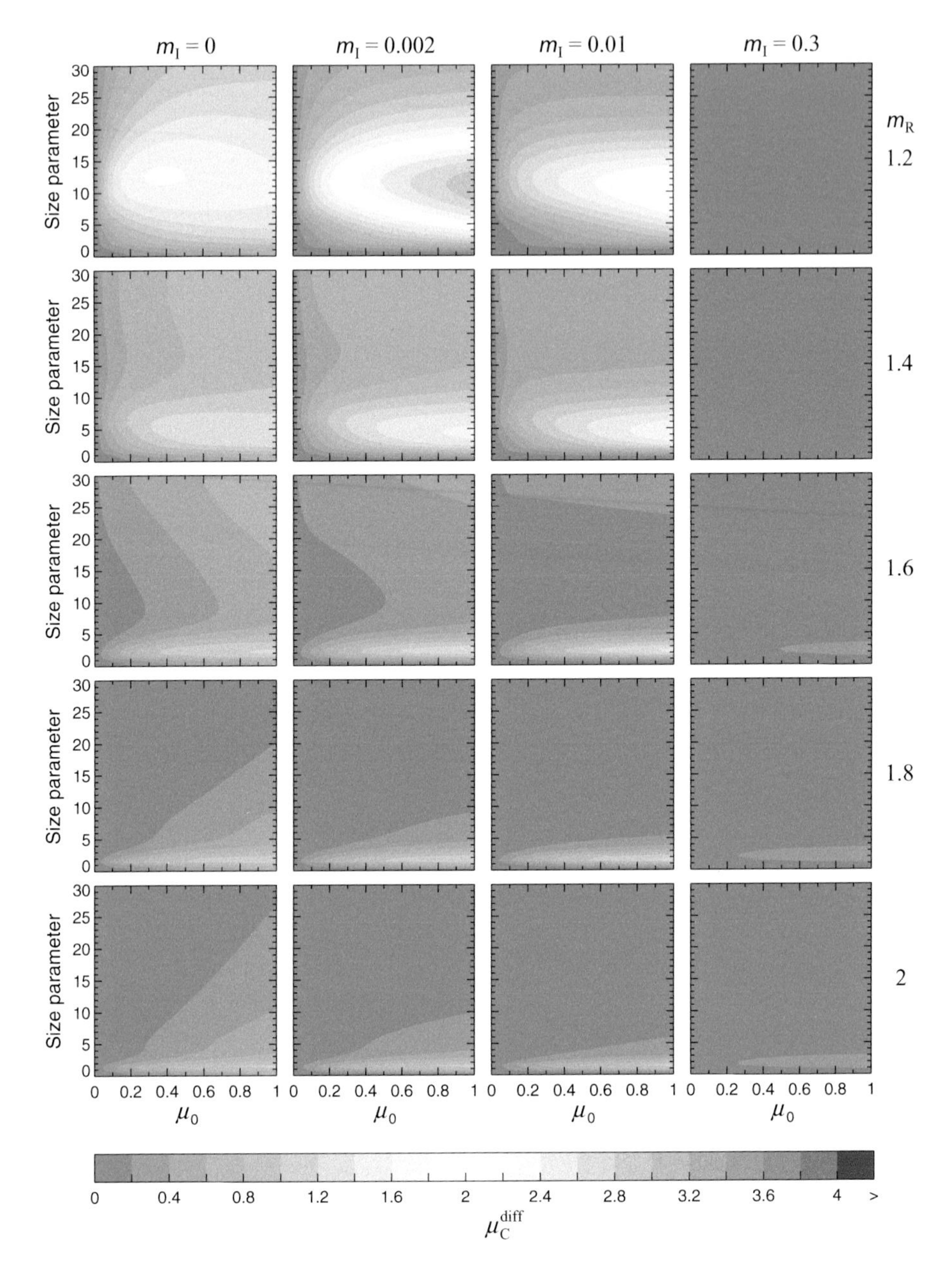

Plate 14.6.8. Diffuse circular polarization ratio μ_C^{diff} versus effective size parameter x_{eff} and μ_0 for a homogeneous semi-infinite slab composed of polydisperse spherical particles with $m_R = 1.2$, 1.4, 1.6, 1.8, and 2 and $m_I = 0$, 0.002, 0.01, and 0.3.

Printed in the United States
By Bookmasters